SpringerWienNewYork

Acta Neurochirurgica
Supplements

Editor: H.-J. Steiger

Intracranial Pressure and Brain Monitoring XIII

Mechanisms and Treatment

Edited by
G. Manley, C. Hemphill, S. Stiver

Acta Neurochirurgica
Supplement 102

SpringerWienNewYork

Geoffrey T. Manley, M.D., Ph.D.
Claude Hemphill, M.D., Ph.D.
Shirley Stiver, M.D., Ph.D.
San Francisco General Hospital
USA

Printed in Germany

Springer-Verlag Wien New York is a part of Springer ScienceþBusiness Media
springer.at

Typesetting: SPI, Manila, Phillipines
Printing and Binding: Druckerei Stürtz, 97080 Würzburg

Printed on acid-free and chlorine-free bleached paper

SPIN: 12110117

Library of Congress Control Number: applied for

With 160 Figures (thereof 1 coloured)

ISSN 0065-1419
ISBN 978-3-211-85577-5 SpringerWienNewYork

ICP 13 preface

The XIII International Symposium on Intracranial Pressure and Brain Monitoring was held July 22–26, 2007, in San Francisco, CA, USA. The international advisory board reviewed and selected a number of highly qualified submissions. The scientific sessions included invited speakers, oral presentations, poster sessions, and panel discussions. The symposium was highlighted by an honorary lecture by Dr. Anthony Marmarou and a consensus building session on the clinical applications of brain tissue oxygen monitoring. ICP XIII adopted a new format in which oral presentations were presented in multidisciplinary sessions, rather than topic oriented groups. We believe that this was a successful approach as it fostered lively cross-disciplinary discussions and stimulated exchange of new ideas.

The symposium was well attended by the basic science and clinical communities, both of which contributed to an outstanding scientific program. Presentations included new work in the areas of some of the topics which were presented include traumatic brain injury, intracerebral hemorrhage, brain ischemia, hydrocephalus, brain edema, and the rapidly growing fields of advanced neuromonitoring, bioinformatics and neuro-imaging. There was a particular focus on the increasing use of decompressive craniectomy for the treatment of brain edema following both traumatic and spontaneous ischemic acute brain injury. It is also apparent that neuromonitoring has extended beyond ICP measurement and that these newer modalities will continue to be part of future meetings. This volume includes manuscripts presented at the symposium. They have been carefully reviewed, edited, and organized thematically.

We were pleased to announce at the conclusion of the meeting that the International Advisory Board selected Tübingen, Germany as the site for the next Intracranial Pressure and Brain Monitoring XIV under the direction of Dr. Martin Schumann and his colleagues. Symposium attendees enthusiastically look forward to a successful meeting in 2010.

The editors wish to thank Ms. Silvia Schilgerius and the staff at Springer-Verlag, as well as Professor Steiger for their commitment and expertise in preparation of this volume. The editors are especially indebted to the expertise and tireless efforts of Michele Meeker, RN, for her help in the organization of the manuscript review and the completion of this book.

Contents

PART 1:
Brain injury: ICP management and cerebral physiology

PART 2:
Hydrocephalus and cerebrospinal fluid dynamics

PART 3:
Advanced neuromonitoring

PART 1:
Brain injury: ICP management and cerebral physiology

Acta Neurochir Suppl (2008) 102: 3–7

Cerebrovascular reactivity and autonomic drive following traumatic brain injury

Andrea Lavinio · Bogdan Ene-Iordache · Ilaria Nodari · Alan Girardini · Elena Cagnazzi · Frank Rasulo · Piotr Smielewski · Marek Czosnyka · Nicola Latronico

Abstract
Introduction The autonomic nervous system exerts tonic control on cerebral vessels, which in turn determine the autoregulation of cerebral blood flow. We hypothesize that the impairment of cerebral autoregulation following traumatic brain injury might be related to the acute failure of the autonomic system.
Methods This prospective, observational study included patients with severe traumatic brain injury requiring mechanical ventilation and invasive monitoring of intracranial pressure (ICP) and arterial blood pressure (ABP). Pressure reactivity index (PRx), a validated index of cerebrovascular reactivity, was continuously monitored using bedside computers. Autonomic drive was assessed by means of heart rate variability (HRV) using frequency domain analysis.
Findings Eighteen TBI patients were included in the study. Cerebrovascular reactivity impairment (PRx above 0.2) and autonomic failure (low spectral power of HRV) are significantly and independently associated with fatal outcome (P=0.032 and P<0.001, respectively). We observed a significant correlation between PRx and HRV spectral power (P<0.001). The high frequency component of HRV (HF, 0.15–0.4Hz) can be used to predict impaired autoregulation (PRx>0.2), although sensitivity and specificity are low (ROC AUC=0.67; P=0.001).
Conclusion Following traumatic brain injury, autonomic failure and cerebrovascular autoregulation impairment are both associated with fatal outcome. Impairment of cerebrovascular autoregulation and autonomic drive are interdependent phenomena. With some refinements, HRV might become a tool for screening patients at risk for cerebral autoregulation derangement following TBI.

Keywords Traumatic brain injury · Autoregulation · Autonomic nervous system · Autonomic failure

A. Lavinio · I. Nodari · A. Girardini · E. Cagnazzi · F. Rasulo · N. Latronico
Institute of Anaesthesiology and Intensive Care Medicine, Ospedale Civile,
Pzz.le Spedali Civili, 1,
25100 Brescia, Italy

B. Ene-Iordache
Department of Biomedical engineering, Mario Negri Institute,
Via Gavazzeni, 11,
24125 Bergamo, Italy

A. Lavinio (✉) · P. Smielewski · M. Czosnyka
Department of Clinical Neurosciences,
Academic Neurosurgical Unit, Addenbrooke's Hospital,
Cambridge, UK
e-mail: andrea.lavinio@gmail.com

Introduction

The autonomic nervous system is thought to modulate the autoregulation of cerebral blood flow in humans [19]. Cerebral blood vessels are innervated by autonomic fibres originating mostly from the superior cervical sympathetic ganglion, and the sphenopalatine and otic parasympathetic ganglia [10].

It is known that autonomic disorders can severely affect cerebral vasoreactivity. Diabetic autonomic failure was reported to affect cerebral autoregulation [3], and represents a significant independent risk factor for the occurrence of stroke in diabetic patients [5]. Autoregulation can also be impaired in Shy–Drager syndrome [14], and the loss of sympathetic innervations of cerebral vessels is thought to contribute to the impairment of cerebral autoregulation in patients with end-stage liver disease [8].

However, the relationship between autonomic disorders and the disturbances of cerebrovascular reactivity remains controversial. Cerebral autoregulatory vasodilation may still occur in patients with autonomic failure [12], and cerebral autoregulation (CA) was reported to be normal even after hypercapnic challenge in a series of patients affected by preganglionic autonomic failure [11]. Moreover, it was pointed out that in some diabetic patients cerebrovascular autonomic regulation may be intact even when cardiac autonomic function is severely compromised [7].

Data regarding autonomic failure and cerebral autoregulation impairment in acute critical brain injury are absent. However, it was independently reported that following traumatic brain injury the derangement of cerebrovascular reactivity [4] and poor heart rate variability (HRV) [17] are both predictors of fatal outcome. Since HRV is predominantly controlled by autonomic neural activity [18], it could be argued that the loss of cerebrovascular autoregulation following brain injury can be the manifestation of acute autonomic failure.

We therefore explored the relationship between autonomic failure and cerebrovascular reactivity impairment following severe traumatic brain injury.

Materials and methods

This is a prospective, observational study approved by the local Ethics Committee of the University Hospital of Brescia. Patients' next of kin granted written informed consent for publishing recorded data.

Adult (>18 years) comatose mechanically ventilated patients with a diagnosis of traumatic brain injury, requiring invasive intracranial pressure (ICP) and invasive arterial blood pressure (ABP) monitoring were eligible for this study. Comatose patients were defined according to current criteria as those patients not obeying to simple commands, not opening their eyes, and not uttering words, with a GCS score <8. Exclusion criteria were age <18 years, GCS score >8, non-sinus rhythm or electrical pacing, and history of diabetes or autonomic disease.

ICP was continuously monitored according to clinical indication with Codman parenchymal probes (Johnson & Johnson Medical, Raynham, MA, USA) or via a fluid-filled intraventricular catheter connected to an electrical transducer positioned and zeroed at the external auditory meatus. Arterial blood pressure was invasively monitored through a catheter positioned in the radial artery. Lead II ECG was also continuously monitored. All data were digitalized and captured using a bedside computer with a sampling rate of 125 Hz. Artefacts were manually removed.

Monitoring cerebrovascular reactivity Cerebral autoregulation was continuously monitored by means of the pressure reactivity index (PRx), calculated on bedside computers running ICM+ software for multimodal brain monitoring [13] (Intensive Care Monitor, University of Cambridge, UK; www.neurosurg.cam.ac.uk/icmplus). The pressure reactivity index is defined as the moving correlation coefficient between spontaneous changes in averaged ICP and ABP [15]. PRx measures the ability of the intracranial resistive vessels to cope with spontaneous changes in ABP, and it is a quantitative index of cerebral vasoreactivity. When autoregulation is intact, a spontaneous decrease in ABP will trigger a physiological dilation of cerebral arterioles in order to reduce vascular resistance and to maintain cerebral blood flow constant. The arteriolar dilation will lead to an increase in cerebral blood volume and, consequently, to an increase in ICP. Hence, in the case of preserved autoregulation, ABP and ICP are negatively correlated, and PRx is negative (i.e. when ABP falls, ICP increases, and vice versa, see Fig. 1a). On the contrary, when CA is impaired, cerebral blood volume will increase or decrease passively with changes in ABP, and PRx will become positive (i.e. when ABP increases, so does ICP, see Fig. 1b). To summarize, positive average PRx (above 0.2) signifies disturbed pressure-reactivity, while negative PRx (below 0) implies good reactivity. Pressure-reactivity has previously been demonstrated to correlate strongly with CA assessed using transcranial Doppler ultrasonography in head injured patients [6], and with static autoregulation assessed using PET-CBF [16].

From the continuously recorded data of ICP and ABP we calculated 48 hour-averaged PRx for the first 2 days after admission period (PRx) and PRx averaged over 1 h-long window (PRx_{1h})—matching 30 min before and after the 5-min ECG recordings at 0600, 1200, 1800, and 2400 hours (see below).

Monitoring cardiac autonomic function ECG signal (lead II) was continuously monitored and recorded on bedside computers with a sampling rate of 125 Hz. Five minute-long ECG recordings, free of ectopy, missing data, noise, and free of active patient stimulation (such as suctioning, nursing, etc.) were manually selected four times a day at 6 h intervals (0600, 1200, 1800, and 2400 hours respectively) covering the first 48 h from ICU admission for every patient. Heart rate variability analysis was performed off-line from raw ECG data, exported as ASCII text files from ICM+ recordings, for eight ECG recordings for each patient. Text ECG signal data were processed to obtain inter-beat (RR) intervals using the freely available PhysioToolkit software (PhysioNet, http://www.physionet.org) [9]. Briefly, QRS fiducial points were automatically detected by means of a single-channel QRS detector based on length transform algorithm. Successively, the five-minute ECG were visually revised for correction of

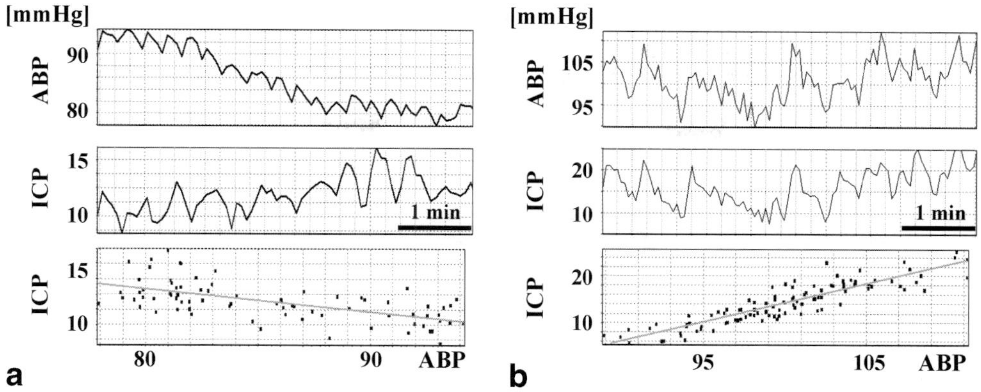

Fig. 1 Time trends and scatter plots with regression lines for ICP and ABP in a case of good pressure reactivity **(a)**, and disturbed pressure reactivity **(b)**. PRx is calculated as the correlation coefficient between slow waves of ICP and ABP from a period of 4 to 6 minutes. In **a** ABP spontaneously falls from 95 to 80 mmHg, while ICP increases from 10 to 15 mmHg. ABP and ICP are thus negatively correlated and PRx is negative (PRx=−0.61). In **b** ICP replicates the changes in ABP, thus demonstrating passive behaviour of cerebral vessels and positive PRx (PRx=0.90)

eventually missing or QRS errors and then the RR intervals were computed. Time and frequency domain HRV analysis on RR time series files was performed using the freely distributed programs HRV Analysis v1.1 (Biomedical Signal Analysis Group, University of Kuopio, Finland, http://it.uku.fi/biosignal/winhrv.shtml).

According to the international guidelines on HRV analysis, frequency domain analysis is to be preferred to the time domain methods, especially when short-term recordings are investigated [1]. We therefore performed a spectral power analysis of the fast Fourier transform in the low frequency range (LF: 0.04–0.15 Hz), the high frequency range (HF: 0.15–0.4 Hz), the total power (TP: 0.04–0.4 Hz), and the LF/HF ratio for five-minute recordings.

Data analysis The relationship between HRV parameters and PRx was evaluated with mixed linear model analysis to account for repeated measures and for the covariates [13]. The model covariates were the SAPSII score [2], age and diagnosis at admission. A receiver operating (ROC) curve was plotted in order to describe sensitivity and specificity of the HRV parameters to predict impairment in cerebral vasoreactivity (PRx>0.2). We considered mortality during ICU stay as the outcome measure. Differences between alive and dead patients groups were evaluated using the Mann-Whitney test. All tests were two-tailed and $P<0.05$ was assumed as statistically significant. All statistical analyses were performed using SPSS v12 software.

Results

Eighteen TBI patients were included in the study. Patients demographics are presented in Table 1.

A total ICP and ABP monitoring time of 864 h, and 144 ECG recordings were analysed respectively for cerebral vasoreactivity and heart rate variability.

Average PRx was significantly lower in patients who survived, demonstrating better cerebrovascular reactivity in patients with favourable outcome [mean PRx difference= 0.32 (95% CI: 0.04–0.81); $t=2.355$, $P=0.032$). Average HF component of HRV variability was significantly higher in patients who survived, demonstrating better parasympathetic drive in patients with favourable outcome [log HF mean difference=2.9 s^2 (95% CI: 1.44–4.39); $t=4.189$, $P<0.001$].

Maximum spectral power of HRV in the high frequency range (HF: 0.15–0.4 Hz) significantly correlates with average PRx ($R=0.630$, $P=0.005$; Fig. 2).

The sensitivity and specificity of a low HF-HRV spectral power to predict impaired autoregulation (defined as a PRx>0.2) were evaluated using a receiver operated curve analysis [ROC area under curve=0.669 (95% CI 0.577–0.761), $P<0.001$; Fig. 3].

Discussion

We demonstrated a significant correlation between the high frequency (parasympathetic) component of the heart rate variability and the status of cerebrovascular reactivity following traumatic brain injury.

Although these preliminary results might be interpreted as a cross correlation between two independent indexes of severity of injury, we hypothesize that cerebrovascular autoregulation impairment might be an aspect of autonomic failure complicating traumatic brain injury. If this hypothesis will be confirmed in a larger cohorts of patients, we will be able to reinterpret the pathogenesis of cerebral

Table 1 Study population

Patient	Age	Sex	SAPS II	ICU stay	GCS(d)
1	21	M	40	31	E4 M5 V1
2	29	M	49	18	E4 M6 V5
3	47	M	47	31	E4 M6 V1
4	24	M	40	30	E1 M6 V1
5	35	M	49	9	Deceased
6	31	M	30	43	E4 M3 V1
7	67	M	58	23	E4 M6 V1
8	18	M	40	29	E4 M5 V1
9	48	M	22	20	E4 M6 V1
10	33	M	46	14	E2 M3 V1
11	55	M	26	15	E4 M6 V5
12	30	M	30	9	E4 M6 V1
13	38	M	40	18	E3 M6 V1
14	34	M	43	27	E4 M6 V1
15	43	M	32	13	E4 M6 V5
16	29	M	39	43	E1 M2 V1
17	49	M	41	20	E4 M5 V1
18	29	M	45	14	E3 M3 V1

SAPS Simplified acute physiology score. ICU stay: days. *GCS(d)* Glasgow Coma Score at ICU discharge—*E* eyes, *M* motor, *V* verbal score

autoregulation derangement following traumatic brain injury.

Also, HRV might acquire a new significance as a non invasive tool for cerebral autoregulation screening following traumatic brain injury. Continuous monitoring of cerebrovascular reactivity is relatively practical, as PRx monitoring requires invasive monitoring of ICP and ABP. On the other hand, HRV testing is completely non invasive, and several devices for automated frequency domain analysis are commercially available. We suggest that, after further investigation and refinement of the technique, HRV analysis might become a practical tool for screening patients at risk for CA derangement. However, at present, sensitivity and specificity are unacceptably low, and further research is needed.

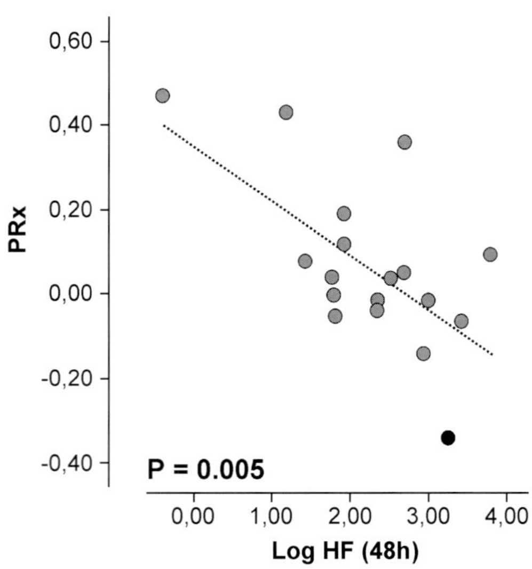

Fig. 2 *PRx* Average pressure reactivity index over 48 h of monitoring. *HF* Maximum high frequency component of heart rate variability. *Dotted line* Regression line

Conclusion

Cerebrovascular autoregulation impairment and autonomic drive are related phenomena. Following traumatic brain injury, parasympathetic failure and cerebrovascular autoregulation impairment are both associated with fatal out-

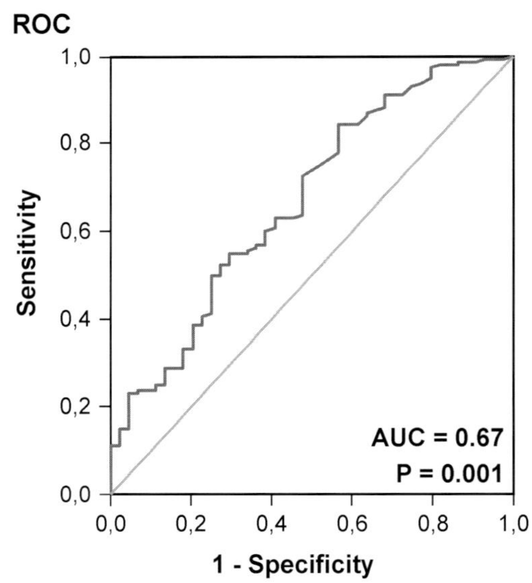

Fig. 3 *ROC* Receiver operated curve. *AUC* Area under curve. The *curve* depicts the performance of HF-HRV as a diagnostic test for impaired cerebrovascular autoregulation (as defined by a PRx greater than 0.2). Ideally, the AUC should approximate the unity, and the curve should be close to the *top left corner of the square* (sens and spec=100%)

come. With some refinements, HRV might become a tool for screening patients at risk for cerebral autoregulation derangement. Further research on the topic is needed.

Conflict of interest statement We declare that we have no conflict of interest.

References

1. Task Force of the European Society of Cardiology and the North American Society of Pacing and Electrophysiology (1996) Heart rate variability. Standards of measurement, physiological interpretation, and clinical use. Eur Heart J 17:354–381
2. Aegerter P, Boumendil A, Retbi A, Minvielle E, Dervaux B, Guidet B (2005) SAPS II revisited. Intensive Care Med 31:416–423
3. Asil T, Utku U, Balci K, Uzunca I (2007) Changing cerebral blood flow velocity by transcranial Doppler during head up tilt in patients with diabetes mellitus. Clin Neurol Neurosurg 109:1–6
4. Balestreri M, Czosnyka M, Steiner LA, Hiler M, Schmidt EA, Matta B, Menon D, Hutchinson P, Pickard JD (2005) Association between outcome, cerebral pressure reactivity and slow ICP waves following head injury. Acta Neurochir Suppl 95:25–28
5. Cohen JA, Estacio RO, Lundgren RA, Esler AL, Schrier RW (2003) Diabetic autonomic neuropathy is associated with an increased incidence of strokes. Auton Neurosci 108:73–78
6. Czosnyka M, Smielewski P, Kirkpatrick P, Laing RJ, Menon D, Pickard JD (1997) Continuous assessment of the cerebral vasomotor reactivity in head injury. Neurosurgery 41:11–17 (discussion 17–19)
7. Daffertshofer M, Diehl RR, Ziems GU, Hennerici M (1991) Orthostatic changes of cerebral blood flow velocity in patients with autonomic dysfunction. J Neurol Sci 104:32–38
8. Frokjaer VG, Strauss GI, Mehlsen J, Knudsen GM, Rasmussen V, Larsen FS (2006) Autonomic dysfunction and impaired cerebral autoregulation in cirrhosis. Clin Auton Res 16:208–216
9. Goldberger AL, Amaral LA, Glass L, Hausdorff JM, Ivanov PC, Mark RG, Mietus JE, Moody GB, Peng CK, Stanley HE (2000) PhysioBank, PhysioToolkit, and PhysioNet: components of a new research resource for complex physiologic signals. Circulation 101:E215–220
10. Gulbenkian S, Uddman R, Edvinsson L (2001) Neuronal messengers in the human cerebral circulation. Peptides 22:995–1007
11. Hetzel A, Reinhard M, Guschlbauer B, Braune S (2003) Challenging cerebral autoregulation in patients with preganglionic autonomic failure. Clin Auton Res 13:27–35
12. Horowitz DR, Kaufmann H (2001) Autoregulatory cerebral vasodilation occurs during orthostatic hypotension in patients with primary autonomic failure. Clin Auton Res 11:363–367
13. Littell RC, Pendergast J, Natarajan R (2000) Modelling covariance structure in the analysis of repeated measures data. Stat Med 19:1793–1819
14. Ogawa M, Fukuyama H, Harada K, Kimura J (1998) Cerebral blood flow and metabolism in multiple system atrophy of the Shy-Drager syndrome type: a PET study. J Neurol Sci 158:173–179
15. Smielewski P, Czosnyka M, Steiner L, Belestri M, Piechnik S, Pickard JD (2005) ICM+: software for on-line analysis of bedside monitoring data after severe head trauma. Acta Neurochir Suppl 95:43–49
16. Steiner LA, Coles JP, Johnston AJ, Chatfield DA, Smielewski P, Fryer TD, Aigbirhio FI, Clark JC, Pickard JD, Menon DK, Czosnyka M (2003) Assessment of cerebrovascular autoregulation in head-injured patients: a validation study. Stroke 34:2404–2409
17. Winchell RJ, Hoyt DB (1997) Analysis of heart-rate variability: a noninvasive predictor of death and poor outcome in patients with severe head injury. J Trauma 43:927–933
18. Zhang R, Iwasaki K, Zuckerman JH, Behbehani K, Crandall CG, Levine BD (2002) Mechanism of blood pressure and R–R variability: insights from ganglion blockade in humans. J Physiol 543:337–348
19. Zhang R, Zuckerman JH, Iwasaki K, Wilson TE, Crandall CG, Levine BD (2002) Autonomic neural control of dynamic cerebral autoregulation in humans. Circulation 106:1814–1820

Acta Neurochir Suppl (2008) 102: 9–13

A comparison between the transfer function of ABP to ICP and compensatory reserve index in TBI

Sima Shahsavari · Tomas McKelvey · Thomas Skoglund · Catherine Eriksson Ritzèn

Abstract
Background The transfer functions which map the arterial blood pressure to the intracranial pressure and the compensatory reserve index have been investigated by various groups to evaluate the brain compliance of patients with traumatic brain injury. The focus of this study has been to assess the capability of both the above mentioned methods to monitor the intracranial compliance in patients suffering from brain swelling.
Materials and methods Clinical data was collected from sixteen traumatic brain injury patients and split into 4 min segments. For each segment, both the magnitude of the empirical transfer function at the fundamental cardiac frequency and the compensatory reserve index were extracted.
Findings The mean values of the compensatory reserve index and the magnitude of the transfer function which scored higher than 0.7 and 0.1 respectively were recorded for all patients suffering from brain swelling. By comparing the histogram of the magnitude of the transfer function at the fundamental cardiac frequency with the histogram of the compensatory reserve index for all patients, a positive correlation between the mean values and a negative correlation among their variances were observed. The linear correlation between the mean values was estimated at $r=0.82$ ($p<0.0001$).
Conclusions These observations suggest that to evaluate the intracranial compensatory reserve, the magnitude of 0.1 could be a useful threshold for the transfer function at the fundamental cardiac frequency.

Keywords Arterial blood pressure (ABP) · Compensatory reserve index (RAP) · Transfer function · Intracranial pressure (ICP) · Traumatic brain injury (TBI)

S. Shahsavari (✉) · T. McKelvey
Department of Signals and Systems,
Chalmers University of Technology,
SE-412 96 Gothenburg, Sweden
e-mail: shahsava@chalmers.se

T. McKelvey
e-mail: mckelvey@chalmers.se

T. Skoglund · C. Eriksson Ritzèn
Neuroscience Clinic, Sahlgrenska University Hospital,
SE-413 45 Gothenburg, Sweden

T. Skoglund
e-mail: thomas.skoglund@vgregion.se

C. Eriksson Ritzèn
e-mail: catherine.eriksson-ritzen@vgregion.se

Introduction

Head injuries are important causes of disability and death in people of all ages. In the neurointensive care of patients with traumatic brain injury or other cerebrovascular diseases such as stroke or subarachnoid hemorrhage, a principal objective is to minimize the risk of secondary injury. Intracranial hypertension or brain swelling which can follow severe traumatic brain injury are life threatening complications which can cause serious permanent damage to the brain and even lead to death. This is due to the fact that the brain is enclosed in a non-expanding cavity and swelling leads to an increased pressure which at high levels can initiate cerebral ischemia or cause structural damage to brain tissues. If such incidents of brain swelling could be better predicted, the treatment of secondary injury could be made more selective and provide better overall results. Today, brain swelling is mainly monitored by measuring the intracranial pressure (ICP) and the cerebral perfusion pressure (CPP) which is defined as the difference between the mean arterial pressure (MAP) and ICP. However, clinical experiences show that the monitored data cannot give enough information regarding the intracranial regulatory processes. At the moment, the only reliable way to detect changes of the brain parenchyma is to perform computer assisted tomography (CT) or magnetic resonance imaging (MRI). These investigations give a momentary picture of the brain and often have to be repeated, sometimes several times a day, and in most instances require the patient to be transported. Several investigations have attempted to find a solution to the demand of reliable and continuous brain monitoring and as a result different concepts have been proposed and applied to this area. In the time domain, the pressure–volume curve and the term "compensatory reserve" which have been introduced by Langefitt et al. [5] and Lofgren et al. [6] are foundations of many studies. More recently much attention has been paid to the volume–pressure compensatory reserve index (RAP) which determines the level of linear correlation between the amplitude of the ICP wave (AMP) and the mean of ICP (ICP_{mean}) [2–4]. The RAP index is not able to define the steepness of the volume–pressure curve as is done by the pressure–volume index (PVI) [7], but it is able to classify the volume–pressure curve into three distinct regions corresponding to: good pressure–volume compensatory reserve, poor pressure–volume compensatory reserve and deranged cerebrovascular reactivity [4]. A RAP coefficient close to 0 indicates an asynchrony between the simultaneous changes in AMP and ICP_{mean} and hence it implies a good compensatory reserve. A RAP index around 1 indicates that the changes in AMP and ICP_{mean} are synchronized and is therefore the indicator of a poor compensatory reserve. In the frequency domain, the study of the pulse transmission between ABP and ICP has a long history [8]. In this approach the brain is modeled as a dynamic system with ABP wave as the input and ICP wave as the output. The idea behind this method is that any change in the transfer function between input and output is indicative of a change in the system itself. Piper and his group defined four different curves for the magnitude of the transfer function (MTF), and classified them into two classes of low and high ICP [8]. According to their observations, elevated ICP was associated to the larger MTF at the fundamental cardiac frequency. This study is an attempt to utilize the above mentioned time and frequency domain methods to explore their potentials to monitor the brain swelling in TBI patients. During this study, patients were clinically classified according to their injury type and later on divided into two groups corresponding to with and without brain swelling.

Materials and methods

Clinical data was collected from sixteen patients admitted to the Neurointensive Care Unit at Sahlgrenska University Hospital in Gothenburg, Sweden. The underlying intracranial pathology for each patient was classified according to the clinical information which is summarized in Table 1. Five patients had brain swelling and four of them underwent decompressive craniectomy surgery to give the brain more

Table 1 Clinical profile of studied group

Patient	Age/sex	CT classification	Mechanism of injury	Surgical treatment	Data length (h)
1	41F	Contusion	Fall	Evacuation of contusion	13
2	26M	DAI	TA	–	24
3	60F	SAH	–	–	26
4	66F	EDH, CCF	Fall	Evacuation of EDH, Embolization	168
5	42M	SDH, Contusion, Brain swelling	WA	Evacuation of SDH and contusion, Craniectomy	43
6	51M	–	Fall	–	342
7	65M	SDH	Fall	Evacuation of SDH	23
8	56F	SAH	–	–	70
9	20F	DAI, Contusion, Brain swelling	Fall	–	193
10	55M	Contusion, Brain swelling	Fall	Evacuation of contusion, Craniectomy	306
11	38M	DAI	TA	–	29
12	69F	SDH, Contusion	–	Evacuation of SDH and contusion	376
13	21M	EDH, Brain Swelling	TA	Evacuation of EDH, Craniectomy	12
14	36M	Brain Swelling	TA	Craniectomy	18
15	50M	SAH, DAI, Contusion	TA	–	306
16	18F	SDH	Fall	Evacuation of SDH	175

CCF carotid cavernous fistula, *DAI* diffuse axonal injury, *EDH* epidural hematoma, *SAH* subarachnoidal hematoma, *SDH* subdural hematoma, *TA* traffic accident, *WA* work accident space to swell and compensate for the high intracranial pressure. ICP was monitored continuously using either a parenchymal fiber optic pressure transducer (ICP express, Codman) or a ventricular catheter (Medtronic). A Datex-Ohmeda S/5 critical care monitor collected both ICP and ABP measurements. These signals were collected from an S/5 network, using S5-collecting software at the sampling rate of 300 Hz and stored on CD-ROM.

The 300 Hz sampled data was downsampled to 30 Hz. The entire data for each patient was broken up into 4 min segments. In case of craniectomy, only the recorded data before craniectomy was used. Length of data for each patient can be seen in Table 1. Both ABP and ICP segments were first smoothed by a Blackman-Tukey window and then using a 4096 point FFT transformed to the frequency domain (frequency resolution 0.007 Hz). The fundamental frequency (F_C) of the cardiac components for each segment was estimated using a search algorithm to find the bin with the highest magnitude value in ABP. Although the amplitude of the fundamental component appears to be resilient to noise, a small error in estimation of the fundamental frequency will turn to the large deviations in higher harmonics. Therefore, to compensate for the effect of noise and make a robust estimation of the fundamental frequency and its magnitude, a quadratic function was fitted to the magnitude of ABP at the frequency range of $f=[F_C -0.3 \text{ Hz}: F_C +0.3 \text{ Hz}]$. The desired quadratic function was linearly modeled as

$$\boldsymbol{X} = \mathbf{H}\theta + W$$

where $\boldsymbol{X}$ is the vector of magnitudes, $\mathbf{H}$ is a known observation matrix dependent on f, $\boldsymbol{\theta}$ is a vector containing curve parameters to be estimated and W is the unknown noise. The least square (LS) estimation of $\boldsymbol{\theta}$ was simply found using:

$$\theta = \left(\mathbf{H}^{\mathrm{T}}\mathbf{H}\right)^{-1}\mathbf{H}^{\mathrm{T}}X$$

Knowing the curve parameter, the maximum of the quadratic function and the respective frequency were used as an updated estimation for the fundamental frequency and its magnitude in ABP. Fitting another quadratic function to the magnitude of ICP around the estimated fundamental frequency, the magnitude of ICP was found and subsequently MTF at the fundamental frequency was calculated as the ratio between the ICP and ABP magnitudes.

In the next step the amplitude of ICP was calculated using a time domain method which measures the peak-to-peak value of ICP waveform directly from the time domain samples. ICP beats were detected for each ICP segment employing an automatic pressure waveform components detector algorithm [1]. AMP and ICP_{mean} were extracted for each beat, averaged over a 6 s interval and subsequently used to calculate the RAP index of each segment by determining the linear correlation between them.

Results

Histograms of MTF and RAP coefficients were plotted for each of the sixteen patients. It was recognized that whenever MTF has a sharp and concentrated distribution around small values, the RAP index shows a broad and smooth distribution stretched from the values close to 1 toward −1. In a similar way, the wider distribution of MTF, extended toward the larger values, was observed for the sharp and concentrated distribution of the RAP index around 1. The

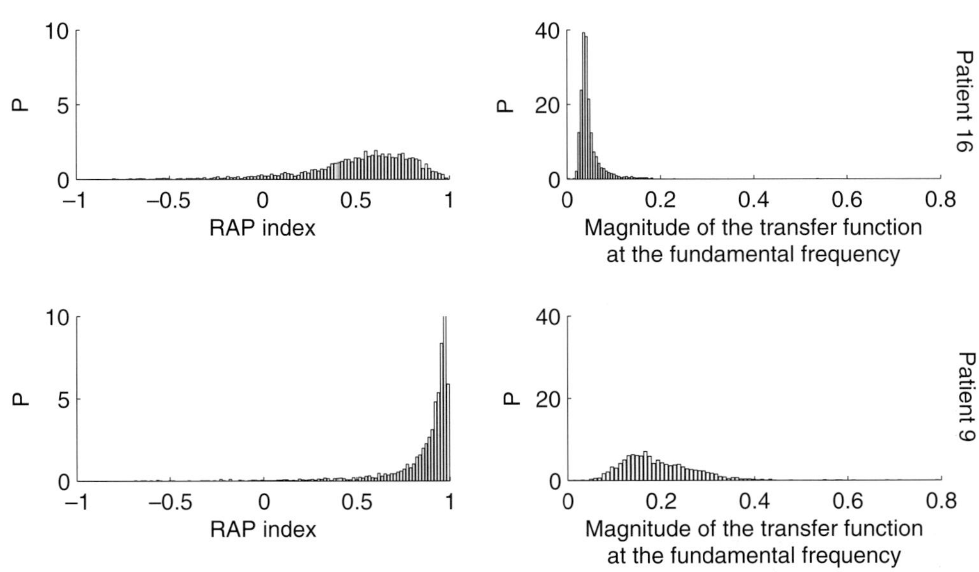

Fig. 1 Histograms of MTF and RAP coefficients are shown for patient 16 (*top panels*) and patient 9 (*bottom panels*). In *top panels* MTF has a sharp and concentrated distribution below 0.1, while RAP index indicates a smooth distribution between −0.3 and 1. The opposite situation can be recognized in bottom panels. The wider distribution of MTF extended between 0.005 and 0.4 is observed for the sharp and concentrated distribution of RAP index around 1

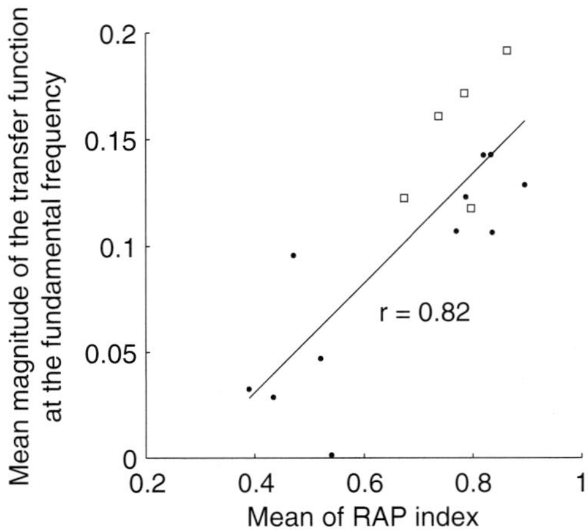

Fig. 2 Mean of MTF versus mean of RAP index. Patients with brain swelling are highlighted by the *square marks*. For mean RAP index less than 0.7, mean of MTF below 0.1 is observed, while for mean RAP index greater than 0.7, mean MTF above 0.1 is the most probable. Note that all brain swelling patients are positioned in the region with the high mean values for both MTF and RAP coefficient

case of concentrated MTF can be seen in the top panels of Fig. 1, which show the distributions of MTF and RAP coefficients from patient 16. The bottom panels of Fig. 1, represent the case of concentrated RAP index and are from patient 9 (who was a patient with brain swelling.).

The mean value of MTF versus the mean value of RAP coefficients are shown in Fig. 2. A linear correlation of $r=0.82$ ($p<0.0001$) has been estimated to be exist between these values. For a RAP index scored lower than 0.7, MTF below 0.1 are most probable while for the larger values of RAP index, MTF values larger than 0.1 are observed. In this figure patients who suffered from brain swelling have been marked with squares. Note how all these values are concentrated in the right side of the window with high mean values of both MTF and RAP coefficient, with three of them having the largest mean values of MTF. Although five members of the non-brain swelling patients have low mean values concentrated in the left side of the window, the other six show mean values of MTF and RAP coefficient above 0.7 and 0.1 respectively, which could be suggestive of a poor pressure–volume compensatory reserve in these patients. However, there is no clinical proof of brain swelling for this group.

Figure 3 represents the box plot of MTF and RAP coefficients over only 1 h time segment for two groups of patients with and without brain swelling (left and right panels respectively). In this part of study, only the data from five patients without brain swelling who had the least amount of MTF and RAP index (see Fig. 2) were used. For patients with brain swelling who had decompressive craniectomy, MTF and RAP index were extracted just before the surgery. As observed in the left panels, all the patients with brain swelling had the median of RAP index above 0.8 and the median of MTF above 0.13, with the

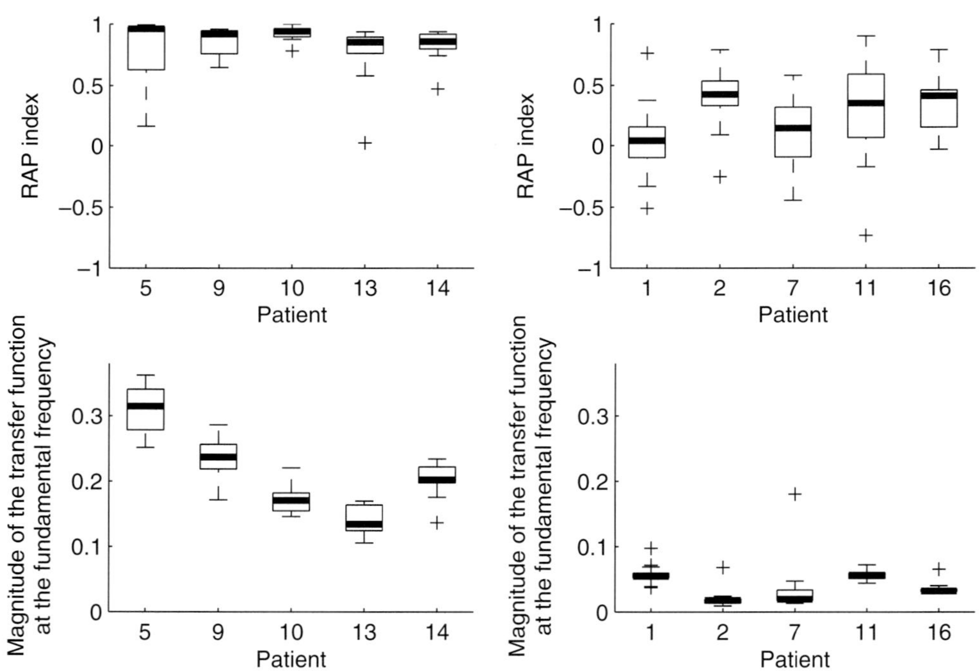

Fig. 3 Boxplot of MTF and RAP coefficients for two groups of swelling and non-swelling brain patients over 1 h time segment. Each box has lines at the lower quartile, median (*highlighted line within the box*), and upper quartile values. The *lines extending from each end of the box* are whiskers to show the extent of the rest of the data. The data shown with + are outliers with values beyond the ends of the whiskers. In patients with brain swelling RAP index and MTF show the median above 0.8 and 0.13 consequently and even the range of MTF is positioned above 0.1 (*left panels*). Note how concentrated the RAP coefficients are (except patient no. 5). Patient without brain swelling have median of RAP index and MTF below 0.5 and 0.1 respectively (*right panels*). Rap index has a broad range of values while MTF is completely concentrated

range of MTF spread above 0.1. However, comparing the box plot of RAP index with the box plot of MTF, it can be recognized that even though all the five patients have a similar median of RAP index, MTF is quite different. For example, both patient 5 and patient 10 have a median RAP index around 0.95, while they have the completely different medians and ranges for MTF. This dissimilarity might be due to the fact that during RAP index calculation, AMP and ICP_{mean} are normalized to their standard deviations within the individual time segment (here 4 min time segment) and therefore there is not any reference to compare the amount of the changes in AMP and ICP_{mean}, and consequently RAP index, over the different time intervals. On the other hand, RAP index just indicates the synchronization degree among AMP and ICP_{mean} and does not care about their actual values. Therefore, it is possible to have the same RAP index for two different time segments, while ICP and AMP are not the same. In such cases depending on the simultaneous changes in ABP waveform, MTF may change. Coming back to the right panels which refer to the patients without brain swelling, all patients show the median and range of MTF below 0.1 and the median of RAP below 0.5. Note how RAP coefficients are distributed over a wide range, while MTF parameters are completely concentrated.

Discussion

Both MTF and RAP coefficients could experience large variations during a short time interval. Today, the only way to use the information conveyed by RAP coefficients is averaging over a reasonable time window, which means one RAP coefficient by itself does not provide any reliable evaluation of the pressure–volume compensatory reserve. Furthermore, the raw Fast Fourier Transform is not a consistent estimation of the spectrum either. Under this circumstance, it is not possible to find a way to directly describe the underlying connection between each RAP coefficient and the corresponding MTF. However, the above observations suggest that highly correlated changes occur in both the distributions of MTF and RAP coefficients as the brain working point moves along the pressure–volume curve. The positive correlation between the mean values of MTF and RAP coefficients, which emphasizes the increase in the magnitude of the transfer function of ABP to ICP at the fundamental cardiac frequency during the poor compensatory reserve state of the brain, is in complete agreement with previous findings [8]. Furthermore, the negative correlation among the variances of MTF and RAP index seems promising for evaluation of brain compliance by relying more on the parameter with less variability.

Assuming a threshold of 0.7 for RAP index to describe the intracranial states as a good or poor compensatory reserve, the current data suggests the magnitude of 0.1 as the corresponding threshold for MTF at the fundamental cardiac frequency. However, consideration of the contribution of higher harmonics and their relationship to the fundamental component might provide more information regarding the brain compliance.

Conclusions

It is obvious that both RAP index and MTF are the results of a heavy data reduction which can lead to the loss of a significant part of the information contained in the ICP waveform. A reliable evaluation of the intracranial compliance requires more features to be found, extracted and included. The transfer function of ABP to ICP in conjunction with time has an intrinsic ability to compare the ICP waveform with a reference such as ABP to provide a detailed sort of information regarding the amount and rate of changes in each component and even specify the relationship between different components to take the advantage of ICP waveform study. Considering this capability besides the results of current study, the transfer function of ABP to ICP seems promising to provide a multi-dimensional feature space to evaluate brain compliance.

Acknowledgements We thank Professor Bertil Rydenhag for his support and contribution. This study is supported by the Swedish Research Council.

Conflict of interest statement We declare that we have no conflict of interest.

References

1. Aboy M, McNames J, Wakeland W, Goldstein B (2005) Pulse and mean intracranial pressure analysis in pediatric traumatic brain injury. In: Poon WS, Avezaat CJJ, Chan MTV, Czosnyka M, Goh KYC, Hutchinson PJA, Katayama Y, Lam JMK, Marmarou A, Ng SCP, Pickard JD (eds) Intracranial pressure and brain monitoring XII. Springer, Wien
2. Czosnyka M, Pickard JD (2004) Monitoring and interpretation of intracranial pressure. J Neurol Neurosurg Psychiatry 75(6):813–821
3. Czosnyka M, Steiner L, Balestreri M, Schmidt E, Smielewski P, Hutchinson PJ, Pickard JD (2005) Concept of "true ICP" in monitoring and prognostication in head trauma. In: Poon WS, Avezaat CJJ, Chan MTV, Czosnyka M, Goh KYC, Hutchinson PJA, Katayama Y, Lam JMK, Marmarou A, Ng SCP, Pickard JD (eds) Intracranial pressure and brain monitoring XII. Springer, Wien
4. Czosnyka M, Smielewski P, Timofeev I, Lavinio A, Guazzo E, Hutchinson P, Pickard JD (2007) Intracranial pressure: more than a number. J Neurosurg Focus 22(5):1–7
5. Langfitt TW, Weinstein JD, Kassell NF, Gagliardi LJ (1964) Transmission of increased intracranial pressure. II. Within the supratentorial space. J Neurosurg 21:998–1005
6. Lofgren J, von Essen C, Zwetnow NN (1973) The pressure–volume curve of the cerebrospinal fluid space in dogs. Acta Neurol Scand 49:557–574
7. Marmarou A (1973) A theoretical and experimental evaluation of the cerebrospinal fluid system. PhD thesis, Drexel University
8. Piper IR, Miller JD, Dearden NM (1990) System analysis of cerebrovascular pressure transmission: an observational study in head-injured patients. J Neurosurg 73:871–880

Acta Neurochir Suppl (2008) 102: 15–20

Cranioplasty effect on the cerebral hemodynamics and cardiac function

Yoo-Dong Won · Do-Sung Yoo · Ki-Tae Kim · Suck-Gu Kang · Sang-Bock Lee · Dal-Soo Kim · Seong-Tai Hahn · Pil-Woo Huh · Kyung-Suck Cho · Chun-Kun Park

Abstract

Background Cranioplasty is usually performed for aesthetic, protective and patient comfort reasons. The objective of this study is to examine the effects of cranioplasty on the cerebral hemodynamics and cardiovascular system.

Methods Twenty-seven patients who had undergone cranioplasty after extensive skull bone removal to prevent uncontrollable intracranial hypertension were included in this study. Arterial blood flow velocities in the middle cerebral artery (MCA) and internal carotid artery (ICA) were assessed by transcranial doppler (TCD). The cardiac functions were evaluated using the echocardiogram. And cerebral blood flow were measured by perfusion CT.

Findings The blood flow velocity at the MCA ipsilateral to the cranioplasty was decreased from 50.5±15.4 cm/s preoperative to 38.1±13.9 cm/s following cranioplasty ($p<$ 0.001) and from 33.1±8.3 cm/s to 26.4±6.6 cm/s at the ICA ($p<0.001$). The stroke volume was increased from 64.7±18.3 ml/beat, to 73.3±20.4 ml/beat ($p<0.001$), while the cardiac output and mean arterial blood pressure were unchanged. The cerebral blood flow was increased from 39.1±7.2 ml/100g/min to 44.7±8.9 ml/100g/min on the cranioplasty side (P=0.05).

Conclusions Cranioplasty can get rid of the atmospheric pressure on the brain and increase the cerebral blood flow as well as improve the cardiovascular functions. A skull defect should be corrected, because cranioplasty has not only aesthetic or protective effects but also improves the cardiovascular functions.

Keywords Cranioplasty · Stroke volume · Atmospheric pressure · craniectomy

D.-S. Yoo (✉) · S.-G. Kang · S.-B. Lee · D.-S. Kim · P.-W. Huh · K.-S. Cho · C.-K. Park
Department of Neurosurgery, Uijeongbu St. Mary's Hospital, College of Medicine, The Catholic University of Korea, Seoul, Korea
e-mail: yooman@catholic.ac.kr

Y.-D. Won · K.-T. Kim · S.-T. Hahn
Department of Radiology, Uijeongbu St. Mary's Hospital, College of Medicine, The Catholic University of Korea, Seoul, Korea

Introduction

Cranioplasty is usually indicated for cosmetic or protective, but recent reports have suggested that its also has therapeutic effects [2, 4, 7, 12, 14, 16, 19]. Most neurosurgeons agree that the skull defect should be corrected after the offending pathologic problems had been resolved. But in patients with a poor neurological status, neurosurgeons should consider the surgical risks.

We hypothesized that, in patients with skull defect, the atmospheric pressure compresses the brain and this may have effects on the cardiovascular functions as well as the cerebral hemodynamics. This study was designed to investigate the cerebral blood flow velocity, cerebral blood flow and the cardiac function before and after cranioplasty.

Patients and methods

Patient population

Twenty-seven patients with skull bone defect larger than 100 cm^2 were included in this study. Clinical data on the patients are presented in Table 1. All patients underwent bilateral (17 cases) or unilateral (ten cases) craniectomy. After the offending pathologic problems had been resolved, cranioplasty with freeze autologous bone were performed.

Table 1 Clinical data and outcome obtained in 27 patients who underwent cranioplasty

Case No.	Age (years), sex	Reason for craniectomy	Laterality of craniectomy	Time Lapse (weeks)	GCS score cranioplasty		Changed neurologic symptoms
					Before	After	
1	44, F	Massive brain swelling due to Rt. contused ICH	Bilateral	6	6	6	No change in neurologic condition
2	43, M	Rt. acute SDH with brain swelling	Bilateral	6	14	15	Improved headache and dizziness
3	61, F	Rt. acute SDH with brain swelling	Bilateral	8	8	10	Brisk eye opening
4	65, M	Brain swelling post-SAH, Lt. A-comm. aneurysm	Bilateral	8	8	10	Brisk eye opening
5	26, M	Rt. acute SDH with brain swelling	Right	6	15	15	Improved restlessness
6	52, F	Lt. putaminal H-ICH	Left	6	15	15	Improved right hemiparesis, paresthesia
7	57, F	Vasospasm post-SAH, Lt. A-comm. aneurysm	Bilateral	16	8	8	No change in neurologic condition
8	40, M	Rt. acute SDH with brain swelling	Bilateral	5	6	8	Brisk eye opening
9	16, M	Lt. EDH	Left	12	6	8	Brisk eye opening
10	62, F	Vasospasm post-SAH, Lt. MCA aneurysm	Left	5	12	14	Improved headache, orientation
11	78, F	Rt. acute SDH with brain swelling	Bilateral	3	6	6	No change in neurologic condition
12	25, M	Massive brain swelling due to Lt. contused ICH	Bilateral	5	12	15	Improved orientation
13	63, M	Massive brain swelling due to Lt. contused ICH	Left	4	12	14	Improved orientation
14	42, M	Rt. acute SDH with brain swelling	Bilateral	14	10	12	Improved verbal response, orientation
15	46, F	Rt. putaminal H-ICH	Bilateral	13	6	6	No change in neurologic condition
16	41, F	Rt. putaminal H-ICH	Bilateral	8	8	13	Improved verbal response, orientation
17	48, M	Rt. acute SDH with brain swelling	Right	5	15	15	Improved headache, fine movement
18	57, F	Rt, putaminal H-ICH	Right	10	8	12	Improved mentality, orientation
19	62, F	Lt. acute SDH with brain swelling	Left	7	15	15	Improved restlessness
20	41, M	Vasospasm post-SAH, Rt. A-comm. aneurysm	Bilateral	8	6	6	No change in neurologic condition
21	60, F	Massive brain swelling due to Rt. contused ICH	Bilateral	8	8	13	Improved verbal response, orientation
22	66, F	Brain swelling post-SAH, Rt. MCA aneurysm	Bilateral	6	6	6	No change in neurologic condition
23	69, F	Rt. putaminal H-ICH	Bilateral	4	13	15	Improved orientation, restlessness
24	69, F	Rt. putaminal H-ICH	Bilateral	6	13	14	Improved orientation, headache
25	45, M	Brain swelling post-SAH, Lt. A-comm. aneurysm	Bilateral	10	9	12	Improved verbal response, orientation
26	19, F	Massive ICH due to Lt. parietal AVM	Left	3	14	15	Improved right hemiparesis, dizziness
27	72, M	Cerebral infarction due to Lt. ICA occlusion	Left	5	12	13	Improved right hemiparesis, orientation

Time Lapse = duration from craniectomy to crnioplasty
EDH epidural hematoma; *GCS* Glasgow coma scale; *H-ICH* hypertensive intracerebral hematoma; *ICH* intracerebral hematoma; *SDH* subdural hematoma;

Table 2 Mean values in echocardiogram before and after the cranioplasty

Case no.	Cardiac output (ml/min)		Stroke volume (ml/beat)		Heart rate (beat/min)	
	Before	After	Before	After	Before	After
1	3,942	5,154	41.1	50	96	103
2	5,849	8,789	77.9	113	75	78
3	5,601	5,081	64.4	69.6	87	73
4	6,749	7,058	61.4	65.4	110	108
5	5,111	5,512	74.1	75.5	69	73
6	7,201	6,276	84.7	89.7	85	70
7	6,588	6,334	64.6	67.9	97	98
8	4,414	4,193	58.1	67.6	76	62
9	3,590	3,281	34.9	38	103	86
10	5,627	4,415	64.6	62.6	87	66
11	8,675	9,165	92.3	119	94	77
12	6,080	6,337	78	84.5	77	75
13	8,610	7,200	106	112	81	64
14	6,328	5,390	64	67.9	99	79
15	3,169	3,690	47.3	61.7	67	60
16	5,160	5,560	57.9	69	89	80
17	4,290	5,550	47.7	68.3	90	81
18	3,340	3,650	33.9	44.9	99	81
19	7,895	7,880	84.9	98.5	93	80
20	6,800	6,500	62.1	56.6	120	104
21	6,960	6,070	80.3	74.8	93	76
22	8,750	8,520	97.2	99	90	86
23	6,150	8,240	57.8	67.6	106	122
24	8,240	6,030	65.2	67.2	126	90
25	5,228	5,450	61.5	67.3	85	81
26	3,370	3,770	55.9	49.3	68	67
27	4,580	5,110	52.4	44.5	103	98
Mean±SD	5,862±1,705	5,933±1,621	65.6±17.9	72.3±21.2	91.4±14.7	82.2±15.1

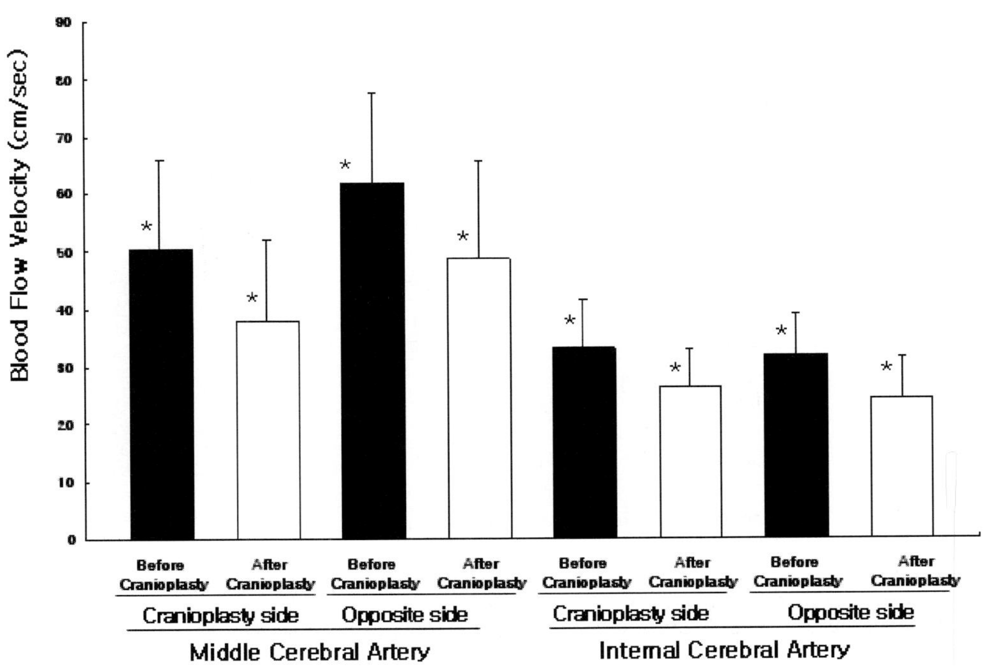

Fig. 1 Bar graph depicting the influence of cranioplasty on the cerebral blood flow velocity obtained in 27 patients (mean ±SD). The *symbols on the left of the bars* show the significance (*$p<0.05$) compared with before cranioplasty

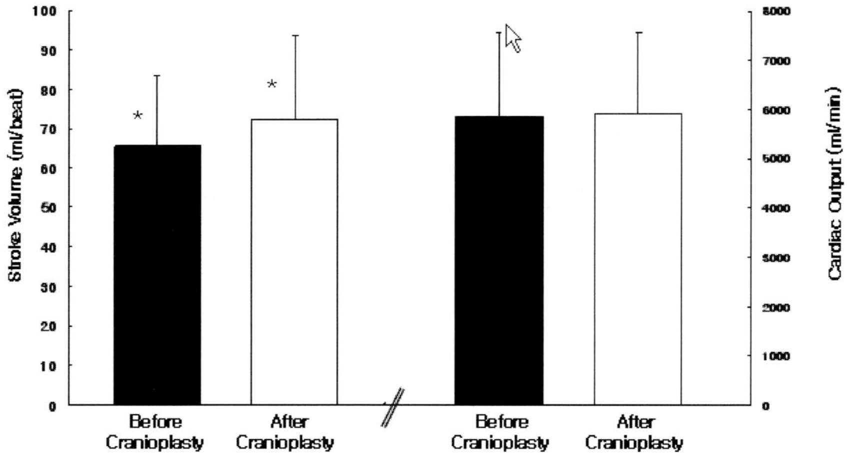

Fig. 2 Bar graph depicting the influence of cranioplasty on the stroke volume and the cardiac output (mean±SD). The *symbols on the left of the bars* show the significance (*$p<0.05$) compared with before cranioplasty

In all patients, echocardiography, cerebral blood fow velocity by transcranial Doppler and Ct perfusion cerebral blood flow studies were performed before and after the cranioplasty.

Data collection

Transcranial doppler ultrasonography (TC2020; Eden Medical Electronics, Lake Constance, Germany) was performed to determine the blood flow velocities in the both middle cerebral arteries (2mm distal from the ICA bifurcation site) and both extracranial intracranial carotid arteries (depth 46 mm from the both madibular angle). From the echocardiography (Sonos 5500; Hewlett Packard, Andover, MA, USA) routine echocardiographic values, cardiac output and stroke volume were measured (Table 2). Perfusion CT (Somatom plus 4; Siemens, Forcheim, Germany) was performed to evaluate the cerebral blood flow changes. All of these studies were performed before and 14±2 days after the cranioplasty operation.

Statistical analysis

All values are presented as the mean ± standard deviation. Comparisons between data groups were computed using Student's t-test. Statistical significance was defined as a probability value of less than 0.05.

Results

Changes of the cerebral blood flow velocity

The blood flow velocity in the MCA ipsilatral to the cranioplasty was decreased from 50.5±15.4 cm/s pre-operatively to 38.1±13.9 cm/s ($p<0.001$) post-operatively and from 33.1±8.3 to 26.4±6.6 cm/s ($p<0.001$) in the ICA. Blood flow velocity on the contralateral side was decreased from 61.9±15.7 to 48.7±16.9 cm/s ($p=0.002$) at the MCA and from 31.8±7.3 to 24.5±7.1 cm/s ($p<0.001$) at the ICA (Fig. 1).

Effect on the cardiac function and systemic blood pressure

The cardiac function evaluation with echocardiogram revealed that the stroke volume was increased from 64.7± 18.3 to 73.3±20.4ml/beat ($p<0.001$). The cardiac output showed no statistical significance ($p=0.731$) before and after the cranioplasty (Fig. 2).

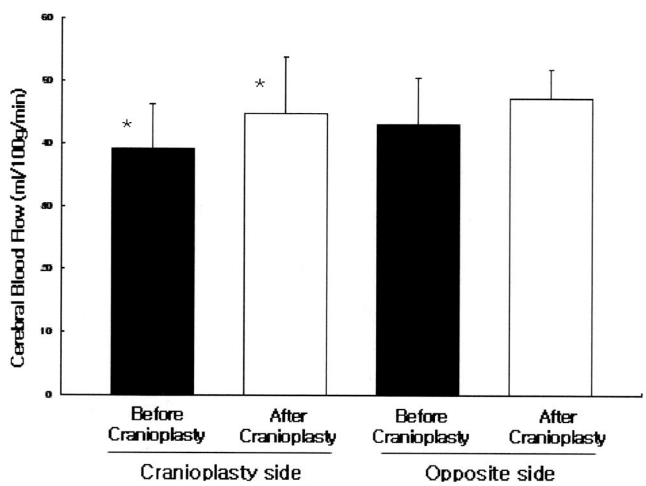

Fig. 3 Bar graph depicting the influence of cranioplasty on the cerebral blood flow (mean±SD). The *symbols on the left of the bars* show the significance (*$p<0.05$) compared with before cranioplasty

Cerebral blood flow changes

The cerebral blood flow, evaluated with perfusion CT, in terms of the hemispheric values were increased on both sides (Fig. 3). On the cranioplasty side it was increased from 39.1±7.2 to 44.7±8.9 ml/100 g/min (p=0.05), and on the contralateral side, from 42.9±7.5 to 47.2±4.4 ml/100 g/min (p=0.137).

Discussion

Accepted indications for cranioplasty are for aesthetic and for physical protection of intracranial structures. Recently, the potential improvement of neurological function should be added as an acceptable indication for cranioplasty [6–8, 12, 15, 16, 19, 24].

Gardner reported a syndrome characterized by headaches, dizziness, irritability, epilepsy, discomfort, and psychiatric symptoms that he had observed in patients with large cranial defect. He called it the "syndrome of the trephined" and was the first to describe an improvement in the neurological function of some patients who underwent cranioplasty with tantalum [4, 6–8, 10, 15]. The mechanism of "syndrome of trephined" remains controversial. These mechanisms include, cicatrical changes occurring in the cortex, dura, and skin that may exert pressure on the skull contents [6, 8, 15, 24], cerebrospinal fluid hydrodynamic changes [7–9, 15–17], the effects of atmospheric pressure on the brain cerebral blood flow changes [7, 11, 12, 16, 18, 21, 24] and disturbed cerebral energy metabolism [15, 22, 24].

Previous reports have shown, a 15 to 30% increase in cerebral blood flow in the area of cortex adjacent to the cranioplasty [5, 13, 15, 16, 21, 24]. Furthermore cranioplasty after decompressive craniectomy may increase cerebral blood flow in the ipsilateral hemisphere, as well as the contralateal hemisphere [11, 21, 24]. In our study, the cerebral blood flow on the cranioplasty side was increased as observed by others in these previous studies.

Scar tissue and atmospheric pressure produced by skull defect may increase pressure on the brain cortex and subarachnoid space [8, 16, 18, 19]. Consequently, vessels under the skull defect, with added atmospheric compression may alter their cerebral blood flow hemodynamics [5–7, 12, 15, 20, 22, 24]. Under such circumstances, cerebral blood flow velocity of the compressed vessel may increase. Following the law of Laplace, we suggest that the diameter of cerebral blood vessels may be dilated after cranioplasty, based on our observations of a decrease in blood flow velocities and an increase in cerebral blood flow following the cranioplasty procedure [23].

Cardiac output is mathematically equal to the stroke volume multiplied by the heart rate [2, 3]. Stroke volume is influenced by peripheral vascular resistance [1–3]. In echocardiogram studies, stroke volume was increased and the heart rate was decreased after cranioplasty while the cardiac output and systemic blood pressure were unchanged.

Based on the results of our study, cranioplasty may remove the compression that arise from the atmospheric pressure and scar tissue [4, 12]. As a result, the cerebral blood flow was increased and cardiac function improved following cranioplasty. Therefore cranioplasty should be performed even on the poor neurologic patients, because cranioplasty has not only aesthetic and skull protective effects but it also improves the systemic cardiovascular and cerebral hemodynamic functions.

Conflict of interest statement We declare that we have no conflict of interest.

References

1. Anton H (1980) Function of the heart. In: Schmidt RF, Thews G (eds) Human physiology. Springer, New York, pp 358–96
2. Beumer J, Firtell DN, Curtis TA (1979) Current concept in cranioplasty. J Prosthet Dent 42:67–77
3. Carter BS, Ogilvy CS, Candia GJ et al (1997) One-year outcome after decompressive surgery for massive nondominant hemispheric infarction. Neurosurgery 40:1168–1176
4. Delashaw JB Jr, Persing JA (1994) Repair of cranial defects. In: Youmans JR (ed) Neurosurgical surgery. 4th edn. WB Saunders, Philadelphia, pp 1853–1864
5. Dujovny M, Agner C, Aviles A (1999) Syndrome of the trephined: theory and facts. Crit Rev Neurosurg 9:271–278
6. Dujovny M, Aviles A, Agner C et al (1997) Cranioplasty: cosmetic or therapeutic? Surg Neurol 47:238–241
7. Dujovny M, Fernandez P, Alperin N et al (1997) Post-cranioplasty cerebrospinal fluid hydrodynamic changes: magnetic resonance imaging quantitative analysis. Neurol Res 19:311–316
8. Fodstad H, Ekstedt J, Fridén H (1979) CSF hydrodynamic studies before and after cranioplasty. Acta Neurochir Suppl (Wien) 28:514–518
9. Fodstad H, Love JA, Ekstedt H et al (1984) Effect of cranioplasty on cerebrospinal fluid hydrodynamics in patients with the syndrome of the trephined. Acta Neurochir (Wien) 70:21–30
10. Grantham EG, Landis HP (1948) Cranioplasty and the post-traumatic syndrome. J Neurosrug 15:19–22
11. Maekawa M, Awaya S, Teramoto A (1999) Cerebral blood flow (CBF) before and after cranioplasty performed during the chronic stage after decompressive craniectomy evaluated by xenon-enhanced computerized tomography (Xe-CT) CBF scanning. No Shinkei Geka 27:717–7 22, (In Japanese)
12. Matsumura H, Shigehara K, Ueno T et al (1996) Cranial defect and decrease in cerebral blood flow resulting from deep contact burn of the scalp in the neonatal period. Burns 22:560–565
13. Richaud J, Boetto S, Guell A et al (1985) Effects of cranioplasty on neurological function and cerebral blood flow. Neurochirugie 31:183–188
14. Sanan A, Haines SJ (1997) Repairing holes in the head: a history of cranioplasty. Neurosurgery 40:588–603
15. Schiffer J, Gur R, Nisim U et al (1997) Symptomatic patients after craniectomy. Surg Neurol 47:231–237

16. Segal DH, Oppenheim JS, Murovic JA (1994) Neurological recovery after cranioplasty. Neurosurgery 34:729–31
17. Shapiro K, Fried A, Takei F et al (1985) Effect of the skull and dura on neural axis pressure–volume relationships and CSF hydrodynamics. J Neurosurg 63:76–81
18. Stula D (1985) Intracranial pressure measurement in large skull defects. Neuorchirurgia (Stuttg) 28:164–169, (In Germany)
19. Stula D, Muller HR (1980) Cranioplasty after extensive decompressive craniotomy with displacement of the cerebral hemisphere. CT analysis Neurochirurgia (Stuttg) 23:41–46
20. Suga H (1991) Physiological interpretation of negative circumferential tension in vascular walls. Jpn Heart J 32:473–80
21. Suzuki N, Suzuki S, Iwabuchi T (1993) Neurological improvement after cranioplasty. Analysis by dynamic CT scan. Acta Neurochir (Wien) 122:49–53
22. Winkler PA, Stummer W, Linke R et al (2000) Influence of cranioplasty on postural blood flow regulation, cerebrovascular reserve capacity, and cerebral glucose metabolism. J Neurosurg 93:53–61
23. Witzleb E (1980) Function of the vascular system. In: Schmidt RF, Thews G (eds) Human physiology. Springer, New York, pp 397–455
24. Yoshida K, Furuse M, Izawa A et al (1996) Dynamics of cerebral blood flow and metabolism in patients with cranioplasty as evaluated by 133Xe CT and 31P magnetic resonance spectroscopy. J Neurol Neurosurg Psychiatry 61:166–171

Acta Neurochir Suppl (2008) 102: 21–24

Hyperbaric oxygen therapy for consciousness disturbance following head injury in subacute phase

Takehiro Nakamura · Yasuhiro Kuroda · Susumu Yamashita · Kenya Kawakita · Nobuyuki Kawai · Takashi Tamiya · Toshifumi Itano · Seigo Nagao

Abstract

Background Hyperbaric oxygen (HBO) therapy has been shown to improve outcome after brain injury, however its mechanisms are not understood. The purpose of the present study was to investigate the effect of hyperbaric oxygen (HBO) therapy on the cerebral circulation and metabolism of patients with disturbances in consciousness after head injury in the subacute phase.

Methods Seven head injury patients underwent HBO treatment after leaving the intensive care unit. Oxygen (100%O_2, 2.7 atm absolute) was delivered to patients in a hyperbaric chamber for 60 min every 24 h (total five treatments/patient). Cerebral circulation monitoring (mean flow velocity: mFV, and pulsatility index: PI at horizontal portion of middle cerebral artery by transcranial Doppler) and cerebral metabolism monitoring (arterio-jugular venous difference of oxygen: $AJDO_2$ and jugular venous lactate: lac-JV) before and after the series of treatments were evaluated.

Findings Both PI and lac-JV were significantly decreased after HBO theatment, while there were no significant changes in mFV and $AJDO_2$. The decreased PI and lac-JV after HBO therapy might indicate that this treatment couples cerebral circulation and metabolism.

Conclusions The measurement of cerebral circulation and metabolism parameters, especially PI and lac-JV, is useful for estimation of effect of HBO therapy in patients with distubances in consciousness after head injury in the subacute phase.

Keywords Hyperbaric oxygen therapy · Consciousness disturbances · Subacute phase · Jugular venous lactate

T. Nakamura (✉) · T. Itano
Department of Neurobiology, Faculty of Medicine, Kagawa University, 1750-1 Ikenobe, Miki, Kita, Kagawa 761-0173, Japan
e-mail: tanakamu@kms.ac.jp

T. Itano
e-mail: toshi@kms.ac.jp

T. Nakamura · N. Kawai · T. Tamiya · S. Nagao
Department of Neurological Surgery, Faculty of Medicine, Kagawa University, 1750-1 Ikenobe, Miki, Kita, Kagawa 761-0173, Japan

N. Kawai
e-mail: nobu@kms.ac.jp

T. Tamiya
e-mail: tamiya@kms.ac.jp

S. Nagao
e-mail: snagao@kms.ac.jp

Y. Kuroda · S. Yamashita · K. Kawakita
Department of Emergency and Critical Care Medicine, Faculty of Medicine, Kagawa University, 1750-1 Ikenobe, Miki, Kita, Kagawa 761-0173, Japan

Y. Kuroda
e-mail: kuroday@kms.ac.jp

S. Yamashita
e-mail: sum-ygc@umin.ac.jp

K. Kawakita
e-mail: kenfact@kms.ac.jp

Introduction

Evidence on hyperbaric oxygen (HBO) therapy for severe head injury is insufficient to prove whether it is effective or ineffective [4]. Furthermore, there have been few studies in which the mechanisms underlying HBO therapy have been investigated in patients with disturbed consciousness following severe head injury. The purpose of the present study was to investigate the effect of HBO therapy on cerebral circulation and metabolism in patients with consciousness disturbance after head injury in the subacute phase.

Materials and methods

Patients

Between October 2001 and September 2002, seven head injury patients who suffered prolonged disturbances in consciousness even after acute therapy in the intensive care unit in the Kagawa Medical University Hospital were considered for this study (Table 1). All patients had a severe head injury and had a Glasgow Coma Scale (GCS) of less than or equal to 8 at admission (four patients with diffuse axonal injury and three patients with acute subdural hematoma). Mean patients age was 37 years, with a range of 9–76 years. As an acute therapy, hypothermia therapy was induced in all patients as soon as possible after admission. If necessary, patients underwent craniotomy for hematoma evacuation. After acute treatment, all patients were able to spontaneously breathe and did not require artificial ventilation. However, there were prolonged disturbances in consciousness. Patient outcome was judged at leaving hospital according to Glasgow Outcome (GOS).

Hyperbaric oxygen therapy

The patients entered into the study received HBO treatments while in a monoplace HBO chamber (Sechrist Industries, Anaheim, CA, USA). During the entire HBO treatment, oxygen (100% O_2, compression to 2.7 atm absolute) was delivered to patients in a hyperbaric chamber for 60 min every 24 h (total five treatments/patient).

Table 1 Summary of patients

	Age	Gender	Injury	Cons level of pre-HBO	Outcome (GOS) at leaving hospital
Case 1	66	M	ASDH	GCS 9	Unfavorable (SD)
Case 2	76	F	DAI	GCS 11	Unfavorable (SD)
Case 3	33	M	DAI	GCS 11	Favorable (MD)
Case 4	32	F	ASDH	GCS 8	Unfavorable (SD)
Case 5	19	M	DAI	GCS 10	Favorable (MD)
Case 6	24	F	DAI	GCS 13	Favorable (GR)
Case 7	9	F	ASDH	GCS 9	Favorable (MD)

HBO Hyperbaric oxygen, *F* female, *M* male, *ASDH* acute subdural hematoma, *DAI* diffuse axonal injury, *GCS* Glasgow Coma Scale, *GOS* Glasgow Outcome Scale, *SD* severe disability, *MD* moderate disability, *GR* good recovery

Monitoring

Cerebral circulation monitoring (mean flow velocity: mFV, and pulsatility index: PI) and cerebral metabolism monitoring (arterio-jugular venous difference of oxygen: $AJDO_2$ and jugular venous lactate: lac-JV) before and after a series of treatments was evaluated. A transcranial Doppler system (Neuroguard, Medasonics, Fremont, CA, USA) was used to insonate the horizontal portion of the middle cerebral artery via the transtemporal window. The mFV was registered, and the PI was determined according to the formula [PI=(systolic velocity−diastolic velocity)/mean velocity]. Arterial and jugular venous blood samples were obtained at the same time for determination of oxygen saturation and lactate level. The $AJDO_2$ was calculated by multiplying the difference between arterial (SaO_2) and jugular venous (SjO_2) by the hemoglobin concentration (Hb) and 1.34, divided by 100: [$AJDO_2=1.34Hb(SaO_2-SjO_2)/100$].

Statistical analysis

Differences between the levels of each monitoring parameter (mFV, PI, $AJDO_2$ and lac-JV) before and after the set of HBO treatments were analysed by unpaired *t*-test. The relationship between values of PI and levels of lac-JV was examined using linear regression. Differences were considered statistically significant at $P<0.05$.

Results

Cerebral circulation monitoring

To assess the effects of hyperbaric oxygen therapy on the cerebral circulation, mFV and PI were measured before and after therapy (Fig. 1a,b). After HBO treatment, there was a significant decrease in PI (1.07±0.26 vs. 0.73±0.19, $P<0.05$), whilst there were no significant changes in mFV.

Cerebral metabolism monitoring

To assess the effects of HBO therapy on cerebral metabolism, $AJDO_2$ and lac-JV were measured before and after therapy (Fig. 1c,d). The levels of lac-JV were significant decreased after HBO treatment (9.5±3.1 mg/dl vs. 6.3±2.5 mg/dl, $P<0.05$). There were no significant changes in $AJDO_2$.

Relation of cerebral circulation and metabolism

Figure 2 shows a linear regression of PI (cerebral circulation) versus lac-JV (cerebral metabolism), including

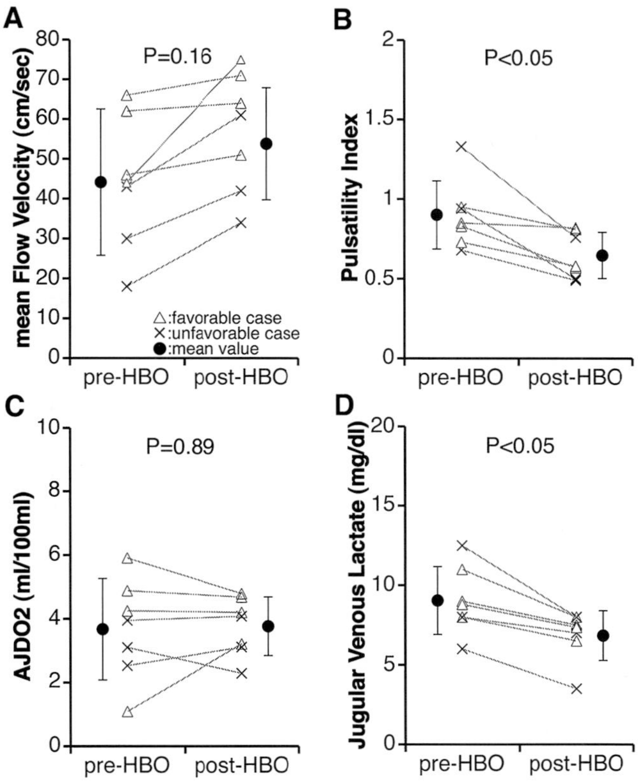

Fig. 1 Evaluation of the effects of hyperbaric oxygen therapy on the cerebral circulation and metabolism. Mean flow velocity (**a**) and pulsatility index (**b**) at horizontal portion of the middle cerebral artery using transcranial Doppler before and after HBO treatment. Arterio-jugular venous difference of oxygen ($AJDO_2$, **c**) and jugular venous lactate (**d**) before and after HBO treatment

values both before and after HBO treatments. The PI and lac-JV were significantly related (r=0.768, P<0.01).

Discussion

The precise mechanisms by which HBO therapy may affect severe head injury have not been fully elucidated. Cerebral monitoring is one means to clarify the mechanism and pathophysiology [7, 8]. In the present study, mFV and PI were performed to monitor cerebral circulation, while $AJDO_2$ and lac-JV were measured to monitor cerebral metabolism.

Effect of HBO on cerebral circulation

A previous study showed that HBO treatment leads to improvements in cerebral blood flow (CBF) as measured by SPECT [2]. Another study, using the nitrous oxide saturation method, demonstrated that the effects of HBO therapy on CBF is complicated, with low CBF increasing and high CBF decreasing [10]. Instead of these measurements of CBF, the current study performed cerebral circulation monitoring using TCD. In general, TCD is easily performed at the patient's bedside allowing for determination of mFV and PI. Moreno et al. showed that PI may be predictive of outcome and correlates significantly with intracranial pressure and cerebral perfusion pressure in patients with severe head injury [6]. In the present study, there was significant correlation of PI and lac-JV. This result might suggest that HBO treatment couples cerebral circulation and metabolism.

Effect of HBO on cerebral metabolism

When cerebral metabolic rate of oxygen ($CMRO_2$) and CBF are normally coupled, the ratio between them ($AJDO_2=CMRO_2/CBF$) does not change [9]. In the present study, there were no significant changes in $AJDO_2$ before and after HBO treatment. Rockswold et al. [10] also showed that $AJDO_2$ was unaffected by HBO therapy in patients with severe head injury. These findings suggest that HBO might normalize the coupling of cerebral circulation and metabolism in severe head injury. This is also supported by the correlation between PI and lac-JV in the present study.

Most patients with severe head injury have increased lactate production, as measured by microdialysis [3]. In our study, high levels of lac-JV were observed before HBO treatment. Increased lactate production indicates an anaerobic metabolic status caused by a lack of oxygen and mitochondria damage [1]. Cerebrospinal fluid lactate levels were consistently decreased by HBO treatment [10]. Mean brain tissue lactate levels decreased by 40% in oxygen-treated patients who suffered from head injury [5]. The present study demonstrates that HBO treatments significantly reduced lac-JV levels, regardless of the

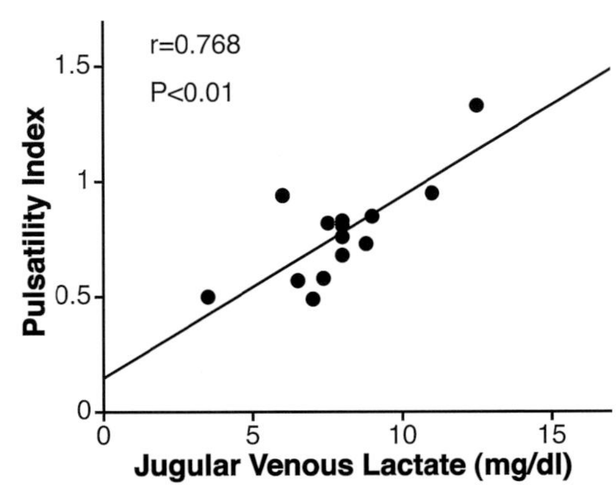

Fig. 2 Linear regression of pulsatility index versus levels of jugular venous lactate, including values both before and after hyperbaric oxygen therapy

patient outcome. The decreased lactate levels after HBO treatment might indicate that HBO could improve aerobic metabolism.

Conclusions

The decreased PI and lac-JV after HBO therapy might indicate that this treatment couples cerebral circulation and metabolism. The measurement of cerebral circulation and metabolism parameters, especially PI and lac-JV, may be useful for estimation of the effects of HBO therapy in patients with disturbed consciousness after head injury in subacute phase.

Acknowledgments The authors thank Prof. Shinji Ogra (Gifu University Hospital) for excellent technical assistance. We also thank Prof. Richard F. Keep (University of Michigan) for providing useful suggestions.

Conflict of interest statement We declare that we have no conflict of interest.

References

1. DeSalles AAF, Muizelaar JP, Young HF (1987) Hyperglycemia, cerebrospinal fluid lactatic acidosis, and cerebral blood flow in severely head-injured patients. Neurosurgery 21:45–50
2. Golden ZL, Neubauer R, Golden CJ, Greene L, Marsh J, Mleko A (2002) Improvement in cerebral metabolism in chronic brain injury after hyperbaric oxygen therapy. Int J Neurosci 112:119–131
3. Goodman JC, Valadka AB, Gopinath SP, Uzura M, Robertson CS (1999) Extracellular lactate and glucose alterations in the brain after head injury measured by microdialysis. Crit Care Med 27:1965–1973
4. McDonagh M, Helfand M, Carson S, Russman BS (2004) Hyperbaric oxygen therapy for traumatic brain injury: a systematic review of the evidence. Arch Phys Med Rehabil 85:1198–1204
5. Menzel M, Doppenberg EM, Zauner A, Soukup J, Reinert MM, Bullock R (1999) Increased inspired oxygen concentration as a factor in improved brain tissue oxygenation and tissue lactate levels after severe human head injury. J Neurosurg 91:1–10
6. Moreno JA, Mesalles E, Gener J, Tomasa A, Ley A, Roca J, Fernandez-Llamazares J (2000) Evaluating the outcome of severe head injury with transcranial Doppler ultrasonography. Neurosurg Focus 8:e8
7. Nakamura T, Nagao S, Kawai N, Honma Y, Kuyama H (1998) Significance of multimodal cerebral monitoring under therapeutic hypothermia severe head injury. Acta Neurochir (Suppl) 71:85–87
8. Nakamura T, Tatara N, Morisaki K, Kawakita K, Nagao S (2002) Cerebral oxygen metabolism monitoring under hypothermia for severe subarachnoid hemorrhage: report of eight cases. Acta Neurol Scand 106:314–318
9. Robertson CS, Narayan RK, Gokaslan ZL, Pahwa R, Grossman RG, Caram P Jr, Allen E (1989) Cerebral arteriovenous oxygen difference as an estimate of cerebral blood flow in comatose patients. J Neurosurg 70:222–230
10. Rockswold SB, Rockswold GL, Vargo JM, Erickson CA, Sutton RL, Bergman TA, Biros MH (2001) Effects of hyperbaric oxygenation therapy on cerebral metabolism and intracranial pressure in severely brain injured patients. J Neurosurg 94:403–411

Acta Neurochir Suppl (2008) 102: 25–28

Gender-related differences in intracranial hypertension and outcome after traumatic brain injury

Marek Czosnyka · Danila Radolovich · Marcella Balestreri · Andrea Lavinio · Peter Hutchinson · Ivan Timofeev · Peter Smielewski · John D. Pickard

Abstract

Background We have previously studied possible gender-related differences in intracranial hypertension and outcome following head injury. The results were always close to the limit of significance so that to achieve greater statistical power we have continued recruitment of patients for further 5 years.
Methods Head injury patients (612) who were sedated and ventilated were studied from 1992–2007. All had intracranial pressure (ICP), arterial blood pressure (ABP) and cerebral perfusion pressure (CPP) monitored continuously. Patients' outcome was assessed at 6 months post-injury (469 were available for follow-up).
Results This retrospective analysis enrolled 98 females and 371 males. Males and females were well matched for age (mean±standard deviation: 33±17 and 33±16 years respectively) and the initial median Glasgow Coma Score (GCS) [females and males 6]. The difference in mortality rate between sexes was age-related. In the subgroup of patients younger than 50 years mortality was 17% in males and 29% in females (p=0.026), whereas there was no difference above 50 years (around 40% both males and females). Mean ICP, CPP and ABP were not different between males and females. However, cerebrovascular pressure reactivity was found to be significantly worse in females than in males in the age group below 50 years (PRx; males: 0.044±0.031; females 0.11±0.047; p<0.05).
Conclusion Following head injury females younger than 50 have a significantly worse pressure reactivity and greater mortality rate than males of the same age.

Keywords Traumatic brain injury · Outcome · Intracranial pressure · Gender · Head injury · Pressure reactivity

Recent clinical studies investigating gender-related differences in patients following traumatic brain injury (TBI) have been inconsistent [5, 8], particularly in reference to the relatively simple question: do women have the same, better, or worse outcome compared to men? Should such a difference exist and be attributable to a specific cause, this might translate into gender-dependant management strategies.

The possible neuroprotective effect of oestrogen and progesterone have been highlighted in the past [10] with numerous experimental studies conducted in the 1990's in rodent models of stroke [1], and head injury [4]. However, in clinical practice, studies show conflicting results [5, 8]. Although it has been recently demonstrated that hormonally active women have a better physiologic and haemodynamic response to shock and trauma than do their male counterparts [7], other factors such as the peculiar predisposition to brain swelling of young females [8] might outweigh the theoretical neuroprotective effects of oestrogen.

We have recently published a short letter [3] commenting on the paper by Farin et al. [8] who demonstrated gender-related differences in head-injured patients. We have now

Source of support: UK Government Technology Foresight Initiative, and the Medical Research Council (Grant No G9439390 ID 65883).

M. Czosnyka (✉) · D. Radolovich · M. Balestreri · P. Hutchinson · I. Timofeev · P. Smielewski · J. D. Pickard
Academic Neurosurgical Unit, Box 167, Addenbrooke's Hospital, Cambridge CB20 QQ, UK
e-mail: MC141@medschl.cam.ac.uk

D. Radolovich · M. Balestreri
Department of Anaesthesiology, University of Pavia, Policlinica San Matteo, 27100 Pavia, Italy

A. Lavinio
Department of Anaesthesia, University of Brescia, Spedali Civili, 25100 Brescia, Italy

explored the effect of gender in a larger group of 612 head injured patients monitored over the past 15 years in a UK regional neurosurgical unit with special attention paid to intracranial pressure (ICP), cerebral perfusion pressure (CPP), cerebrovascular pressure-reactivity (PRx) and outcome following TBI.

Materials and methods

This retrospective analysis is based on 612 TBI patients admitted to the Neurosciences Critical Care Annexe and Neurosciences Critical Care Unit (NCCU) at Addenbrooke's Hospital, Cambridge, UK between January 1992 and April 2007. Only patients with invasive monitoring of ICP and ABP for more than 12 h and connected to a bedside computerized system were included in the study. It is important to emphasize that the studied group is not representative of all NCCU head injured patients but includes approximately 44% of all head trauma admissions. Patients who were admitted and discharged promptly, who died soon after admission or in whom ICP was not monitored were not included in the analysis.

Patients were sedated, paralysed and mechanically ventilated in order to maintain the ICP below 20–25 mmHg. Systemic hypotension was treated with fluids and inotropes. CPP was maintained above 70 mmHg to avoid secondary ischaemic injury. Episodes of intracranial hypertension were treated with mild hyperventilation ($PaCO_2$ 4.0–4.5 kPa), boluses of mannitol, moderate hypothermia, thiopentone and CSF drainage depending on the size of ventricles. A CPP-oriented therapy protocol for head injury introduced in 1997 included a more aggressive management of intracranial hypertension and a stricter control of ABP aiming to minimize the detrimental effect of hypoperfusion on brain tissue [9].

ICP was monitored using an intraparenchymal probe (Camino ICP transducer: 12 patients or Codman ICP MicroSensors: 566 patients) or through a ventricular drain and an external pressure transducer (Baxter: 34 patients, pre-1994). ABP was monitored invasively (indwelling arterial catheter in the radial artery). Signals were sampled from the analogue output of the monitors, digitised and subsequently analysed as 8-s time averages [11]. From each 40 samples of mean ICP and ABP, the moving correlation coefficient was calculated (named PRx-pressure-reactivity index). A positive correlation between ABP and ICP reveals a passive, non-reactive cerebrovascular bed. A negative correlation is specific for a well reactive bed. A positive value of PRx has been previously demonstrated to be a strong predictor of fatal outcome following head injury [2].

Time trends were analysed and global averaged ICP, ABP, CPP and PRx for each patient were recorded. Post-resuscitation Glasgow Coma Scale Score (GCS) was used for analysis. The Glasgow Outcome Score (GOS) was determined at 6 months (either by follow-up clinic or by telephonic questionnaire). Patients (469) were available for follow up. For the data analysis, subjects were subdivided into two age categories with a cut-off point of 50 years. This follows recently published work [8] which suggested that gender-related differences disappear after the onset of menopause.

Routine clinical and brain monitoring data, collected and approved by the multidisciplinary local Neuro Critical Care Users Group, were retrospectively analysed as a part of a routine clinical audit.

Results

Ninety-eight females and 371 males with known outcome were included in this analysis. Gender subgroups were matched for age (mean age±SD females: 34±16.5 years; males: 34±17 years) and GCS at admission (median GCS was 6 both in males and females).

In patients younger than 50 years, the mortality rate (at 6 months) was 17% in males and 29% in females ($N=344$; $p=0.026$). In patients above 50 mortality rate was around 43% for both males and females.

There was no obvious difference in males and females above and below 50 years of age in mean ICP, ABP and CPP. However, cerebrovascular pressure-reactivity was found to be worse in females than in males (PRx; males: 0.044±0.031; females 0.11±0.047; $p<0.05$) in the age group <50 years. No such difference was observed in patients older than 50.

In younger patients with disturbed reactivity (PRx>0.3) gross intracranial hypertension (mean ICP>25 mmHg) was observed in 60% of females and only in 20% of males ($p<0.05$).

Discussion

In the group of TBI patients younger than 50 years of age, females had a significantly greater rate of fatal outcome than males. This was associated with a worse pressure reactivity in females compared to males. These differences disappeared in older patients. One practical implication may be that any possible management oriented on restoration of pressure reactivity should be administered more readily in young females than in males, as disturbed reactivity may lead to refractory intracranial hypertension and implicate fatal outcome in pre-menopausal females.

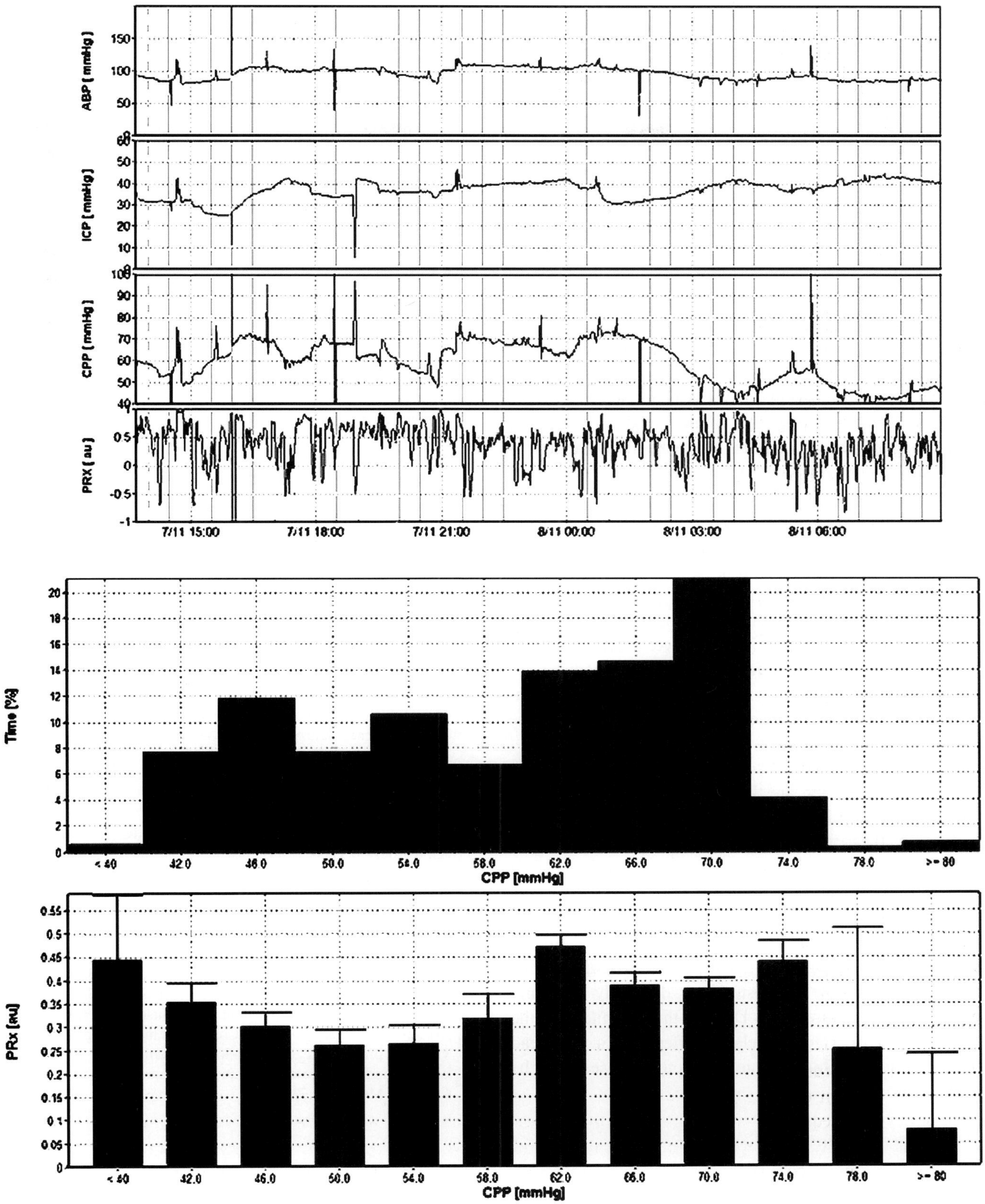

Fig. 1 Typical monitoring of ICP, ABP, CPP and pressure-reactivity (PRX) in a young girl with severe head injury (GCS 3) who was admitted with initial ICP close to 30 mm Hg, CPP around 60 mmHg. Pressure reactivity was from the very beginning disturbed (PRx>0.2). Although CPP was stabilized at 70 mm Hg soon after admission, PRx did not improve and patient died 15 h later. In the bottom part of the figure, so called 'optimal CPP' [12] graph was plotted. It consists of a histogram of CPP and distribution of PRx versus CPP. In this particular case 'optimal' CPP did not exist. PRx plotted against mean CPP shows almost flat distribution with all values greater than 0.2

In spite of the postulated neuroprotective role of oestrogen in experimental studies, the susceptibility to brain swelling reported in the recently published clinical audit [8] is very likely to be the predominant factor in determining outcome following traumatic brain injury. This study also confirms that, after the successful introduction of a CPP-oriented therapy, cerebrovascular reactivity may be an important target in the management of head injured patients.

Our data support the previously proposed hypothesis [6] that mortality rate is age-related both in males and females. Worse cerebrovascular reactivity in younger females seems to be the only pathophysiological factor which may explain the differences between younger males and females.

Apart from statistical analysis, close monitoring of ICP, CPP and PRx in individual cases is helpful in identifying patients prone to secondary brain insults. Real time monitoring of cerebrovascular reactivity allows individualised cerebral haemodynamic management in the majority of cases, and it triggers prompt correction of the physiological variables determining cerebral blood flow (Fig. 1).

Limitations

Unfortunately, our series included only 27 females older than 50 so that our study is less powerful than the data presented by the San Diego group [8]. Also, we have considered only patients included in our multimodal monitoring program whose data were available. These patients do not represent the entire group of severely head injured patients admitted to our neurocritical care unit. Missing in this group are the patients who either died or who displayed a favourable clinical course shortly after admission. It is important to note that not all patients in our group can be classified as 'severe head injury' since 21% of males and 17% females had a GCS above 8. However their clinical condition required sedation, paralysis and mechanical ventilation for at least 24 h. We did not check menopausal state in females. Therefore the age of 50 years cannot be understood as definitely pre- and post-menopausal threshold. Despite these limitations the main finding of a gender-related difference in post-head injury intracranial hypertension and outcome seems to be statistically robust.

Conclusion

Among patients younger than 50, in comparison to males, females tend to develop worse pressure-reactivity following TBI and this is associated with a higher mortality.

Conflict of interest statement We declare that we have no conflict of interest.

References

1. Alkayed NJ, Harukuni I, Kimes AS, London ED, Traystman RJ, Hurn PD (1998) Gender-linked brain injury in experimental stroke. Stroke 29:159–165
2. Balestreri M, Czosnyka M, Steiner LA, Schmidt E, Smielewski P, Matta B, Pickard JD (2004) Intracranial hypertension: what additional information can be derived from ICP waveform after head injury? Acta Neurochir (Wien) 146:131–141
3. Balestreri M, Steiner LA, Czosnyka M (2003) Sex-related differences and traumatic brain injury. J Neurosurg 99(3):616
4. Bramlett HM, Dietrich WD (2001) Neuropathological protection after traumatic brain injury in intact female rats versus males or ovariectomized females. J Neurotrauma 18:891–900
5. Coimbra R, Hoyt DB, Potenza BM, Fortlage D, Hollingsworth-Fridlund P (2003) Does sexual dimorphism influence outcome of traumatic brain injury patients? The answer is no!. J Trauma 54:689–700
6. Czosnyka M, Balestreri M, Steiner L, Smielewski P, Hutchinson PJ, Matta B, Pickard JD (2005) Age, intracranial pressure, autoregulation, and outcome after brain trauma. J Neurosurg 102(3):450–454
7. Deitch EA, Livingston DH, Lavery RF, Monaghan SF, Bongu A, Machiedo GW (2007) Hormonally active women tolerate shock-trauma better than do men: a prospective study of over 4000 trauma patients. Ann Surg 246(3):447–455
8. Farin A, Deutsch R, Biegon A, Marshall LF (2003) Sex-related differences in patients with severe head injury: greater susceptibility to brain swelling in female patients 50 years of age and younger. J Neurosurg 98:32–36
9. Patel HC, Menon DK, Tebbs S, Hawker R, Hutchinson PJ, Kirkpatrick PJ (2002) Specialist neurocritical care and outcome from head injury. Intensive Care Med 28:547–553
10. Singer CA, Rogers KL, Strickland TM, Dorsa DM (1996) Estrogen protects primary cortical neurons from glutamate toxicity. Neurosci Lett 212:13–16
11. Smielewski P, Czosnyka M, Steiner L, Belestri M, Piechnik S, Pickard JD (2005) ICM+: software for on-line analysis of bedside monitoring data after severe head trauma. Acta Neurochir Suppl; 95:43–49
12. Steiner LA, Czosnyka M, Piechnik SK, Smielewski P, Chatfield D, Menon DK, Pickard JD (2002) Continuous monitoring of cerebrovascular pressure reactivity allows determination of optimal cerebral perfusion pressure in patients with traumatic brain injury. Crit Care Med 30(4):733–738

Acta Neurochir Suppl (2008) 102: 29–31

Long term outcomes following decompressive craniectomy for severe head injury

U. Meier · S. Ahmadi · T. Killeen · F. T. Al-Zain · J. Lemcke

Abstract
Background Severe head injury is one of the commonest indications for neurosurgical intervention. For the neurosurgeon, the operative last resort in cases of generalised brain oedema of traumatic origin is the decompressive craniectomy. Is it possible to use predictive factors to ascertain what degree of success, in terms of both the acute and long-term outcome, is to be expected in patients who undergo this treatment?
Methods The clinical records of 131 patients treated with decompressive craniectomy for severe head injury were evaluated. All patients were operated on between September 1997 and September 2005 in the neurosurgical department of the Unfallkrankenhaus Berlin. A follow-up examination was carried out 49±25 months after the initial trauma. The clinical outcome was compared with several patient and radiographic factors to establish if any of these showed a relationship to the long-term outcome.
Findings A significant relationship was demonstrated between quality of outcome and the Glasgow Coma Scale score on admission. Quality of outcome was similarly related to the age of the patient, the condition of the basal cisterns and the degree of midline shift in the initial cranial computed tomography. Factors which correlated with poor outcome included pupil reactivity on admission, established clotting disorders and post-traumatic hydrocephalus internus. Hyperglycaemia and initial acidosis were also associated with a poor outcome.
Conclusions The clinical outcome in patients with a severe head injury is to a great degree determined by the extent and type of the primary injury. When considering decompressive hemicraniectomy as a treatment for raised intracranial pressure following traumatic brain injury, the predictive factors detailed here should be taken into consideration.

U. Meier (✉) · S. Ahmadi · T. Killeen · F. T. Al-Zain · J. Lemcke
Department of Neurosurgery, Unfallkrankenhaus Berlin,
Warener Strasse 7,
12683 Berlin, Germany
e-mail: ullrich.meier@ukb.de

Keywords Severe head injury · Decompressive craniectomy · Outcomes · Predictive factors

Introduction

Primary brain injury in severe head trauma occurs at the moment of the mechanical insult. With this in mind, the neurosurgeon must decide how to intervene therapeutically in order to at least minimize secondary brain injury. Decompressive craniectomy represents the neurosurgeon's *ultima ratio*—the last resort—in terms of operative therapy. The decision for or against this measure depends largely on the prognostic indications of the individual patient. Do factors exist that can more reliably predict the quality of outcome in such patients?

Patients and methods

All 131 patients who underwent decompressive craniectomy for severe head injury in our department between September 1997 and September 2005 were included for study. Their clinical records were evaluated and a number of preoperative factors recorded. These included basic epidemiological information as well as Glasgow Coma Scale (GCS) score on admission, the initial computed tomography (CT) scan results, pupil reactivity on admission and the position of the midline shift on cranial CT (cCT) on admission. Visibility of basil cisterns was also assessed and cisterns were scored as absent if there was no CSF present in the cisterns at the level of the midbrain. Follow-up examination was conducted at least 12 months after discharge from the Unfallkrankenhaus Berlin. The condition of the patient was determined, retrospectively from the clinic records, using the Glasgow Outcome Scale (GOS). Poor GOS was defined as grades 1 (dead) and 2 (persistent vegetative) and good GOS as grades 3 (severely disabled),

4 (moderately disabled), and 5 (good recovery). Statistical comparisons between patients good and poor GOS scores were made using student *T*-test, spearman, and Mann–Whitney analyses. Statistical significance considered for $p \leq 0.05$.

Results

The mean age of the 99 men (76%) and 32 women (24%) was 36±20 years at the time of the operation. The commonest causes of head injury were as follows; road traffic accident (53%; *n*=69), falls (26%; *n*=34) and physical violence (including suicide attempts; 14%; *n*= 18). Other causes accounted for the remaining ten (7%) patients' injuries. It was possible to follow-up 95% of patients (*n*=124). The mean period between initial trauma and follow-up was 49±25 months. At the time of follow-up, 75 patients (57%) had died (GOS 1) and nine patients (7%) remained in a persistent vegetative state (GOS 2). Fourteen patients (11%) were severely disabled (GOS 3), while 13 patients (10%) were moderately disabled and a further 13 patients (11%) had made a good recovery (GOS 5).

When the specific indicators and the GOS score at follow-up were compared, a number of significant relationships were demonstrated (Table 1). Patients with an initial GCS of less than eight were significantly more likely to go on to have a GOS of 1–3 at follow-up. Pathological pupil reactivity on admission similarly led to a significantly higher chance of a poor outcome. A marked midline shift at the time of admission was also highly predictive of an unsatisfactory GOS (1–3) at follow-up. Pre-existing and perioperatively apparent clotting disorders as well as post traumatic hydrocephalus internus also revealed themselves as predictors of a poor outcome. Hyperglycaemia and/or initial acidosis were likewise associated with undesirable outcomes at follow-up.

Discussion

At the present time, no large prospective studies of decompressive craniectomy for severe head injury are to be found in the literature. The neurosurgeon therefore, when faced with deciding for or against decompressive

Table 1 Correlation analysis and determination of single-factor statistically significant potential predictors of the Glasgow Outcome Scale (GOS) score at the time of discharge to rehabilitation services

Factor	Test	Significance (p)	Comments
Age	Spearman SpC=−0.208 *n*=131	0.017	Older patients had poorer GOS scores
GCS	*T*-test and MWT *n*=123	0.032 and 0.041	An initial GCS value of less than 8 showed a significant association with worse GOS scores
Cisterns discernible on cCT	*T*-test and MWT *n*=118	0.001	Obliterated cisterns preoperatively were associated with significantly worse GOS scores
Pupil status (initial)	T-Test and MWT *n*=128	0.002	Initial pathological findings on pupil examination were associated with significantly worse GOS scores
Midline Shift	Spearman SpC=−0.186 *n*=121	0.041	Patients with a large midline shift had significantly poorer GOS scores
Clotting disorders	*T*-test and MWT *n*=131	0.000	Clotting disorders were associated with significantly poorer outcomes
Glucose (from 2001)	Spearman SpC=−0.382 *n*=74	0.001	Preoperative hyperglycaemia (>6 mmol/L) was associated with significantly worse GOS scores
	T-test and MWT Cut off=(>6.10 mmol/L)	0.000	
pH (from 2000)	Spearman SpC=0.285 *n*=75	0.015	Low pH values were associated with significantly poorer GOS scores, while alkalosis appeared to have a protective effect.
	T-test and MWT Cut off=(<7.34 mmol/L)	0.000 and 0.002	
Post-traumatic hydrocephalus internus	X^2 *n*=2	0.001	Significantly poorer outcomes were observed in cases of post-traumatic hydrocephalus internus

Spearman correlation analysis was used for non-parametric values and the *T*-test and Mann–Whitney test were used for parametric data. *GCS* Glasgow Coma Scale, *cCT* cranial computer tomography, *SpC* Spearman coefficient, χ^2 Chi-square test, *MWT* Mann–Whitney test craniectomy, has only a combination of generally accepted opinion and his/her own personal experience to draw upon. The demographics of the population in the current study is comparable to those reported in the international literature [3–6]. The results of the Rescue ICP Study, which began in 2006 with a target population of 500 patients, are not expected until 2008 or 2009.

The findings of this study add to the consensus in the international literature regarding predictors of outcome in craniectomy. In this study, the demonstration of a relationship between a low GCS score at the scene of the injury or on admission and a poor prognosis—the principal finding of various other studies [1, 3]—is replicated. Pathological pupil reactivity in terms of an initial anisocoria or an absent reaction to light correlated with poor outcome. This has also been proposed as a negative predictor of outcome by other authors [2]. In this study, shift of the midline on cCT was significantly correlated with the quality of outcome. Patients with a poor outcome (GOS 1–3) had a mean midline shift of 9.76± 6.7 mm whereas patients with a good outcome (GOS 4–5) showed mean midline shift of 6.69±5.0 mm. The important predictive value of this factor has been confirmed by Marshall et al. [7], among others. The visibility of the basal cisterns is also shown in this study to be very significantly connected to the quality of outcome at follow-up.

Conclusion

The initial GCS score, pupil reactivity, visibility of the basal cisterns and the degree of midline shift on cCT are highly significant predictors of outcome in patients undergoing decompressive craniectomy for severe head injury. Clotting disorders and hyperglycaemia are similarly significant as predictors of outcome. These predictive factors should be taken into account when deciding for or against craniectomy.

Conflict of interest statement We declare that we have no conflict of interest.

References

1. Arabi B, Hesdorffer DC, Ahn ES, Aresco C, Scalea TM, Eisenberg HM (2006) Outcome folling decompressive craniectomy for malignant swelling due to severe head injury. J Neurosurg 104 (4):469–479
2. Bullock MR, Chesnut RM, Clifton GL, Ghajar J, Marion DW, Narayan RK, Newel DW (2000) Part I: Guidelines for the management of severe head injury, Management and prognosis of severe traumatic brain injury. Brain Trauma Foundation, New York, pp 7–159
3. Gaab MR, Rittierodt M, Lorenz M, Heissler HE (1990) Traumatic brain swelling and operative decompression: a prospective investigation. Acta Neurochir Supp (Wien) l51:326–328
4. Guerra WK, Gaab MR, Dietz H, Mueller JU, Pieck J, Fritsch MJ (1999) Surgical decompression for traumatic brain swelling: indications and results. J Neurosurg 90(2):187–196
5. Meier U, Lemcke J, Reyer T, Gräwe A (2006) Decompressive craniectomy for severe head injury in patients with major extracranial injuries. Acta Neurochir 96(Suppl):373–376
6. Ransohoff J, und Benjamin V (1971) Hemicraniectomy in the treatment of acute subdural haematoma. J. Neurolog Neurosurg Psychiat 34(1):106
7. Marshall LF, Gautille T, Klauber MR, Eisenberg HM, Jane JA, Luerssen TG, Marmarou A Foulkes MA (1991) The outcome of severe closed head injury. J Neurosurge 75:28–36

Acta Neurochir Suppl (2008) 102: 33–36

Relationship of a cerebral autoregulatory index with outcome in head injured patients

Martin Shaw · Ian Piper · Michael Daley

Abstract
Background Cerebral autoregulation is the process by which cerebral blood flow (CBF) is maintained constant over a specific cerebral perfusion pressure (CPP) range. We have reworked a version of the Ursino and Lodi autoregulation model to derive an index of autoregulation (G), and compared it to a number of other autoregulatory models as well as a gold standard measure of autoregulation obtained from an animal model study (6 piglets with a cranial window preparation and ICP, ABP sampled at 250 Hz). The results of that study have shown that this index G correlates well with the "Bouma" index of autoregulation.
Methods In this study this new autoregulatory index has been calculated for a sample of 12 head injury patient's data over multiple time points and then used to firstly investigate if this index in conjunction with other clinical prognostic factors may give a better indication of outcome and then analyse its trend with time to quantify how the level of autoregulation changes post-injury.
Findings The index correlates well with dichotomised GOSe outcome (p=0.03) and the trend in the result between middle and late time periods shows early signs of being predictive of outcome as well.
Conclusions Though more work is needed these results warrant further investigation with larger numbers of patients

Keywords Cerebral autoregulation · Mathematical modelling · Outcome prediction

M. Shaw (✉) · I. Piper
Clinical Physics, Institute of Neurological Sciences,
Southern General Hospital,
1345 Govan Road,
Glasgow G514TF, UK
e-mail: martin.shaw@nhs.net

M. Daley
Department of Electrical and Computer Engineering,
University of Memphis,
Memphis, TN 38152, USA

Introduction

Cerebral autoregulation is the process by which cerebral blood flow (CBF) is maintained relatively constant over a specific cerebral perfusion pressure (CPP) range. There are a number of models of cerebral autoregulation of which the Ursino Lodi model is an example of a mathematical modelling approach [6]. We used a reworked version of the Ursino and Lodi autoregulation model to derive an index of autoregulation (G) and compared it with a number of other autoregulatory models as well as a gold standard measure of autoregulation obtained from an animal model dataset. This dataset was obtained from 6 piglets with a cranial window preparation with intracranial pressure (ICP) and arterial blood pressure (ABP) monitoring sampled at 250 Hz. The results of that study have shown that this modified Ursino index correlates well with the "Bouma" index [1] of autoregulation computed as the ratio of the % change in CPP divided by the % change in CVR. In this study this new autoregulatory index has been calculated for a sample of patient data obtained from the BrainIT dataset [2] over multiple time points and then used to:

1) Investigate if this index in conjunction with other clinical prognostic factors may give a better indication of outcome as assessed by 6 month dichotomised extended Glasgow outcome scale (GOSe).
2) Analyse its trend with time to quantify how the level of autoregulation changes post-injury.

Fig. 1 Block diagram describing action of autoregulatory mechanisms on arterial-arteriolar compliance in response to a given percent change in CBF

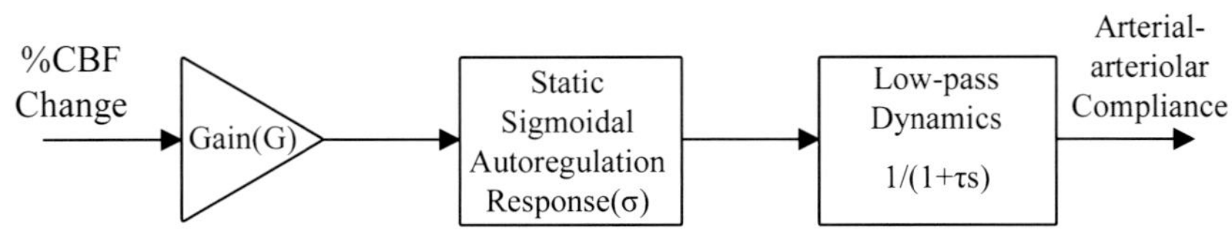

Materials

We used data from 12 patients extracted from the BrainIT database [2] and the algorithms developed in the stats package R [3].

Methods

Reworked predictor from Ursino and Lodi's model

Autoregulation in the original Ursino–Lodi model can be thought of as a combination of three processes affecting arterial-arteriolar compliance for a given percentage change in CBF. (Fig. 1)

It is from this continuous index for cerebral autoregulation G that the new model is derived. With this in mind the mathematical representation of Ursino's original physical model has been rearranged to output G instead of using it as an input. In the validation of the new model it has been found that the variance of the G parameter is a closer predictor of autoregulation [5].

As with previous testing of the model, the relative change in arterial arteriolar volume has to be considered. This has been done in the past with the measurement of the arterial diameters and estimation of the volume using the physics of a venturi flow meter [4]. In the current study, as there was no arterial diameter parameters available that would trend in the correct direction a scaled version of intracranial pressure (ICP) pulse amplitude was used as an estimate for this parameter.

Analysis

12 patients were sampled from the BrainIT database [2] with measurable ICP pulse amplitude and had a good spread of clinical outcome as measured by the GOSe. Using the new model, the absolute value of G was calculated for the full patients monitoring time course. The data for each of the patients was then split into 3 equal time periods (early, middle and late) and the variance of G in each period calculated (Table 1). Patients were then grouped by a dichotomised GOSe ("good outcome" GOSe 4–8). The outcome sets for the late time period were then compared to each other time period using a one-sided student's *t*-test. This analysis was used to test if impaired autoregulation was associated with bad outcome i.e. the mean variance was greater for bad outcome than good.

We also looked at the time course of autoregulation over the duration of stay in 2 patients, one sampled from the bad clinical outcome set (GOSe=1), the other from the good (GOSe=5).

Results

The 12 patients are evenly dichotomised on GOSe (7 bad, 5 good). The results of the one sided *t*-test, presented in

Table 1 Initial results from the reworked Ursino model

Patient ID	Early	Middle	Late	GOSe	Outcome
04027262	2183.343	10.404	142.808	1	bad
04816150	8.493	0.002	1.039	8	good
16137372	20.398	152.872	2.558	7	good
26248362	2.827	163.958	304.209	4	bad
27240484	0.003	0.001	0.003	1	bad
38351504	21.641	80.444	82.968	1	bad
48355050	5.680	0.015	208.945	1	bad
51573727	11538.370	1060.402	19.082	1	bad
61683383	22.941	19.093	0.000	8	good
73684838	0.017	10.719	1.220	1	bad
83815161	84.795	232.755	46.522	5	good
84861515	1.928	1.659	0.282	8	good

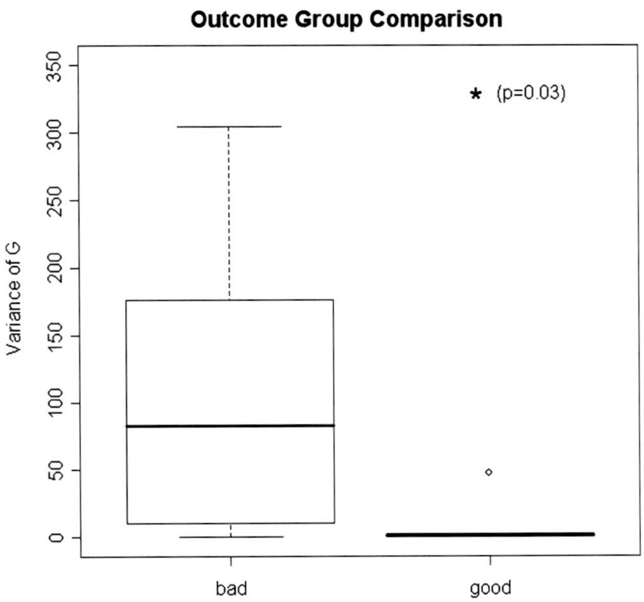

Fig. 2 Box plot representation of var(G) comparing both outcome groups

Fig. 2, have shown that there was a significantly greater mean variance in G in the bad outcome set compared with the good (p=0.033).

Figure 3 shows the pilot data trending var(G) for a patient with good outcome and one with bad outcome, these are representative of the other patients dichotomised in their respective groups. Of course these autoregulatory time courses are not conclusive but it is interesting to note that the overall trending may also be predictive of outcome.

Discussion

As the results of the t-tests were statistically significant there is a good correlation of the reworked Ursino model output G with clinical outcome over the later stages of stay in the ICU. In the pilot analysis considering the trend in G over time between patients with good and bad outcome the patients in the "good" group the trend from "middle" to "late" is intuitively what would be expected for a patient that is recovering; going from non functional autoregulation to functional, and the exact opposite can be seen in the patient with bad outcome.

Limitations

As it stands there are three main limitations to this study which we hope to address in a future analysis. Firstly as previously mentioned the study numbers need to be larger. Secondly to make the study more clinically relevant we should redefine the time periods "early", "middle" and "late" as a standardised measure of days post injury instead

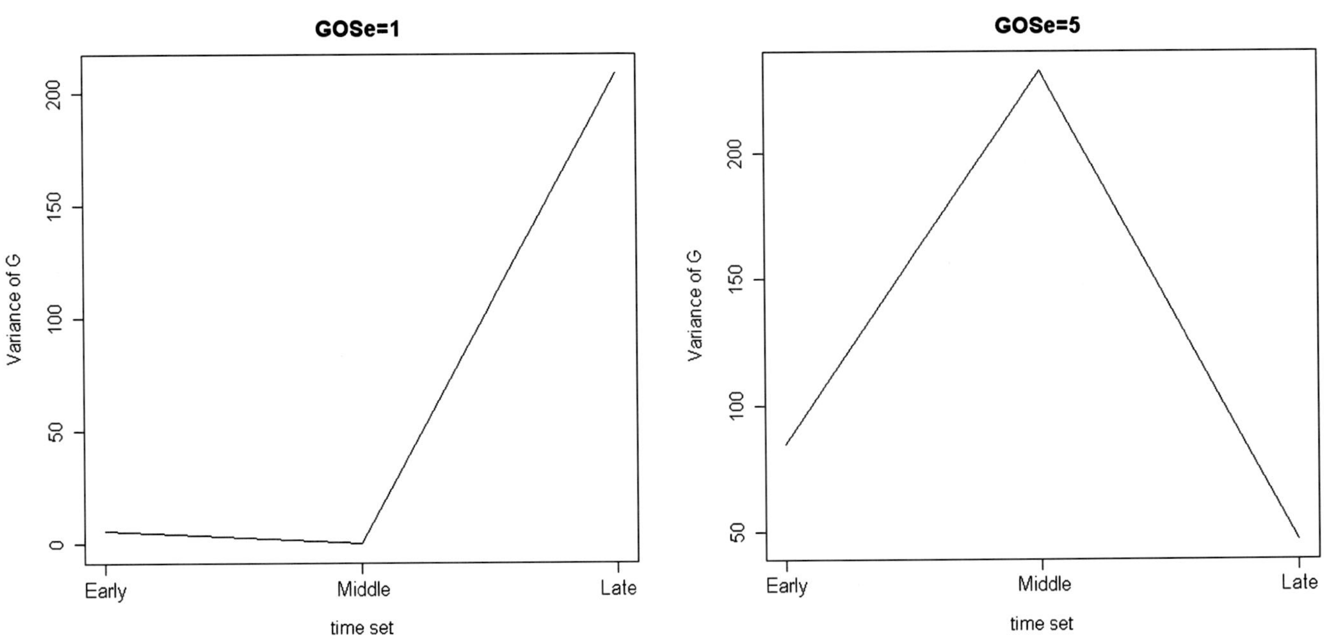

Fig. 3 Representation of autoregulation over time in one patient (*left*) Bad group (*right*) Good group

of the proportional of time monitored. Lastly we need to look for a more accessible measure of arterial volume or arterial compliance so that ICP wave amplitude shouldn't be needed.

Conclusion

Though the numbers of patients analysed were small it is felt that these results warrant further investigation with larger numbers of patients.

Conflict of interest statement We declare that we have no conflict of interest.

References

1. Bouma GJ, Muizelaar JP (1992) Blood pressure and intracranial pressure-volume dynamics in severe head injury relationship with cerebral blood flow. J Neurosurg 77:15–19
2. http://www.BrainIT.org
3. R Development Core Team (2007) R: A language and environment for statistical computing. R Foundation for Statistical Computing, Vienna, Austria. ISBN 3-900051-07-0, URL http://www.R-project.org.
4. Shaw M, Piper IR (2006) Improving comparison of models of pressure autoregulation by adaption of existing models to provide normalised outputs. Proceedings BNRG
5. Shaw M, Piper IR, Dayley M (2006) Comparison of two models of cerebral pressure autoregulation. J Neurotrauma 23(5):798 May
6. Ursino M, Lodi CA (1997) A simple mathematical model of the interaction between intracranial pressure and cerebral hemodynamics. J Appl Physiol 82(4):1256–1269 Apr

Acta Neurochir Suppl (2008) 102: 37–41

Assessment of cerebrovascular resistance with model of cerebrovascular pressure transmission

Nithya Narayanan · Charles W. Leffler · Marek Czosnyka · Michael L. Daley

Abstract

Background A two step modeling method of cerebrovascular pressure transmission, the dynamic relationship between arterial blood pressure (ABP) and intracranial pressure (ICP) has been developed as a means to continuously assess cerebrovascular regulation and resistance. Initially, system identification modeling was used to construct a numerical model of cerebrovascular pressure transmission. Next, the modal frequencies of the numerical model and the actual ABP recording were used to manipulate the parameters of a physiologically-based biomechanical model such that: (1) the actual and simulated ICP; and (2) the numerical and biomechanical model modal frequencies match.

Materials and methods This study was designed to compare changes of cerebrovascular resistance of the biomechanical model with the expected changes of cerebrovascular resistance associated with the occurrence of either a plateau wave or refractory intracranial hypertension. Pressure recordings from five patients with plateau waves and five patients with intracranial hypertension were used.

Findings Vascular resistance decreased significantly during the plateau wave and was inversely related to CPP, indicating active vasoreactivity. In contrast, vascular resistance increased significantly during intractable intracranial hypertension and was directly related to CPP, indicating impaired cerebrovascular regulation.

Conclusions Such results support the use of the modeling method as a means to continuously assess changes of cerebrovascular regulation and resistance.

N. Narayanan
Department of Electrical and Computer Engineering,
The University of Memphis,
Engineering Science Building, Rm. 209A,
Memphis, TN 38152-3180, USA

C. W. Leffler
Department of Physiology,
The University of Tennessee Health Science Center,
301 Nash Building,
Memphis, TN 38163, USA

M. Czosnyka
Academic Neurosurgical Unit, Addenbrooke's Hospital,
Cambridge, Box 167 CB2 2QQ, UK

M. L. Daley (✉)
Department of Electrical and Computer Engineering,
The University of Memphis,
Engineering Science Building, Rm. 208B,
Memphis, TN 38152-3180, USA
e-mail: mdaley@memphis.edu

Introduction

Clinical methods designed to provide continuous assessment of regulation of cerebral blood flow (CBF) and further insight into evolving pathophysiological processes during the intensive care management of patients with brain injury are needed. The approach of this study to address these needs is based on serial modeling of cerebrovascular pressure transmission as defined by the dynamic relationship between arterial blood pressure (ABP) and intracranial pressure (ICP). Biomechanical models of intracranial pressure dynamics based on the Monroe–Kellie doctrine (which assumes the total volume within the craniospinal sac remains constant) have been developed to simulate ICP dynamics and hemodynamic alterations [1, 9, 10]. In contrast to these biomechanical models which require physiological compatibility, the use of numerical system identification modeling has been proposed recently as a method to continuously assess regulation of cerebral blood flow [4]. This procedure is a strictly numerical one in which the digitized paired input/output values are coupled to a generalized dynamic description of a process to produce a simulated output that matches the actual output by minimization of the least squares difference between the actual and simulated digitized output series [6]. The two step modeling method of cerebrovascular pressure transmission has been developed to continuously assess cerebrovascular regulation and resistance. Initially, system identification modeling is used to construct a numerical black-box model of cerebrovascular pressure transmission. Next, the modal frequencies of the numerical model and the actual ABP recording are used to manipulate the parameters of a physiologically-based biomechanical model such that: (1) the actual and simulated ICP; and (2) the numerical and biomechanical model modal frequencies are matched.

In this study two distinct clinical conditions associated with traumatic brain-injured patients, the plateau wave and intractable intracranial hypertension were studied. The plateau wave is a pathophysiological event that is characterized by a sudden rapid elevation of ICP followed by a sustained period of elevation that is terminated by a rapid decrease in ICP [2, 7, 8]. The onset of the wave has been described by a vasodilatory cascade phase which links active vasodilation, increased cerebral blood volume, elevated ICP and decreased cerebral perfusion pressure (CPP) [8]. The relatively rapid termination of the wave has been described by a vasoconstriction cascade phase in which active vasoconstrictive events lead to decreases in cerebral blood volume and ICP and an increase of CPP [8]. Intractable intracranial hypertension is a pathophysiological condition that is characterized by an uncontrollable increase of ICP and a decrease in CPP which leads to compression of the vascular bed and ischemic levels of cerebral blood flow (CBF). The aim of this study was to use the two-step modeling method on pressure recordings obtained during plateau waves and during intractable intracranial hypertension as a means to derive values of simulated arterial/arteriolar bed resistance (SimR) and determine whether changes in this resistance were consistent with those expected to occur for each pathophysiological condition.

Materials and methods

Pressure recordings sampled at a rate of 30 or 50 Hz from two groups of five patients each were used in this study. One group demonstrated plateau waves; while the other group demonstrated a prolonged increase of ICP characteristic of intractable intracranial hypertension. Neurocritical Care Users Committee, Addenbrooke's Hospital, Cambridge, UK approved the fact that physiological signals were recorded numerically and since this did not introduce any additional risk or burden to routine patients care, individual consent was not required at the time of recording. Each set of recordings was labeled in such a manner that the patients could not be identified by the Memphis group. The Institutional Review Board at The University of Memphis approved the data analysis protocol.

The first step of the two step modeling procedure, system identification modeling, has been previously reported in detail [4, 5]. Briefly a modified biomechanical model proposed by Czosynka et al. [1] was used to provide a generalized dynamic differential equation of cerebrovascular pressure transmission. A system identification technique was applied to each 500 paired samples of digitized pressure recordings to obtain the modal frequencies of cerebrovascular pressure transmission. For the brief period on which each numerical model is based, perturbations of ABP and ICP are assumed to be small and cerebrovascular pressure transmission is considered to be linear and time-invariant. In the second step, mean values of the modal frequencies derived by the first step numerical model for each three consecutive 500 sample modeling intervals of ABP recording are applied to the physiologically-based biomechanical model to simulate an ICP recording for the recording interval based on 1500 samples. Using the physiologically-based biomechanical model, a computational method was constructed to simulate ICP for the ABP recording and the corresponding numerically derived modal frequencies for each modeling interval. The computation was designed to manipulate the resistance and compliance elements of the biomechanical model such that the minimum square error for the 1,500 sample interval between: (1) the actual and simulated ICP recordings; (2) the numerical model value of the highest modal frequency (HMF) and the biomechanical value of HMF; and (3) the numerical model value of a dampening factor (DF) of the lower modal frequencies and the DF of the biomechanical model were minimized. Compliance and resistance elements were allowed to range over values consistent with those reported in previously published modeling studies of intracranial pressure dynamics [1, 9, 10]. The parameters of the biomechanical model were estimated by recursive computation to obtain a best fit set which produced a minimum least square error and maximum correlation value for each segment of the pressure recording.

Results

ICP recordings obtained from a patient during the demonstration of a plateau wave and of a patient with intractable intracranial hypertension are shown (see Fig. 1). Throughout the three phases of the plateau wave ABP did not change significantly while ICP increased and CPP decreased relative

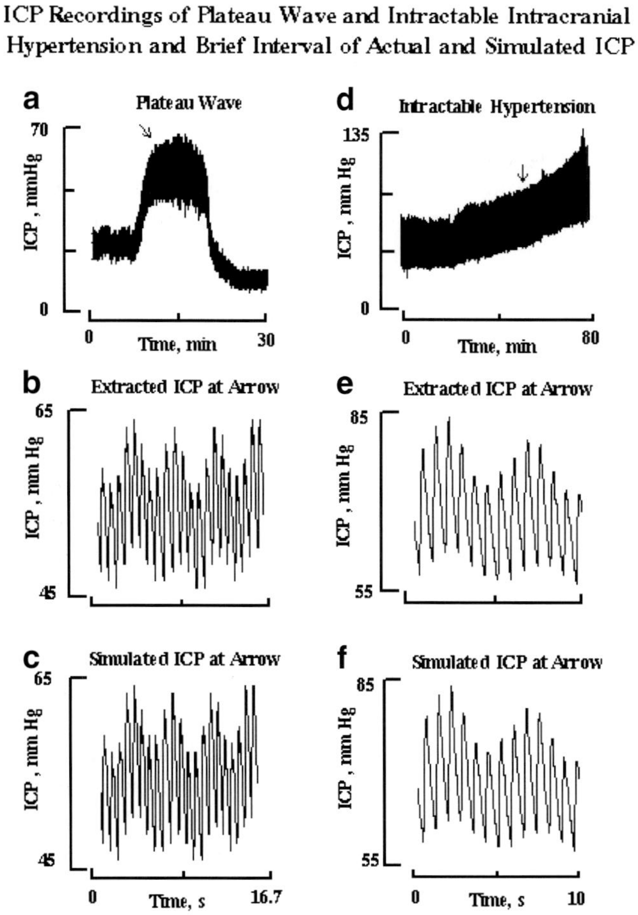

Fig. 1 ICP recordings o f plateau wave and intractable intracranial hypertension and brief interval of actual and simulated ICP. **a** ICP recording during plateau wave. *Arrow* indicates the location of wave used for extracting 16.7 s interval of ICP recording shown below. **b** Extracted ICP recording at arrow located in recording of plateau wave. **c** Simulated ICP recording of biomechanical model. **d** ICP recording during intractable intracranial hypertension. *Arrow* indicates location used to extract 10.0 s interval of ICP recording. **e** Extracted ICP recording at arrow located on recording of intractable intracranial hypertension. **f** Simulated ICP recording by two step modeling method of 10 s extracted ICP recording

to baseline during the wave; at the termination of the wave, ICP decreased and CPP increased (see Table 1). Throughout intracranial hypertension ABP did not change while ICP increased and CPP decreased (see Table 1).

For the ICP recordings of a plateau wave and intractable intracranial hypertension, changes in the values of HMF of both models and the corresponding changes in simulated arterial–arteriolar bed resistance are shown in Fig. 2. The derived HMF values of the numerical system identification model and the corresponding HMF values of the biomechanical model increased at the onset of the plateau wave during vasodilation and decreased at termination during the constrictive phase of the wave (see Fig. 2a). Consistent with the active changes in the cerebrovasculature during the wave, the corresponding simulated values of arterial–arteriolar bed resistance decreased at the onset of the wave and increased at the termination of the wave (see Fig. 2b). Correlations between simulated bed resistance and CPP overall, prior to, during, and after the plateau wave were 0.97 (±0.01), 0.97 (±0.01), 0.98 (±0.02), and 0.97 (±0.02) respectively with a level of significance for each individual correlation at $p<0.001$. The HMF values derived by the first step numerical model and the corresponding HMF values of the biomechanical model decreased during intractable intracranial hypertension (see Fig. 2c). Consistent with vasocompression at the periphery of the arterial tree during intractable intracranial hypertension, simulated resistance of the arterial–arteriolar bed progressively increased (see Fig. 2d). Mean correlations (±SD) between simulated bed resistance and CPP overall, at the start, and end of the intractable intracranial hypertension recording were −0.98 (±0.02), −0.98 (±0.04) and −0.98 (±0.04) respectively with a level of significance for each individual correlation of $p<0.001$.

Overall mean values and corresponding standard deviations of ABP, ICP, CPP, HMF, and simulated resistance of

Table 1 Overall mean values (±SD) values of ABP, ICP, CPP, HMF, simulated resistance of the arterial–arteriolar bed and correlation of HMF value with bed resistance

Condition	Number	ABP (±SD, mmHg)	ICP (±SD, mmHg)	CPP (±SD, mmHg)	HMF (±SD, Hz)	Sim. resist. (±SD, mmHg ml^{-1} s^{-1})	SimR vs CPP r value (±SD)
Plateau wave							
Prior	5	100.7 (±13.9)	27.1 (±2.4)	73.6 (±2.5)	7.5 (±1.1)	8.6 (±1.1)	0.97 (±0.01)[c]
During	5	108.5 (±9.6)	44.4 (±4.4)[a]	67.1 (±3.1)[b]	15.3 (±3.0)[a]	4.1 (±1.3)[a]	0.98 (±0.02)[c]
After	5	100.4 (±7.1)	25.7 (±3.1)[a]	76.1 (±3.2)[a]	6.9 (±2.1)[a]	9.8 (±1.3)[a]	0.97 (±0.02)[c]
Intractable ICP							
Start	5	129.2 (±6.3)	65.6 (±3.8)	64.1 (±3.1)	16.4 (±0.3)	5.1 (±0.62)	−0.98 (±0.01)[c]
End	5	135.3 (±5.2)	98.1 (±2.5)[a]	43.9 (±3.8)[a]	5.4 (±0.8)	12.4 (±1.4)[1]	−0.98 (±0.04)[c]

[a] Level of significance between mean values at $p<0.001$
[b] Level of significance between mean values at $p<0.005$
[c] Level of significance of each individual correlation ($n=5$) at $p<0.001$

Changes of HMF Values of Numerical and Biomechanical Models and Simulated Arterial-Arteriolar Bed Resistance During Plateau Wave and Intractable Hypertension

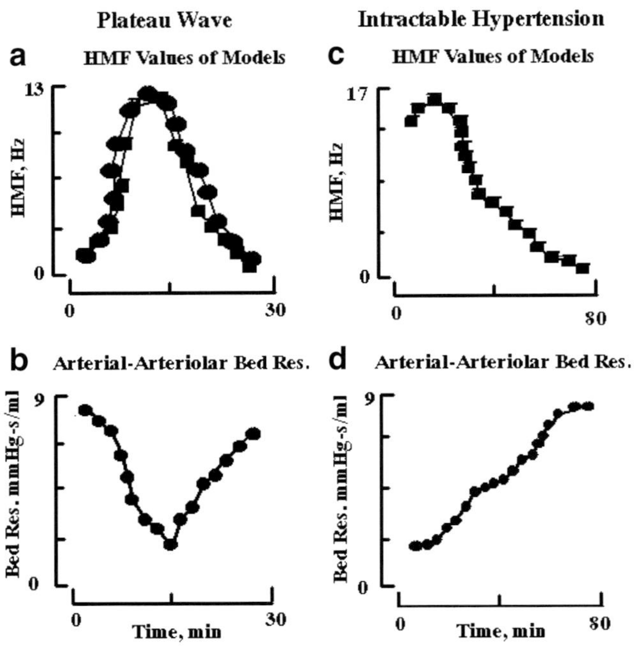

Fig. 2 Changes of HMF values of numerical and biomechanical models and simulated arterial–arteriolar bed resistance during plateau wave and intractable intracranial hypertension. **a** Derived HMF values for ICP recording with plateau wave. For each three consecutive values of HMF, the mean value and standard deviation was computed and are plotted as *closed squares* with corresponding error bars. The corresponding HMF values of the biomechanical model are plotted as *closed circles*. Correlation between model HMF values was strong and significant at 0.98. and $p<0.001$. **b** Simulated values of arterial–arteriolar bed resistance for ICP recording with plateau wave. HMF values and simulated arterial–arteriolar bed resistance were inversely related with a slope of the regression line relationship of 1.38 mmHg ml^{-1} s^{-1}. **c** Derived HMF values for ICP recording of intractable intracranial hypertension. Mean and standard deviation of three consecutive values of HMF of numerical model (closed squares) with corresponding error bars and corresponding HMF values of the biomechanical model (*closed circles*) with error bars are plotted. Correlation between model HMF values was strong and significant at 0.97 and $p<0.0001$. **d** Simulated values of arterial–arteriolar bed resistance for ICP recording with intractable intracranial hypertension. HMF values and simulated arterial–arteriolar bed resistance were inversely related with a slope of the regression line relationship of −2.08 mmHg ml^{-1} s^{-1}

the arterial–arteriolar bed and correlation of HMF with simulated cerebrovascular resistance of both groups of patients demonstrated results similar to those described by the above individual examples. For the five patients with plateau waves, mean ABP did not change during the recording period. However during the plateau wave, ICP increased and CPP decreased. Also during the plateau wave, HMF increased while simulated vascular resistance decreased. Prior to, during, and after the plateau wave, simulated resistance of the arterial–arteriolar bed was strongly correlated to CPP, which is indicative of active cerebrovascular regulation. In contrast, for the group of patients with intractable intracranial hypertension, mean ICP at the end of the recording period was higher than at the start of recording; while mean CPP decreased over the period. Furthermore, mean HMF decreased while simulated bed resistance increased. Finally, the correlations between simulated vascular resistance and CPP at the start and the end of the recording were strongly inverse, indicating impaired cerebrovascular regulation of cerebral blood flow.

Discussion

The major finding of study is that a novel two step modeling method based on ICP and ABP recordings available in the intensive care setting can be used to continuously assess changes in cerebrovascular resistance. For the five patients in the plateau wave group, simulated arterial–arteriolar vascular resistance decreased during the vasodilatory phase of the wave and increased during the vasoconstrictive phase of the wave. Two findings based on the recordings obtained from these patients indicate that regulation of CBF was intact during the recording period. First, simulated vascular resistance was directly correlated with CPP. And second, HMF was inversely related to CPP which is also consistent with intact cerebrovascular regulation as reported in laboratory study of percussive brain injury [4]. In contrast the group of five patients with intractable intracranial hypertension had findings indicative of impaired cerebrovascular regulation. For all patients in this group, simulated arterial–arteriolar bed resistance steadily increased as CPP decreased. The correlation between simulated arterial–arteriolar bed resistance and CPP was strongly inverse and the HMF was directly correlated with CPP. While the change in the resistance of arterial–arteriolar bed is different for these distinct pathophysiological events, the correlation between HMF and simulated arterial–arteriolar bed resistance was strongly inverse for all patients.

Previous black-box numerical modeling of cerebrovascular pressure transmission based on clinical pressure recordings obtained during the occurrence of a plateau wave or during intractable intracranial hypertension have demonstrated that values of HMF are inversely related to CPP when cerebrovascular regulation is thought to be intact and directly related when cerebrovascular regulation is likely lost [3]. In this study of clinical recordings of ABP and ICP obtained during the two distinct pathophysiological conditions, the described two-step modeling method of cerebrovascular pressure transmission was used to obtain simulated values of arterial–arteriolar bed resistance. Indicative of active vasoreactivity, in addition to demonstrating the inverse relationship between HMF and CPP, simulated bed resistance was shown to increase with CPP during the plateau wave. In contrast, during intractable intracranial hypertension a direct relationship between HMF and CPP existed and simulated bed resistance was shown to decrease with increasing CPP, indicative of impaired cerebrovascular regulation.

Possible limitations of the proposed methodology with respect to the development of a clinical application should be considered. First, the modal frequencies of cerebrovascular pressure transmission are not solely dependent on arterial–arteriolar bed resistance, but on all the resistance and compliance elements of the model. However, except for extremely high values of ICP, because arteriolar resistance is nearly ten times the magnitude of venous resistance, the simulated bed resistance has a dominant influence on the modal frequencies of cerebrovascular pressure transmission. A second limitation is that the method requires artifact free continuous dynamic recordings of ICP and ABP. Alteration of the dynamic pulsatile pressure recordings either by the inherent design of the recording instrumentation or by sensor alteration during monitoring will corrupt the modeling process. In such cases an otherwise highly pulsatile ICP recording with sharp peaks might be recorded as a rounded pulse and the computations based on cerebrovascular pressure transmission would be erroneous.

In conclusion, independent of whether the two step process modeling method of cerebrovascular pressure transmission was based on pressure recordings obtained either during a plateau wave or during intractable hypertension, the relationship between HMF and simulated resistance of the arterial–arteriolar bed was an inverse one. Pressure recordings obtained during the plateau wave demonstrated: (1) a positive correlation between CPP and simulated arterial–arteriolar vascular resistance; and (2) an inverse relationship between HMF and CPP, indicative of intact cerebrovascular regulation [4]. In contrast, pressure recordings obtained during intracranial hypertension demonstrated: (1) a negative correlation between simulated vascular resistance and CPP; and (2) a direct relationship between HMF and CPP, indicative of impaired cerebrovascular regulation [4]. The results of this study provide a preliminary basis for the use of the proposed modeling method of cerebrovascular pressure transmission as a means to continuously assess cerebrovascular regulation and interpret changes of cerebrovascular resistance of the arterial–arteriolar bed.

Acknowledgements This project was partially (MC and JDP) supported by the UK Government Technology Foresight Initiative, and the Medical Research Council (Grant No G9439390 ID 65883). This research was also supported in part by the NHLB and NINDS of National Institutes of Health and the Southeast Affiliate of the A.H. A.

Conflict of interest statement We declare that we have no conflict of interest.

References

1. Czosnyka M, Piechnik P, Richards HK, Kirkpatric P, Smielewski P, Pickard JD (1997) Contribution of mathematical modeling to the interpretation of bedside tests of cerebrovascular autoregulation. J Neurol Neurosurg Psychiatry 63:721–731
2. Czosnyka M, Smielewski P, Piechnik S, Schmidt E, Al-Rawi PG, Kirkpatric P, Pickard JD (1999) Hemodynamic characterization of intracranial pressure plateau waves in head-injured patients. J Neurosurg 92:11–19
3. Daley ML, Leffler CW, Czosnyka M, Pickard JD (2006) Intracranial pressure monitoring: modeling cerebrovascular pressure transmission. Acta Neurochir [Suppl] 96:103–107
4. Daley ML, Pourcyrous M, Timmons SD, Leffler CW (2004) Assessment of cerebrovascular autoregulation: changes of highest modal frequency of cerebrovascular pressure transmission with cerebral perfusion pressure. Stroke 35:1952–1956
5. Daley ML, Pourcyrous M, Timmons SD, Leffler CW (2007) Mode changes of cerebrovascular pressure transmission induced by cerebral vasodilation. J Neurotrauma 24(3):559–566
6. Ljung L (1991) The basic steps of system identification. System identification toolbox user's guide. The MathWorks, Natick, MA, pp 8–9
7. Lundberg N (1960) Continuous recording and control of ventricular fluid pressure in neurological practice. Acta Psychiatrica et Neurologica Scandinavica, Copenhagen (thesis)
8. Rosner MJ (1987) Cerebral perfusion pressure: link between intracranial pressure and systemic circulation. In: Wood JH (ed) Cerebral blood flow. McGraw-Hill, New York, pp 425–448
9. Ursino M (1988) A mathematical study of human intracranial hydrodynamics. Part I—The cerebrospinal fluid pulse pressure. Ann Biomed Eng 16:379–401
10. Ursino M, Lezzi M, Stocchetti N (1995) Intracranial pressure dynamics in patients with acute brain damage: a critical analysis with the aid of a mathematical model. IEEE Trans Biomed Eng 42:529–540

Acta Neurochir Suppl (2008) 102: 43–47

Generation of very low frequency cerebral blood flow fluctuations in humans

Malgorzata Turalska · Miroslaw Latka · Marek Czosnyka · Krystyna Pierzchala · Bruce J. West

Abstract

Background Slow oscillations of cerebral blood flow induced by synchronous variations of arterial blood pressure (ABP) are often used for clinical assessment of cerebral autoregulation. In the alternative scenario, spontaneous cerebral vasocycling may produce waves in cerebral blood flow that are, to a large extent, independent of ABP fluctuations. We use wavelet analysis to test the latter hypothesis.

Methods The wavelet variability *V(f)*, defined as the time averaged moduli of frequency dependent wavelet coefficients, is employed to analyze the relation between dynamics of arterial blood pressure and that of cerebral blood flow velocity in middle cerebral artery (MCA).

Findings In the very low frequency (VLF, 0.02–0.07 Hz) band the variability in traumatic brain injury (TBI) patients with low intracranial pressure (V_{ABP}=0.36±0.28) is significantly smaller than that of the volunteers (V_{ABP}=0.70±0.25) with $p=7\times10^{-5}$. Interestingly, the corresponding variabilities of MCA flow velocity for both cohorts are comparable. V_{MCA}=0.83±0.65 of the brain injury patients is not statistically different from that of the volunteers V_{MCA}=1.06±0.41 (p=0.11).

Conclusions In TBI patients without cerebral hypertension, the VLF oscillations must have been spontaneously generated within intracranial volume to compensate for the reduced ABP variability. Vasomotion is identified as a plausible physiological mechanism underlying such oscillations. We argue that vasomotion may be beneficial for brain tissue oxygenation especially during periods of critically low perfusion.

Keywords Cerebral autoregulation · Brain injury · Vasomotion · Brain tissue oxygenation

M. Turalska
Physics Department, University of North Texas,
P.O. Box 311427, Denton, TX 76203, USA

M. Latka (✉)
Institute of Biomedical Engineering,
Wroclaw University of Technology,
Wybrzeze Wyspianskiego 27,
50–370 Wroclaw, Poland
e-mail: miroslaw.latka@pwr.wroc.pl

M. Czosnyka
Department of Neurosurgery, University of Cambridge,
Addenbrooke's Hospital,
Cambridge CB2 1QQ, UK

K. Pierzchala
Department of Neurology, Silesian Medical University,
ul. 3 Maja 13/15,
41–800 Zabrze, Poland

B. J. West
Mathematical & Information Science Directorate,
Army Research Office,
P.O. Box 12211, Research Triangle, NC27709–2211, USA

Introduction

Under normal physiological conditions changes in cerebral perfusion pressure can be compensated by local adjustments of the tone of the cerebrovascular bed – the mechanism known as cerebral autoregulation [13]. Adjustment of the tone of cerebral arteriolas underlies not only pressure autoregulation but also vasomotor reactivity and metabolic regulation. Diverse physiologic stimuli such as varying partial pressures of oxygen and CO_2 exert their influence on vascular smooth muscle cells either through elevated production of nitric oxide, resulting in vasodilation, or through an overproduction of the vasoconstrictor endothelin. Until recently, autoregulation, metabolic regulation, and vasomotor reactivity seemed to be closely interrelated [14]. However, Lavi et al [8] recently demonstrated that patients with endothelial dysfunction preserve pressure dependent autoregulation despite having impaired CO_2 vasomotor reactivity.

The application of transcranial Doppler ultrasonography (TCD) [2] has enabled study of the dynamic aspects of cerebral autoregulation. Cerebral autoregulation is a vital protective mechanism that has come to be perceived as a physiological high-pass filter transmitting rapid changes in arterial blood pressure (ABP), but dampening low-frequency perturbations [5]. However, some studies raised the possibility of spontaneous generation of very low frequency cerebral blood flow oscillations [4]. Herein, we use TCD to explore the nature of very low frequency variability of cerebral blood flow.

Methods

Patients

The study comprised 38 brain injury patients treated at the Department of Neurosurgery of Addenbrooke's Hospital whose intracranial pressure during the monitoring interval did not exceed 15 mmHg (mean ICP in the group was 10.9±3.6 mmHg). The median GCS was 5 (range 3–8). The control group was made up of 17 volunteers (7 men and 10 women, mean age of 24±3 years) free of cardiovascular, pulmonary, and cerebrovascular disorders. Flow velocity in middle cerebral artery (MCAfv) was recorded by transcranial Doppler ultrasonography (TCD). The arterial blood pressure was concurrently measured by either the radial artery cannulation (patients) or noninvasively by finger photoplethysmography (volunteers). It is worth pointing out that the differences between aortic and finger blood pressure fluctuations can lead to erroneous results for very short segments of data [11]. Herein we average the calculated hemodynamic parameters over relatively long recordings which minimizes a potentially significant source of error.

Data analysis

For each patient or volunteer a single 15 min monitoring interval was chosen for analysis. Beat-to-beat average values of arterial blood pressure and flow velocity in middle cerebral artery were calculated via waveform integration of the corresponding signals. In numerical calculations non-uniformly spaced time series were resampled at 2 Hz by cubic spline interpolation.

We use time averaged moduli of complex Morlet wavelet coefficients $W_s(f, t_0)$ as a frequency-dependent measure of variability of the signal $s(t)$:

$$V_s(f) = \langle |W_s(f, t_0)| \rangle_{t_0}$$

hereafter the angular brackets denote averaging over time [6]. The interdependence of ABP and MCAfv spontaneous fluctuations may be quantified by the frequency dependent synchronization index:

$$\gamma_{ABP}^{MCA}(f) = \langle \sin \Delta\phi \rangle^2 + \langle \cos \Delta\phi \rangle^2,$$

where $\Delta\phi = \phi_1 - \phi_2$ is the instantaneous phase difference between the two time series calculated using the complex Morlet continuous wavelet transform [7]. The synchronization index lies in the interval $0 \leq \gamma \leq 1$. A vanishing index $\gamma=0$ corresponds to a uniform distribution of the phase differences (no synchronization) while $\gamma=1$ corresponds to perfect synchronization (phase locking of the two processes).

To quantify the amplification or attenuation of arterial blood pressure fluctuations we introduce the scale-dependent wavelet gain η defined as the ratio of time averaged wavelet coefficients of the corresponding signals:

$$\eta_{ABP}^{MCA}(f) \equiv \frac{\langle |W_{MCA}(f, t_0)| \rangle}{\langle |W_{ABP}(f, t_0)| \rangle}.$$

Results

Figure 1 shows the group averaged wavelet variability of both ABP (Fig. 1a) and MCAfv (Fig. 1b) plotted as a function of frequency. The prominent peak in Fig. 1a, centered at approximately 0.23 Hz is associated with the strong periodic component in ABP generated by the mechanical ventilation in brain injured patients. The respiration peak of the control cohort is less pronounced indicating that spontaneous breathing has much weaker influence on ABP dynamics. In fact, the mean arterial blood pressure variability for brain injury patients in the RF band (0.17–0.38 Hz) V_{ABP}=0.72±0.31 is more than a factor of two larger than that of healthy individuals (0.30±0.10). For lower frequencies the relation between ABP variability for the TBI patients and controls is reversed. In the very low frequency (VLF, 0.02–0.07 Hz) band the variability of the brain injured patients (V_{ABP}=0.36±0.28) is a factor of two smaller than that of the volunteers (V_{ABP}=0.70±0.25). Surprisingly, the differences in the patient's ABP variability are not transferred to the dynamics of flow velocity in middle cerebral artery. In fact, in both the RF and VLF intervals the mean values of the MCAfv wavelet variability between patients and controls *are not* statistically significant. For example the VLF wavelet variability of the brain injury patients V_{MCA}=0.83±0.65 is

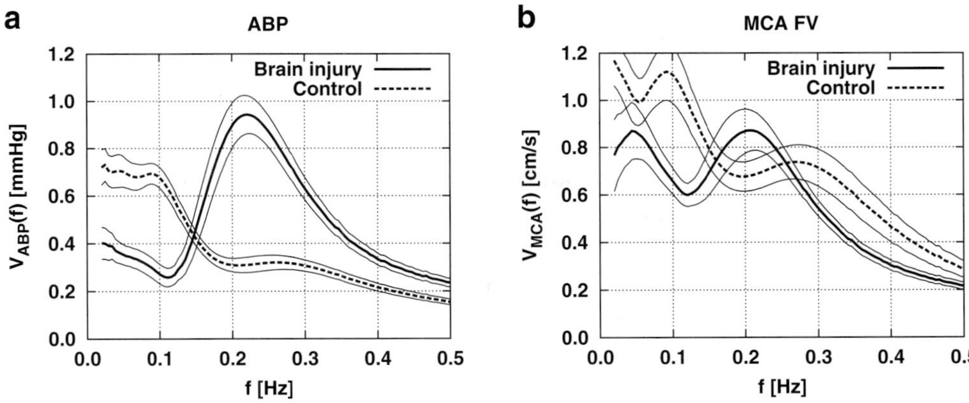

Fig. 1 Group averaged wavelet variability of time series of: (**a**) ABP, (**b**) MCAfv is plotted as a function of frequency. The solid line corresponds to the cohort of patients with low intracranial pressure and the dashed line to the control group. The thin lines denote the error of the mean

not statistically different from that of the volunteers 1.06± 0.41 (p=0.11).

The frequency dependence of the wavelet gain η^{MCA}_{ABP} and synchronization parameter γ^{MCA}_{ABP} is depicted in Fig. 2. One can see that in the brain injury patients the strong ABP fluctuations related to respiration are *attenuated* below that of the healthy cohort (Fig. 2a). Consequently, the patient group average RF wavelet gain $\eta^{MCA}_{ABP} = 0.99 \pm 0.43$ is more than a factor of two smaller than that of the controls ($\gamma^{MCA}_{ABP} = 0.18 \pm 0.14$). The opposite effect is observed for the VLF frequencies. The low ABP variability of the patients does not lead to reduced fluctuations of MCAfv. This is distinctly manifested in the values of the wavelet gain for the TBI patients: $\eta^{MCA}_{ABP} = 3.09 \pm 2.41$and the controls: $\eta^{MCA}_{ABP} = 1.61 \pm 0.70$. It is worth pointing out that in the VLF part of the spectrum the fluctuations of ABP are essentially statistically *independent* of MCAfv fluctuations as indicated by the low values of the synchronization index (Fig. 2b) for both the brain injury patients $\gamma^{MCA}_{ABP} = 0.18 \pm 0.14$ and the healthy individuals $\gamma^{MCA}_{ABP} = 0.20\pm$ 0.13.

Discussion

In the VLF range, the mean wavelet ABP variability of brain injury patients is *half* of that of the control group. Nevertheless, as clearly seen in Fig. 1b, the corresponding cohort averaged variabilities of cerebral flow velocity are not statistically different. *Thus, the VLF MCAfv oscillations must have been generated within the intracranial volume.* This active generation of cerebral flow oscillations in turn explains the lack of synchronization with ABP fluctuations (reflected by the low value of synchronization index $\gamma^{MCA}_{ABP} = 0.20 \pm 0.13$). It goes without saying that spontaneous generation of VLF flow oscillations cannot be reconciled with a passive control function of cerebral autoregulation.

The suppressed ABP variability of the patients in the VLF part of the spectrum was instrumental in demonstrating the spontaneous generation of cerebral flow velocity oscillations. However, an intravenous infusion of trimethaphan can block both sympathetic and parasympathetic nerve activity effectively inhibiting regulatory mechanisms controlled by autonomous nervous systems, e.g. the baroreflex. In a pioneering experiment [16] such a ganglion blockade was used to decrease mean arterial blood pressure variability of healthy volunteers in the 0.02–0.07 Hz frequency range by as much as 82% (measured as a fraction of the baseline Fourier spectrum). Despite this significant decrease, the MCAfv variability persisted leading to the 81% increase of the VLF transfer function gain. Zhang et al point out “Interestingly, although transfer function gain increased and phase decreased at the very low frequencies, they remained unchanged at higher frequencies. Thus, the frequency-dependent nature of dynamic autoregulation was abolished by ganglion blockade, suggesting that the cerebral circulation was unable to buffer even slow changes in arterial pressure.” Finally, these authors conclude that autonomic neural control of the cerebral circulation is tonically active and plays a significant role in the regulation of beat-to-beat cerebral blood flow in humans.

Two aspects of the ganglion experiment lead us to suggest an alternative interpretation. Zhang et al demonstrated that the application of oscillating lower body negative pressure (LBNP) did restore low-frequency ABP variability but it did not significantly modify the hemodynamic effects of the blockade. This is a strong argument supporting the autonomic neuronal control hypothesis. However, the analysis of the representative spectra of mean blood pressure and MCAfv during LBNP (Fig. 4 of reference [16]) shows that the cerebral blood flow velocity spectrum is dominated by a distinct peak centered at a frequency different than that of the LBNP oscillations. The appearance of such a peak is a clear manifestation of spontaneous generation of oscillations – precisely the same effect observed in our brain injury cohort. We suggest that the generation of oscillations observed both in the ganglion blockade and in our current study does not imply impairment of pressure autoregulation. In fact in a closely related recent study Lavi et al [8] confirmed the integrity of pressure dependent autoregulation in a group of patients with impaired autonomic cardiovascular control caused by diabetes mellitus and/or hypertension. Lavi et al conclude that their research precludes the involvement of autonomic nervous system in mechanoregulation of cerebral blood flow. They point out the work of Brevan et al [3] showing that isolated sections of human pial arteries have poor innervations and attenuated responsiveness to adrenergic vasoconstrictors. Taking into account both the experimental results and the controversy concerning the influence of the autonomous nervous system on cerebral circulation we sought to identify a plausible physiological mechanism, which could account for the observed generation of spontaneous VLF oscillations.

Vasomotion is an oscillation of vascular tone that is generated from within the vascular wall and is not a consequence of heartbeat, respiration, or neuronal input. The frequency of these oscillations, observed in vivo and in vitro in all vascular beds, span the broad range from 0.02 Hz to 0.3 Hz [1, 10]. Even though this phenomenon was first observed more than 150 years ago, its physiological significance still remains elusive. From the viewpoint of cerebral hemodynamics, one of the most interesting theoretical hypothesis is that vasomotion is beneficial for tissue oxygenation. The results of modeling [15] suggest that low-frequency vasomotion characteristic of proximal arterioles enhances long-range diffusion of oxygen to the capillaries. Consequently, during periods of critically low perfusion oxygen becomes more homogeneously distributed reducing the amount of surrounding tissue exposed to hypoxia. Certainly, vasomotion may play a pivotal protective role in tissues with low oxygen buffering capacity. Brain tissue is one of them.

The high value of the synchronization parameter γ_{ABP}^{MCA} in the RF band is a strong indication that for these frequencies the variability of cerebral blood flow may, to large extent, be attributed to the ABP dynamics (cf. Fig. 2b). This interpretation is consistent with previous studies [9, 12]. With decreasing frequency the connection between the statistical properties of the ABP and MCAfv time series becomes less obvious. We can see that for healthy volunteers the synchronization parameter γ_{ABP}^{MCA} has a distinct peak at 0.12 Hz which coincides with the operating frequency of baroreflex. This peak is conspicuously missing in the traumatic brain injury patients. We consider this evidence that autonomic cardiovascular control is impaired in TBI patients. On the other hand, the low synchronization in the VLF part of the spectrum, observed in both TBI patients and controls, may result from either the lack of causal relation between ABP and MCAfv fluctuations or more intricate nonlinear interdependence. Using different methodologies, other investigators [9, 12] have found that the VLF end-tidal CO_2 fluctuations significantly influences MCAfv dynamics. In addition, the work of Mitsis et al [9] emphasizes the considerable effect of nonlinear interactions between ABP and CO_2 on cerebral blood flow variability.

Spontaneous vasomotion, which could naturally account for weak synchronization between ABP and MCA velocity at low frequency, is physiologically a second order effect. However, when the ABP variability is abolished, either by pharmacological intervention or as a result of brain injury, this second order mechanism may in fact determine cerebral hemodynamics. We can see in Fig. 2a that for the brain injury cohort the wavelet gain begins to rapidly increase at 0.2 Hz. We hypothesize that this frequency delineates a frequency region, in which active spontaneous generation of oscillations due to vasomotion is significant. The broad range of vasomotion frequencies (from 0.02 Hz to 0.3 Hz) is compatible with such an interpretation.

We conclude that spontaneous vasomotion may be beneficial for brain tissue oxygenation especially during

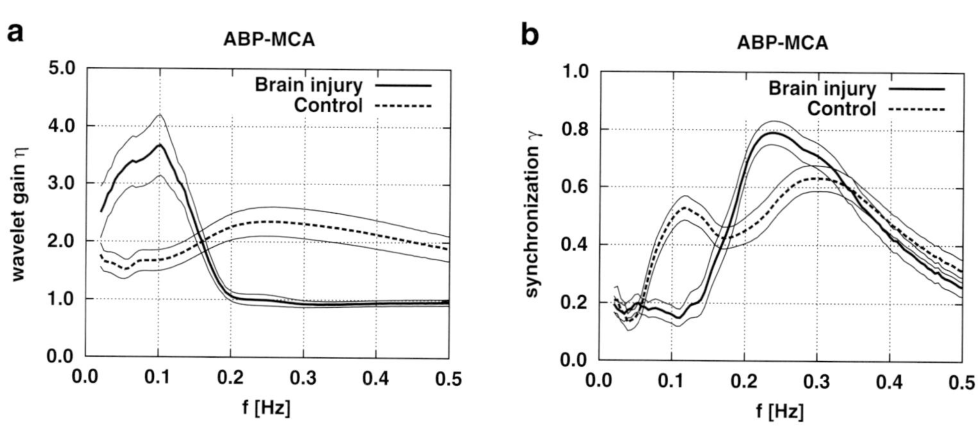

Fig. 2 Group averaged wavelet gain (**a**) and synchronization parameter (**b**) are plotted as a function of frequency. The solid line corresponds to the cohort of patients with low intracranial pressure and the dashed line to the control group. The thin lines denote the error of the mean. The wavelet gain is given in (cm/s)/mmHg

periods of critically low perfusion and consequently may play a significant role in cerebral hemodynamics of TBI patients.

Acknowledgments Malgorzata Turalska, Miroslaw Latka, and Krystyna Pierzchala acknowledge the financial support of the Polish Ministry of Science and Information Society Technologies (Grant 3 T11E 023 29). Bruce J. West was supported by the U.S. Army Research Office.

Conflict of interest statement We declare that we have no conflict of interest.

References

1. Aalkaer C, Nilsson H (2005) Vasomotion: cellular background for the oscillator and for the synchronization of smooth muscle cells. Br J Pharmacol 144:605–616
2. Aaslid R, Lindegaard KF, Sorteberg W, Nornes H (1989) Cerebral autoregulation dynamics in humans. Stroke 20:45–52
3. Bevan RD, Dodge J, Nichols P, Penar PL, Walters CL, Wellman T, Bevan JA (1998) Weakness of sympathetic neural control of human pial compared with superficial temporal arteries reflects low innervation density and poor sympathetic responsiveness. Stroke 29:212–221
4. Giller CA, Hatab MR, Giller AM (1999) Oscillations in cerebral blood flow detected with a transcranial Doppler index. J Cereb Blood Flow Metab 19:452–459
5. Giller CA, Mueller M (2003) Linearity and non-linearity in cerebral hemodynamics. Med Eng Phys 25:633–646
6. Latka M, Kolodziej W, Turalska M, Latka D, Zub W, West BJ (2007) Wavelet assessment of cerebrospinal compensatory reserve and cerebrovascular reactivity. Physiol Meas 28:465–479
7. Latka M, Turalska M, Glaubic-Latka M, Kolodziej W, Latka D, West BJ (2005) Phase dynamics in cerebral autoregulation. Am J Physiol Heart Circ Physiol 289:H2272–H2279
8. Lavi S, Gaitini D, Milloul V, Jacob G (2006) Impaired cerebral CO2 vasoreactivity: association with endothelial dysfunction. Am J Physiol Heart Circ Physiol 291:H1856–H1861
9. Mitsis GD, Poulin MJ, Robbins PA, Marmarelis VZ (2004) Nonlinear modeling of the dynamic effects of arterial pressure and CO2 variations on cerebral blood flow in healthy humans. IEEE Trans Biomed Eng 51:1932–1943
10. Nilsson H, Aalkjaer C (2003) Vasomotion: mechanisms and physiological importance. Mol Interv 3:79–89, 51
11. Panerai RB, Sammons EL, Smith SM, Rathbone WE, Bentley S, Potter JF, Samani NJ (2007) Transient drifts between Finapres and continuous intra-aortic measurements of blood pressure. Blood Press Monit 12:369–376
12. Panerai RB, Simpson DM, Deverson ST, Mahony P, Hayes P, Evans DH (2000) Multivariate dynamic analysis of cerebral blood flow regulation in humans. IEEE Trans Biomed Eng 47: 419–423
13. Paulson OB, Strandgaard S, Edvinsson L (1990) Cerebral autoregulation. Cerebrovasc Brain Metab Rev 2:161–192
14. Settakis G, Molnar C, Kerenyi L, Kollar J, Legemate D, Csiba L, Fulesdi B (2003) Acetazolamide as a vasodilatory stimulus in cerebrovascular diseases and in conditions affecting the cerebral vasculature. Eur J Neurol 10:609–620
15. Tsai AG, Intaglietta M (1993) Evidence of flowmotion induced changes in local tissue oxygenation. Int J Microcirc Clin Exp 12:75–88
16. Zhang R, Zuckerman JH, Iwasaki K, Wilson TE, Crandall CG, Levine BD (2002) Autonomic neural control of dynamic cerebral autoregulation in humans. Circulation 106:1814–1820

Acta Neurochir Suppl (2008) 102: 49–52

Accuracy of non-invasive ICP assessment can be increased by an initial individual calibration

B. Schmidt · M. Weinhold · M. Czosnyka · S. A. May · R. Steinmeier · J. Klingelhöfer

Abstract

Objective In a formerly introduced mathematical model, intracranial pressure (ICP) could be non-invasively assessed using cerebral blood flow velocity (FV) and arterial blood pressure (ABP). The current study attempts to check whether the accuracy of the non-invasive ICP assessment (nICP) improves after an initial individual calibration by implanted ICP probes.

Methods Thirteen patients with brain lesions (35–77 years, mean: 58±13 years) were studied. FV, ABP and ICP signals were recorded at days 1, 2, 4 and 7. nICP was calculated and compared to ICP. In the first recording of each patient the (invasively assessed) ICP signal was used to calibrate the nICP calculation procedure, while the follow-up recordings were used for its validation.

Findings In 11 patients 22 follow-up recordings were performed. The mean deviation between ICP and the original nICP (±SD) was 8.3±7.9 mmHg. Using the calibrated method this deviation was reduced to 6.7±6.7 mmHg ($P<0.005$).

Conclusions Initial individual calibration of nICP assessment method significantly improves the accuracy of nICP estimation on subsequent days. This hybrid method of ICP assessment may be used in intensive care units in patients with initially implanted ICP probes. After removal of the probes, ICP monitoring can be continued using the calibrated nICP assessment procedure

Keywords Intracranial pressure · Cerebral blood flow · Transcranial Doppler ultrasonography · Arterial blood pressure

B. Schmidt (✉) · M. Weinhold · J. Klingelhöfer
Department of Neurology, Medical Centre Chemnitz,
Dresdner Str.178,
09131 Chemnitz, Germany
e-mail: B.Schmidt@skc.de

M. Czosnyka
Academic Neurosurgical Unit, Addenbrooke's Hospital,
P.O. Box 167, Hills Road,
Cambridge CB2 2QQ, UK

S. A. May · R. Steinmeier
Department of Neurosurgery, Medical Centre Chemnitz,
Flemmingstr. 2,
09116 Chemnitz, Germany

Introduction

Various approaches have been made in the past to analyse the relationship between ICP and parameters derived from cerebral blood flow velocity (FV) and arterial blood pressure (ABP) [1–5]. The authors formerly introduced a mathematical model in which selected hemodynamic parameters (TCD characteristics) were used for a transformation of the ABP into the ICP signal. [8, 9]. Since the introduction of this method of non-invasive ICP (nICP) assessment its accuracy and potential benefit for clinical application have been demonstrated in various clinical studies [10–12]. It could be shown that the procedure was generally able to assess ICP pulse waves, B waves, plateau waves [6, 7] and ICP trends, while the mean deviation between non-invasively and invasively assessed ICP was between 4 and 8 mmHg depending on the studied population. Although being generally valid, some results suggested that this model might show individual deviations. Several attempts have been made by the authors in order to contribute to this fact and reduce the error in nICP assessment even in heterogeneous patient groups. One approach was the introduction of an nICP assessment model which was self-adaptive to dynamic changes of cerebral autoregulatory state [12], another was the use of fuzzy pattern classification method for model construction [13], a non-linear method that allowed the integration of a large amount of influential parameters. The objective of the current study was to verify whether the accuracy of the nICP assessment procedure could be improved if the procedure was individually calibrated by an implanted ICP probe.

Materials and methods

Patient population Thirteen acute neurocritical care patients (35–77 years, mean age: 58±13 years, 8 male, 5 female) were included. They suffered either from traumatic brain injury (N=5), subarachnoidal hemorrhage (N=2), MCA infarction (N=2), intracranial hemorrhage, sinus venous thrombosis, hypoxic encephalopathy or encephalitis. Patients were sedated and ventilated.

Monitoring FV measurements were taken using 2 MHz pulsed Doppler devices (TC 2-64B, EME, Überlingen, Germany). Flow patterns of the MCA were continuously recorded in the hemisphere ipsilateral to the ICP measurement. Blood pressure was measured with a standard manometer line inserted into the radial artery. ICP was measured using either implanted intraventricular or intraparenchymal microsensors (Raumedic AG, Rehau, Germany).

Monitoring was a routine clinical practice and did not require separate Local Ethical Committee approvals. Data were processed anonymously as a part of internal clinical audit.

Computer-assisted recording A personal computer fitted with data acquisition system (Daq 112B, Iotech, Inc., Cleveland, OH, USA) was used for recording and analyzing FV, ABP and ICP signals. The sampling frequency was 25 Hz. Signal data consisting of FV, ABP and (invasively assessed) ICP were initially recorded at day 1, and if possible repeatedly recorded at days 2, 4 and 7, for the duration of one hour each. The day 1 recording of each patient was taken to calibrate the nICP calculation procedure while the follow-up recordings were used for its validation. Patients with day 1 recording only could not be evaluated. Data recording software written for this purpose was used [9].

Non-invasive ICP assessment In the procedure of non-invasive ICP assessment the ABP signal is transformed into the ICP signal by means of a signal transformation called Impulse Response, a function closely related to the more familiar transfer function. During nICP assessment the ABP→ICP impulse response is controlled and continuously updated by parameters called TCD characteristics. The TCD characteristics again are continuously recalculated from the currently measured FV and ABP signals. They substantially consist of coefficients of an ABP→FV impulse response and additional ICP related parameters such as pulsatility index. The model assumes linear relationship between TCD characteristics and the ABP→ICP impulse response. This relationship has been computed by means of multiple regression analysis on data recordings of FV, ABP and (invasively assessed) ICP from a well-defined group of reference patients. In Fig. 1 a flow chart of the nICP assessment is presented.

Calibration of non-invasive ICP assessment

In order to individually calibrate non-invasive ICP assessment to a certain patient the signal data of this patient was added to the above-mentioned reference patient data. Using this enriched data the nICP assessment procedure was reconstructed (Fig. 2). In this way the peculiarity of this patient with regard to the relationship between TCD characteristics and ABP→ICP impulse response was integrated into the model.

After calibration of nICP assessment using the patient's day 1 recording, nICP signals were calculated in the subsequent days using both the calibrated method as well as the standard method.

Results

In two of 13 patients follow-up data could not be assessed because the patients died. In 11 patients 22 follow-up recordings were performed (in three patients: one, in five patients two, and in three patients: three follow-up recordings). On average of the follow-up days 2, 4 and 7 the invasively assessed ICP (±SD) was 15.3±17.2 mmHg, the standard nICP was 19.6±9.6 mmHg, and the calibrated nICP assessed was 15.8±11.3 mmHg. The mean deviation

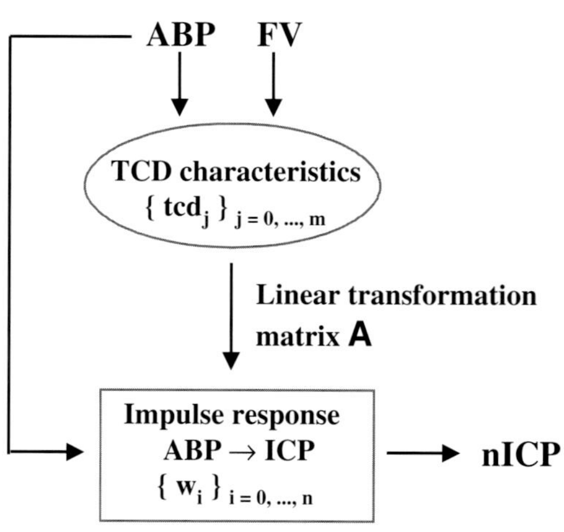

Fig. 1 nICP assessment procedure. From FV and ABP signals, TCD characteristics are computed. Applying a linear transformation matrix **A** to the TCD characteristics yields the ABP→ICP impulse response function, which transforms ABP into the nICP signal. The matrix **A** have been formerly calculated by a multiple regression analysis on patients reference data

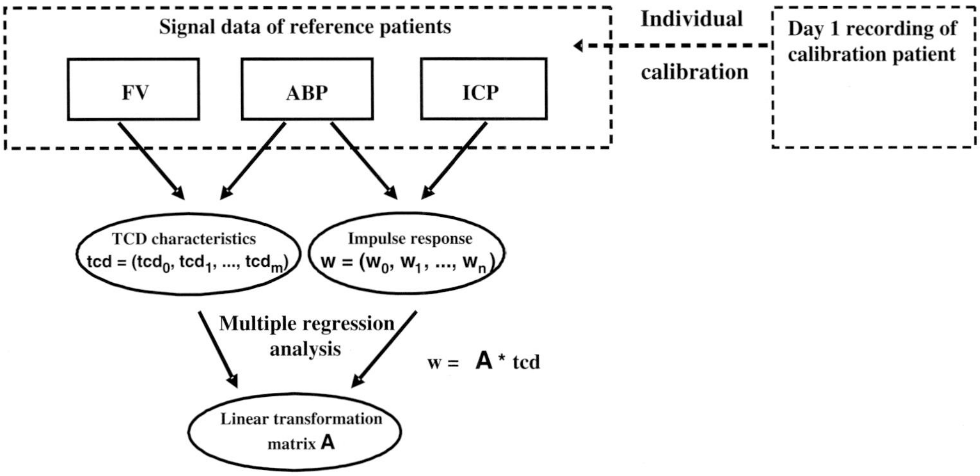

Fig. 2 Construction of nICP assessment procedure. Signal data of reference patients is used for the calculation of transformation matrix **A**. Samples of TCD characteristics are computed at different time points from FV and ABP recordings. Simultaneously ABP→ICP impulse response parameters are calculated from ABP and ICP recordings. Both parameter groups are then related by multiple regression analysis in terms of matrix **A**. Individual calibration to a certain patient is achieved by previous inclusion of this patient's signal data to the reference data

between ICP and in the standard way assessed nICP (±SD) was 8.3±7.9 mmHg, the nICP median was 6.5 mmHg. Using the calibrated method this deviation was significantly ($P<0.005$) reduced to 6.7±6.7 mmHg, the nICP median was 4.4 mmHg (Fig. 3).

Discussion

The results showed that initial individual calibration of nICP assessment method could significantly improve the accuracy of follow-up nICP assessments. The patients included so far suffered from a great variety of cerebral diseases. This was in contrast to former studies of the authors [11, 12] where homogeneity of the patient population was an important point for accurate nICP assessment results. This heterogeneity of population in the current study may be seen as one of the reasons for relatively high deviations between measured and calculated ICP values. However, being able to broaden its applicability was another intention for the introduction of initial calibration of nICP assessment procedure. In fact, stenosis of intra- or extracranial vessels needs no longer be an exclusion criterion [8, 12] because of the initial adaptation

Fig. 3 Comparison between ICP, standard nICP and calibrated nICP in a patient with subarachnoidal hemorrhage during day 2 (*left*) and day 4 (*right*) recordings. In both recordings similarity in terms of mean values as well as shapes of pulse waves was visibly higher between calibrated nICP and ICP than it was between standard nICP and ICP. In the day 4 recording mean FV has been increased above 110 cm/s indicating the development of vasospasm in this patient. Vasospasm was an exclusion criterion for the application of the standard nICP method in former studies [8, 12]. This might explain the high deviation between standard nICP and ICP in this recording

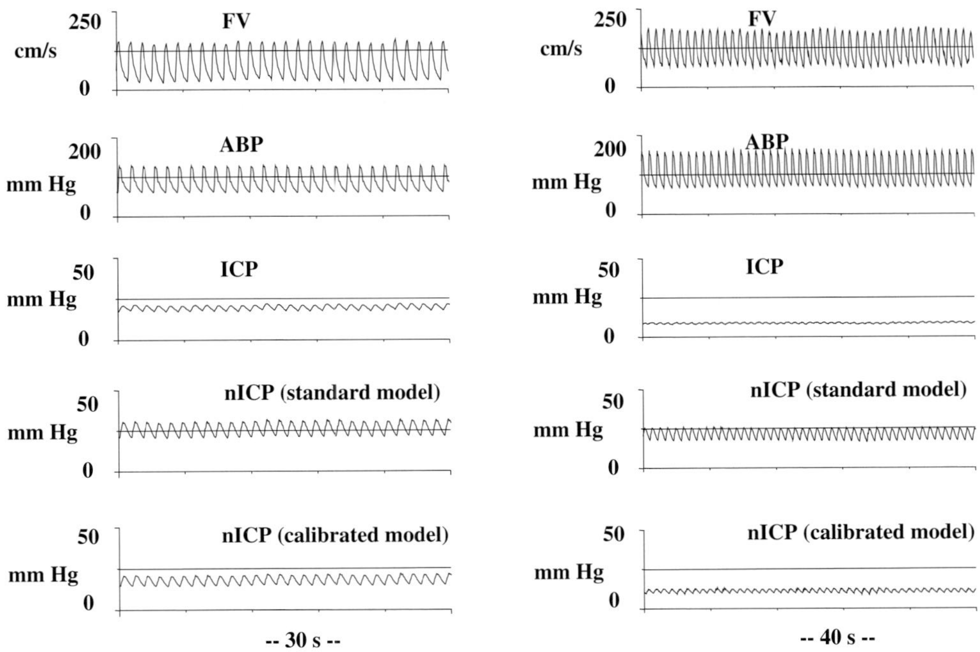

of the nICP procedure to the possibly occurring individual differences in hemodynamics. A slightly different point is the inclusion of patients with vasospasm. The degree of vasospasm may rapidly change in time and abolish the validity of former procedure adaptations. Although in the presented patient with vasospasm (Fig. 3) clearly improved results could be achieved by calibration, this question remains a subject for further investigations.

Strictly speaking, the introduced method cannot be called non-invasive, it rather provides a hybrid procedure and includes the usage of invasive ICP assessment. However, in many cases, i.e. in patients with acute severe traumatic brain injuries, implantation of pressure probes is currently an act of clinical routine and will probably remain in future. Such a situation may be used for a calibration of the nICP procedure in order to increase its accuracy or even to enable its applicability. The intention of this study was not to replace nICP assessment by the hybrid method rather than to search for improvements of accuracy whenever it is possible. Other studies to improve the accuracy of the "pure" nICP assessment are in progress. There are still open questions: It is not clear how the benefit depends on the time of follow-up assessment. It seems plausible that this benefit might decrease with increment of time passed since the calibration. However. the population is yet to small to quantify or even to confirm this effect.

Conclusion and outlook

Initial individual calibration of nICP assessment method significantly improves the accuracy of follow-up nICP assessment and provides a broader applicability of this method. Such a hybrid method of ICP assessment may be used in intensive care units in patients with temporarily implanted ICP probes. After removal of the probes, whether planned or as a result of any technical/clinical complication ICP monitoring can be continued using the calibrated nICP assessment procedure. The study is ongoing and a minimum of 30 included patients is intended.

Conflict of interest statement We declare that we have no conflict of interest.

References

1. Aaslid R, Lundar T, Lindegaard KF, Nornes H (1993) Estimation of cerebral perfusion pressure from arterial blood pressure and transcranial Doppler recordings. In: Miller JD, Teasdale GM, Rowan JO, Galbraith SL, Mendelow AD (eds) Intracranial pressure VI. Springer, Berlin, pp 226–229
2. Chan KH, Miller JD, Dearden NM, Andrews PJ, Midgley S (1992) The effect of changes in cerebral perfusion pressure upon middle cerebral artery blood flow velocity and jugular bulb venous oxygen saturation after severe brain injury. J Neurosurg 77:117–130
3. Czosnyka M, Matta BF, Smielewski P, Kirkpatrick P, Pickard JD (1998) Cerebral perfusion pressure in head-injured patients: a noninvasive assessment using transcranial Doppler ultrasonography. J Neurosurg 88(5):802–808
4. Klingelhöfer J, Conrad B, Benecke R, Sander D (1987) Relationship between intracranial pressure and intracranial flow patterns in patients suffering from cerebral diseases. J Cardiovasc Ultrasonogr 6:249–254
5. Klingelhöfer J, Conrad B, Benecke R, Sander D, Markakis E (1988) Evaluation of intracranial pressure from transcranial Doppler studies in cerebral disease. J Neurol 235:159–162
6. Lundberg N (1960) Continuous recording and control of ventricular fluid pressure in neurosurgical practice. Acta Psych Neurol Scand (Suppl) 149:1–193
7. Newell DW, Aaslid R, Stooss R, Reulen HJ (1992) The relationship of blood flow velocity fluctuations to intracranial pressure B waves. J Neurosurg 76:415–421
8. Schmidt B, Klingelhöfer J, Schwarze JJ, Sander D, Wittich I (1997) Noninvasive prediction of intracranial pressure curves using transcranial Doppler ultrasonography and blood pressure curves. Stroke 28:2465–2472
9. Schmidt B, Schwarze JJ, Czosnyka M, Sander D, Wittich I, Klingelhöfer J (1998) A method for a simulation of continuous intracranial pressure curves. Comp Biomed Res 31(4):231–243
10. Schmidt B, Czosnyka M, Schwarze JJ, Sander D, Gerstner W, Lumenta CB, Pickard JD, Klingelhöfer J (1999) Cerebral vasodilatation causing acute intracranial hypertension—a method for noninvasive assessment. J Cereb Blood Flow Metab 19:990–996
11. Schmidt B, Czosnyka M, Schwarze JJ, Sander D, Gerstner W, Lumenta CB, Klingelhöfer J (2000) Evaluation of a method for noninvasive intracranial pressure assessment during infusion studies in patients with hydrocephalus. J Neurosurg 92:793–800
12. Schmidt B, Czosnyka M, Raabe A, Yahya H, Schwarze JJ, Sackerer D, Sander D, Klingelhöfer J (2003) Adaptive noninvasive assessment of cerebral autoregulation and ICP. Stroke 34:84–89
13. Schmidt B, Bocklisch SF, Päßler M, Czosnyka M, Schwarze JJ, Klingelhöfer J (2005) Fuzzy pattern classification of hemodynamic data can be used to determine noninvasive intracranial pressure. Acta Neurochir [suppl] 95:345–349

Acta Neurochir Suppl (2008) 102: 53–55

The impact of ventricular catheter impregnated with antimicrobial agents on infections in patients with ventricular catheter: interim report

George K. C. Wong · W. S. Poon · Stephanie C. P. Ng · Margaret Ip

Abstract

Introduction Previous prospective study in our unit had shown that the use of dual antibiotic prophylaxis in patients with external ventricular drain was associated with decreased incidence of cerebrospinal fluid infection but complicated with opportunistic extracranial infection. In recent years, cerebrospinal fluid shunt catheters impregnated with antimicrobial agents have become available. Theoretically, these catheters provide antibiotic prophylaxis locally without the associated complications of systemic opportunistic infection.

Methods We carried out a prospective randomized, controlled clinical trial in a regional neurosurgical center in Hong Kong. We recruited patients admitted for emergency neurosurgical operation after informed consent was obtained from next-of-kin. Eligible patients were randomized to receive an antibiotic-impregnated ventricular catheter or plain ventricular catheter Dual prophylactic antibiotic coverage was given to the patients randomized for plain ventricular catheter only. Patients who received antibiotic impregnanted catheters were not treated with systematic prophylactic antibiotics. Here we present the analysis of 110 patients, recruited over a 2-year period, to receive antibiotic-impregnanted ventricular catheters versus non-impregnated ventricular cathethers with prophylactic antibiotic coverage.

Findings Fifty-two patients were randomized to antibiotic-impregnated ventricular catheter with no systemic antibiotic prophylaxis (Group A) and 58 patients were randomized to plain ventricular catheters with prophylactic dual antibiotics (Group B). There was no ventriculostomy-related infection in either groups of patients. There was also no statistical significant difference in incidences of extracranial infections between the two groups, p=0.617.

Conclusions In this analysis, antibiotic-impregnation of ventricular catheters was as effective as systemic antibiotics in the prevention of ventriculostomy infections, with the potential advantage of avoiding the systemic side-effects of prophylactic antibiotics.

Keywords Ventriculostomy · Infection · Cerebrospinal fluid · Antibiotic

G. K. C. Wong · W. S. Poon (✉) · S. C. P. Ng
Division of Neurosurgery, Prince of Wales Hospital,
The Chinese University of Hong Kong,
Shatin, New Territories, Hong Kong SAR, China
e-mail: wpoon@surgery.cuhk.edu.hk

M. Ip
Department of Microbiology, Prince of Wales Hospital,
The Chinese University of Hong Kong,
New Territories, Hong Kong SAR, China

Introduction

External ventricular catheters are used for intracranial pressure monitoring and temporary cerebrospinal fluid drainage in neurosurgery. An incidence of ventriculostomy-related cerebrospinal fluid infection has been quoted to be between 2.2% and 10.4% [3–6, 10–12] in the more recent literature. A previous prospective study in our unit [7] had shown that the use of dual antibiotics prophylaxis in patients with external ventricular drainage was associated with decreased incidence of cerebrospinal fluid (CSF) infection but was complicated by opportunistic extracranial infections. The current practice is to cover all patients with external ventricular drains with prophylactic dual antibiotics, unless microbiology results guide more selective antibiotic regimens. In recent years, cerebrospinal fluid shunt catheters impregnated with antimicrobial agents have become available. The antibiotic-impregnated catheter is a silicone catheter impregnated with 0.15% clindamycin and 0.054% rifampicin. Experimental studies [2, 7] have shown that these antibiotic impregnanted catheters provide protection against Staphylococcus aureus and coagulase-negative staphylococci strains for at least 42 days. Theoretically, this provides antibiotic prophylaxis locally without the associated complications of systemic antibiotics. The beneficial effect in reducing CSF infection was shown with reduction in positive CSF cultures by Zabramski [14], in which they were not employing the effective policy of systemic dual antibiotics for prophylaxis as long as the ventricular catheter was in situ, and no data was available in terms of nosocomial infection was available in the literature.

Method

We carried out a prospective randomized controlled clinical trial in a regional neurosurgical center in Hong Kong to investigate whether antibiotics-impregnated ventricular catheters could replace prophylactic systemic antibiotics prophylaxis.

The pre- and post-operative care was in accordance with standard protocols currently used in the center. The ventricular catheter might be inserted as a separate procedure or in the setting of craniotomy as determined clinically. Prophylactic antibiotic coverage was used during the period of external ventricular drain placement. This was based on the results of a randomized controlled trial performed in our unit [8], which demonstrated a reduction in CSF infections with peri-procedural prophylactic antibiotic therapy. Subsequent analyses by Rebuck et al. [9] and Alleyne et al. [1] have questioned this effect, but the retrospective nature of their studies and the choice of antibiotics used makes comparison difficult. We performed all our ventricular drain insertion under stringent aseptic techniques in the operating theatre with a subgaleal tunneling of 5 cm. We had a standard guideline and standard instrument sets. Cerebrospinal fluid was collected under aseptic technique from the three-way connector just distal to the ventricular catheter every 5 days or on evidence of clinical sepsis. We did not perform regular exchanges of external ventricular drain catheters according to the published randomized controlled clinical trial of our unit [13].

Patients who were likely to have ventricular catheter in-situ for 5 days or more would be recruited before emergency or elective neurosurgical operation. They included patients with head injury, haemorrhagic stroke or hydrocephalus. Randomization would be taken care of by the research assistant through a computer-generated randomization number. Consent was obtained from a guardian or next-of-kin if the patient was mentally unfit for consent. Patients were randomized into one of the two groups:

1. Periprocedural antibiotics only ie Unasyn and Rocephin and insertion of the antibiotics-impregnated (0.15% Clindamycin and 0.05% Rifampicin) ventricular catheter;
2. Periprocedural antibiotics and prophylactic dual antibiotics ie Unasyn and Rocephin and insertion of ventricular catheter without impregnation of antibiotics.

"Prophylactic" means to use the antibiotics named above, as long as the ventricular catheter was in-situ. Changes in antibiotic treatment were directed by the clinical status of the patient, as for example, by the presence of concurrent, extracranial infection.

Primary outcomes include ventriculostomy-related infection and extracranial infection. Statistical analysis was carried out with SPSS for Window Version 14.0. Categorical outcome measures, such as CSF infection, were compared between groups using Fisher Exact Test. Continuous outcome measures, such as hospital stay, were compared between groups using unpaired t-test.

Results

Here we presented the interim analysis of the 110 patients, which were recruited over a 2 year period. 52 patients were randomized to antibiotic-impregnated ventricular catheter (Group A) and 58 patients were randomized to plain ventricular catheters with prophylactic dual antibiotics (Group B). Age (mean ± SD) was 50.1±14.1 years. Male to female ratio was 1:1. The neurosurgical indications were stroke (74%), trauma (17%) and others such as tumor and hydrocephalus (9%). Risk factors for ventriculsotomy-related infection included diabetes (11%), steroid consumption (4%), skull base fracture (10%) and craniotomy/craniectomy (57%). The duration of ventricular catheter in-situ (mean ± SD) was not different between the two groups, 9.9±5.4 days for Group A and 10.1±5.4 for Group B, p= 0.823. There were no ventriculostomy-related infections in either group of patients. There was also no statistical significant difference in the incidence of extracranial infections between the two groups, 51% for Group A and 46% for Group B, p=0.617. The incidence of resistant organisms (MRSA or fungi) in cases of extracerebral infection was also not different between the two groups.

Discussion

External ventricular catheter insertion through burr hole and ventriculostomy represented an advance in the management of neurosurgical patients. Through a simple procedure, neurosurgeons could have a simple way to (a) monitor change in intracranial pressure, which per sec would be a good indicator for new intracranial events such as oedema or rebleed; (b) provide a reliable mean for drainage of cerebrospinal fluid for hydrocephalus or intracranial pressure control. In patients with severe head injury, the application of a threshold intracranial pressure leads to an improvement of management outcome. The downside is that as a foreign body and conduit to the cerebrospinal fluid compartment, it carries a risk of infection, which may result in neurological morbidity and mortality. Systemic antibiotic prophylaxis has been prescribed, following a variety of different protocols, to reduce the incidence of catheter-induced CSF infection. Being systemic in nature, this has the risk of systemic side-effects such as allergic reactions and pseudomembranous colitis. Moreover, prophylactic antibiotic treatment has been associated with nosocomial infections with resistant strains.[7] The other way to give antibiotics prophylaxis would be topical. One simple way would be to give the antibiotics locally through the ventricular catheter. With the advance of technology, antibiotics impregnation of ventricular catheter became feasible. These catheters have the advantage of sustained release throughout the catheter-in-situ period. In our interim analysis, our results showed that the strategy of antibiotics-impregnated catheter insertion is as effective as systemic dual antibiotics prophylaxis in terms of cerebrospinal fluid infection and clinical outcome.

Conclusion

In this analysis, antibiotic-impregnated ventricular catheter was as effective as systemic antibiotics in prevention of ventriculostomy infection, with the potential advantage of avoiding the systemic side-effects of prophylactic antibiotic treatment.

Conflict of interest statement We declare that we have no conflict of interest.

Reference

1. Alleyne CH, Hassan M, Zabramski JM (2000) The efficacy and cost of prophylactic and periprocedural antibiotics in patients with external ventricular drains. Neurosurgery 47:1124–1129
2. Bayston R, Lambert E (1997) Duration of protective activities of cerebrospinal fluid infection shunt catheters impregnated with antimicrobial agents to prevent bacterial catheter-related infection. J Neurosurg 87:247–251
3. Gaskin PR, St John MA, Cave CT, Clarke H, Bayston R, Levett PN (1994) Cerebrospinal fluid shunt infection due to corynebacterium xerosis. J Infect 28:323–325
4. Holloway KL, Barnes T, Choi S, Bullock R, Marshall LF, Eisenberg HM, Jane JA, Ward JD, Young HF, Marmarou A (1996) Ventriculostomy infections: the effects of monitoring duration and catheter exchange in 584 patients. J Neurosurg 85:419–426
5. Mayhall CG, Archer NH, Lamb VA, Spadora AC, Baggett JW, Ward JD, Narayan RK (1984) Ventriculostomy-related infections: a prospective epidemiological study. N Engl J Med 310:553–559
6. Paramore CG, Turner DA (1994) Relative risks of ventriculostomy infection and morbidity. Acta Neurochir (Wien) 127:79–84
7. Pattavilakom A, Kotasnas D, Korman TM, Xenos C, Danks A (2006) Duration of in vivo antimicrobial activity of antibiotic-impregnated cerebrospinal fluid catheters. Neurosurgery 58:930–935
8. Poon WS, Ng S, Wai S (1998) CSF antibiotic prophylaxis for neurosurgical patients with ventriculostomy: a randomized study. Acta Neurochir Suppl (Wien) 71:146–148
9. Rebuck JA, Murry KR, Rhoney DH, Michael DB, Coplin WM (2000) Infection related to intracranial pressure monitors in adults: analysis of risk factors and antibiotic prophylaxis. J Neurol Neurosurg Psychiatry 69(3):381–389
10. Rosner MJ, Becker DP (1976) ICP monitoring: complications and associated factors. Clin Neurosurg 23:494–519
11. Rossi S, Buzzi F, Paparella A, Mainini P, Stocchetti N (1998) Complications and safety associated with ICP monitoring: a study of 542 patients. Acta Neurochir Suppl (Wien) 71:91–93
12. Winfield JA, Rosenthal P, Kanter RK, Casella G (1993) Duration of intracranial pressure monitoring does not predict daily risk of infective complications. Neurosurgery 33:424–431
13. Wong GK, Poon WS, Wai S, Yu LM, Lyon D, Lam JM (2002) Failure of regular external ventricular drain exchange to reduce cerebrospinal fluid infection: result of a randomized controlled trial. J Neurol Neurosurg Psychiatry 73:759–761
14. Zabramski JM, Whiting D, Darouiche RO, Horner TG, Olson J, Robertson C, Hamilton JA (2003) Efficacy of antimicrobial-impregnated external ventricular drain catheters: a prospective, randomized, controlled trial. J Neurosurg 98:725–730

Acta Neurochir Suppl (2008) 102: 57–61

Assessment of mitochondrial impairment and cerebral blood flow in severe brain injured patients

Gunes A. Aygok · Anthony Marmarou · Panos P. Fatouros · Birgit Kettenmann · Ross M. Bullock

Abstract

Background We believe that in traumatic brain injury (TBI), the reduction of *N*-acetyl aspartate (NAA) occurs in the presence of adequate cerebral blood flow (CBF) which would lend support to the concept of mitochondrial impairment. The objective of this study was to test this hypothesis in severely injured patients (GCS 8 or less) by obtaining simultaneous measures of CBF and NAA.

Methods Fourteen patients were studied of which six patients presented as diffuse injury at admission CT, while focal lesions were present in eight patients. CBF using stable xenon method was measured at the same time that NAA was measured by magnetic resonance proton spectroscopy (^{1}HMRS) in the MR suite. Additionally, diffusion weighted imaging (DWI) and maps of the apparent diffusion coefficient (ADC) were assessed.

Findings In diffuse injury, NAA/Cr reduction occurred uniformly throughout the brain where the values of CBF in all patients were well above ischemic threshold. In focal injury, we observed ischemic CBF values in the core of the lesions. However, in areas other than the core, CBF was above ischemic levels and NAA/Cr levels were decreased.

Conclusions Considering the direct link between energy metabolism and NAA synthesis in the mitochondria, this study showed that in the absence of an ischemic insult, reductions in NAA concentration reflects mitochondrial dysfunction.

Keywords *N*-acetyl aspartate (NAA) · Cerebral blood flow (CBF) · Apparent diffusion coefficient (ADC) · Mitochondrial impairment

This work was supported in part by National Institutes of Health Grant NS12587, M. Ross Bullock, P.I./Anthony Marmarou Project and by National Institutes of Health Grant NS19235, Anthony Marmarou, P.I.

G. A. Aygok (✉) · A. Marmarou · R. M. Bullock
Department of Neurosurgery,
Virginia Commonwealth University Medical Center,
1001 East Broad Street, Suite 235, P.O. Box 980508, Richmond,
VA 23298-0508, USA
e-mail: gaaygok@vcu.edu

P. P. Fatouros · B. Kettenmann
Department of Radiology,
Virginia Commonwealth University Medical Center,
Richmond, VA 23298, USA

P. P. Fatouros
Division of Radiation Physics and Biology,
Virginia Commonwealth University Medical Center,
Sanger Hall, B3-021A,
Richmond, VA 23298, USA

B. Kettenmann
Division of Radiation Physics and Biology,
Virginia Commonwealth University Medical Center,
Sanger Hall, Room B3-020,
Richmond, VA 23298, USA

R. M. Bullock
The Miami Project to cure Paralysis,
1095 NW 14 th Terrace,
Miami, FL 33136, USA

Introduction

Traumatic brain injury (TBI) remains a significant clinical entity that may result in death and severe disability among a predominantly young population. Investigators have found that in addition to the structural damage to the brain tissue, a series of complex pathophysiological mechanisms are initiated by the trauma [9]. These results in less ATP production and disturbance of the ion gradients essential for functional neurons which lead to changes at cellular levels such as swelling in astrocytes and neurons described as cytotoxic edema [8]. Furthermore, glutamate is released and causes a state of excitotoxicity [9] and changes at sub cellular levels such as mitochondrial impairment [7]. Neuronal function ceases, although the neurons may remain viable and recover.

These complex processes involved in energy crisis are not visualized by conventional imaging and one measure of energy crisis secondary to mitochondrial impairment is the amount of NAA reduction which can be measured quantitatively by ^{1}HMRS. NAA is synthesized in mitochondria and found to be almost exclusively located within the nervous system with a concentration which is only second to glutamate [9].Therefore, NAA reduction is a reflection of mitochondrial impairment and/or neuronal death. However, studies indicate that while 33% of severely head injured patients suffer an ischemic insult in the first 4 h, the proportion of patients with ischemia drops to 5% in the first 24 h [3]. Therefore, we hypothesize that, with an exception of the very first few hours following injury, there is a profound energy crisis even in the presence of adequate CBF. Our objective was to measure NAA and CBF simultaneously and establish the degree to which CBF confounds the interpretation of NAA. To better understand the mechanisms responsible for water movement into the brain, in vivo DWI and ADC were also utilized.

Materials and methods

Patient population and management

After obtaining informed consent, severely head-injured patients with an admission GCS of 8 or less were enrolled. All patients were treated with a standard TBI protocol and received ICP monitoring. Patients were transported to the CT scanner for measurement of CBF by Xenon technique and to the MR suites (Sigma 3.0 T, GE Medical Systems) for measurement of ^{1}HMRS, DWI and ADC and returned to the NICU without complication. All studies were performed within the first two weeks post injury (mean 9 days, range 3–18 days). CBF examinations reported here were performed sequentially on the same day of the ^{1}HMRS and ADC studies.

MRI-^{1}HMRS acquisition technique and data analysis

After stabilization into the magnet, T1 and T2 weighted pulse sequences were used to produce images in the axial and sagittal planes and semi-quantitative analysis of NAA, creatine containing compounds (Cr/PCr), and choline (Cho) was performed using the point resolved spectroscopy (PRESS) pulse sequence (TE=135 ms TR=1500 ms) Following localized shimming and water suppression, a spectrum from 8 cm^3 single voxel (SV) was obtained. The NAA, Cho and Cr peak areas were measured and results were reported as ratios. For diffuse injury, six different regions of interest (ROI) were selected including frontal, parietal and occipital lobes. White matter (WM) and gray matter (GM) were identified separately. For focal injury, regions of interest were core, perilesional and the symmetrically corresponding area in the contralateral hemisphere. All these voxels were placed on the white matter regardless of the ROI. Spectroscopy with 3-T magnet has an increased signal to noise ratio that allowed us to obtain spectra with higher resolution. In addition, 3-T magnet reduced the shim time which reduces the time that the patient is away from the neuroscience ICU.

Stable xenon-enhanced CT CBF technique and data analysis

CBF studies were performed using a CT scanner (Siemens, Erlangen, Germany) equipped with Xe-CT CBF imaging (Xe/TC system-2TM, Diversified Diagnostic Products, Inc., Houston, TX). Our technique required the acquisition of four head CT slices, each 10 mm thickness and separated from one another by 5 mm. Two baseline scans were performed at each level followed by multiple enhanced scans during inhalation of 30% xenon and 70% oxygen. CBF maps were calculated by means of the Kety–Schmidt equation using a commercially available package (Diversified Diagnostic Products, Inc., Houston, Texas). ROI's were positioned on CBF maps which were corresponding to the location and volume of spectral voxel. For purposes of analysis, CBF values below 18 ml/100 g/min were considered ischemic [2].

ADC acquisition technique and data analysis

DWI was performed using SE-EPI sequences. These pulse sequences generated an ADC trace image using a single shot technique with b value of 1,000 s/mm^2. Twenty five slices were generated with 5 mm slice thickness, 2 mm gap, 26×26 cm FOV, 96×132 matrix.

Statistical analysis

The NAA/Cr, CBF and ADC values were compared with controls for each ROI with an independent t test. Differences were regarded as statistically significant at $p<0.05$.

Results

Patient characteristics

This prospective study was composed of 14 patients (13 males and one female) that sustained serious head injury resulting from motor vehicle accidents (MVA; n=8), assault (n=4) and falls (n=2). The mean age of the entire study population was 30±10 years and ranged from 18 to 54 years. Six patients were presented as diffuse injury at admission CT, while eight evidenced a lesion, in which six patients had either surgical evacuation or decompressive craniectomy.

^{1}HMRS, CBF and ADC in healthy controls

After obtaining informed consent, ^{1}HMRS studies carried out in 12 healthy, age-matched controls previously screened to exclude prior head injuries. Figure 1 shows an axial MR image used to identify the spectroscopic voxel location and a normal spectra. The ratio of NAA/Cr from controls were calculated and averaged, equaling, 1.40±0.12 in gray matter and 1.75±0.17 in white matter. In mid 1990's, while stable xenon studies were first conducted in our institution, Bouma et al. [2] obtained CBF values in healthy volunteers using the imaging modalities and CBF protocols identical to those utilized in our head injured patients. Therefore, we referenced their result for the normal values which was about 50–55 ml/100 g of brain tissue/min in the adult. However, these values represent the global CBF and did not specify the gray and white matter separately. The ADC computed from hemispheric white matter ROI's in normal volunteers averaged 0.96±0.045 S.D.

^{1}HMRS in TBI patients

Of the 14 patients studied with SV technique, six patients were classified as diffuse injury and they had significantly lower white matter NAA/Cr ratios (1.46±0.24) than normal (p=0.0001). Gray matter NAA/Cr ratios were also significantly reduced (1.28±0.16; p=0.04). In focal injury, brain tissue was distorted in the core of the lesion, and therefore we did not measure the NAA/Cr ratios in this affected area. Perilesional NAA/Cr levels were decreased (1.20±0.38) as well as contra lateral NAA/Cr (1.32±0.15; p=0.0001 and p= 0.0001, respectively).

CBF in TBI patients

Despite the uniform NAA/Cr reduction in diffuse injury, the CBF values corresponding to each SV were above ischemic levels averaging 37.07±14 ml/min/100 g in white matter and 51±16 ml/min/100 g in gray matter. In diffuse injury, the global CBF of the left hemisphere was 46.3±10 ml/min/100 g and the right hemisphere was 45.3±10 ml/min/100 g which were normal for comatose patients. However, in focal injury, CBF levels in the core were ischemic (11.82±3.68 ml/min/ 100 g). The remaining brain tissue including perilesional and contralateral regions has non-ischemic CBF values (40.15± 17.63 ml/min/100 g and 37.85±14.01 ml/min/100 g respectively). In focal injury, the global CBF of the injured hemisphere was 47.3±12 ml/min/100 g and the non-injured hemisphere averaged.51.2±10 ml/min/100 g s.

ADC in TBI patients

In diffuse injury, ADC values corresponding to each SV equaled to 0.66±0.07 (p=0.0001) for white matter indicating

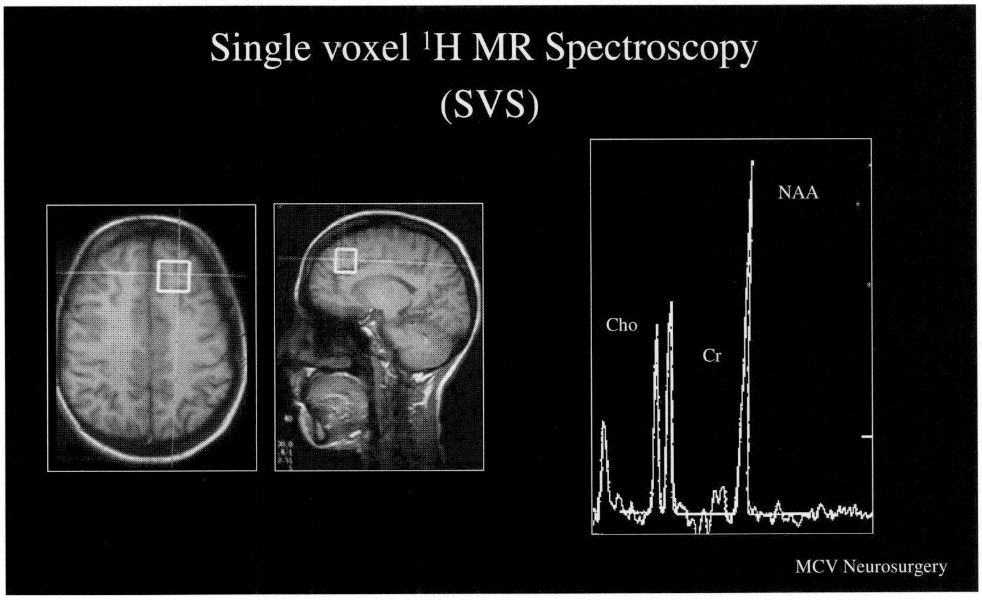

Fig. 1 Axial and sagittal MR image of a normal volunteer showing SV located in the right frontal lobe along with the corresponding proton spectrum. The tallest peak on the right represents NAA, the middle peak Cr and the leftmost peak is Cho

that in the presence of NAA/Cr depletion, ADC values were below normal indicating a cellular swelling. For focal injury, we did measure the ADC in the core and as expected ADC levels were high signifying vasogenic edema (1.05±0.03; p=0.08). Contrary, in the perilesional and contralateral areas, the ADC was below normal and averaged 0.71±0.05 (p=0.0001) and 0.69±0.05 (p=0.0001), respectively.

Discussion

This report provides supportive evidence that at time of study, when CBF measures were made in conjunction with ^{1}HMRS and ADC studies, the blood flow was well above ischemic thresholds. In six diffuse injury patients, NAA/Cr reduction occurred uniformly throughout the brain where the values of CBF in all patients were well above ischemic threshold. In focal injury, we observed ischemic CBF values in the core of the lesions which is a result of a direct damage to the neurons after trauma. In areas other than the core, CBF was above ischemic levels. An ischemic event at the time of injury or prior to our studies which recovered at the time of the study cannot be excluded. However we have to consider, based on previous work [3], that majority of these patients do not survive after a severe ischemic insult. For the patients who do survive, our data provide simultaneous NAA/Cr depletion and reduced ADC values coupled with non-ischemic CBF levels in patients. The results suggest a mitochondrial injury rather than a neuronal death.

The importance of *N*-acetylaspartate reduction in TBI

^{1}HMRS, if performed in the acute phase of the trauma would give an insight for the long term outcome which remains uncertain during the coma status. Garnet et al. [6] presented 26 patients, in which the frontal white matter, that appeared normal on conventional MRI, showed significant NAA/Cr and NAA/Cho reduction, significantly correlating with poor clinical outcome. Therefore the survival chance of impaired neurons and patient outcome would improve if we identify the underlying reason for the NAA loss.

The concept of neuronal loss resulting in reduced NAA

To date, studies have indicated that one third of severe head injured patients suffer from an ischemic insult in the very early hours after injury [3] which leads to a deprivation of oxygen and nutrients thereby causing a reduction in NAA and subsequent neuronal death. This hypothesis is also supported by the stroke literature where they found a close correlation between the decline in NAA and total number of neurons following permanent focal ischemia in mice [10]. However, there are several limitations with these studies and ischemia induced neuronal death may not be the only explanation accounting for NAA alteration. Firstly, the reported NAA values in stroke studies were mostly selected from a necrotic region thereby reflecting an obvious neuronal death and decline in NAA. Secondly, ^{1}HMRS and CBF studies were not performed sequentially; making correlation between metabolic and perfusion findings in addition, permanent decline or a possible recovery in NAA levels were not assessed with follow up studies.

The concept of mitochondrial impairment resulting in reduced NAA

Our laboratory findings indicated that [11] following moderate experimental injury, NAA reduced gradually and eventually recovered to reach baseline levels. Recovery of ATP was coupled with NAA recovery in moderately injured animals, while in severe injuries, with sustained ATP depletion, the NAA reduction was prolonged and did not recover. Our clinical studies as well as work by others are consistent with these experimental findings [5]. The association between mitochondrial impairment and subsequent decline in ATP and NAA can be explained by the pathological cascades triggered following TBI, initiated by activation of NMDA glutamate receptors and resulting in mitochondrial dysfunction and impaired ATP production [1, 4]. Considering the direct link between energy metabolism and NAA synthesis in the mitochondria, we conclude that NAA reduction found in head injured patients, in the absence of ischemia as confirmed by our measures, may reflect this energy crisis due to TBI-induced mitochondrial impairment.

Conflict of interest statement We declare that we have no conflict of interest.

References

1. Bernardi P, Broekemeier KM, Pfeiffer DR (1994) Recent progress on regulation of the mitochondrial permeability transition pore; a cyclosporin-sensitive pore in the inner mitochondrial membrane. J Bioenerg Biomembr 26:509–517
2. Bouma GJ, Muizelaar JP (1995) Cerebral blood flow in severe clinical head injury. New Horiz 3:384–394
3. Bouma GJ, Muizelaar JP, Choi SC, Newlon PG, Young HF (1991) Cerebral circulation and metabolism after severe traumatic brain injury: the elusive role of ischemia. J Neurosurg 75:685–693
4. Bullock R, Fujisawa H (1992) The role of glutamate antagonists for the treatment of CNS injury. J Neurotrauma 9(Suppl 2):S443–462
5. Friedman SD, Brooks WM, Jung RE, Chiulli SJ, Sloan JH, Montoya BT et al (1999) Quantitative proton MRS predicts outcome after traumatic brain injury. Neurology 52:1384–1391
6. Garnett MR, Corkill RG, Blamire AM, Rajagopalan B, Manners DN, Young JD et al (2001) Altered cellular metabolism

following traumatic brain injury: a magnetic resonance spectroscopy study. J Neurotrauma 18:231–240

7. Marmarou A, Signoretti S, Fatouros P, Aygok GA, Bullock R (2005) Mitochondrial injury measured by proton magnetic resonance spectroscopy in severe head trauma patients. Acta Neurochir Suppl 95:149–151
8. Marmarou A, Signoretti S, Fatouros PP, Portella G, Aygok GA, Bullock MR (2006) Predominance of cellular edema in traumatic brain swelling in patients with severe head injuries. J Neurosurg 104:720–730
9. Reilly PL, Bullock R (eds) (2005) Head injury pathophysiology and management. Great Britain: Arnold
10. Sager TN, Hansen AJ, Laursen H (2000) Correlation between *N*-acetylaspartate levels and histopathologic changes in cortical infarcts of mice after middle cerebral artery occlusion. J Cereb Blood Flow Metab 20:780–788
11. Signoretti S, Marmarou A, Tavazzi B, Lazzarino G, Beaumont A, Vagnozzi R (2001) *N*-acetylaspartate reduction as a measure of injury severity and mitochondrial dysfunction following diffuse traumatic brain injury. J Neurotrauma 18:977–991

Acta Neurochir Suppl (2008) 102: 63–69

Mathematical models of cerebral hemodynamics for detection of vasospasm in major cerebral arteries

Federico S. Cattivelli · Ali H. Sayed · Xiao Hu · Darrin Lee · Paul Vespa

Abstract
Background Vasospasm is a common complication of aneurismal subarachnoid hemorrhage (SAH) that may lead to cerebral ischemia and death. The standard method for detection of vasospasm is conventional cerebral angiography, which is invasive and does not allow continuous monitoring of arterial radius. Monitoring of vasospasm is typically performed by measuring Cerebral Blood Flow Velocity (CBFV) in the major cerebral arteries and calculating the Lindegaard ratio. We describe an alternative approach to estimate intracranial arterial radius, which is based on modeling and state-estimation techniques. The objective is to obtain a better estimation than that offered by the Lindegaard ratio, that might allow for continuous monitoring and possibly vasospam prediction without the need for angiography.
Methods We propose two new models of cerebral hemodynamics. Model 1 is a more general version of Ursino's 1991 model that includes the effects of vasospasm, and Model 2 is a simplified version of Model 1. We use Model 1 to generate Intracranial Pressure (ICP) and CBFV signals for different vasospasm conditions, where CBFV is measured at the middle cerebral artery (MCA). Then we use Model 2 to estimate the arterial radii from these signals.
Findings Simulations show that Model 2 is capable of providing good estimates for the radius of the MCA, allowing the detection of the vasospasm. These changes in arterial radius are being estimated from measurements of CBFV, and CBF is never being measured directly. This is the main advantage of the model-based approach where several interrelations between CBFV, ABP and ICP are taken into account by the differential equations of the model.
Conclusions Our results indicate that arterial radius may be estimated using measurements of ABP, ICP and CBFV, allowing the detection of vasospasm.

Keywords Vasospasm · State estimation · System identification · Cerebral hemodynamics

F. S. Cattivelli (✉) · A. H. Sayed
Electrical Engineering Department,
University of California Los Angeles,
44-123 Engineering IV Building,
Los Angeles, CA 90095, USA
e-mail: fcattiv@ee.ucla.edu

A. H. Sayed
e-mail: sayed@ee.ucla.edu

X. Hu · D. Lee
Division of Neurosurgery, Geffen School of Medicine,
University of California Los Angeles,
Suite 219, MP 100,
Los Angeles, CA 90095, USA

X. Hu
e-mail: xiaohu@ucla.edu

D. Lee
e-mail: DarrinLee@mednet.ucla.edu

P. Vespa
Division of Neurosurgery, Geffen School of Medicine,
University of California Los Angeles,
10833 Le Conte Ave, CHS 18-218,
Los Angeles, CA 90095, USA
e-mail: PVespa@mednet.ucla.edu

Introduction

Vasospasm is a common complication of aneurysmal subarachnoid hemorrhage (SAH) that may lead to cerebral ischemia and death. Vasospasm may have a predictable time course and several treatment interventions exist [5]. The "gold standard" method for detection of vasospasm is conventional cerebral angiography. Angiography is invasive and continuous monitoring of arterial radius is not possible through this technique.

A physiological quantity that is closely related to vasospasm is Cerebral Blood Flow Velocity (CBFV). CBFV through a vessel of (inner) radius r is equal to the ratio of Cerebral Blood Flow (CBF) through the vessel, and its area, as follows:

$$CBFV = \frac{CBF}{\pi r^2} \quad (1)$$

CBFV may be measured non-invasively and continuously using the Transcranial Doppler ultrasonography (TCD). CBFV is typically measured at the Middle Cerebral Artery (MCA) and Internal Carotid Artery (ICA), though measurements at the Anterior Cerebral Artery (ACA), and various posterior circulation vessels are also possible. However, knowledge of CBFV is not sufficient for the correct prediction of arterial radius r. An alternative method that is used in clinical practice is calculation of the Lindegaard ratio [3] which is a ratio of the CBFV of an intracranial artery to an extracranial artery (usually the ICA). While arterial radius may be measured from conventional angiography, current TCD methods do not allow this determination.

We describe an alternative approach to estimation of intracranial arterial radius. It constitutes a model-based approach where state-estimation is applied to estimate physiological variables of interest such as arterial radii. The objective is to obtain a better estimation than that offered by the Lindegaard ratio, that might allow for continuous monitoring and possibly vasospam prediction without the need for angiography. In its current form, this work represents a simulation-based approach whose aim is to assess the possibility of using model-based state estimation to estimate arterial radii, since in general a model will likely never be able to replicate the actual complex physiological system of cerebral hemodynamics.

Methodology

The methodology used for arterial radii estimation is a model-based State-Estimation approach [1]. We use a two-step approach, consisting of Model Training and State Estimation. All the variables used correspond to time domain signals, sampled at 1 Hz. The mathematical models used in this work have inputs, outputs, state variables and parameters. The input in this case is Arterial Blood Pressure (ABP), and the outputs are Intracranial Pressure (ICP) and Cerebral Blood Flow Velocity (CBFV). We assume measurements of all inputs and outputs are available. The models have several parameters which are in general unknown. An example of a parameter is the nominal value of vessel resistance (see Section 3). Since these parameters are unknown, it is necessary to estimate them using available measurements. This is the first step of the methodology, and is called Model Training.

Figure 1a shows the Model Training scenario. A model is used to generate artificial outputs (ICP and CBFV), and the measurements of these outputs are subtracted to generate an error signal. An optimization block is used to select the set of parameters that minimizes some cost function that depends on the error. For instance, in our case we use the cost function

$$J(\theta) = \sum_{l=1}^{L} \sum_{i=1}^{N} w_l(i)[y_l(i) - \hat{y}_l(i,\theta)]^2 \quad (2)$$

where N is the total number of measurements, L is the total number of outputs, θ is the unknown parameter vector, $y_l(i)$ is the ith measurement of output l, $\hat{y}_l(i,\theta)$ is the ith output l generated by the model using parameter θ, and $w_l(i)$ is some weighting function. In our case, we use the weighting function that weights every variable y_l inversely proportional to the energy of the signal $y_l(i)$.

The models considered in this work are nonlinear, and hence Eq. 2 will in general be a non-convex function of θ. As such, algorithms based on gradient descent are not guaranteed to converge to a global optimum. Hence, the optimization is done in two steps as proposed in [1]. First, a global search is performed using a genetic algorithm known as Differential Evolution (DE) [7], which has low complexity and good convergence. After the global search, a local search is performed using a standard gradient descent algorithm through the MATLAB Optimization Toolbox.

The second step of the methodology is called State Estimation. The states typically represent some physiolog-

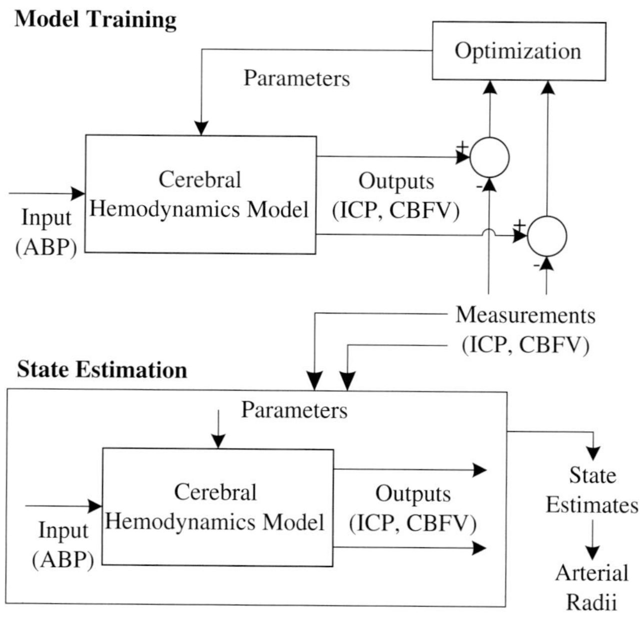

Fig. 1 Model training and state estimation

ical variables which may not be measured directly, such as arterial radii of the vessels, compartment compliances, etc, and therefore need to be estimated. After the model has been trained, and a good value of L is known, the estimation is performed, as shown in Fig. 1b. This stage relies on models of the form shown in Eq. 3 where $x(t)$ and $x(t)$ are vectors corresponding to the state of the system at time t and its derivative with respect to time, respectively, $y(t)$ is its output vector, $u(t)$ is the input, $v(t)$ and $w(t)$ are process noise and measurement noise, respectively, and f and g are some nonlinear functions that may change with time.

$$\begin{aligned} \dot{x}(t) &= f(x(t), u(t), v(t), t) \\ y(t) &= g(x(t), w(t), t) \end{aligned} \tag{3}$$

Let $\widehat{x}(t|t)$ denote the minimum mean-square error (MMSE) estimate of $x(t)$ given all observations $y(t)$ up to time t. It is well known that for linear systems in white Gaussian noise, the MMSE estimate can be obtained recursively using the Kalman Filter [2]. For non-linear systems, however, this is not the case, and a typical approach to solve the problem is to use the Extended Kalman Filter (EKF), which has the disadvantage of requiring the Jacobian matrix of the system, its calculation being error prone. Derivative-free state estimation approaches in non-linear systems have also been proposed, for example, the Unscented Kalman Filter [11] and the DD1 and DD2 filters [6], which have been shown to provide better performance than the EKF. In this work we use DD1 and DD2 filters.

Mathematical models

In Section 2 we introduced a methodology for the estimation of arterial radii based on continuous time measurements of CBFV, ABP and ICP. This methodology relies heavily on mathematical models that relate these quantities, together with the desired arterial radii. For our purpose, a good mathematical model should provide good correlation with observed quantities, and at the same time have low complexity to allow fast training and state estimation, and avoid possible instability. In general, these two characteristics will contradict each other, i.e., a less complex model will be less able to capture the interrelations between all the variables.

Another limitation of the approach is that even if we have a good model that closely matches the observed variables, it is virtually impossible to obtain continuous measurements of the actual arterial radii to compare them with the estimates. Hence, in this work we propose a simulation-based approach as follows: we develop a mathematical model of cerebral hemodynamics that is more general than previous models, and takes into account mechanisms such as autoregulation and vasospasm. We will denote this model as Model 1. Then, we will use Model 1 to generate artificial data for different values of spasm severity. Next, we will develop a second model, denoted as Model 2, to estimate the arterial radii from Model 1 based on its outputs. As mentioned before, we want Model 2 to be simple, in order to reduce the complexity of the parameter and state estimation. This simulation-based approach will give us good insight into how capable simple models are in predicting states from more complex ones, and is the first step towards the application of the state estimation on actual patient data.

The mathematical models derived in this work are based on the models proposed by Ursino *et al.* These models were first introduced in [8, 9] and [10]. Our work is based on the model of [8]. One inconvenience of the model in [8] is that it does not model vasospasm, which makes it inappropriate for the generation of data at different levels of spasm. Vasospasm was modeled in the work by Lodi and Ursino [4], but several simplifications were introduced to the original Ursino model, such as a much simpler autoregulation mechanism, and collapsing of the small and large arterial sections into one single section. Hence, we combined the two aforementioned models into one more general model that takes into account vasospasm, has a detailed autoregulation mechanism, and has four sections: namely those corresponding to the large arteries (MCA, ACA, PCA), followed by the large pial arteries, small pial arteries and capillaries, and finally the venous compartment. We refer to this model as Model 1, and present it in the form of an electrical circuit in Fig. 2 (left circuit).

Next we introduced several simplifications to Model 1, namely collapsing small and large pial arterial sections into one, a simpler autoregulation mechanism, and assuming $P_v = P_{ic}$. We also added one capacitance at the large arteries to obtain a state variable that allowed us to obtain the desired MCA radius. We refer to this model as Model 2, and present it in the form of an electrical circuit in Fig. 2 (right circuit).

Details of model 1

Model 1 has one input (ABP), two outputs (CBFV at the MCA and ICP) and 10 state variables. The state r_{jk} represents the radii of the arteries at compartment i, and branch k. The index $i=1$ represents the proximal (medium arteries) and $i=2$ represents the distal (small arteries and capillaries). The index $k=\mathrm{v}$ represents the top branch of the compartment, which is in spastic state, and $k=n$ is the bottom branch, which is in normal state (see Fig. 2).

According to the Hagen-Poiseuille law, the hydraulic resistance of several parallel tubes of equal caliber is inversely proportional to the fourth power of the inner

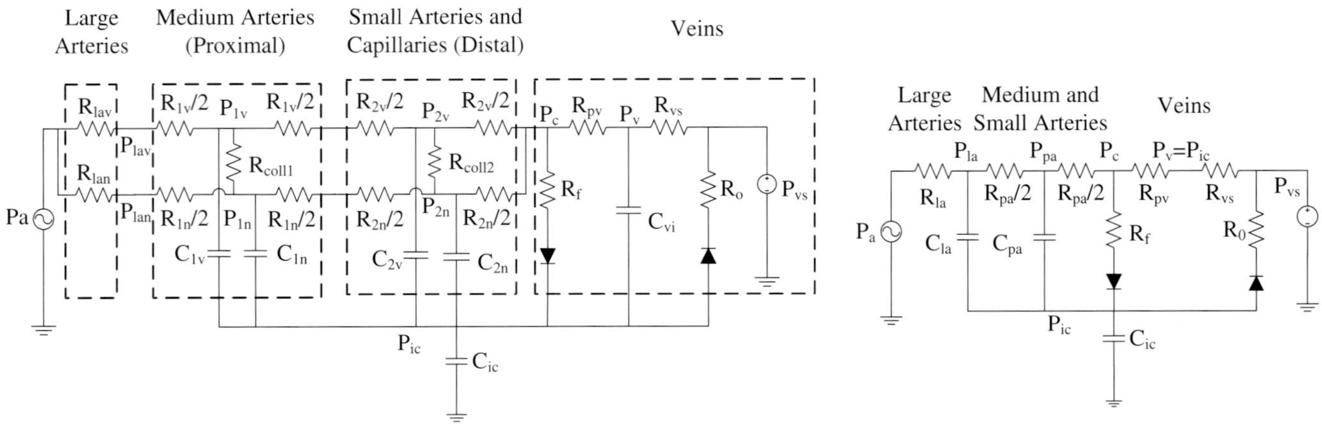

Fig. 2 The more general model 1 (*left*) and the simpler model 2 (*right*)

radius. The four states r_{jk} are related to the corresponding resistances via

$$R_{jk} = R_{jk,0}\frac{r_{jnom}^4}{r_{jk}^4}, \qquad j \in \{1,2\}, \qquad k \in \{v,n\}$$

where the sub-index "zero" indicates nominal values. The resistances between the nominal and vasospastic branches have an equivalent resistance of $R_{j,tot}$ in the absence of vasospasm. This is modeled through the following equations

$$R_{jv,0} = \frac{R_{j,tot}}{k_{pv}}, \qquad R_{jn,0} = \frac{R_{j,tot}}{1-k_{pv}}, \qquad j \in \{1,2\}$$

where k_{pv} is a parameter that depends on the artery where the vasospasm is present (MCA, PCA or ACA).

In the absence of vasospasm, the resistances at the large arteries are given by

$$R_{la,n} = \frac{R_{la,tot}}{1-k_{pv}} \qquad R_{la,v} = \frac{R_{la,tot}}{k_{p,v}} = \frac{8\eta}{\pi r_v^4} l_v$$

where l_v and r_v are the length and radius of the vessel, respectively. These values are shown in Table 1 for different arteries, and will depend on the artery that has vasospasm, which also corresponds to the one where CBFV is being measured. When vasospasm is present, the vasospastic radius is $\widetilde{r}_v$ over a length $k_d l_v$, where k_d is the coefficient of diffusion of the vasospasm (a number between 0 and 1, 0 being no spasm). In this case, the resistance at the large arteries is given by

$$R_{la,v} = \frac{8\eta}{\pi r_v^4}(1-k_d)l_v + \frac{8\eta}{\pi \widetilde{r}_v^4} k_d l_v + \frac{k_t}{2}\frac{\rho q_v}{\pi^2 r_v^4}\left(\frac{r_v^2}{\widetilde{r}_v^2} - 1\right)^2$$

where q_v is the flow through the vasospastic arteries (through $R_{la,v}$). The CBFV at the affected arteries is $CBFV_v = q_v / (\pi \widetilde{r}_v^2)$.

The vasospastic radius $\widetilde{r}_v$ is assumed to fluctuate over its nominal value according to the following equation

$$\widetilde{r}_v = \widetilde{r}_{v0}\left[\frac{1}{k_f}\ln\left(\frac{P_a - P_{ic}}{P_{an} - P_{icn}}\right) + 1\right]$$

In order to calculate state equations for the inner radius r_{jk} and calculating P_{jk}, from Laplace's law we obtain

$$P_{jk} r_{jk} - P_{ic}(r_{jk} + h_{jk}) = T_{ejk} + T_{mkj} + T_{vjk}$$
$$j \in \{1,2\},\ k \in \{v,n\}$$

where h_{jk} is the thickness of the vessel and is given by

$$h_{jk} = -r_{jk} + \sqrt{r_{jk}^2 + 2r_{j0}h_{j0} + h_{j0}^2} \quad j \in \{1,2\},\ k \in \{v,n\}$$

and r_{j0}, h_{j0} are the corresponding values in unstressed conditions. The elastic, muscle and viscous tensions, respectively, are

$$T_{ejk} = h_{jk}\left\{\sigma_{0j}\left[\exp\left(k_{ej}\frac{r_{jk} - r_{j0}}{r_{j0}}\right) - 1\right] - \sigma_{collj}\right\}$$

$$T_{vjk} = \frac{\eta_j}{r_{j0}}\frac{dr_{jk}}{dt} h_{jk}$$

$$T_{mjk} = T_{\max,j}(1 + M_{jk})\exp\left(-\left|\frac{r_{jk} - r_{mj}}{r_{tj} - r_{mj}}\right|^{n_{mj}}\right), \quad \text{with}$$

$$M_{jk} = \frac{M_{\min} + M_{\max}\exp(x_{jk}/k_m)}{1 + \exp(x_{jk}/k_m)}$$

Volume is related to radius according to the following equation

$$V_{jk} = K_{vjk} r_{jk}^2 \quad j \in \{1,2\}z,\ j \in \{1,2\}$$

Table 1 Parameters for model 1

Parameters	
Fixed parameters	
P_{vs}=6.07437 mmHg	r_{m1}=0.027 cm
P_{an}=100 mmHg	r_{m2}=0.0128 cm
q_n=12.5 ml s^{-1}	r_{t1}=0.018 cm
P_{icn}=9.5 mmHg	r_{t2}=0.0174 cm
P_{lan}=92.5 mmHg	n_{m1}=1.83
$P_{1normal}$=85 mmHg	n_{m2}=1.75
P_{cn}=25 mmHg	η_1=232 mmHg s
$R_{latot}=(P_{an}-P_{lan})/q_n$ mmHg s ml^{-1}	η_2=47.8 mmHg s
$R_{1,tot}=2(P_{lan}-P_{1normal})/q_n$ mmHg s ml^{-1}	σ_{01}=0.1425 mmHg
$R_{2,tot}=(P_{lan}-P_{cn})/q_n-R_{1tot}$ mmHg s ml^{-1}	σ_{02}=11.19 mmHg
ρ =7.87563e-4 mmHg s^2/cm^2	$k_{\sigma 1}$=10
k_t=1	$k_{\sigma 2}$=4.5
k_f=12	σ_{coll1}=62.79 mmHg
k_m=0.5	σ_{coll2}=41.32 mmHg
R_{pv}=0.875 mmHg s ml^{-1}	G_1=0.02 $mmHg^{-1}$
R'_{vs}=0.3656 mmHg s ml^{-1}	τ_1=10 s
P_{v1}=−2.5 mmHg	G_2=5.2 $mmHg^{-1}$
R_f=2.38e3 mmHg s ml^{-1}	τ_2=20 s
R_0=0.526e3 mmHg s ml^{-1}	r_{10}=1.5e−2 cm
M_{min}=−1	r_{20}=7.5e−3 cm
M_{max}=1	h_{10}=3e−3 cm
$T_{max,1}$=2.16 mmHg cm	h_{20}=2.5e−3 cm
$T_{max,2}$=1.50 mmHg cm	r_{1nom}=0.023435 cm
k_{v1}=4640 cm	r_{2nom}=0.007346 cm
k_{v2}=154320 cm	
Vessel parameters for different arteries	
MCA: k_p=0.3, r_v=0.14 cm, l_v=10.87 cm	
ACA: k_p=0.1, r_v=0.09 cm, l_v=5.57 cm	
PCA: k_p=0.1, r_v=0.095 cm, l_v=6.92 cm	
Trained parameters (nominal)	
k_e=0.11 ml^{-1}	k_{ven}=0.31 ml^{-1}
C_{max}=0.2 ml $mmHg^{-1}$	R_{coll1}=56 mmHg s ml^{-1}
k_d=0.368	R_{coll2}=56 mmHg s ml^{-1}
Initial values of states (nominal)	
r_{1v}=0.023435 cm	P_{ic}=9.5 mmHg
$r_{1n}=r_{1v}$ cm	$x_{1v}=x_{1n}=x_{2v}=x_{2n}=0$
r_{2v}=0.007346 cm	P_v=14.0682 mmHg
$r_{2n}=r_{2v}$ cm	

where $K_{vjv}=k_{pv}K_{vj}$ and $K_{vjn}=(1-k_{pv})K_{vj}$, for $j\in\{1,2\}$, from which we obtain

$$\frac{dV_{jk}}{dt}=2K_{v_{jk}}r_{jk}\frac{dr_{jk}}{dt}$$

Since dV_{jk}/dt represents the current (or flow) from node P_{jk} to node P_{ic}, we readily obtain four equations for dV_{jk}/dt through conservation of flow at the four nodes P_{jk}.

In order to compute state equations dr_{jk}/dt, we need to compute first the value of $R_{la,v}$, which at the same time depends on P_{1v}, which can be computed from dr_{1v}/dt. Solving for dr_{1v}/dt, we obtain a third order equation in $R_{la,v}$ from which we can compute its value. Then we can compute all the state equations of the system, and also the intermediate pressures P_{jk}.

The remaining states are the pressure on the venous bed P_v, the intracranial pressure P_{ic}, and the four Autoregulation variables x_{jk}, $j\in\{1,2\}$, $k\in\{v,n\}$. The state equations for these variables are

$$\tau_{1k}\frac{dx_{1k}}{dt}+x_{1k}=G_1(P_a-P_{ic}-(P_{an}-P_{icn})),\quad k\in\{v,n\}$$

$$\tau_{2k}\frac{dx_{2k}}{dt}+x_{2k}=G_2(q_{2k}-q_{2kn})/q_{2kn},\qquad k\in\{v,n\}$$

where q_{2k} is the flow through R_{2k}, $k\in\{v,n\}$, and q_{2kn} is the nominal value of q_{2k}, and is given by $q_{2vn}=k_{pv}q_n$ and $q_{2nn}=(1-k_{pv})q_n$. Finally, state equations for P_{ic} and P_v may be computed from the conservation of flow at their corresponding nodes, and noting that the diode in the circuit only allows current to flow from P_c to P_{ic} and from P_{ic} to P_{vs}. We also need the resistance $R_{vs}=R'_{vs}(P_v-P_{vs})/(P_v-P_{ic})$ and the capacitances $C_{vi}=1/k_{ven}(P_v-P_{ic}-P_{v1})$ and $C_{ic}=1/k_E P_{ic}$.

The fixed parameters, trained parameters (nominal values) and initial values of states (nominal values) for Model 1 are shown in Table 1. For both the trained parameters, and initial values of states, the nominal values are provided, though the model training of Section 2 selects a better set of parameters which are close to the nominal ones.

Details of model 2

Model 2 has one input (ABP), and two outputs (ICP and CBFV at the MCA). It has four states, namely, pressure at the large arteries, P_{la}, pressure at pial arteries, P_{pa}, intracranial pressure, P_{ic}, and capacitance at pial arteries, C_{pa}.

One simplification of the model is that it assumes a linear relation between volume and pressure at the large and pial arteries (recall from Model 1 that this relation is of exponential nature). Thus, we have for the volumes (V_{ka} and $V_{ka,0}$, $k\in\{l,p\}$)

$$V_{ka}=C_{ka}(P_{ka}-P_{ic})\quad k\in\{l,p\} \tag{4}$$

From these volumes we can readily compute the resistances at the large and pial arteries as follows

$$R_{ka}=R_{ka,0}V_{ka,0}^2\Big/V_{ka}^2\qquad k\in\{l,p\}$$

Another simplification of the model is the assumption that $P_v=P_{ic}$, which eliminates one state variable. This assumption was introduced in [9]. State equations for P_{la}, and P_{pa}, can now be obtained by differentiating (4). Care must be exercised while differentiating (4) since C_{pa} also depends on time due to autoregulation. Using $C_{ic}=1/k_E P_{ic}$, a state equation for state P_{ic} may be obtained through the conservation of flow at node P_{ic}.

A third approximation of the model is a much simpler autoregulation mechanism at the pial arteries as in [4]. This is accomplished using a state variable for the compliance at the pial arteries

$$\tau_{aut}\frac{dC_{pa}}{dt} + C_{pa} = C_{pa0}\sigma(x), \quad x = G_{aut}\left(q_{pa} - q_0\right)/q_0$$

$$\sigma(x) = \frac{(1 + \Delta\sigma/2) + (1 - \Delta\sigma/2).\exp(4x/\Delta\sigma)}{1 + \exp(4x/\Delta\sigma)}$$

where q_{pa} is the flow through R_{pa}, and $\Delta\sigma=\Delta\sigma_{\max}$ if $x<0$ and $\Delta\sigma_{\max}$ if $x>0$. Finally, the radius at the large arteries is given by $r_{la}=(k_{rla}/R_{la})^4$. The total CBFV at the large arteries is given by $v_{la}=q_{la}/\pi r_{la}^2$, where q_{la} is the flow through R_{la}. CBFV at the MCA is approximated by a sixth of the total, i. e., $CBFV_{MCA}=v_{la}/6$. The fixed parameters, trained parameters (nominal values) and initial values of states (nominal values) for Model 2 are shown in Table 2.

Table 2 Parameters for model 2

Parameters	
Fixed parameters	
P_{vs}=6.07437 mmHg	$\Delta\sigma_{max}$=6
q_n=12.5 ml s^{-1}	$\Delta\sigma_{\min}$=0.6
P_{ic0}=9.5 mmHg	R_{pv}=0.875 mmHg s ml^{-1}
P_{pa0}=58.75 mmHg	R'_{vs}=0.3656 mmHg s ml^{-1}
G_{aut}=2 mmHg^{-1}	R_f=2.38e3 mmHg s ml^{-1}
τ_{aut}=20 s	R_0=0.526e3 mmHg s ml^{-1}
k_{rla}=3.04e–4 mmHg s cm	
Trained parameters (nominal)	
k_e=0.11 ml^{-1}	V_{la0}=2.5 cm^3
$C_{la}=2.5/(P_{la}-P_{ic})$ ml mmHg^{-1}	R_{la0}=0.6 mmHg s ml^{-1}
C_{pa0}=0.202 ml mmHg^{-1}	R_{pa0}=5.4 mmHg s ml^{-1}
Initial values of states (nominal)	
P_{la}=92.5 mmHg	P_{ic}=9.5 mmHg
P_{pa}=58.75 mmHg	C_{pa}=0.202 ml mmHg^{-1}

Simulation results and discussion

The two models (Model 1 and Model 2) were implemented in C code and the Differential Equation solver CVODE was used for the simulations. The parameter and state-estimation algorithms were implemented in MATLAB. Figure 3 shows the radius at the MCA versus time in seconds. The dashed curve corresponds to the actual MCA radius of Model 1. This radius was gradually decreased during simulation from about 0.14 cm to 0.11 cm. The solid curve shows the Estimate obtained using Model 2. It can be noted that even though the estimated radius is slightly off by about 0.01 cm, it correctly tracks the dashed curve and allows estimation of the variation in radius.

It is interesting to note that the changes in arterial radius are being tracked based on measurements of CBFV only. From Eq. 1, we recall that these two variables are related also to Cerebral Blood Flow (CBF), but CBF is never being measured directly. Although this may seem counterintuitive, this is the main advantage of the model-based approach. This approach takes into account several interrelations between CBF, CBFV, ABP and ICP, which are captured by the differential equations of the model.

We also note that the simpler Model 2 is able to estimate the MCA radius from Model 1, even though these two models have several differences. This is a first step towards the application of the estimation framework using Model 2, to the much more relevant problem of estimating vasospasm from real patient data.

Acknowledgement The work of F.S. Cattivelli and A. H. Sayed was supported in part by the awards NIH/NINDS NS055045 and NSF ECS-0601266. X. Hu is supported by NIH/NINDS NS055045 and NS055998.

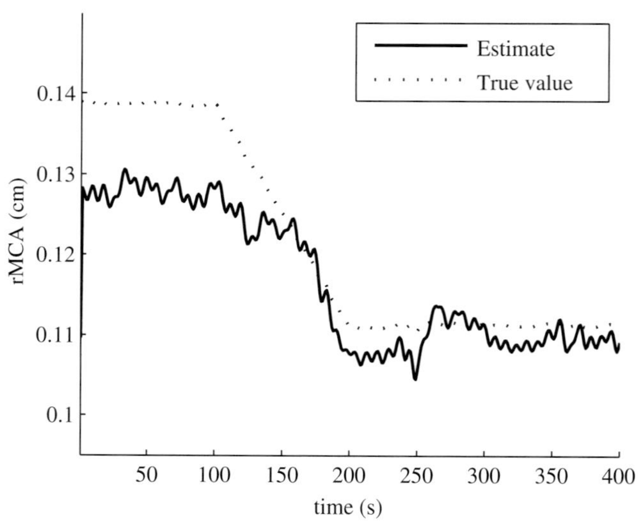

Fig. 3 Actual (*dashed*) and estimated (*solid*) MCA arterial radius versus time

Conflict of interest statement We declare that we have no conflict of interest.

References

1. Hu X, Nenov V, Bergsneider M, Glenn TC, Vespa P, Martin N (2006) Estimation of hidden state variables of the intracranial system using constrained nonlinear Kalman filters. IEEE Trans Biomed Eng 54(4):597–610
2. Kailath T, Sayed AH, Hassibi B (2000) Linear Estimation. Prentice Hall, NJ
3. Lindegaard KF et al (1988) Cerebral vasospasm after subarachnoid hemorrhage investigated by means of transcranial Doppler ultrasound. Acta Neurochir (Vienna) 24:81–84
4. Lodi CA, Usino M (1999) Hemodynamic effect of cerebral vasospasm in humans: a modeling study. Ann Biomed Eng 27:257–273
5. Macdonald RL, Weir B (2001) Cerebral vasospasm. Academic, San Diego
6. Norgaard M, Poulsen N, Ravn O (2000) New developments in state estimation for nonlinear systems. Automatica 36(11):1627–1638
7. Storn R, Price K (1997) Differential evolution: a simple and efficient adaptive scheme for global optimization over continuous spaces. J Global Optimization 11:341–359
8. Ursino M, Di Giammarco P (1991) A mathematical model of the relationship between cerebral blood volume and intracranial pressure changes: the generation of plateau waves. Ann Biomed Eng 19:15–42
9. Ursino M, Lodi CA (1997) A simple mathematical model of the interaction between intracranial pressure and cerebral hemodynamics. J Appl Physiol 82:1256–1269
10. Ursino M, Lodi CA (1998) Interaction among autoregulation, CO2 reactivity, and intracranial pressure: a mathematical model. Am J Physiol Heart Circ Physiol 274:H1715–H1728
11. Wan EA, Van Der Merwe R (2000) The unscented Kalman filter for nonlinear estimation. In Proc. IEEE 2000 Adaptive Systems for Signal Processing, Communications, and Control Symposium, pp. 153–158

Acta Neurochir Suppl (2008) 102: 71–75

Intracranial pressure in patients with sepsis

D. Pfister · B. Schmidt · P. Smielewski · M. Siegemund · S. P. Strebel · S. Rüegg · S. C. U. Marsch · H. Pargger · L. A. Steiner

Abstract

Introduction In sepsis the brain is frequently affected although there is no infection of the CNS (septic encephalopathy). One possible cause of septic encephalopathy is failure of the blood-brain barrier. Brain edema has been documented in animal models of sepsis. Aggressive fluid resuscitation in the early course of sepsis improves survival and is standard practice. We hypothesized that aggressive fluid administration will increase intracranial pressure (ICP) and may cause critical reductions in cerebral perfusion pressure (CPP).

Materials and methods Patients with sepsis were investigated daily on up to four consecutive days in the intensive care unit. Mean arterial blood pressure (MAP) and blood flow velocity in the middle cerebral artery were monitored for one hour each day. ICP was calculated non-invasively from MAP and flow velocity data. S-100β was determined daily.

Findings Fifty-two measurements were performed in 16 patients. ICP could be determined in 45 measurements in 15 patients. Seven patients had an ICP>15 mmHg and 11 patients had a CPP<60 mmHg on at least 1 day. We found no significant correlation between ICP and fluid administration, but low CPP was significantly correlated with elevated S-100β ($r=-0.47$, $p=0.001$).

Conclusions Further research is needed to determine the role of ICP/CPP monitoring in patients with sepsis.

Keywords Intracranial pressure · Cerebral perfusion pressure · Sepsis · S-100β

D. Pfister · M. Siegemund · S. P. Strebel · H. Pargger · L. A. Steiner (✉)
Department of Anaesthesia, Operative Intensive Care Unit, University Hospital Basel,
Spitalstrasse 21,
CH-4031 Basel, Switzerland
e-mail: lsteiner@uhbs.ch

B. Schmidt
Department of Neurology, Chemnitz Medical Center,
Dresdner Straße 178,
09131 Chemnitz, Germany

P. Smielewski
Department of Clinical Neurosciences, University of Cambridge,
Neurosurgery Unit, Addenbrooke's Hospital,
Box 167, Level 4, A Block, Hills Road,
Cambridge CB2 2QQ, UK

S. Rüegg
Department of Neurology, University Hospital Basel,
Petersgraben 4,
CH-4051 Basel, Switzerland

S. C. U. Marsch
Medical Intensive Care Unit, University Hospital Basel,
Petersgraben 4,
CH-4051 Basel, Switzerland

Introduction

In sepsis the brain is frequently affected although there is no infection of the central nervous system, a condition referred to as septic encephalopathy. Reported incidences vary widely from 8% to 70%. This wide range is most probably due to differences in diagnostic criteria. Nevertheless, septic encephalopathy is a common organ dysfunction in sepsis. The importance of septic encephalopathy is reflected by the fact that it is associated with an increase in mortality [17]. Yet, the pathophysiology of septic encephalopathy remains poorly understood. Among several proposed mechanisms, a disturbance of the blood-brain barrier is thought to play a major role [11]. Disruption of astrocytic end-feet and perimicrovascular edema could be demonstrated in animal models of sepsis [3, 10], but have not been demonstrated in humans. However, a recent magnetic resonance imaging (MRI) study of the brain in patients with septic shock contributes to the concept of brain oedema in septic encephalopathy

[15]. Early aggressive fluid resuscitation has been shown to improve outcome in sepsis [12] and has become an integral part of treatment guidelines for septic patients [6]. In severe disease large amounts of fluid are often required in order to stabilize hemodynamics. We hypothesized that aggressive fluid administration in conjunction with a disturbed blood-brain barrier will lead to cerebral edema and to an increase in intracranial pressure (ICP), causing a critical reduction in cerebral perfusion pressure (CPP) as mean arterial pressure (MAP) levels are typically low in sepsis.

Materials and methods

This study was approved by the regional ethics committee. Informed consent was obtained from all patients or their closest relatives. Patients admitted to the intensive care unit were eligible if they were aged 18 years or older and had sepsis, severe sepsis or septic shock according to the criteria of the ACCP/SCCM consensus conference [1]. Patients with an intracranial focus of infection, with a relevant preexisting central neurological disorder or with delirium attributable to another cause than sepsis were excluded. Patients were included within 24 to 36 h of admission to the intensive care unit.

Patients were investigated daily on up to four consecutive days. Routine monitoring included electrocardiography, pulse oximetry and MAP measured directly in the radial or femoral artery. S-100β was determined at each monitoring session (Roche Diagnostics GmbH, D-68298 Mannheim, Germany). The manufacturer proposes a cutoff of 0.105 μg/l for patients with possible cerebral damage. Fluid balance was calculated for the time interval between two measurements, including every type of fluid given and all fluid losses. During the examination, patients were in the supine position with head elevation of no more than 30°. Using transcranial Doppler sonography (TCD) with a 2-MHz probe (Multidop T, DWL, Germany), blood flow velocity in the middle cerebral artery on both sides was monitored for one hour. Values from the right and left side were averaged for analysis. Analogue outputs from arterial pressure monitoring and TCD were transferred to a laptop computer via an analogue-to-digital converter and stored by using the 'ICM^{+} software', version 6.1, from the University of Cambridge, UK [16]. ICP was assessed non-invasively offline by an observer (BS) blinded to the clinical course of the patients. The TCD characteristics were used to calculate a dynamic transformation formula connecting ICP and MAP as described previously [13]. CPP was then calculated as MAP–ICP. All presented ICP and CPP data are therefore non-invasive estimates. All parameters were averaged over the 60 min recording period for analysis. Statistical analysis was performed with SPSS 14.0 for Windows (SPSS Inc. Chicago, Illinois, USA). Data are shown as mean ± standard deviation unless otherwise specified. A p value <0.05 was considered significant.

Results

Fifty-two measurements were performed in 16 patients. Due to signal quality ICP could not be calculated from five recordings in four patients. In one patient nitroglycerine was administered during two measurements. These two measurements were excluded because of the cerebrovascular effects of this drug. Analysis is therefore based on 45 datasets of 15 patients. Mean patient age was 67±17 years, 44% were female, mean APACHE II score was 21±7, and 30-day mortality was 25%. Source of sepsis was pneumonia in 11, abdominal in three, necrotizing myositis, and prosthetic joint infection in one patient each. ICP> 15 mmHg was measured on at least 1 day in seven of the 15 patients (47%). However, the increases in ICP were moderate and never exceeded 20 mmHg (Fig. 1a). ICP correlated strongly with MAP (r=0.63, p<0.0001). Patients who died did not have higher peak ICP than patients who survived: (16±3 and 16±3 mmHg, respectively, p=0.89, Mann–Whitney U test). Reductions in CPP below 60 mmHg and below 50 mmHg on at least 1 day were found in 11 (73%) and three (20%) patients respectively (Fig. 1b). Low CPP correlated significantly with high S-100β levels (r=−0.47, p=0.001; Fig. 2). The lowest recorded CPP values in patients who died were lower than in survivors: 50±8 vs. 58±6 mmHg, respectively. However, this difference did not reach statistical significance (p=0.07, Mann–Whitney U test). There was no significant relationship between ICP and S-100β. The maximal S-100β levels were higher in non-survivors (0.21±0.11 μg/l) than in survivors (0.12±0.06 μg/l) but again this difference did not reach statistical significance (p=0.12 Mann–Whitney U test).

Fluid administration between measurements (i.e. over approximately 24 h) ranged from 370 to 11,840 ml (mean, 4,600±2,500 ml). Daily fluid balances ranged from −2,000 to 8,400 ml (mean, 2,400±2,500 ml). We found no significant correlations between ICP, daily change in ICP or relative change in ICP and overall or daily fluid administration or balance (Fig. 3).

Discussion

Although moderate elevations in ICP seem to occur in a relevant number of patients with sepsis, we were not able to

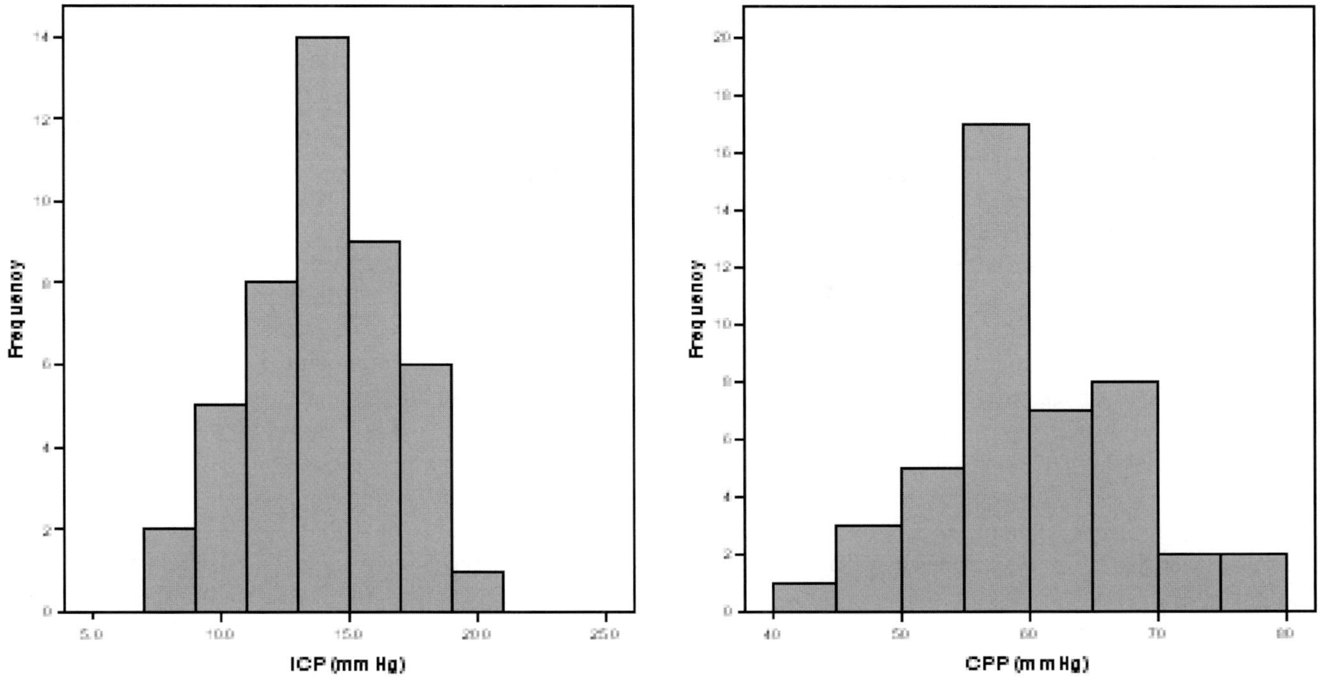

Fig. 1 **a** Distribution of daily values of intracranial pressure (*ICP*). **b** Distribution of daily values of cerebral perfusion pressure (*CPP*)

find a significant correlation between ICP and fluid administration. In contrast, we found a strong correlation between low CPP and elevated S-100β. There are several possible interpretations of our data.

First, our methods have several limitations. The accuracy of the non-invasive method used may be insufficient, as calculation of ICP with the procedure described has a mean deviation of 7.1 mmHg in patients with traumatic brain injury and 4.3 mmHg in patients with stroke [13]. Hence, the values shown should be regarded as estimations rather than exact measurements. However, in the study cited [13] a relevant proportion of patients with ICP>30 mmHg caused an increase in the difference between measured and calculated ICP. As such high ICP did not occur in this study, the estimation in our patients can be expected to be more accurate than in head injured patients. Furthermore, as the methodological error in our calculations would be two-tailed, it is, in our opinion, unlikely that the elevated ICP values are merely a product of chance. It is unknown whether ICP waves occur in patients with sepsis. As we performed measurements once daily rather than continuously, we cannot exclude that we missed ICP peaks or waves. However, there were only small differences between day-to-day results within patients, with measurements

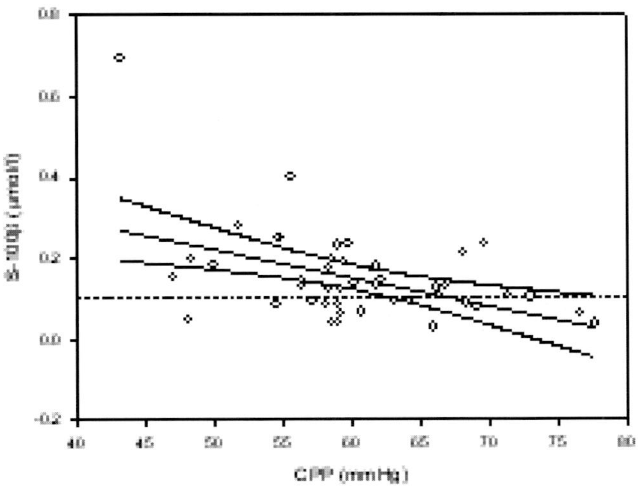

Fig. 2 Cerebral perfusion (*CPP*) plotted against S-100β. *Dotted line* indicates upper normal limit for S-100β at our institution (0.105 μg/l). *Solid lines*: regression line and 95% confidence intervals

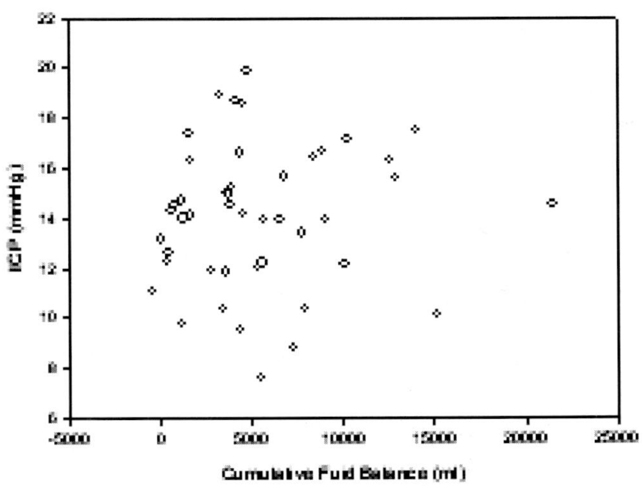

Fig. 3 Intracranial pressure (*ICP*) and cumulative fluid balance

performed randomly regarding time of day, and ICP was very stable during the measurements (mean SD±3.0). We therefore think it is unlikely that we missed relevant ICP peaks or waves.

Since we did not find a correlation between fluid administration and ICP—and in view of the blood flow data [4]—it could be speculated that, if brain edema occurs in sepsis, it may be cytotoxic rather than vasogenic oedema. In contrast to animal models, brain oedema has not yet been demonstrated in humans with sepsis. A recent MRI study found evidence of blood-brain barrier breakdown in patients with septic shock [15], but in an autopsy study of two patients who died of septic shock no brain edema was found [19]. In another pathologic study of 23 patients who died of septic shock, neither existence of cerebral edema nor brain weight were reported [14]. A recent case report described necrosis but not edema in a patient with septic encephalopathy who died [5].

We found a significant correlation between low CPP and high S-100β. This suggests that brain injury in sepsis may at least partly be caused by inadequate cerebral perfusion. Earlier work using the ^{133}Xe clearance technique showed reduced cerebral blood flow in patients with sepsis [4], supporting this concept. An increase in S-100β, a calcium-binding protein that is found predominately in astrocytes and in Schwann cells, is not absolutely specific for brain damage [18], but can also occur from a disturbance of the blood-brain barrier [7]. It has been suggested that low values reflect blood–brain barrier dysfunction whereas higher values reflect brain damage. A cut-off value has been suggested based on a pharmacokinetic model [8]. However, S-100β cutoff values depend on the used kit and comparisons can only be made when identical kits have been used. In our patients we found moderate elevations of S-100β but we cannot differentiate between blood–brain barrier dysfunction and glial damage. We did not measure neuron-specific enolase (NSE), another possible marker of brain damage. However, in a large study including 170 patients with severe sepsis and septic shock a similar proportion of patients showed increased S-100β and NSE levels, with S-100β being a better predictor of disease severity [9]. Extracranial sources of S-100β including heart, skeletal muscle, and kidneys have been described [2]. In the cited study [9] there was no increase in S-100β in a group of postoperative control patients, and acute renal failure which often accompanies sepsis did not significantly influence S-100β levels.

A significant association between mortality and low CPP, high ICP, or elevated S-100β would have strengthened our data. However, the investigated group of patients is very small. Furthermore, ICP/CPP data are only available in three of the four patients who died. A larger group of patients would be needed to investigate the relationship between CPP and outcome in sepsis. In a group of 170 patients with severe sepsis or septic shock a significant association between elevated S-100β and mortality has been reported [9]. It would also be very interesting to investigate not only mortality but also functional outcomes such as late cognitive dysfunction.

In conclusion, moderate elevations in ICP seem to occur quite frequently in patients with sepsis, independent of the amount of fluid administered. As MAP levels are typically low in sepsis, even small increases in ICP may negatively affect cerebral perfusion pressure. This hypothesis may be supported by our finding of a significant correlation between CPP and S-100β. However, the interpretation of S-100β values is difficult and as this was a strictly observational study it remains speculative whether therapeutic manipulation of CPP would lead to a benefit. Cerebral pathophysiology in sepsis is a field that has not been intensively investigated so far and further research is required to determine the role of ICP and CPP monitoring in patients with sepsis and septic encephalopathy.

Conflict of interest statement We declare that we have no conflict of interest.

References

1. American College of Chest Physicians/Society of Critical Care Medicine (1992) American College of Chest Physicians/Society of Critical Care Medicine Consensus Conference: definitions for sepsis and organ failure and guidelines for the use of innovative therapies in sepsis. Crit Care Med 20:864–874
2. Anderson RE, Hansson LO, Nilsson O, Dijlai-Merzoug R, Settergren G (2001) High serum S100B levels for trauma patients without head injuries. Neurosurgery 48:1255–1258, discussion 1258–1260
3. Ari I, Kafa IM, Kurt MA (2006) Perimicrovascular edema in the frontal cortex in a rat model of intraperitoneal sepsis. Exp Neurol 198:242–249
4. Bowton DL, Bertels NH, Prough DS, Stump DA (1989) Cerebral blood flow is reduced in patients with sepsis syndrome. Crit Care Med 17:399–403
5. Finelli PF, Uphoff DF (2004) Magnetic resonance imaging abnormalities with septic encephalopathy. J Neurol Neurosurg Psychiatry 75:1189–1191
6. Hollenberg SM, Ahrens TS, Annane D, Astiz ME, Chalfin DB, Dasta JF, Heard SO, Martin C, Napolitano LM, Susla GM, Totaro R, Vincent JL, Zanotti-Cavazzoni S (2004) Practice parameters for hemodynamic support of sepsis in adult patients: 2004 update. Crit Care Med 32:1928–1948
7. Kapural M, Krizanac-Bengez L, Barnett G, Perl J, Masaryk T, Apollo D, Rasmussen P, Mayberg MR, Janigro D (2002) Serum S-100beta as a possible marker of blood–brain barrier disruption. Brain Res 940:102–104
8. Marchi N, Rasmussen P, Kapural M, Fazio V, Kight K, Mayberg MR, Kanner A, Ayumar B, Albensi B, Cavaglia M, Janigro D (2003) Peripheral markers of brain damage and blood–brain barrier dysfunction. Restor Neurol Neurosci 21:109–121

9. Nguyen DN, Spapen H, Su F, Schiettecatte J, Shi L, Hachimi-Idrissi S, Huyghens L (2006) Elevated serum levels of S-100beta protein and neuron-specific enolase are associated with brain injury in patients with severe sepsis and septic shock. Crit Care Med 34:1967–1974
10. Papadopoulos MC, Lamb FJ, Moss RF, Davies DC, Tighe D, Bennett ED (1999) Faecal peritonitis causes oedema and neuronal injury in pig cerebral cortex. Clin Sci (Lond) 96:461–466
11. Papadopoulos MC, Davies DC, Moss RF, Tighe D, Bennett ED (2000) Pathophysiology of septic encephalopathy: a review. Crit Care Med 28:3019–3024
12. Rivers E, Nguyen B, Havstad S, Ressler J, Muzzin A, Knoblich B, Peterson E, Tomlanovich M (2001) Early goal-directed therapy in the treatment of severe sepsis and septic shock. N Engl J Med 345:1368–1377
13. Schmidt B, Czosnyka M, Raabe A, Yahya H, Schwarze JJ, Sackerer D, Sander D, Klingelhofer J (2003) Adaptive noninvasive assessment of intracranial pressure and cerebral autoregulation. Stroke 34:84–89
14. Sharshar T, Annane D, de la Grandmaison GL, Brouland JP, Hopkinson NS, Francoise G (2004) The neuropathology of septic shock. Brain Pathol 14:21–33
15. Sharshar T, Carlier R, Bernard F, Guidoux C, Brouland JP, Nardi O, de la Grandmaison GL, Aboab J, Gray F, Menon D, Annane D (2007) Brain lesions in septic shock: a magnetic resonance imaging study. Intensive Care Med 33:798–806
16. Smielewski P, Czosnyka M, Steiner L, Belestri M, Piechnik S, Pickard JD (2005) ICM+: software for on-line analysis of bedside monitoring data after severe head trauma. Acta Neurochir Suppl 95:43–49
17. Sprung CL, Peduzzi PN, Shatney CH, Schein RM, Wilson MF, Sheagren JN, Hinshaw LB (1990) Impact of encephalopathy on mortality in the sepsis syndrome. The Veterans Administration Systemic Sepsis Cooperative Study Group. Crit Care Med 18: 801–806
18. Stocchetti N (2005) Brain and sepsis: functional impairment, structural damage, and markers. Anesth Analg 101:1463–1464
19. Talvik R, Liigant A, Tapfer H, Tamme K, Metsvaht T (1998) Septic shock with disseminated microfoci in multiple organs in humans. Intensive Care Med 24:73–76

Acta Neurochir Suppl (2008) 102: 77–80

Intracranial pressure and cerebral oxygenation changes after decompressive craniectomy in children with severe traumatic brain injury

A. A. Figaji · A. G. Fieggen · A. C. Argent · P. D. Le Roux · J. C. Peter

Abstract

Introduction There has been a resurgence of interest in decompressive craniectomy for traumatic brain injury (TBI), but the impact of craniectomy on intracranial pressure (ICP) and cerebral oxygenation has not been well described for diffuse injury in children.

Methods ICP and brain tissue oxygenation (PbtO2) changes after decompressive craniectomy for diffuse brain swelling after TBI in children were analysed.

Findings Decompressive craniectomy was performed for diffuse brain swelling in 18 children under 15 years old. For 8 patients, craniectomy was performed as an emergency for malignant brain swelling, and in 10, for sustained ICP >25 mmHg refractory to conventional medical treatment. In 6 of these patients, PbtO2 was also monitored. Median ICP was reduced from 40 mmHg before craniectomy to 16 mmHg for 24 hours thereafter, and PbtO2 improved from a median of 17.4 to 43.4 mmHg. Clinical outcome was favourable in 78%.

Conclusions In selected pediatric patients with TBI, craniectomy for diffuse brain swelling can significantly improve ICP and cerebral oxygenation control. The use of the procedure in appropriate settings does not appear to increase the proportion of disabled survivors.

Keywords Decompressive craniectomy · Intracranial pressure · Cerebral oxygenation · Brain injury

A. A. Figaji (✉) · A. G. Fieggen · J. C. Peter
Division of Neurosurgery, Institute for Child Health,
School of Child and Adolescent Health, University of Cape Town,
Red Cross Childrens Hospital,
Cape Town, South Africa
e-mail: Anthony.Figaji@uct.ac.za

A. G. Fieggen
e-mail: gfieggen@gmail.com

J. C. Peter
e-mail: Jonathan.Peter@uct.ac.za

A. C. Argent
Division of Pediatric Critical Care, Institute for Child Health,
School of Child and Adolescent Health, University of Cape Town,
Red Cross Childrens Hospital,
Cape Town, South Africa
e-mail: Andew.Argent@uct.ac.za

P. D. Le Roux
Department of Neurosurgery,
The Hospital of the University of Pennsylvania,
330 S. 9th Street, 4th floor,
Philadelphia, PA 19107, USA
e-mail: Peter.LeRoux@uphs.upenn.edu

Introduction

There has been renewed interest in decompressive craniectomy for the treatment of elevated intracranial pressure (ICP) in traumatic brain injury (TBI). Although many large studies have been reported in adults [4], there are few reports limited to children, and these reports are generally small studies [3, 6, 7, 11, 16, 17, 20]. Some pediatric studies also include patients up to 21 years of age in their definition of children and combine craniectomy performed for diffuse brain swelling and for the removal of mass lesions. Changes in cerebral oxygenation after decompressive craniectomy have been described in adult patients [9, 15, 18], but only in one case report for pediatric TBI [8].

Currently there are 2 multi-centred randomised trials of decompressive craniectomy being conducted in adults (RESCUE-ICP and DECRAN). There has however been a single-centre randomised study of decompressive craniectomy versus medical treatment reported as a pilot study in children [20]. The surgical procedure in this trial was relatively conservative: small bitemporal (3-4 cm) craniectomies without durotomy were performed. As would be expected from the size of the craniectomy, the difference in ICP reduction after craniectomy, although significant, was relatively small: 8.98 mmHg in the operative group versus 3.69 mm Hg in the medically treated group. Fewer episodes of intracranial hypertension and a trend towards better outcome were observed in the craniectomy group. Their operative approach, though, is very different to the large bifrontal or hemi-craniectomy with dural expansion that is more frequently used in adults. In this study, we report the ICP and cerebral oxygenation changes in children with diffuse traumatic brain swelling after decompressive craniectomy using a large craniectomy and dural expansion.

Materials and methods

Data for consecutive patients less than 15 years old who underwent decompressive craniectomy for TBI at Red Cross Childrens Hospital, a University-affiliated dedicated Pediatric Hospital were analysed. Only craniectomy for diffuse swelling was considered; procedures where mass lesions were simultaneously removed were excluded. Craniectomy was performed in 2 circumstances: 1) for elevated ICP refractory to medical treatment, or 2) as an initial emergency intervention before any ICP monitoring in a patient who has experienced an early secondary neurological deterioration accompanied by clinical and radiographic signs of cerebral herniation.

All patients were treated in a standard manner and intracranial hypertension was managed in accordance with established practice [2]. Brain tissue oxygen tension (PbtO2) was monitored with Licox catheters (Integra Neurosciences, Plainsboro, NJ) placed in relatively uninjured frontal white matter, based on computed tomography (CT) findings, usually on the right side, but occasionally on the side demonstrating greater swelling or focal contusions. PbtO2 <20 mmHg was considered compromised oxygenation and <10 mmHg, critical hypoxia [19]. PbtO2 <20 mmHg was treated using a stepwise approach to optimise ICP, CPP, haemoglobin (Hb) and arterial carbon dioxide tension (PCO2). The inspired fraction of oxygen (FiO2) was increased only as an emergency temporary measure or if PbtO2 remained low despite optimisation of the above parameters.

Data analysis

Comparisons were made between values for ICP, MAP and PbtO2 (where available) for the 4 hour period before craniectomy (or part thereof) and the 24 hour period after surgery. FiO2 before surgery and 24 hours after were also compared. Data are presented as means ± standard deviation (SD) or median and range. Differences were defined as statistically significant if $p<0.05$. The non-parametric Wilcoxon signed-rank test was used to compare pre- and post-intervention readings. Outcome was assessed with the Glasgow Outcome Score (GOS), dichotomised into favourable (GOS 4-5) and unfavourable (GOS 1-3) outcome.

Results

Mean age was 7.8±3.4 years (range 11 months to 14 years old). Median GCS before craniectomy was 5 (range 3-8). At the time of craniectomy, a unilateral unreactive pupil was documented in 7 patients and bilaterally unreactive pupils in 5. Craniectomy was performed in 8 patients as an emergency intervention for secondary acute neurological deterioration (secondary drop in GCS with development of pupillary abnormalities and obliterated cisterns on CT) and in 10 children for medically refractory ICP. Of the latter 10 patients: individual median ICP values for the 4-hour period before craniectomy were between 20 and 30 mmHg in 1 patient, between 30 and 40 mmHg in 5 patients, and ≥40 mmHg in 4. Six patients also had PbtO2 monitoring before and after craniectomy. PbtO2 before craniectomy deteriorated to between 10-20 mmHg in 1 patient, and to <10 mmHg in 4 patients, requiring increased FiO2. Pooled data for ICP and PbtO2 before and after craniectomy are summarised in Table 1. Median ICP for all patients was 40 mmHg before craniectomy and 16 mmHg after craniectomy. Only one patient required the addition of barbiturate therapy after craniectomy to control ICP. Pooled PbtO2 improved from a median of 17.4 mmHg before surgery to 43.4 mmHg after surgery. Graphic comparisons of data are presented in Figs. 1 and 2. Statistically significant differences were found for ICP before and after craniectomy (z= 2.803, p=0.005) and for PbtO2 before and after craniectomy (z=-2.201, p=0.03). There was also a trend towards lower MAP (91±13 mmHg versus 78±11 mmHg) and FiO2 (73± 21 versus 48±6) after craniectomy.

Table 1 ICP and PbtO2 data for 4 hours before and 24 hours after decompressive craniectomy

	Before DC	After DC	*p*-value
ICP (10 patients)	40±14.9	17±9.02	0.0051
PbtO2 (6 patients)	18±12.8	43±15.4	0.027

Data are presented as means ± standard deviation.
ICP, intracranial pressure; **PbtO2**, partial pressure of brain tissue oxygen. **Before DC**, data over the 4 hours, or part thereof recorded, preceding craniectomy. **After DC**, data for the 24 hour period after the procedure.

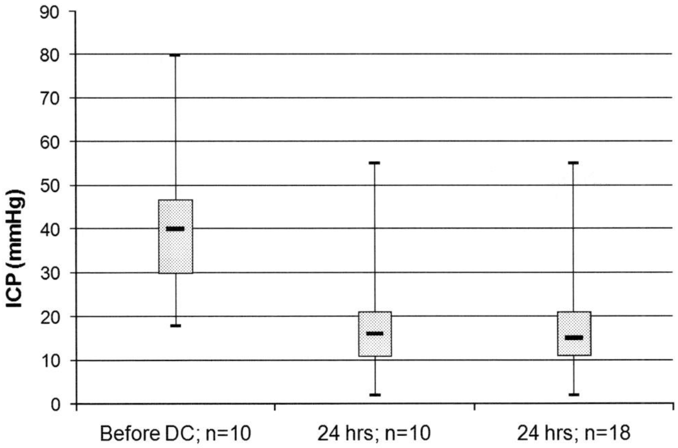

Fig. 1 Box-and-whisker plot of ICP averaged over 4 hours before craniectomy (first column – before DC; n=10) compared with ICP averaged over 24 hours after craniectomy for the 10 patients with pre- and postoperative ICP data (second column – 24 hrs; n=10). The third column (24 hours; n=18) shows the 24-hour period ICP data for all 18 patients who underwent craniectomy

Clinical outcome

Mean time of follow-up was 36±25 months (range 3-84). Outcome was favourable in 14 (78%) and unfavourable in 4 (22%); there were no vegetative survivours. There was one death (5.6%); all survivors were eventually discharged home. One patient required bone flap removal for sepsis after it had been replaced. Two patients had ventriculoperitoneal shunts placed for hydrocephalus.

Discussion

There is resurgent interest in decompressive craniectomy to treat elevated ICP or brain swelling today. This has come about because the limitations of earlier reports have been identified [6] and rather than wait for deterioration more recent reports emphasize ICP monitoring and early control of ICP [1, 3, 5–7, 10, 11, 20, 21]. The use of decompressive craniectomy is well described in adults but studies that describe children with diffuse TBI generally include few patients (on average 2-12), and ICP changes often are not reported [3, 6, 7, 11, 16, 17, 20]. Jagannathan et al [10] described a larger group of 23 patients but since patients up to the age of 19 were included, the mean age of almost 12 years is relatively high. At the time of surgery 4 of their patients had extra-axial mass lesions removed (subdural or extradural haematomas). Kan et al [12] described a mortality of 31% in pediatric patients after decompressive craniectomy. However, non-accidental trauma accounted for 23.5% of the cases and only 6 cases underwent surgery for elevated ICP with diffuse swelling; the rest underwent craniectomy in conjunction with removal of a mass lesion. In contrast, the present study represents a relatively homogenous population of children less than 15 years old with diffuse brain swelling.

There is a single case report that describes PbtO2 after craniectomy in children [8]. In our study, PbtO2 was compromised (PbtO2 <20 mmHg) prior to craniectomy in 5 of 6 patients who underwent PbtO2 monitoring. These patients also required a higher inspired fraction of oxygen (FiO2) to augment brain oxygenation. Four had critical hypoxia, with PbtO2 readings below 10 mmHg. There was an immediate and significant change in PbtO2 in these patients after craniectomy. This increase was sustained, and allowed a reduction of FiO2 settings.

In this report, craniectomy significantly reduced ICP in all but one of the patients, in whom barbiturate therapy also failed. Group median ICP in the 4 hours before craniectomy was 40 mmHg whereas for the 24 hour period after craniectomy it was 16 mmHg. PbtO2 was also significantly improved after craniectomy: median PbtO2 improved from 17.4 to 43.4 mmHg. Outcome was favourable in 78% of patients, despite a relatively high incidence of preoperative pupillary abnormalities and low median GCS. In a number of patients these clinical features occurred as a secondary insult from elevated ICP, which was then effectively treated, rather than as a manifestation of irreversible primary brain injury. The fact that these insults were secondary not primary and that they were promptly treated may have contributed to the good outcome.

The pilot study of Taylor et al [20] is noteworthy because it was the first attempt at a randomised trial of decompressive surgery versus medical treatment. However, it has not provided a definitive answer about the role of craniectomy in pediatric TBI. The effectiveness of craniectomy for the management of ICP and brain tissue

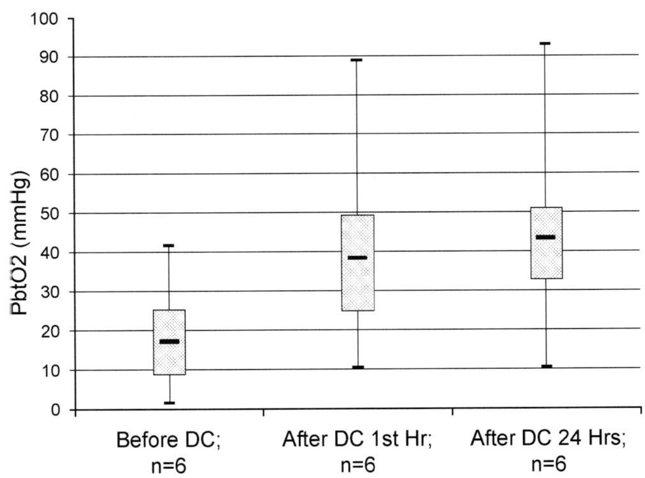

Fig. 2 Box-and-whisker plot demonstrating PbtO2 for 4 hours before craniectomy (first column: Before DC) compared with PbtO2 immediately after craniectomy (second column: After DC 1st Hr) and over the 24 hour period after craniectomy (third column: After DC 24 hrs)

oxygenation compared with continued medical management, therefore, is not established yet. If outcome is improved in the randomised trials in adults, it may be reasonable to extrapolate these results to children, as many investigators feel that craniectomy is more likely to be of benefit in children [13, 14, 20]. If not, a multi-centred randomised controlled trial should be conducted to finally provide class 1 evidence for the role of decompressive craniectomy versus medical management in the treatment of elevated ICP refractory to first tier medical therapy in children with diffuse brain swelling after TBI. The data from this study provide useful information about the cerebral physiologic effects of decompressive craniectomy in pediatric patients.

Conflict of interest statement We declare that we have no conflict of interest.

References

1. Aarabi B, Hesdorffer DC, Ahn ES, Aresco C, Scalea TM, Eisenberg HM (2006) Outcome following decompressive craniectomy for malignant swelling due to severe head injury. J Neurosurg 104:469–479
2. Adelson PD, Bratton SL, Carney NA et al (2003) Guidelines for the acute medical management of severe traumatic brain injury in infants, children, and adolescents. Chapter 17. Critical pathway for the treatment of established intracranial hypertension in pediatric traumatic brain injury. Pediatr Crit Care Med 4(3):S65–S67
3. Berger S, Schwartz M, Huth R (2002) Hypertonic saline solution and decompressive craniectomy for treatment of intracranial hypertension in pediatric severe traumatic brain injury. J Trauma 53: 558–563
4. Bullock MR, Chestnut R, Ghajar J et al (2006) Guidelines for the surgical management of traumatic brain injury. Chapter 5: Surgical management of traumatic parenchymal lesions. Neurosurgery 58:S25–S46
5. Coplin WM, Cullen NK, Policheria PS, Vinas FC, Wilseck JM, Zafonte RD, Rengachary SS (2001) Safety and feasibility of craniectomy with duraplasty as the initial surgical intervention for severe traumatic brain injury. J Trauma 50:1050–1059
6. Figaji AA, Fieggen AG, Argent A, Peter JC (2006) Surgical treatment for "brain compartment syndrome" in children with severe head injury. S Afr Med J 96:969–975
7. Figaji AA, Fieggen AG, Peter JC (2003) Early decompressive craniotomy in children with severe traumatic brain injury. Childs Nerv Syst 19:666–673
8. Figaji AA, Fieggen AG, Sandler SJ, Argent AC, Le Roux PD, Peter JC (2007) Intracranial pressure and cerebral oxygenation changes after decompressive craniectomy in a child with traumatic brain swelling. Childs Nerv Syst 23:1331–1335
9. Jaeger M, Soehle M, Meixensberger J (2003) Effects of decompressive craniectomy on brain tissue oxygen in patients with intracranial hypertension. J Neurol Neurosurg Psych 74:513–515
10. Jagannathan J, Okonkwo DO, Dumont AS, Ahmed H, Bahari A, Prevedello DM, Jane JA Sr, Jane JA Jr (2007) Outcome after decompressive craniectomy in children with severe traumatic brain injury: a 10-year single-center experience with long term follow-up. J Neurosurg 106(4 Suppl Pediatrics):268–275
11. Josan VA, Sgouros S (2006) Early decompressive craniectomy may be effective in the treatment of refractory intracranial hypertension after traumatic brain injury. Childs Nerv Syst 22: 1268–1274
12. Kan P, Amini A, Hansen K, White GL, Brockmeyer DL, Walker ML, Kestle JR (2006) Outcomes after decompressive craniectomy for severe traumatic brain injury in children. J Neurosurg 105 (5 Suppl):337–342
13. Polin RS, Ayad M, Jane JA (2003) Decompressive craniectomy in pediatric patients. Crit Care 7(6):409–410
14. Polin RS, Shaffrey ME, Bogaev CA, Tisdale N, Germanson T, Bocchicchio B, Jane JA (1997) Decompressive bifrontal craniectomy in the treatment of severe refractory posttraumatic cerebral edema. Neurosurgery 41(1):84–92
15. Reithmeier T, Lohr M, Pakos P, Ketter G, Ernestus R-I (2005) Relevance of ICP and ptiO2 for indication and timing of decompressive craniectomy in patients with malignant brain edema. Acta Neurochir 147:947–952
16. Ruf B, Heckmann M, Schroth I, Hugens-Penzel M, Reiss I, Borkhardt A, Gortner L, Jodicke A (2003) Early decompressive craniectomy and duraplasty for refractory intracranial hypertension in children: results of a pilot study. Crit Care 7(6):R133–R138
17. Rutgliano D, Egnor MR, Priebe CJ, McCormack JE, Stong N, Scriven RJ, Lee TK (2006) Decompressive craniectomy in pediatric patients traumatic brain injury with intractable elevated intracranial pressure. J Ped Surg 41:83–87
18. Stiefel MF, Heur GG, Smith MJ, LeRoux PD et al (2004) Cerebral oxygenation following decompressive hemicraniectomy for the treatment of refractory intracranial hypertension. J Neurosurg 101: 241–247
19. Stiefel MF, Spiotta A, Gracias VH, Garuffe AM, Guillamondegui O, Maloney-Wilensky E, Bloom S, Grady MS, LeRoux PD (2005) Reduced mortality rate in patients with severe traumatic brain injury treated with brain tissue oxygen monitoring. J Neurosurg 103:805–811
20. Taylor A, Butt W, Rosenfeld J, Shann F, Ditchfield M, Lewis E, Klug G, Wallace D, Henning R, Tibbals J (2001) A randomized trial of very early decompressive craniectomy in children with traumatic brain injury and sustained intracranial hypertension. Child's Nerv Syst 17:154–162
21. Whitfield PC, Patel H, Hutchinson PJA, Czosnyka M, Parry D, Menon D, Pickard JD, Kirkpatrick PJ (2001) Bifrontal decompressive craniectomy in the management of posttraumatic intracranial hypertension. Br J Neurosurg 15:500–507

Acta Neurochir Suppl (2008) 102: 81–84

Are head injury guidelines changing the outcome of head injured children? A regional investigation

P. A. Jones · I. R. Chambers · R. A. Minns ·
T. Y. M. Lo · L. M. Myles · A. J. W. Steers

Abstract

Background Secondary pathophysiological CPP insult is related to outcome after head injury, and improved management would be expected to reduce secondary brain insult. Paediatric head injury management guidelines have been published in recent years, by SIGN (2000), RCPCH (2001), NICE (June 2003), and jointly by Critical/Intensive Care Societies (C/ICS July 2003). We investigated whether outcome of children's head injury (and total burden of secondary CPP insult) has changed (1) annually; (2) before and after the introduction of any HI guidelines, and (3) following other service changes.

Methods Seventy-six children (aged 1–14 years with severe HI) were admitted to the Edinburgh Regional Head Injury Service between 1989 and 2006, and dichotomised at various time points and compared in terms of: demographic factors, intracranial pressure (ICP), cerebral perfusion pressure (CPP) insults [e.g. age-banded pressure–time index (PTI)], and Glasgow Outcome Scale (GOS) score (assessed at 6 months post injury).

Findings When dichotomised around the SIGN guidelines, there were no statistically significant differences between the two group's demography or in primary brain injury, but the outcomes were different ($p=0.03$), with 6 vs 4 GOS1 (died), 2 vs 4 GOS3 (severely disabled), 5 vs 16 GOS4 (moderately disabled) and 23 vs 14 GOS5 (good recovery), when comparing before and after year 2000. GOS4 was significantly different (chi-square=7.99, $p<0.007$). There was a (non-significant) trend for the later years to have longer insult durations of ICP, hypertension, CPP, hypoxia, pyrexia, tachycardia and bradycardia, greater PTI for both CPP and ICP, and more CPP insults ($p=0.003$). There was, however, significantly less CPP insult ($p=0.030$) after the introduction of the more management-oriented C/ICS guidelines.

Conclusions The most recent paediatric HI guidelines appear to have reduced the burden of secondary insult, but more time is required to determine if this will be reflected in improved outcomes.

Keywords Paediatric head injury · Secondary brain insult · Outcome · Guidelines

Introduction

Secondary pathophysiological CPP insult has been consistently shown to be related to outcome after head injury in both adults [4–6, 9, 11, 13, 16] and children [3, 2, 7, 8], and improved management would be expected to reduce secondary brain insult in both duration and intensity. Pediatric head injury management guidelines have been published in recent years, by the Scottish Intercollegiate Guidelines Network (SIGN; 2000) [15], the Royal College of Paediatrics and Child Health (RCPCH; 2001) [14], the NHS National Institute for Clinical Excellence (NICE; June 2003) [12], and jointly by the Society of Critical Care Medicine, the World Federation of Pediatric Intensive and Critical Care Societies, and the Paediatric intensive Care Society UK (C/ICS; July 2003) [1]. We investigated whether

P. A. Jones · R. A. Minns (✉)
Child Life and Health, University of Edinburgh,
20 Sylvan Place,
Edinburgh EH9 1UW Scotland, UK
e-mail: Robert.Minns@ed.ac.uk

R. A. Minns · T. Y. M. Lo · L. M. Myles · A. J. W. Steers
Royal Hospital for Sick Children, Edinburgh,
9 Sciennes Road,
Edinburgh EH9 1LF Scotland, UK

I. R. Chambers
Regional Medical Physics Department,
The James Cook University Hospital,
Marton Road,
Middlesbrough TS4 3BW England, UK

outcome after children's head injury and total burden of secondary brain insult, (particularly CPP insult) has changed (1) annually; (2) before and after the introduction of any HI guidelines, and (3) following other service changes.

Materials and methods

Seventy-six children (aged 1–14 years with severe HI) were admitted to the Edinburgh Regional Head Injury Service for adults and children between 1989 and 1996 (Western General Hospital), and between 2000 and 2006 to a new Paediatric ICU (Royal Hospital for Sick Children). Demographic (age, gender, cause of injury, GCS, ISS, Marshall CT score, pupil response etc.) and physiological data from the ICU bedside monitors were collected prospectively and analysed.

Outcome at 6-months post injury was assigned from responses to a questionnaire sent to all parent/carers. The groups were dichotomised at various time points, and compared in terms of: (1) demographic factors, (2) secondary brain insults including intracranial pressure (ICP) and cerebral perfusion pressure (CPP) insults, and (3) Glasgow Outcome Scale (GOS—paediatric modification) score. We used a previously developed age-banded pressure–time index (PTI) [10] to give a measure of the amount of 'brain insult' which occurred during the ICU management period.

The pressure–time index is a two-dimensional cumulative measure combining intensity and duration of secondary insult found in both ICP and CPP, calculated from data recorded every minute from the bed-side monitors in the Intensive Care Unit, using the example formula below for the cPTI for CPP: $\text{cPTI} = \sum(\text{CPP}_{\text{threshold}} - \text{CPP}) \times t_{\text{sample}}$ mmHg min, where cPTI is the cumulative pressure time index, and t is the time over which the data was sampled.

The data set was divided so that cases admitted before *any* of the mentioned guidelines were published, were compared with those of later years (i.e. division point immediately pre SIGN guidelines—2000). Data was also analysed on an annual basis to look for trends over time, and finally, the data set was split by admission date before and after July 2003, when the treatment specific guidelines of the C/ICS became widely available.

The statistical package of SPSS© for Windows 14.0 (SPSS Inc. USA), was used for the analysis.

Results

Pre- and post-SIGN guidelines

When dichotomised around the SIGN guidelines, the groups were comparable with no statistically significant differences between the demographic features (age, sex, cause of injury, GCS, ISS, ICP monitoring characteristics etc) or in primary brain injury. The outcomes however, were different (chi-square 9.11, $p=0.028$), with 6, 2, 5 and 23 having GOS 1 (died), GOS 3 (severely disabled), GOS 4 (moderately disabled) and GOS 5 (good recovery) before, compared to 4, 4, 17 and 15 respectively post-2000. In particular, the change in relative positions of the GOS 4 and 5 outcomes was highly significant (chi-square=7.99, $p<0.007$) (Fig. 1)

There was a (non-significant) trend for the later years to have longer mean insult durations of ICP, hypertension, CPP, hypoxia, pyrexia, tachycardia and bradycardia, greater mean cPTI for ICP, and a significantly greater mean number of episodes of CPP insults ($p=0.005$). i.e. a less optimal trend.

Fluctuations annually

When these head injury cases were analysed year-by-year, there were no overall significant differences found but a closer look at the referral pattern (Fig. 2) indicated that a change had taken place, with more children being referred from tertiary centres from 2001 onwards. There were 17, 1, and 18 admitted from the hospital Accident and Emergency department, a GP, and from tertiary referral centre before 2000, compared to 6,1, and 33 respectively after 2000 (chi-square=9.49, $p=0.009$).

The PTIicp and PTIcpp by year (see Fig. 3) for all 76 patients from 1989 to 2006, and the median amount of

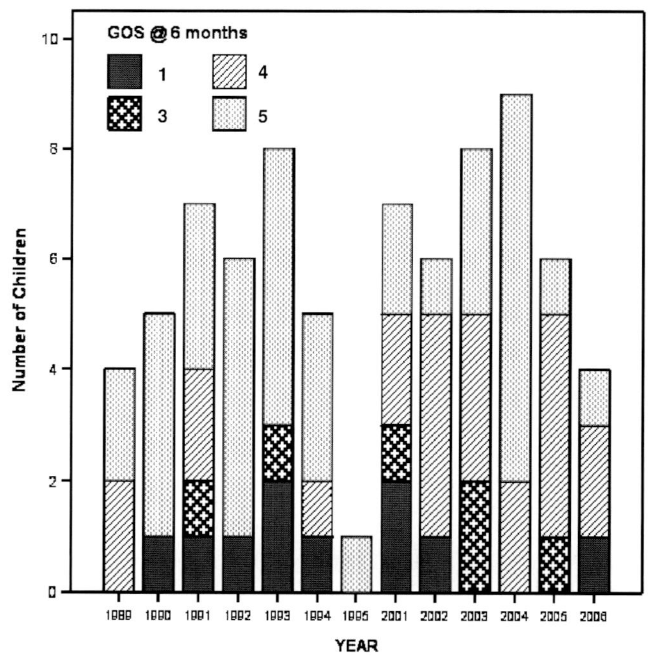

Fig. 1 Outcome at 6 months post-injury, illustrating that there were more GOS4 outcomes in later years compared to earlier years of the study. *GOS* Glasgow Outcome Score, where *GOS1* dead, *GOS3* severe disability, *GOS4* moderate disability, and *GOS5* good recovery (note there were no cases of *GOS2* vegetative state)

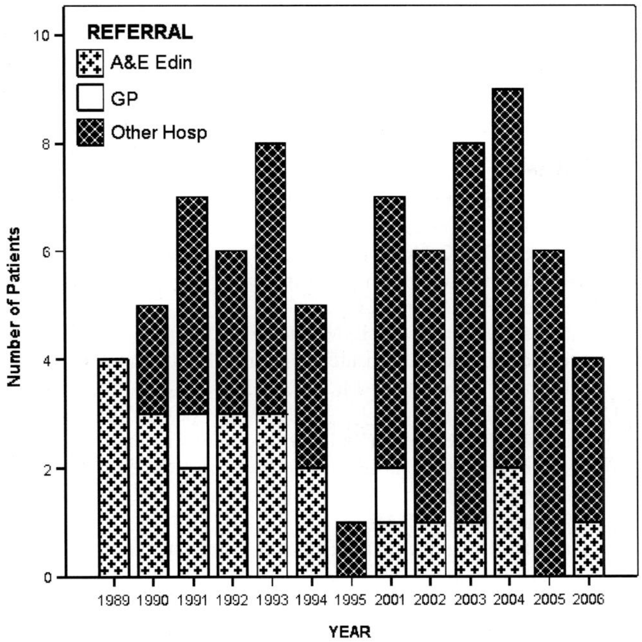

Fig. 2 Referral pattern year by year. Since 2001 the referral pattern has changed, with more children being admitted from peripheral hospitals

measured secondary insult (cPTI) per patient, independent of outcome, was similar.

Pre- and post C/ICS guidelines

When dichotomised around the time point of July 2003, when the paediatric C/ICS Head Injury guidelines were published, the outcomes were 9, 5, 14 and 29, compared to 1, 1, 8 and 9, for GOS1, 3, 4, and 5 respectively, for the pre- and post-July 2003 groups. While overall this was not a significant difference, there was a trend for fewer deaths and poor (GOS 3) outcomes.

The mean cPTI for CPP was 10,000.53 mmHg min (the product of duration and intensity) compared to 4,218.37 mmHg min respectively when the cohort was split pre-and post-July 2003. Although this gave an unequal distribution of cases (57 vs. 19), there was still significantly less CPP insult overall ($p=0.030$) after the introduction of the more management-oriented C/ICS guidelines, with a decrease of almost 60%. The mean cPTI for ICP was 35,186.95 mmHg min after July 2003, having fallen from 58,355.07 mmHg min, a decrease of about 40%. While this shows a trend in the desired direction, it was not significant.

Discussion

Pre- and post-SIGN guidelines

One of the notable guidelines advocated in the SIGN guidelines publication was for the transfer of head injured patients to a Neurosurgical Centre. We showed there was a significant change in referral pattern, with almost twice as many cases coming from other hospitals, largely in the eastern half of Scotland. However, an unexpected finding was an increase in the burden of ICP insult, as measured by the cPTI. We speculate that this could have been due to more insult occurring before admission to our unit, or

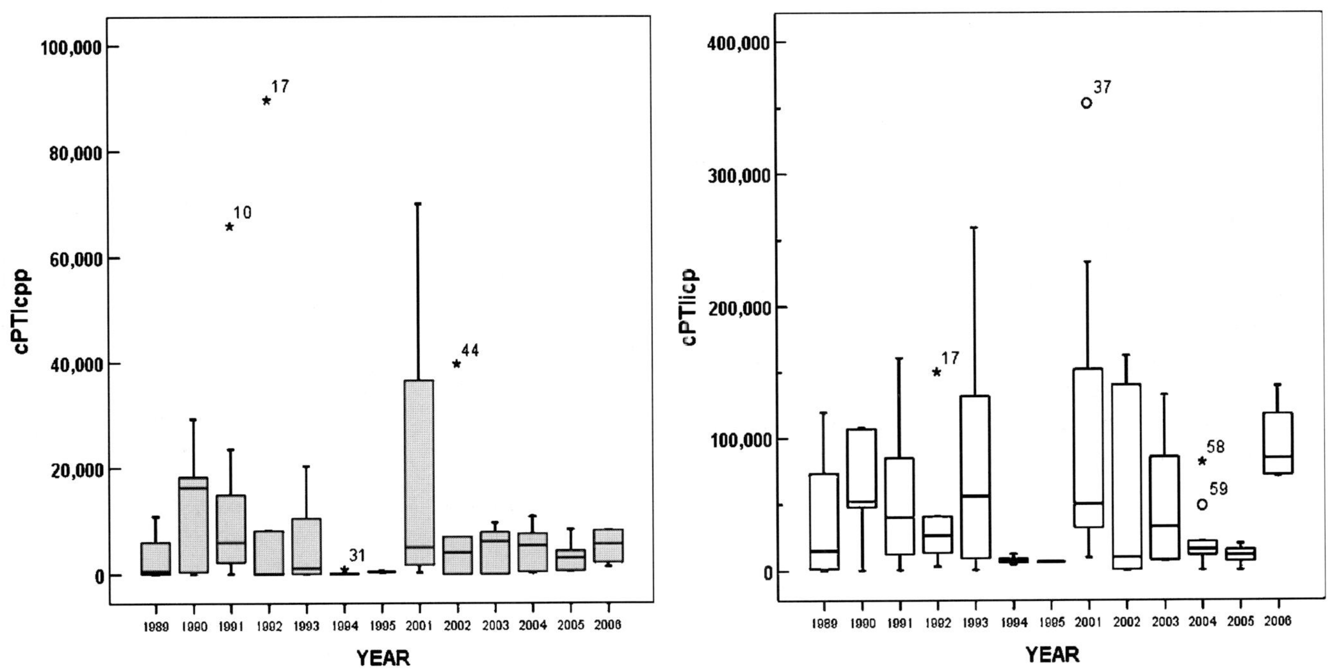

Fig. 3 *Box plots* of the yearly distribution of secondary brain insult assessed by the cumulative pressure time index (for both CPP and ICP), with all outcomes included in each year group

different treatment routines employed before patient transfer. The difference however, was not due to the time interval from injury to the instigation of intracranial pressure monitoring in these two groups and the later group were actually monitored on average, slightly more speedily (17.9 h, compared to 13.8 h). The mean duration of monitoring once at the PICU was 86.6 h compared to 92.0 h, but again this was not statistically significant.

As there were no demographic factors or GCS differences, it is difficult to explain the subsequent increase in numbers achieving only a moderate (GOS 4) recovery, compared to GOS5 (good recovery). The same questionnaire was used throughout the whole study period, and the same personnel were responsible for assigning the GOS score at 6 months, so internal bias is unlikely. We explored the change in the outcome pattern of those with GOS4 and GOS 5 more closely, and found that after 2000, the mean cPTI for CPP increased 3 fold, with a 4 fold increase in mean cPTI ICP in the same period, while those with GOS5 outcome had a 10% and 17% reduction in cPTI CPP and ICP respectively.

Annual evaluation

The amount of cPTI for both ICP and CPP on a year-to-year basis, independent of outcome, was found to be not significantly different for the year groups as a whole. There were however relatively small number of cases per year.

Unsurprisingly, those with the poorest outcome had the greatest burden of secondary brain insult, whether considered annually, before and after 2000, or before and after the July 2003 dividing point.

Pre-and post-C/ICS guidelines

After the publication of the more management directed head injury C/ICS guidelines in July 2003, and despite the declining prevalence of paediatric head injury cases there was significantly less secondary 'pressure' insult: cPTI, for CPP (p= 0.030) and a trend to less ICP. This was accompanied by a trend to fewer deaths and GOS3 outcomes.

Clearly guidelines may be implemented completely or partially and will require some time to show an effect. Additional time will also ensure larger study numbers, however the trend is for a definite improvement in outcome and less secondary brain insult which may reach significance in the future.

Conflict of interest statement We declare that we have no conflict of interest.

References

1. Adelson PD, Bratton SL, Carney NA, Chesnut RM, Du Coudray HEM, Goldstein B, Kochanek PM, Miller HC, Partington MD, Selden NR, Warden CR, Wright DW (2003) Guidelines for the acute medical management of severe traumatic brain injury in infants children and adolescents. J Trauma Inj Infect Crit Care 54 (Suppl 6):S237–S239
2. Chambers IR, Kirkham FJ (2003) What is the optimal cerebral perfusion pressure in children suffering from traumatic coma? Neurosurg Focus 15(6):E3
3. Chambers IR, Treadwell L, Mendelow AD (2000) The cause and incidence of secondary insults in severely head-injured adults and children. Br J Neurosurg 14(5):424–431
4. Chambers IR, Treadwell L, Mendelow AD (2001) Determination of threshold levels of cerebral perfusion pressure and intracranial pressure in severe head injury by using receiver-operating characteristic curves: an observational study in 291 patients. J Neurosurg 94(3):412–416
5. Huang SJ, Hong WC, Han YY, Chen YS, Wen CS, Tsai YS, Tu YK (2006) Clinical outcome of severe head injury using three different ICP and CPP protocol-driven therapies. J Clin Neurosci 13(8):818–822
6. Huang SJ, Hong WC, Han YY, Chen YS, Wen CS, Tsan YS, Tu YK (2007) Clinical outcome of severe head injury in different protocol-driven therapies. J Clin Neurosci 14(5):449–454
7. Jackson S, Piper IR, Wagstaff A, Souter M (2000) A study of the effects of using different cerebral perfusion pressure (CPP) thresholds to quantify CPP "secondary insults" in children. Acta Neurochir Suppl 76:453–456
8. Jones PA, Andrews PJ, Easton VJ, Minns RA (2003) Traumatic brain injury in childhood: intensive care time series data and outcome. Br J Neurosurg 17:29–39
9. Jones PA, Andrews PJ, Midgley S, Anderson SI, Piper IR, Tocher JL, Housley AM, Corrie JA, Slattery J, Dearden NM (1994) Measuring the burden of secondary insults in head-injured patients during intensive care. J Neurosurg Anesthesiol 6 (1):4–14
10. Jones PA, Chambers IR, Lo TY, Andrews PJ, Chaudhry W, Clark A, Croft J, Forsyth R, Fulton B, Mendelow AD, Wilson G, Minns RA (2005) Quantification of secondary CPP insult severity in paediatric head injured patients using a pressure–time index. Acta Neurochir Suppl 95:29–32
11. Marmarou A, Saad A, Aygok G, Rigsbee M (2005) Contribution of raised ICP and hypotension to CPP reduction in severe brain injury: correlation to outcome. Acta Neurochirurgica Suppl 95:277–280
12. NHS National Institute for Clinical Excellence (NICE) (2003) Head injury. Triage, assessment, investigation and early management of head injury in infants, children and adults. Clinical Guideline 4
13. Rosner MJ, Rosner SD, Johnson AH (1995) Cerebral perfusion pressure: Management protocol and clinical results. J Neurosurg 83(6):949–962
14. Royal College of Paediatrics and Child Health (2001) Good practice guidelines. Early management of patients with a head injury
15. Scottish Intercollegiate Guideline Network (SIGN) (2000) Early management of patients with a head injury. No 46
16. Young JS, Blow O, Turrentine F, Claridge JA, Schulman A (2003) Is there an upper limit of intracranial pressure in patients with severe head injury if cerebral perfusion pressure is maintained? Neurosurg Focus 15(6):E2

Acta Neurochir Suppl (2008) 102: 85–88

Low frequency pressure waves of possible autonomic origin in severely head-injured children

R. A. Minns · P. A. Jones · I. R. Chambers

Abstract

Background Useful information (both clinical and pathophysiological) which may be extracted from intracranial pressure (ICP) recordings include: (1) the mean level of ICP (and CPP), (2) cerebrovascular autoregulation status, (3) the intracranial pulse pressure (the pulse wave index, ICPpp/ICPm) or the pressure-volume compensatory reserve index (RAP) and (4) the presence of any abnormal ICP waveform. This paper describes a slow frequency ICP waveform in children with TBI and postulates the pathophysiological basis and whether it contains clinically useful detail.

Methods Children admitted to the Regional Head Injury Service in Edinburgh with TBI have continuously monitored ICP, MAP, CPP, and other physiological data (stored at a 1-min resolution). Slow frequency waveforms were noted, prompting a review of the stored monitoring from all cases over a 10 year period.

Findings Episodic slow pressure waves were detected in 11 of 122 severely head-injured (HI) children The waveforms were detected in children of all ages (1.6–15 years) in the ICP signal, which were in phase with similar fluctuations in the MAP, CPP, and HR signals Their mean periodicity was 1 per 7 min (range 1 per 5–10 min), with a mean ICP pulse wave amplitude of 5.45 mmHg (range 4–7.5), and mean MAP pulse wave amplitude (pulse pressure) of 10.4 mmHg (range 4–15 mmHg). The duration was variable (range approx 2 h to 4.5 days). They were detected in the pre-terminal phase after serious HI, as well as in those children who made an independent recovery (GOS 4/5). The waves were not related to the mean levels of ICP, CPP, MAP, temperature or the state of cerebrovascular autoregulation.

Conclusions We postulate that these previously unreported slow waveforms may reflect the very low frequency (VLF) and ultra low frequency (ULF; $\leq$1 per 5 min) components of heart rate and arterial blood pressure variability.

Keywords Pressure waves · ICP · Ultra low frequency · Head injury

R. A. Minns (✉) · P. A. Jones
Child Life and Health, University of Edinburgh,
20 Sylvan Place,
Edinburgh EH9 1UW Scotland, UK
e-mail: Robert.Minns@ed.ac.uk

R. A. Minns
Royal Hospital for Sick Children,
Edinburgh, 9 Sciennes Road,
Edinburgh EH9 1LF Scotland, UK

I. R. Chambers
Regional Medical Physics Department,
The James Cook University Hospital,
Marton Road,
Middlesbrough TS4 3BW England, UK

Introduction

Various ICP waveforms superimposed on the ICP signal have been described, such as the well recognised A (plateau), B and C waves. We have detected very slow episodic pressure waves, which to our knowledge have not been previously reported in the paediatric head-injured population.

In Edinburgh and Newcastle, when children with serious head injuries are admitted to the Intensive Care Units, continuous physiological data from the bedside monitors including intracranial pressure (ICP), mean arterial pressure (MAP), cerebral perfusion pressure (CPP), heart rate (HR), and oxygen saturations etc., are stored and analysed (with consent) as part of various ethically approved studies on

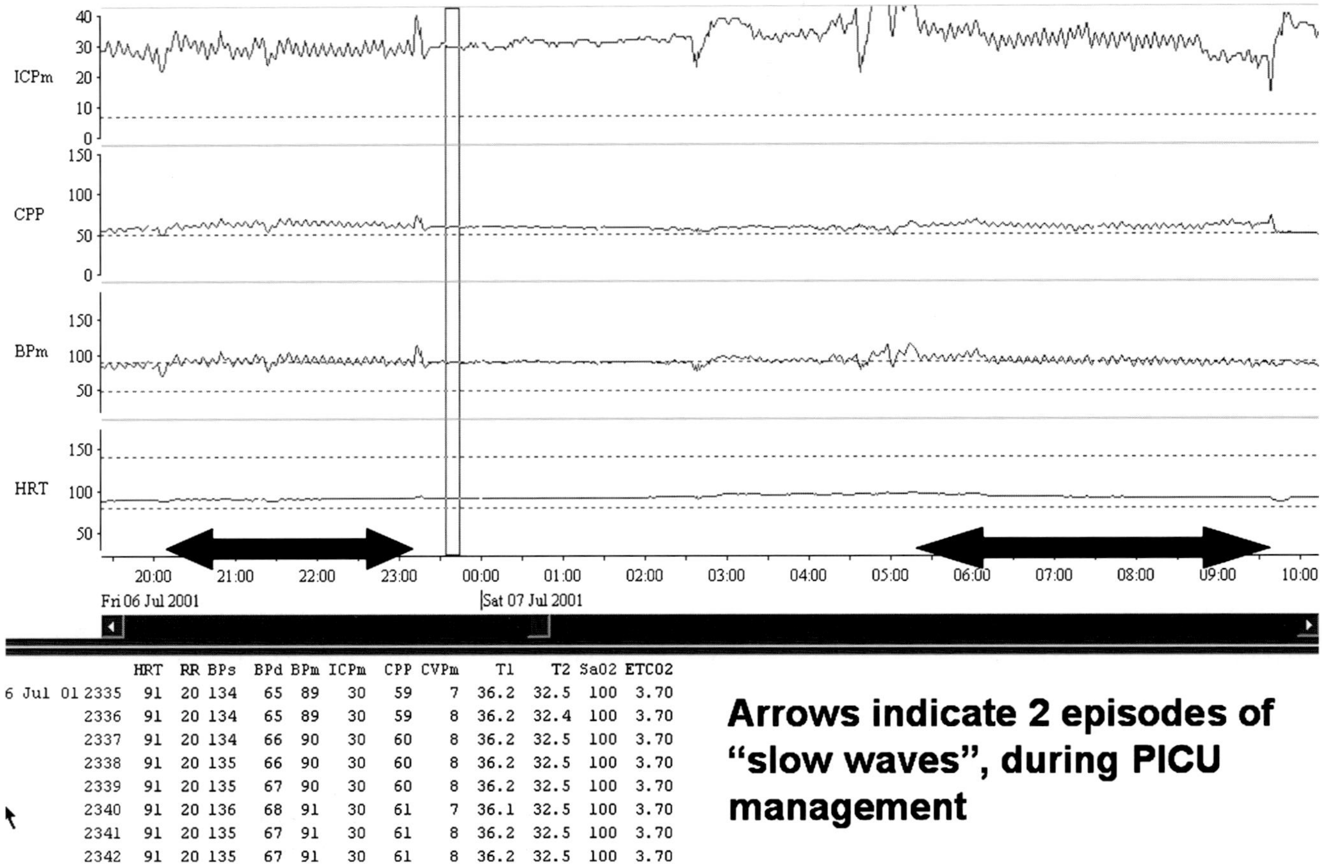

Fig. 1 illustrates two episodes of 'slow waves' seen in ICP, CPP, MAP and heart rate, in a 9 year old child with severe head injury, as viewed using the Odin Browser Software. *ICPm* mean intracranial pressure; *CPP* cerebral perfusion pressure; *BPm* mean arterial pressure; *HRT* heart rate. The *arrows* indicate two episodes of slow wave phenomena

"secondary brain insults". This time series data can be viewed in various ways, using the Edinburgh Monitor/Browser© Software [5], including a graphical display, where several variables can be viewed concurrently on the computer screen (Fig. 1).

Materials and methods

Having noticed an unusual ICP wave pattern in a number of cases, we reviewed all the electronically stored physiological data from a 10 year period from 122 seriously head injured (HI) children who had had ICP monitoring. Each record was scanned manually to identify episodes of the slow waves, and the characteristics of each episode recorded.

Results

Slow ICP pressure waves were detected in 11 of 122 children, and occurred in children of all ages (1.6–15 years) in the ICP signal. These appeared in phase with similar fluctuations in the MAP, CPP and HR signals. Several children had a number of episodes of slow wave activity while in the acute treatment phase. Five of the 11 children died. The mean periodicity of the ICP wave, the ICP pulse wave amplitude, the MAP pulse pressure, duration of slow wave activity and mean background ICP level during these slow wave episodes are seen in Table 1.

An example of these slow waves over a 2 h period is seen in Fig. 2. This record is from a seriously head injured 14 year old child who died in intensive care 2 weeks following a road traffic accident. It can be seen that the ICP

Table 1

Features	Value	Range
Mean periodicity	1 per 7.39 min	1 per 5–10 min
Mean ICP amplitude	5.45 mmHg	4–7.5 mmHg
Mean MAP amplitude	10.4	5–15 mmHg
Duration	Variable	2 h–4.5 days
Mean background ICP during slow wave episodes	22.3 mmHg	11–44 mmHg

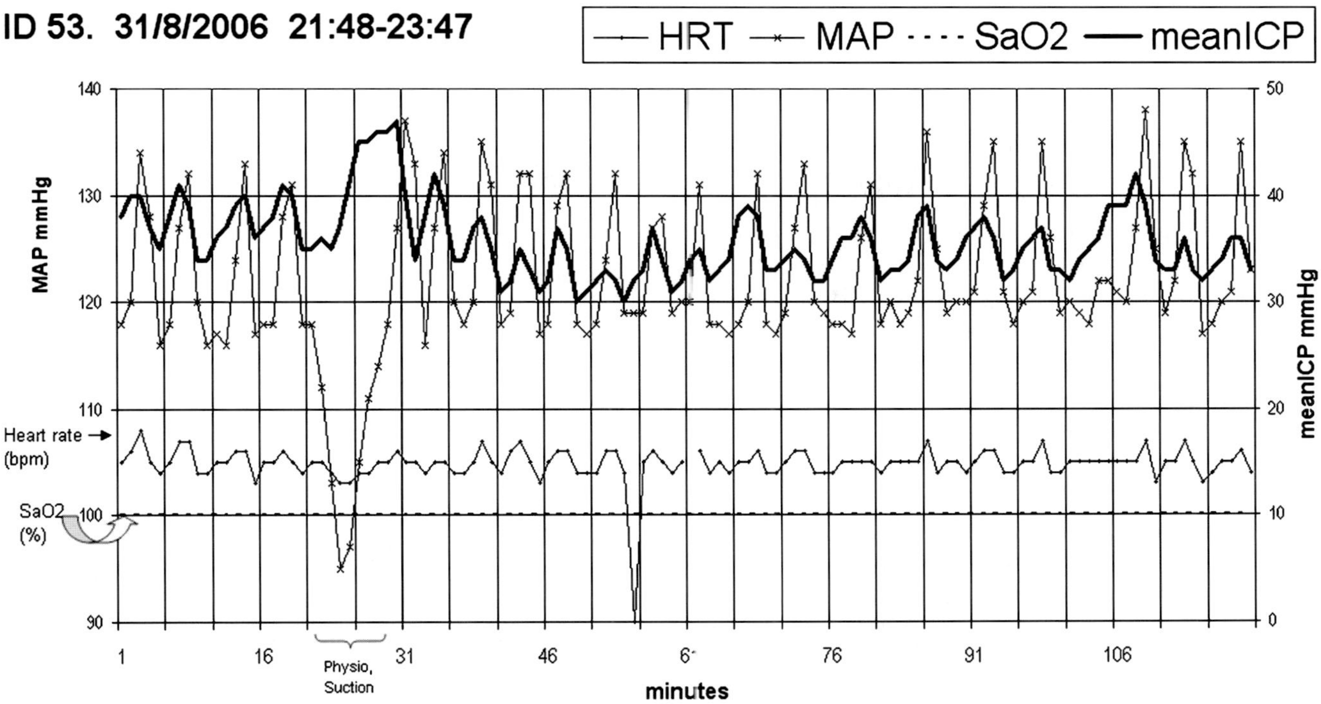

Fig. 2 illustrates these waves from a 2 h time period (in a 14 year old child who died following a RTA) showing fluctuations in both ICP and MAP, which were also reflected in the systolic and diastolic pressure, and in the pulse, but not in the O_2 saturation recording

fluctuations are also recorded in the mean arterial pressure and heart rate.

These slow ICP waveforms were detected in the pre-terminal stage after serious head injury, as well as in the recordings of children who made an independent recovery (GOS 4 and 5). They were not confined to any particular level of ICP, CPP, MAP and appeared independent of the state of cerebrovascular autoregulation.

Discussion

We describe very slow waveforms in ICP, BP and HR signals of children having physiological monitoring on account of serious traumatic brain injury. Investigations of the clinical circumstances surrounding these episodes showed that there was no obvious artifactual cause for these. In one case similar fluctuations occurring in the respiratory signal occurred irregularly in relation to the ICP fluctuations, and were deemed to be due to inadequate ventilation or fluid in the ventilator tubing. It can be seen in Fig. 2 that an interruption of these episodic wave forms occurs during physiotherapy and airway suction although the rhythmicity returns immediately after this activity. There was no change in the child's state or alteration of treatment modalities which could be responsible.

Our networked system of monitoring intracranial pressure and other physiological variables depends on a minute-to-minute sampling of signals. For this reason we are unable to precisely define the phase relationship of the peaks in the ICP, BP and heart rate channels and although the pressure pulses appear simultaneous and totally in phase, the sampling rate of our recording system does not allow that assumption, and hence we cannot deduce if they originate in the intracranial compartment and are reflected systemically or vice versa.

We postulate that the periodicity of these waveforms i.e. mean of 1 per 7 min, with apparent synchrony in BP and heart rate may reflect the very low frequency (VLF) and ultra low frequency (ULF; ≤1 per 5 min, or ≤0.003 Hz) components of heart rate and blood pressure variability (HRV and R–R interval) assessed by frequency domain [11]. These peaks most approximate the periodicity of the observed intracranial pressure waveforms we describe. HRV indicates disordered autonomic regulation of circulatory function and is known to be disordered in SAH [6], Intracranial hypertension and head injury [1, 10]. HRV is normally mediated via the negative feedback loop (baroreflex) through Nucleus Tractus Solitarius to coupled sympathetic and parasympathetic efferents from the vasomotor centre and vagal nucleus.

Wide band spectral analysis allows reliable information of ULF and VLF components of BP and RRI variability and one explanation for the genesis for the VLF fluctuations has included thermoregulation [7]. However, throughout all of our 11 cases with these slow frequency waveforms there were no temperature changes reported at any time.

For the most part the origin of these fluctuations in the circulation are not known although fluctuations in baroreceptor sensitivity [3], brain stem activity [9] and membrane potential in the sinus node [8] have been causally suggested. It is known that even in steady state conditions, human vagal 'baroreflex sensitivity' fluctuates in a major way at very low frequency [4].

Waveforms similar to our description above have been seen in the blood pressure and heart rate records of neonates with and without asphyxial brain damage [2]. The limitations of our sampling of pressure and heart rate signals does not allow a calculation of serial coefficients of the sequences but manual calculation of samples of these episodes suggest a very small variability over short time periods, e.g. the mean peak-to-peak interval in the illustrated case of 46 peaks was 5.02 with a standard deviation of 1.15 pointing to a possible generator (central or peripheral) in their origin.

Conflict of interest statement We declare that we have no conflict of interest.

References

1. Biswas AK, Scott WA, Sommerauer JF, Luckett PM (2000) Heart rate variability after acute traumatic brain injury in children. Crit Care Med 28(12):3907–3912
2. Cunningham S, Deere S, McIntosh N (1993) Cyclical variation of blood pressure and heart rate in neonates. Arch Dis Child 69 (1 Spec No):64–7
3. Di Rienzo M, Castiglioni P, Parati G, Frattola A, Mancia G, Pedotti A (1992) Role of arterial baroreflex in producing the 1/f shape of systolic blood pressure and heart rate spectra. 283–286
4. Eckberg DL, Kuusela TA (15-9-2005) Human vagal baroreflex sensitivity fluctuates widely and rhythmically at very low frequencies. J Physiol 567(Pt 3):1011–1019
5. Howells TP (1994) Edinburgh monitor and browser. Computer Program (v1.4)
6. Kawahara E, Ikeda S, Miyahara Y, Kohno S (2003) Role of autonomic nervous dysfunction in electrocardio-graphic abnormalities and cardiac injury in patients with acute subarachnoid hemorrhage. Circ J 67(9):753–756
7. Kitney RI (1974) The analysis and simulation of the human thermoregulatory control system. Med Biol Eng 12(1):57–65
8. Kobayashi M, Musha T (1982) 1/f fluctuation of heartbeat period. IEEE Trans Biomed Eng 29(6):456–7
9. Marsh DJ, Osborn JL, Cowley AW Jr. (1990) 1/f fluctuations in arterial pressure and regulation of renal blood flow in dogs. Am J Physiol 258(5 Pt 2):F1394–400
10. Rapenne T, Moreau D, Lenfant F, Vernet M, Boggio V, Cottin Y, Freysz M (2001) Could heart rate variability predict outcome in patients with severe head injury? A pilot study [see comment]. J Neurosurg Anesthesiol 13(3):260–268
11. Togo F, Kiyono K, Struzik ZR, Yamamoto Y (2006) Unique very low-frequency heart rate variability during deep sleep in humans. IEEE Trans Biomed Eng 53(1):28–34

Acta Neurochir Suppl (2008) 102: 89–91

Controlled lumbar drainage in medically refractory increased intracranial pressure. A safe and effective treatment

Ali Murad · Samer Ghostine · Austin R. T. Colohan

Abstract
Background A prospective study of lumbar CSF drainage in the setting of raised intra-cranial pressure refractory to medical management and ventriculostomy placement is presented. There have been no controlled trials of its use reported in the literature, to the best of our knowledge.
Method An IRB approved prospective study was conducted. 8 patients with increased intracranial pressure secondary to traumatic brain injury or aneurysm rupture were initially managed with sedation, ventriculostomy placement, mild hyperventilation (pCO_2=30–35), and hyperosmolar therapy (Na=150–155). A lumbar drain was placed if ICP continued to be above 20 mmHg despite optimization of medical therapy.
Findings After lumbar drain placement, ICP was reduced from a mean of 27±7.8 to 9±6.3, an average decrease of 18 mm H_2O ($p<0.05$). Requirements for hypertonic saline and/or mannitol boluses and sedation to control ICP were also decreased. There were no complications noted.
Conclusions We have shown that controlled lumbar drainage is a safe, efficacious and minimally invasive method for treatment of elevated ICP refractory to medical management. Ventriculostomies are always placed before utilizing lumbar drains to minimize the risk of cerebral herniation. We would advocate making controlled lumbar drainage a standard part of ICP control protocols.

A. Murad (✉) · S. Ghostine · A. R. T. Colohan
Department of Neurosurgery,
Loma Linda University Medical Center,
11234 Anderson Street, Rm. 2562-B,
Loma Linda, CA 92354, USA
e-mail: amurad@llu.edu

Keywords Head injury · Intracranial pressure · Intracranial hypertension · Lumbar drain

Introduction

After brain injury the intra-cranial pressure (ICP) may remain normal at first, because the brain swelling can be compensated by displacing blood and cerebrospinal fluid (CSF). When no more blood and CSF (~100 to 150 ml of intra-cranial volume) can be pushed out of the cranial vault, the intra-cranial pressure starts rising, which can lead to secondary brain injury. Ventriculostomies are commonly used for monitoring and treating high ICP. However, external drainage of CSF is not limited to the ventricular route. Up to 30% of the total compliance of the CSF system is in the spinal axis [1], and even more if basal cisterns are taken into account. Drainage of some of this CSF by use of lumbar drains can help reduce ICPs further than ventriculostomies alone. However, lumbar CSF drainage has not been a 1st tier tool in treatment of high ICP because of concern about cerebral herniation. This concern mostly stems from reports of such complications in publications several decades ago [2, 3]. There has been renewed interest recently in the use of lumbar drainage for treatment of high ICP refractory to ventriculostomy placement and medical treatment [4–6].

Methods

A prospective study was designed with the aim of enrolling patients with brain injury—traumatic, aneurysm rupture, brain tumor (post resection)—and high ICP refractory to medical management and ventriculostomy. A protocol approved by Loma Linda University Medical Center (LLUMC) Institutional Review Board was used to enroll patients.

Inclusion criteria:

- Patient at LLUMC
- Increased intra-cranial pressure (ICP): higher than 20 mm H_2O
- Ventriculostomy catheter placed
- Medical therapy has been refractory
- Patients age 18–99

Exclusion criteria:

- Presence of surgical pathology (LD could be placed after removal of such pathology)
- Patient transferred with lumbar drains
- Patients without ventriculostomy catheters
- Patients age 0–18

Eight patients with increased intracranial pressure secondary to brain injury (six patients with traumatic injury and two with aneurysm rupture) have so far been included in the study. Three of the trauma patients had emergent surgery for hematoma evacuation on arrival; one of the two aneurysm patients had endovascular coiling of aneurysm, while the other had clipping of aneurysm along with evacuation of hematoma. Their management included ventriculostomy placement and medical measures such as hypertonic saline (Na>150), mild hyperventilation (pCO_2= 30–35), and sedation. A lumbar drain was placed if ICP

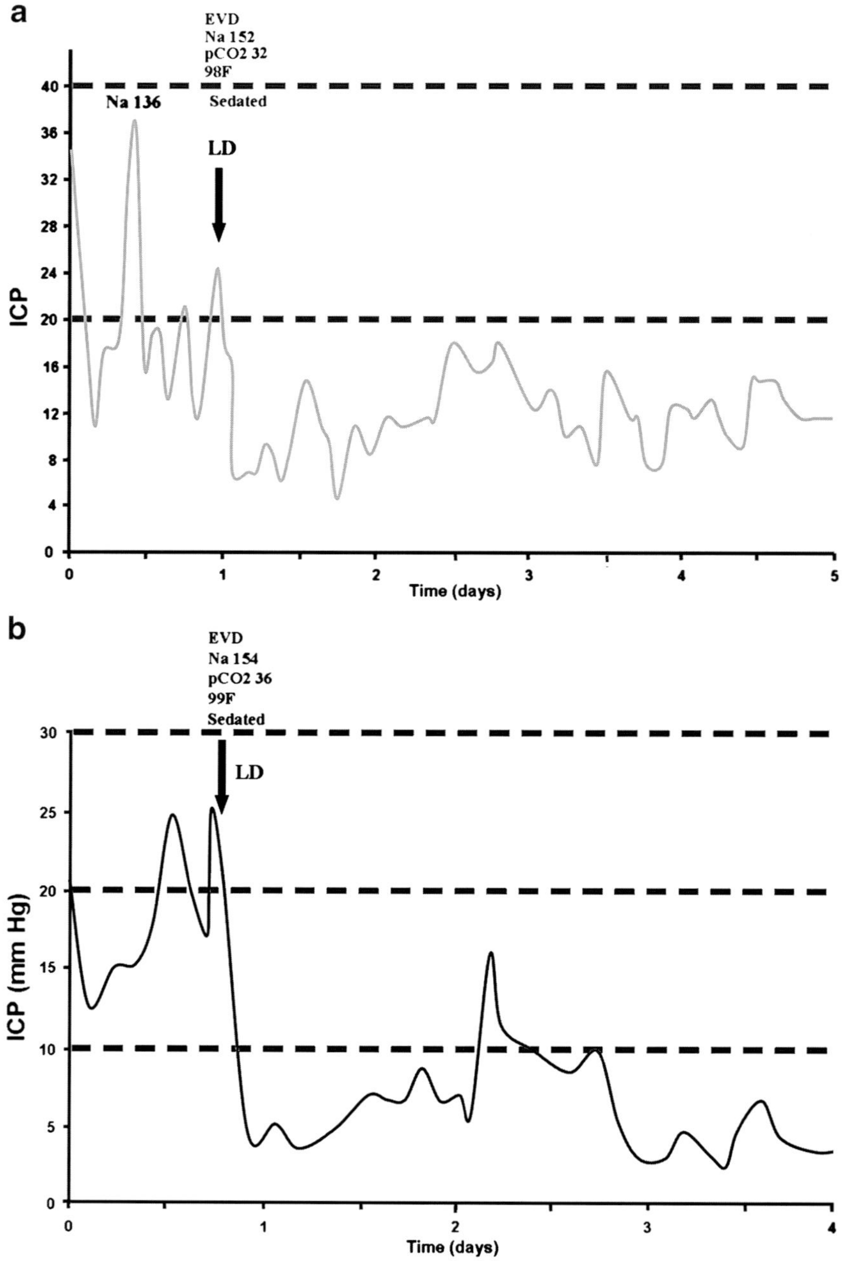

Fig. 1 a, b Pattern of ICP before and after lumbar CSF drain (*LD*) placement in two patients. Parameters such as serum sodium (*Na*), pCO_2, patient temperature (*F*) at the time of LD placemen are shown (*bold arrow* indicates when LD was placed)

trended above 20 mmHg for an average of 3 h despite optimization of parameters mentioned above. In some patients, the lumbar drain and ventriculostomy were both drained at the same level (0 cm H_2O above the Foramen of Monroe) and in others, the lumbar drain was kept clamped but 10 ml of CSF were drained each hour by unclamping the lumbar drain to gravity intermittently until 10 ml of CSF were obtained in the reservoir chamber.

While all patients had several head CTs during their admission, they were not done routinely immediately before or after placement of lumbar drains. After placement of the lumbar drain, there was no worsening of neurological examination or ICP.

Results

Eight patients were enrolled in the study. Mean age was 36 years (range 20–51 years). After lumbar drain placement, ICP was reduced from a mean of 27±7.8 to 9±6.3 (mean ICP values given are 3 h pre and post LD placement). This was an average decrease of 18 mm H_2O ($p<0.05$). The reduction of ICP was seen in a clear and sustained manner in all patients, as illustrated in two patients shown (Fig. 1a,b). In the 24-h period preceding lumbar drain placement, an average of 2.75 boluses of hypertonic saline and Mannitol per patient were given to control ICP. In the 24-h period following lumbar drain insertion, no hypertonic saline or mannitol boluses were needed. Sedation requirements were also decreased. There were no intracranial complications noted due to lumbar drain placement.

Discussion

We demonstrate the safety of lumbar CSF drainage in the setting of high ICP. We would advocate making controlled lumbar drainage a standard part of ICP protocols when medical management and ventriculostomy placement have failed to control ICP. In our series of eight patients, use of lumbar drainage was used effectively to significantly lower ICP.

The risk of herniation is particularly high when the ICP is high and the basal cisterns are already at least partially effaced at the time of drain insertion. In the setting of high ICPs, minimizing CSF leakage at the time of needle insertion for lumbar drain placement is of great importance. Once the lumbar drain is placed, CSF can be drained in a controlled manner, with continued vital signs and neuro-checks monitoring.

Outcome data for these patients is being deferred at this time. In this current paper, the goal of the authors is to evaluate the safety and effectiveness of lumbar CSF drainage in patients suffering from raised ICPs refractory to medical management. The study is ongoing and analysis of patient's outcome and changes in therapy intensity level (TIL) [7] as a result of lumbar drain placement is planned.

Conflict of interest statement We declare that we have no conflict of interest.

References

1. Shapiro K, Marmarou A, Shulman K (1980) Characterization of clinical CSF dynamics and neural axis compliance using the pressure–volume index: I. The normal pressure–volume index. Ann Neurol 7:508–514
2. Nash CS (1937) Cerebellar herniation as a cause of death. Ann Otol Rhinol Laryngol 46:673–680
3. Verbrugghen A (1946) Spinal puncture. Surg Clin North Am 26:78–90
4. Levy DI, Baldwin HZ et al (1995) Controlled lumbar drainage in pediatric head injury. J Neurosurg 83:453–460
5. Munch EC, Vajkoczy P et al (2001) Therapy of malignant intracranial hypertension by controlled lumbar cerebrospinal fluid drainage (23 patients). Crit Care Med 29(5):976–981
6. Willemse RB, Egeler-Peerdeman SM (1998) External lumbar drainage in uncontrollable intracranial pressure in adults with severe head injury: a report of 7 cases. Acta Neurochir Suppl (Wien) 71:37–39
7. Maset AL, Young HF et al (1987) Pressure volume index in head injury. J Neurosurg 6:832–840

Acta Neurochir Suppl (2008) 102: 93–97

Decompressive craniotomy: durotomy instead of duroplasty to reduce prolonged ICP elevation

Ralf Burger · David Duncker · Naureen Uzma · Veit Rohde

Abstract

Background Usually, decompressive craniectomy (DC) in patients with increased intracranial pressure (ICP) is combined with resection of the dura and large-scale duroplasty. However, duroplasty is cumbersome, lengthens operation time and requires heterologous or autologous material. In addition, the swelling brain could herniate into the duroplasty with kinking of the superficial veins at the sharp cutting edges and subsequent ICP exacerbation. Several longitudinal durotomies avoid these limitations, but it remains a matter of discussion if durotomies reduce ICP sufficiently.

Methods DC was performed in ten patients (mean age 45 years) with increased ICP after head trauma or subarachnoid hemorrhage. After craniectomy, the dura was opened by three to four durotomies from midline to the temporal base. Duration of surgical procedure and ICP during each surgical step and postoperatively were recorded.

Findings Mean duration of surgery was 90±10 min. ICP prior to skin incision was 39±12 mmHg and dropped to 22± 9 mmHg after craniectomy. During durotomy ICP decreased stepwise and reached stable values of 12±6 mmHg at the end of surgery. On days 1–10 after surgery, ICP values ranged between 12–17 mmHg.

Conclusion This study showed that durotomy is a fast and easy, but likewise effective method to lower ICP further after craniectomy.

Keywords Intracranial pressure · Decompressive craniectomy · Traumatic brain injury · Durotomy

R. Burger (✉) · D. Duncker · N. Uzma · V. Rohde
Department of Neurosurgery, Georg-August University, Robert–Koch Str. 40, 37075 Goettingen, Germany
e-mail: ralf.burger@med.uni-goettingen.de

D. Duncker
e-mail: David.Duncker@stud.uni-goettingen.de

N. Uzma
e-mail: uzma@med.uni-goettingen.de

V. Rohde
e-mail: veit.rohde@med.uni-goettingen.de

Introduction

Initially, decompressive craniectomy (DC) was introduced to lower the intracranial pressure (ICP) in patients with inoperable brain tumors [8]. Since the late sixties DC has been performed in uncontrollable ICP after traumatic brain injury [2, 5]. It was implemented in the Guidelines of the Brain Trauma Foundation [3] as one of the "second tier therapies" of increased ICP when first level options have failed. The European Brain Injury Consortium considered DC as "ultima ratio" after failure of other treatment options [22]. In the literature a wide variety of surgical techniques could be found either for location and extent of craniectomy, dura opening and duroplasty had been reported. Unilateral or bilateral [6, 7, 10, 11, 18, 20, 21, 24, 27, 31, 32], bifrontal [14, 15, 19, 26, 34, 35], subtemporal [8, 12, 33] or circumferential [4] bony decompressions had been proposed. However, unilateral hemicraniectomies and bifrontal craniectomies have gained the widest acceptance [29]. Dura opening after bone removal is considered to be mandatory [8]. The majority of centers prefer extensive dura resection and subsequent duroplasty. However, duroplasty is cumbersome, lengthens operation time in these critically ill patients and requires heterologous or autologous material. Longitudinal durotomies avoid these limitations. However, it is a matter of discussion if mere durotomy indeed lowers the ICP over time. Thus, it was the aim of the present study to investigate the effect of simple longitudinal durotomies on intra- and postoperative ICP.

Materials and methods

DC was performed in ten patients (traumatic brain injury [TBI] n=8; subarachnoid hemorrhage [SAH], n=2; six males, four females; mean age 45±5 yrs.) with ICP>30 mmHg for 30 min or longer despite ventricular drainage. Glasgow coma scale score at admission [34], TBI classification [3], Marshall computed tomography (CT) grade [23], Hunt and Hess grade [13] and Fisher CT classification [9] in SAH patients, pre- peri- and postoperative ICP and Glasgow outcome scale (GOS) scale were assessed [17], (Table 1).

All patients were fixed in a Mayfield clamp. A question mark shaped skin incision from the preauricular to the frontal region was performed. Galea, periostium, and temporal muscle were mobilised from the skull. A bone flap of a size of 10×12 cm was removed, bone resection was extended to the temporal base. The dura was opened by straight incisions (n=4; lenght 6–8 cm) from parasagitally to the lower inferior temporal gyrus (Fig. 1).

Duration of surgical procedure, mean arterial blood pressure (MABP), endexpiratory $etpCO_2$ and ICP were recorded during each surgical step with possible effect on ICP (beginning of surgery, during burrhole placement, removal of the bone flap, each durotomy and skin closure). In the postoperative period, ICP was recorded until day 10. ICP, MABP, $etpCO_2$, and duration of surgery are presented as absolute values with mean and standard deviation. One-Way Repeated Measures Analysis of Variance was employed to compare intracranial pressure, MABP and

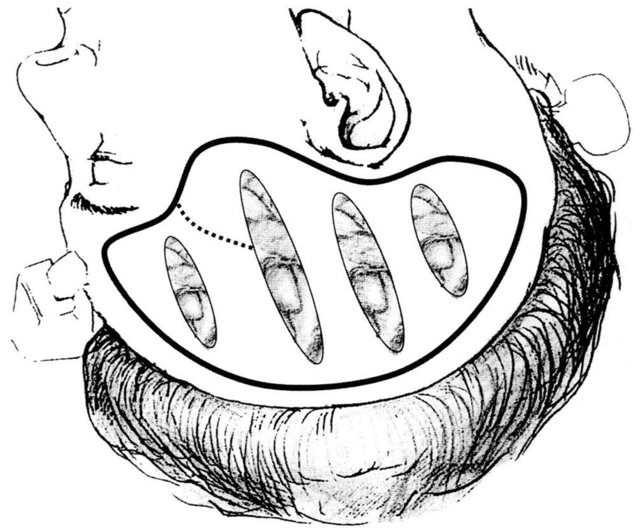

Fig. 1 Schematic illustration of bone flap size (*bold line*) and contour of sylvian fissure (*scattered line*). After decompressive craniectomy intracranial pressure was significantly decreased by widening of durotomies without bulging of edemateous brain (modified illustration from Yarsagil M.G., Microneurosurgery Part I, Thieme Verlag, Stuttgart, New York, 1984)

Table 1 Demographic data of study patients

	Impact	TBI grading	H & H scale	GCS	Marshall CT grading	CT Fisher scale	Level of ICP [indication for DC] (mmHg)	Surgery (DC) [days after impact]	GOS [3 months]	Comments
TBI	fall from stairway	III		4	DI 3		~50	8	4	
TBI	fall from stairway	III		4	DI 3		~35	2	3	
TBI	car accident	III		3	DI 4		~35	1	2	Anisocoria before DC
TBI	public bus accident	III		3	DI 4		~58	1	3	
TBI	bicycle accident	III		4	DI 3		~70	0	2	Anisocoria before DC
TBI	fall from stairway	III		3	DI 3		~21	3	3	Early DC due to CAT scan
TBI	fall from scaffolding	III		3	DI 3		~42	0	1	
TBI	car accident	III		4	DI 3		~25	4	3	
SAH	ICA–aneurysm		IV	3		IV	~38	3	3	No ischemia, massive brain edema
SAH	PCA–aneurysm		IV	3		III	~41	5	3	PCA infarction, brain edema

TBI Traumatic brain injury, *GCS* Glasgow coma scale [34], *DC* decompressive craniectomy, *DI* diffuse injury, *H & H* Hunt and Hess grading [13], *GOS* Glasgow outcome scale [17]

TBI grading by Bullock et al. [3]; Fisher grading I–IV post SAH [9]; Marshall CT grading after TBI [23].

$etpCO_2$ values over time in the study group. The Kolmogorov–Smirnov test (with Lilliefors' correction) was used to test data for normality. The significance level was $p<0.05$. Statistical analysis was performed with SigmaStat 2.0 (Jandel Scientific®, SPSS Inc., Chicago, IL, USA).

Results

Before surgery, the mean ICP was 42±16 mmHg with a range between 21–70 mmHg. Baseline ICP immediately prior to skin incision was 39±12 mmHg. ICP showed high or sometimes slightly increased values during burrhole placement and dropped significantly for the first time after removal of the bone flap to 22±9 mmHg ($p<0.001$). The first durotomy reduced the mean ICP to 16±6 mmHg ($p=0.018$). After the third durotomy, an ICP of 10±3 mmHg ($p=0.003$) was reached and remained stable until skin closure (mean value of 12±6 mmHg ($p>0.05$). Mean duration of operative procedure was 90±10 min. Mean arterial blood pressure was 80±18 mmHg and the mean endexpiratory pCO_2 was 34±4 mmHg. During surgery, these values did not show statistically significant differences. In the postoperative period (days 1 to 10 after DC) ICP values ranged between 12±9 and 17±5 mmHg ($p=0.031$).

The outcome 3 months after admission was assessed by the GOS scale. One patient made a good recovery (GOS 4), 6 patients remained moderately disabled (GOS 3), one patient remained in a vegetative status (GOS 2) and one patient died 10 days after DC due to sepsis and multiorgan failure (GOS 1).

Discussion

Decompressive craniectomy is considered as a second tier therapy in the Guidelines of the Brain Trauma Foundation [3] and as last tier therapy in the European Brain Injury Consortium [22]. A well-defined ICP trigger for DC does not exist. ICP thresholds between 20 mmHg [26, 32] and 40 mmHg [31] were reported. However, many authors propose to perform DC in patients with medically uncontrollable ICP of 30 mmHg and higher or with a cerebral perfusion pressure below 70 mmHg with further neurological detoriation [7, 20, 21]. In accordance with these data, we proceed to DC in medically refractory ICP of 30 mmHg and higher. In selected patients with acute subdural hematoma, massive brain swelling and signs of herniation, we performed DC even if the registered ICP did not reach 30 mmHg.

The technical details of DC are not well defined. Uni- and bilateral, subtemporal, bifrontal and circumferential craniectomies had been performed. However, unilateral and bilateral hemicraniectomies [6, 7, 10, 11, 18, 20–22, 27, 31, 32] as well as bifrontal craniectomies [14, 15, 19, 26, 35, 36] have gained the broadest acceptance. Several studies have indicated that opening of the dura is crucial for effective ICP reduction but again, a variety of methods had been described, Stellate-, X-, Z -, question mark- or fishmouth-shaped dura openings [1, 21, 24–27, 30, 36] had been proposed followed by large-scale duraplasty with heterologous or autologous material [1, 6, 10, 20, 21, 24–27, 30, 35, 36]. Duroplasty has some major disadvantages: (1) prolongation of the operative procedure, which might have a negative effect on the outcome in a severely ill patient; (2) use of heterologous or autologous material; (3) herniation of the edematous brain into the duroplasty with kinking of brain arteries and veins at the sharp cutting edges of the resected dura which might result in ischemia or venous congestion, thereby further contributing to brain swelling. Therefore some authors [7] recommended to create a vascular channel with small pillars on each side of crossing vessels at the dural margin made of absorbable gelatin (hemostatic sponge) combined with polyglycol acid (absorbable suture); (4) Dura resection and duroplasty is a cumbersome procedure for the neurosurgeon.

The present study shows that mere durotomy by dural incisions avoids the disadvantages of dural resection and duroplasty without having a diminished effect on ICP control. ICP dropped by 44% after bone flap removal and by another 26% after durotomy. This observation is in line with other studies which monitored ICP during DC. In these studies, ICP prior to surgery ranged between 29.2–40.4 mmHg [20, 25, 30, 31, 35, 36] and dropped to values below 20 mmHg [1, 20, 25, 28, 30, 31, 35, 36] after bone flap removal. Some authors found a clear correlation between the extent of bone removal and decline of ICP [30]. However, a strong focus on the size of the bone flap is not justifed as dura opening after DC likewise has a strong effect on ICP [18]. Authors who monitored ICP and brain $p(ti)O_2$ showed that ICP continuously dropped to low pathological values during DC and duroplasty whereas brain $p(ti)O_2$ tremendeously improved to normal values [16, 27, 32]. Interestingly, brain $p(ti)O_2$ was not related to bone flap removal but to dura opening, which underlines again that dura opening is the crucial factor in DC [27]. Within the next 24–72 h after DC ICP remained stable below 20 mmHg in our and other studies [1, 25, 30, 36]. We measured ICP until day 10 post surgery, and did not see any ICP rebound despite reduction of the sedation and anti-edema drugs.

The lack of any ICP rebound suggests that secondary brain swelling does not play a role in our modification of DC, possibly because the dura stripes between the durotomies avoid extensive brain herniation and, thereby, any kinking of arteries and veins at the sharp cutting edges of the incised dura. It has to be pointed out that this "protective" effect is not guaranteed with too many dura incisions respectively with too small dural openings: In one patient in whom several dural incisions had been performed and who was re-operated due to a subgaleal hematoma, we observed brain herniation between the small dural openings. Thus, we believe that the number of durotomies should be limited to three or four. This is supported by the fact that in our study a further decrease of the elevated ICP was not seen if more than three durotomies had been performed.

In our study mean duration of surgery was 90 min which is very short in comparison to DC surgery with duroplasty, supporting again the concept of the modification of DC presented here.

In conclusion, these preliminary results clearly indicate, that unilateral craniectomy with durotomy has the potential to overcome the disadvantages of craniectomy and duroplasty: (1) the operation is shorter and less cumbersome than conventional DC which might have a positive effect on the outcome of the patient with increased ICP after brain injury and subarachnoid haemorrhage; (2) heterologous or autologous material for duroplasty after dura resection is not required; (3) it is assumed that brain herniation with vessel kinking and secondary brain swelling is avoided by the dura stripes between the durotomies. However, these positive preliminary results have to be further evaluated in a larger series which is currently under way in our institution.

Conflict of interest statement We declare that we have no conflict of interest.

References

1. Aarabi B, Hesdorffer DC, Ahn ES, Aresco C, Scalea TM, Eisenberg HM (2006) Outcome following decompressive craniectomy for malignant swelling due to severe head injury. J Neurosurg 104(4):469–479
2. Britt RH, Hamilton RD (1978) Large decompressive craniectomy in the treatment of acute subdural hematoma. Neurosurgery 2 (3):195–200
3. Bullock R, Chesnut RM, Clifton G, Ghajar J, Marion DW, Narayan RK et al (2000) Management and prognosis of severe traumatic brain injury. Part 1: Guidelines for the management of severe traumatic brain injury. Part 2: Early indicators of prognosis in severe traumatic brain injury. J Neurotrauma 17(6–7):451–627
4. Clark K, Nash TM, Hutchison GC (1968) The failure of circumferential craniotomy in acute traumatic cerebral swelling. J Neurosurg 29(4):367–371
5. Cooper PR, Rovit RL, Ransohoff J (1976) Hemicraniectomy in the treatment of acute subdural hematoma: a re-appraisal. Surgical Neurol 5:25–28
6. Coplin WM, Cullen NK, Policherla PN, Vinas FC, Wilseck JM, Zafonte RD, Rengachary SS (2001) Safety and feasibility of craniectomy with duraplasty as the initial surgical intervention for severe traumatic brain injury. J Trauma 50(6):1050–1059
7. Csokay A, Egyud L, Nagy L, Pataki G (2002) Vascular tunnel creation to improve the efficacy of decompressive craniotomy in post-traumatic cerebral edema and ischemic stroke. Surg Neurol 57(2):126–129
8. Cushing H (1905) The establishment of cerebral hernia as a decompressive measure for inaccessible brain tumors; with the description of intramuscular methods making the bone defect in temporal and occipital regions. Surg Gyn Obstet 1:297–314
9. Fisher CM, Kistler JP, Davis JM (1980) Relation of cerebral vasospasm to subarachnoid hemorrhage visualized by computerized tomographic scanning. Neurosurgery 6(1):1–9
10. Gaab MR, Rittierodt M, Lorenz M, Heissler HE (1990) Traumatic brain swelling and operative decompression: a progressive investigation. Acta Neurochir [Suppl.] 51:326–328
11. Gerl A, Tavan S (1980) Bilateral craniectomy in the treatment of severe traumatic brain edema. Zentralbl Neurochir 41(2):125–138
12. Gower DJ, Lee KS, McWhorter JM (1988) Role of subtemporal decompression in severe closed head injury. Neurosurgery 23 (4):417–422
13. Hunt WE, Hess RM (1968) Surgical risk as related to time of intervention in the repair of intracranial aneurysms. J Neurosurg 28(1):14–20
14. Hutchinson PJ, Kirkpatrick PJ (2004) Decompressive craniectomy in head injury. Curr Opin Crit Care 10(2):101–104
15. Hutchinson PJ, Corteen E, Czosnyka M, Mendelow AD, Menon DK, Mitchell P, Murray G, Pickard JD, Rickels E, Sahuquillo J, Servadei F, Teasdale GM, Timofeev I, Unterberg A, Kirkpatrick PJ (2006) Decompressive craniectomy in traumatic brain injury: the randomized multicenter RESCUEicp study (www.RESCUEicp.com). Acta Neurochir Suppl 96:17–20
16. Jaeger M, Soehle M, Meixensberger J (2003) Effects of decompressive craniectomy on brain tissue oxygen in patients with intracranial hypertension. J Neurol Neurosurg Psychiatry 74 (4):513–515
17. Jennett B, Bond M (1975) Assessment of outcome after severe brain damage. Lancet 1(7905):480–484
18. Jourdan C, Convert J, Mottolese C, Bachour E, Gharbi S, Artru F (1993) Evaluation of the clinical benefit of decompression hemicraniectomy in intracranial hypertension not controlled by medical treatment. Neurochirurgie 39(5):304–310
19. Kjellberg RN, Prieto A Jr (1971) Bifrontal decompressive craniotomy for massive cerebral edema. J Neurosurg 34(4):488–493
20. Kontopoulos V, Foroglou N, Patsalas J, Magras J, Foroglou G, Yiannakou-Pephtoulidou M, Sofianos E, Anastassiou H, Tsaoussi G (2002) Decompressive craniectomy for the management of patients with refractory hypertension: should it be reconsidered? Acta Neurochir (Wien) 144(8):791–796
21. Kunze E, Meixensberger J, Janka M, Sorensen N, Roosen K (1998) Decompressive craniectomy in patients with uncontrollable intracranial hypertension. Acta Neurochir Suppl 71:16–18
22. Maas AI, Dearden M, Teasdale GM, Braakman R, Cohadon F, Iannotti F, Karimi A, Lapierre F, Murray G, Ohman J, Persson L, Servadei F, Stocchetti N, Unterberg A (1997) EBIC-guidelines for management of severe head injury in adults. European Brain Injury Consortium. Acta Neurochir (Wien) 139(4):286–294
23. Marshall LF, Marshall SB, Klauber MR, Van Berkum Clark M, Eisenberg H, Jane JA, Luerssen TG, Marmarou A, Foulkes MA (1992) The diagnosis of head injury requires a classification based on computed axial tomography. J Neurotrauma 9(Suppl 1):287–292
24. Munch E, Horn P, Schurer L, Piepgras A, Paul T, Schmiedek P (2000) Management of severe traumatic brain injury by decompressive craniectomy. Neurosurgery 47(2):315–322

25. Olivecrona M, Rodling-Wahlstrom M, Naredi S, Koskinen LO (2007) Effective ICP reduction by decompressive craniectomy in patients with severe traumatic brain injury treated by an ICP-targeted therapy. J Neurotrauma 24(6):927–935
26. Polin RS, Shaffrey ME, Bogaev CA, Tisdale N, Germanson T, Bocchicchio B, Jane JA (1997) Decompressive bifrontal craniectomy in the treatment of severe refractory posttraumatic cerebral edema. Neurosurgery 41:84–92
27. Reithmeier T, Lohr M, Pakos P, Ketter G, Ernestus RI (2005) Relevance of ICP and p(ti)O_2 for indication and timing of decompressive craniectomy in patients with malignant brain edema. Acta Neurochir (Wien) 147(9):947–952
28. Rutigliano D, Egnor MR, Priebe CJ, McCormack JE, Strong N, Scriven RJ, Lee TK (2006) Decompressive craniectomy in pediatric patients with traumatic brain injury with intractable elevated intracranial pressure. J Pediatr Surg 41(1):83–87
29. Sahuquillo J, Arikan F (2006) Decompressive craniectomy for the treatment of refractory high intracranial pressure in traumatic brain injury. Cochrane Database Syst Rev 25(1): CD003983
30. Skoglund TS, Eriksson-Ritzen C, Jensen C, Rydenhag B (2006) Aspects on decompressive craniectomy in patients with traumatic head injuries. J Neurotrauma 23(10):1502–1509
31. Schneider GH, Bardt T, Lanksch WR, Unterberg A (2002) Decompressive craniectomy following traumatic brain injury: ICP, CPP and neurological outcome. Acta Neurochir Suppl 81:77–79
32. Stiefel MF, Heuer GG, Smith MJ, Bloom S, Maloney-Wilensky E, Gracias VH, Grady MS, LeRoux PD (2004) Cerebral oxygenation following decompressive hemicraniectomy for the treatment of refractory intracranial hypertension. J Neurosurg 101(2):241–247
33. Taylor A, Butt W, Rosenfeld J, Shann F, Ditchfield M, Lewis E, Klug G, Wallace D, Henning R, Tibballs J (2001) A randomized trial of very early decompressive craniectomy in children with traumatic brain injury and sustained intracranial hypertension. Childs Nerv Syst 17(3):154–162
34. Teasdale G, Jennett B (1976) Assessment and prognosis of coma after head injury. Acta Neurochir (Wien) 34(1–4):45–55
35. Whitfield PC, Patel H, Hutchinson PJ, Czosnyka M, Parry D, Menon D, Pickard JD, Kirkpatrick PJ (2001) Bifrontal decompressive craniectomy in the management of posttraumatic intracranial hypertension. Br J Neurosurg 15(6):500–507
36. Yoo DS, Kim DS, Cho KS, Huh PW, Park CK, Kang JK (1999) Ventricular pressure monitoring during bilateral decompression with dural expansion. J. Neurosurg. 91(6):953–959

Acta Neurochir Suppl (2008) 102: 99–104

Ventriculostomy for control of raised ICP in acute traumatic brain injury

I. Timofeev · C. Dahyot-Fizelier · N. Keong · J. Nortje · P. G. Al-Rawi · M. Czosnyka · D. K. Menon · P. J. Kirkpatrick · A. K. Gupta · P. J. Hutchinson

Abstract

Summary The aim of this study was to evaluate the effect of ventriculostomy on intracranial pressure (ICP), and related parameters, including cerebrospinal compensation, cerebral oxygenation ($PbtO_2$) and metabolism (microdialysis) in patients with traumatic brain injury (TBI).

Materials and methods Twenty-four patients with parenchymal ICP sensors were prospectively included in the study. Ventriculostomy was performed after failure to control ICP with initial measures. Monitoring parameters were digitally recorded before and after ventriculostomy and compared using appropriate tests.

Results In all patients ventriculostomy led to rapid reduction in ICP. Pooled mean daily values of ICP remained <20mmHg for 72h after ventriculostomy and were lower than before ($p < 0.001$). In 11 out of 24 patients during the initial 24-h period following ventriculostomy an increase in ICP to values exceeding 20mmHg was observed. In the remaining 13 patients ICP remained stable, allowing reduction in the intensity of treatment. In this group ventriculostomy led to significant improvement in craniospinal compensation (RAP index), cerebral perfusion pressure and $PbtO_2$. Improvement in lactate/pyruvate ratio, a marker of energy metabolism, was correlated with the increase in $PbtO_2$.

Conclusion Ventriculostomy is a useful ICP-lowering manoeuvre, with sustained ICP reduction and related physiological improvements achieved in >50% of patients.

Keywords Ventriculostomy · Head injury · Intracranial pressure · Monitoring · Microdialysis · Brain tissue oxygen · Compliance · Pressure-reactivity · Hydrocephalus

Introduction

Intracranial hypertension is common after traumatic brain injury (TBI) and treatment is aimed at controlling intracranial pressure (ICP) and maintaining adequate cerebral perfusion pressure (CPP) [4]. Drainage of cerebrospinal fluid (CSF) via ventriculostomy may be considered as one of the ICP reducing options; however the efficacy of this manoeuvre has not been extensively studied. The aim of the current study was to evaluate the effects of protocol-driven ventriculostomy on ICP and related physiological parameters including indices of cerebrospinal compensation and cerebrovascular pressure-reactivity, cerebral oxygenation and metabolism and to assess the duration of such effects.

Materials and methods

Twenty-four patients were prospectively included in this observational study. All patients had TBI that required

I. Timofeev (✉) · N. Keong · P. G. Al-Rawi · M. Czosnyka · P. J. Kirkpatrick · P. J. Hutchinson
Academic Neurosurgery Unit,
Department of Clinical Neurosciences, University of Cambridge,
Addenbrooke's Hospital,
P.O. Box 167, Hills Road,
Cambridge CB2 0QQ, UK
e-mail: it227@cam.ac.uk

J. Nortje · D. K. Menon · A. K. Gupta
Division of Anaesthesia, Department of Medicine,
University of Cambridge, Addenbrooke's Hospital,
P.O. Box 93, Hills Road,
Cambridge CB2 0QQ, UK

C. Dahyot-Fizelier
Department of Anaesthesiology and Intensive Care,
INSERM ERI-23, University Hospital,
Poitiers, France

management in a neurointensive care unit with mechanical ventilation and neuromonitoring. ICP was recorded with a parenchymal ICP monitor (Codman, Raynham, MA). In all patients continuous recording of physiological parameters (ICP, CPP, mean arterial blood pressure [MAP]) and ICP-waveform derived indices (PRx and RAP) was performed. The PRx index is a moving correlation coefficient between MAP and ICP and represents the response of cerebral vasculature to changes in arterial blood pressure, serving as a measure of autoregulation. The RAP index represents correlation between mean ICP and the pulse component of the ICP waveform (changes in intracranial volume with each heart beat) and as such allows assessment of craniospinal compensation (compliance). More detailed description of these indices and their clinical utility can be found in previous publications [3, 5].

Other monitoring modalities included cerebral microdialysis in all patients (CMA71 catheters, placed via a cranial access device (Technicam, Newton Abbot, UK) [12] and perfused with CNS fluid at a standard rate of 0.3μL/min with analysis performed hourly on a bedside analyser for concentrations of glucose, lactate, pyruvate, glutamate and glycerol; all microdialysis equipment by CMA Microdialysis, Solna, Sweden) and measurement of cerebral oxygenation ($PbtO_2$), extracellular pH (pHb), PCO_2 ($PbCO_2$) and temperature in 13 patients (Neurotrend™ sensor, Codman, Raynham, MA).

To assess ventricular size prior to ventriculostomy the frontal horns/inner [skull] diameter (FH/ID) ratio from the most recent CT scan prior to insertion of ventricular drain (Fig. 3a) was calculated. Neurological outcome was assessed at 6months using the Glasgow Outcome Scale (GOS) dichotomised into favourable (GOS = 4–5) and unfavourable (GOS = 1–3) outcome groups.

The decision to perform ventriculostomy was driven by the local ICP/CPP driven management protocol [18] and CSF drainage was considered when initial measures (sedation, analgesia, myorelaxation, head elevation, optimisation of ABP and $PaCO_2$) failed to maintain adequate levels of ICP and CPP (<20 and >60–70mmHg, respectively) and when ventricular size was considered sufficient by the neurosurgeon. In most instances ventriculostomy was performed in the intensive care unit. Following insertion of the ventricular catheter continuous free drainage of CSF was allowed, limited only by the height of the collecting reservoir (≈15mmHg above the external projection of foramen of Monro).

Statistical analysis was performed using SPSS 15.0 (SPSS Inc., Chicago, IL) and p value of ≤0.05 was considered significant for all tests. Monitoring parameters were continuously digitally captured at the bedside for at least 24 h before and after ventriculostomy and patient averaged mean values were compared using ANOVA or Wilcoxon signed-rank test, depending on data distribution. The degree of change in each parameter from the mean value over 24-h period prior to ventriculostomy to corresponding mean over 24 h post ventricular catheterisation was estimated (Δ). Nonparametric statistics were used to determine if correlations were significant (Spearman's rank correlation coefficient) as distribution of several variables deviated from normal. Demographic parameters relating to patients and injuries were compared using appropriate tests, depending on data type (Mann–Whitney or chi-square tests). All values are presented as mean (±SD) unless stated otherwise.

Results

Ventriculostomy was successful in all patients and led to a rapid instant reduction in intracranial pressure (Fig. 1). Pooled mean daily values of ICP remained <20mmHg for at least 72h after ventriculostomy and were significantly lower than before the procedure ($p < 0.001$, ANOVA, Bonferroni corrections). However, further analysis of the initial 24-h period following the insertion of ventricular drain demonstrated that in 11 out 24 patients initial ICP reduction was followed by gradual or rapid increase to the values exceeding 20mmHg by the end of 24-h period, whereas in the remaining 13 patients ICP remained stable and below 20mmHg, allowing subsequent reduction in the intensity of treatment (Fig. 1). In the latter group ventriculostomy also led to statistically significant and sustained improvement in craniospinal compensation (the RAP index) (Fig. 2a), increase in $PbtO_2$ (Fig. 2b) and CPP despite lower values of MAP. There was no significant change in cerebrovascular pressure-reactivity (the PRx index), pHb, $PbCO_2$, brain temperature, cerebral glucose, glycerol and glutamate levels in either group. Summary of changes in physiological and biochemical parameters between 24-h periods before and after external ventricular drain (EVD) insertion, based on ICP response to ventriculostomy is presented in Table 1.

In addition, in the cohort of patients with brain tissue oxygen monitoring, a significant correlation ($r = 0.512$, $p = 0.031$) was observed between the degree of decrease in ICP (ΔICP) and corresponding increase in $PbtO_2$ ($\Delta PbtO_2$). This increase in $PbtO_2$ did not depend on the $PbtO_2$ level, prior to ventriculostomy ($r = 0.03$, $p = 0.92$).

A marker of cerebral oxidative metabolism and mitochondrial function—the lactate/pyruvate ratio (LP ratio)—exhibited highly variable changes in individual patients in relation to the time of EVD insertion. The response of LP ratio (ΔLP) did not depend on ΔICP or improvement in CPP (ΔCPP) and overall mean values were not statistically different before and after EVD insertion in either group (Fig. 2c). However, there was a significant correlation ($r =$

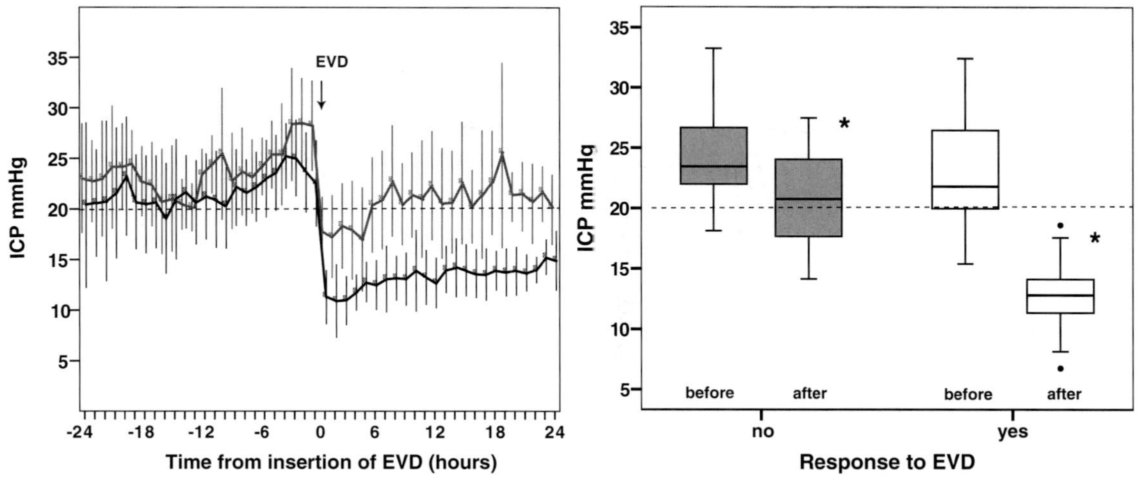

Fig. 1 Hourly dynamics (**a**) and mean values (**b**) of ICP before and after EVD, demonstrating two types of ICP response to ventriculostomy

0.560, $p = 0.019$) between $\Delta PbtO_2$ and decrease in LP ratio, indicating that patients in whom $PbtO_2$ increased more also sustained larger improvement in LP ratio (Fig. 2d). A significant relationship between the initial level of LP ratio and ΔLP ($r = 0.45$, $p = 0.34$) and the length of time from injury to EVD insertion and ΔLP ($r = -0.487$, $p = 0.019$) was also observed.

The demographic and injury type comparison of the two "ICP response" groups is presented in Table 2. There was no difference between these groups, with the exception of

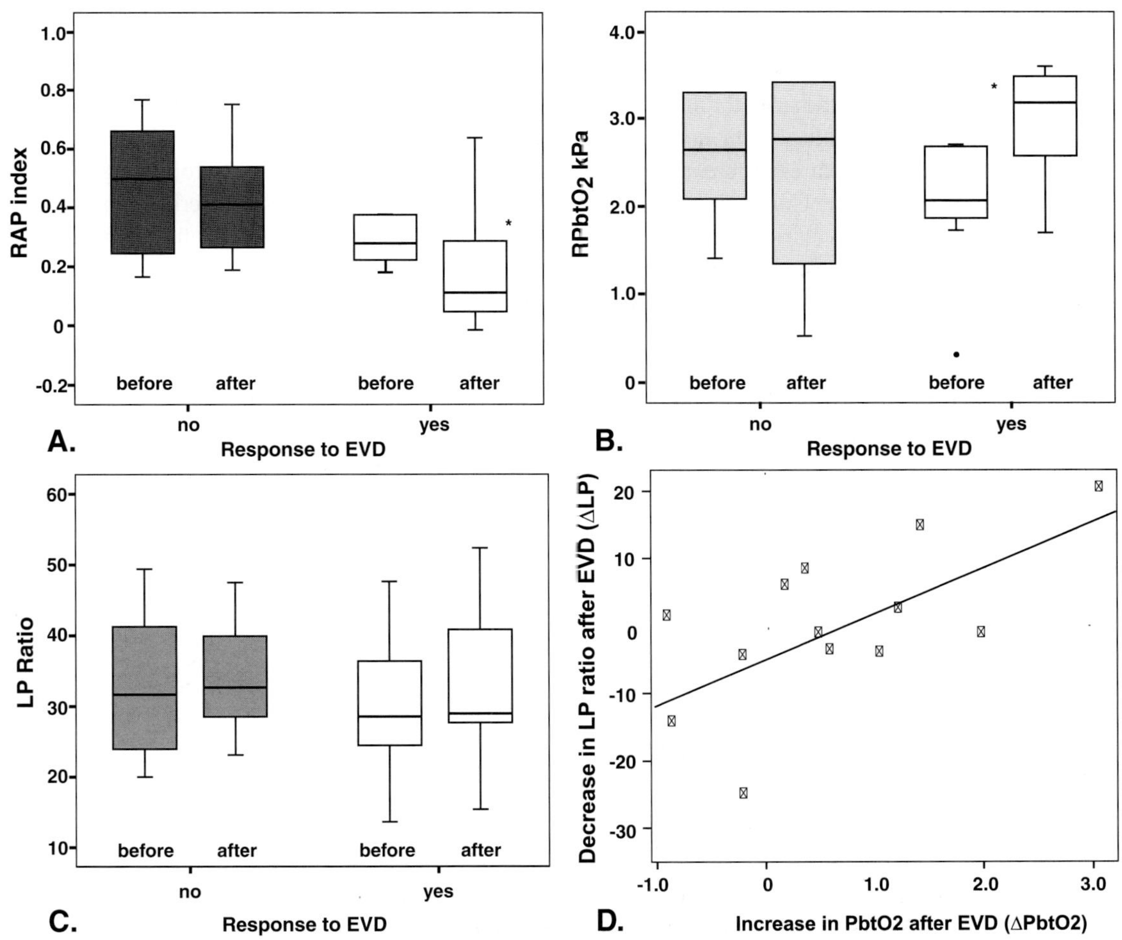

Fig. 2 Changes in RAP (**a**), $PbtO_2$ (**b**), and L/P ratio (**c**). **d** The relationship between changes in LP ratio and $PbtO_2$

Table 1 Physiological and biochemical parameters before and after ventriculostomy, depending on response to ventriculostomy

Parameter	Responders			Non-responders		
	Before	After	*p* value	Before	After	*p* value
ICP (mmHg)	22.7±5.2	12.5±3.4	0.001	24.7±4.4	20.8±4.0	0.026
CPP (mmHg)	75.5±8.4	79.6±6.5	0.046	79.4±7.4	78.7±6.9	0.328
MAP (mmHg)	98.1±10.0	92.1±6.0	0.016	104.0±9.3	99.5±7.2	0.05
RAP	0.20±0.21	0.07±0.11	0.002	0.42±0.2	0.01±0.12	0.477
$PbtO_2$ (kPa)	2.3±1.2	3.2±1.1	0.036	3.4±2.5	3.4±3.2	0.917
Lactate (mmol/L)	3.2±1.4	3.7±1.5	0.055	3.1±1.2	3.8±1.4	0.008

the GOS based neurological outcome at 6months. We found no statistically significant difference in FH/ID ratio values between the groups, despite some trend towards higher values in the positive response group (Fig. 3b).

No haemorrhagic complications or CSF infections were observed in this series.

Discussion

The results of this study suggest that ventriculostomy leads to rapid reduction in ICP and that and in a large proportion of patients this reduction is sustained and associated with improvement in other physiological parameters. Whilst many units use ventricular catheters primarily for monitoring ICP in patients with traumatic brain injury, with an option of intermittent CSF drainage, in other centres different methods of ICP measurement, such as parenchymal or subdural probes are chosen. In this situation, ventriculostomy can be considered as a later stage therapeutic intervention, aimed at ICP reduction and used for continuous or interrupted drainage of CSF to achieve clinical targets.

Despite the long history of using ventricular drains for ICP monitoring [16] and treatment of posttraumatic brain edema [6] there is a relative paucity of clinical information on physiological effects of this procedure. Fortune et al. [7] have studied the immediate effect of ventriculostomy on intracranial pressure and jugular bulb venous saturation (SjO_2) as a surrogate measure of cerebral blood flow (CBF). They used only brief (3min) period of CSF drainage and found that this led to modest changes in SjO_2 despite pronounced reduction in ICP. More recently, Kerr et al. [13], also found that drainage of CSF via ventriculostomy caused immediate reduction in ICP, minimal changes in CPP and had no significant effect on cerebral blood flow velocity or regional oxygen saturation assessed with near-infrared spectroscopy.

The present study confirmed the immediate ICP lowering effect of ventriculostomy in all patients. We also aimed to evaluate the longer term effects of this intervention and found that overall mean intracranial pressure was significantly lower for at least 72h following ventriculostomy. More detailed analysis was reserved for the 24-h epochs before and after insertion of the ventricular drain on assumption that beyond these periods many other clinical variables and treatments may have increasing influence on ICP and related physiological parameters, making interpretation difficult. Within these time intervals we observed a sustained decrease in ICP after ventriculostomy in one group of patients ("responders") and a return of ICP to

Table 2 Patients, injury and outcome characteristics

	Overall *n*=24	Response—"no" *n*=11	Response—"yes" *n*=13	Differences between the response groups (*p* value)
Age	41±16	41±13	41±18	0.839
Male/female	Male=18 (75%) Female=6 (25%)	Male=9 (82%) Female=2 (18%)	Male=9 (69%) Female=4 (31%)	0.649
GCS	Severe=17 971%) Moderate=4 (17%) Mild=3 (12%)	Severe=7 (64%) Moderate=1 (9%) Mild=3 (27%)	Severe=10 (77%) Moderate=3 (23%) Mild=0	0.111
Type of injury (CT)	Diffuse=12 (50%) Mass lesion=12 (50%)	Diffuse=4 (36%) Mass lesion=7 (64%)	Diffuse=8(62%) Mass lesion=5 (38%)	0.414
Days from injury to EVD	5.1±3.4	5.4±3.4	4.9±3.5	0.620
Outcome	Favourable=8 (33%) Unfavourable=16 (67%)	Favourable=1 (9%) Unfavourable=10 (91%)	Favourable=7 (54%) Unfavourable=6 (46%)	0.033

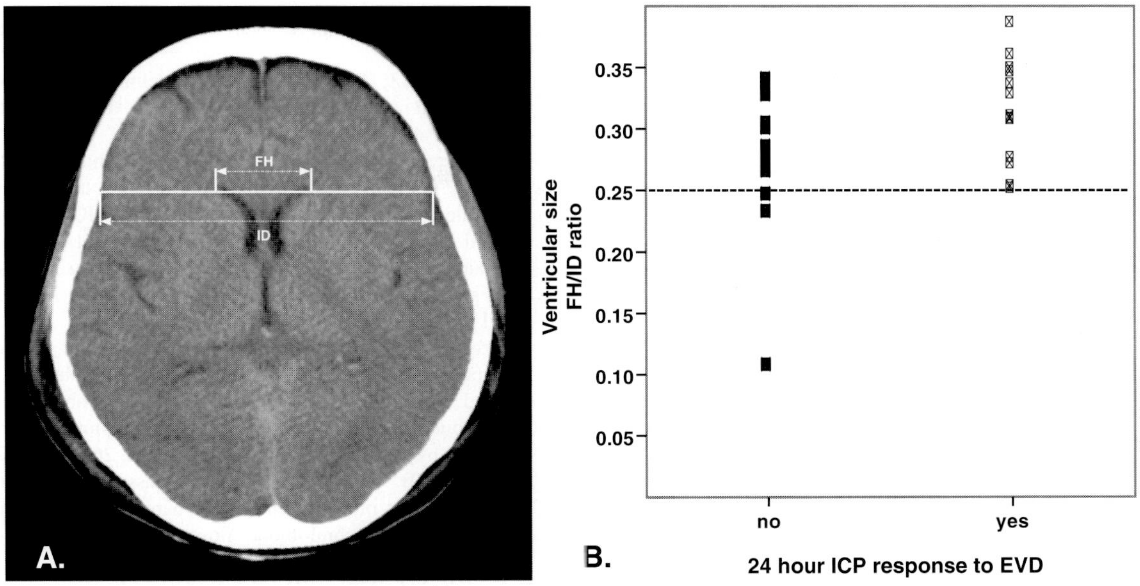

Fig. 3 **a** Calculation of frontal horns/inner diameter (FH/ID) ratio as an estimate of ventricular size **b** Ventricular size (FH/ID ratio) in two different response groups

levels exceeding 20mmHg within 24 h of EVD insertion in the second group ("non-responders"). In the former group, due to substantial ICP reduction, a small but statistically significant improvement in CPP was observed and was associated with decreased levels of MAP, suggesting reduced intensity of cardiovascular support required following ventriculostomy. Patients in this group also demonstrated improvements in cranio-spinal compensation (assessed with RAP index, surrogate measure of compliance) and cerebral oxygenation. In the "non-responders" group, mean ICP was also lower in the post-insertion 24-h period; however there was no associated improvement in CPP or other clinically significant changes. Increase in the cerebral lactate level, seen in both groups, reached statistical significance only in the second group.

Although we did not record arterial PaO_2 for each patient, the significant correlation that we observed between $\Delta PbtO_2$ and ΔICP, suggests that reduction in ICP by ventriculostomy played at least some role in increasing $PbtO_2$ levels. Improvement in cerebral metabolism, reflected by decrease in the LP ratio, was in turn coupled to an improvement in tissue oxygenation and depended on the initial value of the LP ratio (patients with the higher initial values of LP ratio exhibiting larger improvement). In view of these changes, one could hypothesize that improvement in ICP by ventriculostomy leads to increase in cerebral blood flow and oxygen delivery (CPP and $PbtO_2$) with associated improvement in cerebral metabolism, when the latter is deranged. The size of this study is, however, too small to firmly support this suggestion.

In a search for pre-ventriculostomy features that could predict the ICP response we examined pre-operative demographic, physiological and radiological variables. Patients with larger ventricles prior to ventriculostomy may have a larger contribution of CSF compartment to the overall intracranial hypertension and therefore may be more likely to better respond to ventriculostomy. However, despite a trend towards larger ventricular size in patients who responded to EVD (Fig. 3), there was no significant difference between the two groups. The lack of difference may be partly due to the preselection bias, i.e. only patients with prominent ventricles had the option of ventriculostomy and all three patients with relatively 'small' (FH/ID ratio < 0.25) ventricles in this study had a rapid increase in ICP back to baseline values. The two groups did not significantly differ in age, gender, injury characteristics, preoperative values of physiological parameters or timing of the EVD insertion. Interestingly, the only difference between the groups was neurological outcome at 6months with more patients in the 'responders' group achieving a favourable outcome. This study was not designed or powered to evaluate the influence of EVD on neurological outcome and the detected difference may reflect different disease course [17] in two groups, despite similar original characteristics, with predominance of tissue oedema in the "non-responders" group and hydrocephalus in the "responders". However, this finding agrees with the previous reports of Ghajar et al. [9], who demonstrated in their non-randomised study that improved outcome was seen in patients in whom ventriculostomy was used as a method of ICP monitoring and control. Further randomised prospective investigation is warranted.

Limitations of this study include its observational nature, which makes cause–effect interpretations difficult. The sample size is small and represents a selection of patients, based on the subjective assessment of ventricular size prior to insertion of the drain. This limits generalisation of the finding and precludes firm assumptions about outcome and the rate of complications. Furthermore, many monitoring techniques used in this study are based on very "focal" measurements, thus representing only a small portion of brain tissue.

In summary, ventriculostomy, being relatively minimally invasive procedure, which can be performed at the bedside, offers benefits of rapid and sustained ICP reduction in a large proportion of patients with raised ICP. However these benefits need to be weighed against a risk of known complications, particularly infection [11] and haemorrhage, which may exceed those of parenchymal ICP monitor [1, 10, 14], and the appropriate indications for and timing of this procedure in patients with parenchymal ICP monitoring are yet to be defined. The choice of an intermittent or continuous CSF drainage is also open to discussion. Whilst the former may minimize the risk of "overdraining" the ventricles and their subsequent collapse with EVD obstruction, it may subject the patients to potentially dangerous elevations of ICP between drainage periods, whereas continuous drainage may achieve more stable ICP control and allow weaning of the supportive therapy. In patients with small ventricular size guided EVD insertion [19] or controlled lumbar drainage [2, 8, 15] may be considered.

Conclusion

The results of this study support the use of ventricular drainage of CSF as a secondary ICP-lowering manoeuvre, with sustained ICP reduction and improvement in cerebral compliance and oxygenation achieved in >50% of patients. Nevertheless, in a substantial proportion of patients ICP fails to be controlled by ventriculostomy and other treatment methods may need to be considered. Predicting the likely response to EVD insertion may help with clinical management decisions and requires further research.

Acknowledgements: Ivan Timofeev is supported by Codman & Shurtleff, Inc., The Evelyn Trust Grant, Cambridge Overseas Trusts BP-TNK Kapitza Scholarship and the Medical Research Council RESCUEicp trial grant. Nicole Keong is supported by the Royal College of Surgeons of England fellowship. Peter Hutchinson is supported by the Academy of Medical Sciences/Health Foundation senior fellowship.

Conflict of interest statement We declare that we have no conflict of interest.

References

1. Anderson RC, Kan P, Klimo P, Brockmeyer DL, Walker ML, Kestle JR (2004) Complications of intracranial pressure monitoring in children with head trauma. J Neurosurg 101: 53–58
2. Baldwin HZ, Rekate HL (1991) Preliminary experience with controlled external lumbar drainage in diffuse pediatric head injury. Pediatr Neurosurg 17:115–120
3. Balestreri M, Czosnyka M, Steiner LA, Schmidt E, Smielewski P, Matta B, Pickard JD (2004) Intracranial hypertension: what additional information can be derived from ICP waveform after head injury? Acta Neurochir (Wien) 146:131–141
4. Brain Trauma Foundation (2007) Guidelines for the management of severe traumatic brain injury (3rd edition). J Neurotrauma 24 (s1):S1–S106
5. Czosnyka M, Pickard JD (2004) Monitoring and interpretation of intracranial pressure. J Neurol Neurosurg Psychiatry 75:813–821
6. Ford R, Spatz EL (1960) Post-traumatic obstruction of the aqueduct of Sylvius and postdecompression cerebral edema treated by ventriculostomy. N Engl J Med 263:263–267
7. Fortune JB, Feustel PJ, Graca L, Hasselbarth J, Kuehler DH (1995) Effect of hyperventilation, mannitol, and ventriculostomy drainage on cerebral blood flow after head injury. J Trauma 39:1091–1097 (discussion 1097–1099)
8. Fritsch MJ, Moss SD, Beyda DH, Manwaring KH (1997) Controlled external lumbar drain as treatment for therapy resistant intracranial hypertension—case report. Zentralbl Neurochir 58:192–195
9. Ghajar JBG, Hariri RJ, Patterson RH (1993) Improved outcome from traumatic coma using only ventricular cerebrospinal fluid drainage for intracranial pressure control. Adv Neurosurg 21:173–177
10. Guyot LL, Dowling C, Diaz FG, Michael DB (1998) Cerebral monitoring devices: analysis of complications. Acta Neurochir Suppl 71:47–49
11. Holloway KL, Barnes T, Choi S, Bullock R, Marshall LF, Eisenberg HM, Jane JA, Ward JD, Young HF, Marmarou A (1996) Ventriculostomy infections: the effect of monitoring duration and catheter exchange in 584 patients. J Neurosurg 85:419–424
12. Hutchinson PJ, Hutchinson DB, Barr RH, Burgess F, Kirkpatrick PJ, Pickard JD (2000) A new cranial access device for cerebral monitoring. Br J Neurosurg 14:46–48
13. Kerr ME, Weber BB, Sereika SM, Wilberger J, Marion DW (2001) Dose response to cerebrospinal fluid drainage on cerebral perfusion in traumatic brain-injured adults. Neurosurg Focus 11: E1
14. Khan SH, Kureshi IU, Mulgrew T, Ho SY, Onyiuke HC (1998) Comparison of percutaneous ventriculostomies and intraparenchymal monitor: a retrospective evaluation of 156 patients. Acta Neurochir Suppl 71:50–52
15. Levy DI, Rekate HL, Cherny WB, Manwaring K, Moss SD, Baldwin HZ (1995) Controlled lumbar drainage in pediatric head injury. J Neurosurg 83:453–460
16. Lundberg N, Troupp H, Lorin H (1965) Continuous recording of the ventricular-fluid pressure in patients with severe acute traumatic brain injury. A preliminary report. J Neurosurg 22:581–590
17. Marmarou A, Maset AL, Ward JD, Choi S, Brooks D, Lutz HA, Moulton RJ, Muizelaar JP, DeSalles A, Young HF (1987) Contribution of CSF and vascular factors to elevation of ICP in severely head-injured patients. J Neurosurg 66:883–890
18. Menon DK (1999) Cerebral protection in severe brain injury: physiological determinants of outcome and their optimisation. Br Med Bull 55:226–258
19. Ruchholtz S, Waydhas C, Muller A, Lewan UM, Nast-Kolb D, Euler E, Pfeiffer KJ, Schweiberer L (1998) Percutaneous computed tomographic-controlled ventriculostomy in severe traumatic brain injury. J Trauma 45:505–511

Acta Neurochir Suppl (2008) 102: 105–108

Intracranial pressure variability and long-term outcome following traumatic brain injury

Catherine J. Kirkness · Robert L. Burr · Pamela H. Mitchell

Abstract

Background Research suggests that intracranial pressure (ICP) dynamics beyond just absolute ICP level provide information reflecting intracranial adaptive capacity. Specifically, evidence indicates that physiologic variability provides information about system functioning that may reflect dimensions of adaptive capacity. The purpose of this study was to examine the association between ICP variability in patients following moderate to severe traumatic brain injury (TBI) and outcome at hospital discharge and 6 months post-injury.

Methods ICP was monitored continuously for 4 days in 147 patients (78% male; mean (SD) age=37 years (18 years)). ICP variability indices were calculated for four time scales (24 h, 60 min, 5 min and 5 s). Functional outcome was assessed using the Extended Glasgow Outcome Scale (GOSE). Logistic regression was used to estimate odds of survival or favorable outcome, and ordinal regression was used to estimate odds for outcome above versus below GOSE thresholds, predicted by ICP variability, controlling for age, gender, Glasgow Coma Scale motor score, craniectomy, and ICP level.

Findings ICP variability indices were better predictors of 6-month outcome than mean ICP. Survival was significantly associated with greater 5-s ICP variability ($p<0.001$). Higher ICP variability on shorter time scales was associated with better functional outcome (5-s RMSSD, 5-min SD: $p<0.002$; 60-min SD: $p<0.011$).

Conclusions ICP variability may reflect the degree of intactness of intracranial adaptive ability.

This study was funded by NIH NINR R01NR004901.

C. J. Kirkness (✉) · R. L. Burr · P. H. Mitchell
School of Nursing, University of Washington,
P.O. Box 357266, Seattle, WA 98195-7266, USA
e-mail: kirkness@u.washington.edu

Keywords Intracranial pressure · Variability · Traumatic brain injury · Outcome

Introduction

Research on intracranial pressure (ICP) monitoring following traumatic brain injury (TBI) has traditionally focused primarily on the relationship between absolute ICP level and outcome. However interest has grown in examining dynamics of the ICP signal as possible indicators of regulatory capacity that may provide additional information beyond absolute ICP level [1, 2, 13]. In addition, research suggests that the dynamic variability of biologic signals provides information about physiologic system functioning, potentially reflecting dimensions of adaptive capacity. A general hypothesis has been proposed that decreased physiologic complexity and greater physiologic regularity are associated with disease, reflecting uncoupling of normal regulatory system components [14]. This uncoupling results in a decreased ability to respond appropriately to internal and external perturbations [11]. Impaired functioning of intracranial adaptive mechanisms, for example, cerebral autoregulation, following TBI have been shown to correlate with poorer outcome [5, 9, 10]. Therefore the purpose of this study was to examine the association between ICP variability, as a potential measure of intracranial adaptive ability, in the first 4 days following moderate to severe TBI and outcome at hospital discharge and 6 months post-TBI.

Materials and methods

This study was a part of the physiologic data analysis of data from a larger randomized trial of a clinical cerebral perfusion pressure display [7, 8]. The local Institutional Review Board approved the trial. ICP data was obtained

Table 1 Mean ICP level and variability over 4 days of monitoring

Variable (mmHg)	Mean (SD)
ICP median	16.36 (5.87)
ICP 5-s RMSSD	1.10 (.56)
ICP 5-min SD	1.52 (.80)
ICP 60-min SD	2.53 (1.25)
ICP 24-h SD	4.13 (1.77)

continuously over 4 days via Camino intraparenchymal ICP monitors (Integra, Plainsboro, NJ). Raw ICP signals were saved as 5-s averages. Demographics and data related to injury severity and hospital management were obtained. Outcome was assessed at hospital discharge and at 6 months post-injury using the Extended Glasgow Outcome Scale (GOSE) (1=dead, 8=upper good recovery).

Variability indices were calculated for four time scales, including 5 s, 5 min, 60 min, and 24 h. Variability at the 5-s time scale was calculated as the root mean square successive difference (RMSSD) between adjacent 5-s averages, and for the 5-min, 60-min, and 24-h time scales was calculated as the standard deviation (SD). Logistic regression was used to predict the odds of survival or favorable outcome (GOSE> 4). Ordinal regression (PLUM) was used to estimate the odds for outcome above versus below GOSE thresholds, predicted by ICP variability. ICP level, age, gender, post-resuscitation Glasgow Coma Scale motor score (GCS-M), and craniectomy were controlled for in the analysis.

Results

Of the 156 subjects with TBI enrolled in the larger clinical trial, adequate ICP data was unavailable for nine subjects, thus data from 147 subjects was included in this analysis. Seventy-eight percent of the subjects were male. The mean (SD) age was 37 years (18 years). Motor vehicle accident was the most common mechanism of injury (46%), followed by falls (19%) and motor bike accidents (12%). The mean (SD) GCS-M score was 4.2 (1.6). One-quarter of the subjects underwent craniectomy. Mortality at hospital discharge was 14% and at 6 months was 17%. The mean (SD) GOSE score at 6 months was 4.28 (2.09).

The mean ICP level and variability indices over the first 4 days of monitoring are presented in Table 1. ICP variability increased at each incremental time scale, from 1.10 mmHg for the 5-s scale to 4.13 mmHg for the 24-h scale. In addition, mean ICP variability for all time scales increased as mean ICP level increased (Spearman's r=0.391 to 0.484, p<0.001). Figure 1 presents the odds ratios for discharge survival by mean ICP and ICP variability indices, controlling only for ICP level for the plot on the left and additionally controlling for age, gender, GCS-M, and craniectomy for the plot on the right. Greater ICP variability for the 5-s, 5-min, and 60-min time scales were all significantly associated with greater odds of discharge survival in both uncontrolled and controlled models. Figure 2 presents the odds of better 6-month outcome (GOSE>4) by ICP level and variability, similarly controlling for ICP level in the plot on the left and the additional variables in the plot on the right. While neither absolute ICP mean level nor 24-h SD were significant predictors of outcome in the uncontrolled or controlled models, 5-s, 5-min, and 60-min ICP variability were all significantly associated with greater odds of better outcome at 6 months post-injury.

The relationship between ICP variability, represented as 5-s RMSSD, and percent survival at 6 months within groups with lower and higher ICP was examined. For the group with the lower mean ICP (≤15 mmHg), survival was greater in those with a higher RMSSD (90% versus 75%). In the higher mean ICP group (>15 mmHg), the difference in survival based on lower or higher ICP RMSSD was

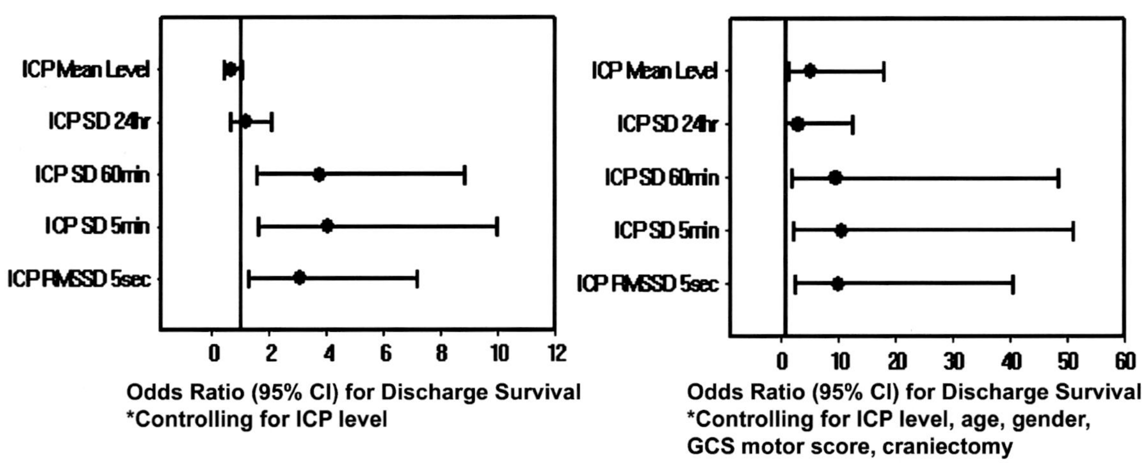

Fig. 1 Odds ratios for discharge survival by mean ICP and ICP variability

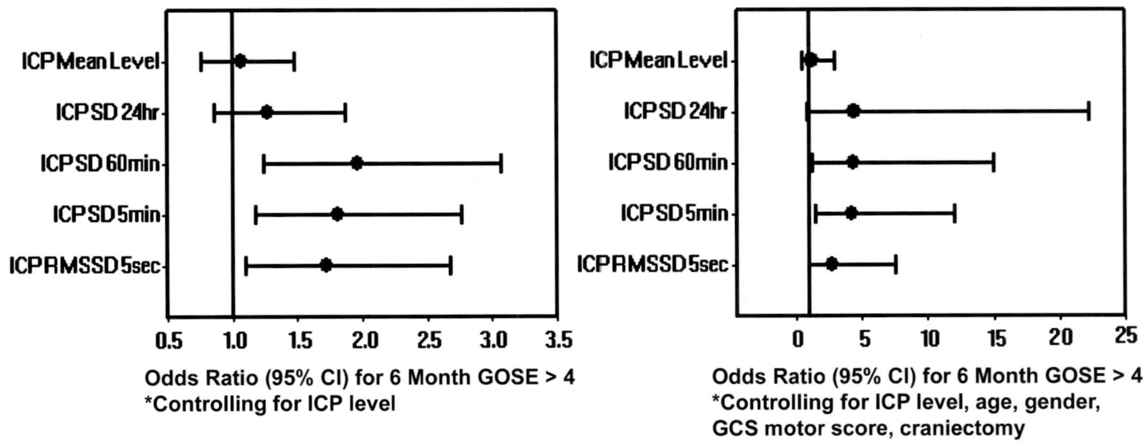

Fig. 2 Odds ratios for 6-month GOSE>4 by mean ICP and ICP variability

particularly pronounced, with a survival rate of 92% for those with higher ICP variability versus 52% for those with lower variability.

Table 2 presents the ordinal regression analysis of ICP variability indices and 6-month GOSE. The changes in chi-square between a base model controlling for ICP level, age, gender, GCS-M, and craniectomy, and models with added individual ICP variability indices were significant for the 5-s, 5-min, and 60-min indices. Each of these three ICP variability indices was individually a significant predictor of outcome in the full models.

Table 3 illustrates the association between lower and higher ICP variability (5-s RMSSD) and GOSE at 6 months. The percentage of those in the higher RMSSD group who were dead at 6 months was considerably higher than that of those in the lower RMSSD group (39% versus 8%, respectively). In addition, the percentage of those in the higher RMSSD group who were in the upper GOSE categories was greater.

Discussion

ICP variability indices were stronger predictors of outcome than absolute ICP level. Higher ICP variability for the 5-s, 5-min, and 60-min time scales was predictive of increased survival and better functional outcome at 6 months. In keeping with the general hypothesis of decreased physiologic variability reflecting greater system component isolation and decreased ability to respond to perturbations, greater ICP variability may reflect better functioning of intracranial adaptive mechanisms, contributing to better outcome.

Few studies have reported simple ICP variability indices following TBI. ICP SD in patients with various disorders (hydrocephalus, head injury, cerebrovascular) was found to differ by disorder, with highest values occurring in the head injury group [4]. Within the cerebrovascular group, ICP SD was lower in those with cerebral vasospasm. The presence of subarachnoid hemorrhage in this study, although traumatic not aneurysmal and with an unknown association with vasospasm, was significantly inversely correlated with mean ICP variability at all time scales ($r=-0.222$ to -0.238, $p=0.004$ to 0.007). A poor relationship between hourly ICP variance and outcome was found in Traumatic Coma Data Bank patients [12]. ICP SD in the first 48 h post-TBI was inversely related to survival and recovery in children [6]. Differing measurement time scales, patient populations, and ICP management may contribute to these different findings.

The use of 5-s averages in the calculation of the variability indices in this analysis removed the structured

Table 2 Ordinal regression: ICP variability and 6-month GOSE

GOSE	RMSSD	5-min SD	60-min SD	24-h SD
$\Delta\chi^{2a}$	12.89	16.24	10.20	2.97
p value	<0.001	<0.001	<0.002	0.064

Controlling for ICP level, age, gender, GCS-M, and craniectomy
[a] $\Delta\chi^2$ between base model (controlling for ICP level, age, gender, GCS-M, and craniectomy) and base model with ICP variability measure added

Table 3 Percentage of lower and higher ICP 5-s RMSSD groups in 6-month GOSE categories

	GOSE category							
	1	2	3	4	5	6	7	8
Percentage of lower RMSSD group in GOSE category	39	3	15	3	26	10	3	3
Percentage of higher RMSSD group in GOSE category	8	1	22	5	36	15	3	11

Numbers do not add to 100% due to rounding

variability related to the primary cardiac cycle and to respiratory cycles of 12 breaths per minute or faster. Potential influences on ICP variability over periods longer than 5 s include arterial blood pressure (ABP) variability, cerebrovascular pressure transmission, and other adaptive mechanisms. We are unable to fully elucidate the degree to which ABP variability drives the ICP variability in this study. Although it does not address the within-subject beat-by-beat ABP to ICP transmission occurring with each cardiac cycle, over 4 days ABP variability explained 10% of the across-subject ICP variability at the 5-s time scale and 3% at the 5-min time scale. Using a within-subjects index of cerebrovascular transmission over the 5- to 200-s range (PRx), [3], on average 5% of the ICP variability was attributable to ABP variability. In this study the 5-min time scale is most comparable to that of PRx, however, PRx and 5-min ICP SD were not significantly correlated. While PRx reflects ABP to ICP transmission, it is affected by the degree of incoming ABP variability and does not address the absolute amount of ICP variability.

The presence of ICP B waves at a frequency of 0.5 to 2 cycles per minute is also a potential source of ICP variability at time scales of 30 s and longer. However the contribution to ICP SD over a 4-day period would likely be minimal given the large number of data points. In addition, B waves in patients with hydrocephalus are not associated with ICP SD despite their relatively frequent occurrence [4]. While averaging the data over 4 days results in a relatively minimal effect of abnormal slow ICP waves and shorter periods of ICP variability, it provides the advantage of a representative composite view of ICP variability over an extended period and shows an association between ICP variability and outcome.

Greater ICP variability is associated with better functional outcome and survival in this study. ICP variability may reflect the degree of intactness of intracranial system component connections and ability to respond to perturbations. Changes in ICP variability may occur with alterations in physiologic state and may thus be of value clinically as indicators of changing adaptive capacity and in monitoring response to intervention strategies following TBI.

Conflict of interest statement We declare that we have no conflict of interest.

References

1. Balestreri M, Czosnyka M, Steiner LA et al (2004) Intracranial hypertension: what additional information can be derived from ICP waveform after head injury? Acta Neurochir (Wien) 146:131–141
2. Contant CF Jr, Robertson CS, Crouch J, Gopinath SP, Narayan RK, Grossman RG (1995) Intracranial pressure waveform indices in transient and refractory intracranial hypertension. J Neurosci Methods 57:15–25
3. Czosnyka M, Smielewski P, Kirkpatrick P, Laing RJ, Menon D, Pickard JD (1997) Continuous assessment of the cerebral vasomotor reactivity in head injury. Neurosurgery 41:11–17 (discussion 17–19)
4. Gaab MR, Ungersbock K, Hufenbeck B (1986) Evaluation of ICP by computerized bedside monitoring: methods and clinical significance. Neurol Res 8:44–52
5. Hiler M, Czosnyka M, Hutchinson P et al (2006) Predictive value of initial computerized tomography scan, intracranial pressure, and state of autoregulation in patients with traumatic brain injury. J Neurosurg 104:731–737
6. Jones PA, Andrews PJ, Easton VJ, Minns RA (2003) Traumatic brain injury in childhood: intensive care time series data and outcome. Br J Neurosurg 17:29–39
7. Kirkness CJ, Burr RL, Cain KC, Newell DW, Mitchell PH (2006) Effect of continuous display of cerebral perfusion pressure on outcomes in patients with traumatic brain injury. Am J Crit Care 15:600–609
8. Kirkness CJ, Burr RL, Cain KC, Newell DW, Mitchell PH (2007) The impact of a highly visible display of cerebral perfusion pressure on outcome in individuals with cerebral aneurysms. Heart Lung 37(3):227–237
9. Lam JM, Hsiang JN, Poon WS (1997) Monitoring of autoregulation using laser Doppler flowmetry in patients with head injury. J Neurosurg 86:438–445
10. Lang EW, Lagopoulos J, Griffith J et al (2003) Noninvasive cerebrovascular autoregulation assessment in traumatic brain injury: validation and utility. J Neurotrauma 20:69–75
11. Lipsitz LA (1995) Age-related changes in the complexity of cardiovascular dynamics: a potential marker of vulnerability to disease. Chaos 5:102–109
12. Marmarou A, Anderson RL, Ward JD et al (1991) Impact of ICP instability and hypotension on outcome in patients with severe head trauma. J Neurosurg 75:S59–S66
13. Mitchell PH, Kirkness C, Burr R, March KS, Newell DW (1998) Waveforms predictors: adverse response to nursing care. In: Marmarou A et al (eds) Intracranial pressure and neuromonitoring in brain injury. Springer, New York
14. Pincus SM (1994) Greater signal regularity may indicate increased system isolation. Math Biosci 122:161–181

Acta Neurochir Suppl (2008) 102: 109–112

Predictive values of age and the Glasgow Coma Scale in traumatic brain injury patients treated with decompressive craniectomy

Matthew B. Potts · John H. Chi · Michele Meeker · Martin C. Holland · J. Claude Hemphill III · Geoffrey T. Manley

Abstract
Background The use of decompressive craniectomy (DC) as an aggressive therapy for traumatic brain injury (TBI) has gained renewed interest. While age and the Glasgow Coma Scale (GCS) are frequently correlated with outcome in TBI, their prognostic values after decompressive craniectomy are ill-defined.
Methods We retrospectively reviewed data from 103 TBI patients treated with DC from 2001 to 2003. Age, preoperative GCS, and injury severity scores were recorded. Outcome at time of discharge was measured with the Glasgow Outcome Scale (GOS). Patients were stratified into the following age groups: <35, 35–49, 50–64, and ≥65 years. Spearman's correlation coefficients between age, GCS, and GOS were calculated for the entire population and each age group.
Findings Mortality rates for each age group were 19.2%, 66.7%, 60%, and 80%, respectively. There was a significant negative correlation between age and GOS ($r=-0.42$, $p<0.0001$) and patients <35 years had significantly better outcomes than patients ≥35 years ($p<0.0001$). The overall correlation between GCS and GOS did not reach significance ($r=0.18$, $p=0.076$). When stratified by age, there was a significant correlation between GCS and GOS only in patients 35–49 years ($r=0.51$, $p=0.011$).
Conclusions This data suggests that in TBI patients treated with DC, age correlates with outcome while the correlation between GCS and outcome is age-dependent.

Keywords Brain injuries · Decompressive craniectomy · Age · Glasgow Coma Scale

M. B. Potts · J. H. Chi · M. Meeker · M. C. Holland · G. T. Manley
Department of Neurological Surgery, University of California, 505 Parnassus Avenue, M-779, P.O. Box 0112, San Francisco, CA 94143-0112, USA

J. C. Hemphill III
Department of Neurology and UCSF Brain and Spinal Injury Center, University of California, 1001 Potrero Avenue, SFGH 5 4M62, P.O. Box 0870, San Francisco, CA 94143-0870, USA

M. Meeker · G. T. Manley (✉)
San Francisco General Hospital, Building 1, Room 101, 1001 Potrero Avenue, UCSF Box 0899, San Francisco, CA 94110, USA
e-mail: manley@itsa.ucsf.edu

J. H. Chi
Department of Neurosurgery, The Johns Hopkins Bayview Medical Center, B Building, Room 122, 4940 Eastern Avenue, Baltimore, MD 21224, USA

M. C. Holland
Naval Medical Center, 34800 Bob Wilson Drive, San Diego, CA 92134, USA

Introduction

Decompressive craniectomy (DC) in traumatic brain injury (TBI) is typically reserved as a second-tier therapy for patients with medically refractory intracranial hypertension. Recently, however, there has been renewed interest in this procedure as an aggressive therapy for head-injured patients [1, 3]. While it appears that younger patients may benefit the most from DC [11], predicting outcome after this procedure remains a challenge at any age. The Glasgow Coma Scale (GCS), initially developed as a rating scale for severity of brain injury [14], strongly correlates with mortality and functional outcome after TBI [10, 15]. Although there is evidence that the motor component of the GCS is a more reliable measure of injury severity [9], the full GCS score remains the standard, especially during the initial assessment of head-injured patients. Evidence also suggests that the predictive value of GCS is age-dependent [4]. The prognostic values and interaction of age and preoperative GCS, however, are not known in patients treated with DC, in whom the initial assessment is often critical in the decision to operative. We present our experience with a large number of TBI patients treated with DC.

Patients and methods

We conducted a retrospective review of 103 consecutive TBI patients treated with DC at San Francisco General Hospital, an urban level I trauma center, between 2001 and 2003. Surgical criteria for TBI patients included evidence of a focal lesion with midline shift or compression of the basal cisterns and the presence of any brainstem function. Patients with only epidural hematomas or with penetrating head wounds were excluded. Complete hospital records for each patient were reviewed and age, injury severity scores (ISS), and GCS scores were recorded. GCS scores were obtained from emergency department records and neurosurgery preoperative reports. The highest of these scores for each patient was used in our analysis. Glasgow Outcome Scale (GOS) scores at time of discharge were determined by review of discharge summaries and physical therapy consultation notes and were obtained blinded to GCS and age group.

Spearman's correlation coefficients were calculated between age, GCS, and GOS. Patients were then stratified by age into the following four groups: <35, 35–49, 50–64, and ≥65 years. The Kruskal–Wallis one-way ANOVA test was used to compare ISS, GCS, and GOS among age groups and Spearman's correlation coefficients between GCS and GOS were then calculated for each group. This study was approved by the Institutional Review Board of the University of California, San Francisco.

Results

During the 3-year study period, we obtained complete data for 95 patients who were treated with DC for non-penetrating head trauma resulting in subdural hematomas, intraparenchymal contusions, or combinations thereof. Ages ranged from 13 to 88 years with an average of 47.7±17.9 (standard deviation). Median preoperative GCS was 8 (interquartile range 4–12) and mean ISS was 29.2±7.9. Overall, there was a significant negative correlation between age and outcome ($r=-0.42$, $p<0.0001$). The correlation between GCS and outcome, however, did not reach significance ($r=0.18$, $p=0.076$).

When stratified by age, there were no significant differences in preoperative GCS among age groups ($p=0.76$). However, a significant difference in ISS was observed between patients aged <35 years (31.9±7.9) and those aged 50–64 years (26.5±3.6, $p<0.05$, Dunn's post test), indicating that the younger age group had more severe injuries. Patient's aged <35 years also had significantly better outcomes than each of the three older age groups ($p<0.0001$). Mortality for each age group was 19.2%, 66.7%, 60%, and 80%, respectively. There was a significant positive correlation between GCS and outcome in patients aged 35–49 years ($r=0.51$, $p=0.011$). In each of the other age groups, this correlation was not significant (Table 1).

Discussion

Accurate prognosis of the head-injured patient is critical for subsequent management, especially when aggressive therapies such as surgical decompression are being considered. In this study of TBI patients treated with DC, we examined the prognostic values of age and preoperative GCS and demonstrate a significant negative correlation between age and outcome. The correlation between GCS and outcome, however, is age-dependent and only significant in patients aged 35–49 years.

Table 1 Age group characteristics and correlation between GCS and GCS

Age group (years)	Number of patients	ISS mean ±SD	GCS median (IQR)	GOS median (IQR)	Mortality (%)	GCS-GOS correlation coefficient	*p* value
<35	26	31.9±7.9	7.5 (3–12)	3 (1–4)	19.2	0.15	0.46
35–49	24	27.5±5.1	7 (3.5–12.5)	1 (1–3)	66.7	0.51	0.011[a]
50–64	30	26.5±3.6	10 (3.5–13)	1 (1–2.5)	60	0.035	0.85
≥65	15	32.7±13.8	7 (4–13)	1 (1–1)	80	−0.11	0.71
Overall	95	29.2±7.9	8 (4–12)	1 (1–3)	53.7	0.18	0.076

GCS Glasgow Coma Scale, *GOS* Glasgow Outcome Scale, *IQR* interquartile range
[a] Significant correlation

The population studied in this report is unique because it is limited to head-injured patients who underwent DC. Variables such as age, preoperative GCS, preoperative ICP, and time to surgical decompression [11, 12] are thought to play a role in outcome after this procedure, but these factors are not yet clearly defined in this population. At our institution we are aggressive in our surgical management of head-injured patients and the vast majority of patients in this study underwent DC within 4 h after injury. Easily obtainable prognostic factors, such as age and GCS, are thus commonly emphasized when deciding to take a patient to the operating room.

Increasing age has long been associated with an increasingly poor outcome after TBI [16]. While systemic complications and different injury mechanisms may account for this, older age remains an independent risk factor, suggesting that the ageing nervous system has an inherently weaker pathophysiologic response to TBI [16]. Our findings agree with this, and indicate that old age predicts poor outcome in TBI patients treated with DC even if they present with a high preoperative GCS.

Although many studies have correlated GCS with outcome in TBI patients [5, 6, 8], there is evidence that GCS may not be as valid a prognostic factor as once thought. Levati et al. [7] reported a significant correlation between patients with GCS ≤5 and mortality but noted that mortality was also high in patients with higher GCS scores. Balestreri et al. [2] recently reported a 10-year retrospective review of TBI patients and found that while GCS correlated with outcome during the first 5 years, that correlation was lost during the latter half of the study. They suggest that this difference may be due to improved prehospital care (including sedation and intubation), which may impair the accurate assessment of GCS once a patient has reached the hospital. For this reason, it has been suggested that the motor component of the GCS may be a more accurate measure of neurologic status and prognosis in the head-injured patient [9].

Several recent series of decompressive craniectomy have reported overall mortality rates of 23–52% [1, 3, 11, 12]. In patients <35 years, our mortality rate was 19.2%, suggesting that decompressive craniectomy is beneficial in this young age group. Although overall mortality for patients 35–49 years was 66.7%, our data indicates that GCS is a valid predictor of outcome in this age group. As such, those patients with a preoperative GCS >8 had only a 42.9% mortality rate compared to 76.5% in patients with a preoperative GCS ≤8. In patients aged ≥50 years, where GCS does not correlate with outcome, mortality rates were high, indicating that DC may not be the choice surgical treatment in older TBI patients. Interestingly, a recent study by Pompucci et al. [13] of TBI patients treated with DC found no differences in outcome among patients <40 years and those aged 40–65 years, suggesting that age is less of a prognostic factor in patients under 65 years. They also reported that pre-operative GCS scores were predictive of outcome. The indications for surgery in this study, however, were not limited to focal intracranial mass lesions as in ours, but also included diffuse brain swelling without a focal intracranial lesion. Further investigation into how these differences affect outcome after TBI is warranted.

We acknowledge that this study has several limitations. The retrospective design prevented us from controlling the timing of GCS assessment and we were unable to assess GCS motor scores alone. While there are many factors associated with outcome after TBI, we have chosen to focus on two that are immediately available in every trauma situation—age and GCS. Overall, we believe these results are compelling because they begin to define the patient population that will most benefit from DC. They also suggest the shortcomings of using GCS to prognosticate TBI patients. Although it is generally understood that older patients tend to have poorer outcomes, the treating physician may impart a false sense of reassurance when basing the prognosis of an older patient on GCS. On the other hand, a low GCS should not necessarily be interpreted as a poor prognostic factor in young patients, and these patients should be treated aggressively.

Conflict of interest statement We declare that we have no conflict of interest.

References

1. Albanese J, Leone M, Alliez JR, Kaya JM, Antonini F, Alliez B, Martin C (2003) Decompressive craniectomy for severe traumatic brain injury: evaluation of the effects at one year. Crit Care Med 31:2535–2538
2. Balestreri M, Czosnyka M, Chatfield DA, Steiner LA, Schmidt EA, Smielewski P, Matta B, Pickard JD (2004) Predictive value of Glasgow Coma Scale after brain trauma: change in trend over the past ten years. J Neurol Neurosurg Psychiatry 75:161–162
3. Coplin WM, Cullen NK, Policherla PN, Vinas FC, Wilseck JM, Zafonte RD, Rengachary SS (2001) Safety and feasibility of craniectomy with duraplasty as the initial surgical intervention for severe traumatic brain injury. J Trauma 50:1050–1059
4. Demetriades D, Kuncir E, Murray J, Velmahos GC, Rhee P, Chan L (2004) Mortality prediction of head Abbreviated Injury Score and Glasgow Coma Scale: analysis of 7,764 head injuries. J Am Coll Surg 199:216–222
5. Gennarelli TA, Champion HR, Copes WS, Sacco WJ (1994) Comparison of mortality, morbidity, and severity of 59,713 head injured patients with 114,447 patients with extracranial injuries. J Trauma 37:962–968
6. Jennett B, Teasdale G, Galbraith S, Braakman R, Avezaat C, Minderhoud J, Heiden J, Kurze T, Murray G, Parker L (1979) Prognosis in patients with severe head injury. Acta Neurochir Suppl (Wien) 28:149–152
7. Levati A, Farina ML, Vecchi G, Rossanda M, Marrubini MB (1982) Prognosis of severe head injuries. J Neurosurg 57:779–783
8. Levin HS, Gary HE Jr., Eisenberg HM, Ruff RM, Barth JT, Kreutzer J, High WM Jr., Portman S, Foulkes MA, Jane JA et al (1990)

Neurobehavioral outcome 1 year after severe head injury. Experience of the Traumatic Coma Data Bank. J Neurosurg 73:699–709
9. Marion DW, Carlier PM (1994) Problems with initial Glasgow Coma Scale assessment caused by prehospital treatment of patients with head injuries: results of a national survey. J Trauma 36:89–95
10. Marshall LF, Klauber GT, Eisenberg HM (1991) The outcome of severe closed head injury. J Neurosurg 75:S28–S36
11. Munch E, Horn P, Schurer L, Piepgras A, Paul T, Schmiedek P (2000) Management of severe traumatic brain injury by decompressive craniectomy. Neurosurgery 47:315–322
12. Polin RS, Shaffrey ME, Bogaev CA, Tisdale N, Germanson T, Bocchicchio B, Jane JA (1997) Decompressive bifrontal craniectomy in the treatment of severe refractory posttraumatic cerebral edema. Neurosurgery 41:84–92
13. Pompucci A, De Bonis P, Pettorini B, Petrella G, Di Chirico A, Anile C (2007) Decompressive craniectomy for traumatic brain injury: patient age and outcome. J Neurotrauma 24:1182–1188
14. Teasdale G, Jennett B (1974) Assessment of coma and impaired consciousness. A practical scale. Lancet 2:81–84
15. Teasdale G, Jennett B (1976) Assessment and prognosis of coma after head injury. Acta Neurochir (Wien) 34:45–55
16. Vollmer DG, Torner JC, Jane JA, Sadovnic B, Charlesbois D, Eisenberg HM, Foulkes MA, Marmarou A, Marshall LF (1991) Age and outcome following traumatic coma: why do older patients fare worse? J Neurosurg S75:37–49

PART 2:
Hydrocephalus and cerebrospinal fluid dynamics

Acta Neurochir Suppl (2008) 102: 115–118

Noninvasive estimation of intracranial compliance in idiopathic NPH using MRI

Mitsuhito Mase · Toshiaki Miyati · Harumasa Kasai · Koichiro Demura · Tomoshi Osawa · Masaki Hara · Yuta Shibamoto · Kazuo Yamada

Abstract
Background The pathophysiology of idiopathic normal pressure hydrocephalus (I-NPH) is still unclear and the diagnosis is sometimes difficult. The aim of this study was to assess the biophysics of I-NPH by measuring intracranial compliance using cine MRI.
Methods The study included patients with I-NPH (I-NPH group, n=13), brain atrophy or asymptomatic ventricular dilation (VD group, n=10), and healthy volunteers (control group, n=13). Net blood flow (bilateral internal carotid and vertebral arteries and jugular veins) and cerebrospinal fluid (CSF) flow in the subarachnoid space at the C2 cervical vertebral level were measured using phase-contrast cine MRI. CSF pressure gradient (PG_{p-p}) and intracranial volume changes (VC_{p-p}) during a cardiac cycle were calculated.
Findings The compliance index ($CI=VC_{p-p}/PG_{p-p}$) in the I-NPH group was significantly lower than in the control and VD groups, whereas no difference was found between the control and VD groups. CI values of I-NPH patients after the tap test were larger than those before. These results clearly show that the intracranial compliance of I-NPH is relatively low compared to that of brain atrophy or normal subject. The increase of CI after a tap test also supports this finding.
Conclusions It is possible to estimate intracranial compliance as CI non-invasively using cine MRI. CI could become a useful method for the diagnosis of I-NPH.

Keywords Intracranial compliance · Idiopathic normal pressure hydrocephalus · MRI · Cerebrospinal flow dynamics

M. Mase (✉) · K. Demura · T. Osawa · K. Yamada
Department of Neurosurgery and Restorative Neuroscience, Graduate School of Medical Sciences, Nagoya City University, 1-Kawasumi, Mizuho-cho, Mizuho-ku, Nagoya 467-8602, Japan
e-mail: mitmase@med.nagoya-cu.ac.jp

T. Miyati
Division of Health Sciences, Graduate School of Medical Science, Kanazawa University, 5-11-80, Kadotsuno, Kanazawa 920-0942, Japan

H. Kasai · M. Hara · Y. Shibamoto
Department of Radiology, Nagoya City University Hospital, 1-Kawasumi, Mizuho-cho, Mizuho-ku, Nagoya 467-8602, Japan

Introduction

Intracranial compliance is one of the most important parameters in estimating the compensatory capacity of the cranial cavity. One method of directly determining the intracranial compliance is to measure intracranial pressure (ICP) response after intrathecal or intraventricular infusion of saline (pressure volume response [PVR]). However, this procedure has some risks of infection or other complications. Recently, we have developed a non-invasive method to measure intracranial compliance as compliance index (CI) using MRI based on Alperin's method [1, 11]. We have also reported on an improved method to measure CI [14]. In the present study, we assess the biophysics of idiopathic normal pressure hydrocephalus (I-NPH) by measuring intracranial compliance using this method.

Materials and methods

MR imaging conditions

Retrospective cardiac gated phase-contrast (PC) cine MRI was performed on a 1.5-T MR system (Gyroscan Intera;

Philips Medical Systems International, Best, Netherlands). A gradient echo pulse sequence (T1-FFE) was used with the shortest echo time, 25° (for CSF flow and spinal cord displacement) and 20° (for blood flow) flip angles, 6 mm slice thickness, two signals averaged, 140×140 mm rectangular field of view, and 256×128 acquisition matrix. We set the vertical slice plane at the dense (C2) level against the cerebrospinal axis, and obtained velocity-mapped phase images with different velocity encoding gradients (7 cm/s for CSF flow and cord motion, and 80 cm/s for blood flow).

Compliance index

The compliance index (CI) was defined as the ratio of the maximal intracranial volume change to the maximal craniospinal CSF pressure gradient change during the cardiac cycle, which was calculated from the net transcranial blood flow, CSF flow, and displacement of spinal cord measured with phase-contrast (PC) cine [1, 11, 14]. CSF flow [$V_c(t)$], displacement of cord [$V_s(t)$], arterial inflow (the sum of both internal carotid arteries and both vertebral arteries) [$V_a(t)$], and venous outflow (the sum of both internal jugular veins) [$V_v(t)$] were measured with correction of the baseline offset due to eddy currents by a subtraction process. Then, to match the difference in inflow and outflow capacity to the cranium in a cardiac cycle, the flow wave of the venous outflow was scaled up, and the net intracranial volume change in each cardiac phase was obtained from Eq. 1:

$$\mathrm{VC}(t) = \int \{V_a(t) - V_v(t) - V_c(t) - V_s(t)\}\mathrm{d}t. \tag{1}$$

Next, we determined the craniospinal CSF pressure gradient change during the cardiac cycle, which was calculated from the above measured CSF flow velocity using a simplified Navier–Stokes Eq. 2 [16]:

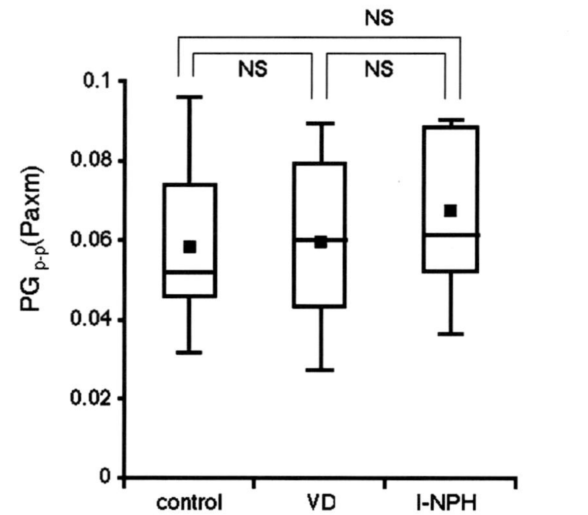

Fig. 2 Peak to peak pressure gradient (PG_{p-p}) of each group is shown in box–whisker plots. A *black square* shows a mean value of PG_{p-p} in each group. There is no significant difference of PG_{p-p} in each group. *NS* not significant

$$\nabla P = \frac{\partial P}{\partial z} = -\rho\frac{\partial w}{\partial t} + \mu\left(\frac{\partial^2 w}{\partial x^2} + \frac{\partial^2 w}{\partial y^2}\right) \tag{2}$$

where x, y, and z are coordinates, and the z-direction is the canal axis. P and w are the pressure and the axial velocity, and ρ and μ are the fluid density (CSF=1.0007 g/cm^3) and viscosity (CSF=1.1 cP), respectively.

Finally, the CI ($CI=VC_{p-p}/PG_{p-p}$) was obtained by dividing the peak-to-peak CSF pressure gradient change (PG_{p-p}) into the peak-to-peak intracranial volume change (VC_{p-p}).

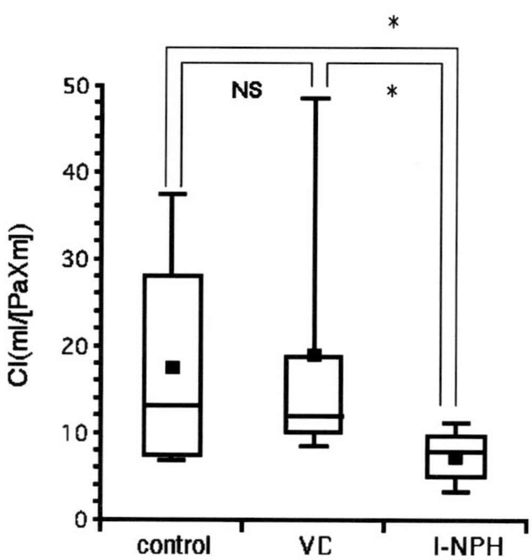

Fig. 1 Compliance index (CI) of each group is shown in box–whisker plots. A *black square* shows a mean value of CI in each group. CI in the I-NPH group is significantly lower than in the control and VD groups (*$p<0.01$). *NS* not significant

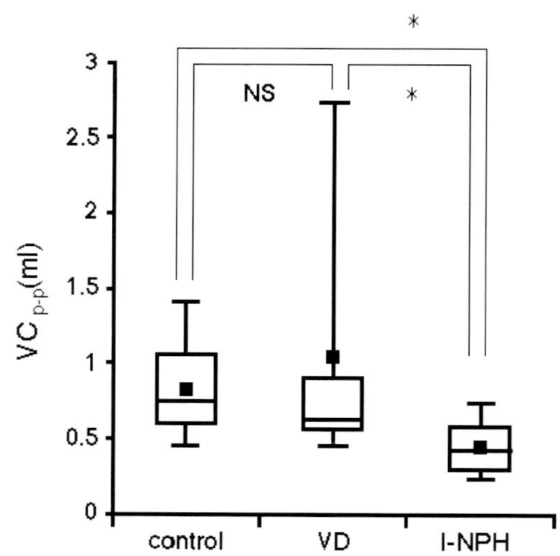

Fig. 3 Peak to peak intracranial volume change (VC_{p-p}) of each group is shown in box–whisker plots. A *black square* shows a mean value of VC_{p-p} in each group. VC_{p-p} in I-NPH group is significantly lower than in the control and VD groups (*$p<0.05$). *NS* not significant

Patients

The study included patients with I-NPH (I-NPH group, $n=13$), brain atrophy or asymptomatic ventricular dilation (VD group, $n=10$), and healthy volunteers (control group. $n=13$).

All patients with I-NPH showed improvement in their symptoms after a CSF tap test (withdrawal of CSF by a lumbar puncture more than 30 mL) and a shunt operation.

Results

The CI in the I-NPH group (7.16±3.00 mL/(Pa m), mean± SD) was significantly lower than in the control (17.57± 12.39, $p=0.0071$) and VD (18.97±16.91, $p=0.0036$) groups, whereas no difference was found between the control and VD groups ($p=0.7565$) (Fig. 1). There was no significant difference in the PG_{p-p} among groups (Fig. 2). In contrast, the VC_{p-p} in I-NPH group was significantly smaller than in the others (Fig. 3).

CI values of I-NPH patients after the tap test were larger than those before (data not shown).

Discussion

Various intracranial pathologies cause the intracranial pressure (ICP) to increase, resulting in reduced compensatory capacity of the cranial cavity. The adequate control of ICP is a very important clinical issue in neurosurgical patients. However, ICP is not always sensitive to detection of the increased elasticity of the cranial cavity. For example, NPH is associated with normal ICP. The intracranial compliance is another parameter to understand the compensatory capacity of the cranial cavity in addition to ICP. However, prior direct measures of intracranial compliance have often come from an invasive test such as the infusion test. In the present study, we were able to directly measure the intracranial compliance as CI non-invasively using cine MRI based on Alperin's method [1]. CI appears to be an accurate and reliable parameter based on prior correlation with the pressure volume response (PVR) in patients with NPH after subarachnoid hemorrhage [11]. In a previous study [14], we improved upon the measurement of the CSF pressure gradient with regard to the normalization of CSF flow area [1]. The present study uses this improved method. Based on this, we believe that CI could become a useful clinical index to estimate intracranial compliance non-invasively.

The diagnosis of I-NPH continues to be difficult in some cases. Even with recent evidenced based clinical guidelines for the diagnosis and management of I-NPH [5, 7], accurate diagnosis remains challenging. While some advocate an infusion test for evaluating intracranial compliance or a CSF drainage tap-test with evaluation of clinical symptoms after withdrawal of CSF [4], these test are invasive. Numerous studies have been conducted in which CSF flow dynamics were measured using MRI to evaluate the changes in intracranial conditions in normal-pressure hydrocephalus (NPH) [3, 6, 8–15]. However, the diagnosis of NPH based on MRI CSF flow study has not been clearly established. In the present study, we showed that patients with I-NPH had significantly lower CI, meaning that compensatory capacity had decreased. This corresponds with our previous papers and the other reports that NPH has low compliance [6, 8, 9, 11, 12, 14]. The CI value increases after a tap test in I-NPH patients [14]. Since the withdrawal of CSF is a volume reduction of intracranial components, the intracranial compliance certainly increases after the tap test. This also supports that I-NPH has lower CI.

The present study also showed that the significant decrease of CI in I-NPH was not due to changes of PG_{p-p} but rather a significant decrease of VC_{p-p}. It appears that the volume change of one cardiac cycle is restricted in I-NPH. Due to the reduced compliance of the intracranial compartment, venous and CSF outflow occur more immediately following the systolic increase in arterial blood flow, resulting in a smaller systolic increase in intracranial volume (VC_{p-p}) [2].

In conclusion, CI (compliance index) analysis measured by phase contrast cine MRI makes it possible to assess intracranial compliance and dynamics non-invasively. Decreased CI and VC_{p-p} could become a new parameter for the diagnosis of I-NPH. Although further investigation is needed, CSF flow study including CI using MRI could become a useful component of preoperative assessment for patients with I-NPH.

Conflict of interest statement We declare that we have no conflict of interest.

References

1. Alperin NJ, Lee SH, Loth F et al (2000) MR-Intracranial pressure (ICP): a method to measure intracranial elastance and pressure noninvasively by means of MR imaging: baboon and human study. Radiology 217:877–885
2. Alperin N, Sivaramakrishnan A, Lichtor T (2005) Magnetic resonance imaging-based measurements of cerebrospinal fluid and blood flow as indicators of intracranial compliance in patients with Chiari malformation. J Neurosurg 103:46–52
3. Bradley WG, Scalzo D, Queralt J et al (1996) Normal pressure hydrocephalus; evaluation with cerebrospinal fluid flow measurements at MR imaging. Radiology 198:523–529

4. Damasceno BP, Carelli EF, Honorato DC (1997) The predictive value of cerebrospinal fluid tap-test in normal pressure hydrocephalus. Arq Neuropsiquiatr 55:179–185
5. Ishikawa M, Guideline Committee for Idiopathic Normal Pressure Hydrocephalus, Japanese Society of Normal Pressure hydrocephalus (2004) Clinical guidelines for idiopathic normal pressure hydrocephalus. Neurol Med Chir (Tokyo) 44:222–223
6. Kim DS, Choi JU, Huh R et al (1999) Quantitative assessment of cerebrospinal fluid hydrodynamics using a phase-contrast cine MR image in hydrocephalus. Child's Nerv Syst 15:461–467
7. Marmarou A, Black P, Bergsneider N et al (2005) Guidelines for management of idiopathic normal pressure hydrocephalus: progress to date. Acta Neurochir Suppl 95:237–240
8. Mase M, Yamada K, Banno T et al (1998) Quantitative analysis of CSF flow dynamics using MRI in normal pressure hydrocephalus. Acta Neurochir Suppl 71:350–353
9. Mase M, Yamada K, Banno T et al (1998) Differential diagnosis of normal pressure hydrocephalus based on CSF flow dynamics study using MRI. Curr Treat Hydroceph 8:13–18
10. Mase M, Banno T, Yamada K et al (1998) Quantitative measurement of CSF flow velocity in the aqueduct using MRI—experimental simulation and normal value. Curr Treat Hydroceph 8:6–12
11. Mase M, Miyati T, Yamada K et al (2005) Non-invasive measurement of intracranial compliance using cine MRI in normal pressure hydrocephalus. Acta Neurochir Suppl 95:303–306
12. Matsumoto T, Nagai H, Kasuga Y et al (1986) Changes in intracranial pressure (ICP) pulse wave following hydrocephalus. Acta Neurochir (Wien) 82:50–56
13. Miyati T, Mase M, Banno T (2003) Frequency analysis of CSF flow on cine MRI in normal pressure hydrocephalus. Eur Radiol 13:1019–1024
14. Miyati T, Mase M, Kasai H et al (2007) Noninvasive MRI assessment of intracranial compliance in idiopathic normal pressure hydrocephalus. J Magn Reson Imaging 26:274–278
15. Ohara S, Nagai H, Matsumoto T (1988) MR imaging of CSF pulsatory flow and its relation to intracranial pressure. J Neurosurg 69:675–682
16. Urchuk SN, Plewes DB (1995) MR measurement of time-dependent blood pressure variations. J Magn Reson Imaging 5:621–627

Acta Neurochir Suppl (2008) 102: 119–123

The role of cerebrospinal fluid flow study using phase contrast MR imaging in diagnosing idiopathic normal pressure hydrocephalus

F. T. Al-Zain · G. Rademacher · U. Meier · S. Mutze · J. Lemcke

Abstract

Background The purpose of this prospective study was to identify the ability of cerebrospinal fluid flow study using phase contrast MR imaging to replace the invasive methods currently used to establish the diagnosis of idiopathic normal pressure hydrocephalus (iNPH).

Materials and methods Between January 2003 and April 2005, 61 patients with clinical symptoms fitting the Hakim triad and a dilated ventricular system on CT underwent a intrathecal infusion test and cerebrospinal tap test. All patients also had a phase contrast MRI to determine the CSF flow rate in the aqueduct. Shunted patients were followed postoperatively up to 12 months. The pre- and postoperative symptomatic condition was evaluated using the clinical Kiefer score. The outcome was calculated by the NPH Recovery Rate.

Findings Patients were classified into 41 with iNPH and 20 patients with brain atrophy. Thirty-nine iNPH patients were shunted and two patients refused surgery. The mean Kiefer score of the shunted patients was statistically significantly lower after surgery. In patients screened for clinical symptoms and ventriculomegaly on CT imaging, an aqueduct-CSF flow rate greater than 24.5 ml/min was found to be statistically specific for a diagnosis of iNPH.

Conclusions The measurement of the CSF flow rate in the aqueduct by using the phase contrast MRI technique is a highly specific pre-selective method for diagnosing iNPH.

F. T. Al-Zain (✉) · U. Meier · J. Lemcke
Department of Neurosurgery, Unfallkrankenhaus Berlin,
Warener Strasse 7,
12683 Berlin, Germany
e-mail: ferass.alzain@ukb.de

G. Rademacher · S. Mutze
Department of Radiology, Unfallkrankenhaus Berlin,
Berlin, Germany

Keywords Idiopathic normal pressure hydrocephalus · Phase contrast MRI · CSF flow

Introduction

Since the observation of the CSF flow in the early forties by O'Connell [11] and the development of the MR imaging in the early 1980s, efforts have been made to regenerate these flow theories and patterns in MR imaging procedures. In 1980, Holland et al. [6] showed that the disappearance of the CSF MR signal was a proof of the oscillatory movements of the molecules. More refined MR studies have shown the signal void in the aqueduct. The first quantitative study of cerebrospinal fluid flow was introduced by Edelman et al. [13] who used a multiphase MR imaging method. In this prospective study, we measured the CSF flow in the aqueduct by phase contrast MR technique and studied its relation to the diagnosis of iNPH and the response to shunting.

Materials and methods

The classical Hakim triad clinical symptoms were the first criterion for the selection of iNPH-suspected patients. All patients either showed one or a combination of symptoms. These clinical features were converted into numerical scores based on the clinical Kiefer score [9] The second line of investigation included a radio-morphological evidence of the dilated ventricles in cranial computer tomography (Evans index>0.3) A total of 61 patients met these clinical and radiographic criteria. The third investigational line was dedicated to the differentiation of iNPH from other diseases with a similar clinical picture. We used the dynamic infusion test [10]. The resistance to CSF outflow is significant for shunt responsiveness at a value ≥13 mmHg/ml/min. Subsequently, a cerebrospinal fluid tap test of 30–70 ml was done to establish the responsiveness to shunting. Based on these results, patients were divided into two groups, those diagnosed to have iNPH were offered shunting and those diagnosed as having cerebral atrophy were followed conservatively.

All 61 patients underwent a phase contrast MRI (1.5 Tesla, Philips Medical systems). To visualize the CSF flow dynamics and to detect flow turbulences, the sagittal 2D T1- weighted fast field echo technique was used in all studies with following parameters: flip angle=10°, TE=11, TR=25, 256×256 matrix, slice thickness=5 mm, flow compensation (Fig. 1a). For accurate measurements of the velocity of CSF flow through the aqueduct, the single slice retrospective cardiac-gated 2D phase-contrast flow-quantification sequence was used with the following parameters: 15° flip angle, TE=9,8, TR=18, 256×256 matrix, 160 mm field of view (FOV), 5 mm slice thickness, flow compensation. The flow velocity encoding values were 10 cm/s for the first and 20 cm/s for the second series. If aliasing was observed at 20 cm/s, the encoding was increased to 30 cm/s. These 2D phase contrast sequences were performed with retrospective cardiac gating using a vector electrocardiograph and arrhythmia-rejection method. Any heartbeat greater or less than 15% of the encoded average R-R interval was rejected. The scan time depended on the patient's heart rate by using the maximum number of cardiac phases for one R-R interval. A phase velocity map was generated for quantitative assessments of CSF volume flow through an imaging plane. By integration of the area

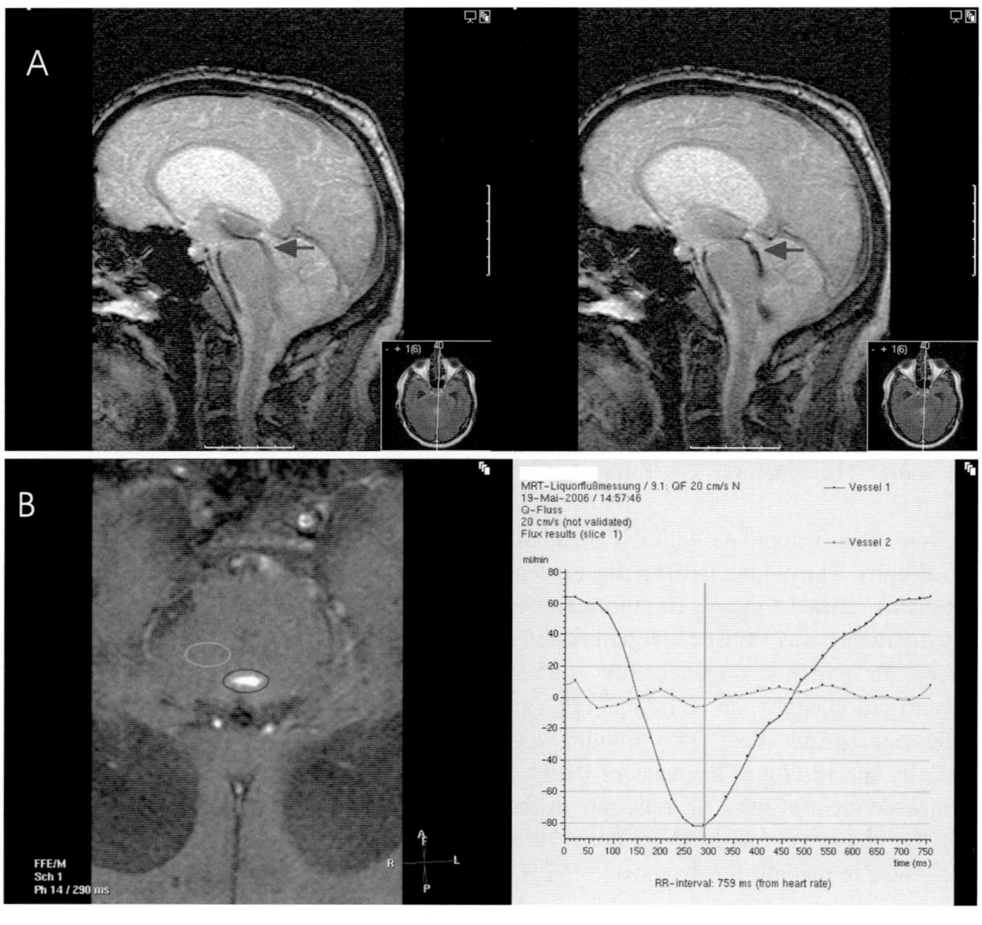

Fig. 1 **a** A 73 year old male with idiopathic normal pressure hydrocephalus. MR images (*midline sagittal* 2D T2-weighted TFE sequence) in different cardiac phases to visualise CSF flow shows prominent flow void within the aqueduct (*arrows*), which extends into the third ventricle, in the fourth ventricle and the cisterna magna. **b** (*left image*): Modulus map of a phase contrast MR sequence across the aqueduct diameter at peak systolic flow. The method of calculating the CSF flow rate showing the ROI (circumscribes the aqueduct) being slightly larger than the cerebral aqueduct. To rule out the partial volume effect and the mass brain movement, an identical ROI shape is set in the brain stem parenchyma and is being subtracted from the net flow to obtain the real flow. By integration flow velocity across the aqueduct, a mean cross sectional flow is calculated and shown as a flow-versus-time curve. (*Right image*): Plot of volumetric flow rate (ml/min) versus time (ms) over the entire cardiac cycle (large deflection with a sinusoidal flow pattern) and the mass brain movement (small deflection). The quantitative analysis demonstrates an average flow rate of 65.1 ml/min

under the curves, one can calculate fluid flow volumes through a tube. The data were stored and analyzed by two independent examiner (MRI radiologists) using a commercial flow analysis software (EasyVision-Workstation, Philips, Netherlands). The region of interest (ROI) was drawn to include all pixels that demonstrated flow, being usually slightly larger than the aqueduct. To reduce the partial volume effect due to brain movement, a drawing of the ROI identical in size and shape was localised in the brain stem parenchyma. The mass brain movement was calculated and subtracted from the flow in the aqueduct to get accurate aqueductal net flow measurements (Fig. 1b). We used the mean flow rate in ml/min as the standard unit of flow. According to the three investigational lines, the patients were differentiated into the iNPH group who were consequently shunted and the group with brain atrophy who were treated conservatively. Postoperatively, the shunted patients were clinically controlled by using the Kiefer score. The outcome was calculated by the NPH recovery rate (NPH-RR).

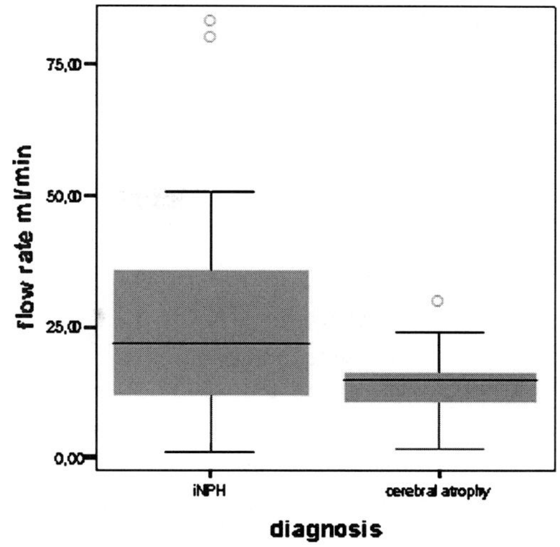

Fig. 2 Using the Wilcoxon test. A significant statistical difference ($p<0.05$) was found between the mean of the CSF flow rate in the cerebral aqueduct between the iNPH group and the cerebral atrophy group. *Box plots* for both iNPH and cerebral atrophy patients demonstrate the distribution of and the median value of the CSF flow-rates

$$\mathrm{NPH-RR} = \frac{\text{Kiefer score}_{\text{pre-operatively}} - \text{Kiefer score}_{\text{post-operatively}}}{\text{Kiefer score}_{\text{pre-operatively}}} \times 10$$

The outcome was classified in excellent (NPH-RR≥7 points), good (NPH-RR≥5 points), fair (NPH-RR≥2 points), poor outcome (NPH-RR<2 points). For the explorative statistics, the Wilcoxon test was used ($p<0.05$). The sensitivity and specificity of the CSF flow rate measured by the phase contrast MR technique were evaluated with the receiver-operator-characteristic (ROC) curve (Fig. 3).

Results

Descriptive statistics The gender distribution was 20 males and 21 females in iNPH group and 14 males and 6 females in cerebral atrophy group. The age distribution in iNPH group was between 27–84 years (mean=67.8) and in cerebral atrophy group between 37–87 years (mean=70). According to the results of the infusion test, the patients were classified into the iNPH group with a mean R-out value of 20.26 mmHg/min/ml and the cerebral atrophy group with a mean R-out value of 8.85 mmHg/min/ml. The mean of the CSF flow rate in the iNPH group was 25.7 ml/min±18.24 and 14.7 ml/min±6.94 in the cerebral atrophy group. Out of the 41 patients with idiopathic normal pressure hydrocephalus, 39 underwent a ventriculo-peritoneal shunt implantation. Two patients refused surgery. We implanted three shunt systems: in 15 patients the Dual Switch®-valve (Miethke-Aesculap, Potsdam, Germany), in 15 patients the proGAV® - The adjustable MIETHKE gravitational valve (Miethke-Aesculap, Potsdam, Germany) and in nine patients the CODMAN® Programmable Valve (Codmann and Johnson & Johnson company, Massachusetts, USA) with ShuntAssistant®. (Miethke-Aesculap, Potsdam, Germany). On admission, the mean value of the Kiefer score in the iNPH group was 8 points and in the cerebral atrophy group was 6.75 points. The follow-up time was between 3 and 12 months with an average of 9.4 months. The mean value of the postoperative Kiefer score in the iNPH group was 3.6 points. Using the NPH recovery rate, we estimated that 18 patients had an excellent outcome, six patients had a good outcome and 11 patients had a fair outcome. Four patients had a poor outcome.

Explorative statistics We surveyed the data of the phase-contrast MR flow study and evaluated it statistically using the Wilcoxon test. A significant statistical difference ($p<0.05$) was found between the mean of the CSF flow rate in the cerebral aqueduct between the iNPH group and the cerebral atrophy group (Fig. 2). We needed to define a quantitative value of CSF flow in the cerebral aqueduct by phase contrast MR at or above which the CSF flow is considered hyperdynamic and specific for iNPH. For that purpose, we used the (ROC) curve. The chosen pair of sensitivity and specificity on the ROC curve was selected to minimize the risk of shunting a patient with atrophy. A SPECIVICITY of 95% and a SENSITIVITY of 46% for establishing the diagnosis of iNPH and to predict a positive shunt response corresponded to a CSF flow rate of 24.5 ml/min in the aqueduct (Fig. 3).

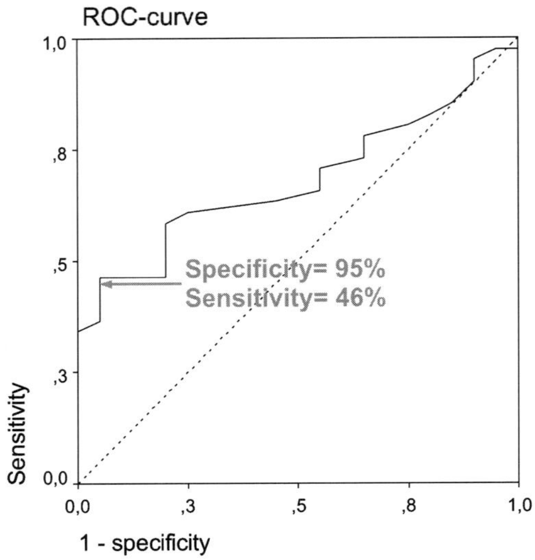

Fig. 3 Conjoined in a curve, all the pairs of specificity and the sensitivity of the CSF flow rate are shown. The chosen pair to establish the diagnosis of idiopathic normal pressure hydrocephalus using the receiver–operator-characteristic (ROC) curve is marked by the *arrow*

Discussion

Since the 1980s, numerous efforts have been made to study the quantitative and qualitative properties of CSF dynamics and its relation to iNPH using MR imaging. Brand-Zawadazki et al. [3] were the first to describe the signal void phenomenon. They found a low signal in the region of the aqueduct on the T2-weighted MRI scans. This low signal was referred to as the pulsatile CSF flow by Sherman and Citrin [12] and was termed “flow void sign”. Bradley et al. [2] tried to subdivide the flow void according to the extent of the signal void in the aqueduct, third ventricle and fourth ventricle. They showed a significant relation between the extent of the signal void and the response to shunt surgery in 20 patients with normal pressure hydrocephalus. Krauss et al. [7] investigated the degree as well as the extent of the signal void in comparison with the large cerebral arteries. These authors found no statistical correlation between the score of the fluid void and the surgical outcome in 37 iNPH patients as compared to 37 age-matched controls. We however, have shown a statistically significant difference in the CSF flow rate between the iNPH and cerebral atrophy groups. Similar results have also been reported by Luetmer et al. [8]. In their study, 43 patients with iNPH were selected out of 236 patients (47 normal elderly patients, 115 patients with cognitive impairment, and 31 with suspected iNPH which ultimately were excluded). Statistically, the CSF flow rate was significantly higher in the iNPH group. In a study by Gideon et al. [5], the CSF flow rates in the aqueduct of 9 patients with iNPH were compared with those of 9 healthy volunteers using a gradient flow sensitive MR-study method. The mean CSF flow rate was calculated to be 21.8 ml/min in the iNPH group. This was statistically significantly higher than the mean CSF flow in the control group (7 ml/min). These data are similar to the results of our study. The quantification of the CSF flow by measuring the stroke volume of the CSF was also used by Bradley et al. [1] in their study evaluating the CSF flow rate of 18 patients diagnosed with NPH. 12 patients had a CSF stroke volume greater than 42 μl/cc and a favourable shunting outcome. Of the group of patients (n=6) having a CSF stroke volume of 42 μl/cc or less, three improved after shunting, while three did not. The relationship between a CSF stroke volume greater than 42 μl/cc and a favourable response to shunting was statistically significant.

We statistically analysed the phase contrast MR CSF flow rate through the aqueduct of these patients and found that a flow rate ≥24.5 ml/min is 95% specific to iNPH patients., Dixon et al. [4, 8] showed a statistically significant difference between the CSF flow rates of 43 patients having the diagnosis of iNPH as compared to patients with other causes of cognitive impairment. They set a limit of 18 ml/min above which the CSF flow rate is defined as hyperdynamic. They further studied 49 patients with a CSF flow rate higher or equal to 18 ml/min but were not able to show a significant statistical association between outcome and the aqueductal CSF flow. However the study did show that flow rates greater than 33 ml/min trended to significant improvement after ventriculo-peritoneal shunting.

Based on the statistically calculated specificity and sensitivity of our study, we can use the phase contrast MR flow studies as a selective non-invasive method for establishing the diagnosis of iNPH. The value of 24.5 ml/min for the CSF flow rate through the aqueduct is the threshold we set to consider the flow hyperdynamic. Returning to our study data, 16 out of 18 shunted patients (having a hyperdynamic CSF flow) had a post-operative reduction of 20–100% in the severity of their symptoms. Only in one patient did the clinical symptoms became worse after shunting, and in one patient the symptoms did not change in the follow-up period. Thus the phase contrast MR method of investigating iNPH and the responsiveness to shunting was successful in 88% of our patients with hyperdynamic CSF flow through the aqueduct. Nevertheless it should be mentioned that the false-negative percentage could not be ignored, or in other words, the sensitivity of this investigational method is low. This will lead to the consensus that patients having a suspected iNPH and a normodynamic CSF flow should undergo additional invasive investigational methods like the dynamic infusion test or continuous ICP monitoring in order to further evaluate a possible diagnosis of iNPH.

Conclusion

The CSF-flow through the aqueduct can be considered a selective non-invasive method for establishing the diagnosis of idiopathic normal pressure hydrocephalus if the flow rate is ≥24.5 ml/min. Due to the low sensitivity of the method, 54% of the iNPH patients have false negative values. Therefore we recommend that patients with symptoms of the Hakim triad and a dilated ventricular system on CT scan (Evans index≥0.3) who have CSF flow rates lower than 24.5 ml/min undergo further invasive testing to better evaluate whether they should be shunted.

Conflict of interest statement We declare that we have no conflict of interest.

References

1. Bradley WG, Scalzo D, Queralt J, Nitz WN, Atkinson DJ, Wong P (1996) Normal-pressure hydrocephalus: evaluation with cerebrospinal fluid flow measurements at MR imaging. Radiology 198 (2):523–529
2. Bradley WG Jr, Whittemore AR, Kortman KE, Watanabe AS, Homyak M, Teresi LM et al (1991) Marked cerebrospinal fluid void: indicator of successful shunt in patients with suspected normal-pressure hydrocephalus. Acta Radiol 178(2):459–466
3. Brant-Zawadzki M, Kelly W, Kjos B, Newton TH, Norman D, Dillon W et al (1985) Magnetic resonance imaging and characterization of normal and abnormal intracranial cerebrospinal fluid (CSF) spaces. Neuroradiology 27(1):3–8
4. Dixon GR, Friedman JA, Luetmer PH, Quast LM, McClelland RL, Petersen RC et al (2002) Use of cerebrospinal fluid flow rates measured by phase-contrast MR to predict outcome of ventriculoperitoneal shunting for idiopathic normal-pressure hydrocephalus. Mayo Clin Proc 77(6):509–514
5. Gideon P, Stahlberg F, Thomsen C, Gjerris F, Sorensen PS, Henriksen O (1994) Cerebrospinal fluid flow and production in patients with normal pressure hydrocephalus studied by MRI. Neuroradiology 36(3):210–215
6. Holland GN, Hawkes RC, Moore WS (1980) Nuclear magnetic resonance (NMR) tomography of the brain: coronal and sagittal sections. J Comput Assist Tomogr 4(4):429–433
7. Krauss JK, Regel JP, Vach W, Jungling FD, Droste DW, Wakhloo AK (1997) Flow void of cerebrospinal fluid in idiopathic normal pressure hydrocephalus of the elderly: can it predict outcome after shunting? Neurosurgery 40(1):67–73 discussion 73–74
8. Luetmer PH, Huston J, Friedman JA, Dixon GR, Petersen RC, Jack CR et al (2002) Measurement of cerebrospinal fluid flow at the cerebral aqueduct by use of phase-contrast magnetic resonance imaging: technique validation and utility in diagnosing idiopathic normal pressure hydrocephalus. Neurosurgery 50(3):534–543 discussion 543–544
9. Meier U (2002) The grading of normal pressure hydrocephalus. Biomed Tech (Berlin) 47(3):54–58
10. Meier U, Bartels P (2002) The importance of the intrathecal infusion test in the diagnosis of normal pressure hydrocephalus. J Clin Neurosci 9(3):260–267
11. O'Connell JEA (1943) Vascular factor in intracranial pressure and maintenance of cerebrospinal fluid circulation. Brain 66: 204–228
12. Sherman JL, Citrin CM (1986) Magnetic resonance demonstration of normal CSF flow. AJNR Am J Neuroradiol 7(1):3–6
13. Wedeen VJ, Meuli RA, Edelman RR, Geller SC, Frank LR, Brady TJ et al (1985) Projective imaging of pulsatile flow with magnetic resonance. Science 230(4728):946–948

Acta Neurochir Suppl (2008) 102: 125–127

Course of disease in patients with idiopathic normal pressure hydrocephalus (iNPH): a follow-up study 3, 4 and 5 years following shunt implantation

U. Meier · J. Lemcke · F. Al-Zain

Abstract

Background In spite of recent advances in the diagnosis and treatment of iNPH, favorable outcomes following CSF diversion continue to be limited by complications, both valve dependent and valve independent, as well as by a reduction, over time, in the response to shunting.

Materials and methods Between September 1997 and December 2006, 148 patients underwent ventriculo-peritoneal shunt surgery in our department. All patients underwent the implantation of gravitational valves. These patients were followed-up 3, 6 and 12 months after surgery and then at annual intervals.

Findings The mean age of the 94 men and 54 women in our study was 68 years. The perioperative mortality was 0.7% (one patient died from a pulmonary embolism). A further 23 patients died during the follow-up period from causes unrelated to iNPH or the surgery. This study reports on groups of patients followed-up for 2 years (n=92), 3 years (n=62), 4 years (n=38) and 5 years (n=21) postoperatively. Valve independent complications occurred postoperatively in 6% of patients (n=10). Of these, five patients (3% of the total) had an infection and catheter displacement was recorded in a further five. Valve dependent complications occurred in 24 patients (16%), with overdrainage found in seven patients (5%) and underdrainage apparent in 17 (11%). Responder rates were 79% at 2 years, 79% at 3 years, 64% at 4 years and 60% at 5 years. The optimal valve opening pressure in programmable valves with a gravitational unit was between 30 and 70 mmHg.

Conclusions Sixty percent of patients with iNPH who underwent a ventriculo-peritoneal shunt using a gravitational valve continue to benefit from surgery 5 years postoperatively.

U. Meier (✉) · J. Lemcke · F. Al-Zain
Department of Neurosurgery, Unfallkrankenhaus Berlin,
Warener Strasse 7,
12683 Berlin, Germany
e-mail: ullrich.meier@ukb.de

Keywords Idiopathic normal pressure hydrocephalus · Gravitational valve · Programmable Medos-Codman-Valve · Miethke proGAV · Outcome · Gravitational valve-shunt operation · Clinical outcome study

Introduction

The successful treatment of normal pressure hydrocephalus continues to represent a challenge to practitioners. Optimal long term outcomes are reached via a lengthy path which begins with the appropriate choice of surgical indication. The selection of individually tailored methods for controlling internal CSF drainage is the next step and the journey ends with careful postoperative management over many years. The aim of this paper is to prospectively investigate outcomes following surgical treatment of idiopathic normal pressure hydrocephalus (iNPH) using gravitational valves (valves in which the opening pressure is posture-dependent) 3, 4 and 5 years after implantation.

Materials and methods

Patients, 148, diagnosed with iNPH at a major acute hospital (Unfallkrankenhaus Berlin) between September 1997 and December 2006 were enrolled into this prospective clinical outcomes study. Treatment took the form of surgical implantation of a gravitational ventriculo-peritoneal shunt. The course of their disease was recorded postoperatively using the Kiefer score [7], initially at 3, 6 and 12 months following implantation and thereafter at yearly intervals. All patients underwent a computed tomography scan of the head (head CT) before surgery, 4 days post-surgery and regularly during the follow-up period to assess the Evan's index [8].

The same diagnostic pathway was used with all patients. To fulfil the requirements for further assessment of possible iNPH, a patient had to demonstrate gait ataxia during an initial clinical examination as well as at least one additional symptom of the Hakim triad. An enlarged ventricular system (as evidenced neuroradiologically with an Evan's index of >0.3) was also required. Patients meeting these criteria then underwent a constant-rate lumbar infusion test followed by a spinal tap test in which 60 ml of CSF was removed. Shunt therapy was indicated when the patient showed a pathological resistance to outflow (R_{out}) in the infusion test and an improvement in symptoms following the spinal tap test.

The normal pressure hydrocephalus recovery rate (NPH-RR) was used to assess the postoperative development of clinical symptoms and to allow comparison between patients with varying grades of symptoms on presentation.

$$\text{NPH Recovery Rate} = \frac{\text{Kiefer Score}_{\text{preoperative}} - \text{Kiefer Score}_{\text{postoperative}}}{\text{Kiefer Score}_{\text{preoperative}}} \times 10$$

The NPH-RR scores obtained were categorised as follows; 7–10 points was classified as excellent, ≥5 as good and ≥2 as fair. NPH-RR scores of less than two points were graded as poor and were deemed to be unsatisfactory. NPH-RR scores equal to or above 2 were considered satisfactory.

Results

During the 9 year study period, 148 patients underwent shunt operations for iNPH at the Unfallkrankenhaus Berlin.

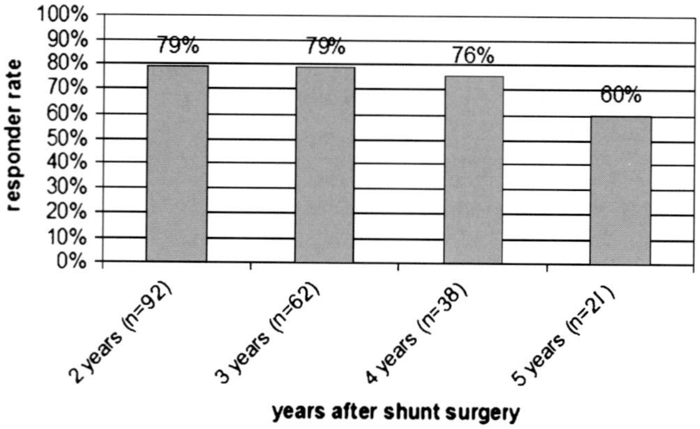

Fig. 1 Responder rate 2, 3, 4 and 5 years after shunt surgery

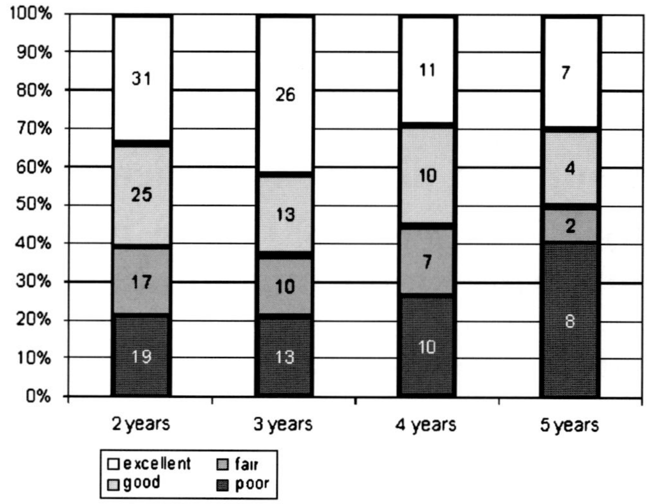

Fig. 2 Clinical outcomes measured using the NPH-RR 2, 3, 4 and 5 years after surgery. *Numbers within the boxes* represent the number of patients in that group

All patients received gravitational valves (63 dual switch valves, 42 programmable Medos® valves with a gravitational assistant valve and 43 proGAVs). Perioperative mortality was 0.7% (n=1; pulmonary embolism). A further 23 patients died during the follow-up period from causes unrelated to surgery or iNPH. The data presented here are from groups of patients followed-up at 2 years (n=92), 3 years (n=62), 4 years (n=38) and 5 years (n=21) postoperatively. The responder rate is presented in Fig. 1.

Clinical outcomes as measured using the NPH-RR (Fig. 2.) reveal that the percentage of patients experiencing a good or excellent outcome remains relatively stable after 2, 3, 4 and 5 years. At 2 years 60% of the patients had a good or excellent outcome, dropping to 50% at 5 years. By contrast, the proportion of patients suffering a poor outcome increases from 20% to 40% during this time period.

Complications 6% of patients (n=10) experienced valve independent complications postoperatively. Of these, five patients (3% of the total) had an infection and catheter displacement was recorded in a further five. Valve dependent complications occurred in 24 patients (16%), with overdrainage found in seven patients (5%) and underdrainage apparent in 17 (11%).

Discussion

The comparison of data for iNPH shunt outcome studies reported in the international literature presents several problems. The lack of distinction between patients with different pathogeneses for their iNPH often renders comparison of patient groups essentially meaningless [3, 10, 11] and the use of a wide range of evaluation measures and systems further complicates critical analysis of the literature [7, 9]. The frequency of follow-up periods in the literature also varies to a significant degree. It is the authors' view that 3, 6, 12 and 36 month intervals are desirable, a suggestion supported by other experts in this field [2, 4, 10, 12].

The available literature concerning iNPH outcomes following shunting reports a broad range of improvement rates. In 1984, Børgesen [2] suggested average improvement rates of 52% (range 42 – 67%) in an analysis of six studies published between 1972 and 1980. More recently, Hebb and Cusimano (2001) [5] analysed 44 publications and reported an average improvement rate of 59%, falling to 29% in the long term. Large single studies provide further figures; Romner and Zemack [12] report an improvement rate of 79% for their 146 iNPH patients at an average of 27 months (range 3 months to 9 years) postoperatively. Boon et al. [1] established average improvement rates of 53% and 47% using low pressure versus medium pressure standard valves respectively at *x* months. Mori et al. [9] determined an average improvement rate of 73% for 120 iNPH patients after 3 years while Kiefer et al. [6] observed an improvement rate of 70% in 122 iNPH patients an average of 26 months after surgery.

This study reports figures similar to those cited in the international literature. When comparing these figures with the work of others, however, it should be noted that in the classification system used in our clinical work and in this study, we consider an improvement of less than 20% (NPH-RR<2) to be a poor clinical outcome. Excellent, good and satisfactory clinical outcomes were recorded in 79% of patients 2 years postoperatively. Five years after surgery this figure had decreased to 60%. Closer examination reveals a trend towards the two extremes; the proportion of patients with good or fair outcomes decreases markedly (from 27% to 19% and from 18% to 10% respectively) whereas the poor outcome group increases considerably (20% to 40%) between year 2 and year 5, postoperatively. The percentage of patients with an excellent outcome meanwhile remains unchanged over this time period.

Conclusion

Sixty percent of patients undergoing an operation to establish a ventriculo-peritoneal shunt using a gravitational valve continue to benefit from surgery 5 years postoperatively. However, the proportion of patients with poor outcomes increases over time.

Conflict of interest statement We declare that we have no conflict of interest.

References

1. Boon AJ, Tans JT, Delwel EJ, Egeler-Peerdeman SM, Hanlo PW, Wurzer HA, Hermans J (1999) Dutch normal-pressure hydrocephalus study: the role of cerebrovascular disease. J Neurosurg 90(2):221–226
2. Børgesen SE (1984) Conductance to outflow of CSF in normal pressure Hydrocephalus. Acta Neurochir (Wien) 71:1–15
3. Decq P, Barat JL, Duplessis E, Leguerinel C, Gendrault P, Keravel Y (1995) Shunt failure in adult hydrocephalus: flow-controlled shunt versus differential pressure shunts—a cooperative study in 289 patients. Surg Neurol 43:333–339
4. Drake JM, Kestle JRW, Milner R (1998) Randomized trial of cerebrospinal fluid shunt valve design in pediatric hydrocephalus. Neurosurgery 43:294–305
5. Hebb AO, Cusimano MD (2002) Idiopathic normal pressure hydrocephalus: a systematic review of diagnosis and outcome. Neurosurgery 49:1166–1184
6. Kiefer M, Eymann R, Meier U (2002) Five years experience with gravitational shunts in chronic hydrocephalus of adults. Acta Neurochir (Wien) 144:755–767
7. Kiefer M, Eymann R, Komenda Y, Steudel WI (2003) A grading system for chronic hydrocephalus. Zentralbl Neurochir 64:109–115
8. Meier U (2004) Outcome of idiopathic normal pressure hydrocephalus after surgery with gravity valves. Neurosurg Q 14:119–126
9. Mori K (2001) Management of idiopathic normal-pressure hydrocephalus: a multiinstitutional study conducted in Japan. J Neurosurg 95:970–973
10. Pollack IF, Albright AL, Adelson PD (1999) A randomized, controlled study of a programmable shunt valve versus a conventional valve for patients with hydrocephalus. Hakim-Medos Investigator Group. Neurosurgery 45:1399–1408
11. Sprung C, Miethke C, Shakeri K, Lanksch WR (1998) Pitfalls in shunting of hydrocephalus–clinical reality and improvement by the hydrostatic dual-switch valve. Eur J Pediatr Surg 8(Suppl 1):26–30
12. Zemack G, Romner B (2002) Adjustable valves in normal-pressure hydrocephalus: a retrospective study of 218 patients. Neurosurgery 51:1392–1400

Acta Neurochir Suppl (2008) 102: 129–130

Chronic hydrocephalus that requires shunting in aneurysmal subarachnoid haemorrhage [a-SAH]: its impact on clinical outcome

W. S. Poon · S. C. P. Ng · G. K. C. Wong · L. Y. C. Wong · M. T. V. Chan

Abstract
Background Chronic hydrocephalus is a common occurrence following aneurismal subarachnoid haemorrhage [a-SAH] but its impact on neurological outcome has not been re reviewed systematically.
Patients and methods One hundred and eleven patients were recruited from a prospectively collected a-SAH registry over a 3-year period between 2002 and 2004. Their 6-month extended Glasgow Outcome Scale [GOSE] scores were correlated with routine clinical data and the need for CSF shunting [chronic hydrocephalus that required shunting, CHS].
Results Thirty patients with CHS were identified and they were associated with an initial poor WFNS grading [median 4 versus 2, p=0.028]. Among patients with poor WFNS grading, CHS was associated with a better GOSE [median 4 versus 2, p=0.041] and among patients with good WFNS grading, CHS paradoxically was associated with a poor GOSE [median 3.5 versus 7, p=0.016].
Conclusion The relationships between CHS and GOSE in a-SAH were complex. Their true clinical significance requires a more in-depth prospective study.

W. S. Poon (✉) · S. C. P. Ng · G. K. C. Wong · L. Y. C. Wong
Division of Neurosurgery, Department of Surgery,
Prince of Wales Hospital, The Chinese University of Hong Kong,
Shatin, Hong Kong
e-mail: wpoon@cuhk.edu.hk

M. T. V. Chan
Department of Anaesthesia and Intensive Care,
Prince of Wales Hospital, The Chinese University of Hong Kong,
Shatin, Hong Kong

Keywords Aneurismal subarachnoid haemorrhage · Hydrocephalus · CSF shunting · Glasgow Outcome Scale

Background

Chronic hydrocephalus occurs in 4.3–33% in patients with aneurysmal subarachnoid haemorrhage (a-SAH) [1, 2] and is associated with complications such as delayed cerebral ischaemia and hyponatraemia, resulting in poor neurological outcome. Its clinical significance and impact on functional neurological outcome have not been systematically studied. The aim of this study was to investigate the clinical impact of chronic hydrocephalus that required shunting (CHS) on the extended Glasgow Outcome Scale score (GOSE) at 6 months in 111 patients with a-SAH.

Patients and methods

From a prospectively collected a-SAH registry in a 3-year period (2002–2004), 111 patients were recruited for a retrospective study, correlating the clinical impact of CHS—chronic hydrocephalus that required shunting with GOSE. All data are presented as mean [range] unless otherwise specified. Difference between the means and medians of two measures was analysed by t test and Mann–Whitney test respectively; and the association between two nominal measures was analysed by Pearson chi-squared test.

Results

This group of 111 patients had a mean age of 59±13 years with female predominance (78 versus 33), median WFNS

Table 1 Demographic and outcome data

Patient demographics and outcome	
Sex	
Male	33 (30%)
Female	78 (70%)
Age (mean±standard deviation)	58.5± 12.9 years
WFNS grading	3 [1–8]
Good grade (1–3, n=66)	1 [1–3]
Poor grade (4–5, n=45)	4 [4–5]
Fisher grading	3 [2–4]
Location of aneurysm	
Anterior communicating artery	32 (29%)
Internal carotid artery	44 (39%)
Middle cerebral artery	23 (21%)
Posterior circulation	12 (11%)
Treatment	
Coiling	62 (56%)
Clipping	38 (34%)
No surgical intervention	11 (10%)
Implantation of shunt	
Implanted	30 (27%)
Not implanted	81 (73%)
GOSE at 6 months	6 [1–8]

All data are presented as median [range] unless otherwise specified

[1–5] and Fisher [2–4] gradings of 3 and a total of 30 patients were shunted (Table 1).

Poor GOSE at 6 months was associated with a poor WFNS grading (median GOSE 3 versus 7, $p<0.0005$) and older age (mean 62 versus 56, $p=0.021$) but not CHS (median GOSE 4 versus 7, $p=0.227$) (Table 2).

Table 2 Results of outcome analysis

	n	GOSE at 6 months	p value
WFNS grading			<0.0005
Good [1–3]	66	7 [1–8]	
Poor [4–5]	45	3 [1–8]	
Age (mean±SD years)			0.021
56.2±13.0	65	Favourable [5–8]	
61.8±11.7	46	Unfavourable [1–4]	
Implantation of shunt			0.227
Implanted	30	4 [2–8]	
Not implanted	81	7 [1–8]	

All data are presented as median [range] unless otherwise specified

Table 3 Relationships between outcome and implantation of shunt regarding the WFNS grading

	Implanted	Not implanted	p value
Implantation of shunt			0.028
n	30	81	
WFNS grading	4 [1–5]	2 [1–5]	
WFNS grading [1–3], n=66			0.016
n	14	52	
GOSE at 6 months	3.5 [2–8]	7 [1–8]	
WFNS grading [4–5], n=45			0.041
n	16	29	
GOSE at 6 months	4 [2–8]	2 [1–8]	

All data are presented as median [range] unless otherwise specified

Those 30 patients of CHS were associated with a poor WFNS grading (median 4 versus 2, $p=0.028$) (Table 3). However, when patients with poor WFNS grading [4 and 5, n=45] were analysed, CHS was associated with a better GOSE (median GOSE 4 versus 2, $p=0.041$), whereas in patients with good WFNS grading [1–3, n=66], CHS was associated with a poor GOSE (median GOSE 3.5 versus 7, $p=0.016$).

Discussion/conclusion

The relationship between CHS (chronic hydrocephalus that requires shunting) in patients with a-SAH and GOSE is complex: unfavourable in patients with good WFNS [1–3] grading, but improved in patients with poor WFNS [4–5] grading. Further correlation is needed to reveal its clinical relevance.

Conflict of interest statement We declare that we have no conflict of interest.

References

1. Demirgil BT et al (2003) Factors leading to hydrocephalus after aneurysmal subarachnoid haemorrhage. Minim Invasive Neurosurg 46(6):344–348
2. Ohwaki K et al (2004) Relationship between shunt-dependent hydrocephalus after subarachnoid haemorrhage and duration of CSF drainage. J Neurosurg 18:403–404

Acta Neurochir Suppl (2008) 102: 131–136

Morphological changes of intracranial pressure pulses are correlated with acute dilatation of ventricles

Xiao Hu · Peng Xu · Darrin J. Lee · Vespa Paul · Marvin Bergsneider

Abstract

Background Potentially useful information may exist in the morphological changes in intracranial pressure pulse therefore their extraction by automated methods is highly desirable.

Methods Long-term continuous recordings of intracranial pressure and electrocardiogram (ECG) signals were analyzed for four patients undergoing intracranial pressure (ICP) monitoring with their implanted shunts externalized and clamped. A novel clustering algorithm was invented to process hours of continuous ICP recordings such that a dominant ICP pulse was extracted every 5 min. Morphological characteristics of dominant ICP pulses were then extracted after detecting characteristics points of a dominant ICP pulse that include the locations of ICP pulse onset, the first (P_1), the second (P_2), and the third peaks (P_3) (or inflection points in the absence of peaks).

Findings It was found that ratios of P_2 amplitude to P_1 amplitude and P_3 amplitude to P_1 amplitude showed a strong increasing trend for a patient whose lateral ventricles were significantly enlarged (bi-frontal distance was 33 cm on day 0 and 37 cm on day 2) while there were no consistent trends in these morphological features of ICP pulse for the three patients whose ventricles size was not altered during the monitoring period.

Conclusion The present work demonstrates the usefulness of this novel ICP pulse analysis algorithm in terms of its potential capabilities of extracting predictive pulse morphological features from long-term continuous ICP recordings that correlate with the development of ventriculomegaly.

X. Hu · P. Xu · D. J. Lee · V. Paul · M. Bergsneider (✉)
Department of Neurosurgery, The David Geffen School of Medicine, University of California, Los Angeles, Los Angeles, CA 90095, USA
e-mail: mbergsneider@mednet.ucla.edu

X. Hu · M. Bergsneider
Biomedical Engineering Graduate Program, The UCLA Henry Samueli School of Engineering and Applied Science, University of California, Los Angeles, Los Angeles, CA, USA

P. Xu
School of Biological Science and Technology, University of Electronic Science and Technology, Chengdu, China

Keywords Intracranial pressure · Pulse morphology · Hydrocephalus · Hierarchical clustering

Introduction

Intracranial pressure (ICP) has a unique role in the management of patients in a neurosurgical environment. Mathematical modeling of ICP dynamics and computerized analysis of ICP signals have gained an increasing popularity and extracted researchers with different backgrounds working towards a better understanding of pathophysiology of ICP and better clinical utilizations of information, beyond the absolute value, in ICP dynamics. The main objective of the present work is to introduce a novel computerized ICP analysis method that is capable of characterizing morphological properties of ICP pulses and tracking their temporal changes. It will be demonstrated in the present work that this new analysis method helped reveal a consistent ICP pulse morphological pattern change associated with the development of acute ventriculomegaly.

Although, description of morphological changes in ICP pulses has been widely used in existing medical literatures [3, 7], e.g., elevation of ICP associated with rounding of ICP pulses, computerized characterization of ICP pulse morphologies has seldom been attempted. A visualization method of ICP morphology change was proposed recently by McNames and colleagues [5]. In addition, Eide developed an automated method analyzing distribution of ICP pulse amplitude and mean ICP [4]. However, no attempt has been made to identify the sub-peaks in an ICP pulse and to characterize subtle changes associated with those peaks. The presence of three distinct peaks in an ICP pulse is well-known in neurosurgical literature with accumulated clinical knowledge and interpretations of different peak configurations [8–11]. Therefore, there is a clear advantage of characterizing the ICP pulse morphology in terms of properties associated with these peaks, i.e., a starting point for the computerized analysis is to identify and assign these peaks in an ICP pulse. This is apparently a demanding requirement for the automated analysis considering the variability of pulse morphological configurations and the inherent noise in typical clinical recordings of physiological signals. To enable a computerized ICP analysis algorithm with the capability of resolving the presence and locations of multiple ICP peaks, we propose a technique called morphological clustering and analysis of intracranial pressure pulse (MOCAIP). MOCAIP is composed of three distinct but integrated processing modules that handle the robust delineation of individual ICP pulses, the clustering of ICP pulses based on their morphological properties into different groups, and the analysis of single averaged ICP pulse for extracting morphological features. In the rest of text, a brief description of the MOCAIP will be given and then will be followed by an introduction of clinical materials we used for a preliminary validation of this technique. Discussion of the analysis results will be presented and conclusion drawn to conclude the paper.

Materials and methods

Basic definitions

An unambiguous definition of landmarks on an ICP pulse is a prerequisite for proceeding with computerized analysis. A typical ICP pulse is shown in Fig. 1 with our interested landmarks and indices marked. In the proposed method, time zero of a pulse signal corresponds to the QRS peak of the ECG beat just preceding the ICP pulse. The landmarks sought for each individual pulse include locations of pulse onset, the first (P_1), the second (P_2), and the third peaks (P_3). Pulse onset is determined, according to the definition used in arterial blood pressure pulse analysis [2], as the intersection of lines A and B in the figure. Line A is a fitted straight line to the rising edge of the ICP pulse and line B is the horizontal line passing the minimum of the ICP pulse within the time period between QRS peak and the first ICP peak. A point k is considered the location of an ICP peak if $ICP_{k-1}<ICP_k<ICP_{k+1}$ is satisfied. However, this condition is often too restrictive as ICP peaks, including the P_1 and the P_3, may become "shoulder" points in many conditions. To accommodate this, the necessary condition for a point k to be considered as P_1 is $\frac{d^2ICP_k}{dt^2} \leq 0$ and $\frac{d^2ICP_{k+1}}{dt^2} > 0$. The necessary condition for a P_3 candidate is $\frac{d^2ICP_{k-1}}{dt^2} > 0$ and $\frac{d^2ICP_k}{dt^2} \leq 0$. These constraints are defined to position shoulder points at locations where portions of pressure pulse contour change their shape from a downward concave one to an upward one (for $P1$) or from an upward concave one to a downward one (for P_3).

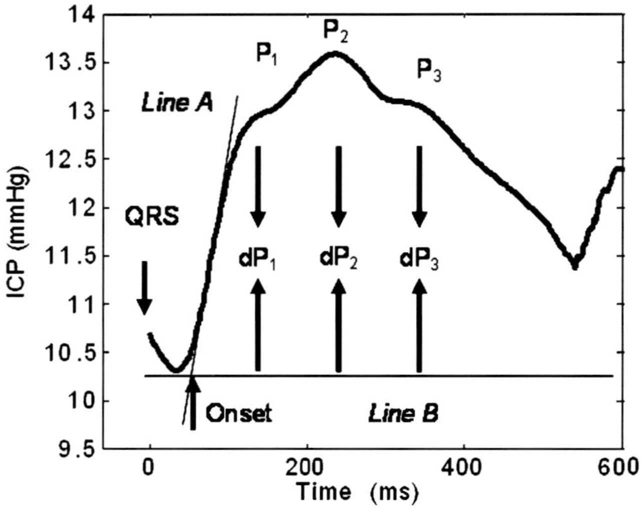

Fig. 1 Illustration of the basic definitions for the proposed morphological analysis of ICP pulses

Block diagram of computerized algorithm

According to the block diagram of the proposed algorithm, there are three major components of MOCAIP. We adopted the well developed ECG QRS detection methods for detecting QRS peak. This information is further exploited in delineating each individual ICP pulse in such a way that the missed or spurious ECG beats will not damage the delineation of ICP pulses. The resultant set of ICP pulses will then be subjected to a clustering process that groups pulses solely based on their morphological similarity by removing mean values from ICP pulses. The dominant ICP pulse will then be extracted, by averaging a subset of pulses belonging to the biggest cluster, for further analysis that involves detecting the aforementioned landmarks and calculating the morphological indices. These indices could be used to represent the average morphological information of the ICP segment under analysis. It is often the case that noises in signals were clustered into non-dominant group and will not affect the characterization of the morphology of ICP by utilizing the dominant pulse.

Beat-by-beat pulse delineation

Three processing steps are involved in beat-by-beat ICP pulse delineation as shown in Fig. 2. R wave could be located using well-established QRS detection methods [1]. Detecting QRS location from clinical recordings, especially for one-lead ECG recording, could not achieve 100% accuracy. Therefore, the process of delineating ICP pulse has to deal with the situations with missing and/or spurious ECG beat detections. Missing beat detection will not create problems for our purpose of delineating ICP pulses because an ICP pulse will only be identified if ECG beat is detected. Given the high sensitivity and specificity of most ECG QRS detection algorithms, mild number (<1%) of missing beats is achievable and hence the missed information can be recovered by interpolation. However, spurious ECG beat detection, if untreated, will create spurious ICP pulses and damage further morphological analysis. We proposed an adaptive beat-by-beat ICP "peak" location algorithm for detecting spurious ECG beat detections. Particularly, the *i*th ICP pulse candidate is considered a spurious ICP pulse if the location of its "global" peak does not satisfy the timing constraint $\beta_i \leq p_i \leq \alpha_i$, where p_i is the location, relative to its corresponding QRS peak, of the global peak of the *i*th ICP pulse candidate, β_i and α_i are time-varying lower and upper bounds of this peak location. This constraint essentially states that the timing difference between an ICP peak and ECG QRS peak should stay within a physiological temporal window. The global peak of an ICP pulse is obtained by a heart-rate dependent band-pass filter, which is used to pre-process raw ICP waveform for two purposes: (1) remove both high-frequency noise and baseline wander in ICP; (2) simplify ICP pulse morphology such that one global peak, which may not correspond to any of its original peaks, is retained per ICP pulse. This peak is termed the global peak of an ICP pulse. This band-pass filter is further designed based on the information in the heart rate such that the cut-off frequencies of the pass-band are 0.45 and 0.55 times mean heart rate, respectively. This essentially achieves a patient specific band-pass filter. Note that both β_i and α_i can change from beat to beat. The adjustment algorithm for β_i and α_i was designed with the objective of maintaining a window of a constant length around p_i for its robust detection, i.e., $p_i - \beta_i \equiv \Delta_0$ and $\alpha_i - p_i \equiv \Delta_1$. Since p_i can change from beat-by-beat, the requirement of a constant window length hence implies that β_i and α_i are adjusted beat-by-beat accordingly. This can be viewed as an adaptive interval determination such that the true but unknown p_i is always trapped within the detection window. An equation for adapting α_i from its value at previous beat α_{i-1} in a beat-by-beat fashion can be implemented as:

$$\alpha_i = \alpha_{i-1} + \lambda_{\text{pos}}[\Delta_1 - (\alpha_{i-1} - p_{i-1})] + \lambda_{\text{pre}}(p_{i-1} - p_{i-2}) \quad (1)$$

where λ_{pos} and λ_{pre} are weighting coefficient within [0, 1]. The correction added to α_{i-1} consists of two parts: (1) the error between the desired window length and the actual window length at the previous beat; (2) the prediction of change of p_i relative to p_{i-1} based on the change from p_{i-2} to p_{i-1}. Similar idea can be applied to derive β_i as $\beta_i = \beta_{i-1} + \lambda_{\text{pos}}[(p_{i-1} - \beta_{i-1}) - \Delta_0] + \lambda_{\text{pre}}\ (p_{i-1} - p_{i-2})$.

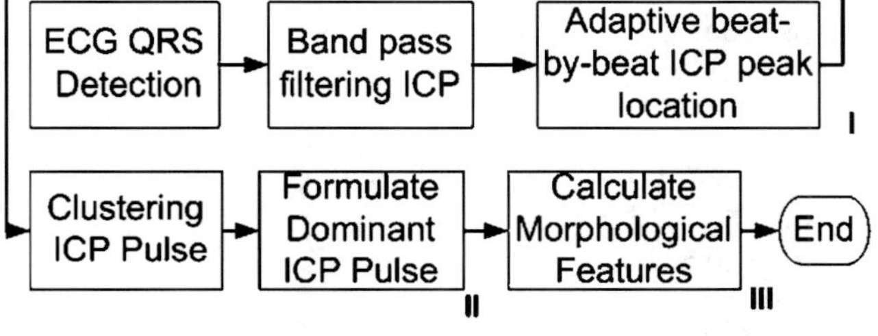

Fig. 2 Diagram of the proposed computerized algorithm of morphological clustering and analysis of intracranial pressure

Clustering of pulses

Each pulse is delineated by taking the preceding QRS peak as the starting point and the next QRS peak (exclusive) as the ending point. We chose the hierarchical clustering as it does not require a prior specification of number of clusters. The clustering algorithm starts with treating each single pulse as a different cluster and then merges two clusters with the smallest dissimilarity at each step until there is only one cluster left. The average linkage [6] is chosen for calculating dissimilarity between two clusters such that it equals the average of distances between all pairs of pulses, where each pair is made up of one pulse from each cluster. The Euclidian distance between two pulses is adopted for determining their closeness after removing mean of each pulse. A 5 Hz low-pass zero-phase filter is used for preliminary noise removal before subjecting pulses to the clustering process.

After the above clustering procedure, the number of optimal clusters is determined by the Silhouette criterion [6]. The average pulse is then extracted for each cluster. The dominant pulse for the signal segment analyzed is taken as the average pulse of the biggest cluster. This dominant pulse is usually well shaped with a highly enhanced signal to noise ratio.

Single pulse analysis

Further analysis of the dominant pulse is performed by first detecting the landmarks on the ICP pulse whose signal to noise ratio should have been significantly improved due to the averaging process. To avoid the issue of arbitrarily assigning a detected peak to one of P_1, P_2, and P_3, only those pulses with all three peaks detected will be processed further for extracting morphological features. As a safeguard for false detection, we implemented two peak sorting criteria. The first criterion is based on an amplitude measure and the second is based on a curvature measure for candidate peaks that satisfy either the peak or the shoulder point criterion. All detected peaks and shoulder points within an ICP pulse will be sorted based on the two criteria, respectively. If the top three peaks resulted from each criteria match with each other, then these three peaks were taken as the P_1, P_2, and P_3 of the ICP pulse. The amplitude measure of a candidate peak is calculated as the summation of the distance from the peak to the starting and the ending points of the identified peak. The amplitude measure of a shoulder point is calculated as the difference between the maximal value and the minimal value of the concave curve where the shoulder point is detected. The curvature is calculated following the equation: $\kappa = \frac{d^2\mathrm{ICP}}{\mathrm{d}t} \Big/ (1+ (\frac{\mathrm{dICP}}{\mathrm{d}t})^2)^{\frac{3}{2}}$.

Table 1 Basic characteristics of the four patients in the present study

Patient	1		2				3			4		
Sex	Female		Female				Female			Female		
Age (years)	51		47				18			28		
CT brain date	D_1	D_2	D_0	D_1	D_2	D_3	D_0	D_2	D_3	D_0	D_1	D_2
Ventricle size (cm)	29	29	20	20	23	25	30	30	33	33	33	37

The ventricle size is approximated as the distance between two front horns of the third ventricles

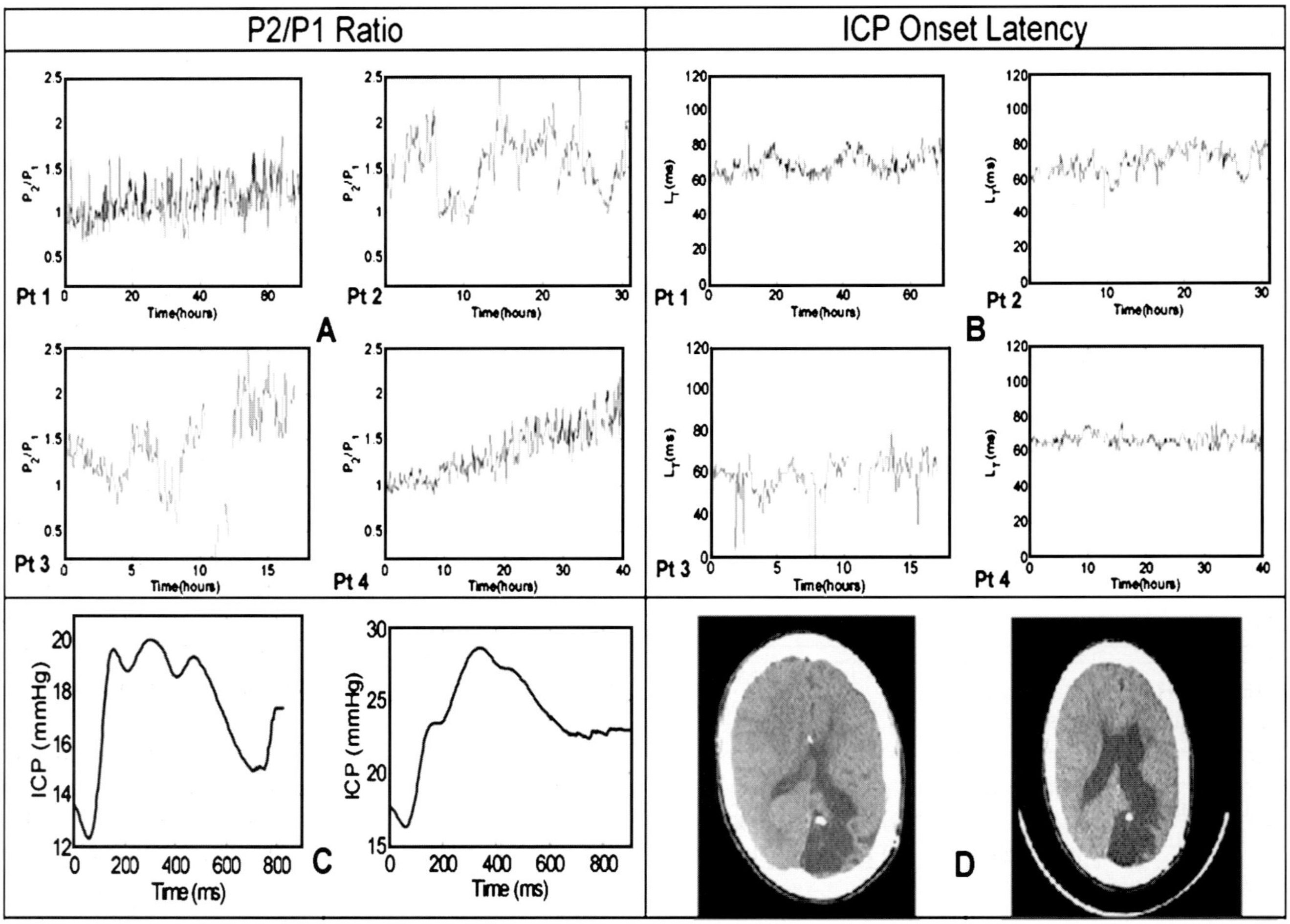

Fig. 3 a–d Summary of results of MOCAIP analysis to long-term ICP from four silt ventricle patients

Patient data

Continuous recordings of ICP and ECG data from four slit ventricle syndrome patients were analyzed using the MOCAIP algorithm. These patients had long-term shunt implantation before the admission to the hospital. To prepare them for an endoscopic third ventriculostomy procedure, their shunts were externalized and clamped during an ICP monitoring period. Only archived waveform data were used in this retrospective study hence informed consents were waived by the IRB committee of the UCLA medical center. Basic patient characteristics are summarized in Table 1. Ventricle size was assessed by one of authors (M.B.) as the maximal distance between two frontal horns of the third ventricles. As shown in Table 1, only patient #4 developed significant hydrocephalus within the ICP monitoring period.

All signals were sampled at 400 Hz with a 12-bit quantization. The MOCAIP was applied to every consecutive 5-min segment and the resulting trend of P_2/P_1 ratio and the ICP onset latency (LT) are reported here.

Results

Panels A and B of Fig. 3 show a summary plot of the P_2/P_1 and the LT trends for all four patients, respectively. The linear increasing trend of P_2/P_1 is significant for patient #4 as it approximately increased from a value of one to two in a period of 41 h. P_2/P_1 trends for other three patients did not show such a conspicuous trending. Panel B of the figure demonstrates the LT trending, which does not show a definite monotonic trending for any patients but shows a diurnal periodic pattern for patient #2. Two dominant ICP pulses, taken at the first and the last 5 min of the recording, for patient #4 are shown in panel C of the figure with two representative CT brain scans taken at the beginning and end of the monitoring sessions shown in panel D.

Discussion

We have presented key elements of a novel computerized algorithm of conducting morphological clustering and analysis of ICP pulses and its preliminary application in analyzing continuous long-term ICP recordings for four patients who underwent a shunt externalization/clamping procedure. The proposed algorithm has two novel aspects compared to several existing attempts at ICP pulse analysis. First, a clustering procedure was applied that is necessary to group similar pulses together for extracting a representative dominant pulse. The dominant pulse usually has a much better signal to noise ratio due to the averaging process that is necessary for the proposed procedure of identifying various landmarks on an ICP pulse. Second, we implemented a two-criteria matching procedure for identifying the locations of all three peaks in one ICP pulse. Although, the existence of three peaks in an ICP pulse is a well-known observation, the direct identification of them in an automated fashion has not been proposed. Various morphological features can be defined once the locations of the three peaks are identified. In this preliminary application, two measures were studied including the P_2/P_1 ratio and latency of ICP pulse onset to the ECG QRS. It is found that the P_2/P_1 ratio showed a significant increasing trend in a 40-h period as the ventricles dilated. On the other hand, the P_2/P_1 ratio did not show a monotonic trend for three cases where no significant ventricle dilatation occurred. This result indicates that the P_2/P_1 ratio may be strongly influenced by the compliance of the intracranial pressure dynamic systems as dilated ventricles decreased the overall compliance of the craniospinal space. However, this needs further validation as it is only demonstrated in one patient.

Future improvement of the proposed technique is desirable in terms of assigning peaks to P_1, P_2, and P_3 at situations when the number of detected peaks is less than three. A validation study is being performed to assess the accuracy of the peak detection algorithm.

Acknowledgements The present work is partially supported by NINDS R21 awards (NS055998 and NS055045) to X.H. and a R01 award (NS054881) to M.B.

Conflict of interest statement We declare that we have no conflict of interest.

References

1. Afonso VX, Tompkins WJ, Nguyen TQ, Luo S (1999) ECG beat detection using filter banks. IEEE Trans Biomed Eng 46(2):192–202
2. Chiu YC, Arand PW, Shroff SG, Feldman T, Carroll JD (1991) Determination of pulse wave velocities with computerized algorithms. Am Heart J 121(5):1460–1470
3. Contant CF Jr, Robertson CS, Crouch J, Gopinath SP, Narayan RK, Grossman RG (1995) Intracranial pressure waveform indices in transient and refractory intracranial hypertension. J Neurosci Methods 57(1):15–25
4. Eide PK (2006) A new method for processing of continuous intracranial pressure signals. Med Eng Phys 28(6):579–587
5. Ellis T, McNames J, Aboy M (2007) Pulse morphology visualization and analysis of intracranial pressure. IEEE Trans Biomed Eng 54(9):1552–1559
6. Kaufman L, Rousseeuw PJ (1990) Finding groups in data : an introduction to cluster analysis. Wiley series in probability and mathematical statistics. Applied probability and statistics. Wiley, New York (xiv, 342 p)
7. Kirkness CJ, Mitchell PH, Burr RL, March KS, Newell DW (2000) Intracranial pressure waveform analysis: clinical and research implications. J Neurosci Nurs 32(5):271

8. Piper IR, Chan KH, Whittle IR, Miller JD (1993) An experimental study of cerebrovascular resistance, pressure transmission, and craniospinal compliance. Neurosurgery 32(5):805–815 (discussion 815–816)
9. Piper IR, Miller JD, Dearden NM, Leggate JR, Robertson I (1990) Systems analysis of cerebrovascular pressure transmission: an observational study in head-injured patients. J Neurosurg 73(6):871–880
10. Portnoy HD, Chopp M (1981) Cerebrospinal fluid pulse wave form analysis during hypercapnia and hypoxia. Neurosurgery 9 (1):14–27
11. Portnoy HD, Chopp M, Branch C, Shannon MB (1982) Cerebrospinal fluid pulse waveform as an indicator of cerebral autoregulation. J Neurosurg 56(5):666–678

Acta Neurochir Suppl (2008) 102: 137–140

Pulse amplitude of intracranial pressure waveform in hydrocephalus

Z. Czosnyka · N. Keong · D. J. Kim · D. Radolovich · P. Smielewski · A. Lavinio · E. A. Schmidt · S. Momjian · B. Owler · J. D. Pickard · M. Czosnyka

Abstract

Background There is increasing interest in evaluation of the pulse amplitude of intracranial pressure (AMP) in explaining dynamic aspects of hydrocephalus. We reviewed a large number of ICP recordings in a group of hydrocephalic patients to assess utility of AMP.

Materials and methods From a database including approximately 2,100 cases of infusion studies (either lumbar or intraventricular) and overnight ICP monitoring in patients suffering from hydrocephalus of various types (both communicating and non-communicating), etiology and stage of management (non-shunted or shunted) pressure recordings were evaluated. For subgroup analysis we selected 60 patients with idiopathic NPH with full follow-up after shunting. In 29 patients we compared pulse amplitude during an infusion study performed before and after shunting with a properly functioning shunt. Amplitude was calculated from ICP waveforms using spectral analysis methodology.

Findings A large amplitude was associated with good outcome after shunting (positive predictive value of clinical improvement for AMP above 2.5 mmHg was 95%). However, low amplitude did not predict poor outcome (for AMP below 2.5 mmHg 52% of patients improved). Correlations of AMP with ICP and Rcsf were positive and statistically significant (N=131 with idiopathic NPH; R= 0.21 for correlation with mean ICP and 0.22 with Rcsf; $p<$ 0.01). Correlation with the brain elastance coefficient (or PVI) was not significant. There was also no significant correlation between pulse amplitude and width of the ventricles. The pulse amplitude decreased ($p<0.005$) after shunting.

Conclusions Interpretation of the ICP pulse waveform may be clinically useful in patients suffering from hydrocephalus. Elevated amplitude seems to be a positive predictor for clinical improvement after shunting. A properly functioning shunt reduces the pulse amplitude.

Keywords Intracranial pressure · Pulse waveform · Shunting · Improvement · Normal pressure hydrocephalus

Z. Czosnyka (✉) · N. Keong · D. J. Kim · D. Radolovich · P. Smielewski · A. Lavinio · E. A. Schmidt · S. Momjian · B. Owler · J. D. Pickard · M. Czosnyka
Academic Neurosurgical Unit, Addenbrooke's Hospital, Box 167, Cambridge CB2 0QQ, UK
e-mail: ZC200@medschl.cam.ac.uk

D. J. Kim
Department of Engineering, University of Cambridge, Trumpington Str, Cambridge CB2 1PZ, UK

D. Radolovich
Department of Anaesthesiology, University of Pavia, Policlinica San Matteo, 27100 Pavia, Italy

A. Lavinio
Department of Anaesthesia, University of Brescia; Spedali Civili, 25100 Brescia, Italy

E. A. Schmidt
Department of Neurosurgery, Hopital Purpan, TSA 40031 31059 Toulouse Cedex 9, France

S. Momjian
Department of Neurosurgery, Ho^pitaux Universitaires de Gene`ve, Geneva, Switzerland

B. Owler
Institute of Clinical Neuroscience, Department of Neurosurgery, Royal Prince Alfred Hospital, Missenden Rd, Camperdown, NSW 2050, Australia

Introduction

Several dynamic components can be recognized in the intracranial pressure (ICP) waveform. Pulsatile changes in arterial cerebral blood volume (CBV) evoke the pulse waveform of ICP [1]. Changes in venous CBV due to varying in intra-thoracic pressures are responsible for the respiratory component of the ICP waveform. Slower, intrinsic vasomotor changes of CBV provoke slow waves, classified as B, C, Meyer or plateau waves [18].

Studies of the intracranial pulse pressure waveform have existed for over three decades [1, 12] and have continued through the present [10, 11, 27].

Relatively early it was postulated [8, 16, 23] that increased pulse pressure may cause ventricular enlargement. With the advent of dynamic MRI, pulsatile flow of CSF has been intensively studied and reported to be increased in hydrocephalus [4, 9, 13, 22]. However, there have been conflicting reports regarding its role in prognostication following shunting, both enthusiastic [4] and highly critical [15]. Some novel theories of the development of communicating hydrocephalus, though still awaiting conclusive documentation [10, 14], are largely based on MRI investigations of pulsatile CSF flow.

Although invasive, direct measurement of the intracranial pressure remains a common tool in the investigation of hydrocephalus. Studies on the pulse pressure ICP waveform [1, 3, 12, 17, 20, 21, 24–26] pioneered the understanding of CSF dynamics. In this short report, we analyzed pulse pressure amplitude during CSF infusion studies [7] in both non-shunted and shunted hydrocephalic patients.

Material and methods

We interrogated a database of around 2100 clinical infusion studies performed in 980 patients suffering from hydrocephalus of various type and aetiology (idiopathic NPH 47%, post SAH NPH 12%, others 19%, non-communicating 22%,). Average age was 65 (range 24 to 94) and male to female ratio was 2:1. All of the patients attended the hydrocephalus clinic following referral by their treating neurosurgeon for ventricular dilatation on brain scan (CT or MRI) and clinical symptoms belonging to Hakim's triad (poor gait, memory problems and urinary incontinence). Due to the nature of patient selection many had complex clinical problems. This group of patients was investigated with a constant rate infusion study (either lumbar 20%, or via pre-implanted Ommaya reservoir 38%, shunt pre-chamber 40%, open EVD 2%) and/or overnight ICP monitoring in addition to routine clinical and imaging assessment. 44% of tests were performed in patients with shunt in-situ, in order to check its performance.

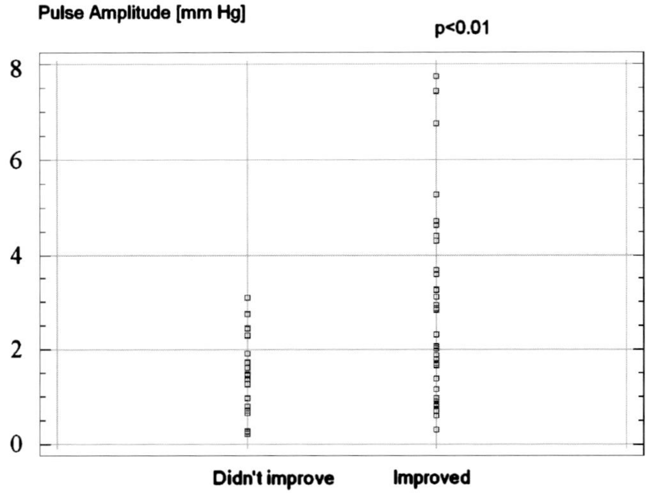

Fig. 1 Sixty patients with typical symptoms of iNPH. 66% improved after shunting and 33% did not. Those with pulse amplitude greater than 3 mmHg all improved. Only about half of those with amplitude less than 3 mmHg improved ($p<0.01$; Kruskall–Wallis test)

The infusion studies were performed through two lumbar needles or the shunt pre-chamber proximal to the valve or a pre-implanted ventricular access device. If lumbar access was used, lumbar needles (usually 21 gauge) were used. With intraventricular access, in both situations two needles (25 gauge butterfly) were inserted.

One needle was connected to a pressure transducer and the other to an infusion pump mounted on a purpose-built trolley containing a pressure amplifier (Simonsen & Will, Sidcup, U. K.) and an IBM-compatible personal computer running ICM+ software (www.neurosurg.cam.ac.uk/icmplus). A strict aseptic technique was used to keep all the pre-filled tubing and

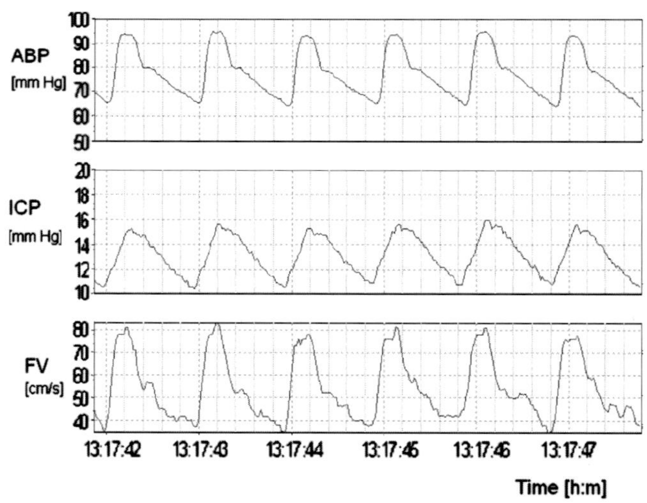

Fig. 2 Amplitude peak-to-peak above 4 mmHg or first harmonic above 2.5–3 mmHg is considered as exceptionally high. However interpretation may be inaccurate without monitoring of arterial pressure (ABP-Finapress) and/or blood flow velocity in the basal cerebral arteries (FV)

the transducer sterile. The skin was very carefully prepared with antiseptic solution.

After 10 min of baseline measurement, the infusion of normal saline or Hartmann's solution was started at a rate of 1.5 ml/min (or 1 ml/min if the baseline pressure was higher than 15 mmHg) and continued until a steady state ICP plateau was achieved. If the mean ICP increased over 40 mmHg, the infusion was stopped immediately. All compensatory parameters were calculated using computer-supported methods based on physiological models of CSF circulation [7]. Baseline ICP and R_{csf} (resistance to CSF outflow; calculated as the plateau ICP reached during the test minus baseline ICP, divided by infusion rate) characterize static conditions of CSF circulation. Elastance coefficient (or elasticity [1]) characterizes the ability of the system to store extra volumes of fluid- a larger coefficient indicates that smaller volume may be stored under the same incremental pressure conditions. During the infusion study, the ICP waveform was processed through a Fourier transform analysis [5, 6] to determine the pulse amplitude of ICP (AMP) as the magnitude (peak-to peak) of the first harmonic component related to the heart rate. This method is an alternative to time-domain analysis [1, 11] and, in our experience, both methods are generally equivalent.

Results

In our clinical practice, most patients who receive a shunt have increased resistance to CSF outflow (Rcsf>13 mmHg/(ml/min). Even in such a pre-selected group, pulse amplitude was a predictor of outcome after shunting. In a subgroup of patients with idiopathic NPH with a stable follow-up assessment (N=60), the relationship between baseline amplitude and improvement (decrease in Stein–Langfitt score) is shown in Fig. 1. Non-parametric statistical testing using the Kruskall–Wallis test indicates that amplitude is greater in patients who improved ($p<0.01$). The plot suggests that when the amplitude is larger (greater than 2.5 mmHg) improvement is very likely (more than 90% patients improve; Fig. 2). However, when the amplitude is less 2.5 mmHg, this predictive ability is lost as about half of these patients improved with shunting while half did not.

Pulse amplitude correlated rather weakly with other compensatory parameters assessed during the infusion studies. Correlations with ICP and Rcsf were positive and statistically significant (N=131 in patients with idiopathic NPH; R=0.21 for correlation with mean ICP and R=0.22 with Rcsf; $p<0.01$). Correlation with brain elastance coefficient (or PVI) was not statistically significant. There was no significant correlation between pulse amplitude and the width of the ventricles.

After shunting, the pulse amplitude decreased. Differences were significant for both baseline values and those obtained during infusion study. In shunted patients, the pulse amplitude was lower in patients with a normally functioning shunt compared to those with a blocked shunt (0.95±0.4 vs 1.95±0.61; P=0.000033).

Discussion

Intracranial pressure recording and infusion studies are routinely performed in our institution in patients with complex hydrocephalus. The usefulness of physiological measurements has been recently summarized in guidelines [19] for the diagnosis and management of NPH. Contrary to the resistance to CSF outflow [2], the clinical value of pulse amplitude is not supported by any randomized clinical trial. Our own experience indicates that knowledge of the pulse amplitude may help in interpretation of recordings in cases of disturbed CSF compensation, shunt blockage, and slit ventricles. It may also help in distinguishing postural changes and vasogenic ICP waves. While low amplitude is not predictive for lack of improvement following shunting, large amplitude is strongly associated with improvement after shunting. In our clinical series, there is no evidence that increased pulse amplitude promotes ventricular dilatation (no correlation between amplitude and the width of the ventricles). It should be emphasized that we do not interpret pulse amplitude alone as a single parameter directing management, but rather consider it in conjunction with the clinical presentation, results of neuroimaging, neurophysiological tests, laboratory data, and other parameters describing CSF compensation (mean pressure, resistance to CSF outflow, vasogenic waves, RAP index, elastance coefficient, etc.). A decrease in amplitude after shunting may signify that diversion of CSF results in improved pressure–volume compensation.

Conflict of interest statement We declare that we have no conflict of interest.

References

1. Avezaat CJ, Eijndhoven JH (1986) Clinical observations on the relationship between cerebrospinal fluid pulse pressure and intracranial pressure. Acta Neurochir (Wien) 79(1):13–29
2. Borgesen SE, Gjerris F (1982) The predictive value of conductance to outflow of CSF in normal pressure hydrocephalus. Brain 105(Pt 1):65–86
3. Bergsneider M, Alwan AA, Falkson L, Rubinstein EH (1998) The relationship of pulsatile cerebrospinal fluid flow to cerebral blood flow and intracranial pressure: a new theoretical model. Acta Neurochir Supp 71:266–268

4. Bradley WG Jr, Scalzo D, Queralt J, Nitz WN, Atkinson DJ, Wong P (1996) Normal-pressure hydrocephalus: evaluation with cerebrospinal fluid flow measurements at MR imaging. Radiology 198(2):523–529
5. Christensen L, Borgesen SE (1989) Single pulse pressure wave analysis by fast Fourier transformation. Neurol Res 11(4):197–200
6. Czosnyka M, Wollk-Laniewski P, Batorski L, Zaworski W (1988) Analysis of intracranial pressure waveform during infusion test. Acta Neurochir (Wien) 93(3–4):140–145
7. Czosnyka Z, Czosnyka M, Owler B, Momjian S, Kasprowicz M, Schmidt EA, Smielewski P, Pickard JD (2005) Clinical testing of CSF circulation in hydrocephalus. Acta Neurochir Suppl 95:247–251
8. Di Rocco C, Pettorossi VE, Caldarelli M, Mancinelli R, Velardi F (1977) Experimental hydrocephalus following mechanical increment of intraventricular pulse pressure. Experientia 33 (11):1470–1472
9. Dombrowski SM, Schenk S, Leichliter A, Leibson Z, Fukamachi K, Luciano MG (2006) Chronic hydrocephalus-induced changes in cerebral blood flow: mediation through cardiac effects. J Cereb Blood Flow Metab 26(10):1298–1310
10. Egnor M, Zheng L, Rosiello A, Gutman F, Davis R (2002) A model of pulsations in communicating hydrocephalus. Pediatr Neurosurg 36(6):281–303
11. Eide PK, Brean A (2006) Intracranial pulse pressure amplitude levels determined during preoperative assessment of subjects with possible idiopathic normal pressure hydrocephalus. Acta Neurochir (Wien) 148(11):1151–1156
12. Foltz EL, Aine C (1981) Diagnosis of hydrocephalus by CSF pulse-wave analysis: a clinical study. Surg Neurol 15(4):283–293
13. Greitz D, Hannerz J, Rahn T, Bolander H, Ericsson A (1994) MR imaging of cerebrospinal fluid dynamics in health and disease. On the vascular pathogenesis of communicating hydrocephalus and benign intracranial hypertension. Acta Radiol 35(3):204–211
14. Greitz D (2004) Radiological assessment of hydrocephalus: new theories and implications for therapy. Neurosurg Rev 27(3):145–165 discussion 166–167
15. Krauss JK, Regel JP, Vach W, Jungling FD, Droste DW, Wakhloo AK (1997) Flow void of cerebrospinal fluid in idiopathic normal pressure hydrocephalus of the elderly: can it predict outcome after shunting? Neurosurgery 40(1):67–73
16. Kuchiwaki H, Misu N, Kageyama N, Ishiguri H, Takada S (1987) Periodic oscillation of intracranial pressure in ventricular dilation: a preliminary report. Neurol Res 9(4):218–224
17. Lenfeldt N, Andersson N, Agren-Wilsson A, Bergenheim AT, Koskinen LO, Eklund A, Malm J (2004) Cerebrospinal fluid pulse pressure method: a possible substitute for the examination of B waves. J Neurosurg 101(6):944–950
18. Lundberg N (1960) Continuous recording and control of ventricular fluid pressure in neurosurgical practice. Acta Psychiatr Scand Suppl 36(149):1–193
19. Marmarou A, Bergsnaider M, Klinge P, Relkin N, Black PM (2005) The value of supplemental prognostic tests for the preoperative assessment of idiopathic normal-pressure hydrocephalus. Neurosurgery 57(3 Suppl):S17–S28
20. Matsumoto T, Nagai H, Kasuga Y, Kamiya K (1986) Changes in intracranial pressure (ICP) pulse wave following hydrocephalus. Acta Neurochir (Wien) 82(1–2):50–56
21. Momjian S, Czosnyka Z, Czosnyka M, Pickard JD (2004) Link between vasogenic waves of intracranial pressure and cerebrospinal fluid outflow resistance in normal pressure hydrocephalus. Br J Neurosurg 18(1):56–61
22. Ohara S, Nagai H, Matsumoto T, Banno T (1988) MR imaging of CSF pulsatory flow and its relation to intracranial pressure. J Neurosurg 69(5):675–682
23. Pettorossi VE, Di Rocco C, Mancinelli R, Caldarelli M, Velardi F (1978) Communicating hydrocephalus induced by mechanically increased amplitude of the intraventricular cerebrospinal fluid pulse pressure: rationale and method. Exp Neurol 59(1):30–39
24. Portnoy HD, Chopp M, Branch C, Shannon MB (1982) Cerebrospinal fluid pulse waveform as an indicator of cerebral autoregulation. J Neurosurg 56(5):666–678
25. Stephensen H, Andersson N, Eklund A, Malm J, Tisell M, Wikkelso C (2005) Objective B wave analysis in 55 patients with non-communicating and communicating hydrocephalus. J Neurol Neurosurg Psychiatry 76(7):965–970
26. Stephensen H, Tisell M, Wikkelso C (2002) There is no transmantle pressure gradient in communicating or noncommunicating hydrocephalus. Neurosurgery 50(4):763–771
27. Wagshul ME, Chen JJ, Egnor MR, McCormack EJ, Roche PE (2006) Amplitude and phase of cerebrospinal fluid pulsations: experimental studies and review of the literature. J Neurosurg 104 (5):810–819

Acta Neurochir Suppl (2008) 102: 141–144

The influence of co-morbidity on the postoperative outcomes of patients with idiopathic normal pressure hydrocephalus (iNPH)

U. Meier · J. Lemcke

Abstract

Background A critical question in the diagnosis and treatment of idiopathic normal pressure hydrocephalus (iNPH) is that of which preoperative factors can most reliably predict outcomes following shunt insertion. The number and type of co-morbidities are increasingly being viewed as important predictive indicators.

Methods Between 1997 and 2004, 95 patients were implanted with a gravitational ventriculo-peritoneal shunt as treatment for iNPH. All coincident disease processes were recorded. Eighty-two of these patients underwent follow-up 2 years postoperatively. The results of this prospective follow-up examination (Kiefer Score, NPH Recovery Rate) were compared with the preoperative Co-Morbidity Index (CMI).

Findings Of the patients with a CMI score of 0–1 (n=18), 67% experienced an excellent outcome, 28% a good outcome and 5% and 0% a fair and poor outcome respectively. A CMI score of 2–3 was associated with markedly poorer outcomes (n=33); 42% excellent, 30% good, 18% fair and 10% poor. A score of 4–5 was related to 14% excellent, 27% good, 23% fair and 36% poor outcomes (n=22). Remarkably few patients scoring between 6 and 8 on the CMI scale experienced a favourable outcome. The outcomes for this latter group were 0% excellent, 10% good, 45% fair and 45% poor (n=9).

Conclusions Co-morbidity is a statistically significant predictor of the quality of clinical outcome for patients with iNPH undergoing shunt therapy.

U. Meier (✉) · J. Lemcke
Department of Neurosurgery, Unfallkrankenhaus Berlin,
Warener Strasse 7,
12683 Berlin, Germany
e-mail: ullrich.meier@ukb.de

Keywords Idiopathic normal pressure hydrocephalus · Co-Morbidity Index (CMI) · Shunt surgery · Outcome · Co-morbidity · iNPH

Introduction

Patients with idiopathic normal pressure hydrocephalus (iNPH) are usually elderly and, as such, often present with multiple co-morbidities. In this prospective audit, we aim to evaluate whether a Co-Morbidity Index (CMI) [7] can be used to provide a prognostic indicator for the quality of clinical outcomes following shunt therapy for iNPH.

Materials and methods

In the neurosurgical unit of a major hospital for acute care (Unfallkrankenhaus Berlin) between September 1997 and September 2004, 95 patients were diagnosed with iNPH and treated surgically. The 62 men and 33 women had an average age of 67 at diagnosis (range 27–83). All patients underwent the implantation of gravitational valves (54x Miethke-Aesculap® Dual switch valve, 20x programmable Codman® Medos valve with Miethke-Aesculap® gravitational assistant valve, 21x Miethke-Aesculap® proGAV). It was possible to follow-up 82 of these patients over a 2-year postoperative period. Eight patients died from causes unrelated to either the shunt operation or their iNPH between 10 and 19 months postoperatively (four from heart disease, two from neoplastic disease, one from pneumonia and one from renal failure). One patient died perioperatively from a pulmonary embolism despite thromboprophylaxis (1% perioperative mortality). Four patients were lost to follow-up.

Table 1 Co-morbidity index (CMI; Kiefer et al. [7])

Risk factors	1 point	2 points	3 points
Vascular	Hypertension	Diabetes mellitus	
	Aortofemoral bypass	Peripheral vascular disease	
	Stent	Vascular occlusion	
Cerebrovascular	Posterior circulatory insufficiency	Vascular encephalopathy	Cerebral infarction
	ICA stenosis	TIA/PRIND	
Cardiac	Arrhythmia		
	Valve disease		
	Heart failure/stent		
	Aortocoronary bypass		
	Myocardial infarction		
Others		Parkinson's disease	

Diagnosis Patients who displayed gait ataxia in addition to other cardinal symptoms of iNPH and who showed neuroradiological evidence of ventricular enlargement were further assessed with an intrathecal infusion test. To determine the individual CSF flow parameters, a dynamic infusion test was performed via lumbar puncture using a computer-assisted constant flow technique with an infusion rate of 2 mL/min. A resistance of 13 mmHg or higher was defined as pathological. Immediately following the dynamic infusion test a diagnostic drainage of at least 60 mL CSF was carried out using the same puncture site. An improvement in the clinical picture over the ensuing 2 or 3 days served as indication for the implantation of a ventriculo-peritoneal shunt. If the patient's symptoms, particularly the gait ataxia, did not initially improve, then 2–3 days of further, external lumbar drainage was commenced. Once more, if symptoms improved over this period, shunting was indicated.

Clinical grading The Black Grading System for Shunt Assessment and the NPH Recovery Rate (based on the clinical grading for NPH by Kiefer) were used to express the results of the clinical examinations. All graded examination results were split into four clinical outcome groups; excellent (restoration of pre-morbid activity levels), good (limited reduction in activity levels), fair (partial improvement) and poor (transient or no improvement). The first group was defined by an NPH recovery rate of ≥7.5 (75–100% improvement), the second by a rate of ≥5 (50–74% improvement), the third by a rate of ≥2 (20–49% improvement) and the fourth by a rate of <2 (partial improvement up to 19% or deterioration):

$$\text{NPH Recovery Rate} = \frac{\text{NPH Grading}_{\text{preoperative}} - \text{NPH Grading}_{\text{postoperative}}}{\text{NPH Grading (Kiefer)}_{\text{preoperative}}} \times 10.$$

Co-morbidity index (CMI) Kiefer et al. [7] introduced an assessment tool for various pathologies and their clinical effects when present in NPH. By summating the scores for each co-morbid disease process found in Table 1, a CMI score of between 0 and 23 can be calculated.

Results

Of the 82 patients studied, 29 (35%) experienced an excellent outcome, 22 (27%) a good outcome and 16 (20%) a fair outcome. 15 patients (18%) experienced a poor outcome. Thus, 67 patients responded to treatment; a responder rate of 82%. Figure 1 graphically relates the preoperative CMI score to the clinical outcomes of all iNPH patients who underwent follow-up examinations postoperatively. Ninety-five percent of patients with a CMI score of 0–1 achieved an excellent or good outcome. A CMI score of 2–3 was associated with a noticeable decrease in the quality of outcomes (n=33; excellent, 42%; good, 30%; fair, 18%; poor, 10%). Patients scoring 4–5 (n= 22; 14%, 27%, 23%, 36%) or 6–8 (n=9; 0%, 10%, 45%,

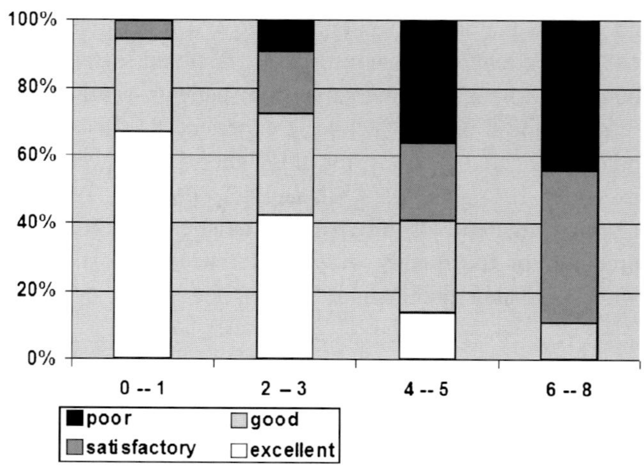

Fig. 1 Responder rate

CMI / Outcome	0 - 3	4 - 8
excellent / good	41	10
satisfactory / poor	10	21

Fig. 2 Kiefer's CMI on postoperative outcomes in patients with iNPH is associated with a sensitivity of 80% and a specificity of 68%

45%) on the preoperative CMI scale experienced remarkably poorer postoperative outcomes.

As a predictive tool, Kiefer's CMI on postoperative outcomes in patients with iNPH is associated with a sensitivity of 80% and a specificity of 68%. Outcomes in patients with a CMI score between 0 and 3 were significantly better than those of patients with a CMI greater than 3 (Fig. 2).

Discussion

In all branches of medicine, a multidisciplinary approach involving various specialities is necessary for two important reasons. Firstly, for an intervention to be effective, it must be ascertained that the observed symptoms are indeed caused by the disease process targeted by the treatment method and not by some other pathology [1–3]. Secondly, it is important to recognise situations in which the detrimental effect of co-morbidity on the probable outcome of a given intervention is such that little or not improvement can be expected. Both of these possibilities represent a potential contraindication to operative therapy [8, 9].

Co-morbidity A review of the international literature yields a mean reported rate of co-morbidity in iNPH of 43% [5]. Cerebrovascular insufficiency is described in 45% of these patients. The commonest co-morbidity (78–100%) is vascular encephalopathy. Parkinson's disease or Parkinsonism were present in 10% of cases, and 10% had histologically unconfirmed but clinically suspected Alzheimer's dementia [11]. In order to distinguish the clinical picture of iNPH from other dementing syndromes, Golomb et al. [4] and Savolainen et al. [10] describe the coincidence of Alzheimer's disease and iNPH. Centrally limited motor disturbances can likewise result in an overlapping entity. In the patient population reported in this study, 8% were found to have Parkinson's disease in addition to iNPH. The exclusion of progressive cerebrovascular dementia as a differential diagnosis in iNPH is an important motivation for the use of invasive diagnostic procedures. Co-morbidities can be clinically quantified in terms of risk factors. Such a system—the Co-morbidity Index—was introduced in 2006 by Kiefer [6, 7]. In it, the most common co-morbid pathologies are assigned between one and three points which, when totalled as a CMI score, can be referred to an empirical threshold value above which the likelihood of a good to excellent outcome significantly decreases. This indirect correlation between outcome and CMI is unequivocally demonstrated in this study—those patients with a CMI score of 0 or 1, 67% had an excellent outcome, while 45% of patients with a CMI of between 6 and 8 experienced an unsatisfactory outcome. Kiefer et al. [7] suggest a CMI of 3 as a threshold value dividing patient groups with a statistically favourable prognosis from those with a tendency towards a poor outcome. The data from this study serve to confirm the value of this cut-off point; of the patients who experienced a good or excellent outcome 80% (41/51) had a CMI ≤ 3. Only 10/31 patients (32%) with a CMI >3 experienced a comparable improvement in symptoms.

Outcome The general improvement rates reported in the literature for patients with NPH undergoing a shunt operation vary around a mean of 53% (range 31–96%). In one meta-analysis, Vanneste [12] gives a figure of between 30% and 50%. Hebb and Cusimano [5] report an immediate improvement rate following iNPH shunt operations of 59%, falling to 29% in the long term. The results of our study show an overall improvement rate at 2 years post-shunt of 82%, in keeping with the international literature.

Conclusion

Co-morbidity is a statistically significant predictor of the quality of the clinical outcome for patients with iNPH undergoing shunt therapy (sensitivity=80%; specificity=68%). A CMI score of less than or equal to 3 points allows for a good postoperative prognosis. A CMI of more than 3 significantly decreases the chance of a good outcome and this should form part of the assessment when the risks and benefits of surgery are considered. Due to factors arising from co-morbidity, a successful outcome in patients with a CMI of 6 or more is unlikely.

Conflict of interest statement We declare that we have no conflict of interest.

References

1. Bech RA, Waldemar G, Gjerris F (1999) Shunting effects in patients with idiopathic normal pressure hydrocephalus: correlations with cerebral and leptomeningeal biopsy findings. Acta Neurochir (Wien) 141:633–639

2. Boon AJ, Tans JT, Delwei EJ, Egeler-Peerdeman SM, Hanlo PW, Wurzer HA, Hermans J (1999) Dutch Normal Pressure Hydrocephalus Study: the role of cerebrovascular disease. J Neurosurg 90(2):221–226
3. Borgesen SE (1984) Conductance to outflow of CSF in normal pressure hydrocephalus. Acta Neurochir (Wien) 71:1–15
4. Golomb J, Wisoff J, Miller DC, Boksay I, Kluger A, Weiner H, Salton J, Graves W (2000) Alzheimer's disease comorbidity in normal pressure hydrocephalus: prevalence and shunt response. J Neurol Neurosurg Psychiatry 68(6):778–81
5. Hebb AO, Cusimano MD (2001) Idiopathic normal pressure hydrocephalus: a systematic review of diagnosis and outcome. Neurosurgery 49:1166–1186
6. Kiefer M, Eymann R, Meier U (2002) Five years experience with gravitational shunts in chronic hydrocephalus of adults. Acta Neurochir (Wien) 144:755–767
7. Kiefer M, Eymann R, Steudel WI (2006) Outcome predictors for normal pressure hydrocephalus. Acta Neurochir Suppl 96:364–367
8. Meier U, Kiefer M, Sprung C (2003) Normal-pressure hydrocephalus: pathology, pathophysiology, diagnostics, therapeutics and clinical course. PVV Science, Ratingen
9. Meier U, Zeilinger FS, Kintzel D (1999) Signs, symptoms and course of normal pressure hydrocephalus in comparison with cerebral atrophy. Acta Neurochir (Wien) 141(10):1039–1048
10. Savolainen S, Paljarvi L, Vapalahti M (1999) Prevalence of Alzheimer's disease in patients investigated for presumed normal pressure hydrocephalus: a clinical and neuropathological study. Acta Neurochir (Wien) 141(8):849–583
11. Tulberg M (2000) CSF sulfatide distinguishes between NPH and subcortical ateriosclerotic encephalopathy. J Neurol Neurosurg Psychiatry 69:74–81
12. Vanneste JA (2000) Diagnosis ad management of normal-pressure hydrocephalus. J Neurol 247(1):5–14

Acta Neurochir Suppl (2008) 102: 145–151

ICM+, a flexible platform for investigations of cerebrospinal dynamics in clinical practice

P. Smielewski · A. Lavinio · I. Timofeev · D. Radolovich · I. Perkes · J. D. Pickard · M. Czosnyka

Abstract

Background ICM+ software encapsulates 20 years of our experience in brain monitoring gained in multiple neurosurgical and intensive care centres. It collects data from a variety of bedside monitors and produces on-line time trends of parameters defined using configurable signal processing formulas. The resulting data can be displayed in a variety of ways including time trends, histograms, cross histograms, correlations, etc. For technically minded researchers there is a plug-in mechanism facilitating registration of third party libraries of functions and analysis tools.

Methods The latest version of the ICM+ software has been used in 162 severely head injured patients in the Neurosciences Critical Care Unit of the Addenbrooke's Cambridge University Hospital. Intracranial pressure (ICP) and invasive arterial blood pressure (ABP) were monitored routinely. Mean values of ICP, ABP, cerebral perfusion pressure (CPP) and various indices describing pressure reactivity (PRx), pressure–volume compensation (RAP) and vascular waveforms of ICP were calculated. Error-bar chart showing reactivity index PRx versus CPP ('Optimal CPP' chart) was calculated continuously.

Findings PRx showed a significant relationship with CPP (ANOVA: $p<0.021$) indicating loss of cerebral pressure-reactivity for low CPP (CPP<55 mmHg) and for high CPPs (CPP>95 mmHg). Examining PRx–CPP curves in individual patients revealed that CPP_{OPT} not only varied between subjects but tended to fluctuate as the patient's state changed during the stay in the ICU. Calculation window of 6–8 h provided enough data to capture the CPP_{OPT} curve.

Conclusions ICM+ software proved to be useful both academically and clinically. The complexity of data analysis is hidden inside loadable profiles thus allowing clinically minded investigators to take full advantage of signal processing engine in their research into cerebral blood and fluid dynamics.

Keywords Multimodal monitoring · Cerebral autoregulation · On-line data analysis · Head injury

Introduction

Contemporary brain monitoring in clinical practice includes multiple global and local modalities such as arterial and intracranial pressures, transcranial Doppler blood flow velocity, laser Doppler flowmetry, brain tissue oxygenation etc. Each of those modalities produce time varying data and some contain complex waveforms. Trends of minute by minute time averages are these days widely used to aid in interpretation of the monitoring data but they largely

P. Smielewski (✉) · A. Lavinio · I. Timofeev · D. Radolovich · I. Perkes · J. D. Pickard · M. Czosnyka
Neurosurgery Unit, Department of Clinical Neurosciences, University of Cambridge, Addenbrookes Hospital, Level 4, A Block, Cambridge CB0 2QQ, UK
e-mail: ps10011@cam.ac.uk

A. Lavinio
e-mail: andrea.lavinio@gmail.com

I. Timofeev
e-mail: it227@cam.ac.uk

D. Radolovich
e-mail: iperdany@hotmail.com

I. Perkes
e-mail: iain.perkes@gmail.com

J. D. Pickard
e-mail: jdpsecretary@medschl.cam.ac.uk

M. Czosnyka
e-mail: mc141@medschl.cam.ac.uk

dispose of information carried by the waveforms. Also, it is often the nature and strength of association between different modalities that provides crucial information rather than the modalities themselves. For example, analysis of correlations between signals such as blood flow velocity, arterial blood and intracranial pressures has been shown to carry information related to regulation of cerebral blood flow [2].

ICM+ software was originally developed some 20 years ago at the Warsaw University of Technology, Poland, and at present encapsulates our 20 years experience in data monitoring and analysis in the neuro-critical care and neurosurgery units. Using previous versions of the software [3, 8] we collected and analysed multimodal data from nearly 600 severely head injured patients, 1000 patients suffering from hydrocephalus, performed over 900 intra-operative monitoring, and over 200 recordings in subarachnoid haemorrhage (SAH) patients. All this experience proved invaluable in fine-tuning the design issues of the new software [7].

The software was first licensed for distribution to other centres in the world in 2004 and officially launched during the ICP 12 conference in Hong Kong. The aim of this paper was to present enhancement to the original software and discuss experiences gained during its use over the past 4 years.

Material

One hundred sixty-two head injury patients (123 males, 39 females) with an average age of 34 years were managed in

A)

B)

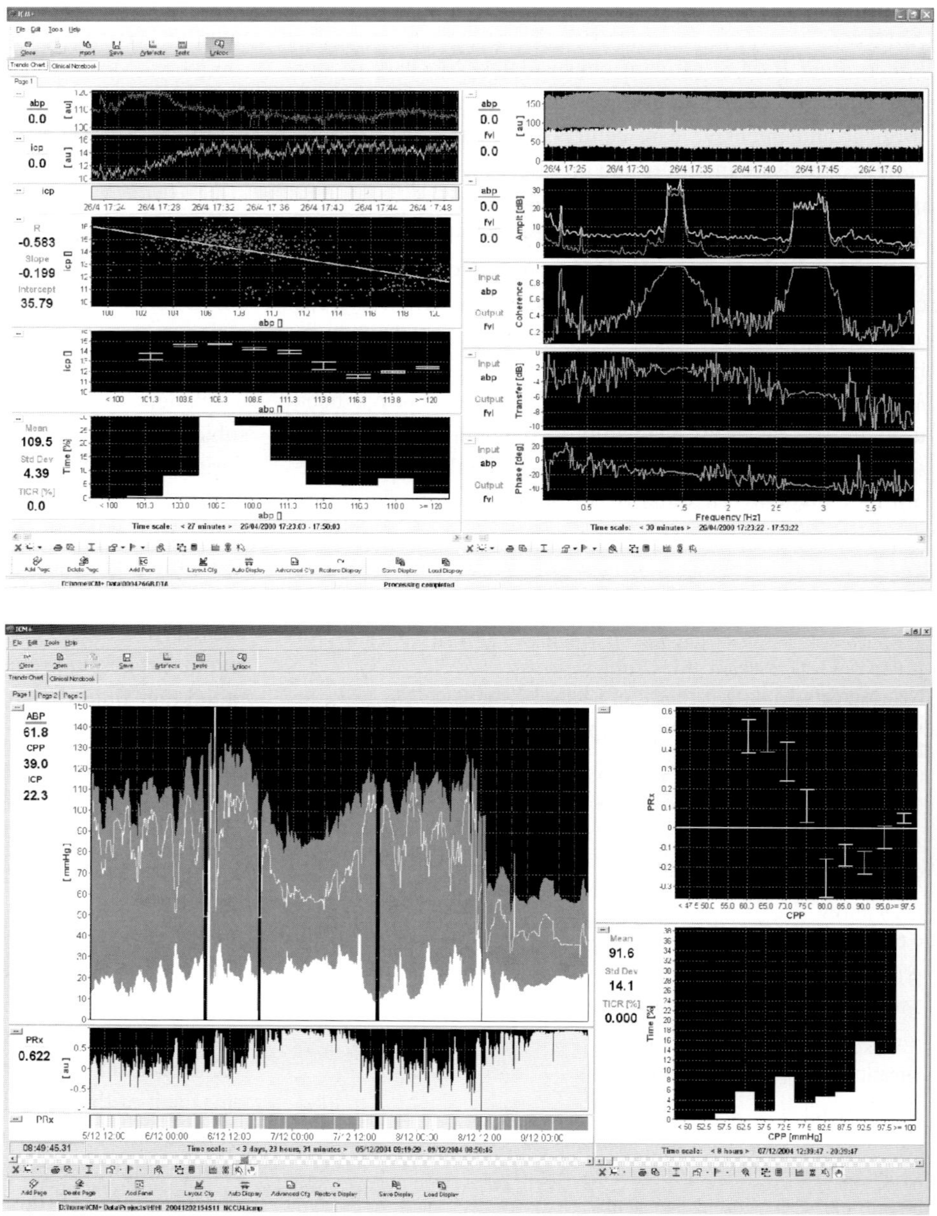

Fig. 1 Example of display configuration for off-line analysis of raw ICP, ABP and TCD FV signals showing variety of different chart types available (**a**) and on-line analysis of ICP and ABP signals aimed to provide direct feedback on the CPP (**b**), state of autoregulation (colour coded at the bottom), the optimal CPP curve and the CPP histogram (for autoregulation guided CPP management the peak of CPP histogram should correspond to the valley of the optimal CPP curve)

the Neurosciences critical Care Unit (NCCU) between 2004 and 2007. Median admission Glasgow Coma Scale (GCS) score was 6 (range 3 to 14), with 20% of patients having a GCS of 9 or greater, but deteriorating later.

Methods

Software details

A detailed description of the software has been published elsewhere [7]. Briefly, ICM+ includes a user configurable signal processing engine that allows real-time trending of complex parameters derived from signals from bedside monitoring devices. Data from multiple monitoring devices is downloaded continuously using analog or digital interfaces. Configuration of on-line data analyses utilizes arithmetic expressions of statistical and signal processing functions. Calculations are performed using multiple levels of analysis with the output of each one providing input to the next. The list of functions available for analysis configuration can now be extended via a plug-in system that allows researchers to write their own functions and register them with the calculation engine. The final data is displayed in a variety of ways including simple time trends, as well as time window based histograms, cross histograms, correlations etc (Fig. 1a). The selection also includes a variety of spectral analysis charts.

In addition to time trends resulting from on-line analysis, the software also stores raw signals acquired from the bedside monitors. Those can be later re-processed using the on-line analysis engine with different configurations, thus providing means of building a data bank for testing novel indices and methods of on-line processing of multimodal monitoring data.

Finally, there is also an extendable toolbox for analysis of controlled interventions. Currently these include: CSF infusion test for identification of a model of cerebrospinal fluid dynamics in hydrocephalic patients and a set of tests of cerebrovascular reactivity: CO_2 reactivity, Transient Hyperaemic Response, and the cuff test.

Data collection

ICP was monitored using Codman intraparenchymal probes and ABP was measured from a peripheral artery. Data was acquired using analog outputs from the monitors through the analog-to-digital converter (Data Translation 9801 USB box) with a sampling frequency of 50 Hz. The data analysis configuration was set to calculate average values of the several derived parameters every minute (Table 1). Particular care was paid to the organization of the front page display, which showed time trends of ABP, ICP, CPP, and pressure reactivity (Fig. 1b). For clarity, the pressure reactivity index (PRx) was also displayed at the bottom of the screen as ‘risk graph’, converting information about the reactivity to colours: green—good, red—impaired. In addition a histogram of CPP and an error bar chart of PRx versus CPP were plotted from a period of the last 8 h and automatically updated with every new data sample.

Table 1 List of parameters calculated from ICP and ABP signals composing a standard monitoring configuration for head injury patients in the Neurosciences Critical Care Unit, Addenbrookes Hospital, Cambridge

List of parameters calculated from ICP and ABP signals	
ABP	Mean arterial blood pressure
CPP	Mean cerebral perfusion pressure
ICP	Mean intracranial pressure
Slow	Power of slow waves of ICP (periods from 20 s to 5 min)
Resp	Amplitude of respiratory waveform
PRx	Pressure reactivity index (moving ABP–ICP correlation)
RAP	Index describing CSF compensatory reserve (moving ICP–AMP correlation)
HR	Mean heart rate
aABP	Pulse amplitude of arterial blood pressure
AMP	Pulse amplitude of ICP waveform
RAC	Moving correlation between pulse waveform of ICP and mean CPP
RR	Mean respiratory rate

Results

When the data from all the head injury patients monitored with ICM+ was pulled together the autoregulation index PRx showed a significant relationship with CPP (ANOVA: $p<0.021$) indicating loss of cerebral pressure-reactivity for low CPP (CPP<55 mmHg) and for high CPPs (CPP> 95 mmHg). The value of CPP at the bottom of the valley was called ‘Optimal’ CPP (CPP_{OPT}) in our original paper from 2002 [9]. Examining PRx–CPP curves in individual patients revealed that CPP_{OPT} not only varied between subjects but tended to fluctuate as the patient's state changed during the stay in the ICU (Fig. 2). Setting the curve calculation window to 6–8 h provided enough data to capture the CPP_{OPT} curve and yet it was short enough to provide useful feedback for the intensivists.

Although the Optimal CPP seems at the moment to be our most important fruit of the ICM+ enabled monitoring of severe head trauma patients, many other phenomena have been discovered and pursued. They include optimization of CPP (or ABP) using other modalities like $PbtO_2$ [5] and NIRS [1], change in pressure-reactivity in hypothermia and rewarming [6], as well as important work in subarachnoid haemorrhage [10] and in diagnostics of hydrocephalus [4].

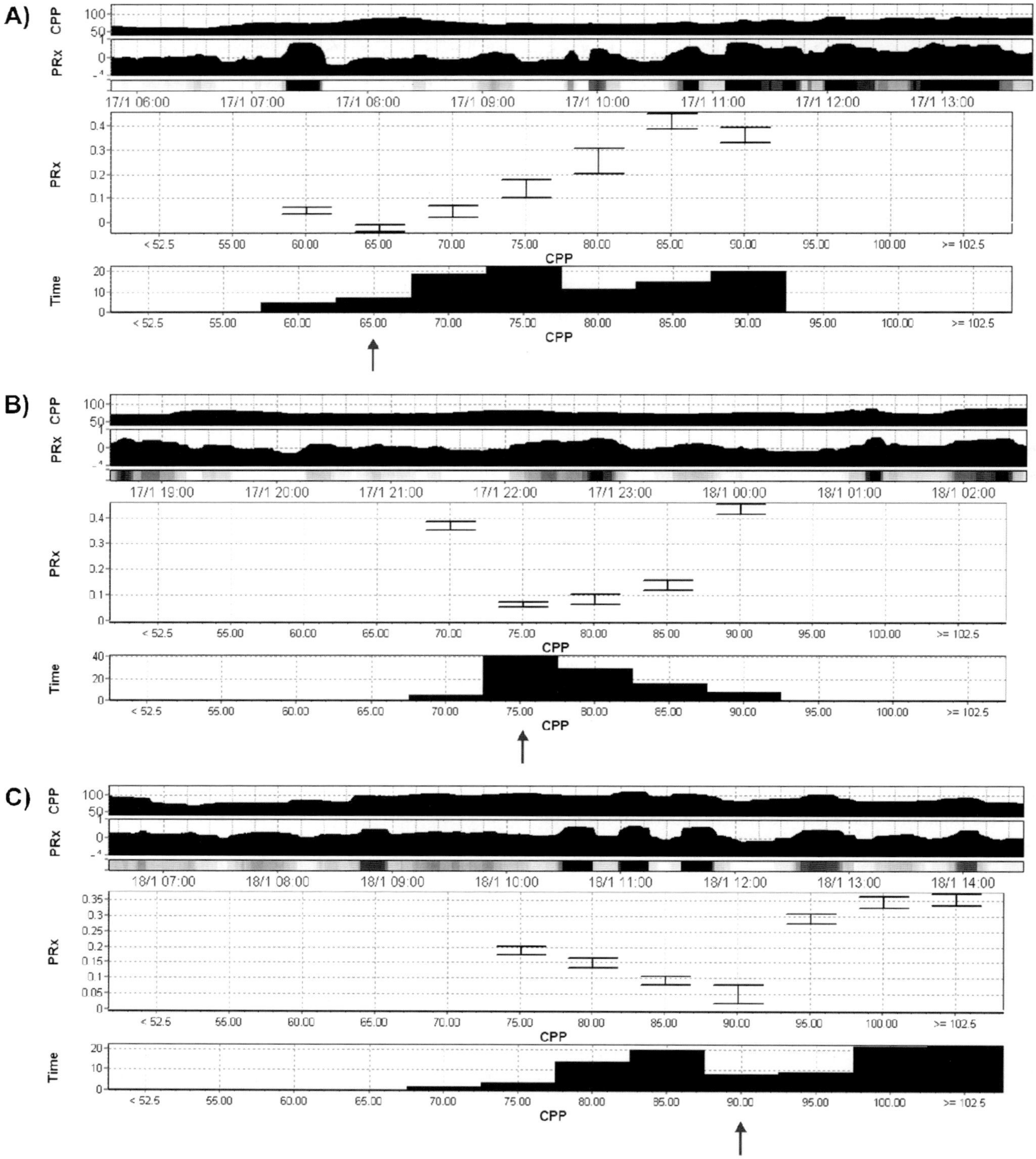

Fig. 2 a–c Screen shots examples showing CPP_{OPT} curves in one patient at different times during his stay on the unit. Note how CPP_{OPT} value (denoted with *vertical arrows* beneath the charts) shifted from 65 to 75 and then to 90 over a period of 24 h

Fig. 3 Example of automatic detection of artifacts associated with arterial line flushing. **a** Raw signals of ABP and ICP, with artifact periods marked with horizontal lines above it. **b** and **c** Result of ABP and ICP trends calculated from signals in **a**, with no artifact detection (**b**) and with (**c**). *Horizontal line* above charts shows periods affected by the artifacts. Note how removal of the corrupted section of the ABP signal changed the quality of the ICP-ABP correlation. The correlation coefficient *R* (PRx) changed from 0.09 to 0.7 ▶

Discussion

Practical application for autoregulation orientated therapy

Many attempts have been made to find an appraisal value for CPP, however, there is no method available currently

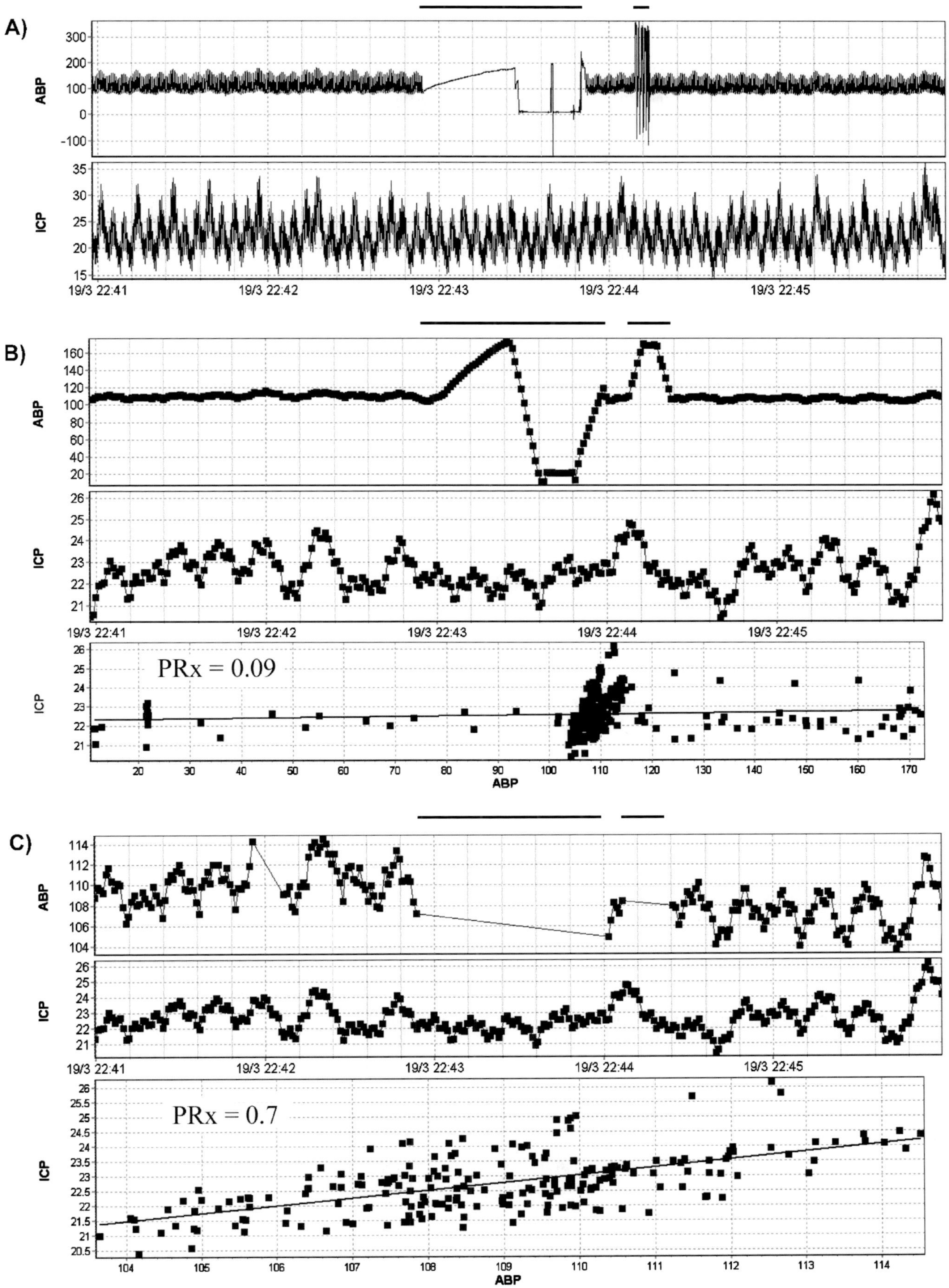
A)
ABP
ICP
19/3 22:41
19/3 22:42
19/3 22:43
19/3 22:44
19/3 22:45
B)
ABP
ICP
PRx = 0.09
ABP
C)
ABP
ICP
PRx = 0.7
ABP

that is accurate enough to be clinically useful. CPP_{OPT}-based management is one possible autoregulation-oriented strategy, as proposed in [9]. In contrast to a simple threshold above which CPP is held, CPP_{OPT}-guided therapy can help to avoid inappropriately high CPP. Excessively high CPP can lead to disruption of the blood brain barrier, development of vasogenic edema and may predispose to cardiac complications. Online, real-time assessment of the CPP_{OPT} may therefore help to optimise patient CPP in order to maintain cerebral perfusion in the most favourable state at all times. A randomised trial of this methodology is already under way in one of our collaborating centres and its results will be published in due course.

Software evolution

ICM+ software is undergoing constant development stimulated by continued multimodal monitoring research in Cambridge as well as, recently, other centres around the world using the software. Since 2004, when it was officially launched, the software has been considerably extended.

The new calculation engine contains many more signal processing functions but most of all it is rewritten in such a way that allows third party libraries to be registered to work with it. This potentially further extends the reach and use of the software as it allows engineers/physicists to experiment with advanced data analysis but at the same time keeps the complexity of it hidden from the more clinically minded researchers and clinicians.

There is now wider support for monitors that do not provide analog output including Phillips Intellivue, Datex-Ohmeda as well as Spiegelberg monitors and Sophysa ICP monitor. That support is growing as more centres are starting to use ICM+.

Automatic artifact treatment has been much improved. The main concept is based on specifying valid ranges for calculated parameters at any stage of processing. For example, in order to detect arterial line flushing, one can look for a period when pulse wave disappears which will manifest itself with the amplitude of the pulse wave dropping below the valid range. That information can then be used to remove the flagged section of the signal in question from further calculations thus potentially improving quality of secondary parameters like PRx (Fig. 3). The detection criterion expressions can be defined in the same way as any other secondary parameter, using expressions of available functions. That means the users can define their own formulas, independently for each signal they monitor, to deal with specific types of artifacts.

Growing interest in waveform analysis

Ever since the software was released on license by the Cambridge University there has been growing interest of neuro-centres around the world to use it. The main attraction for those centres is the fact that it allows them to join into advanced multimodal monitoring research even though they do not have their own support of physicists or engineers capable of providing them with the means to do it. On the other hand, open architecture of the analysis configuration allows more technically minded clinical fellows to modify supplied configurations, to create new analysis strategies or to improve the ones already available. So far there are 28 centres that are already using ICM+, and more are interested to join in. As the data recorded by all those centre is growing in volume, so is also opportunity for collaboration between them. This is particularly promising as most of those centres are very active academically and keen to pursue new concept in multimodal monitoring and diagnosis. Having a common platform for data collection and analysis greatly facilitates multi-centres projects.

Conclusion

ICM+ is a universal tool for clinical and academic purposes. Its flexibility and advanced signal processing features are specialized for the needs of multidisciplinary brain monitoring, and it is particularly well suited for investigations into cerebral haemodynamics.

Conflict of interest statement We declare that we have no conflict of interest.

References

1. Brady KM, Lee JK, Kibler KK, Smielewski P, Czosnyka M, Easley RB, Koehler RC, Shaffner DH (2007) Continuous time–domain analysis of cerebrovascular autoregulation using near-infrared spectroscopy. Stroke 38:2818–2825
2. Czosnyka M, Smielewski P, Piechnik S, Steiner LA, Pickard JD (2001) Cerebral autoregulation following head injury. J Neurosurg 95:756–763
3. Czosnyka M, Whitehouse H, Smielewski P, Kirkpatrick P, Pickard JD (1994) Computer supported multimodal monitoring in neuro intensive care. Int J Clin Monit Comput 11:223–232
4. Czosnyka Z, Czosnyka M, Owler B, Momjian S, Kasprowicz M, Schmidt EA, Smielewski P, Pickard JD (2005) Clinical testing of CSF circulation in hydrocephalus. Acta Neurochir Suppl 95:247–251
5. Jaeger M, Schuhmann MU, Soehle M, Meixensberger J (2006) Continuous assessment of cerebrovascular autoregulation after

traumatic brain injury using brain tissue oxygen pressure reactivity. Crit Care Med 34:1783–1788

6. Lavinio A, Timofeev I, Nortje J, Outtrim J, Smielewski P, Gupta A, Hutchinson PJ, Matta BF, Pickard JD, Menon D, Czosnyka M (2007) Cerebrovascular reactivity during hypothermia and rewarming. Br J Anaesth 99:237–244
7. Smielewski P, Czosnyka M, Steiner L, Belestri M, Piechnik S, Pickard JD (2005) ICM+: software for on-line analysis of bedside monitoring data after severe head trauma. Acta Neurochir Suppl 95:43–49
8. Smielewski P, Czosnyka M, Zabolotny W, Kirkpatrick P, Richards HK, Pickard JD (1997) A computing system for the clinical and experimental investigation of cerebrovascular reactivity. Int J Clin Monit Comput 14:185–198
9. Steiner LA, Czosnyka M, Piechnik SK, Smielewski P, Chatfield D, Menon DK, Pickard JD (2002) Continuous monitoring of cerebrovascular pressure reactivity allows determination of optimal cerebral perfusion pressure in patients with traumatic brain injury. Crit Care Med 30:733–738
10. Tseng MY, Al-Rawi PG, Czosnyka M, Hutchinson PJ, Richards H, Pickard JD, Kirkpatrick PJ (2007) Enhancement of cerebral blood flow using systemic hypertonic saline therapy improves outcome in patients with poor-grade spontaneous subarachnoid hemorrhage. J Neurosurg 107:274–282

Acta Neurochir Suppl (2008) 102: 153–156

Pathophysiology and diagnosis of spontaneous intracranial hypotension

K. Shima · S. Ishihara · S. Tomura

Abstract

Background Spontaneous intracranial hypotension (SIH) has become a well-recognized syndrome. However, diagnosis of SIH is still challenging. The problem with SIH is that the precise mechanism of cerebrospinal fluid (CSF) leakage remains largely unknown and there is no definite diagnostic criterion in the imaging.

Methods The clinical findings of our ten cases and 301 literature reports on SIH were investigated in a retrospective analysis to clarify the pathophysiology of CSF leakage, correlate the findings of imaging studies and determine the most adequate diagnostic criteria.

Results The events precede symptoms of SIH were categorized as traumatic, secondary and strictly spontaneous (62%). The location of the spinal CSF leak remains undetectable in approximately 50% of cases reported. In 93% of patients, the CSF leakage sites were detected at the cervical or thoracic level of the spine. On recent MRI studies, 88% of patients showed spinal epidural fluid collections that most likely represent CSF leakage. MR myelography using heavily T_2-weighted fast-spin-echo sequence can clearly demonstrate the site of CSF leakage. Although numerous treatment options are available, none of the treatments have been evaluated by randomized clinical trials. In 48% of papers, autologous epidural blood patch (EBP) was the treatment of choice in patients who have failed initial conservative management. Forty-nine percent of patients showed relief of symptoms after up to three repeated EBPs.

Conclusion We propose new diagnostic criteria of SIH to avoid misdiagnosis.

Keywords Spontaneous intracranial hypotension · Cerebrospinal fluid leak · CSF hypovolemia · Magnetic resonance imaging

Introduction

Spontaneous intracranial hypotension (SIH) is now defined as a syndrome of low cerebrospinal fluid (CSF) pressure (less than 60 mmH_2O) characterized by postural headaches in patients without any history of dural puncture or penetrating trauma [4, 7]. SIH is thought to result from CSF leak and consequently, in low CSF pressure [8]. The problem with SIH at the present moment is that the precise mechanism of CSF leak remains largely unknown and there is no definite criterion in the diagnostic neuroimaging [10].

The presence of SIH is confirmed by the findings of the cranial and spinal MR imagings and radioisotope (RI) cisternography. Frequent characteristics on the MR neuroimagings used to diagnose SIH include diffuse meningeal enhancement (80%), presence of subdural fluid collections and downward displacement of the cerebral structures. On RI cisternography findings of less radioactivity over the cerebral convexities (90%) and early accumulation of tracer in the bladder (65%) support the diagnostic of SIH [1]. Identification of the actual site of the CSF leak secures the diagnosis of SIH, but spinal MR imaging, RI cisternog-

K. Shima (✉) · S. Ishihara · S. Tomura
Department of Neurosurgery, National Defense Medical College, 3-2 Namiki,
Tokorozawa, Saitama 359-8513, Japan
e-mail: shima@ndmc.ac.jp

S. Ishihara
e-mail: shopar72@saitama-med.ac.jp

S. Tomura
e-mail: tomura@ndmc.ac.jp

K. Shima
Working Group of Intracranial Hypotension,
Japan Society of Neurotraumatology,
Tokyo, Japan

raphy and CT myelography identifies the location of the CSF leak in only about two thirds of patients overall [6].

In Japan, many patients with an intractable headache attributed to cervical spine injury, especially caused by a traffic accident, receive blood patch therapy based on the diagnosis of posttraumatic CSF hypovolemia.

We conducted this study to evaluate the diagnostic criteria for patients managed as a posttraumatic CSF hypovolemia in Japan. More specially, we aimed to propose the most adequate diagnostic criteria for CSF leaks and SIH.

Materials and methods

We studied 301 journal articles retrieved in MEDLINE (1968–2006) using the terms intracranial hypotension, CSF hypovolemia, CSF leak and low-pressure headache by the Working Group for Intracranial Hypotension, Japan society of Neurotraumatology. This group consisted of eight members selected by the past president and secretary, six reviewers selected from the institutions of members and one neuroradiologist. Reference lists of these articles were used. Selected articles were investigated in a retrospective analysis to clarify the existing evidence regarding etiology, pathogenesis, clinical features, methods of diagnosis, and treatment modalities of SIH, and verified the cause of diagnostic confusion for patients with symptoms and imaging findings that can mimic SIH in Japan.

The clinical findings of our ten patients with SIH also were studied.

Results

Of the 301 articles identified, 228 were English journals and 73 were Japanese. Most of the articles are reports on a small number of cases under nine patients. The events precede symptoms of SIH were categorized as traumatic (14%), secondary (24%) and strictly spontaneous (62%). The traumatic group included the patients who had symptoms soon after the history of head injury, cervical injury, fall, skiing, spinal operation, aerobics, epidural block, lumbar puncture or spinal chiropractic manipulation. Secondary group had a history of lumbar puncture, Marfan's syndrome, spondylosis, shunt operation, spinal operation, epidural block, stretch, racket sport or spinal chiropractic care [9].

Although the location of CSF leakage was described in 79% of all articles, only 49% showed the site of CSF leakage. In 96 (49%) of 195 cases demonstrating spinal CSF leakage, the location was detected at the cervical (42%), thoracic (51%) or lumbar (10%) level of the spine by radiological studies. A subgroup of recent spinal MRI studies, however, showed CSF leakage into the epidural space in 35 (88%) of 40 patients. Imaging findings such as subdural fluid collections, diffuse meningeal enhancement and downward displacement of the brain (brain sagging) on MRI and early bladder appearance on RI cisternography have been used as a very specific finding of SIH [10]. However, these MRI features are attributable to compensatory changes related to the low CSF pressure and are indirect findings of SIH. It is a well-known fact that early bladder appearance of RI cisternography can be observed even in normal case [3].

A variety of options from bed rest to surgical repair are available to treat patients with SIH. None of the treatments have been evaluated by randomized clinical trials. Conservative management and bed rest are the first line treatment. In 209 (48%) of 301 journal articles, the injection of autologous blood into the spinal epidural space, the so-called epidural blood patch (EBP) was the treatment of choice. The application of an EBP is the therapy of choice when conservative treatment fails [10]. The first EBP was effective in 58% of patients. Repeated EBP improve the efficacy to 77% (n=504). Forty-nine percent of patients showed relief of symptoms after repeated EBPs until the third time. In Japanese articles, however, although 16% of patients had repeated EBPs over four times, the relief of symptoms remained to 8% of patients. In our ten cases, seven patients were treated conservatively with bed rest and intravenous hydrogen and three patients were treated by intravenous factor XIII administration. The recurrence rate of spinal CSF leakage is low (approximately 10%), regardless of treatment [10]. Long-term outcomes of patients with SIH remain unclear.

Discussion

To date, there are still problems to resolve on SIH, though over 300 studies have described SIH.

Problems with SIH

1. The exact cause and pathogenesis of SIH remain unknown. The generally accepted pathogenesis is that of CSF leakage, which may occur due to the vulnerability of spinal nerve root sleeves [9].
2. Although CSF leakage is the most convincing evidence to diagnose SIH, exact site of the CSF leakage remains undetectable in approximately 50% of patients. Spinal MR imaging for identifying CSF leakage may be most available and least invasive examination for diagnosis

of SIH, but there is no standard sequence to definitely detect the CSF leak. At present, spinal fast-spin-echo MR imaging is the most useful examination for diagnosis of patients suspected of having SIH [5]. Although RI cisternography has been used extensively in the evaluation of SIH, the false-negative rate is 30% [1]. In addition, the early appearance of tracer in the bladder which is used as a typical finding is not pathognomonic sign for SIH [3].

3. SIH diagnostic criteria according to the revised International Classification of Headache Disorders Criteria [2] promote uncertain diagnosis and treatment of SIH in Japan. Although the sensitivity of RI cisternography is not high and the punctured area can be pseudopositive of CSF leak, this diagnostic modality is selected as a preferential imaging method. In addition, the response to treatment with EBP is helpful in confirming the diagnosis of SIH. Most cases of SIH resolve with conservative management and bed rest.

New diagnostic criteria of SIH

We proposed new diagnostic criteria for SIH at the 65th Annual Meeting of the Japan Neurosurgical Society on October 20, 2006 (Table 1). Clinical and imaging characteristic findings of SIH were classified according to the evidence level into three groups: essential symptoms, major criteria and minor criteria. In the classification system, RI cisternography is a minor, not a major, diagnostic criteria. Furthermore, the patient should be diagnosed as a traumatic intracranial hypotension only if the onset of the patient's symptom is within 30 days after trauma and there is no other cause than trauma. We also proposed the algorithm for diagnosis of traumatic intracranial hypotension at the 30th Annual Meeting of the Japan Society of Neurotraumatology on March 16, 2007 (Fig. 1).

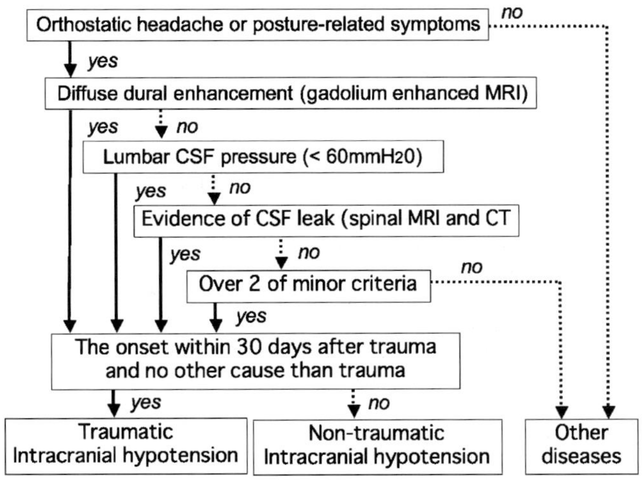

Fig. 1 Algorithm for diagnosis of traumatic intracranial hypotension (Working Group for Intracranial Hypotension, Japan Society of Neurotraumatology 2007)

In conclusion, SIH is a well-recognized syndrome, but consensus on the diagnosis and management of SIH has not been achieved. From our studies, we propose a classification system (Table 1) and an algorithm (Fig. 1) for the diagnosis of SIH. Further research to explore the precise cause of spinal CSF leakage and randomized controlled trials for EBP treatment are required.

Table 1 Diagnostic criteria of spontaneous intracranial hypotension (SIH; Working Group for Intracranial Hypotension, Japan Society of Neurotraumatology 2006)

Diagnostic criteria	
Essential symptoms	Orthostatic headache that worsen within 15 min
	Posture-related associated symptoms[a]
Major criteria	Diffuse dural enhancement on gadolinium-enhanced MRI
	CSF leakage on spinal MRI or CT myelography
	CSF opening pressure less than 60 mmH_2O
Minor criteria	Spinal meningeal diverticula or fluid collections on spinal MRI
	Venous engorgement on cranial and/or spinal MRIs
	Early accumulation of tracer in the bladder or less activity over the cerebral convexities on RI cisternography
	Subdural fluid collections on cranial MRI
	Descent of cerebellar tonsils and flattening of brainstem on cranial MRI
	Enlargement of pituitary gland on cranial MRI
	Decrease in size of ventricles and effacement of basal cisterns on cranial MRI

The diagnosis of SIH is made with certainly when a patient presents with essential symptoms accompanied by at least one major criteria and three of minor criteria

[a] Neck pain, nausea, diplopia, visual blurring, upper limb numbness or pain, etc.

Conflict of interest statement We declare that we have no conflict of interest.

References

1. Chung SJ, Kim JS, Lee MC (2000) Syndrome of cerebral spinal fluid hypovolemia: clinical and imaging features and outcome. Neurology 55:1321–1327
2. Headache Classification Subcommittee of the International Headache Society (2004) The international classification of headache disorders, 2nd edition. Cephalgia 24(Suppl 1):1–160
3. Ishihara S, Otani N, Shima K (2003) Spontaneous intracranial hypotension (SIH): the early appearance of urinary bladder activity in RI cisternography is a pathognomonic sign of SIH? Acta Neurochir 86(Suppl):587–589
4. Ishihara S, Fukui S, Otani N, Miyazawa T, Ohnuki A, Kato H, Tsuzuki N, Nawashiro H, Shima K (2001) Evaluation of spontaneous intracranial hypotension: assessment on ICP monitoring and radiological imaging. Br J Neurosurg 15:239–241
5. Matsumura A, Anno I, Kimura H, Ishikawa E, Nose T (2000) Diagnosis of spontaneous intracranial hypotension by using magnetic resonance myelography. J Neurosurg 92:873–876
6. Mokri B, Piepgras DG, Miller GM (1997) Syndrome of orthostatic headaches and diffuse pachymeningeal gadolinium enhancement. Mayo Clin Proc 72:400–413
7. Rando TA, Fishman RA (1992) Spontaneous intracranial hypotension: report of two cases and review of the literature. Neurology 42:481–487
8. River Y, Schwartz A, Gomori JM, Soffer D, Siegal T (1996) Clinical significance of diffuse dural enhancement detected by magnetic resonance imaging. J Neurosurg 85:777–783
9. Shievink WI (2000) Spontaneous spinal cerebrospinal fluid leaks: a review. Neurosurg Focus 9:8
10. Schievievink WI (2006) Spontaneous spinal cerebrospinal fluid leaks and intracranial hypotension. JAMA 295:2286–2296

PART 3:
Advanced neuromonitoring

Acta Neurochir Suppl (2008) 102: 159–163

Intraoperative infrared brain surface blood flow monitoring during superficial temporal artery–middle cerebral artery anastomosis in a patient with moyamoya disease: clinical implication of the gradation value in postoperative clinical course – A case report

Atsuhiro Nakagawa · Miki Fujimura · Tatsuhiko Arafune · Hideaki Suzuki · Ichiro Sakuma · Teiji Tominaga

Abstract

Background Superficial temporal artery–middle cerebral artery (STA–MCA) anastomosis is a safe and effective treatment for moyamoya disease. Symptomatic cerebral hyperperfusion is a potential complication of this procedure, especially in adult cases. Accurate diagnosis of postoperative hyperperfusion is important because its treatment is contradictory to that for ischemia. Intraoperative techniques to detect hyperperfusion are still lacking.

Methods We performed intraoperative infrared (IR) brain surface monitoring in a 36-year-old man who underwent left STA–MCA anastomosis.

Findings IR monitoring not only detected the patency of bypass, as also confirmed by conventional Doppler sonography and postoperative magnetic resonance angiography, but also delineated the local brain surface hemodynamics after revascularization. Analysis of gradation value disclosed an abnormal increase in brain surface cerebral blood flow (indirectly indicated as a temperature change) after removal of the temporary clip. The patient suffered from transient right upper extremity numbness and dysarthria due to focal hyperperfusion from postoperative days2 through 6. Intensive blood pressure control completely relieved his symptoms, and he was discharged without neurologic deficit.

Conclusions Intraoperative brain surface monitoring by IR imaging may be useful to predict cerebral hyperperfusion after revascularization surgery for moyamoya disease. Further evaluation with a larger number of patients is necessary to validate this technique.

Keywords EC–IC bypass · Intraoperative monitoring · Minimally invasive neurosurgery · Neurocritical care

A. Nakagawa (✉) · M. Fujimura · H. Suzuki · T. Tominaga
Departments of Neurosurgery, Graduate School of Medicine, Tohoku University,
1-1, Seiryo-machi, Aoba-ku,
Sendai, Miyagi 980-8574, Japan
e-mail: tonkan@zc5.so-net.ne.jp

T. Arafune · I. Sakuma
Bio-Medical Precision, Engineering Laboratory,
Institute of Environmental Studies,
Graduate School of Frontier Sciences, University of Tokyo,
7-3-1, Hongo,
Bunkyo-ku, Tokyo 113-8656, Japan

Introduction

Thermography using infrared (IR) imaging is an established technique for studying the surface temperature of human organs. With the advances in IR technology, including detective wavelength, cooling system, filters, as well as data processing methods, intraoperative thermal artery imaging became available. It is a unique method both for morphological evaluation and functional monitoring of superficial vessels [8]. We have developed an IR system with detectable bands located in the range 7–14μm for neurosurgical procedures. In our preliminary experience, we performed intraoperative monitoring of superficial temporal artery–middle cerebral artery (STA–MCA) anastomosis procedures in beagles to determine the patency of the bypass as well as the hemodynamics within small vessels (up to 0.5mm) and cortical cerebral blood flow (CBF) [4].

Moyamoya disease is a chronic, occlusive cerebrovascular disease with unknown etiology characterized by bilateral steno-occlusive changes at the terminal portion of the internal carotid artery and an abnormal vascular network at the base of the brain. Nearly half of the patients with Moyamoya manifest as ischemic attacks during pediatric period and rest of them manifest both as ischemic and hemorrhagic attacks in adults. Surgical revascularization is believed to be beneficial to prevent cerebral ischemic attacks by improving CBF [2, 3]. STA–MCA anastomosis with or without indirect bypass is generally employed. Conventional modalities suitable for comprehensive, visualized evaluation of the hemodynamics of the entire surgical field have not been available. In addition, intraoperative changes of surface CBF and its correlation with postoperative course are not well understood. We present an adult case of moyamoya patient who underwent intraoperative IR surface CBF monitoring during STA–MCA anastomosis. Intraoperative findings are presented and their implications for postoperative course are discussed.

Case report

A 36-year-old man presented with frequent transient ischemic attack (TIA). Magnetic resonance angiography (MRA) satisfied the diagnostic criteria of moyamoya disease. Preoperative *N*-isopropyl-*p*-[^{123}I] iodoamphetamine single-photon emission computed tomography (^{123}I-IMP-SPECT) showed his left CBF and cerebrovascular reserve capacities were markedly impaired, so left bypass surgery was planned.

Intraoperative IR monitoring: After completion of anastomosis, an IRIS V IR imaging system (Sparkling photon Inc., Tokyo, Japan) (Fig. 1), with a high-resolution infrared camera, was set 30cm above brain and surface temperature was continuously monitored during placement and release of the temporary clip. The system has two cameras (IR-band camera and visible-band camera) attached to a head unit with moving arm (whole length = 800mm). IR focal plane array detector (barium strontium titanate; pixel size: 320 × 240pixels) shows the area of 110 × 82mm using F50 IR lens. IR sensor element's sensitivity wavelength is 7 to 14μm. The sensor's recording speed is 30 frames per second and output signal is 8bit (256 gray scale gradation). The range of measuring temperature is confined to 25°C to 45°C (1 gradation value = 0.08°). Visible-band camera has the functions of auto focus and manual zoom. All of the images were stored in the installed computer and recorded with a digital video device. Obtained images were analyzed with imaging software by means of changes in gradation value [5, 6, 9]. *Signal processing*: To capture the image area that had temperature change, the image processing software "Opmap", which was originally developed for optical mapping of the cardiac action potential [1, 10], was modified to enable visualization of surface CBF. In the

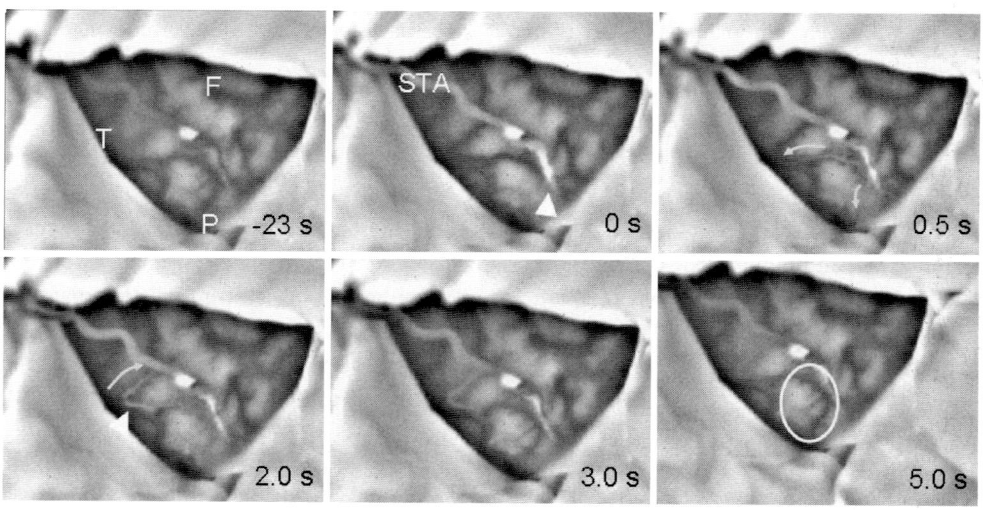

Fig. 1 Intraoperative infrared (IR) monitoring by IRIS V IR imaging system disclosed changes in color of bypass to white after temporary occlusion (indicating decrease in temperature) at 0 s compared to that 23 s before, and then changed to black (indicating increase in temperature) after removal of clip at 0 s, indicating presence of blood flow and patency of bypass. The IR images also disclosed that the white blood flow distributed to both distal and proximal direction of M4 at 0.5 s although backward flow seemed to be dominant. Note that the surrounding brain around anastomosis site slightly changed toward black locally indicating significant temperature increase till 5 s after removal of temporary clip (*circle*). *Arrow* flow direction, *arrowhead* site of anastomosis, *STA* superficial temporal artery, *F* frontal lobe, *P* parietal lobe, *T* temporal lobe

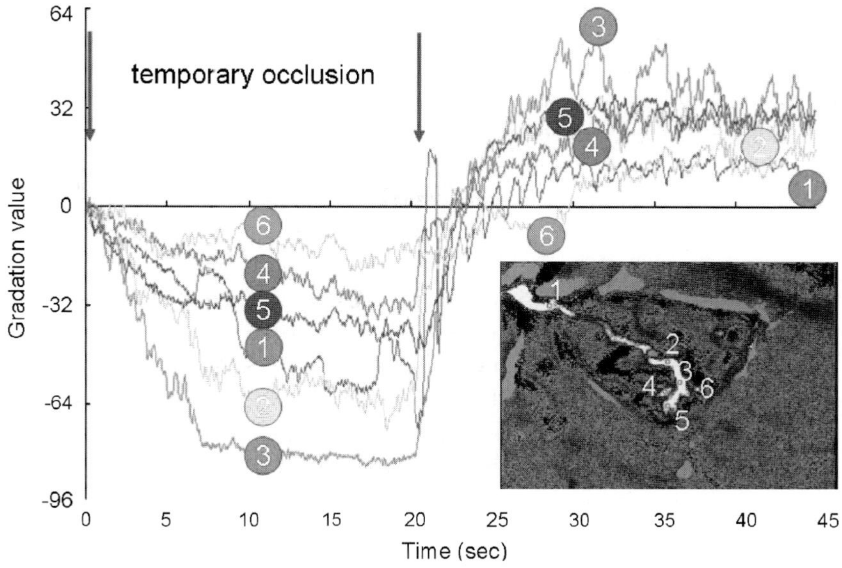

Fig. 2 The changes of gradation value evaluated by imaging software demonstrated different time sequential change of gradation value at various points. The gradation value on bypass graft as well as surrounding brain surface decreased during temporary clip application indicating temperature decrease. However, after removal of temporary clip a significant increase of gradation value indicating an increase in temperature could be observed locally at the brain surface around anastomosis site (*point 5*)

original IR movie recorded by IRIS V, white represents low temperature and black represents high temperature (8bit grayscale). Opmap process software converts the IR movie source data to a color gradation movie that emphasizes the areas with significant temperature changes. *Image process algorithm*: Using Opmap, we normalized and reversed the IR luminance value. Firstly the max variation width of IR luminance (I_{Max}) was determined, and the processed value of the gradient $G(f)$ was given by calculating the following formula. We calculated (*1) in all frame and all pixels:

$$G(f) = (I(f_0) - I(f))/I_{Max} \ldots (*1)$$

f : frame number$/f_0$: frame number of process started/

$I(f)$: luminance value at f[frame]

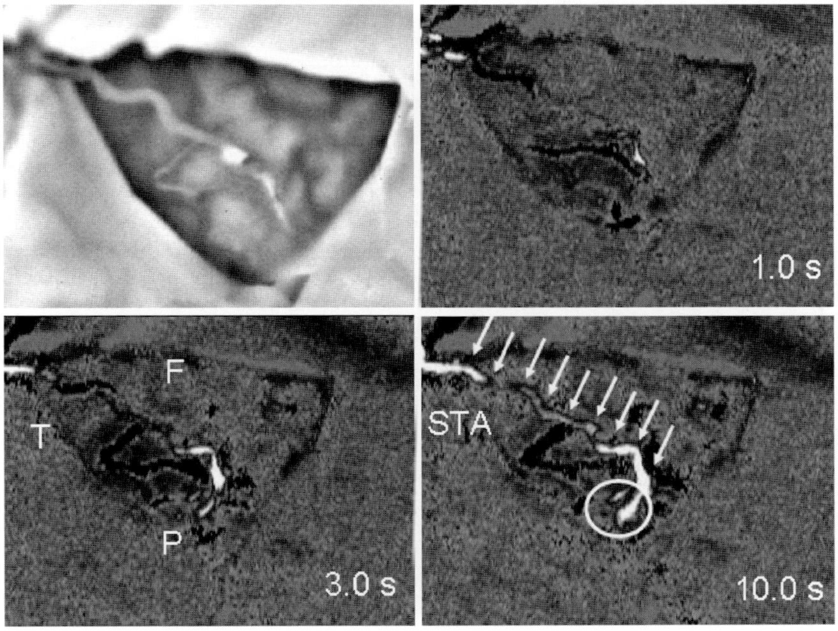

Fig. 3 Imaging analyzed by Opmap showing the area that showed gradation value change over 60 points compared to that at the beginning of temporary clip. In this imaging, area showing increase of gradation value is delineated *white*, and are showing no change is delineated as *black*. *Upper left* original IR imaging taken by IRIS V IR imaging system. After removal of temporary clip, the bypass portion showed increase in gradation value (indicating increase in temperature) at 1 to 3 s. Significant increase in gradation value, indicating increase in temperature and indirectly increase in CBF, was recognized at both vessel (*arrow*) and adjacent brain surface (*circle*) at 10 s after removal of temporary clip

Process application: Moderate I_{Max} values between 45 and 65 were determined to easily show the thermographic change. Opmap can selectively emphasize the area we hope to see by selecting f_0 frame for any purpose. In case of an observation of perfusion after removal of clip at the blood vessel, f_0 should be during clipping frames. On the other hands, f_0 should be before clipping frames for checking the temperature rise after the clip procedure. *Intraoperative IR findings*: The recipient artery at the M4 segment of the anterior parietal branch of the left MCA was explored and anastomosis performed between the stump of the STA and the M4 segment that supplied the parietal lobe. Additional encephalo-duro-myo-synangiosis and dural pedicle insertion [7] were performed. Intraoperative IR monitoring disclosed changes in color in the bypass territory to white after temporary occlusion and placement of a temporary clip (indicating decrease in temperature), and then changes to black (indicating increase in temperature) after removal of the clip, indicating presence of blood flow and patency of the bypass. The images also disclosed that the white blood flow distributed to both proximal and distal directions of M4 at 0.5s. The entire exposed brain progressively changed slightly toward black till the end of monitoring, which was 5s after removal of temporary clip (Fig. 1). Gradation value analysis showed decrease both at bypass and in the surrounding brain surface during placement of temporary clip, but significantly increased after its removal (Fig. 2). Opmap clearly showed areas with significant increase in brain surface temperature both at vessel (arrow) and at adjacent brain surface (circle) following removal of the temporary clip (Fig. 3). *Postoperative course*: Postoperative MRI/MRA showed no ischemic changes, and ^{123}I-IMP-SPECT showed increase in left CBF around the site of anastomosis 1 day after surgery, when the patient showed no evidence of neurological deficit. However, he suffered from fluctuating dysarthria, numbness and fine movement dysfunction on his right upper extremity from postoperative day 2 to 6. The anatomical location and the temporal profile of his CBF increase were in accordance with the neurologic deficits. Taken together with the absence of ischemic changes and brain surface compression, we made diagnosis of symptomatic hyperperfusion. Intensive blood pressure control relieved his symptoms, and he was discharged without neurological deficit at 11 days after surgery.

Discussion

The results of this study suggest that IR monitoring is useful not only for confirming bypass patency, but also for evaluation of local cerebral hemodynamics. The clinical implication of changes in intraoperative brain surface hemodynamics is still obscure, but it might reflect some important pathophysiology during the revascularization to the chronically ischemic brain. In our previous study, a patient who showed an increase in blood flow (i.e. increase in gradation value) by IR imaging had a tendency to suffer symptomatic postoperative hyperperfusion [5]. We suggest that the occurrence of postoperative complication might be predicted by analyzing changes in the gradation value. Extensive explorations still need to be performed to find appropriate parameters and cut off values for analyzing intensity signals, as well as accumulating data on more patients. However, the present system demonstrates the potential of using Opmap analysis in the future to show postoperative risk and important intraoperative information to the surgeon.

Okada and colleagues also applied thermography to evaluate the temperature distribution of cortical surface at the operative field during the temporary occlusion of bypass, and reported that thermography is useful not only to demonstrate the distribution of blood flow through the extracranial–intracranial bypass but also to quantitatively evaluate the regional CBF changes in the operative field [6]. It is now considered that cortical brain temperature is determined by metabolism and blood flow coupling, and sensitive infrared imaging might be a useful measure of ischemia [9]. Although the relationship between gradation value and surface blood flow using Doppler is clearly demonstrated as reported by Watson et al. [9], it is still necessary to further elucidate the exact relationship between the present IR system and conventional tools in the future.

Acknowledgements This work was supported in part by a Grant-in-Aid for Scientific Research (B) (no. 18390388), (no. 19399372), Grant-in-Aid for Young Scientist (A) (no. 19689028) offered by the Japanese Ministry of Education, Culture, Sports, Science, and Technology, and Mitsubishi Pharma Research Foundation. We also acknowledge Etsuko Kobayashi, Ph.D., Takahiro Yamaguchi, Ph.D., and Teruko Sakurai of Bio-Medical Precision, Engineering Laboratory, Institute of Environmental Studies, Graduate School of Frontier Sciences, the University of Tokyo, Tokyo, Japan, for analysis of infrared imaging.

Conflict of interest statement We declare that we have no conflict of interest.

References

1. Arafune T, Mishima A, Sakuma I, Inada H, Shibata N, Nakagawa H, Yamazaki M, Honjo H, Kodama I (2003) Virtual electrode-induced spiral reentry in ventricular myocardium perfused in-vitro. Environ Med 47:72–75
2. Fujimura M, Mugikura S, Shimizu H, Tominaga T (2006) Diagnostic value of perfusion-weighted MRI for evaluating

postoperative alteration of cerebral hemodynamics following STA–MCA anastomosis in patients with moyamoya disease. No Shinkei Geka 34:801–809

3. Fujimura M, Kaneta T, Mugikura S, Shimizu H, Tominaga T (2007) Temporary neurologic deterioration due to cerebral hyperperfusion after superficial temporal artery–middle cerebral artery anastomosis in patients with adult-onset moyamoya disease. Surg Neurol 67:273–282
4. Nakagawa A, Hirano T, Uenohara H, Sato M, Kusaka Y, Shirane R, Takayama K, Yoshimoto T (2003) Intraoperative thermal artery imaging of an EC–IC bypass in beagles with infrared camera with detectable wave-length band of 7–14 micron: possibilities as novel blood flow monitoring system. Minim Invasive Neurosurg 46:231–234
5. Nakagawa A, Fujimura M, Suzuki H, Ohki T, Takayama K, Tominaga T (2007) Prediction of postoperative cerebral hyperperfusion in patients with moyamoya disease by intraoperative monitoring of cerebral hemodynamics using IRIS-V infrared imaging system: two case report. Surgery for Cerebral Stroke 35:136–141
6. Okada Y, Kawamata T, Kawashima A, Hori T (2007) Intraoperative application of thermography in extracranial–intracranial bypass surgery. Neurosurgery 60:362–366
7. Shirane R, Yoshida Y, Takahashi T, Yoshimoto T (1997) Assessment of encephalo-galeo-myo-synangiosis with dural pedicle insertion in childhood moyamoya disease: characteristics of cerebral blood flow and oxygen metabolism. Clin Neurol Neurosurg 99(Suppl 2):S79–S85
8. Suma H, Isomura T, H, orii T, Sato T (2000) Intraoperative coronary artery imaging with infrared camera in off-pump CABG. Ann Thorac Surg 70:1741–1742
9. Watson JC, Gorbach AM, Pluta RM, Rak R, Heiss JD, Oldfield EH (2002) Real-time detection of vascular occlusion and reperfusion of the brain during surgery by using infrared imaging. J Neurosurg 96:918–923
10. Yamazaki M, Honjo H, Nakagawa H, Ishiguro YS, Okuno Y, Amino M, Sakuma I, Kamiya K, Kodama I (2007) Mechanisms of destabilization and early termination of spiral wave reentry in the ventricle by a class III antiarrhythmic agent, nifekalant. Am J Physiol Heart Circ Physiol 292:539–548

Acta Neurochir Suppl (2008) 102: 165–169

Clinical study of craniospinal compliance non-invasive monitoring method

A. Ragauskas · G. Daubaris · V. Petkus · R. Sliteris · R. Raisutis · I. Piper · S. Rocka · E. Jarzemskas · V. Matijosaitis

Abstract

Background The ability to quantify non-invasively the effect of posture on intracranial physiology by using cine phase-contrast MRI may lead to the development of new diagnostic tests to evaluate such functions as regulation of CBF and ICP, and the effect of pathologies on these functions.

Methods Results similar to MRI technology can be obtained using non-invasive ultrasonic method (Vittamed) for intracranial blood volume pulse wave (IBVPW) measurement and intracraniospinal compliance (ICC) monitoring.

Findings IBVPW have been investigated in supine and upright positions of healthy volunteers using Vittamed technology. A group of 13 healthy volunteers (nine females, four males, mean age 25.1±3.4) was studied. More than 3,000 IBVPW were analysed in order to show the difference of shape and amplitude in supine and upright positions. Averaged shape of ten IBVPW waves was presented in the normalized window with dimensions 1.0×1.0.

Conclusions The results show significant difference between averaged IBVPW shapes in upright (highest intracraniospinal compliance) and supine (lower intracraniospinal compliance) body positions. Body posture caused IBVPW subwave P2 and P3 changes $\Delta P2=18\%$ and $\Delta P3=11\%$. Amplitude of IBVPW in upright body position was significantly higher than in the supine one. The value of IBVPW amplitude's ratio in supine and upright positions was 1.55±0.61.

A. Ragauskas (✉) · G. Daubaris
Telematics Scientific Laboratory,
Kaunas University of Technology,
Studentu 50-451a,
LT-51368 Kaunas, Lithuania
e-mail: telematics@ktu.lt

G. Daubaris
e-mail: gdauba@ktu.lt

V. Petkus · R. Sliteris · R. Raisutis
Vittamed Technologijos Ltd.,
V. Putvinskio 47-10,
LT-44243 Kaunas, Lithuania

V. Petkus
e-mail: info@vittamedtechnologijos.lt

R. Sliteris
e-mail: info@vittamedtechnologijos.lt

R. Raisutis
e-mail: info@vittamedtechnologijos.lt

I. Piper
Department Clinical Physics, 5th Floor,
Institute of Neurological Sciences Southern General Hospital,
1345 Govan Road,
Glasgow, UK G514TF
e-mail: ipiper@clinmed.gla.ac.uk

S. Rocka · E. Jarzemskas
Vilnius University Emergency Hospital,
Siltnamiu 29,
LT-04130 Vilnius, Lithuania

V. Matijosaitis
Department of Neurology, Kaunas University of Medicine,
Eiveniu 2,
LT-50009 Kaunas, Lithuania

Keywords ICP · Intracranial blood volume pulse waves · Non-invasive monitoring · Craniospinal compliance

Introduction

Body posture strongly affects intracranial hydrodynamics and cerebral hemodynamics. The link between posture-

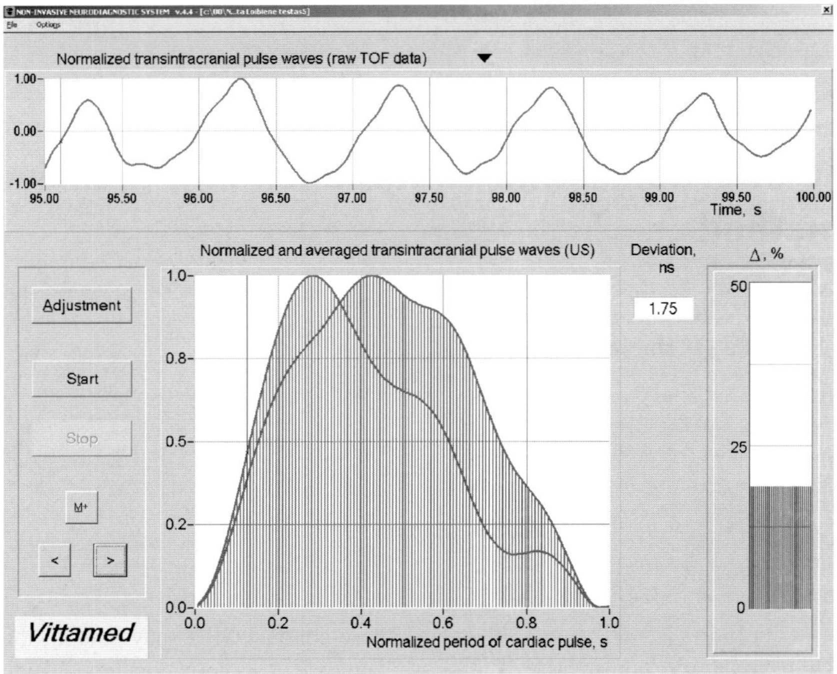

Fig. 1 The display panel of the non-invasive Vittamed monitor for intracranial blood volume pulse wave shape comparison. Two IBV pulse waves are shown in normalised window with dimensions 1.0× 1.0: reference wave (*left wave*) in upright body position, the wave under comparison (*right wave*) in supine body posture when ICP has been elevated

related changes in cerebral hemodynamics, intracranial hydrodynamics, and paterns of venous drainage can be quantified by cine phase-contrast MRI [1, 19]. The ability to non-invasively quantify the effect of posture on intracranial physiology may lead to the development of new diagnostic tests to evaluate such functions as regulation of CBF and ICP, and the effect of pathologies on these functions. It has been shown that in the posture change of

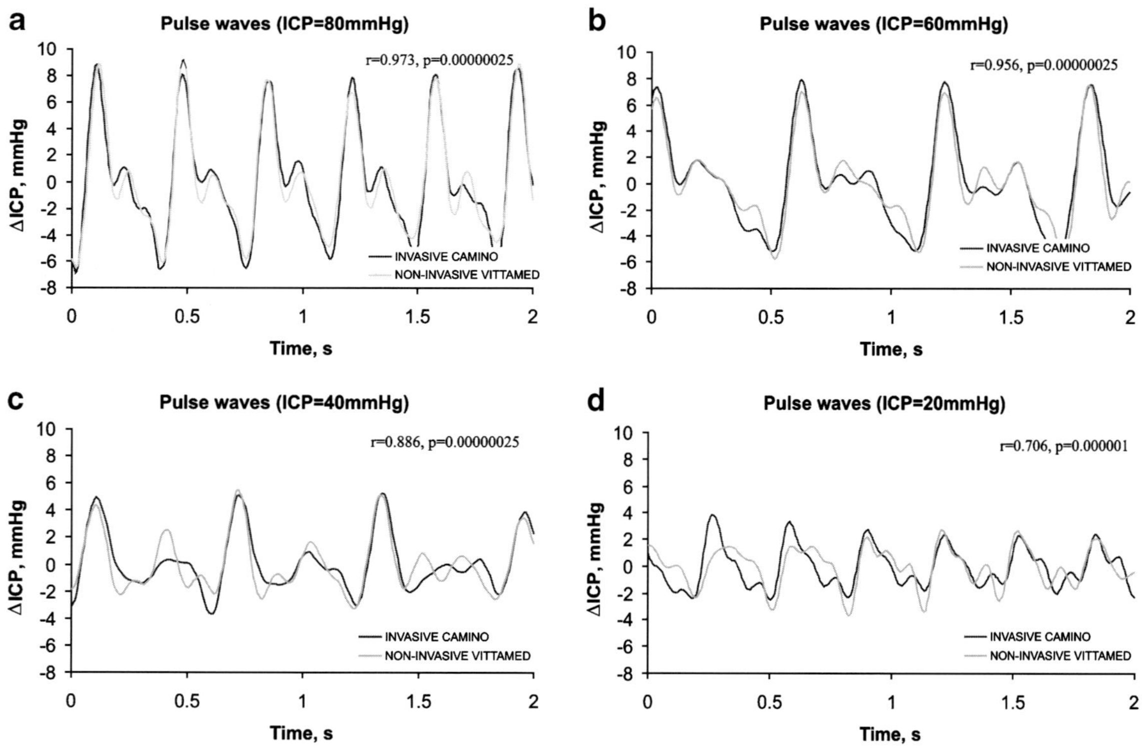

Fig. 2 Simultaneous invasive and non-invasive records of ICP pulse waves when ICP=80 mmHg (**a**), ICP=60 mmHg (**b**), ICP=40 mmHg (**c**) and ICP=20 mmHg (**d**) applying invasive Camino and non-invasive Vittamed monitors (TBI patient in coma)

healthy volunteers from supine to upright postures the intracraniospinal compliance changes up to 2.8 times when ICP changes from 10.6±3.6 to 4.5±1.82 mmHg [1]. The study [1] showed that posture-related changes in ICC and ICP have a great impact on cerebral blood as well as on CSF circulation. It also has been shown in the original MRI study that the shape of IBV pulse waves is strongly related to the intracraniospinal compliance [1].

In our study we found that results similar to those form MRI technology results can be obtained using non-invasive ultrasonic method (Vittamed) for intracranial blood volume pulse wave (IBVPW) measurement [2–18, 20]. The method has been tested using simultaneous invasive ICP and non-invasive IBV wave monitoring of TBI patients. A study on piglets also has been performed, and body posture effect on ICP and IBV pulse wave shape has been investigated [18]. 13 patients with TBI were monitored following Clinical Research Protocol No. 99124006, AIBS No. 990135, HSSRB log No. A-9676. Eighty-seven hours of simultaneous monitoring of invasive and non-invasive data were analyzed [2, 8]. Diagnostic value of IBV pulse waves has been investigated in our previous studies [4–6, 9–14, 20].

A total of 75 patients were examined using Vittamed technology and these included cases of acute, chronic and stabilized hydrocephalus, spinal cord injury and terminal blood flow [2, 6, 20]. These were compared to a control group of 53 healthy volunteers. A detectable change in IBV pulse waveform shape was observed in situations when disturbance in intracranial hydrodynamics was present, e.g., during hypoventilation tests, in cases of terminal blood flow and hydrocephaly, depicting the level of hydrocephalus activity and the patient's compensatory capabilities as well as the effect of treatment [2, 6, 20].

The objective of the present study was to investigate changes in IBV pulse wave shape of healthy volunteers in supine and upright body positions using new ultrasonic Vittamed monitor 105.

Materials and methods

A non-invasive method of intracranial blood volume measurement using ultrasound is based upon the transmission of short ultrasonic pulses from one side of the skull to the other and dynamic measurements of the time-of-flight of ultrasonic pulses. The time-of-flight depends on the acoustic properties of intracranial blood, brain tissue and cerebrospinal fluid. Changes in any of these components' volume will change the time-of-flight [15].

Following previous clinical studies new Vittamed monitor model 105 has been created (Fig. 1). It has been used to measure and record waveforms at a sampling frequency of 25 Hz. In order to decrease the influence of heart rate changes and respiratory modulation on the measured waveforms, the measured IBV pulse waves were averaged over at least three respiratory cycles and normalized to the peak value. What we have termed the normalization window in a 1.0×1.0 dimensionless grid, is shown in Fig. 1 and readily allows comparison of different IBV pulse wave recordings.

A group of 13 healthy volunteers (nine females, four males, mean age 25.1±3.4) were studied in supine and upright body positions. They had no history of headaches, meningitis, head trauma, neurosurgical operations or any other symptoms of elevated ICP. Volunteers were investigated following the protocol beginning in upright (standing) position with 3 min data recording test.

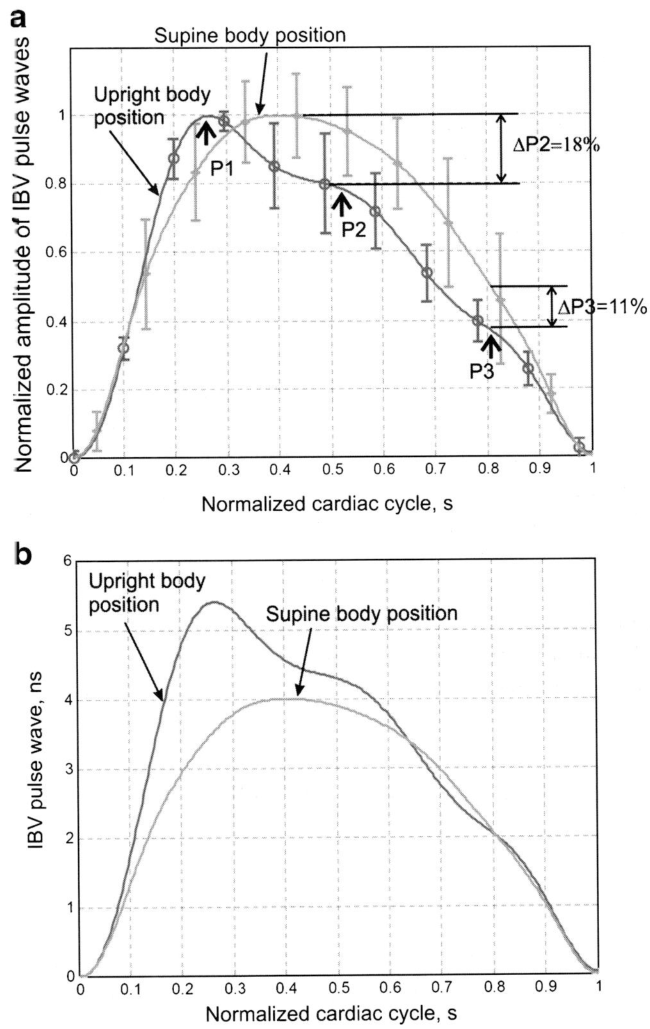

Fig. 3 IBV pulse wave averaged shapes of the group of 13 healthy volunteers in upright and supine body positions: **a** with marked P1, P2 and P3 subwaves and their differences ΔP2=18% and ΔP3=11% which were caused by body posture. *Vertical bars* show physiological fluctuations of pulse wave shape expressed as ±SD for all the group of healthy volunteers; **b** non-normalized IBV pulse waves in upright and supine body postures

A 3-min resting time interval was used after taking supine body position. After that, a 3-min recording was performed.

Before the recording, a mechanical frame with ultrasonic transducers and sonopads has been fixed on the human head in the intraventricular acoustic path position. An electrocardiographic three-lead channel was used for synchronization of each IBV pulse wave recording. Pulse waves which were distorted by artefacts have been automatically excluded from further waveform analysis by software of Vittamed 105 monitor.

Arterial blood pressure was measured in upright and supine positions. More than 3,000 IBV pulse waves were analyzed in order to show the difference of IBV pulse wave shape and amplitude in supine and upright positions.

Results

The results of comparison of simultaneous invasive and non-invasive (after linear conversion) ICP pulse waves monitoring in ICU are shown in Fig. 2. As follows from Fig. 2 correlation factor range is from r=0.973, p= 0.00000025 (ICP=80 mmHg) to r=0.706, p=0.000001 (ICP=20 mmHg). The possibility to obtain information about pulse wave shape non-invasively is evident.

The difference of averaged IBV pulse wave shape and its amplitude of the group of 13 healthy volunteers in supine and upright body positions is shown in Fig. 3a and b. Arterial blood pressure in the group of healthy volunteers was measured in both body positions. It was 119±9 and 74±7 mmHg in supine position and 122±10/78±8 mmHg in upright position.

Results show significant difference of IBV pulse wave shapes in upright (highest intracraniospinal compliance) and supine (lower intracraniospinal compliance) body positions. Body posture caused IBV pulse waves subwaves P2 and P3 changes ΔP2=18% and ΔP3=11% for the group of healthy volunteers (Fig. 3a). The differences are statistically significant with a p value of 0.00001 or smaller.

Discussion

The amplitude of IBV pulse wave shape in upright body position was significantly higher than in supine body position (the averaged value of the amplitude's ratio was 1.55±0.61). Such differences in IBV pulse wave shapes (Fig. 3a) and amplitudes (Fig. 3b) can be explained by physiological changes of arterial/venous blood and CSF volumetric waves in different body positions. It has been discovered in previous studies that total venous outflow in upright position is lower comparing to that of supine position [1]. Slightly lower total cerebral blood flow and smaller CSF volume were found to be in the sitting position [1].

The effect of posture on intracraniospinal physiology can be quantified by ultrasonic Vittamed IBV pulse wave measurement method. It has shown that ultrasonically measured IBV pulse wave shape and amplitude depend on the body posture and intracraniospinal compliance changes. Ultrasonically measured posture-related changes of IBV pulse wave shape reflect the intracraniospinal hydrodynamics and cerebral hemodynamics.

Acknowledgements The research was supported by the US Department of Defence (Agreement DAMD17-00-2-0065) and EU Structural Funds (Project BPD04-ERPF-3.1.7-03-05/0020).

Conflict of interest statement We declare that we have no conflict of interest.

References

1. Alperin N et al (2005) Quantifying the effect of posture on intracranial physiology in humans by MRI flow studies. J Magn Reson Imaging 22:591–596
2. Chambers IR, Daubaris G, Jarzemskas E, Fountas K, Kvascevicius R, Ragauskas A, Rocka S, Robinson JS, Sitkauskas A (2005) The clinical application of non-invasive intracranial blood volume pulse wave monitoring. Physiol Meas 26(6):1019–1032
3. Deltuva V (1999) Intracranial pressure measuring by using non-invasive sonographic technology. Doctoral thesis of Biomedical Sciences, Hospital of Kaunas University of Medicine, Kaunas
4. Fountas KN, Sitkauskas A, Feltes CH, Kapsalaki EZ, Dimopoulos VG, Kassam M, Grigorian AA, Robinson JS, Ragauskas A (2005) Is non-invasive monitoring of intracranial pressure waveform analysis possible? Preliminary results of a comparative study of non-invasive vs. invasive intracranial slow-wave waveform analysis monitoring in patients with traumatic brain injury. Med Sci Monit 11(2):58–63
5. Kalasauskas L (2003) Diagnosis and treatment of traumatic subdural hygromas, possibility of using the new noninvasive ultrasonographic method. Doctoral thesis of Biomedical Sciences, Hospital of Kaunas University of Medicine, Kaunas
6. Kvascevicius R (2003) Transintracranial ultrasonic diagnostics in hydrocephalus. Doctoral thesis of Biomedical Sciences, Vilnius University Emergency Hospital, Vilnius
7. Matukevicius A (2000) Testing of intracranial medium volume dynamics by using new sonographic technology. Doctoral thesis of Biomedical Sciences, Hospital of Kaunas University of Medicine, Kaunas
8. Ragaisis V (2003) Non-invasive optimised monitoring of cerebrovascular autoregulation in patients after severe head injury. Doctoral thesis of Biomedical Sciences, Hospital of Kaunas University of Medicine, Kaunas
9. Ragauskas A et al (1999) Cerebrovasc Dis 9(suppl 2):31–46
10. Ragauskas A et al (2000) Cerebrovasc Dis 10(suppl 1):34
11. Ragauskas A et al (2001) Cerebrovasc Dis 11(suppl 3):44
12. Ragauskas A et al (2002) Cerebrovasc Dis 13(suppl 4):25
13. Ragauskas A et al (2003) Cerebrovasc Dis 16(suppl 4):18
14. Ragauskas A et al (2004) Cerebrovasc Dis 17(suppl 4):4
15. Ragauskas A, Daubaris G, Ragaisis V, Petkus V (2003) Implementation of non-invasive brain physiological monitoring concepts. Med Eng Phys 25:667–678

16. Ragauskas A, Daubaris G (1995) US Patent 5,388,583
17. Ragauskas A, Daubaris G (2002) US Patent 6,387,051
18. Ragauskas A, Kanapienis A (1999) Simultaneous invasive and noninvasive ICP waves study on animals. Medical and Biological Engineering and Computing: Proceedings of the 11th Nordic–Baltic Conference on Biomedical Engineering, June 6–10, Tallin, Estonia. Peter Peregrinus, Herts, UK, 37; 1:330–331
19. Raksin PB, Alperin N et al (2003) Noninvasive intracranial compliance and pressure based on dynamic magnetic resonance imaging of blood flow and cerebrospinal fluid flow: review of principles, implementation, and other noninvasive approaches. Neurosurg Focus 14(4):e4
20. Rocka S (2003) Hydrodynamic peculiarities of the non-invasively acquired ultrasonic intracranial pulse waves. Doctoral thesis of Biomedical Sciences, Vilnius University Emergency Hospital, Vilnius

Acta Neurochir Suppl (2008) 102: 171–175

Cerebral CO_2 reactivity in severe head injury. A transcranial Doppler study

C. Puppo · G. Fariña · L. López Franco · E. Caragna · A. Biestro

Abstract
Background Cerebral circulation is profoundly affected by changes in $PaCO_2$. CO_2 manipulation plays a basic role in the management of intracranial hypertension; CO_2 reactivity (CO_2R) defines the changes in CBF in response to changes in $PaCO_2$. Transcranial Doppler has allowed exploring its effects "on line".
Materials and methods We conducted a prospective clinical trial, with the objective of studying CO_2R in severe head injury patients. Sixteen severe traumatic brain injury patients, mechanically ventilated, were included. Monitoring of MAP, ICP, CPP, SjO_2, $ETCO_2$, and cerebral blood flow velocity (CBFV) was performed. Taking into account basal cerebral hemodynamic pattern, minute ventilation was modified to attain a negative ("A") or positive ("B") ΔPCO_2. CO_2R was calculated as: $CO_2R = \%\Delta CBFV/\Delta ETCO_2$ in mmHg (normal value $3.7 \pm 1\%$/mmHg). CO_2R was compared with $\Delta ICP/\Delta PCO_2$ in each patient.
Findings Three patients were excluded because the change in $ETCO_2$ was too low ($\Delta ETCO_2 < 3$ mmHg). The median value of CO_2R in the total group of 13 patients was 3.38. In "A" the values tended to be lower than in "B". There were four low CO_2R values in "A" and none in "B". There was no significant correlation between CO_2R and $\Delta ICP/\Delta PCO_2$.
Conclusions The different "A" and "B" behavior might be due to dissimilar mechanisms involved in the basis of vasodilatation and vasoconstriction. Changes in ventilation must be performed with caution, avoiding sudden increases in CO_2 that may increase ICP. The absence of correlation between CO_2R and $\Delta ICP/\Delta PCO_2$ is explained, at least partially, by different cranio-cerebral compliance in each patient. Therefore, induced blood volume changes are not directly transmitted to ICP, but their effects depend on the shape of the pressure-volume curve and the position on the curve in which each situation is working.

Keywords CO_2 reactivity · Transcranial Doppler · Intracranial hypertension · Head injury

C. Puppo (✉)
Centro de Tratamiento Intensivo, Hospital de Clínicas,
Facultad de Medicina, Universidad de la República,
Avenida Brasil 3054 ap 401,
11300 Montevideo, Uruguay
e-mail: coripuppo@gmail.com

G. Fariña
Centro de Tratamiento Intensivo, Hospital de Clínicas,
Universidad de la República,
Franzini 865/402,
11300 Montevideo, Uruguay
e-mail: gfarina@adinet.com.uy

L. López Franco
Centro de Tratamiento Intensivo, Hospital de Clínicas,
Universidad de la República,
Magariños Cervantes 1935,
11600 Montevideo, Uruguay
e-mail: llopez@chasque.net

E. Caragna
Centro de Tratamiento Intensivo, Hospital de Clínicas,
Universidad de la República,
Grito de Gloria 1564,
11000 Montevideo, Uruguay
e-mail: dijoe@adinet.com.uy

A. Biestro
Centro de Tratamiento Intensivo, Hospital de Clínicas,
Universidad de la República,
Luis de La Torre 956/403,
11300 Montevideo, Uruguay
e-mail: mapibies@adinet.com.uy

Introduction

CO_2 is a major stimulus for cerebral arteriolar contraction and dilatation. The response of brain vessels to CO_2 has

Table 1 Patients' characteristics

Patient number	Age	Gender	Glasgow coma score	Marshall	Study day	Glasgow outcome score	ICP	MAP	FV	PI	SjO2	Delta $ETCO_2$	CO_2R
3	29	1	8	2	4	Unfavorable	12	80	82	1.11	76	−7	2.00
4	17	2	7	5	3	Favorable	10	129	44	0.73	60	9	3.03
5	56	1	8	4	4	Favorable	30	100	124	0.81	94	−7	5.07
6	52	1	7	2	7	Favorable	12	74	56	1.29	66	11	5.84
7	56	1	7	2	4	Unfavorable	12	101	52	0.91	71	8	5.77
8	27	1	8	5	4	Unfavorable	17	78	50	1.25	51	8	5.00
9	32	1	6	3	4	Favorable	15	96	36	1.42	66	−8	2.78
10	19	1	7	2	4	Favorable	19	77	46	0.93	84	−9	3.38
11	51	1	7	5	3	Unfavorable	20	116	85	0.87	86	−5	5.41
12	60	1	8	2	4	Unfavorable	30	122	33	1.06	63	−7	2.60
13	43	1	3	5	6	Favorable	13	85	64	0.90	99	−8	4.30
14	30	1	3	6	7	Unfavorable	33	79	84	1.86	70	−7	1.02
16	40	2	7	2	4	Favorable	18	101	61	0.66	70	−10	2.13

been used for many decades. Lundberg et al reported its use to lower elevated ICP in 1959 [11]. Many investigators and clinicians have used it since that time [3, 4, 12, 14–16, 19] to diminish brain volume and therefore ICP, through cerebral blood volume (CBV) lowering. However, there is still controversy on specific indications, timing, depth of hypocapnia, and duration. This property of brain vessels has been called CO_2 reactivity (CO_2R). There are other means of diminish CBV, for example by the use of certain drugs, like indomethacin, which also cause microvascular arteriolar contraction. This trial of CO_2 reactivity was performed as the first part of an indomethacin trial in head injury, in which this CO_2 reactivity was performed in order to be compared with indomethacin reactivity. [13].

Patients and methods

The protocol was approved by the Institutional Ethics Committee, and informed consent was obtained from patient's next of kin. Inclusion criteria were: penetrating or closed severe TBI, (post-resuscitation GCS<8) admitted to our ICU, between 16 and 70 years old, abnormal findings on CT scans, and ICP monitoring. Patients with refractory intracranial hypertension, clinical signs of cardiac failure, known renal or hepatic disease or dysfunction, peptic ulcer, or gastric bleeding, and pregnant or nursing women were excluded, because the indomethacin second part of the protocol. Sixteen patients, admitted to our 12 bed adult general intensive care unit (third level university hospital) between October 2001 and November 2003, were included (patients characteristics are shown in Table 1). Management was based on surgical (evacuation of intracranial mass lesions) and medical treatment, including maintenance of ICP≤20 mmHg (ventricular drainage, moderate hyperventilation, and mannitol or hypertonic saline) and cerebral perfusion pressure (CPP)≥60 mmHg after hydrostatic correction. Volume replacement was followed by noradrenaline if it was not enough to reach the required arterial pressure. All patients were intubated and received mechanical ventilation. They were maintained normothermic, sedated, and paralyzed.

A transcranial Doppler (TCD) device, EME TC2-64b was used to measure mean cerebral blood flow velocity (CBFV) at the M1 segment of the middle cerebral artery, through the temporal window, according to the method described by Aaslid. [1]. A 2 MHz pulsed Doppler monitoring probe was fixed in position over the MCA

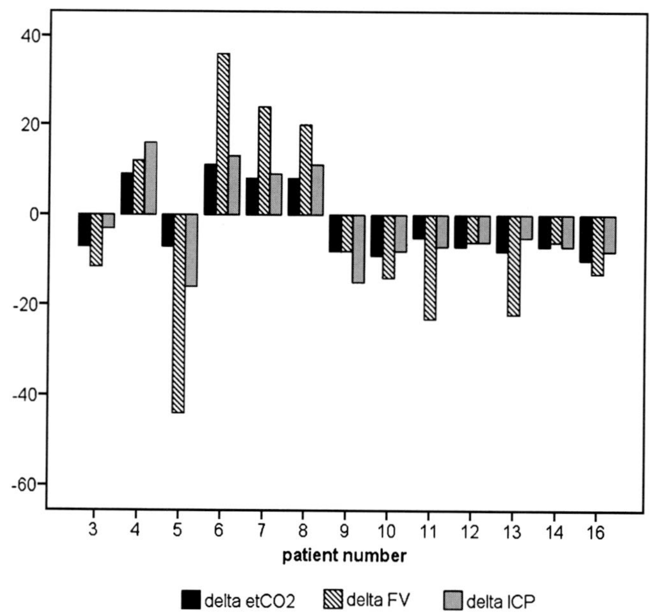

Fig. 1 *Bars* showing the change in cerebral blood flow velocity and intracranial pressure compared to the positive or negative change in CO_2

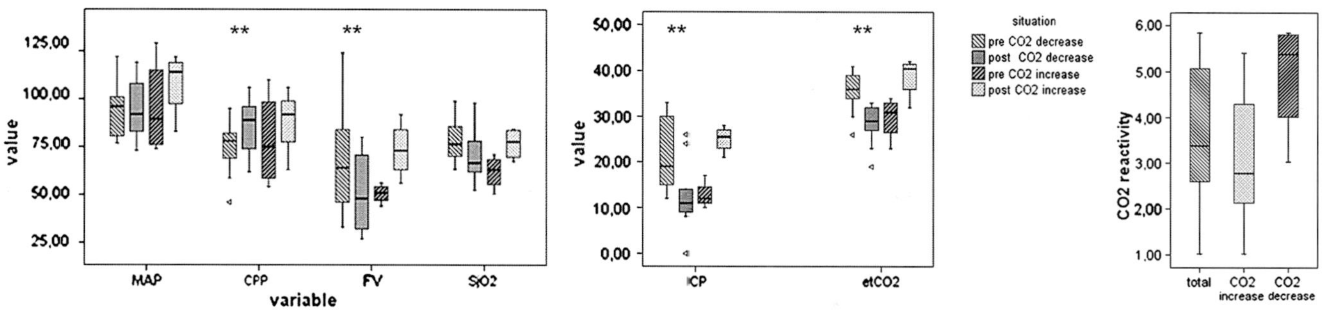

Fig. 2 *Left panel*: *Boxplots* showing the median, interquartile range, minimum and maximum values of the different variables pre and post the change in ventilation. For each variable, the two boxplots at the left show the results of the 9 patients in whom CO_2 was decreased and the two boxplots at the right show the results of the 4 patients in whom CO_2 was increased. *MAP* Mean arterial pressure, *CPP* cerebral perfusion pressure, *FV* mean cerebral flow velocity, *SjO2* jugular bulb oxygen saturation, *ICP* intracranial pressure, $ETCO_2$ end tidal CO_2. *Right panel*: *Boxplots* showing the median, interquartile range, minimum and maximum values of CO_2 reactivity. From left to right: the whole group of 13 patients, the nine hyperventilated patients and the four patients in whom ΔPCO_2 was decreased

through a headband. Heart rate, mean arterial blood pressure (MAP), ICP, CBFV, $ETCO_2$, arterial oxygen saturation (SaO_2) and rectal temperature were continuously monitored, and manually recorded every 15 s during each trial. Data were also recorded with a VCR, and further re-evaluated when needed. ICP monitoring was performed with a Codman intraparenchymal sensor, or a ventricular catheter. Intra-arterial blood pressure was monitored. Arterial and jugular bulb oxymetric values were measured with a co-oxymeter, from blood samples drawn before and after the change in ventilation.

CO_2 reactivity testing CO_2R [5, 7, 9, 10] was performed by modifying minute ventilation, (tidal volume, or respiratory frequency). The change in ventilation was controlled with CO_2 and $ETCO_2$ change. CO_2R was calculated as the percentage change in CBF per mmHg change in CO_2. ICP-CO_2 reactivity was calculated as the mmHg change in ICP per mmHg change in CO_2. CO_2R and ICP-CO_2 reactivity were then compared.

Neurological outcome was assessed at discharge by the Glasgow Outcome Scale [8] for all patients; for statistical purposes, scores of 1 through 3 were defined as an unfavorable outcome and scores of 4 and 5 were defined as a favorable outcome.

Results

Sixteen patients entered the study. Three of them were excluded from further analysis because the change in $ETCO_2$ was too low ($\Delta ETCO_2<3$ mmHg). The median age of the 13 patients included in the analysis was 40 years (range 17–60); there were 11 males out of a total of 13 patients. The median study day was the fourth day (range 3–7). Patients' characteristics are shown in Table 1. The changes in CBFV and ICP in each patient are shown in Fig. 1, compared with the change in $ETCO_2$.

The median value of CO_2R in the total group of 13 patients was 3.38. Fig. 2, right panel. The changes in the different variables are shown in Fig. 2, left panel.

In nine patients ΔPCO_2 was decreased, (group "A"), and it was increased in four (group "B"). There were four low CO_2R values in "A" and none in "B". See reactivity and ICP-CO_2R results in Table 2.

There was no significant correlation between CO_2R and $\Delta ICP/\Delta PCO_2$. (Pearson Correlation 0.338, significance 0.258). When comparing the three groups of CO_2R (normal, low and high) to outcome, the best outcome was found in the normal reactivity group (Fig. 3).

Discussion and conclusions

The present study shows that in this small sample of severe head injury patients, CO_2R was preserved in nine patients (70%), and low in four patients (30%), all of them pertaining to the hyperventilated group. The most novel data of this study is the different behavior of CO_2R in the CO_2 decrease group (group "A"), related to the CO_2 increase group (group "B"). In "A" the values tended to be lower than in "B". Probably due to the small sample of patients, there were no significant differences between both groups.

Table 2 Reactivity results in total group, A and B groups

Parameter	Total group	CO_2 decrease group	CO_2 increase group
CO_2 reactivity	3.38 (1.02–5.84)	2.78 (1.02–5.41)	5.38 (3.03–5.84)
ICP-CO_2R	1.13 (0.43–2.29)	0.89 (0.43–2.29)	1.28 (1.13–1.78)

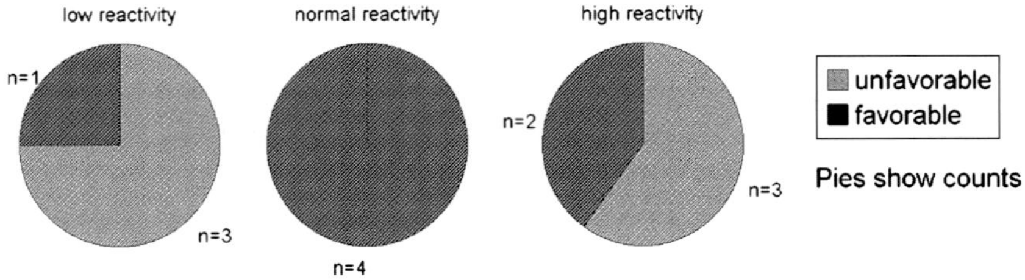

Fig. 3 Relationship between CO_2 reactivity and outcome (GOS) at discharge. CO_2 reactivity has been divided in three: 1) *low*: lower than 2.7%/mmHg, 2) *normal*: between 2.7 and 4.7%/mmHg, and 3) *high*: higher than 4.%/mmHg. The four patients who showed a normal response to CO_2 challenge had a good outcome, three of the four patients with a low CO_2R had an unfavorable outcome, as did three of the five patients with high CO_2 reactivity

The absence of significant changes in systemic hemodynamic variables can be seen in Fig. 2. Systemic effects of changes in CO_2 are multifactorial and interrelated, affecting multiple sites of the body. Substantial differences exist between active changes (when the subject voluntarily increases or decreases the ventilation rate) and passive changes (by means of artificial ventilation), as in our patients. In the former, autonomic flow is markedly affected, while in artificially ventilated patients the effects of CO_2 are combined with those of the complex interaction between artificial ventilation and hemodynamics. Additionally, when hyperventilation is applied for reducing ICP, or in severe head injured patients, it is usually combined with a number of concurrent interventions such as sedation, paralysis, and increased fluid input, interventions that interfere with the pure action of hyperventilation [18].

The different "A" and "B" behavior might be due to different mechanisms involved in the regulation of vasodilatation and vasoconstriction. Moreover, the mild basal hyperventilation used in the management protocol could explain these differences, because the increased basal tone generated by hyperventilation could favor the ability to vasodilate and diminish further vasoconstriction. Similarly, we found in the second part of this trial that indomethacin, which causes vasoconstriction, does not impair vasodilatory capacity, tested by the transient hyperemic response test [6, 17], but, on the contrary, it increases after the administration of the drug [13]. The absence of correlation between CO_2R and $\Delta ICP/\Delta PCO_2$ is explained, at least partially, because each patient has its own cranio-cerebral compliance, represented by a certain pressure volume curve, and at the moment when the patients were studied the place at this curve where dynamic cerebral compliance was working was different. Blood volume changes induced by vasodilation or vasoconstriction are not directly transmitted to ICP, but their effects depend on the shape of the pressure-volume curve and the position on the curve in which each situation is working.

The relationship between the group with preserved CO_2R and favorable outcome, and its difference with the hyper-reactivity and hypo-reactivity groups is also important to underline. Changes in ventilation must be performed with caution, avoiding a sudden increase in CO_2 which may increase ICP.

Conflict of interest statement We declare that we have no conflict of interest.

References

1. Aaslid R, Markwalder TM, Nornes H (1982) Doppler ultrasound recording of flow velocity in basal cerebral arteries. J Neurosurg 57:769–774
2. Ackerman RH, Zilkha E, Bull JWD, du Boulay GH, Marshall J, Russell RWR, Symon L (1973) The relationship of the CO_2 reactivity of cerebral vessels to blood pressure and mean resting blood flow. Neurology 23:21–26
3. Brian JE (1998) Carbon dioxide and the cerebral circulation. Anesthesiology 88:1365–1386
4. Bullock MR, Chesnut R, Clifton G et al (2000) Hyperventilation. J Neurotrauma 17:513–520
5. Cold GE, Jensen FT, Malmros R (1997) The cerebrovascular CO_2 reactivity during the acute phase of brain injury. Acta Anaesth Scand 21:222–231
6. Giller CA (1991) A bedside test for cerebral autoregulation using transcranial Doppler ultrasound. Acta Neurochir (Wien) 108:7–14
7. Imberti R, Bellinzona G, Langer M (2002) Cerebral tissue PO2 and SjvO2 changes during moderate hyperventilation in patients with severe traumatic brain injury. J Neurosurg 96:97–102
8. Jennett B, Bond M (1975) Assessment of outcome after severe brain damage: a practical scale. Lancet 1:480–484
9. Klingelhöfer J, Sander D (1992) Doppler CO_2 test as an indicator of cerebral vasoreactivity and prognosis in severe intracranial hemorrhages. Stroke 23:962–966
10. Lee JH, Kelly DF, Oertel M, McArthur DL, Glenn TC, Vespa P, Boscardin WJ, Martin NA (2001) Carbon dioxide reactivity, pressure autoregulation, and metabolic suppression reactivity after head injury: a transcranial Doppler study. J Neurosurg 95:222–232
11. Lundberg N, Kjallquist A, Bien C (1959) Reduction of increased intracranial pressure by hyperventilation: a therapeutic aid in neurological surgery. Acta Psychiatr Scand 34(suppl):1–64
12. Marion DW, Firlik A, McLaughlin MR (1995) Hyperventilation therapy for severe traumatic brain injury. New Horiz 3:439–447
13. Puppo C, Lopez L, Farina G, Caragna E, Moraes L, Iturralde A, Biestro A (2007) Indomethacin and cerebral autoregulation in severe head injured patients: a transcranial Doppler study. Acta Neurochir (Wien) 149:139–149

14. Raichle ME, Plum F (1972) Hyperventilation and cerebral blood flow. Stroke 3:566–575
15. Reivich M (1964) Arterial PCO_2 and cerebral hemodynamics. Am J Physiol 206:25–35
16. Rossanda M, Vecchi G (1979) Determination of cerebral autoregulatory status and PCO_2 responsiveness. Int Anesthesiol Clin 17:425–438
17. Smielewski P, Czosnyka M, Kirkpatrick P, Pickard JD (1997) Evaluation of the transient hyperemic response test in head-injured patients. J Neurosurg 86:773–778
18. Stocchetti N, Maas A, Chieregato A, van der Plas AA (2005) Hyperventilation in head injury: a review. Chest 127:1812–1827
19. Vender JR (2000) Hyperventilation in severe brain injury revisited. Crit Care Med 28:3361–3362

Acta Neurochir Suppl (2008) 102: 177–183

Acetazolamide vasoreactivity evaluated by transcranial power harmonic imaging and Doppler sonography

T. Shiogai · K. Ikeda · A. Morisaka · Y. Nagakane · T. Mizuno · M. Nakagawa · H. Furuhata

Abstract
Background Cerebral vasoreactivity (CVR) in the major cerebral arteries evaluated by transcranial Doppler sonography has shown some correlation with CVR in the brain tissue measured by other neuroradiological modalities. To clarify vasoreactive differences in the brain tissue and the major cerebral arteries, we have evaluated the relationship of acetazolamide (ACZ) CVR between transcranial ultrasonic power harmonic imaging (PHI) and color Doppler sonography (CDS), in cases of parenchymal pathology with and without occlusive vascular lesions.
Materials and methods The subjects were 31 stroke patients with intraparenchymal pathologies, 15 with (occlusive group) and 16 without (non-occlusive group) occlusive carotid and/or middle cerebral artery lesions. CVR based on values before/after ACZ (angle-collected CDS velocity in the middle and posterior cerebral arteries, PHI contrast area size, peak intensity, time to peak intensity), and correlation of CVR between PHI and CDS were compared between the side with and without lesions in both groups.
Findings (a) PHI CVR tended to be more disturbed than CDS CVR. CVR side differences were not significant. (b) CVR correlations between PHI and CDS were always lower in the pathological sides.
Conclusions CVR in brain tissue evaluated by PHI is susceptible to disturbance in comparison with CDS, due to both parenchymal and vascular occlusive pathologies.

Keywords Cerebral vasoreactivity · Acetazolamide · Transcranial ultrasonic power harmonic imaging · Transcranial color Doppler sonography

T. Shiogai (✉) · K. Ikeda · A. Morisaka
Department of Clinical Neurosciences, Kyoto Takeda Hospital,
Minamikinuta-cho 11, Nishinanajo, Shimogyo-ku,
Kyoto 600-8884, Japan
e-mail: shiogait@pop11.odn.ne.jp

Y. Nagakane · T. Mizuno · M. Nakagawa
Department of Neurology,
Kyoto Prefectural University School of Medicine,
Kajii-cho 465, Kawaramachi-Dori Hirokoji-Agaru, Kamigyo-ku,
Kyoto 602-8566, Japan

Y. Nagakane
e-mail: nagakane@koto.kpu-m.ac.jp

T. Mizuno
e-mail: mizuno@koto.kpu-m.ac.jp

M. Nakagawa
e-mail: mnakagaw@koto.kpu-m.ac.jp

H. Furuhata
Medical Engineering Laboratory,
Tokyo Jikei University School of Medicine,
Nishi-Shinbashi 3-25-8, Minato-ku,
Tokyo 105-8461, Japan
e-mail: furuhata@jikei.ac.jp

Introduction

Cerebral vasoreactivity (CVR) has been evaluated not only by transcranial Doppler sonography (TCD) but also using various neuroradiological perfusion imaging methods including single-photon emission CT (SPECT), Xenon CT (Xe-CT), CT/ magnetic resonance (MR) perfusion imaging, and positron emission tomography [5]. CVR evaluation has been conducted in stroke patients, particularly with carotid occlusive lesions, in relation to risk assessment for cerebral infarction and post-operative hyperperfusion syndrome after carotid endarterectomy, consideration of extra-intracranial bypass surgery, collateral circulation for balloon test occlusion, and selection for other medical interventions [5]. CVR is disturbed not only in carotid occlusive diseases but also in the setting of intracranial parenchymal pathologies [9]. CVR in the major cerebral arteries has been evaluated using conventional TCD, and this has shown some correlation with cerebral blood flow (CBF) measurements in the brain tissue from other neuroradiological modalities [2, 12, 14, 26].

Transcranial power harmonic imaging (PHI) can identify a contrast-perfused area more easily than gray-scale harmonic imaging [20]. Compared to conventional TCD, transcranial color Doppler sonography (CDS) is able to measure more accurately on the basis of angle-collected velocities in the intracranial major vessels [17]. To clarify vasoreactive differences in the brain tissue and the major cerebral arteries, we have evaluated the relationship of acetazolamide (ACZ) CVR between transcranial PHI and CDS [22]. The aim of this study was to evaluate vasoreactive differences in cases of intracranial parenchymal pathology with and without occlusive vascular lesions.

Materials and methods

The subjects were 31 stroke patients with intraparenchymal pathologies in a chronic stage, with open temporal acoustic windows confirmed by transcranial CDS utilizing a SONOS 5500 S4 transducer (Philips). Intraparenchymal pathologies were diagnosed by CT and/or MR imaging (MRI). Cervical and intracranial vascular lesions were evaluated by digital subtraction angiography (DSA) and/or MR angiography (MRA) and cervical and transcranial CDS. Internal carotid and middle cerebral artery occlusions (ICAO and MCAO) were diagnosed by DSA and/or MRA. Diagnosis of cervical ICA stenosis (ICAS) was based on the CDS criteria (>50% narrowing on the longitudinal plain or >70% narrowing on the axial plain). MCA stenosis (MCAS) was diagnosed using transcranial CDS on the basis of increased angle-collected time-averaged maximum velocity (Vm) >120 cm/s. Patients were classified into two groups; one with (Occlusive group) and one without (Non-occlusive group) occlusive lesions in the ICA and/or the MCA. The occlusive and non-occlusive groups consisted of 15 patients (ages 47–92, mean 71) and 16 patients (ages 35–88, mean 65), respectively. Age differences between the two groups were not significant. Details of gender, lesion side, and diagnosis of the two groups are shown in Table 1.

Transient response PHI taken every 2s after a 7ml-bolus Levovist® injection (300 mg/ml) via the antecubital vein was evaluated in an axial diencephalic plane via bilateral temporal windows (Fig. 1). The PHI settings were as previously described [22]. PHI contrast area size was measured by Scion Image (Beta version 4.02, Scion Corporation). Peak intensity (PI) and time-to-peak intensity (TPI) based on time-intensity curve (TIC) analysis were measured by QLAB software (Version 4.1, Philips) (Fig. 2).

Before ACZ (500 mg Diamox®) intravenous injection, Vm in the MCA and posterior cerebral artery (PCA) was measured by CDS and PHI was evaluated on both sides. The PHI was re-evaluated 15 and 30min after ACZ, via right and left temporal windows, respectively (Fig. 1). Vm in the ipsilateral MCA and PCA was measured just before the PHI measurement. The CVR (%Δ) of CDS Vm and PHI parameters were calculated before and after ACZ administration (parameters after ACZ - parameters before ACZ/ parameters before ACZ x 100).

Assessments of CVR on the basis of values before and after ACZ were compared between the side with and the side without pathological lesions on the basis of CDS and other neuroradiological modalities in both groups, in terms of: (a) CDS; Vm in the MCA and PCA; (b) PHI; contrast area size; PI and TPI; and (c) Pearson's correlation coefficients (r) of CVR between PHI and CDS parameters. Utilizing a paired t-test, a non-paired t-test, a chi-square test, and a one-way analysis of variance, statistical significance was set at $p < 0.05$. Informed consent was obtained from patients and/or patients' family members.

Results

(a) CVR evaluated by CDS in occlusive and non-occlusive groups (Table 2)

In the occlusive group, significant Vm increases after ACZ in the MCA and PCA were observed in both lesion and non-lesion sides. Significant Vm increases were more

Table 1 Demographics in occlusive and non-occlusive groups

Parameter	Occlusive group	Non-occlusive group
No. of cases	15	16
Age	71 ± 13	65 ± 15
Gender (male/ female)	14/1	12/4
Lesion side		
Right	5	3
Left	6	8
Bilateral	4	5
Diagnosis[a]		
Cerebral infarction	15	7
ICAO	4	0
ICAS	7	0
MCAO	2	0
MCAS	2	0
Lacunae	0	7
ICH	0	6
SAH	0	3

[a] *ICAO* Internal carotid artery occlusion, *ICAS* internal carotid artery stenosis, *MCAO* middle cerebral artery occlusion, *MCAS* middle cerebral artery stenosis, *ICH* intracerebral hemorrhage, *SAH* subarachnoid hemorrhage

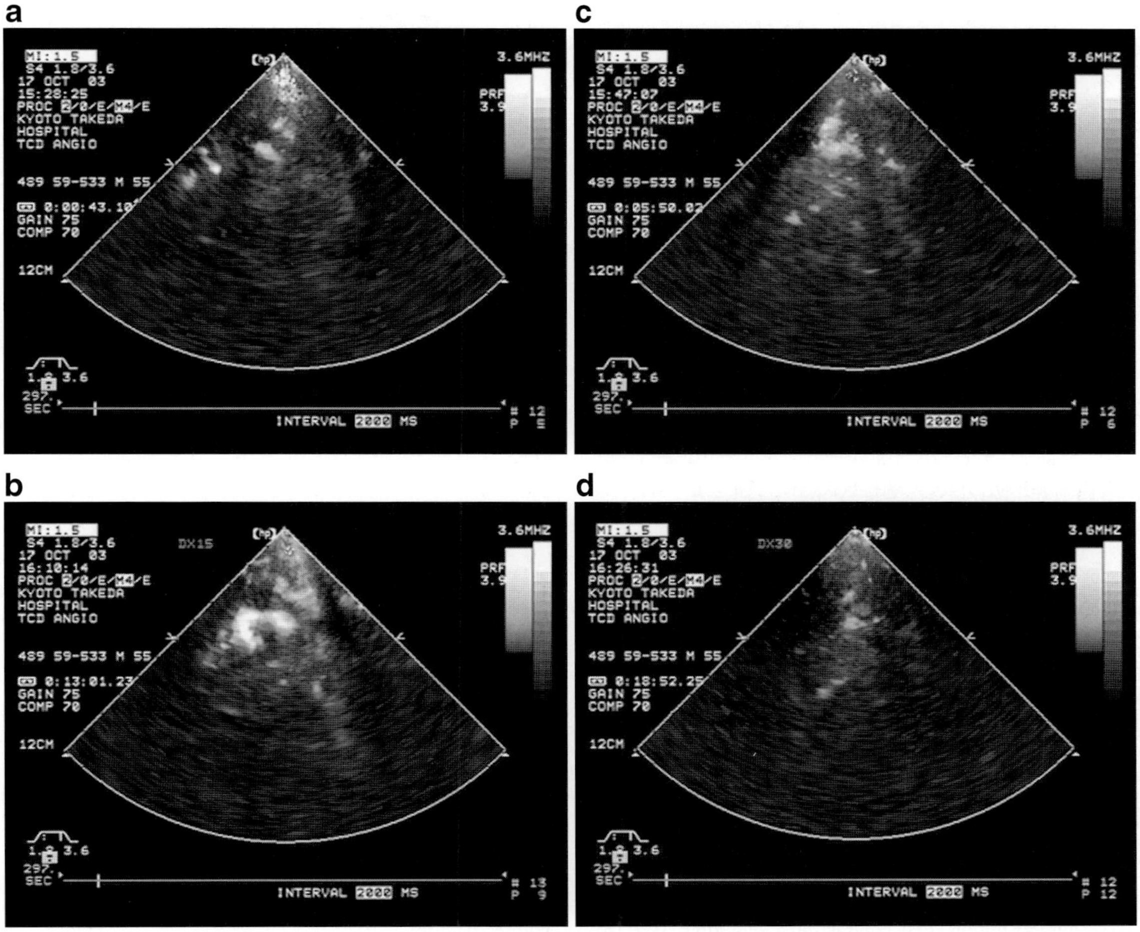

Fig. 1 Acetazolamide (ACZ) CVR tests by transcranial power harmonic imaging (PHI) in a 55-year-old subject were performed 5 months after atherothrombotic infarction caused by the left ICAO. PHI images are represented before (**a** *left*, **c** *right*) and after (**b** *left*, **d** *right*) ACZ bolus intravenous injection. Despite increased contrast area size after ACZ (**b**), decreased size in the left side (**d**) is probably due to a steal phenomenon [22]

obvious in the non-lesion side of the MCA and in the lesion side of the PCA. In the non-occlusive group, significant Vm increases after ACZ were observed in both sides of the PCA and in the lesion side of the MCA. However, there was an increased Vm tendency only in the non-lesion side of the MCA.

Side differences of CVR in the MCA and PCA were not significant in either group.

(b) CVR evaluated by PHI in occlusive and non-occlusive groups (Table 2)

In the occlusive group, significant increases of contrast area size after ACZ were observed in both sides. Significant increases of TPI after ACZ were observed only in the non-lesion side. However, there was no significant increase of PI after ACZ in either side or PHI in the lesion side. In the non-occlusive group, a significant increase after ACZ was observed only in the contrast area size of the lesion side and PI in the non-lesion side. There were no significant increases in other PHI parameters.

Side differences of CVR in all PHI parameters were not significant in either group.

(c) CVR values between CDS and PHI parameters in occlusive and non-occlusive groups (Table 2)

PHI CVR values, except for contrast area size, tended to be lower than CDS CVR values in both groups. CVR based on TPI in both sides of the occlusive group was significantly lower than CVR based on the MCA Vm and PCA Vm. In the non-occlusive group, CVR based on PI in the lesion side was significantly lower than CVR based on MCA Vm, and CVR based on TPI was significantly lower than CVR based on PCA Vm.

(d) Correlation between CDS and PHI CVR in occlusive and non-occlusive groups (Table 3)

In the occlusive group, close and significant correlations between CDS and PHI parameters were observed in the non-lesion side, except for a correlation between PCA Vm and TPI. The correlation coefficients in the lesion side were always lower than those in the non-lesion sides.

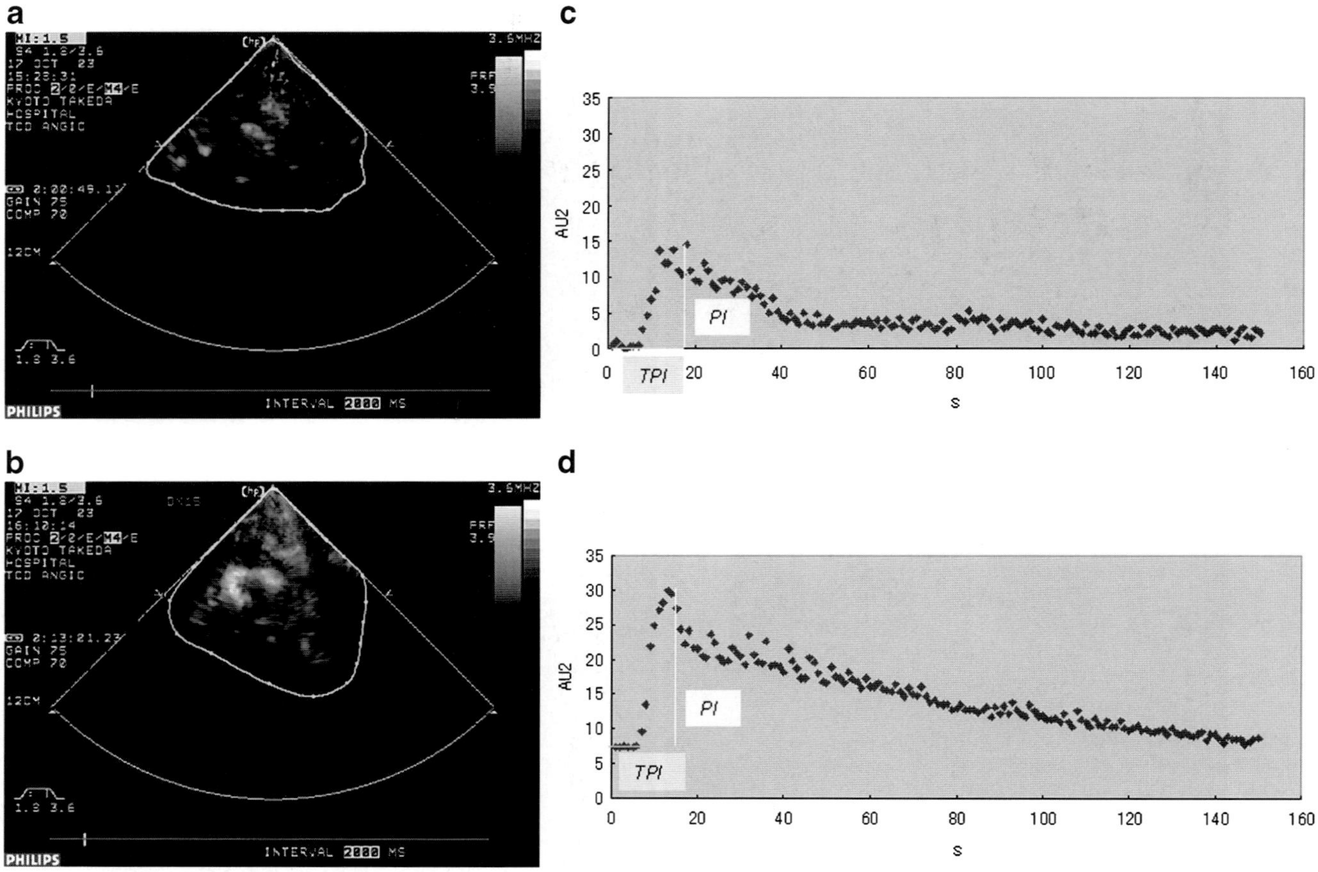

Fig. 2 Time-intensity curve (TIC) analysis before (**a** and **c**) and after (**b** and **d**) ACZ administration. PHI contrast area size after ACZ (b) appears larger than that before ACZ (**a**). On the TIC analysis, after ACZ, PI increased and TPI decreased (**d**) in comparison with state before ACZ (**c**)

Furthermore, there was no significant correlation in the lesion side.

In the non-occlusive group, close and significant correlations between CDS and PHI parameters were observed only between MCA Vm vs. contrast area size and vs. PI in the non-lesion side. There were no significant correlations between the other CDS and PHI parameters in either side. The correlation coefficients in the lesion side were always lower than those in the non-lesion side.

Discussion

(a) CVR evaluated by TCD and PHI

ACZ VMR utilizing TCD has been evaluated in various types of stroke patients [1, 4, 13] including those with lacunar infarction [11] and subarachnoid hemorrhage (SAH) [1, 25], as well as non-stroke patients with and without parenchymal lesions [1, 4]. The lesional side of ICAS or ICAO, e.g. 80% ICAS [15, 16], 90–99% ICAS [12], ICAO [8, 18], or symptomatic side of ICAS or ICAO [12, 14], showed significantly decreased CVR by TCD or CBF measurements in comparison with the contralateral side. However, almost half of asymptomatic severe ICAS (>70%) [7] and non-severe ICAS (≤80%) [16] cases showed good CVR. Furthermore, aside from severe ICAS (>90%), there were no CVR differences between less severe stenosis (70–90%) and ICAO [12]. In our occlusive group, there were no significant CVR side differences between lesion and non-lesion sides (Table 2). These CDS findings in the MCA are probably because our cases of non-severe ICAS, MCAS and MCAO, ICAO or bilateral lesions in the chronic stage were associated with good collateral circulation in both sides of the brain tissue as demonstrated by PHI.

Side differences of CVR were not significant in patients with lacunar infarction [11] and aneurysmal SAH [25]. In our non-occlusive group, four of seven patients with lacunar infarction and one with SAH had bilateral lesions. Furthermore, three patients with ICH (two frontal lobe and one pons) did not have involvement of the MCA or PCA territory. These etiological factors probably explain the absence of significant CVR side differences evaluated by either CDS or PHI between lesion and non-lesion sides in our non-occlusive group (Table 2).

Table 2 Acetazolamide (ACZ) induced cerebral vasoreactivity (CVR) in occlusive and non-occlusive groups

	Occlusive group (n=15)							Non-occlusive group (n=16)						
	Lesion side			Non-lesion side				Lesion side			Non-lesion side			
Parameters[a]	Before ACZ	After ACZ	CVR (%)	Before ACZ	After ACZ	CVR (%)	SD[b]	Before ACZ	After ACZ	CVR (%)	Before ACZ	After ACZ	CVR (%)	SD[b]
MCA Vm	62±28	93±53*	47±61	74±27	106±48***	42±3	ns	41±15	56±18***	43±42	48±12	58±20	21±31	ns
PCA Vm	45±20	62±30**	39±37	42±13	51±14*	31±47	ns	26±7	35±12***	38±33	33±8	45±11**	37±21	ns
Size	10±7	14±5*	65±69	10±3	14±4*	46±61	ns	15±12	18±12*	44±95	10±4	14±5	52±76	ns
PI	13±5	15±4	23±35	15±5	17±10	19±49	ns	17±10	19±11	13±43‡	14±3	17±5*	21±28	ns
TPI	24±5	22±8	8±29‡‡,§§	26±9	22±6*	−12±20‡‡‡,§§	ns	23±10	25±11	17±54	27±11	27±9	20±36§§	ns

$^{*}p<0.05$ in comparison with parameters before ACZ based on paired t-test.
$^{**}p<0.01$ in comparison with parameters before ACZ based on paired t-test.
$^{***}p<0.001$ in comparison with parameters before ACZ based on paired t-test.
$^{‡}p<0.05$ in comparison with MCA Vm based on non-paired t-test.
$^{‡‡}p<0.01$ in comparison with MCA Vm based on non-paired t-test.
$^{‡‡‡}p<0.001$ in comparison with MCA Vm based on non-paired t-test.
$^{§}p<0.05$ in comparison with PCA Vm based on non-paired t-test.
$^{§§}p<0.01$ in comparison with PCA Vm based on non-paired t-test.
[a] *Vm* Time-averaged maximum velocity (cm/s), *MCA* middle cerebral artery, *PCA* posterior cerebral artery, *Size* contrast area size (cm^2), *PI* peak intensity (AU^2), *TPI* time-to-peak intensity (s)
[b] *SD* Side differences, *ns* not significant.

In the studies of ACZ CVR utilizing TCD and CBF measurements, CVR derived from TCD was significantly higher than that from ^{133}Xe SPECT in normal subjects [24] and tended to be higher than that found on Xe-CT in various brain diseases and in occlusive ICA diseases [12]. However, these tendencies were not always seen in normal subjects [3, 26], unselected neurological diseases [4], and obstructive carotid diseases [14]. In our study, CVR based on PHI parameters of PI or TPI tended to be lower than that of PHI contrast area size or CVR based on CDS parameters in both groups. PHI parameters of PI and TPI derived from TIC analysis are not measured in the same units as CBF data. If the dye-dilution principle [10] were applicable for PHI, CDS measured CVR could be compared to calculate CBF data derived from PHI parameters. Since there was data scattering of calculated CBF based on our previous observations utilizing ultrasonic gray scale harmonic imaging in various neurological diseases, the calculated CVR (%ΔCBF) was not always lower than CVR (%ΔVm) in the MCA and PCA [19].

(b) Correlation of CVR between TCD and PHI

Correlations of CVR between TCD and CBF in published studies have not always been consistent. Close

Table 3 Correlation between transcranial CDS and PHI CVR in occlusive and non-occlusive groups

	Correlation Coefficient (p value[b])			
	Occlusive group (n=15)		Non-occlusive group (n=16)	
Parameters[a]	Lesion side	Non-lesion side	Lesion side	Non-lesion side
MCA Vm vs. contrast area size	0.45 (ns)	0.69 (p<0.01)	0.26 (ns)	0.76 (p<0.05)
MCA Vm vs. PI	0.48 (ns)	0.64 (p<0.05)	0.21 (ns)	0.82 (p<0.01)
MCA Vm vs. TPI	−0.17 (ns)	−0.65 (p<0.05)	0.04 (ns)	0.21 (ns)
PCA Vm vs. contrast area size	0.50 (ns)	0.67 (p<0.05)	0.30 (ns)	0.51 (ns)
PCA Vm vs. PI	0.37 (ns)	0.66 (p<0.05)	0.17 (ns)	0.50 (ns)
PCA Vm vs. TPI	−0.36 (ns)	−0.48 (ns)	0.01 (ns)	0.30 (ns)

[a] *MCA* Middle cerebral artery, *Vm* time-averaged maximum velocity, *PCA* posterior cerebral artery, *PI* peak intensity, TPI time-to-peak intensity
[b] Analysis of variance, *ns* Not significant

correlation of TCD was observed with ^{133}Xe CBF in normal subjects ($r = 0.75$, $p < 0.001$) [26], with ^{133}Xe SPECT in cerebrovascular patients ($r = 0.63$, $p < 0.019$) [2] and in obstructive carotid diseases (symptomatic side: $r = 0.59$, $p < 0.004$, asymptomatic side: $r = 0.45$, $p = 0.04$) [14], and with Xe-CT in occlusive ICA diseases ($r = 0.458$, $p < 0.04$) [12]. However, there were no close correlations of TCD with Xe-CT (regional CBF: $r = 0.23$, global CBF: $r = 0.36$) in a vascular stenosis group [1], with ^{133}Xe SPECT (regional CBF: $r = 0.08$, hemispheric CBF $r = 0.111$) in unselected neurological patients [4] or normal subjects ($r = 0.05$) [3].

In our study of both groups, correlations between CDS and PHI parameters were almost always closer in the non-lesion side (occlusive group: $r = 0.48–0.67$, non-occlusive group: $r = 0.21–0.82$) than in the lesion side (occlusive group: $r = 0.17–0.50$, non-occlusive group: $r = 0.01–0.30$). This is probably because parenchymal pathology in the medial temporal lobe, basal ganglia, and thalamus on the lesion side affected the visualized PHI image. Indeed, PHI has technical problems that mean it cannot visualize in the ipsilateral temporal cortex, or frontal and occipital poles outside the sector scan. In order to overcome these PHI problems, new imaging techniques utilizing pulse inversion harmonic imaging [6] or power modulation imaging [23] are able to visualize contralateral hemispheres via unilateral temporal windows. Another approach utilizing a refill kinetic method of transcranial ultrasonic perfusion imaging [21] would be viable for overcoming these problems.

In conclusion, in comparison with transcranial CDS, CVR in brain tissue evaluated by transcranial PHI is easy to disturb due to extra- and intra-cranial vascular occlusive lesions as well as parenchymal pathologies.

Conflict of interest statement We declare that we have no conflict of interest.

References

1. Brauer P, Kochs E, Werner C, Bloom M, Policare R, Pentheny S, Yonas H, Kofke WA, Schulte am Esch J (1998) Correlation of transcranial Doppler sonography mean flow velocity with cerebral blood flow in patients with intracranial pathology. J Neurosurg Anesthesiol 10:80–85
2. Dahl A, Lindegaard KF, Russell D, Nyberg-Hansen R, Rootwelt K, Sorteberg W, Nornes H (1992) A comparison of transcranial Doppler and cerebral blood flow studies to assess cerebral vasoreactivity. Stroke 23:15–19
3. Dahl A, Russell D, Nyberg-Hansen R, Rootwelt K, Mowinckel P (1994) Simultaneous assessment of vasoreactivity using transcranial Doppler ultrasound and cerebral blood flow in healthy subjects. J Cereb Blood Flow Metab 14:974–981
4. Demolis P, Tran Dinh YR, Giudicelli JF (1996) Relationships between cerebral regional blood flow velocities and volumetric blood flows and their respective reactivities to acetazolamide. Stroke 27:1835–1839
5. Eskey CJ, Sanelli PC (2005) Perfusion imaging of cerebrovascular reserve. Neuroimaging Clin N Am 15:367–381
6. Eyding J, Krogias C, Wilkening W, Meves S, Ermert H, Postert T (2003) Parameters of cerebral perfusion in phase-inversion harmonic imaging (PIHI) ultrasound examinations. Ultrasound Med Biol 29:1379–1385
7. Gur AY, Bova I, Bornstein NM (1996) Is impaired cerebral vasomotor reactivity a predictive factor of stroke in asymptomatic patients? Stroke 27:2188–2190
8. Karnik R, Valentin A, Ammerer HP, Donath P, Slany J (1992) Evaluation of vasomotor reactivity by transcranial Doppler and acetazolamide test before and after extracranial-intracranial bypass in patients with internal carotid artery occlusion. Stroke 23:812–817
9. Marion DW, Bouma GJ (1991) The use of stable xenon-enhanced computed tomographic studies of cerebral blood flow to define changes in cerebral carbon dioxide vasoresponsivity caused by a severe head injury. Neurosurgery 29:869–873
10. Meier P, Zierler LK (1954) On the theory of the indicator-dilution method for measurement of blood flow and volume. J Appl Physiol 6:731–744
11. Molina C, Sabin JA, Montaner J, Rovira A, Abilleira S, Codina A (1999) Impaired cerebrovascular reactivity as a risk marker for first-ever lacunar infarction: a case-control study. Stroke 30:2296–2301
12. Müller M, Voges M, Piepgras U, Schimrigk K (1995) Assessment of cerebral vasomotor reactivity by transcranial Doppler ultrasound and breath-holding. A comparison with acetazolamide as vasodilatory stimulus. Stroke 26:96–100
13. Pavics L, Grunwald F, Barzo P, Ambrus E, Menzel C, Schomburg A, Borda L, Mate E, Bodosi M, Csernay L, Biersack HJ (1994) Evaluation of cerebral vasoreactivity by SPECT and transcranial Doppler sonography using the acetazolamide test. Nuklearmedizin 33:239–243
14. Piepgras A, Schmiedek P, Leinsinger G, Haberl RL, Kirsch CM, Einhaupl KM (1990) A simple test to assess cerebrovascular reserve capacity using transcranial Doppler sonography and acetazolamide. Stroke 21:1306–1311
15. Ringelstein EB, Van Eyck S, Mertens I (1992) Evaluation of cerebral vasomotor reactivity by various vasodilating stimuli: comparison of CO_2 to acetazolamide. J Cereb Blood Flow Metab 12:162–168
16. Rosenkranz K, Hierholzer J, Langer R, Hepp W, Palenker J, Felix R (1992) Acetazolamide stimulation test in patients with unilateral internal carotid artery obstructions using transcranial Doppler and 99mTc-HM-PAO-Spect. Neurol Res 14(2 Suppl):135–138
17. Schöning M, Buchholz R, Walter J (1993) Comparative study of transcranial color duplex sonography and transcranial Doppler sonography in adults. J Neurosurg 78:776–784
18. Schreiber SJ, Gottschalk S, Weih M, Villringer A, Valdueza JM (2000) Assessment of blood flow velocity and diameter of the middle cerebral artery during the acetazolamide provocation test by use of transcranial Doppler sonography and MR imaging. AJNR Am J Neuroradiol 21:1207–1211
19. Shiogai T, Koshimura M, Murata Y, Nomura H, Doi A, Makino M, Mizuno T, Nakajima K, Furuhata H (2003) Acetazolamide vasoreactivity evaluated by transcranial harmonic perfusion imaging: relationship with transcranial Doppler sonography and dynamic CT. Acta Neurochir Suppl 86:57–62
20. Shiogai T, Takayasu N, Mizuno T, Nakagawa M, Furuhata H (2004) Comparison of transcranial brain tissue perfusion images between ultraharmonic, second harmonic, and power harmonic imaging. Stroke 35:687–693
21. Shiogai T, Morisaka A, Takayasu N, Yoshikawa K, Mizuno T, Nakagawa M, Furuhata H (2005) Quantitative evaluation of cerebrovascular reactivity in brain tissue by a refill kinetic method of transcranial ultrasonic perfusion imaging: a comparison with Doppler sonography. Acta Neurochir Suppl 95:183–190

22. Shiogai T, Morisaka A, Arima Y, Ikeda K, Takayasu N, Nagakane Y, Yoshikawa K, Mizuno T, Nakagawa M, Furuhata H (2006) Acetazolamide vasoreactivity evaluated by transcranial ultrasonic power harmonic imaging and Doppler sonography. In: Kanno T, Kato Y (eds) Minimally invasive neurosurgery and multidisciplinary neurotraumatology. Springer-Verlag, Tokyo, pp 360–367
23. Shiogai T, Ikeda K, Matsumoto M, Morisaka A, Nagakane Y, Mizuno T, Nakagawa M, Furuhata H (2007) Transcranial brain tissue perfusion images by power modulation imaging in comparison with second harmonic imaging. Cerebrovasc Dis 23 (Suppl 1):15–16, (Abstract)
24. Sorteberg W, Lindegaard KF, Rootwelt K, Dahl A, Nyberg-Hansen R, Russell D, Nornes H (1989) Effect of acetazolamide on cerebral artery blood velocity and regional cerebral blood flow in normal subjects. Acta Neurochir (Wien) 97(3–4):139–145
25. Szabo S, Sheth RN, Novak L, Rozsa L, Ficzere A (1997) Cerebrovascular reserve capacity many years after vasospasm due to aneurysmal subarachnoid hemorrhage. A transcranial Doppler study with acetazolamide test. Stroke 28:2479–2482
26. Ulrich PT, Becker T, Kempski OS (1995) Correlation of cerebral blood flow and MCA flow velocity measured in healthy volunteers during acetazolamide and CO_2 stimulation. J Neurol Sci 129:120–130

Acta Neurochir Suppl (2008) 102: 185–188

A new semi-invasive method for two dimensional pO_2 measurements of cortical structures

Jan Warnat · Gregor Liebsch · Eva-Maria Stoerr · Alexander Brawanski · Chris Woertgen

Abstract
Background Measuring brain oxygenation in patients with TBI or SAH is of major interest. We present a new semi-invasive method for two dimensional measurements of cortical pO_2.
Methods For this feasibility study, a porphyrin containing sensor foil was placed directly on the cortex of intubated and variably ventilated Wistar rats. The sensor was excited with a light pulse and pictures of the foil's pO_2 dependant emissions were captured with a CCD camera. After online data processing, two-dimensional maps of cortex oxygenation were displayed and analyzed using ROIs (here: arteriole, vein, parenchyma) with a display rate of 7 Hz. The size of one single measurement pixel was $0.03 \times 0.03\ mm^2$.
Findings The mean pO_2 over cortex arterioles was 20.3 ± 0.69, over veins 17.1 ± 0.5 and over parenchyma 9.1 ± 0.6 (mmHg±SD). The arterial pO_2 showed a good correlation to the pO_2 in the ROIs ($r=0.46$–0.72, $p<0.0001$, $n=198$). Comparing groups with different p_aO_2 and p_aCO_2 we found significantly different pO_2 values in the ROIs of the cortex.
Conclusions This prototype is capable of obtaining cortical pO_2 maps with excellent temporal and spatial resolution and provides simultaneous imaging of the cortex structures.

Keywords Brain oxygen · Partial oxygen pressure · Time-resolved luminescence imaging · Neuromonitoring

Introduction

The continuous measurement of cerebral tissue oxygen tension is an important part of advanced neuromonitoring in patients with SAH or TBI. There is evidence that periods with low $p_{bt}O_2$ lead to an increase in secondary brain damage and deterioration in outcome [2, 10, 11]. In clinical practice, Clarke type electrodes or optical sensors are inserted in the parenchyma of the white matter. The measured $p_{bt}O_2$ reflects the regional oxygen concentration in the surrounding area of the probe, probably equating to the end-capillary pO_2 [7]. It can be assumed that a volume of only a few cubic millimeters is sampled [3, 7]. Furthermore, the possible influence of nearby vessels is difficult to control.

We present a new method for two-dimensional measurements of the partial pressure of oxygen of the cortical surface structures. The technique is based on the detection of oxygen-dependent quenching of the sensor luminescence [8, 9]. Although previously used for measurements on skin surfaces [1], this method is for the first time evaluated for cortical pO_2 detection. The setup provides a time-resolved two-dimensional luminescence lifetime image [4, 8, 9] and the online calculation of two dimensional maps of partial pressure of oxygen distribution over the cortex under the sensor.

The feasibility of the system was tested with rodent animals and the effects of variation of arterial pO_2 and pCO_2 on cortical pO_2 were investigated.

J. Warnat (✉) · E.-M. Stoerr · A. Brawanski · C. Woertgen
Department of Neurosurgery, University of Regensburg, Franz-Josef-Strauß-Allee 11, 93053 Regensburg, Germany
e-mail: Jan.Warnat@klinik.uni-regensburg.de

C. Woertgen
e-mail: chris.woertgen@klinik.uni-regensburg.de

G. Liebsch
BIOCAM GmbH, Friedenstraße 30, 93053 Regensburg, Germany

Materials and methods

Animal preparation The experimental protocol was approved by the animal care committee and included ten Wistar rats (376±33 g, Charles River), which were sedated, endotracheally intubated and ventilated with a gas mixture of isoflurane (1.2–1.5%), oxygen and nitrous oxide. The left femoral artery was catheterized and the animals were positioned in a stereotactic frame. Rectal body temperature was kept stable between 36.5°C and 37.5°C using a warming lamp. A craniectomy was performed on the right hemisphere and the dura was carefully opened and removed.

Setup for cortical oxygen measurement A light conducting sterilized polymethylmethacrylate (PMMA) cylinder (diameter 12/14 mm, length 50 mm) with the oxygen sensor foil (polysulfone containing a platinum (II)-octaethyl-porphyrin, surface area 110 mm^2) was placed gently on the cortex and was fixed with a holder. The colour CCD camera (AVT, Germany; resolution 780×580, 10 bit=1,024 grey scale values) was placed on axis over the probe. A ring of LED light sources (λ=405 nm) was mounted on the camera objective allowing light pulses entering the PMMA cylinder to be conducted to the sensor foil. Vice versa light emissions of the sensor foil were optically filtered by a 455 nm longpass filter (GG435; Schott, Germany) and detected by the camera through the cylinder. The LEDs and the camera were precisely controlled by a custom made trigger box (Biocam, Germany). A software program provided the online view of the camera pictures, control of the trigger box and the storage of the data (Biocam acquisition software). A special evaluation-software was used for online calculation and display of the resulting oxygen maps (IDL, Creaso). For offline analysis the program provided functions to read out the pO_2-dependent grey scale values of each pixel of the oxygen maps as well as of freely to determine rectangular regions of interest (ROIs). Moreover the grey scale oxygen maps can be converted into pseudocolour maps or three dimensional profiles.

Measurement protocol and data acquisition The p_aO_2 was varied by altering the FiO_2 between ~10% and ~100%. Additionally, the p_aCO_2 was modified by the ventilation settings while the FiO_2 remained constant at 30%. A blood gas was drawn and afterwards a measurement cycle was started. For each measurement cycle five subsequent sets of images were obtained and stored on hard disk. Each image set consisted of a normal colour image, a background luminescence image, an image of the excitation and one of the emission phases. The mean arterial blood pressure, heart rate and rectal body temperature were recorded for each measurement.

Data processing and analysis of pO_2 maps A decay time dependent parameter R was calculated by rationing the image containing the luminescence intensities of the excitation phase by the image containing the luminescence intensities measured in the emission phase. A calibration function was established for R versus pO_2 using an adapted Stern–Volmer relationship and R values were converted into the corresponding values of partial pressure. The transfer function was obtained by calibrating the sensor in a chamber under defined oxygen partial pressure:

$$pO_2[\mathrm{mmHg}] = \frac{\frac{0.95}{\frac{1.225}{R}-0.05} - 1}{0.0615}.$$

Regions of interest containing 10×10 pixels were set over an arterial structure, a vein and a parenchymal area without apparent vessels. Vessels were identified in the colour image of each measurement set as well as in the corresponding pO_2 maps. The average values for R of each ROI with its standard deviation were determined for every measurement. Statistical analysis was performed with Sigma Stat (SPSS inc.) using the Spearman rank order correlation.

Results

In all experiments colour pictures and luminescence pictures of the measurement region were obtained and displayed simultaneously at approximately 7 Hz. The image quality appeared to be good. The overall spatial resolution was 30×30 μm^2 per measurement pixel resulting in an acceptable picture resolution. Structures of the cortex were well recognizable (Fig. 1). Changes of cortical pO_2 due to variation of the FiO_2 were immediately and well noticeable throughout the experiments.

Altogether 225 measurement cycles with 63,000 single data points were collected. The mean arterial blood pressure ranged from 88.2±5.9 to 96.9±8.1 mmHg. There was no correlation between changes in arterial blood pressure and cortical pO_2 (in all ROIs, $p>0.05$, $n=198$). The arterial pO_2 changed throughout the experiments between 13.0 and 424.7 mmHg. The average arterial pO_2 was 157.5± 102.2 mmHg. The p_aCO_2 ranged from 33.5 to 46.0 mmHg except for the pCO_2 reactivity studies for which p_aCO_2 was altered between 25 and 68 mmHg. The body core temperature remained relatively constant (36.8±0.6°C).

The mean cortical pO_2 over arterial vessels was 20.3± 10.8 mmHg, over veins 16.9±8.0 mmHg and over the parenchyma 9.4±7.7 mmHg. These groups were significantly different ($p<0.0005$). The relationship of p_aO_2 and the cortex pO_2 over an arteriole and a vein is shown in

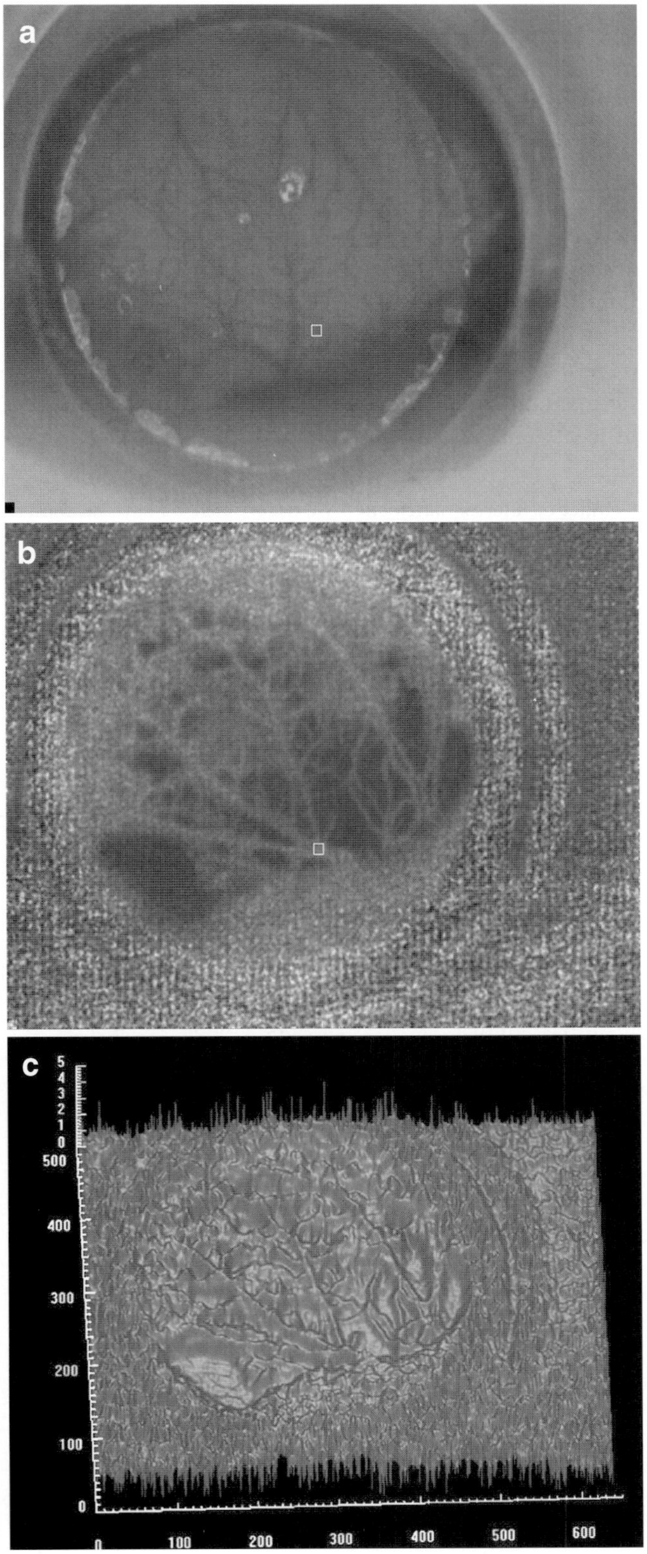

Fig. 1 **a** Conventional camera view through the PMMA cylinder. Venous vessels can be recognized. **b** Calculated online pO_2 map of the same measurement. Bright pixels correspond to high pO_2 values indicating e.g. arterial vessels. *Square* ROI defined over an arterial vessel. **c** Three-dimensional representation. The artefacts surrounding the central measurement area may be excluded by further processing

Fig 2. Arterial systemic pO_2 correlated well with the cortical pO_2 over arteries (correlation coefficient 0.72, $p<0.001$). Smaller, but still strong correlations ($p<0.001$) were found between p_aO_2 and pO_2 over cortical veins (coeff.= 0.58) and over cortical parenchyma (coeff.=0.46). The ROIs over a vein were used for an exemplary analysis of p_aCO_2 and p_aO_2 effects. An increase of either p_aO_2 or p_aCO_2 was followed by an increase of cortical vein pO_2. The comparison of groups with a p_aO_2 below 80 mmHg, between 80 and 120 mmHg and above 120 mmHg showed significant ($p<0.001$) differences in mean cortical pO_2 values: 8.1, 15.7 and 20.3 mmHg respectively. Groups with a p_aCO_2 below 30 mmHg, between 30 and 40 mmHg and above 40 mmHg also showed an increasing and significantly different cortical pO_2 of 12.9, 14.3 and 16.9 mmHg ($p<0.009$).

In order to evaluate the reproducibility we analyzed 106 randomly chosen sets of five subsequent measurements. The average standard deviation for the pO_2 was 1.78% of

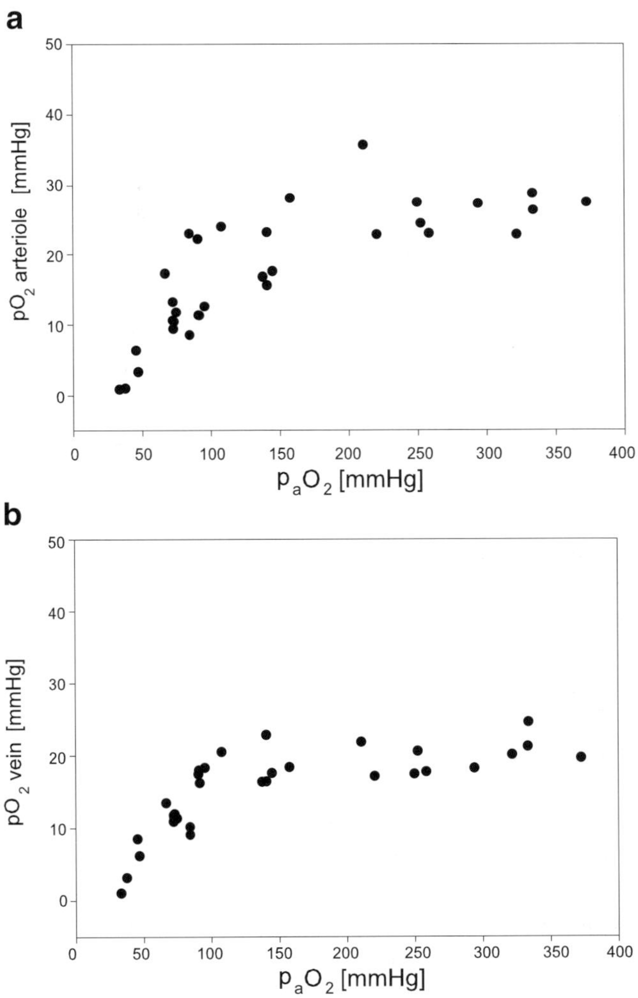

Fig. 2 Relationship of arterial oxygen tension (p_aO_2) with cortical pO_2 over an arteriole (**a**) and a vein (**b**). Individual data points of one representative experiment

the average value for the sequence of five consecutive measurements. The average variance for sets of five subsequent measurements was 0.002. The maximum deviation of one single measurement compared to the average of the five consecutive measurements was 5.88%, although this value was below 3% for 89% of the single measurements.

The visualisation of the pO_2 maps as grey scale images appeared to be sufficient. Three dimensional representations may be more intuitive (Fig. 1); moreover the maps can also be colour coded (not shown).

Discussion

This study shows the suitability of a new technique for the in vivo measurement of cortical pO_2 and its capability of simultaneous measurements of distinct anatomical structures such as supplying and draining vessels. Despite the technical limitations at this stage, the prototype produced sufficient pO_2 image maps of the cortex. The fast reaction of the system even to smaller changes in FiO_2 was recognisable in the online maps and is resembled by the strong correlation of the systemic arterial pO_2 to the obtained cortical measurements. The resolution and the signal to noise ratio, although quite sufficient in the present study, can certainly be increased by the integration of a microscope and by optimisation of the light source and the light conduction.

Since this kind of online imaging of the cortex oxygenation is novel, there are no standard pO_2 values available. Data from intraparenchymal probes (e.g. Licox, Rehau) may not be comparable, because these devices measure an average pO_2 in their surroundings and moreover, are usually placed in the white matter with unclear relation to the cortical vessels.

Single consecutive pO_2 measures with polarographic microelectrodes have been utilized to investigate the pO_2 near to and in vessels of the rat cortex [5, 6, 12] and appear to be consistent with our measures. However, true simultaneous measurements require a two dimensional system.

The correlation coefficients and p values show clearly that systemic and cortical pO_2-values are closely coupled parameters and indicate the good reactivity of the measurement system. The findings may not reflect normal physiological conditions due to the relatively invasive approach in this feasibility study. Possible effects of the invasive craniectomy, pressure effects by the cylinder on the cortex and effects the variation of the N_2O content of the ventilation gas are not considered. Since the sensor foil is placed very close to the cortical vessels or parenchyma, measurements over these regions should represent the true local pO_2. A different local pO_2 may occur in some distance to the sensor foil which then will not be detected. This limitation applies for all local pO_2 measurement devices. The possibility to measure a bigger region of cortex with the presented method may partly overcome this limitation.

At this stage we can only speculate that measuring tissue oxygenation of a whole cortex area might be advantageous for monitoring patients and complementary to the detection of local white matter oxygenation.

Conflict of interest statement We declare that we have no conflict of interest.

References

1. Babilas P, Liebsch G, Schacht V, Klimant I, Wolfbeis OS, Szeimies RM, Abels C (2005) In vivo phosphorescence imaging of pO_2 using planar oxygen sensors. Microcirculation 12:477–487
2. Dings J, Jager A, Meixensberger J, Roosen K (1998) Brain tissue pO_2 and outcome after severe head injury. Neurol Res 20(Suppl 1):S71–S75
3. Hoffman WE, Charbel FT, Abood C, Ausman JI (1997) Regional ischemia during cerebral bypass surgery. Surg Neurol 47:455–459
4. Holst G, Kohls O, Klimant I, König B, Kühl M, Richter T (1998) A modular luminescence lifetime imaging system for mapping oxygen distribution in biological samples. Sens Actuators 51:163–170
5. Ivanov KP, Derry AN, Vovenko EP, Samoilov MO, Semionov DG (1982) Direct measurements of oxygen tension at the surface of arterioles, capillaries and venules of the cerebral cortex. Pflugers Arch 393:118–120
6. Ivanov KP, Sokolova IB, Vovenko EP (1999) Oxygen transport in the rat brain cortex at normobaric hyperoxia. Eur J Appl Physiol Occup Physiol 80:582–587
7. Kett-White R, Hutchinson PJ, Czosnyka M, Boniface S, Pickard JD, Kirkpatrick PJ (2002) Multi-modal monitoring of acute brain injury. Adv Tech Stand Neurosurg 27:87–134
8. Liebsch G, Klimant I, Frank B, Holst G, Wolfbeis OS (2000) Luminescence lifetime imaging of oxygen, pH, and carbon dioxide distribution using optical sensors. Appl Spectrosc 54:548–559
9. Liebsch G, Klimant I, Krause C, Wolfbeis OS (2001) Fluorescent imaging of pH with optical sensors using time domain dual lifetime referencing. Anal Chem 73:4354–4363
10. Meixensberger J, Vath A, Jaeger M, Kunze E, Dings J, Roosen K (2003) Monitoring of brain tissue oxygenation following severe subarachnoid hemorrhage. Neurol Res 25:445–450
11. Nortje J, Gupta AK (2006) The role of tissue oxygen monitoring in patients with acute brain injury. Br J Anaesth 97:95–106
12. Vovenko E (1999) Distribution of oxygen tension on the surface of arterioles, capillaries and venules of brain cortex and in tissue in normoxia: an experimental study on rats. Pflugers Arch 437:617–623

Acta Neurochir Suppl (2008) 102: 189–192

The relationship between intracranial pressure and brain oxygenation following traumatic brain injury in sheep

Robert Vink · Kartik D. Bahtia · Peter L. Reilly

Abstract

Background While it is understood that raised intracranial pressure (ICP) after traumatic brain injury (TBI) may negatively impact on brain tissue oxygenation ($P_{bt}O_2$), few studies have characterized the inter-relationship between these two variables, particularly in a large animal model that replicates the human gyrencephalic brain. The current study uses an ovine model to examine the dynamics of ICP and $P_{bt}O_2$ after TBI.

Materials and methods Five 2-year-old male Merino sheep were anesthetized with isoflurane and impacted in the left temporal region using a humane stunner. ICP and $P_{bt}O_2$ were then monitored over the following 4 h using a Codman ICP Express monitoring system and a LICOX brain tissue oxygen monitoring system, respectively. Two additional sheep were anesthetized and monitored as sham (uninjured) controls.

Findings Mean ICP 60 min following TBI was over 25 mmHg ($p<0.05$ versus controls) and by 4 h, values were consistently greater than 30 mmHg ($p<0.001$). With respect to $P_{bt}O_2$, values fell from mean control values of 52± 11 to 20±4 mmHg by 60 min ($p<0.001$) and by 4 h to 14± 3 mmHg ($p<0.01$). The sigmoidal relationship between the two variables included a negative linear correlation when ICP was between 13 to 27 mmHg.

Conclusions Our results suggest that TBI results in early changes in ICP that are associated with profound declines in $P_{bt}O_2$, and may indicate the need for earlier management of ICP after TBI.

Keywords Neurotrauma · Brain swelling · Intracranial pressure · Brain oxygen · Sheep

Supported, in part, by a Motor Accident Commission of South Australia Grant to RV.

R. Vink (✉)
Discipline of Pathology, School of Medical Sciences,
University of Adelaide,
Adelaide, SA 5005, Australia
e-mail: Robert.Vink@adelaide.edu.au

R. Vink · K. D. Bahtia · P. L. Reilly
Discipline of Neurosurgery, School of Medicine,
University of Adelaide,
Adelaide, SA 5005, Australia

K. D. Bahtia
e-mail: Kartik.Bahtia@adelaide.edu.au

P. L. Reilly
e-mail: Peter.Reilly@adelaide.edu.au

Introduction

Raised intracranial pressure (ICP) is widely thought to account for a significant proportion of the resultant mortality and morbidity after traumatic brain injury (TBI). Indeed, a number of studies have supported an association between raised ICP and neurological outcome [1, 2, 6], with a recent systematic review of the literature demonstrating that absolute ICP values are in fact predictive of neurological outcome, with refractory ICP and response to treatment being the most accurate predictors [14]. Despite the importance of raised ICP to outcome, there have been few significant improvements in the pharmacological management of ICP since the implementation of mannitol therapy, in part because of the lack of appropriate experimental models that can duplicate the changes in ICP that occur in clinical TBI.

A number of experimental studies have investigated the dynamics of brain swelling after experimental TBI [3, 4, 10], although few have incorporated measurement of ICP, and none have measured both ICP and concurrent brain oxygenation ($P_{bt}O_2$). Those studies that have investigated ICP after TBI have generally used rodent models [5, 11, 13] that in most cases incorporate a mass lesion effect due to the severity or focal nature of the induced injury. Rodent models also have inherent deficiencies including a small, lissencephalic brain, small head size relative to body size and a baseline ICP that is significantly different from that in humans. Thus it is impossible to correlate these rodent findings with what might occur in a gyrencephalic human brain after injury.

In contrast to the rodent models, the use of large animal models of TBI would permit the direct determination of ICP and $P_{bt}O_2$ in a gyrencephalic brain with substantial white matter domains. A large animal model has been previously used for the investigation of ICP after TBI [9]. This porcine study also induced a focal injury of a high severity to induce significant changes in ICP, presumably related, in part, to mass lesion effects. $P_{bt}O_2$ was not measured in this particular study. An alternative large animal model of TBI using sheep has been developed at the University of Adelaide, Australia [8] and uses a humane stunner to produce an acceleration type of injury with significant ICP changes. In the absence of skull fracture, there is no development of significant mass lesions in this model. This ovine model has also been well characterized in terms of physiological and neuropathological responses after TBI, and has been successfully used in a number of pharmacological studies [15, 16]. We have therefore used this sheep model of TBI to determine changes in ICP and levels of $P_{bt}O_2$ following traumatic brain injury.

Materials and methods

Subjects and surgery

All studies were performed according to the guidelines established by the National Health and Medical Research Council for the use of animals in experimental research and were approved by the Animal Ethics Committees of the Institute of Medical and Veterinary Science and the University of Adelaide.

Six 2-year-old male Merino sheep were impacted in the left temporal region by a humane stunner as described in detail elsewhere [8, 15]. Briefly, sheep were continuously anesthetized and ventilated using 2.5% isoflurane in oxygen (4 L/min). Animals were then placed into a prone sphinx position, and restrained to the table, leaving the neck and head mobile relative to the body. Impact injury was induced at the midpoint between the left supraorbital process and the left external auditory meatus using a Captive humane bolt stunner armed with a number 7 red charge (model KML, Karl Schermer and Co., Germany). We have previously shown that this impact causes severe diffuse axonal injury [8, 15]. A further two animals were used as surgically prepared but not injured controls.

ICP and brain oxygen monitoring

After injury, animals were stabilized and the heads restrained to the operating table to facilitate insertion of ICP and $P_{bt}O_2$ probes between 15 and 30 min after trauma. Following exposure of the skull, a 5.8 mm burr hole was performed at a point 4 cm lateral to the sagittal midline on the ipsilateral side, the dura matter opened and a calibrated Codman Microsensor ICP transducer inserted such that the tip of the sensor was 1.5 cm into the parenchyma of the left parietal lobe. The probe was attached to a Codman ICP Express monitoring system (Codman and Shurtleff Inc., USA) for digital recording. A second burr hole 1 cm lateral to the sagittal midline and over the left fronto-parietal suture allowed insertion of the distal end of a LICOX® $P_{bt}O_2$ probe to a depth of 2 cm. The probe was attached to a LICOX® brain tissue oxygen monitoring system (Integra, USA) for digital recording. After insertion of the probes, both burr holes were sealed using bone wax. ICP and $P_{bt}O_2$ was recorded every 30 min for a period of 4 h, after which animals were euthanized and their brains removed for assessment of possible mass lesions.

Statistical analyses

Statistical analyses were performed by repeated-measures, two-way analysis of variance using PRISM (Graphpad, San Diego, CA, USA). When the analysis of variance showed a significant effect, the Bonferroni post hoc test was used to establish group differences. All data are expressed as mean± SD. A p value of ≤ 0.05 was considered significant.

Results

Of the six animals that were injured, one incurred a skull fracture with dural tearing and significant intracranial hemorrhage. This animal was eliminated from the study. The remaining five animals did not show any significant hemorrhage throughout the brain during post mortem neuropathological examination and were thus considered to be without mass lesions that could have contributed to changes in ICP.

In control, non-injured animals, the mean ICP over the 4 h assessment period was 13±4 Hg m g (Fig. 1). Insertion of the ICP transducer in control animals was associated with a slightly higher value being recorded initially, but this resolved within the next 30 min and no further changes

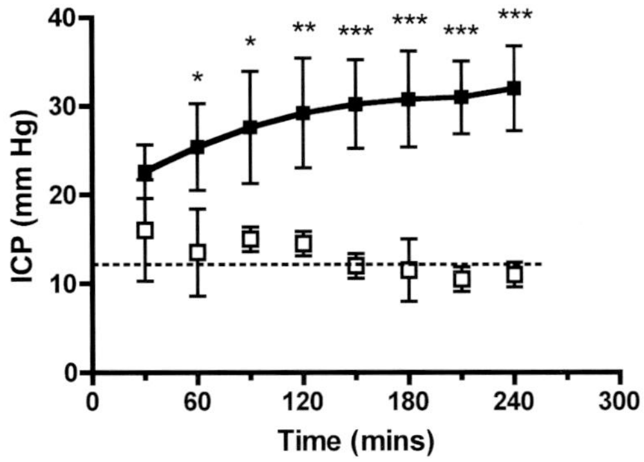

Fig. 1 Changes in ICP following traumatic brain injury in the sheep. *Empty squares*: sham controls; *filled squares*: injured animals. *Dotted line* represents the mean ICP value after probe stabilization. *$p<0.05$, **$p<0.01$, ***$p<0.001$ versus sham controls

were apparent for the remainder of the 4 h monitoring period. In contrast, induction of TBI resulted in a highly significant increase in ICP values relative to controls (ANOVA, $p<0.001$). An immediate increase in ICP to values over 20 mmHg occurred within the first 30 min, and by 60 min, mean ICP was over 25 mmHg ($p<0.05$ versus controls). ICP continued to increase with time such that by 4 h after injury, values were consistently greater than 30 mmHg ($p<0.001$ versus controls).

The mean $P_{bt}O_2$ in control animals over the 4 h monitoring period was 52±11 mmHg (Fig. 2). The values recorded in controls immediately following insertion of the oxygen probe were initially high, but stabilized within 2 h of insertion to a value of 44±6 mmHg. This extended time necessary for the stabilization of the probe is a known feature of such $P_{bt}O_2$ sensors. Despite the time necessary for stabilization of the probe, there was a highly significant (ANOVA, $p<0.001$) and early decrease in $P_{bt}O_2$ recorded following TBI. By 30 min after injury, $P_{bt}O_2$ was approximately 35 mmHg, while by 60 min, it had decreased to 20±4 mmHg ($p<0.001$ versus controls). The $P_{bt}O_2$ continued to decline with time such that by 4 h, the value was 14±3 mmHg ($p<0.01$ versus controls), or 27% of the mean control values.

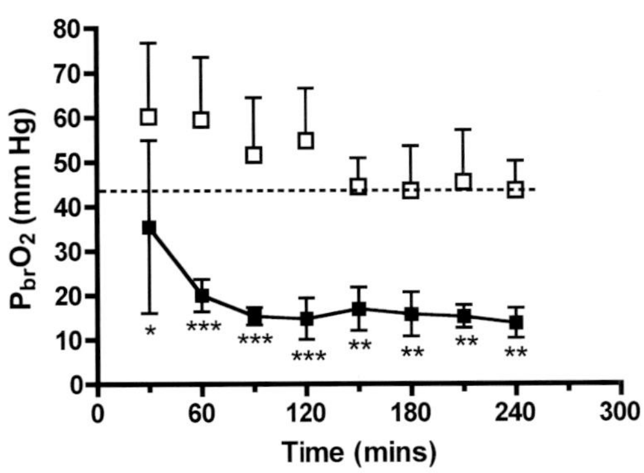

Fig. 2 Changes in brain tissue oxygen levels (mmHg) after traumatic brain injury in sheep. *Empty squares*: sham controls; *filled squares*: injured animals. *Dotted line* represents the mean $P_{bt}O_2$ value after probe stabilization. *$p<0.05$, **$p<0.01$, ***$p<0.001$ versus sham controls

The interrelationship between the ICP and $P_{bt}O_2$ changes are summarized in Fig. 3. The data was best fitted with a sigmoidal curve ($r^2=0.66$) showing oxygen saturation at 45 mmHg, which was similar to the normal values recorded in sham, control animals. Minimum oxygen levels were observed when ICP was above 27 mmHg. A linear relationship between the two variables was noted when ICP was between 13 and 27 mmHg ($r^2=0.58$; $p<0.001$), suggesting that brain oxygen values begin to fall as soon as ICP rises.

Discussion

In the current study, we have demonstrated the utility of a sheep model of TBI to investigate the interrelationship between ICP and $P_{bt}O_2$ after brain injury. Prior to injury, the normal ICP value in sham, control animals was 13 mmHg, while the normal $P_{bt}O_2$ value was 44 mmHg. Both values are similar to normal ICP and $P_{bt}O_2$ values previously reported in clinical studies [7]. Similarly, the temporal kinetics and absolute values of ICP and $P_{bt}O_2$ changes after injury were similar to that previously reported after clinical TBI [7], suggesting that the ovine model of TBI used in the current study is ideal as an approximation of the clinical situation, and may be used to study the dynamic interrelationship between these two variables.

Increases in ICP were noted immediately after TBI demonstrating the rapidity of events after trauma. By 30 min after trauma, the mean ICP value had increased to above 20 mmHg, a value previously shown in itself to be associated with increased mortality [2, 6]. Similarly, brain tissue oxygen levels of less than 20 mmHg have been associated with an increased mortality [12], and these

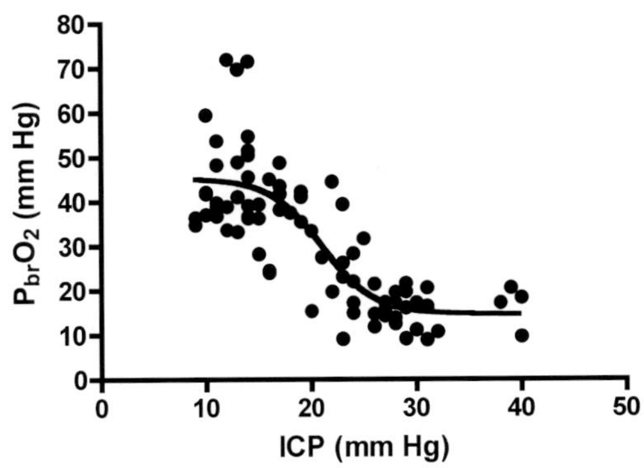

Fig. 3 Relationship between ICP and $P_{bt}O_2$ after traumatic brain injury in sheep. $r^2=0.66$

values were observed within 60 min after trauma. In combination, such values both occur within 60 min after trauma, and continued to deteriorate further with time. ICP values of less than 20 mmHg have been widely considered a treatment threshold in severe TBI patients since mortality increases sharply with ICP values above this threshold [2]. Our findings support that 20 mmHg may be considered a threshold given the interrelationship between ICP and $P_{bt}O_2$. Indeed, in the current study, $P_{bt}O_2$ had already declined by approximately 33% by the time ICP has risen to this value. As ICP increased to 25 mmHg, $P_{bt}O_2$ declined to only 45% of baseline values (20 mmHg). Clearly, any sustained increase in ICP may render the brain hypoxic for prolonged periods of time. Notably an ICP of 27 mmHg or higher was associated with the very low $P_{bt}O_2$ (14 mmHg). Whether this occurs clinically in severe TBI patients needs to be further investigated as it may impact management decisions, including the need for decompressive craniectomy.

In conclusion, our results suggest that changes in ICP and $P_{bt}O_2$ occur very rapidly after TBI and reach critical threshold values within the first hour after trauma. They also highlight the need for early and effective management of ICP given the linear relationship between a rise in ICP and a decline in $P_{bt}O_2$ when ICP is between 13 and 27 mmHg. Given the ability of the ovine TBI model to reproduce the clinical profile for ICP and $P_{bt}O_2$, further studies using this model will be extremely useful in the characterization of current and future ICP management therapies.

Acknowledgements We thank Dr John Finnie, Dr Tim Kuchel and Ms Jodie Dyer of the Hanson Institute Veterinary Services Division for assistance with the sheep experiments.

Conflict of interest statement We declare that we have no conflict of interest.

References

1. Aarabi B, Hesdorffer DC, Ahn ES, Aresco C, Scalea TM, Eisenberg HM (2006) Outcome following decompressive craniectomy for malignant swelling due to severe head injury. J Neurosurg 104:469–479
2. Balestreri M, Czosnyka M, Hutchinson P, Steiner LA, Hiler M, Smielewski P, Pickard JD (2006) Impact of intracranial pressure and cerebral perfusion pressure on severe disability and mortality after head injury. Neurocrit Care 4:8–13
3. Barzo P, Marmarou A, Fatouros P, Hayasaki K, Corwin F (1997) Contribution of vasogenic and cellular edema to traumatic brain swelling measured by diffusion-weighted imaging. J Neurosurg 87:900–907
4. Cernak I, Vink R, Zapple DN, Cruz MI, Ahmed F, Chang T, Fricke ST, Faden AI (2004) Characterization of a highly adaptable, new model of diffuse traumatic brain injury in rodents. Neurobiol Dis 17:29–43
5. Engelborghs K, Verlooy J, Van Reempts J, Van Deuren B, Van de Ven M, Borgers M (1998) Temporal changes in intracranial pressure in a modified experimental model of closed head injury. J Neurosurg 89:796–806
6. Jagannathan J, Okonkwo DO, Dumont AS, Ahmed H, Bahari A, Prevedello DM, Jane JA Sr., Jane JA Jr (2007) Outcome following decompressive craniectomy in children with severe traumatic brain injury: a 10-year single-center experience with long-term follow up. J Neurosurg 106:268–275
7. Korsic M, Jugovic D, Kremzar B (2006) Intracranial pressure and biochemical indicators of brain damage: follow-up study. Croat Med J 47:246–252
8. Lewis SB, Finnie JW, Blumbergs PC, Scott G, Manavis J, Brown C, Reilly PL, Jones NR, McLean AJ (1996) A head impact model of early axonal injury in the sheep. J Neurotrauma 13:505–514
9. Manley GT, Rosenthal G, Lam M, Morabito D, Yan D, Derugin N, Bollen A, Knudson MM, Panter SS (2006) Controlled cortical impact in swine: pathophysiology and biomechanics. J Neurotrauma 23:128–139
10. O'Connor C, Cernak I, Vink R (2006) The temporal profile of edema formation differs between male and female rats following diffuse traumatic brain injury. Acta Neurochir (Suppl) 96:121–124
11. Prough DS, Kramer GC, Uchida T, Stephenson RT, Hellmich HL, Dewitt DS (2006) Effects of hypertonic arginine on cerebral blood flow and intracranial pressure after traumatic brain injury combined with hemorrhagic hypotension. Shock 26:290–295
12. Stiefel MF, Udoetuk JD, Spiotta AM, Gracias VH, Goldberg A, Maloney-Wilensky E, Bloom S, Le Roux PD (2006) Conventional neurocritical care and cerebral oxygenation after traumatic brain injury. J Neurosurg 105:568–575
13. Thomale UW, Griebenow M, Kroppenstedt SN, Unterberg AW, Stover JF (2006) The effect of *N*-acetylcysteine on posttraumatic changes after controlled cortical impact in rats. Intensive Care Med 32:149–155
14. Treggiari MM, Schutz N, Yanez ND, Romand JA (2007) Role of intracranial pressure values and patterns in predicting outcome in traumatic brain injury: a systematic review. Neurocrit Care 6:104–112
15. Van Den Heuvel C, Donkin JJ, Finnie JW, Blumbergs PC, Kuchel T, Koszyca B, Manavis J, Jones NR, Reilly PL, Vink R (2004) Downregulation of amyloid precursor protein (APP) expression following post-traumatic cyclosporin-A administration. J Neurotrauma 21:1562–1572
16. Van Den Heuvel C, Finnie JW, Blumbergs PC, Manavis J, Jones NR, Reilly PL, Pereira RA (2000) Upregulation of neuronal amyloid precursor protein (APP) and APP mRNA following magnesium sulphate (MgSO4) therapy in traumatic brain injury. J Neurotrauma 17:1041–1053

Acta Neurochir Suppl (2008) 102: 193–196

Dual microdialysis probe monitoring for patients with traumatic brain injury

Hiroyasu Koizumi · Hirosuke Fujisawa · Sadahiro Nomura · Shoichi Kato · Koji Kajiwara · Masami Fujii · Michiyasu Suzuki

Abstract

Background In vivo microdialysis can be used to examine the focal metabolism by measuring the extracellular concentration of amino acids and other biochemicals in the focal brain tissue in which the probe is placed. We report two patients with traumatic brain injury who underwent placement of dual microdialysis probes.

Materials and methods One probe was placed in the penumbra zone and another in a region remote to the injury. Multiple measurements of several biochemical's were made.

Findings In the first case, a 54-year-old man with right acute subdural hematoma and right fronto-temporal contusion was monitored. Quantitative analysis of in vivo microdialysis was performed. While the extracellular concentration of lactate and glycerol were higher in the penumbra zone than in the region remote to the injury, glutamate remained lower in the penumbra zone than in the remote region. In the second case, dual microdialysis probes were placed in a 56-year-old woman with left acute subdural hematoma and left fronto-temporal lobe contusion. Also in this case, the extracellular glutamate remained lower in the penumbra zone than in the remote region.

Conclusions The reason why the extracellular glutamate value remained lower in the penumbra zone than in the remote region is unclear. The position of each microdialysis probe was ensured by CT scan after the operation. It is important to be aware of the limitations of performing microdialysis in brain injured patients with a single probe.

H. Koizumi (✉) · H. Fujisawa · S. Nomura · S. Kato · K. Kajiwara · M. Fujii · M. Suzuki
Department of Neurosurgery, Yamaguchi University School of Medicine, 1-1-1 Minami-kogushi, Ube, Yamaguchi 755-8505, Japan
e-mail: hiroyasu@yamaguchi-u.ac.jp

Keywords In vivo microdialysis · Traumatic brain injury · Lactate/pyruvate ratio · Glutamate

Introduction

In vivo microdialysis methods have been established for monitoring patients with traumatic brain injury (TBI) [1–4]. In this unique manner, it is possible to quantify the interstitial levels of biochemicals in contused human brains. However, the data obtained from this monitoring reflects cerebral metabolisms only in a small area of the brain tissue where the catheter is placed, being unlike other monitoring modalities such as ICP and $SjvO_2$. Since in the clinical setting, an evaluation of cerebral metabolism in a contused human brain should be included in clinical decision making, it is important to be aware of differences in focal brain metabolisms between the penumbra zone [5] and the area remote to the injury. Although in Europe and the United States, many clinical institutions have introduced this monitoring method for patients with TBI, most of them have practiced with single microdialysis probe insertion. In this report, we describe patients who underwent dual microdialysis probe monitoring following TBI. Engström et al. reported a study of 22 consecutive patients with TBI using the in vivo microdialysis method, which showed that interstitial glutamate levels were higher in the penumbra zone than in the remote area [1]. Their findings are consistent with our expectations; however they experienced two exceptional cases showing gradual normalization for glutamate levels in the penumbra zone. Elevated interstitial lactate and glutamate levels observed during the microdialysis monitoring period may be explicable to be following an occurrence of ischemic insult in the brain, however it is difficult to account for the cases to the contrary.

Therefore it is also important to be aware of the limited capability of the measured area of a single probe.

Materials and methods

Patients

Case 1 is a 54-year-old man who sustained a traumatic brain injury after falling at a wharf. A head CT scan on admission showed an acute subdural hematoma on the right side, a right frontal lobe contusion and 12 mm midline shift from the right to the left. This patient eventually underwent decompressive hemicraniectomy. Microdialysis probes were inserted, during this open surgery, into the right frontal lobe as the penumbra zone and into the right temporal lobe as the region remote to the injury (Fig. 1 *upper*).

Case 2 is a 56-year-old woman who sustained a traumatic brain injury by falling off a moving bicycle. A head CT scan on admission showed an acute subdural hematoma on the left side, a left temporal lobe contusion and 5 mm midline shift from the left to the right. This patient also eventually underwent decompressive hemicraniectomy. Microdialysis probes were implanted into the left temporal lobe cortex as the penumbra zone and into the left frontal lobe as the region remote to the injury (Fig. 1 *lower*).

Microdialysis method

Microdialysis probes (CMA 70, CMA Microdialysis, Stockholm, Sweden) implanted into each of the targeted regions were perfused with Ringer's solution at a flow rate of 0.3 μl/min. Dialysates were collected at 2-h intervals during the study period. Extracellular levels of glucose, lactate, pyruvate, glycerol, glutamate were analyzed using conventional enzymatic techniques (ISCUS, a compact portable microdialysis analyzer, CMA Microdialysis, Stockholm, Sweden).

Results

In this investigation, microdialysis probes implanted into the penumbra zone and a region remote to the injury. Placement was confirmed by CT scans during the monitor-

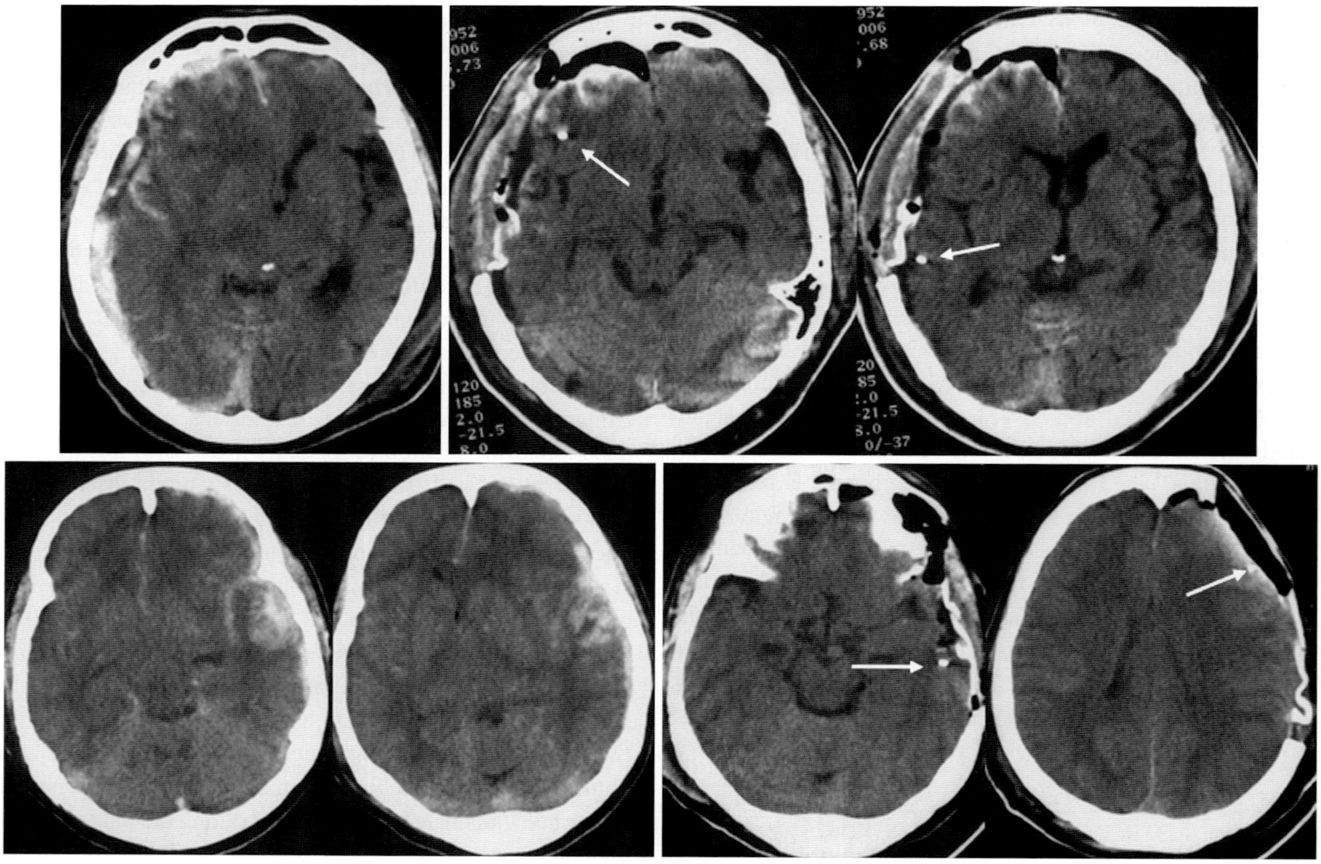

Fig. 1 In Case 1, a head CT scan on admission showed an acute subdural hematoma and a right frontal lobe contusion (*upper right*), and microdialysis probes were confirmed to be implanted into right frontal lobe and right temporal lobe with a review of the CT scans performed after the open surgery (*upper left*). In Case 2, an acute subdural hematoma and a left temporal lobe contusion were revealed on a head CT scan (*lower right*), and microdialysis probes were identified to be implanted into the left temporal lobe and frontal lobe correctly (*lower left*)

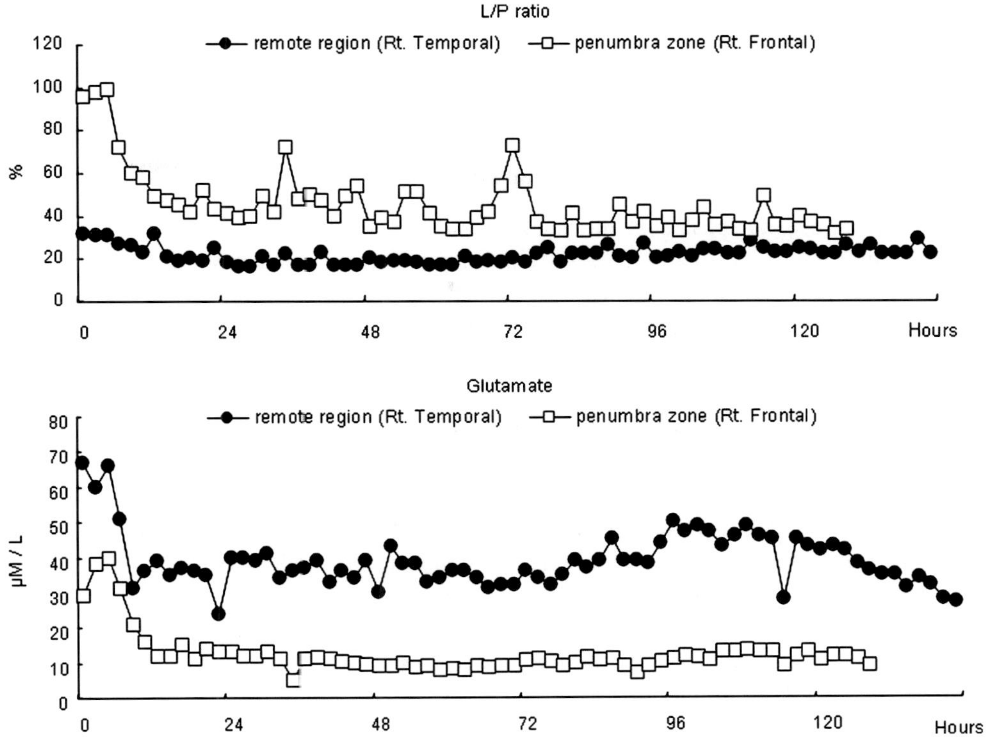

Fig. 2 Extracellular levels of lactate were higher in the penumbra zone than in the remote region and that of pyruvate had no serious difference between the two areas; therefore the lactate/pyruvate ratio was inevitably higher in the penumbra zone than in the remote region (*upper*). Extracellular levels of glutamate were higher in the remote region than in the penumbra zone (*lower*)

ing period after the open surgery in both cases (Fig. 1). Line graphs showing the time course changes in extracellular levels of glucose, lactate, pyruvate, and glycerol both in the penumbra zone and in the remote region have been omitted, because of space limitation.

In Case 1, the extracellular levels of glucose were lower in the penumbra zone than in the remote region. Although, interstitial levels of pyruvate were similar in the penumbra zone and the remote region, interstitial levels of lactate were higher in the penumbra zone than in the remote region. That is why the lactate/pyruvate ratios were higher in the penumbra zone than in the remote region (Fig. 2 *upper*). Similar to the changes in extracellular levels of lactate, glycerol levels were higher in the penumbra zone than in the remote region. Contrary to our expectations, extracellular levels of glutamate were higher in the remote region than in the penumbra zone (Fig. 2 *lower*).

In Case 2, the changes in the extracellular levels of glucose, lactate, pyruvate, and glycerol were the same as in Case 1. As shown in Fig. 3, also in Case 2, the changes in interstitial glutamate levels were higher in the remote region than in the penumbra zone (Fig. 3).

Discussion

Although in vivo microdialysis monitoring is useful for quantifying the extracellular levels of biochemicals in the cerebral cortex of human brain during the period of intensive care following TBI. However, neuromonitoring with only a single microdialysis probe might not fully represent the state of extracellular levels of biochemicals the entire brain. In fact, the results of this investigation show that interstitial levels of lactate and lactate/pyruvate

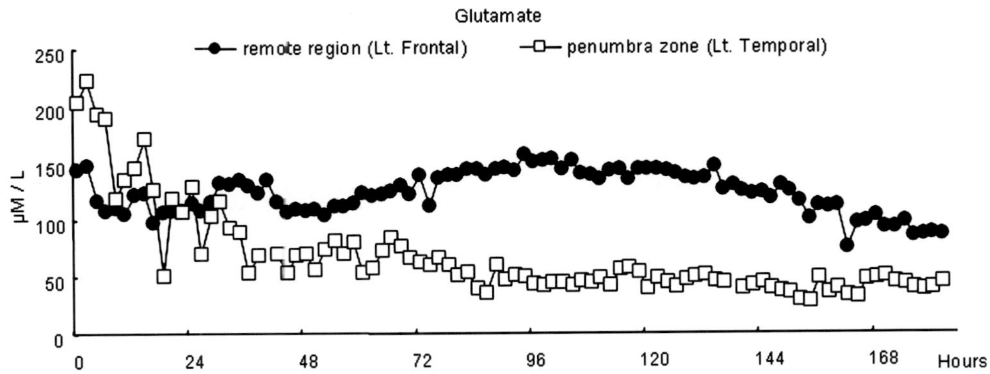

Fig. 3 Also in Case 2, similar to Case 1, the changes in extracellular levels of glutamate were higher in the remote region than in the penumbra zone

ratios were higher in the penumbra zone, which indicates ischemic insult in the penumbra zone, while levels of glutamate were lower in the same lesion. As the data obtained from the microdialysis monitoring should be included in clinical decision making, in the two cases shown in this study, if we monitored only with a single microdialysis catheter placed in the region remote to the injury, interpretation of the current condition of the contused brain would be unknown and the plan for treatment of the patients might be changed. Therefore, in vivo microdialysis reflects the focal metabolism in only the small area of the brain tissue in which the probe is placed, but does not reflect the metabolism of the global cerebral hemisphere. It may be risky to base the decision for treatment of patients only on the results of single probe monitoring for in vivo microdialysis. In order to avoid these risks, we are continuing dual microdialysis monitoring for patients with TBI.

The reason why interstitial levels of glutamate were lower in the penumbra zone, where the lactate/pyruvate ratio was higher, than in the remote region, is unclear. The in vivo microdialysis study by Engström et al. included 22 consecutive patients with traumatic brain injury [1]. They inserted microdialysis catheters into the penumbra zone and outside the penumbra zone, and in some cases, into the contralateral normal hemisphere. In their study, consistent with our expectations, the glutamate levels were within the normal range in ipsilateral and contralateral normal brain tissue, however they were remarkably high in the penumbra zone. The time course of changes in calculated lactate/pyruvate ratios were similar to changes in the glutamate levels. These results are in agreement with our expectations, however it was notable that 22% of the catheters were not located in the sites expected, although these catheters had been inserted during open surgery. This misplacement of microdialysis catheters means that it is sometimes difficult to maintain the precise location of a microdialysis catheter under control, even if it were placed under direct visualization during open surgery. Furthermore, it is also difficult to know the biological condition of the brain tissue where the catheter is placed only with CT scan images. These issues might be related to the discrepancy found in the present investigation. In addition, Engström et al. experienced two exceptional cases in which -a gradual normalization for glutamate levels and lactate/pyruvate ratios in the penumbra zone were observed. Unfortunately, in relation to the exceptional two cases, comparison with the time course changes in glutamate levels outside the penumbra zone were not obtained in their study. These exceptional cases in their study might correspond to our two cases presented here. One reason may be that the extracellular levels of glutamate become exhausted in the penumbra zone following traumatic brain injury. In any case, the question may arise as to whether a single microdialysis catheter monitoring is reliable. Therefore, dual microdialysis catheter monitoring, which enables us to compare the time courses of simultaneous changing in extracellular levels of biochemicals between the penumbra lesion and the remote region, is appropriate for the management of patients with severe TBI in our institution.

Conflict of interest statement We declare that we have no conflict of interest.

References

1. Engström M, Polito A, Reinstrup P, Romner B, Ryding E, Ungerstedt U, Nordström CH (2005) Intracerebral microdialysis in severe brain trauma: the importance of catheter location. J Neurosurg 102:460–469
2. Reinstrup P, Ståhl N, Mellergard P, Uski T, Ungerstedt U, Nordström CH (2001) Intracerebral microdialysis in clinical practice: baseline values for chemical markers during wakefulness, anesthesia, and neurosurgery. Neurosurgery 47:701–710
3. Ståhl N, Mellergård P, Hallström A, Ungerstedt U, Nordström CH (2001) Intracerebral microdialysis and bedside biochemical analysis in patients with fatal traumatic brain lesions. Acta Anaesthesiol Scand 45:977–985
4. Ståhl N, Ungerstedt U, Nordström CH (2001) Brain energy metabolism during controlled reduction of cerebral perfusion pressure in severe head injuries. Intensive Care Med 27:1215–1223
5. Ståhl N, Schalén W, Ungerstedt U, Nordström CH (2003) Bedside biochemical monitoring of the penumbra zone surrounding an evacuated acute subdural haematoma. Acta Neurol Scand 108:211–215

Acta Neurochir Suppl (2008) 102: 197–202

In vitro comparison of two generations of Licox and Neurotrend catheters

Iain Haitsma · Guy Rosenthal · Diane Morabito · Mark Rollins · Andrew I. R. Maas · Geoffrey T. Manley

Abstract

Background Clinical reports on brain tissue oxygen tension differ in suggested threshold values for defining cerebral ischemia using the Licox and Neurotrend/Paratrend system. We evaluated in vitro performance of both first and second generation devices.

Materials and methods Response rate and accuracy in solutions with oxygen tensions from 0 to 150 mm Hg were measured.

Findings Ninety-five percent Response times were 102±13 seconds for first generation Licox probes and 135±24 s for Paratrend (n=6, each probe), with second generation probes at 134±4 and 116±16 s respectively. At pO_2 150 mmHg Licox and Paratrend probes were accurate with 2.2% and 2.1% error, respectively and 2.6% and 4.1% for later generation. At pO_2 18 mmHg, Paratrend overestimated by 16.5% (absolute error range 2.18 to 4.18 mmHg), 7.4% for Neurotrend, Licox underestimated by 1.8% (absolute error range 0.08 to 0.52 mmHg) with 3.6% for the second generation probe.

Conclusions Differences between the first generation probe types, while statistically significant (p<0.001), may not be clinically relevant. Overestimation of pO_2 by Neurotrend and small underestimation by Licox partially explain differences in published thresholds for cerebral ischemia. The Neurotrend was slightly more accurate and faster than the Paratrend system.

Keywords Licox · Neurotrend · Brain tissue oxygen monitors · Neuromonitoring · Accuracy · In vitro

I. Haitsma
Department of Neurosurgery, Erasmus University, Rotterdam, Netherlands
e-mail: i.haitsma@erasmusmc.nl

I. Haitsma
Department of Neurosurgery, Erasmus MC Rotterdam, P.O. Box, 2040, 3000 CA, Rotterdam, The Netherlands

G. Rosenthal · D. Morabito · G. T. Manley (✉)
Department of Neurosurgery, University of California, San Francisco,
1001 Potrero Ave., Bldg 1, Room 101,
San Francisco, CA, 94110, USA
e-mail: manleyg@neurosurg.ucsf.edu

D. Morabito · G. T. Manley
San Francisco Injury Center, University of California, San Francisco, CA, USA

M. Rollins
Department of Anesthesiology, University of California, San Francisco, CA, USA
e-mail: rollinsm@anesthesia.ucsf.edu

M. Rollins
Department of Anesthesia and Perioperative Care, University of California, San Francisco,
533 Parnassus Ave, Room U-280,
San Francisco, CA, 94143-0648, USA

A. I. R. Maas
Department of Neurosurgery, University Hospital Antwerp, Wilrijkstraat 10,
2650, Edegem, Belgium
e-mail: andrew.maas@uza.be

Introduction

Brain tissue oxygen tension ($PbtO_2$) measurement is available clinically since 1996 [17]. Depth and duration of low $PbtO_2$ values are related to poorer outcome [16]. There are differences in the published oxygen values for cerebral hypoxia using the two most commonly devices used: Licox and Neurotrend/Paratrend. To resolve whether accuracy of these systems is a cause of this finding, we evaluated and compared the in vitro performance of both first generation devices and later available FDA approved probes under a range of temperature and oxygen levels frequently encountered in brain tissue.

Materials and methods

Device description

Licox

The Licox device is a Clark-type polarographic oxygen probe [2] developed by GMS mbH (Kiel, Germany). First introduced in Europe in the early 1990's, was the "Micro-catheter pO_2 probe", model C1. Tissue temperature is simultaneously monitored every 2 seconds with a separate thermocouple (C8 temperature probe) or, less accurately, central body temperature can be manually entered in the measuring computer. Calibration was performed in the factory and prior to use the relevant values had to be entered in the measuring computer or, it could be calibrated at room oxygen level in its sterile housing. In 2000, a second generation probe, the C1.SB, was introduced with a chip card containing the calibration information. In early 2001 the Licox system received FDA 510(k) clearance and GMS was acquired by Integra Neurosciences. All Licox oxygen catheters should be stored at 2°C to 10°C. The accuracy according to the manufacturers was a sensitivity error<20% and a zero error<0.7 mmHg for the first generation probes. With the C1.SB probes this was changed to a difference < 2 mmHg with values from 0 to 20 mmHg, 10% difference at 21–50 mmHg and a maximum 12% difference at 51 to 150 mmHg, temperature was within 0,2%. These values were valid for 5 days of use.

Paratrend/Neurotrend

The Paratrend 7 monitoring system (Diametrics Medical Incorporated, Buckinghamshire, England) was originally designed for continuous intra-arterial blood gas measurement. It used a polarographic Clark electrode for pO_2 measurement, optical sensors for measurement of pH and hydrogen ions, and a thermocouple for temperature measurement. The monitor contains a barometric pressure transducer. Each sensor is stored at room temperature and calibrated in its tonometer filled with buffer solution with gas mixtures of O_2, CO_2 and N_2 just prior to use. In 1999 a second generation sensor designed for intracranial tissue monitoring, the Neurotrend, was introduced. The oxygen sensor was changed to a fiber optode [6]. As of October 2004 production of Neurotrend and Paratrend catheters has ceased.

In vitro testing

Unused first generation sensors (six each of C1, C8 and Paratrend) were compared at 37°C with each of the following four gas mixtures in random order: a (0% O_2, 2% CO_2, 98% N_2), b (2.5% O_2, 5% CO_2, 92.5% N_2), c (10% O_2, 10% CO_2, 80% N_2), d (21% O_2, 2% CO_2, 77% N_2)

After calibration, probes were inserted through the cover of a 500 mL glass tube filled with normal saline, held at a constant temperature using a circulating water bath, continuously bubbled with a gas mixture. After ≥20 min, the probes and cover were transferred to a second container being bubbled with another gas mixture. All data was down loaded to a PC at 5 s intervals. The effects of temperature (33°C, 35°C, 37°C, and 39°C) on response time and accuracy were also measured using gas mixtures b and c. Thermocouple accuracy was compared to a glass mercury thermometer.

The equilibrium value under each condition was measured twice with each probe. The differences between actual pO_2 and measured pO_2 at each condition were determined, and mean values calculated. The actual partial pressure is calculated from the difference in atmospheric pressure and water vapor pressure multiplied by the gas fraction in the mixture.

Later, the second generation sensors (six C1SB, six Neurotrend) were compared under conditions as described above. This time the catheters were placed in a buffer mixture as used in the neurotrend tonometers, instead of saline. The temperature accuracy of each system was not evaluated as the thermocouple technology used in each system had not changed from the first generation.

Data are presented as means±SD. Results are tabulated for comparison of the measured pO_2 for each probe generation. Error is taken as an absolute number, with means of both actual and percentage pO_2 error shown with standard deviation. For statistical analysis two tailed *t*-tests were performed, a *p*-value of <0.05 was considered to be statistically significant.

Results

At 37°C and 0% O_2 the C1 and Paratrend devices measured 0.03±0.08 and 0.33±0.5 mmHg O_2 (Table 1). At 2.5% O_2 (pO_2 calculated is 17.8 mmHg) the C1 measured 17.6± 0.4 mm (−1.8%) Hg and Paratrend devices slightly over read the oxygen at 19.8±2.5 mm Hg (16.5%). The 95% response times were 102±13 s for the first generation Licox

Table 1 Absolute and relative errors of pO_2 measurements at different pO_2 levels at 37°Celsius for first generation catheters

Oxygen %	Calculated pO_2 (mmHg)	pO_2 error±SD				Measured pO_2	
		Absolute error (mmHg)		Percent error		mm Hg	
		C1	Paratrend	C1	Paratrend	C1	Paratrend
0	0	0.0±0.1	0.3±0.5	0	–	0.03	0.33
2.5	17.8	0.3±0.2	3.0±0.8	1.8±1.0	16.6±4.2	17.7	19.8
10	71.2	1.4±1.0	3.7±0.5	2.0±1.4	5.1±0.7	70.8	74.3
21	149.6	3.1±2.8	3.2±3.7	2.1±1.9	2.1±2.5	147.9	148.2

probes and 135±24 s for the Neurotrend probes. The accuracy of the catheters is not significantly influenced at temperatures of 33°C to 39°C (Fig. 1).

With the later generation devices results are comparable to the earlier results (Table 2). The measurement errors of both catheter types are plotted in Fig. 2. At 37°C and 2.5% O_2 Licox measured 17.5±0.88 mmHg against 18.3± 1.87 mmHg for the Neurotrend. Combining the values at different temperatures, the absolute differences from calculated values were significantly smaller for Licox then Neurotrend at 0 mmHg (p=0.002) and 18 mmHg (p= 0.004). These differences between probe types were not significant at 71 and 149 mmHg. The second generation Licox probe had a 95% response time at 37°C of 134±4 s. The Neurotrend was faster at 116±16 seconds respectively. At lower temperatures the response time is significantly longer for both Licox (170 s at 33°C vs. 133 s at 39°C) and Neurotrend (134 s at 33°C vs. 109 s at 39°C).

We also examined the effect of recalibrating Licox catheters before use in room air, this did not result in improved accuracy compared to using the factory calibration settings from the smart card. Finally we examined three expired Licox catheters (5 months beyond expiry date). With smart card calibration the catheters under read by as much as 15%. This improved after recalibration but remained up to 6.5%.

The Neurotrend PCO_2 measurements were always within 4.5% of the calculated values. The absolute error at a PCO_2 of 35 mmHg and 37°C was 0.3±0.4 mmHg. There was no effect of temperature on accuracy.

The pH was always within 2% of calculated values. At a pH of 7.04 and 37°C the absolute error was 0.02±0.01. At lower temperatures the Paratrend slightly under read the pH and at higher temperatures it slightly over read. The difference at a pH of 7.04 and 33°C was −0.079±0.023 and at 39°C was +0.001±0.012.

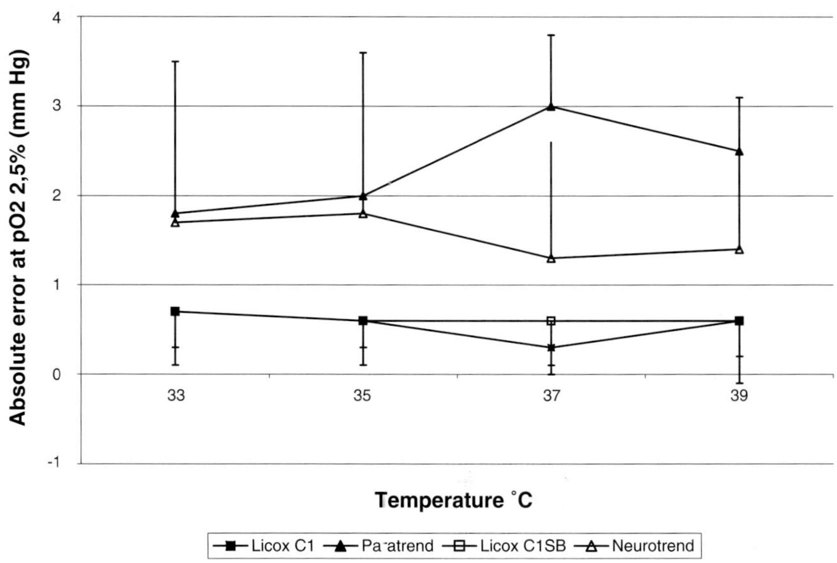

Fig. 1 Absolute measurement error at pO_2 of 18 mmHg for the four catheter types at different temperature levels from 33°C to 39°C. The Y-axis represents the difference between measured and calculated values

Table 2 Absolute and relative errors of pO_2 measurements at different pO_2 levels at 37°Celsius for second generation catheters

Oxygen %	Calculated pO_2 (mmHg)	pO_2 error±SD				Measured pO_2	
		Absolute error mmHg		Percent error		mmHg	
		C1SB	Neurotrend	C1SB	Neurotrend	C1SB	Neurotrend
0	0	0.4±0.8	0.7±1.1	–	–	0.4	0.7
2.5	17.7	0.6±0.6	1.3±1.3	3.6±3.4	7.4±7.4	17.5	18.3
10	70.8	1.5±1.1	2.0±2.2	2.1±1.6	2.9±3.1	71.3	71.6
21	148.8	3.9±2.8	6.0±4.2	2.6±1.9	4.1±2.8	152	154.1

Discussion

Several studies have tried to define a $PbtO_2$ value that is critical for ischemia. However it is probably not possible to have a clear cut off since $PbtO_2$ is influenced by a great number of factors, including $PaCO_2$, PaO_2 and CPP, but also local characteristics as hematomas or contused tissue near the probe [10, 12, 15]. Published critical values range from 8.5 to 20 mmHg. There is a remarkable difference in reported critical values for the Paratrend and Licox system. Two Paratrend studies cite values of 20 mmHg [4, 8]. The reported values for the Licox system range from 8.5 to 15 mmHg [1, 9, 16]. This was a major incentive for our study to examine if this is due to a calibration difference between the two probe types. A study by Valadka et al suggested this, as the Paratrend showed values of 7.0± 1.4 mmHg in a zero solution (Table 3) [14]. Our study did show slightly higher values for the Para/Neurotrend system but not to such an extent as to fully explain the reported difference in critical values. The accuracy of the Licox system was slightly better then the Para/Neurotrend system at low O_2 values. This has improved with the new catheters with a mean error at approximately 18 mmHg decreasing from 3.0 to 1.3 mmHg for the Para/Neurotrend system. For the Licox this remained stable at 0.3 to 0.6 mmHg. One study performed an in vivo comparison of both systems and did not show a significant difference in measured values [13]. Perhaps the difference in reported critical values can be further explained by localization of the catheters, as the oxygen sensing part is located proximally in the Paratrend probe, several centimeters from the tip. This could have caused the oxygen sensing part of these catheters to be located in the grey matter which has higher pO_2 values. Conversely, the oxygen sensing area of the Licox catheter is

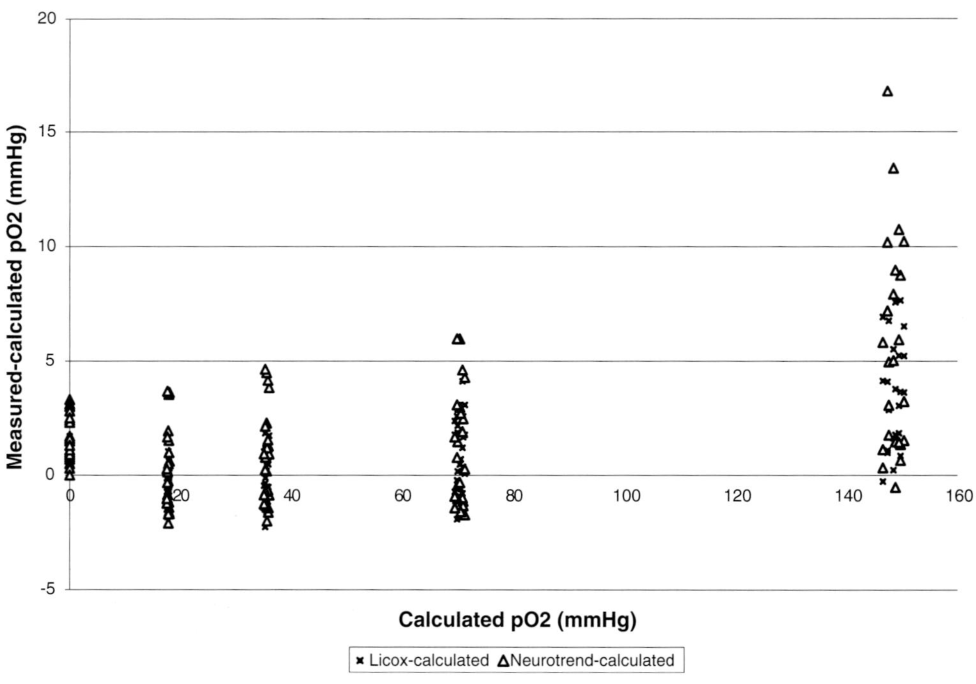

Fig. 2 Measurement error of Licox C1.Sb and Neurotrend probes at 37°C at five different O_2 levels, individual results are plotted against calculated values

Table 3 Previous in vitro studies with Licox and Para/Neurotrend catheters

Study	Year	Catheter type	Number	90% response time	Sensitivity error %	Zero error
Valadka [14]	1998	Paratrend	4	–	±8	7.0±1.4
Valadka [14]	1998	C1	37	–	±6	0.3±0.3
Dings [3]	1998	C1	12	–	<3.9	−0.21±0.25
Hoelper [7]	2005	C1.Sb	12	129 (7 to 56 mmHg)	−4.5 to 9	–
Hoelper [7]	2005	Neurotrend	12	55 (7 to 56 mmHg)	4.8 to 26	–

located near the tip, typically located in the white matter which has lower pO_2 values than grey matter. This effect can also be enhanced by the smaller O_2 sensing area of the neurotrend catheter (1.4 mm) vs. the Licox (5 mm). It can lead to greater standard deviation in measured values and leads to extra concerns on placement of the catheters. It is well known that there is significant heterogeneity in regional cerebral blood flow and brain tissue tension. Thus, the larger sensing area of the Licox catheter serves to average these differences. We have also observed that, when measuring $PbtO_2$ in white matter of uninjured swine, small changes in depth of placement leads to greater variability in Neurotrend values than using the Licox system (unpublished results). However this could be judged as a possible advantage by the experts advocating placement of focal probes as close as possible to the penumbra of lesioned tissue to monitor secondary neurological damage [5].

Clinically our observed measurement error at low O_2 values should pose no problems. The current largest error of less then 8% at levels around 15 to 20 mmHg should cause no extra difficulty in interpreting results.

The 95% response time of both systems is around 2 min. With the first models, the Licox was modestly faster and now the Neurotrend is slightly quicker to respond to changes. In patients increases in arterial pO_2 lead to a plateau phase in $PbtO_2$ within 15 min in 40% of patients [18]. Thus for all current clinical situations the response times seem ample. A recent similar in vitro study by Hoelper et al. [7] reported 90% response times from 1% to 5% O_2 of 129±27 and 55±19 for C1Sb and Neurotrend probes respectively at 37°C. These are comparable to our values of 102±2 (Licox) and 71±8 s (Neurotrend) for a change of 0% to 21% O_2.

The accuracy we found with the Neurotrend catheters was within 3.5% for CO_2 and within 0.5% for pH at these levels. These values are comparable with reports on the accuracy for the intravascular Paratrend catheters [11, 19] and Neurotrend probes [7].

With the second generation devices, the Licox system has not changed much in in vitro measurement characteristics. The Para-/Neurotrend has become more accurate and faster in O_2 measurements using its fluorescent technology.

In 2004 Diametrics Medical Inc. suffered severe financial problems, ultimately leading to restructuring which resulted in the immediate discontinuation of manufacturing of the Neurotrend and Paratrend catheters in October 2004.

Integra developed a new catheter in the Licox range. In April 2004 they received 510k FDA clearance for their PMO probe. It combines the polarographic oxygen catheter with an incorporated thermocouple. The technical specifications for the oxygen sensing part are identical to those for the C1.Sb catheters.

Possible future refinements/developments could be the incorporation of oxygen sensing elements with intraparenchymous ICP transducers and/or thermal diffusion CBF probes.

Conclusion

In vitro oxygen measurements at clinically relevant low oxygen tensions are slightly different between these devices. With regard to in vitro accuracy/reliability both systems perform well. These differences are not clinically significant and do not fully explain the discrepancies in critical brain tissue oxygen thresholds previously reported for the Neurotrend and Licox monitors. The differences in the O_2 sensing areas of the devices may contribute to discrepancy. The fact that the Licox catheter measures O_2 near the tip of the device which is typically located in the white matter versus the Paratrend/Neurotrend catheter that measures O_2 more proximately nearer, or even in, the gray matter which has higher reported O_2 values is the most likely explanation for these differences in reported critical thresholds.

Acknowledgements This study was supported by R49/CCR903697-07. The work of Dr. Haitsma is supported by NWO grant: 920-03-130. The Department of Neurosurgery, EMC, has received in natura support of research in neurocritical care from GMS mbH. (Kiel, Germany) and Codman/Johnson & Johnson (Raynham, MA, USA). All probes were donated by their respective companies.

Conflict of interest statement We declare that we have no conflict of interest.

References

1. Bardt TF, Unterberg AW, Hartl R, Kiening KL, Schneider GH, Lanksch WR (1998) Monitoring of brain tissue PO_2 in traumatic brain injury: effect of cerebral hypoxia on outcome. Acta Neurochir Suppl 71:153–156
2. Clark LC (1956) Monitor and control of blood and tissue oxygen tensions. Trans Am Soc Artif Int Org 2:41–45
3. Dings J, Meixensberger J, Jager A, Roosen K (1998) Clinical experience with 118 brain tissue oxygen partial pressure catheter probes. Neurosurgery 43:1082–1095
4. Doppenberg EM, Zauner A, Watson JC, Bullock R (1998) Determination of the ischemic threshold for brain oxygen tension. Acta Neurochir Suppl 71:166–169
5. Gopinath SP, Valadka AB, Uzura M, Robertson CS (1999) Comparison of jugular venous oxygen saturation and brain tissue pO_2 as monitors of cerebral ischemia after head injury. Crit Care Med 27:2337–2345
6. Henze D, Bomplitz M, Radke J, Clausen T (2004) Reliability of the NeuroTrend sensor system under hyperbaric conditions. J Neurosci Methods 132:45–56
7. Hoelper BM, Alessandri B, Heimann A, Behr R, Kempski O (2005) Brain oxygen monitoring: in-vitro accuracy, long-term drift and response-time of Licox- and Neurotrend sensors. Acta Neurochir (Wien) 147:767–774
8. Hoffman WE, Charbel FT, Edelman G (1996) Brain tissue oxygen, carbon dioxide, and pH in neurosurgical patients at risk for ischemia. Anesth Analg 82:582–586
9. Kiening KL, Unterberg AW, Bardt TF, Schneider GH, Lanksch WR (1996) Monitoring of cerebral oxygenation in patients with severe head injuries: brain tissue pO_2 versus jugular vein oxygen saturation. J Neurosurg 85:751–757
10. Manley GT, Pitts LH, Morabito D, Doyle CA, Gibson J, Gimbel M, Hopf HW, Knudson MM (1999) Brain tissue oxygenation during hemorrhagic shock, resuscitation, and alterations in ventilation. J Trauma 46:261–267
11. Menzel M, Soukup J, Henze D, Engelbrecht K, Senderreck M, Scharf A, Rieger A, Grond S (2003) Experiences with continuous intra-arterial blood gas monitoring: precision and drift of a pure optode-system. Intensive Care Med 29:2180–2186
12. Reinert M, Barth A, Rothen HU, Schaller B, Takala J, Seiler RW (2003) Effects of cerebral perfusion pressure and increased fraction of inspired oxygen on brain tissue oxygen, lactate and glucose in patients with severe head injury. Acta Neurochir (Wien) 145:341–349 discussion 349–350
13. Sarrafzadeh AS, Kiening KL, Bardt TF, Schneider GH, Unterberg AW, Lanksch WR (1998) Cerebral oxygenation in contusioned vs. nonaligned brain tissue: monitoring of PtiO2 with Licox and Paratrend. Acta Neurochir Suppl 71:186–189
14. Valadka AB, Gopinath SP, Contant CF, Uzura M, Robertson CS (1998) Relationship of brain tissue pO_2 to outcome after severe head injury. Crit Care Med 26:1576–1581
15. van den Brink WA, Haitsma IK, Avezaat CJ, Houtsmuller AB, Kros JM, Maas AI (1998) Brain parenchyma/pO_2 catheter interface: a histopathological study in the rat. J Neurotrauma 15:813–824
16. van den Brink WA, van Santbrink H, Steyerberg EW, Avezaat CJ, Suazo JA, Hogesteeger C, Jansen WJ, Kloos LM, Vermeulen J, Maas AI (2000) Brain oxygen tension in severe head injury. Neurosurgery 46:868–876 discussion 876–868
17. van Santbrink H, Maas AI, Avezaat CJ (1996) Continuous monitoring of partial pressure of brain tissue oxygen in patients with severe head injury. Neurosurgery 38:21–31
18. van Santbrink H, vd Brink WA, Steyerberg EW, Carmona Suazo JA, Avezaat CJ, Maas AI (2003) Brain tissue oxygen response in severe traumatic brain injury. Acta Neurochir (Wien) 145:429–438 discussion 438
19. Venkatesh B, Clutton Brock TH, Hendry SP (1994) A multiparameter sensor for continuous intra-arterial blood gas monitoring: a prospective evaluation. Crit Care Med 22:588–594

Acta Neurochir Suppl (2008) 102: 203–206

Cerebral metabolism monitoring during hypothermia following resuscitation from cardiopulmonary arrest

Takehiro Nakamura · Yasuhiro Kuroda · Natsuyo Torigoe · Yuko Abe · Susumu Yamashita · Kenya Kawakita · Nobuyuki Kawai · Takashi Tamiya · Toshifumi Itano · Seigo Nagao

Abstract

Background The aim of the present study was to evaluate cerebral metabolism monitoring during therapeutic hypothermia for global ischemic brain damage after cardiopulmonary resuscitation (CPR).

Methods Jugular venous sampling and positron emission tomography (PET) were used. Seven comatose patients with cardiopulmonary arrest underwent hypothermia treatment as soon as possible after CPR. The body temperature of these patients was maintained at 34°C for 72 h. Rewarming was performed at a rate of 1°C/day. To monitor jugular venous saturation (SjO_2) and lactate (lac-JV), a fiberoptic catheter was inserted into the jugular bulb. Oxygen extraction fraction (OEF) was calculated using the difference between arterial oxygen saturation (SaO_2) and SjO_2. ^{18}F-fluorodeoxyglucose (FDG) PET was performed to investigate cerebral glucose metabolism at the end of therapeutic hypothermia.

Findings The OEF was significantly increased at the end of hypothermia in four patients with favorable outcome on the Glasgow Outcome Scale (hypothermia onset 15.3±2.0% vs. hypothermia end 30.3±2.8%, $P<0.05$). In three patients with unfavourable outcome (severe or worse on the Glasgow Outcome Scale), end hypothermia OEF tended to be low. There was also a reduction in FDG uptake in these three patients with unfavourable outcome. The lac-JV was significantly decreased at the end of hypothermia treatment compared with hypothermia onset (27.7±7.4 vs. 6.0±3.0 mg/dL, $P<0.05$).

Conclusions The measurement of cerebral metabolism parameters, especially OEF, might be useful for estimation of hypothermia therapy in patients with unconsciousness after resuscitation after cardiac arrest.

T. Nakamura (✉) · T. Itano
Department of Neurobiology, Faculty of Medicine,
Kagawa University,
1750-1 Ikenobe, Miki,
Kita, Kagawa 761-0173, Japan
e-mail: tanakamu@kms.ac.jp

T. Itano
e-mail: toshi@kms.ac.jp

T. Nakamura · N. Kawai · T. Tamiya · S. Nagao
Department of Neurological Surgery, Faculty of Medicine,
Kagawa University,
1750-1 Ikenobe, Miki,
Kita, Kagawa 761-0173, Japan

N. Kawai
e-mail: nobu@kms.ac.jp

T. Tamiya
e-mail: tamiya@kms.ac.jp

S. Nagao
e-mail: snagao@kms.ac.jp

Y. Kuroda · N. Torigoe · Y. Abe · S. Yamashita · K. Kawakita
Department of Emergency and Critical Care Medicine,
Faculty of Medicine, Kagawa University,
1750-1 Ikenobe, Miki,
Kita, Kagawa 761-0173, Japan

Y. Kuroda
e-mail: kuroday@kms.ac.jp

N. Torigoe
e-mail: osakanach46@yahoo.co.jp

Y. Abe
e-mail: abekulo@kms.ac.jp

S. Yamashita
e-mail: sum-ygc@umin.ac.jp

K. Kawakita
e-mail: kenfact@kms.ac.jp

Keywords Cardiopulmonary arrest · Hypothermia therapy · Cerebral metabolism · Oxygen extraction fraction

Introduction

Hypothermia induction has proven useful for the treatment of global cerebral ischemia following cardiopulmonary arrest (CPA) [1]. However, comatose survivors from CPA do not always regain consciousness after hypothermia treatment. Furthermore, the exact mechanisms for this effect cerebral resuscitation are still uncertain. In order to better understand the pathophysiological effects of this therapy, cerebral metabolic monitoring may be use.

The aim of the present study was to undertake cerebral metabolism monitoring during therapeutic hypothermia for global ischemic brain damage after cardiac arrest. Jugular venous sampling and positron emission tomography (PET) were used.

Materials and methods

Patients

Between March 2004 and June 2006, seven patients admitted to the Kagawa University Hospital who remained comatose following resuscitation from CPA were considered for this study. The criteria for inclusion were an estimated interval of 5 to 15 min from the patient's collapse to the first attempt at resuscitation and an interval of no more than 60 min from collapse to restoration of spontaneous circulation [4]. Patient outcome was judged at leaving hospital according to Glasgow Outcome Scale.

Hypothermia therapy

The patients were paralyzed by pancuronium and artificially ventilated. Anesthesia was induced and maintained by continuous intravenous injection of midazolam. Jugular venous temperature, used as a substitute for brain temperature, was monitored during hypothermia therapy. The body temperature of these patients was reduced by the use of cooling water blankets until the jugular venous temperature decreased to 34°C. Hypothermia (34°C) was maintained for 72 h. Rewarming was performed at the rate of 1°C/day.

Monitoring

To monitor jugular venous temperature, jugular venous oxygen saturation (SjO_2) and jugular venous lactate levels, a fiberoptic catheter (Abbott Laboratories, North Chicago, IL, USA) was inserted percutaneously into the internal jugular bulb in all patients. The location of the fiberoptic catheter was confirmed radiologically in all patients in order to avoid the contamination of blood from an extracranial source. Arterial and jugular venous blood samples were obtained at the same time for determination of oxygen saturation and lactate level. A normal range of the SjO_2 is from 55% to 75%. The OEF was calculated as follows: $[OEF = (1 - SjO_2/SaO_2) \times 100]$. The OEF is used as a monitor of adequacy of cerebral oxygen delivery and is normally between 25% and 45%. A normal range of jugular venous lactate is between 4 and 16 mg/dL.

PET procedure

Dynamic ^{18}F-fluorodexoxyglucose (FDG) PET was performed using an ECAT EXACT HR scanner (Siemens/CTI, Knoxville TN, USA) in three-dimensional acquisition mode. The in-plane reconstructed resolution was 4.7 mm full-width at half-maximum. Patients had no caloric intake for at least 5 h before administration of FDG. Serum glucose levels were examined before FDG injection with dynamic acquisition of PET started immediately after an intravenous tracer injection. Dynamic emission data in three-dimensional mode were acquired for 60 min following bolus injection of FDG [5].

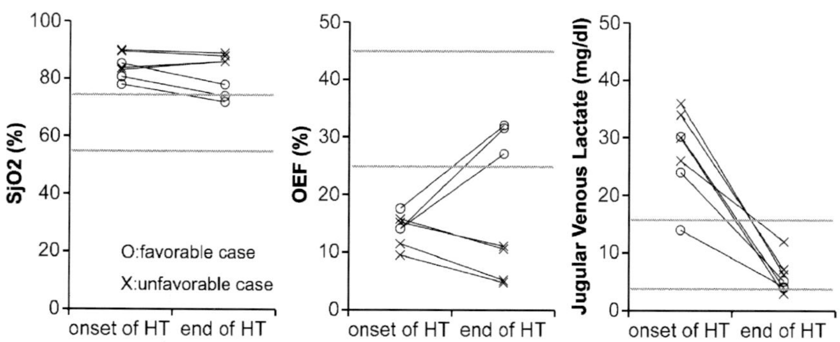

Fig. 1 Jugular venous oxygen saturation (*SjO*$_2$, *left panel*), oxygen extraction fraction (*OEF*, *middle panel*), and jugular venous lactate (*right panel*) at onset and end of hypothermia therapy (*HT*). The dashed lines represent normal values for these parameters. Patient's outcome was judged at leaving hospital according to Glasgow Outcome Scale (favorable: GR and MD; unfavorable: SD, VS, and D)

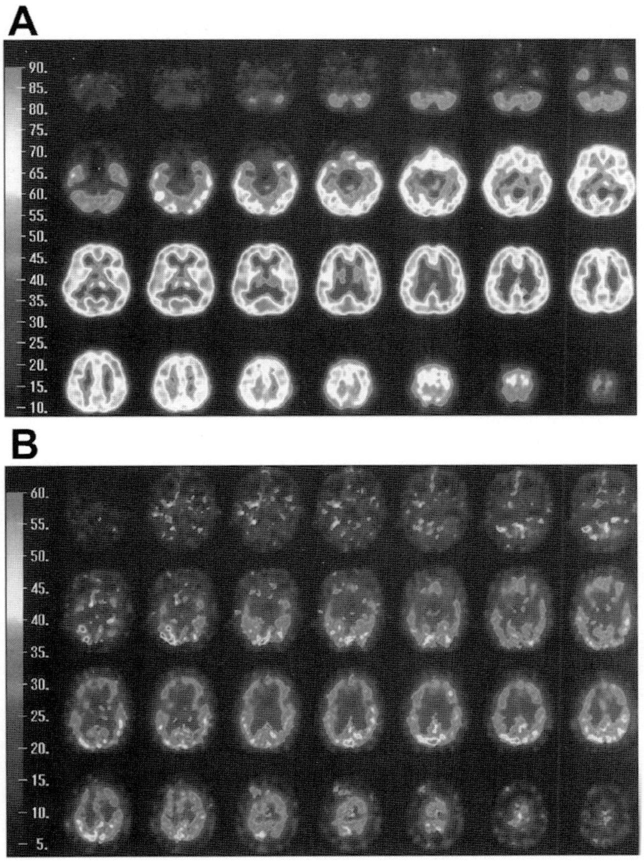

Fig. 2 Cerebral glucose metabolism (CMRGlc) evaluated by ^{18}F-fluorodexoxyglucose positron emission tomography. **a** Favorable outcome case: 9-year-old male patient with cardiopulmonary arrest (CPA) due to near-drowning (case 5 in Table 1). **b** Unfavorable outcome case: 42-year-old female patient with CPA due to hanging (case 6 in Table 1). Patient's outcome was judged at leaving hospital according to Glasgow Outcome scale (favorable: GR and MD unfavorable: SD, VS, and D)

Statistical analysis

Differences between the values of each parameter (SjO_2, OEF and lac-JV) at the start of and the end of hypothermia treatment were analysed by unpaired *t* test. Differences were considered statistically significant at $P<0.05$.

Results

In three cases of favorable outcome (defined as moderate, or good on the Glasgow Outcome Scale), the SjO_2 and OEF were normalized by hypothermia. Although there were no significant SjO_2 changes in patients with favorable outcome, the OEF was significantly increased at the end of hypothermia in these patients (15.3±2.0% vs. 30.3±2.8%, $P<0.05$; Fig. 1, left and middle panels).

In four cases of unfavorable outcome (severe or vegetative on the Glasgow Outcome Scale), high levels of SjO_2 and low levels of OEF were found even if hypothermia therapy was performed (Fig. 1, left and middle panels).

There were significant decreases in lac-JV at the end of hypothermia treatment (27.7±7.4 vs. 6.0±3.0 mg/dL, $P<0.05$), regardless of the patient's outcome (Fig. 1, right panel).

Figure 2a shows a PET scan from a patient who regained consciousness (case 5). The whole brain CNRGlc was maintained at the end of hypothermia. In contrast, there was reduction in FDG uptake in a patient with an unfavourable outcome (Fig. 2b, case 6; Table 1).

Discussion

Previous clinical and experimental studies have shown that hypothermia therapy may have beneficial effects on severe disturbances in consciousness caused by CPA, cerebral ischemia, head injury and subarachnoid hemorrhage [1, 4, 7–9]. The mechanisms of the beneficial effects of hypothermia therapy are still uncertain. Cerebral metabolic monitoring might lend insight into intracranial pathophysiology during this treatment [7, 9].

Continuous SjO_2 measurement allows real time assessment of cerebral oxygen delivery. In general, global cerebral hyperemia and ischemia are defined as an SjO_2 of greater than 75% and less than 55%, respectively. A previous study reported that extreme hyperoxia of internal jugular venous blood is a signal of irreversible brain damage and near brain death [6]. High SjO_2 values are probably

Table 1 Summary of patients

Case	Age (years)	Gender	Etiology of CPA	ECG on arrival	Witness	Outcome (GOS)
1	53	F	Heart disease	VF	(+)	Unfavorable (VS)
2	76	M	FBAO	Asystole	(+)	Unfavorable (VS)
3	52	M	Crush injury	Asystole	(+)	Unfavorable (VS)
4	45	M	Heart disease	VF	(+)	Favorable (GR)
5	9	M	Near-drowning	Sinus	(+)	Favorable (GR)
6	42	F	Hanging	Asystole	(−)	Unfavorable (VS)
7	54	M	Brugada syndrome	VF	(+)	Favorable (GR)

F female, *M* male, *CPA* cardiopulmonary arrest, *FBAO* foreign-body airway obstruction, *ECG* electrocardiogram, *VF* ventricular fibrillation, *GOS* Glasgow Outcome Scale, *VS* vegetative state, *GR* good recovery primarily a consequence of a decrease in cerebral oxygen consumption. High values of SjO_2 in patients who experienced CPA might be a sign of severe brain damage [10].

In the present study, the values of OEF in patients with favorable outcome were significantly normalized. However, there were no significant changes in SjO_2 in these cases. The difference between OEF and SjO_2 is that the OEF is also related to SaO_2. Patients with severe consciousness disturbance following resuscitation from CPA tend to be in poor clinical condition, especially with regard to respiratory and circulatory function. Thus, monitoring parameters that reflect systemic oxygen delivery may help to more adequately estimate cerebral oxygen delivery [2].

The present study shows that hypothermia treatment significantly reduced lac-JV levels, regardless of patient outcome. Lactate clearance is associated with improved outcome in post-cardiac arrest patients [3]. The decreased lac-JV levels after hypothermia treatment might indicate that this therapy could improve aerobic metabolism following CPA.

In the present study, there was a reduction in FDG uptake in patients with an unfavourable outcome. However, it is difficult to perform PET studies during hypothermia treatment. The present study did not show any clear early glucose metabolism in the brain during hypothermia for CPA. Further studies are required to clarify glucose metabolism in the early phase of hypothermia treatment following CPA.

Conclusions

The measurement of cerebral metabolism parameters, especially OEF, might be useful for estimation of the effectiveness of hypothermia therapy in patients with unconsciousness after resuscitation from cardiac arrest.

Acknowledgments The authors thank Prof. Shinji Ogra (Gifu University Hospital) for excellent technical assistance. We also thank Prof. Richard F. Keep (University of Michigan) for providing useful suggestions.

Conflict of interest statement We declare that we have no conflict of interest.

References

1. Bernard SA, Gray TW, Buist MD, Jones BM, Silvester W, Gutteridge G, Smith K (2002) Treatment of comatose survivors of out-of-hospital cardiac arrest with induced hypothermia. N Engl J Med 346:557–563
2. Cruz J, Raps EC, Hoffstad OJ, Jaggi JL, Gennarelli TA (1993) Cerebral oxygenation monitoring. Crit Care Med 21:1242–1246
3. Donnino MW, Miller J, Goyal N, Loomba M, Sankey SS, Dolcourt B, Sherwin R, Otero R, Wira C (2007) Effective lactate clearance is associated with improved outcome in post-cardiac arrest patients. Resuscitation 75(2):229–234
4. The Hypothermia After Cardiac Arrest Study Group (2002) Mild therapeutic hypothermia to improve the neurologic outcome after cardiac arrest. N Engl J Med 346:549–556
5. Kawai N, Miyake K, Nishiyama Y, Yamamoto Y, Sasakawa Y, Haba R, Kushida Y, Tamiya T, Nagao S (2006) FDG-PET findings of the brain in lymphomatoid granulomatosis. Ann Nucl Med 20:683–687
6. Minima T, Ogawa M, Sugimoto T, Katsurada K (1973) Hyperoxia of internal jugular venous blood in brain death. J Neurosurg 39:442–444
7. Nakamura T, Nagao S, Kawai N, Honma Y, Kuyama H (1998) Significance of multimodal cerebral monitoring under therapeutic hypothermia severe head injury. Acta Neurochir Suppl 71:85–87
8. Nakamura T, Miyamoto O, Yamagami S, Hayashida Y, Itano T, Nagao S (1999) Influence of rewarming conditions after hypothermia in gerbils with transient forebrain ischemia. J Neurosurg 91:114–120
9. Nakamura T, Tatara N, Morisaki K, Kawakita K, Nagao S (2002) Cerebral oxygen metabolism monitoring under hypothermia for severe subarachnoid hemorrhage: report of eight cases. Acta Neurol Scand 106:314–318
10. Takasu A, Yagi K, Ishihara S, Okada Y (1995) Combined continuous monitoring of systemic and cerebral oxygen metabolism after cardiac arrest. Resuscitation 29:189–194

Acta Neurochir Suppl (2008) 102: 207–213

Nitric oxide in acute brain injury: a pilot study of NO_x concentrations in human brain microdialysates and their relationship with energy metabolism

Keri L. H. Carpenter · Ivan Timofeev · Pippa G. Al-Rawi · David K. Menon · John D. Pickard · Peter J. Hutchinson

Abstract
Introduction This pilot microdialysis study in acute brain injury patients assessed the relationship between nitric oxide products (total nitrite plus nitrate, termed NO_x) and energy-related molecules: glucose, lactate, pyruvate, glutamate and glycerol.
Methods Twelve patients (11 major head-injury and one subarachnoid haemorrhage) were studied, 11 of which had single catheters and one had two catheters, in the cerebral cortex. Catheters were perfused at 0.3 μL/min with CNS perfusion fluid. Collection vials were changed hourly. Microdialysates were analysed for energy-related molecules on a CMA600 or ISCUS analyser, and for NO_x using a purge vessel (VCl_3 plus HCl at 95°C, purged with nitrogen) connected to a Sievers NOA 280i analyser.
Findings The mean of mean NO_x concentration (±SD) for the 13 catheters was 32.7±16.8 μmol/L, and the lactate/pyruvate ratio was 38.6±20.1. Increasing NO_x concentrations correlated significantly with decreasing lactate/pyruvate ratio (Spearman $r=-0.79$, $p=0.0065$), with decreasing lactate concentration ($r=-0.59$, $p=0.042$), and with increasing glucose concentration ($r=0.71$, $p=0.018$).
Conclusions These pilot data suggest that in injured brains, higher concentrations of nitric oxide are associated with more favourable metabolism. Nitric oxide may act beneficially by increasing blood flow and delivery of oxygen and glucose. Further patients are being recruited.

Keywords Microdialysis · Nitric oxide · Acute brain injury (human) · Energy metabolism

K. L. H. Carpenter (✉) · I. Timofeev · P. G. Al-Rawi · J. D. Pickard · P. J. Hutchinson
Division of Neurosurgery,
Department of Clinical Neurosciences,
University of Cambridge,
Box 167, Addenbrooke's Hospital, Hills Road,
Cambridge CB2 0QQ, UK
e-mail: klc1000@wbic.cam.ac.uk

D. K. Menon
Division of Anaesthesia, Department of Medicine,
University of Cambridge,
Box 93, Addenbrooke's Hospital, Hills Road,
Cambridge CB2 0QQ, UK

K. L. H. Carpenter · D. K. Menon · J. D. Pickard · P. J. Hutchinson
Wolfson Brain Imaging Centre,
Department of Clinical Neurosciences,
University of Cambridge,
Box 65, Addenbrooke's Hospital, Hills Road,
Cambridge CB2 0QQ, UK

Introduction

Nitric oxide (NO) is well known for its ability to relax blood vessels and in addition has many other biological effects [4, 9, 12]. NO is synthesised from L-arginine, by nitric oxide synthase (NOS), which requires NADPH and molecular oxygen [4]. Constitutional forms of NOS are endothelial (eNOS) and neuronal (nNOS), whilst in inflammation an inducible form (iNOS) is produced in activated macrophages and other cell types. NO itself is a short-lived, highly reactive species but its major products nitrate and nitrite (collectively termed NO_x) can readily be measured retrospectively in biological samples as markers of nitric oxide production. NO is auto-oxidised (by oxygen) to nitrite and this process is catalysed by metals or metalloproteins [15]. In blood, the half-life of nitrite is about 10 min, as nitrite reacts rapidly with hemoglobin to form nitrate under oxygenated conditions or iron-nitrosylhemoglobin under deoxygenated conditions [15].

Neuroglobin, a heme protein in brain can also oxidise nitrite, as can cytoglobin [15].

Recently, in TBI patients, Hlatky et al. [9] found a significant relationship between increasing NO_x concentrations in brain microdialysates and increasing regional cerebral blood flow, suggesting that the microdialysate NO_x values were reflecting NO production by eNOS. The present study's aim was to assess the relationship (if any) between NO_x and energy-related molecules.

Materials and methods

Patient selection

The study was approved by the Cambridge Local Research Ethics Committee and assent obtained from the next of kin. Patients over the age of 16 years with acute brain injury (traumatic brain injury (TBI) or subarachnoid haemorrhage (SAH)) requiring ventilation and intracranial pressure monitoring were eligible for the study. The major exclusion criterion was patients with deranged clotting and/or low platelets, which precluded the placement of a microdialysis catheter. None of the patients had any significant previous neurological conditions (disease or injury) or family history of neuro-degenerative disease.

Microdialysis technique

Microdialysis catheters (one per patient) were inserted into the cerebral cortex (not close to a focal lesion), in conjunction with an ICP transducer (Codman, Raynham, MA, USA) using a triple lumen cranial access device (Technicam, Newton Abbot, UK). In one TBI patient an additional catheter was inserted close to a focal lesion via a bone flap. The catheters were CMA71 (100 kDa molecular weight cut-off), or in one case (SAH) CMA70 (20 kDa molecular weight cut-off). The catheters were perfused with CNS Perfusion Fluid (NaCl 147 mmol/L, KCl 2.7 mmol/L, $CaCl_2$ 1.2 mmol/L, $MgCl_2$ 0.85 mmol/L) at 0.3 μL/min with hourly vial changes. The vials were analysed at the bedside using a CMA600 or ISCUS analyser (see below) and then stored at −80°C for subsequent NO_x analysis. Microdialysis monitoring was started as soon as possible following admission to the Neuro-critical care unit.

Microdialysate analysis

Energy related molecules The bedside CMA600 or ISCUS analysers (CMA Microdialysis AB) were used to determine the extracellular concentrations of glucose, lactate, pyruvate and glycerol (and/or glutamate) at the time of collection. The lactate/pyruvate ratio was calculated as a marker of anaerobic metabolism.

Nitric oxide products (NO_x) Total nitrate and nitrite were analysed by injecting 5 μL of each microdialysate sample into a purge vessel containing a solution of vanadium (III) chloride (50 mmol/L) in hydrochloric acid (1 mol/L) at 95°C, continuously purged with a stream of nitrogen gas, connected to a Sievers 280i nitric oxide analyser (GE Analytical Instruments, Boulder, CO, USA). The principle of the assay is that vanadium (III) chloride converts nitrate and nitrite into nitric oxide, which is then reacted with ozone inside the nitric oxide analyser to produce nitrogen dioxide, which sensitively detected by virtue of its chemiluminescence.

Statistical analysis

Data for microdialysate parameters, namely NO_x, glucose, lactate, pyruvate, glycerol, glutamate and the lactate/pyruvate (L/P) ratio were expressed as mean (±standard deviation) for each catheter, and relationships between parameters were assessed by Spearman's rank correlation, using StatView (version 5) software.

Results

Brain microdialysis was carried out in 12 patients (11 traumatic brain injury (TBI), one subarachnoid haemorrhage (SAH)). Each patient had a microdialysis catheter inserted via a cranial access device into the frontal cortex, or in the SAH case into the parietal cortex, not close to a focal lesion. One of the TBI patients had an additional catheter (frontal) inserted via a bone flap, close to a focal lesion. The total number of catheters evaluated was thus 13. Details of patient demography are shown in Table 1. Mean concentrations (±SD) of NO_x and energy-related molecules for each patient are shown in Table 2. The mean of mean concentration of NO_x (±SD) for n=13 catheters was 32.67± 16.82 μmol/L. Mean of mean concentrations of the energy-related molecules were: glucose 1.02±0.51 mmol/L (n= 12), lactate 3.95±1.87 mmol/L (n=13), pyruvate 121.0± 46.5 μmol/L (n=13), glutamate 36.94±55.42 μmol/L (n= 4), glycerol 120.9±60.0 μmol/L (n=11), and the lactate/pyruvate (L/P) ratio 38.62±20.11 (n=13).

Illustrative case examples (four patients) of time-courses of microdialysate NO_x concentrations are shown in Fig. 1, and the corresponding concentrations of energy-related molecules (glucose, lactate, pyruvate, glycerol and L/P ratio) for one of these patients are illustrated in Fig. 2. The general qualitative pattern of behaviour (with quantitative inter-

Table 1 Patient demography

Patient number	Age (years)	Sex	GCS	Injury (CT)	Craniotomy performed	Microdialysis start post-injury day	Microdialysate vial sequence[a] evaluated for NO_x	Outcome
1	35	M	7	2	No	2	1–26 (n=25)	MD
2	19	M	7	2	No	5	1–27 (n=23)	GR
3	17	F	3	3	No	2	1–27 (n=23)	GR
4	51	F	11	5	No	1	21–51 (n=20)	Dead
5	42	M	3	2	No	2	10- 42 (n=18)	Dead
6	44	M	3	5	Yes	1	3–26 (n=21)	Dead
7 (A)[b]	35	M	3	6	Yes	1	1–47 (n=15)	Dead
7 (B)							22–47 (n=25)	
8	51	M	12	6	No	4	1–26 (n=26)	MD
9	17	M	3	2	No	1	2–28 (n=25)	SD
10	42	M	5	5	Yes	3	2–18, 93–108 (n=30)	SD
11	70	F	8	2	No	2	4–24 (n=20)	Dead
12[c]	41	M	8	SAH	No	2	249–285 (n=30)	Dead

GR good recovery, *MD* moderately disabled, *SD* severely disabled

[a] Microdialysate vials were collected hourly after start of microdialysis, and analysed at the bedside (on a CMA600 or ISCUS) for energy-related molecules. Not all vials had sufficient microdialysate volume remaining for NO_x analysis, so n (the number of vials analysed) is less than the total number of vials in the sequence

[b] Patient 7 had an additional microdialysis catheter (A) inserted via a bone flap, close to a focal lesion. In Patients 1–11 all other microdialysis catheters (including patient 7's catheter B) were via a cranial access device, not close to a focal lesion, and were located frontally, except in patient 12 where the catheter was parietal

[c] Patient 12 was subarachnoid haemorrhage (SAH); all the rest were head-injury (TBI)

patient variations) was that NO_x concentrations within-patient were relatively higher at earlier time-points compared with later time-points, in agreement with the findings of Hlatky et al. [9] in TBI, and those of Sakowitz et al. [18] in SAH. Patient 12 (Fig. 1d) died soon after the collection of the last microdialysate vial, by which time, following a period of fluctuation, NO_x concentrations had declined to low levels (6.2–12.1 μmol/L for the last five vials).

In order to establish the overall relationships (taking together all the patients in the study) between NO_x and each

Table 2 Mean (±SD) microdialysis results for NO_x and energy-related parameters in the 12 patients (13 catheters)

Patient number	NO_x (μmol/L)	*L*/*P* ratio	Glucose (mmol/L)	Lactate (mmol/L)	Pyruvate (μmol/L)	Glutamate (μmol/L)	Glycerol (μmol/L)
1	26.9±4.1	25.2±4.7	1.4±0.2	3.5±0.8	140.1±29.3	–	66.2±9.9
2	71.1±12.6	23.4±3.1	1.0±0.1	2.5±0.3	108.3±14.5	–	44.0±9.7
3	61.3±36.0	26.0±8.3	1.2±0.3	1.8±0.6	75.9±19.8	–	61.2±29.6
4	21.7±9.0	34.9±5.6	0.6±0.3	2.7±1.2	77.8±36.2	–	85.5±33.6
5	20.1±4.1	52.2±10.6	0.8±0.3	8.0±0.5	159.1±28.4	–	209.7±39.5
6	27.3±6.5	33.7±7.8	0.5±0.1	3.3±0.3	101.4±15.4	–	195.3±84.1
7(A)	46.8±42.8	29.2±16.3	1.7±0.7	3.2±1.7	118.9±61.2	–	96.0±150.1
7(B)	22.1±7.2	39.6±5.7	0.9±0.2	2.3±0.3	59.1±10.2	–	154.8±55.8
8	27.9±6.2	21.8±3.0	–	5.2±0.9	241.6±44.9	18.0±4.1	116.1±24.3
9	33.9±7.5	29.4±2.8	1.6±0.4	2.9±0.6	98.2±13.9	–	99.3±25.5
10	20.0±6.2	88.4±167.2	0.4±0.3	4.8±2.1	118.3±64.7	119.6±152.1	–
11	28.4±6.7	27.9±2.1	1.8±0.7	4.1±0.8	148.2±29.1	4.5±3.4	202.2±35.7
12[a]	17.2±9.2	70.4±76.6	0.4±0.2	7.0±3.5	126.4±38.4	5.7±22.1	–

For patient 7, (A) denotes catheter A inserted via bone-flap, (B) denotes catheter B inserted via cranial access device. The microdialysates were analysed at the bedside (on a CMA600 or ISCUS) for energy-related molecules, then the same specimens, if sufficient volume remained, were analysed in the laboratory for NO_x. CMA600/ISCUS data were then matched to NO_x data on a vial-by-vial basis, and any vials with CMA600/ISCUS data but without NO_x analysis were excluded from the data analysis

NO_x total nitrate plus nitrite, *L*/*P* lactate/pyruvate ratio, – not analysed

[a] SAH patient, the rest of the patients were TBI

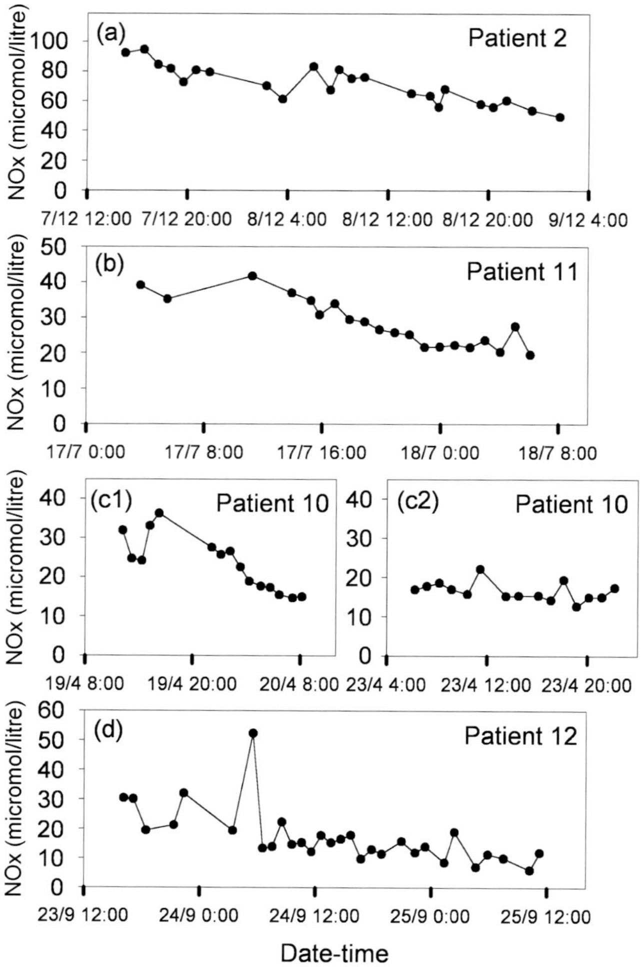

Fig. 1 Illustrative case examples of time-courses of brain microdialysate NO_x concentrations (micromole per litre) following acute brain injury. For the respective days on which microdialysis commenced post-injury, see Table 1. The graphs illustrate **a** patient 2 (TBI) for the first day of microdialysis; **b** patient 11 (TBI) for the first day of microdialysis; **c** *1* and *2* patient 10 (TBI) for the first day and fourth day of microdialysis respectively; **d** patient 12 (SAH) for the final 2 days of microdialysis, 10 days after the start

of the energy-related parameters, Spearman's rank correlation was performed, using the mean concentrations in Table 2. Increasing concentrations of NO_x correlated significantly with increasing glucose concentration ($r=0.71$, $p=0.018$, $n=12$), with decreasing lactate concentration ($r=-0.59$, $p=0.042$, $n=13$), and with decreasing L/P ratio ($r=-0.79$, $p=0.0065$, $n=13$) within the brain microdialysates (Fig. 3a,b,e). There was no significant correlation of NO_x with pyruvate ($r=-0.18$, $p=0.53$, $n=13$) or with glycerol ($r=-0.55$, $p=0.0845$, $n=11$) (Fig. 3c,d). Only four of the patients had glutamate measurements, so correlation between NO_x and glutamate was (as expected) non-significant ($r=-0.4$, $p=0.49$, $n=4$).

The outcomes of the patients (Table 1) were assessed at 6 months post-injury. Figure 3f shows the graph of mean L/P ratio vs NO_x with the data-points differentiated for outcome. Interestingly, the two individuals (patients 2 and 3) with the highest mean NO_x (71.1 and 61.3 μmol/L respectively; Table 2) were the only two who made good recovery within this group of 12 patients. Whether there is any true statistically significant relationship between high NO_x and favourable outcome remains to be determined, by studying more patients. Patients 2 and 3 had mean L/P ratios of 23.4 and 26.0 respectively, which were near the low end of the group (Table 2), but even so, L/P ratios of about 25 or more are generally considered "pathological". NO might thus conceivably exert other beneficial influences on outcome, besides effects on energy metabolism.

Discussion

This pilot study has shown, in acute brain injury patients, that as brain microdialysate NO_x (total nitrate plus nitrite)

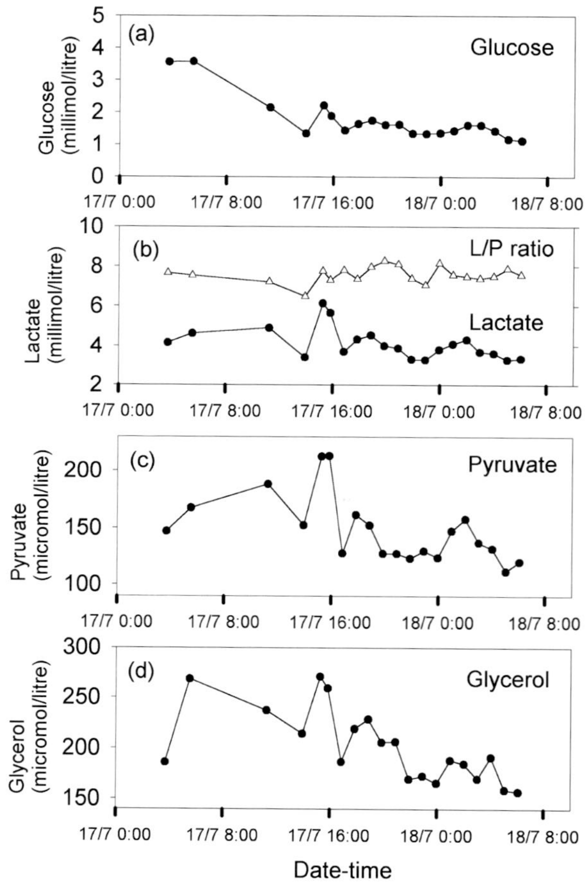

Fig. 2 Illustrative case example of time-courses of molecules related to energy metabolism in brain microdialysates following acute brain injury (patient 11, TBI). For this patient, microdialysis commenced 2 days post-injury, and the graphs illustrated represent the first day of microdialysis. The graphs illustrate **a** glucose (millimol per litre), **b** lactate (millimol per litre) and the lactate/pyruvate (L/P) ratio, **c** pyruvate (micromole per litre) and **d** glycerol (micromole per litre). The microdialysates were the same samples that were analysed for NO_x, illustrated in Fig. 1b

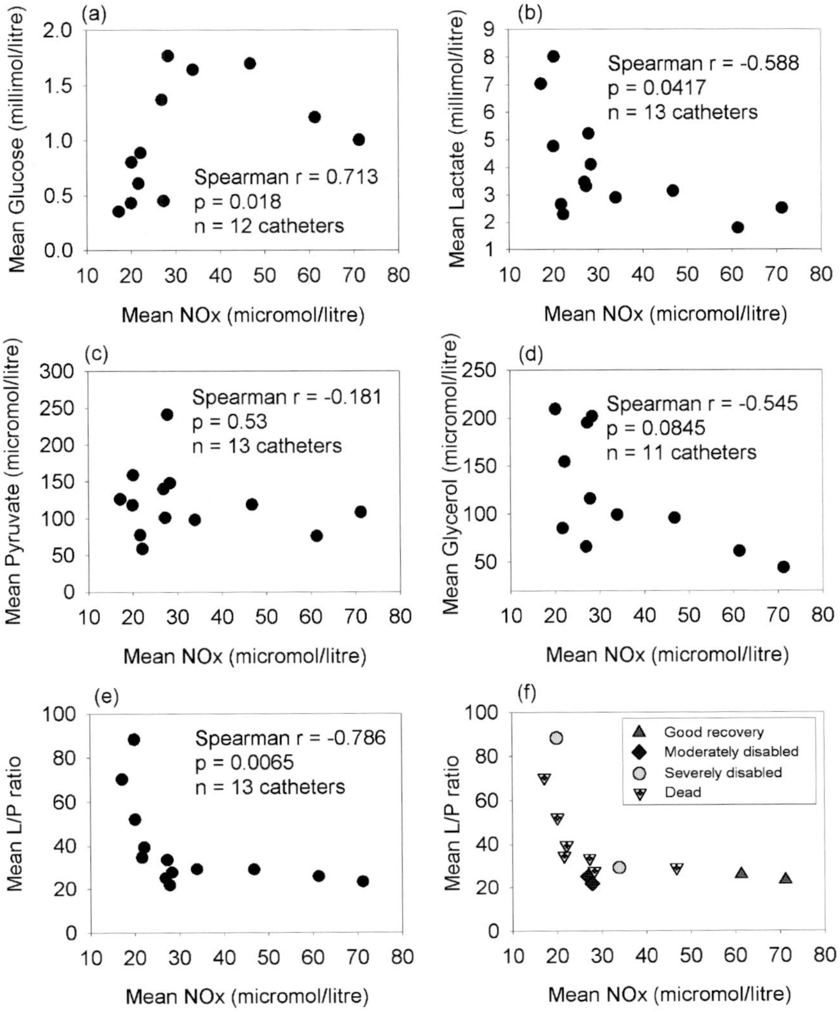

Fig. 3 Bivariate scattergrams of NO_x concentration (micromole per litre) plotted against **a** glucose (millimol per litre), **b** lactate (millimol per litre), **c** pyruvate (micromole per litre), **d** glycerol (micromole per litre), **e** the lactate/pyruvate ratio and **f** the lactate/pyruvate ratio with data-points differentiated for outcome, measured in human brain microdialysates (acute brain injury). Each data-point represents the mean value for each of the microdialysis catheters evaluated (see Table 2). Statistical evaluation was by Spearman's rank correlation

increases, glucose increases, lactate decreases, and the *L*/*P* ratio decreases. These data suggest that in injured brains, higher concentrations of NO are associated with more favourable metabolism. NO may act to increase blood flow and delivery of oxygen and glucose, shifting metabolism towards the TCA cycle. This suggests a beneficial role for NO, discussed in more detail later. Also, nitrite can act as an in vivo "store", which can be mobilised back to NO under the appropriate biological conditions [15]. However, NO is potentially able to exert adverse effects as well as beneficial effects. NO can react with thiols, amines and heme proteins to form nitrosothiols, nitrosamines, nitrosohemoglobin and iron-nitrosyl-hemoglobin [15]. NO can likewise react with neuroglobin and cytoglobin [15], heme proteins found in brain. Notably, NO can react with superoxide radical anion, a product of NAD(P)H oxidase of phagocytes or from electron leakage from mitochondrial electron transport chains, forming highly reactive peroxynitrite, which can nitrate proteins and damage cells [1, 12]. Besides vasodilation, NO has anti-inflammatory effects, at least partly due to its ability to reduce NFkappaB activation, e.g. leading to down-regulation of chemokine MCP-1 [19]. NFkappaB regulates NOS transcription [10, 13], thus NO production constitutes a self-limiting negative feedback loop [6–8, 11]. In contrast, peroxynitrite sustains NFkappaB [8], potentially allowing NO production to escalate.

NO synthesis appears to play an important role in surviving the initial few hours after TBI, evidenced by an experimental study in which inhibition of NO synthesis by pre-treatment with L-NAME (L-nitro-arginine methyl ester) led to more than 67% mortality within 3 h following injury, versus 1% in untreated animals [2]. L-NAME does not discriminate between the three isoforms of NOS. There is evidence, from experimental TBI studies with selective inhibitors, for deleterious roles of nNOS and iNOS in the early and later stages respectively [4]. eNOS is considered to have a beneficial role in preserving cerebral blood flow, at least in the early stages after TBI [4].

A known effect of NO is that it can potentially inhibit mitochondrial respiration, by binding to cytochrome c oxidase and other proteins of the electron transport chain, thereby shifting cellular energy metabolism towards non-oxidative metabolism (glycolysis) [12]. However, the overall behaviour of NO in the context of cellular respiration is multifactorial and complex [12, 14]. Our findings suggest that the dominant effect of NO is to shift energy metabolism away from glycolysis, evidenced by the falling *L*/*P* ratio with increasing NO_x seen in human brain microdialysates. This improvement in energy metabolism may be a result of increased blood supply, probably due to the activity of NO generated by eNOS, thereby relaxing blood vessels. This is consistent with our findings, in human brain microdialysates, of (a) increasing glucose concentrations and (b) falling lactate concentrations, in relation to increasing NO_x concentrations. Our findings are also consistent with those of Hlatky et al. [9] who found that increasing NO_x in human brain microdialysates was significantly related to increasing regional cerebral blood flow. Besides our present study, we are only aware of only one other study of human brain microdialysate NO_x in relation to cerebral energy metabolism. This was in ten SAH patients, by Sakowitz et al. [17]. Like us, they found significant correlations of increasing NO_x with increasing glucose and with decreasing lactate in brain microdialysates. However, unlike us, they found no significant relationship between NO_x and the *L*/*P* ratio. The reason for this apparent discrepancy is unknown, but might be at least partly due to differences in pathology as well as to the evident differences in NO_x assay methods and statistical methods employed in the two studies.

A limitation to brain microdialysis in patients is that of variation in the timing of the microdialysis, in relation to the time of injury. This is unavoidable, as it is dictated by the circumstances of the injury and subsequent clinical considerations. Another constraint was availability of sufficient volume of microdialysate for NO_x analysis, since part of each sample had already been used up at the bedside, for the analysis of energy-related molecules. The patients' microdialysates available for analysis in this study thus of necessity represent sampling for different periods of time, which undoubtedly influences the results. Brain biochemistry following acute brain injury is a dynamic process, as exemplified in Figs. 1 and 2, and expressing the concentrations as averages for each patient (Table 2) clearly masks the time course. However this approach is justifiable as our aim was to establish the overall relationships, taking together all the patients in the study, between NO_x and each of the energy-related parameters. An assumption of the present study is that NO_x is a surrogate marker for NO. Whilst NO can be measured directly using an appropriate electrode in experimental and in vitro systems, there is (to our knowledge) no NO electrode available for use in human brain, thus no alternative to measuring NO_x. In support of this approach, Cherian et al. [2] found that in experimental TBI studies, changes in microdialysate NO_x were similar to those of NO measured directly with an electrode, though the latter was more effective at measuring rapid transient changes. Another assumption is that we are measuring "genuine" levels of NO_x rather than insertion artefacts. Cherian et al. [2] found that careful removal and re-insertion of an NO electrode into rat brain produced minimal changes in NO levels, but whether the micro-dialysis catheter's insertion is similarly non-perturbing for NO_x levels is as yet unknown. Another limitation is that we do not know to what degree the distance of the catheter from lesions is likely to influence levels of NO_x and of the other microdialysate parameters. In the present study, all but one of our patients had one catheter each, not close to a focal lesion, and only one of our patients had two catheters, respectively perilesional (A) right frontal and non-perilesional (B) left frontal, precluding any firm conclusions to be drawn at this stage. In an experimental study, TBI-induced gene expression, including IL-1β and iNOS, far exceeded the primary impact site and was also seen at distant sites in brains [16]. Monitoring at a distant site is thus potentially able to yield useful information.

In conclusion, our present small pilot study of 12 patients shows that as microdialysate NO_x increases, glucose increases, lactate decreases, and the *L*/*P* ratio decreases. These findings suggest that in human acute brain injury, NO may improve blood supply, delivering more oxygen and glucose to brain tissue, and shifting metabolism in favour of the TCA cycle. Whilst we do not yet have enough data to make a proper statistical evaluation of NO in relation to outcome, it is interesting to note that patients 2 and 3, who had with the two highest mean microdialysate NO_x concentrations (71.1 and 61.3 μM respectively) were the only individuals within the group of 12 to progress to good recovery (Tables 1 and 2; Fig. 3f). This raises the question of whether pharmacological augmentation of NO production would be a useful therapy in acute brain injury. Administration of L-arginine, tetrahydrobiopterin and erythropoietin each produced an increase in brain NO together with improved cerebral blood flow, in experimental TBI [3, 5]. Whether analogous treatments in patients after acute brain injury, using these or other NO-increasing agents, would have a similar effect, and most importantly improve outcome, has yet to be investigated.

Acknowledgements We gratefully acknowledge the following. *Study support*: the Medical Research Council (Grant No G9439390 ID 65883) and the Academy of Medical Sciences/Health Foundation. *Authors' support*: KLHC, the MRC (Acute Brain Injury Programme Grant); IT, Codman and the Evelyn Trust; PGA, the Stroke Association, UK; PJH, Academy of Medical Sciences/Health Foundation Senior Surgical Scientist Fellowship. *Technical assistance*: M Higgins (Analytix) and L Maskell.

Conflict of interest statement We declare that we have no conflict of interest.

References

1. Babior BM (2000) Phagocytes and oxidative stress. Am J Med 109:33–44
2. Cherian L, Goodman JC, Robertson CS (2000) Brain nitric oxide changes after controlled cortical impact injury in rats. J Neurophysiol 83:2171–2178
3. Cherian L, Hlatky R, Robertson CS (2004) Comparison of tetrahydrobiopterin and L-arginine on cerebral blood flow after controlled cortical impact injury in rats. J Neurotrauma 21:1196–1203
4. Cherian L, Hlatky R, Robertson CS (2004) Nitric oxide in traumatic brain injury. Brain Pathol 14:195–201
5. Cherian L, Goodman JC, Robertson C (2007) Neuroprotection with erythropoietin administration following controlled cortical impact injury in rats. J Pharmacol Exp Ther 322:789–794
6. Davis ME, Grumbach IM, Fukai T, Cutchins A, Harrison DG (2004) Shear stress regulates endothelial nitric-oxide synthase promoter activity through nuclear factor kappa-B binding. J Biol Chem 279:163–168
7. Grumbach IM, Chen W, Mertens SA, Harrison DG (2005) A negative feedback mechanism involving nitric oxide and nuclear factor kappa-B modulates endothelial nitric oxide synthase transcription. J Mol Cell Cardiol 39:595–603
8. Hattori Y, Kasai K, Gross SS (2004) NO suppresses while peroxynitrite sustains NF-kappaB: a paradigm to rationalize cytoprotective and cytotoxic actions attributed to NO. Cardiovasc Res 63:31–40
9. Hlatky R, Goodman JC, Valadka AB, Robertson CS (2003) Role of nitric oxide in cerebral blood flow abnormalities after traumatic brain injury. J Cereb Blood Flow Metab 23:582–588
10. Hobbs AJ, Higgs A, Moncada S (1999) Inhibition of nitric oxide synthase as a potential therapeutic target. Annu Rev Pharmacol Toxicol 39:191–220
11. Katsuyama K, Shichiri M, Marumo F, Hirata Y (1998) NO inhibits cytokine-induced iNOS expression and NF-kappaB activation by interfering with phosphorylation and degradation of IkappaB-alpha. Arterioscler Thromb Vasc Biol 18:1796–1802
12. Moncada S, Bolanos JP (2006) Nitric oxide, cell bioenergetics and neurodegeneration. J Neurochem 97:1676–1689
13. Nakata S, Tsutsui M, Shimokawa H, Yamashita T, Tanimoto A, Tasaki H, Ozumi K, Sabanai K, Morishita T, Suda O, Hirano H, Sasaguri Y, Nakashima Y, Yanagihara N (2007) Statin treatment upregulates vascular neuronal nitric oxide synthase through Akt/NF-kappaB pathway. Arterioscler Thromb Vasc Biol 27:92–98
14. Palacios-Callender M, Hollis V, Frakich N, Mateo J, Moncada S (2007) Cytochrome c oxidase maintains mitochondrial respiration during partial inhibition by nitric oxide. J Cell Sci 120:160–165
15. Pelletier MM, Kleinbongard P, Ringwood L, Hito R, Hunter CJ, Schechter AN, Gladwin MT, Dejam A (2006) The measurement of blood and plasma nitrite by chemiluminescence: pitfalls and solutions. Free Radic Biol Med 41:541–548
16. Rooker S, Jander S, Van Reempts J, Stoll G, Jorens PG, Borgers M, Verlooy J (2006) Spatiotemporal pattern of neuroinflammation after impact-acceleration closed head injury in the rat. Mediators Inflamm 2006:90123
17. Sakowitz OW, Wolfrum S, Sarrafzadeh AS, Stover JF, Dreier JP, Dendorfer A, Benndorf G, Lanksch WR, Unterberg AW (2001) Relation of cerebral energy metabolism and extracellular nitrite and nitrate concentrations in patients after aneurysmal subarachnoid hemorrhage. J Cereb Blood Flow Metab 21:1067–1076
18. Sakowitz OW, Wolfrum S, Sarrafzadeh AS, Stover JF, Lanksch WR, Unterberg AW (2002) Temporal profiles of extracellular nitric oxide metabolites following aneurysmal subarachnoid hemorrhage. Acta Neurochir Suppl 81:351–354
19. Tsao PS, Wang B, Buitrago R, Shyy JY, Cooke JP (1997) Nitric oxide regulates monocyte chemotactic protein-1. Circulation 96:934–940

PART 4:
Biomedical informatics

Acta Neurochir Suppl (2008) 102: 217–221

The brain monitoring with Information Technology (BrainIT) collaborative network: data validation results

Martin Shaw · Ian Piper · Iain Chambers · Giuseppe Citerio · Per Enblad · Barbara Gregson · Tim Howells · Karl Kiening · Julia Mattern · Pelle Nilsson · Arminas Ragauskas · Juan Sahuquillo · YH Yau · on behalf of the BrainIT Group (www.brainit.org)

Abstract
Background The BrainIT group works collaboratively on developing standards for collection and analyses of data from brain injured patients towards providing a more efficient infrastructure for assessing new health technology.
Materials and methods Over a 2 year period, core dataset data (grouped by nine categories) were collected from 200 head-injured patients by local nursing staff. Data were uploaded by the BrainIT web and random samples of received data were selected automatically by computer for validation by data validation (DV) research nurse staff against gold standard sources held in the local centre. Validated data was compared with original data sent and percentage error rates calculated by data category.
Findings Comparisons, 19,461, were made in proportion to the size of the data received with the largest number checked in laboratory data (5,667) and the least in the surgery data (567). Error rates were generally less than or equal to 6%, the exception being the surgery data class where an unacceptably high error rate of 34% was found.

M. Shaw · I. Piper
Clinical Physics, Southern General Hospital,
1345 Govan Road,
Glasgow, UK

I. Chambers · B. Gregson
Regional Medical Physics Department,
Newcastle General Hospital,
Newcastle, UK

G. Citerio
Neurorianimazione, Hospital San Gerardo,
Via Donizetti,
Monza, Italy

P. Enblad · T. Howells · P. Nilsson
Neurosurgery, Uppsala University Hospital,
Uppsala, Sweden

K. Kiening · J. Mattern
Neurosurgery, Ruprecht-Karls-Universitat Hospital,
Im Neuenheimer Feld,
Heidelberg, Germany

A. Ragauskas
Kaunas University Hospital,
Studentu,
Kaunas, Lithuania

J. Sahuquillo
Neurosurgery, Vall d'Hebron Hospital,
Paseo Vall d'Hebron,
Barcelona, Spain

Y. Yau
Neurosurgery, Royal Adelaide Hospital,
North Terrace,
Adelaide, Australia

I. Piper (✉)
Intensive Care Monitoring, Dept. Clinical Physics, 5th Floor,
Institute of Neurological Sciences Southern General Hospital,
1345 Govan Road,
Glasgow, UKG514TF
e-mail: ipiper@clinmed.gla.ac.uk

I. Chambers
The James Cook University Hospital,
Marton Road,
Middlesbrough, UK

B. Gregson
Neurosurgery, Newcastle General Hospital,
Westgate Road,
Newcastle, UK

Conclusions The BrainIT core dataset (with the exception of the surgery classification) is feasible and accurate to collect. The surgery classification needs to be revised.

Keywords Clinical network · TBI · Methodology · Internet

Background

The BrainIT group works collaboratively on developing standards for collection and analyses of data from brain injured patients towards providing a more efficient infrastructure for assessing new health technology (http://www.brainit.org). The group have defined a core dataset collected using PC based tools as part of an EC funded study (QLGT-2000-00454). A series of meetings spread over one year enabled the group to define a minimum set of data that can be collected from all patients with traumatic brain injury (TBI), which would be useful in most research projects conducted in this population of patients. The core-dataset includes nine categories:

1. *Demographic* and one-off clinical data (e.g.: pre-neurosurgical hospital data, first and worst CT scan data etc.),
2. *Daily management* data (e.g.: use of sedatives, analgesics, vasopressors, fluid input/output balance etc.),
3. *Laboratory* data (e.g.: blood gas, haematology, biochemistry data etc),
4. *Event* data (e.g.: nursing manouevres, physiotherapy, medical procedures (line insertion), calibrations etc.),
5. *Surgical* procedures,
6. *Monitoring* data summary (e.g.: type and placement location of ICP sensor, BP lines, etc.),
7. *Neuro Event* Summary (e.g.: GCS scores, pupil size and reactivity),
8. *Targeted Therapies* (e.g.: mannitol given for raised ICP, pressor given for arterial hypotension etc.),
9. *Vital monitoring* data (e.g.: minute by minute BP, ICP, SaO_2 etc.).

A three year follow up EC funded study (QLGC-2002-00160) enabled the group to develop IT methods to collect the core dataset and to assess the feasibility and accuracy for collection of this core-dataset from 22 neuro-intensive care centres [1]. Data validation research nurse staff were hired on a country by country basis to check samples of the collected data against gold standard clinical record sources in order to quantify the accuracy and therefore the feasibility for collection of the BrainIT core-dataset using the group IT based data collection methods. This paper describes the results of analysis of 200 patients data in whom validation data was also acquired independently by data validation research nurses. The error rates classed by data category are presented and discussed. These validation results calculated on a subset of patients provides an estimate of the data quality for future analyses on the full patient cohort of 350 patients collected as part of the EEC funded study.

Materials and methods

Over a 2 year period, core dataset data (grouped by nine categories: as presented in the background section) were collected from 200 head-injured patients by local nursing staff.

Clinical data is entered by bedside nursing staff on hand held PDA's which, in collaboration with Kelvin Connect Ltd [2], implemented the BrainIT core dataset definition in a flexible and easy to use hand-held PDA based system. A training course was held for the data validation nursing staff in Glasgow on the use of this data collection instrument which supported data entry in six European languages. An anonymisation routine removed patient identification elements from the collected data and labelled the patient data file with a unique BrainIT study code generated from the BrainIT web-site. A local database held in each centre linked the anonymised data to local centre patient id information which was needed during the data checking stage of the study. Anonymised data was uploaded via the BrainIT web upload services. A server side data converter tool converts data from centre based format into BrainIT data format generating nine data category files which are imported into the BrainIT database. A validation request tool samples 20% of the data sent for each data category and generates a validation request file listing the time-stamps and data items to be checked by local data validators. Data validators enter into a *data validation tool* the requested data items for checking from source documentation held in each local centre. Validation data is uploaded to the BrainIT data coordinating centre via the website and using data validation checking software tools, the validated data is checked against the data items originally sent from which percentage accuracy data is calculated. As part of this validation process, in addition to the categorical and numeric clinical data being checked for accuracy, we also assessed the minute by minute monitoring data too. Random samples of monitoring data channels uploaded (e.g.: ICP, SaO_2) were selected and validation staff asked to manually enter the hourly recorded values from the nurses chart (or local gold standard data source) for the first and last 24 h periods of bedside monitoring for a given patient for a given channel. These "validation" values could then be compared with a range of summary measures (e.g.: mean, median) from the computer based monitoring data acquired from the patient.

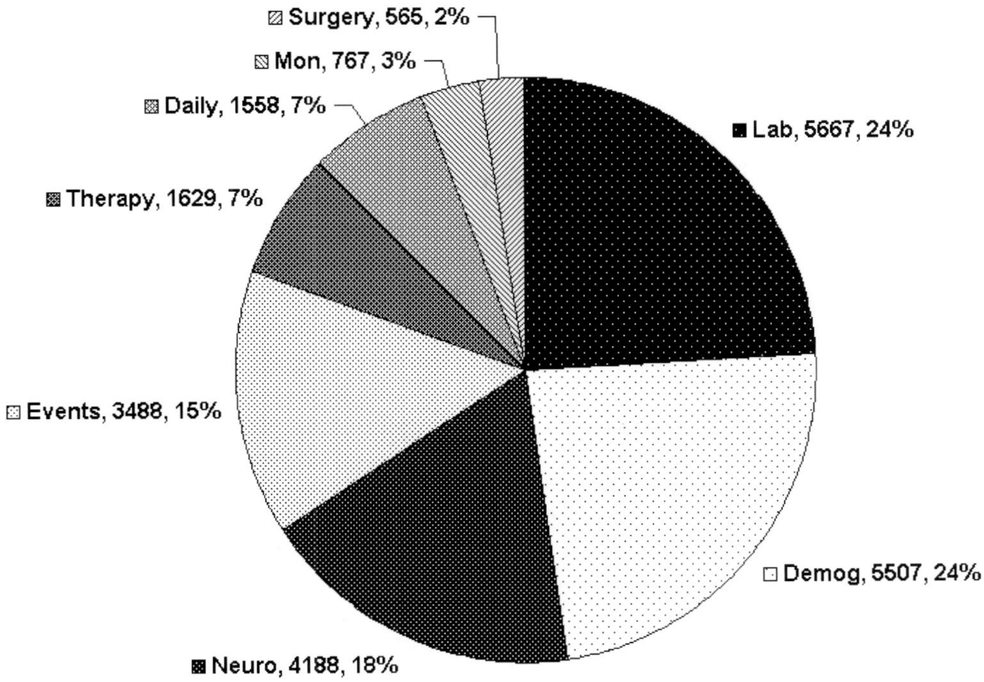

Fig. 1 Pie chart showing the distribution of the 19,461 data validation comparisons which were made in proportion to the size of the data received with the largest number checked in laboratory data (5,667) and the least in the surgery data (567)

Results

In total, 19,461 comparisons were made between collected data elements and source documentation data. The number of comparisons made per data category was in proportion to the size of the data received for that category with the largest number checked in laboratory data (5,667) and the least in the surgery data (567) (Fig. 1). Table 1 summarises error rates by data class. Error rates were generally less than or equal to 6%, the exception being the surgery data class where an unacceptably high error rate of 34% was found.

In the surgery data category, nursing staff had to choose surgical procedures from a fixed list of procedure types: *(1) ICP placement, (2) Evacuation of Mass Lesion, (3) Elevation of depressed skull fracture, (4) Removal of foreign body,(5) Anterior Fossa repair for CSF Leak, (6) Placement of Extra Ventricular Drain, (7) Active external decompression (with bone removal and duroplastia), (8) Other*. This classification system was used in an attempt at simplification and reducing the burden of data entry. However, through discussions with local nursing and data validation staff it was found that there was particular confusion over *when* to record ICP sensor placement and the presence of skull fractures as the primary surgical procedure. Typically, these procedures occur during the same operative procedure as for example "evacuation of mass lesion". Confusion over coding these two procedures accounted for the majority of errors in this data category.

We also checked the detection rate of acute events (e.g.: nursing management, physiotherapy, blood samples etc.). It was found that short duration events were rarely missed but longer duration events such as transfer to CT or theatre were more likely to be not recorded. Through discussions with local nursing and data validation staff it is believed that the intense nursing activity just prior to and following a transfer is more likely to lead to omissions in recording these events on research systems.

Figure 2 shows a scatter plot of computer monitored minute by minute ICP data averaged over 60 min (ICPavg) plotted against nurses chart end hour values (ICPvalid) collected by the data validation nurses. There is a good correlation between the two sets of data with a linear

Table 1 Percentage error rate by data type class with description of common error types

Data class	Error rate (%)	Common errors
Laboratory	2	pCO_2, FiO_2 value
Demographic	4	Monitoring on arrival at neurosurgery, intubation on arrival at neurosurgery
Neuro observations	5	Pupil Size, GCSv (code 1 Vs Unknown)
Monitoring summary	5	ICP type, ICP Location
Daily management summary	5	Infusion type (bolus vs infusion or both), drug number (1, > 1)
Targeted therapy	6	Non-standard target, no Target specified
Surgeries	34	ICP placement, Skull no., mass lesion

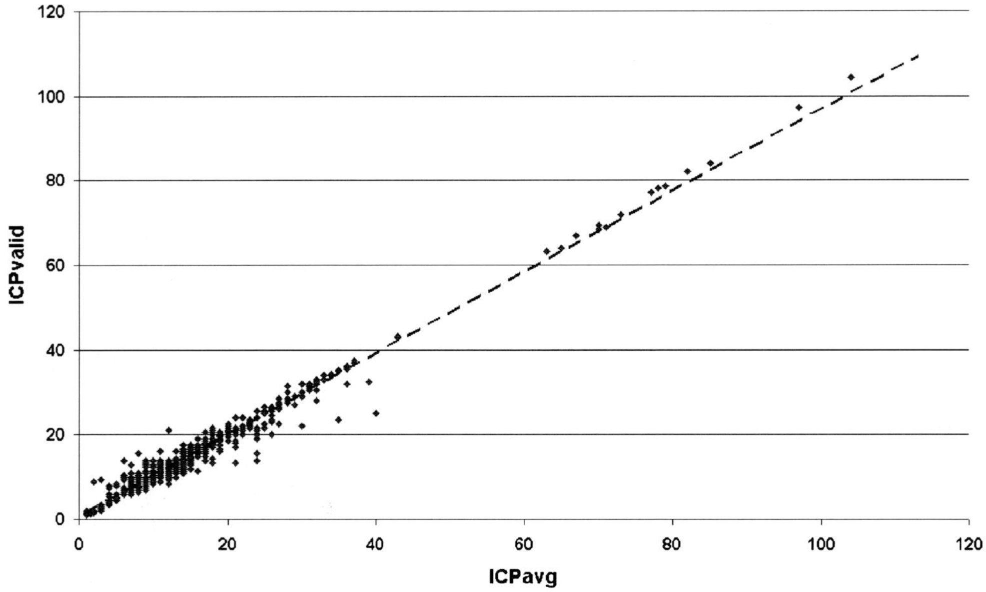

Fig. 2 Scatter plot of computer monitored minute by minute ICP data averaged over 60 min (ICPavg) plotted against nurses chart end hour values (ICP-valid). Linear regression best fit R^2 value=0.9773

regression best fit R^2 value of 0.9773. Figure 3 is an Altman and Bland plot showing the average bias (−0.15 mmHg) and 95% confidence limits (0.12, −0.45) for the computer monitored end hour averaged data Vs the nurses chart end hourly recorded values collected by the validation nurses.

Discussion

Good clinical practice dictates that as part of clinical trial design, acquired data must be checked for accuracy against gold standard data sources. This is often implemented through either employing a contract research organisation or independent research nurse staff to perform this duty. In large multi-centre clinical trials, costs to hire research nurse data validation staff can become prohibitively expensive and feasible only if significant industry or research council funding support is provided. Now with the adoption of the new medical device standard ISO-14155, even small medical device studies are expected to provide some form of check on the accuracy of data.

To our knowledge, this study conducted by the BrainIT group is one of only a few projects to attempt to prospectively assess the data capture error rate within an academic environment [4]. We have shown that it is feasible to collect the BrainIT dataset from multiple centres in an international setting with IT based methods and the accuracy of the data collected is greater than or equal to 94%, with the exception of the surgery data type which

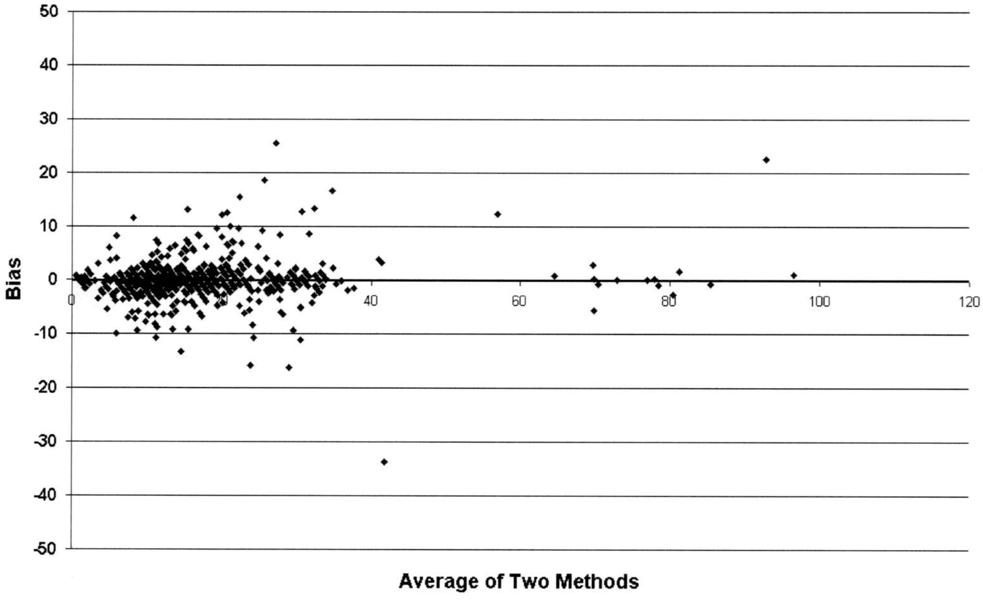

Fig. 3 Altman and Bland plot showing the average bias (−0.15 mmHg) and 95% confidence limits (0.12, −0.45) for the computer monitored end hour averaged data Vs the nurses chart end hourly recorded values collected by the validation nurses

must be revised. We have also shown that computer collected minute by minute vital signs data, summarised as end hour averages, correlated well with nursing chart end hour recordings. This allows the end hour averaged computer records to be used in database analyses assessing nurses chart recorded detection of events with computer based sampling. These validation results calculated on a subset of patients provides an estimate of the data quality for future analyses on the full patient cohort of 350 patients collected as part of the EEC funded study which was conducted over the same time period by the same staff using the same data methods. Clearly though, future data collection projects will generate datasets under differing data collection conditions and will require a separate validation stage if we wish to maintain our confidence in the level of data accuracy. However, the costs of maintaining such a data validation network is prohibitively high. To maintain a full time data validation nurse within each participating country costs in excess of 1 Million Euro's per year. Such large running costs for an academic network is not sustainable in the long term and a more cost-effective solution for data validation must be found.

One promising approach being adopted by the BrainIT group is developing collaborative research with experts in Grid based secure access technology. Grid technology covers more than just access to networks of high end servers in order to solve computationally intensive problems. There is a considerable amount of expertise and middleware software solutions now available that provide secure access to distributed medical datasets so that the right people see the correct data in the appropriate context [3]. Such an approach, once local and national IT policy staff are satisfied with the security, will enable remote data validation systems to directly query hospital based gold standard data sources for data checking. Clearly some data validation staff will still be required to support system queries but increased use of automatic data validation procedures should significantly reduce the cost burden to conduct multi-centre clinical trials. Towards this end, the BrainIT group as part of an EEC funded framework VII project plan to assess such an approach in 6 neuro-intensive care centres equipped with the latest Grid technology.

Acknowledgements We wish to acknowledge the contribution of all data contributing members of the BrainIT group (http://www.brainit.org) who supported the EEC project: QLGC-2002-01160.

Investigators and participating centres Barcelona Spain, Prof J Sahuquillo; Cambridge UK., Prof JD Pickard; Edinburgh UK, Prof R Mins, Prof I Whittle; Glasgow UK, Mr L Dunn; Gothenburg Sweden, Dr B Rydenhag; Heidelberg, Germany, Dr K Kiening; Iasi Romania, Dr S Iencean; Kaunas Lithuania, Prof D Pavalkis; Leipzig Germany, Prof J Meixensberger; Leuven Belgium, Prof J Goffin; Mannheim Germany, Prof P Vajkoczy; Milano Italy, Prof N Stocchetti; Monza Italy, Dr G Citerio; Newcastle upon Tyne UK, Dr IR Chambers; Novara Italy, Prof F Della Corte; Southampton UK, Dr J Hell; Uppsala Sweden, Prof P Enblad; Torino Italy, Dr L Mascia; Vilnius Lithuania, Prof E Jarzemaskas; Zurich Switzerland, Prof R Stocker,

Conflict of interest statement We declare that we have no conflict of interest.

References

1. Piper I, Citerio C, Chambers I et al (2003) The BrainIT Group: Concept and Core Dataset Definition. Acta Neurochir 145:615–629
2. http://www.kelvinconnect.com/
3. http://www.nesc.ac.uk/hub/projects/votes/
4. Beretta L, Aldrovandi V, Grandi E, Citerio G, Stocchetti N (2007) Improving the quality of data entry in a low-budget head injury database. Acta Neurochir (Wien). Jul 31

Acta Neurochir Suppl (2008) 102: 223–227

BrainIT collaborative network: analyses from a high time-resolution dataset of head injured patients

Iain Chambers · Barbara Gregson · Giuseppe Citerio · Per Enblad · Tim Howells · Karl Kiening · Julia Mattern · Pelle Nilsson · Ian Piper · Arminas Ragauskas · Juan Sahuquillo · Y. H. Yau · on behalf of the BrainIT Group

Abstract

Background The BrainIT project was conceived in 1997 and has grown into an international collaboration with the purpose of gathering high time resolution data from head injured patients utilising standardised methodologies.

Materials and methods From 1998, 22 participating neuroscience centres collected three main types of information: demographic, physiological data and clinical treatment information. A data collection solution was provided for each centre dependent on their existing facilities and data

I. Chambers (✉)
Regional Medical Physics Department,
The James Cook University Hospital,
Marton Road,
Middlesbrough, UKTS4 3BW
e-mail: iain.chambers@stees.nhs.uk
url: www.brainit.org

B. Gregson
Ward 31 (North Wing),
Newcastle General Hospital,
Westgate Road,
Newcastle upon Tyne,
UKNE4 6BE
e-mail: barbara.gregson@ncl.ac.uk

G. Citerio
Neurorianimazione,
Hospital San Gerardo,
Via Donizetti Monza,
Monza, Italy
e-mail: gciterio@gmail.com

P. Enblad · T. Howells · P. Nilsson
Neurosurgery,
Uppsala University Hospital,
Uppsala, Sweden

P. Enblad
e-mail: Per.Enblad@neurokir.uu.se

T. Howells
e-mail: timothy.howells@lul.se

P. Nilsson
e-mail: Pelle.Nilsson@neurokir.uu.se

K. Kiening · J. Mattern
Neurosurgery, Ruprecht-Karls-Universitat Hospital,
Im Neuenheimer Feld,
Heidelberg, Germany

K. Kiening
e-mail: Karl.Kiening@med.uni-heidelberg.de

J. Mattern
e-mail: Julia.Mattern@med.uniheidelberg.de

I. Piper
Department Clinical Physics,
5th Floor Institute of Neurological Sciences Southern General Hospital,
1345 Govan Road,
Glasgow, UK
e-mail: ipiper@clinmed.gla.ac.uk

A. Ragauskas
Kaunas University Hospital,
Studentu,
Kaunas, Lithuania
e-mail: Telematics@tef.ktu.lt

J. Sahuquillo
Neurosurgery, Vall d'Hebron Hospital,
Paseo Vall d'Hebron,
Barcelona, Spain
e-mail: sahuquillo@neurotrauma.net

Y. H. Yau
Neurosurgery, Royal Adelaide Hospital,
North Terrace,
Adelaide, Australia
e-mail: yhyau@doctors.org.uk

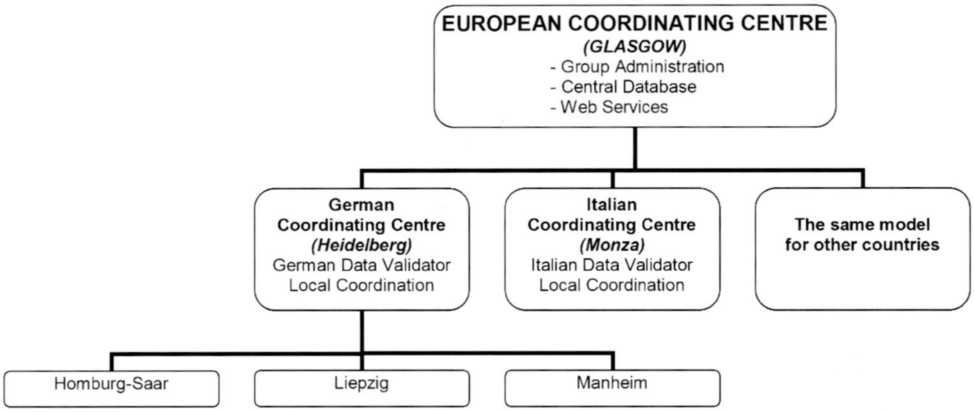

Fig. 1 Structure of the BrainIT group

were collected for the duration of monitoring as defined by the routine care in each centre. On completion of ICP monitoring all personal information was removed and then transferred to Glasgow via the internet where it was converted into a standard format and entered into a central database. Outcome was measured using the extended Glasgow Outcome Score using an interview questionnaire.

Findings Data has been obtained from a total of 349 patients (277 male and 72 female) The age of these patients ranged from 1 to 87 years (median 31); 145 had been involved in a traffic accident and 32 were pedestrians; 78 had suffered a fall; 24 were assaulted and the remaining 70 of other causes. A large amount of physiological data was collected (e.g. BP 2,531 days, ICP 2,212 days in total). This dataset has provided the opportunity to perform unique analysis and these include the statistical features of blood pressure, diurnal variations in ICP, optimal sampling rate determination and a comparison of summary measures of secondary insults.

Conclusions This challenging collaboration has brought together a large number of centres and developed a successful clinical research network focussed on improving the treatment of head injured patients. It has successfully collected a vast quantity of high quality data that provides a rich source for analysis and hypothesis testing.

Keywords Brain injury · Monitoring · Collaborative network · Information technology

Introduction

Head injured patients provide very rich but diverse data from physiological monitoring, patient demographics, treatment and imaging sources and with advances in technology the measurement and recording of high time resolution data as paper based methods is readily available. In comparison, paper based methods, often used in pharmacological studies, may underestimate the severity, length and frequency of secondary insults [7].

The increasing need to ensure that clinical practice is based upon a sound evidence base requires methods and analyses that are capable of replication. Research studies also need to be completed in a timely manner so that changes or improvements in aspects of care do not introduce confounding factors that might affect the analysis and results. In addition data are collected in different ways, formats and time resolution and can be described by different summary measures (e.g. 1 h reading, average over 1 h, rolling means etc.) and this can make it very difficult to compare studies from different centres. There is therefore a need to work collectively to ensure common standards and methods are employed for the collection of data.

In 1997 the BrainIT network evolved from discussions within a group of multi-disciplinary researchers working in the field of head injury. From the outset the underlying ethos was one of openness and collaboration: anyone can join, organisation was non-profit making and time was given voluntarily. All BrainIT studies contribute to a common database to which contributing members have free access. The primary objectives of the BrainIT group were defined as:

1. To develop and disseminate standards for the collection, analysis and reporting of intensive care monitoring data collected from brain injured patients.
2. To provide an efficient multi-centre infra-structure for generating evidence on the utility of new forms of invasive and non-invasive intensive care monitoring and methods for improving the care and outcome of brain-injured patients.
3. To develop and use a standardised database as a tool for hypothesis generation and the development, testing and validation of new data analysis methodologies.

Materials and methods

Three preliminary meetings brought European collaborators together to define the core dataset [6]. This described the

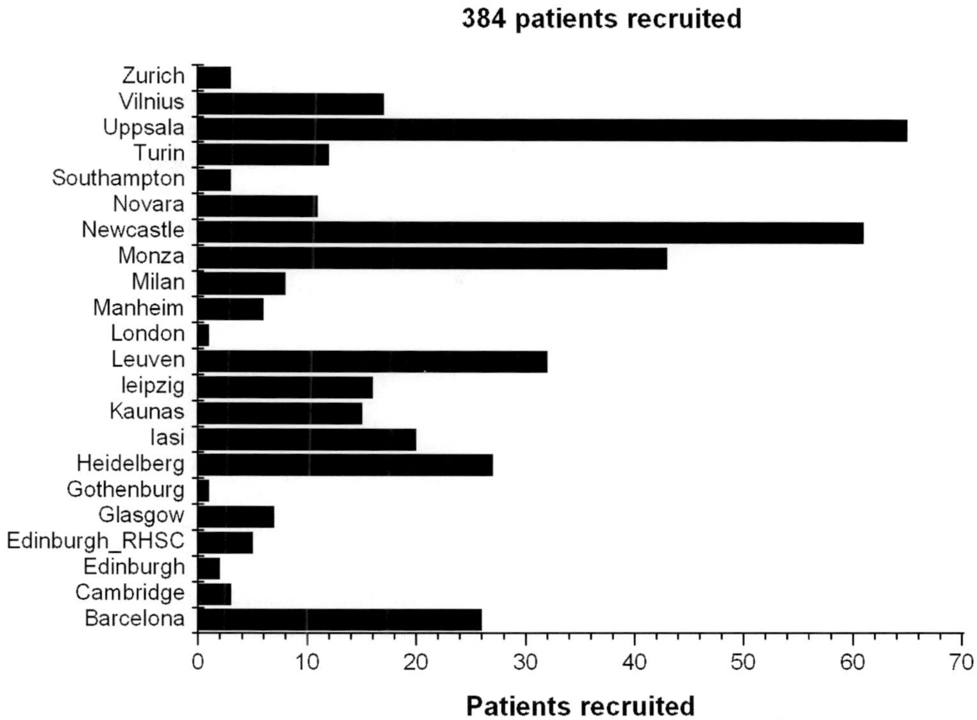

Fig. 2 Patient recruitment to date

format, frequency and units in which commonly measured variables should be stored. The minimum requirements were for invasive blood pressure and intracranial pressure to be monitored and recorded at a one-minute time resolution.

A European Union Framework V grant [3] provided the resources for a multi-centre pilot study to collect high resolution data from head injured patients. This required that equipment was deployed to participating sites for the

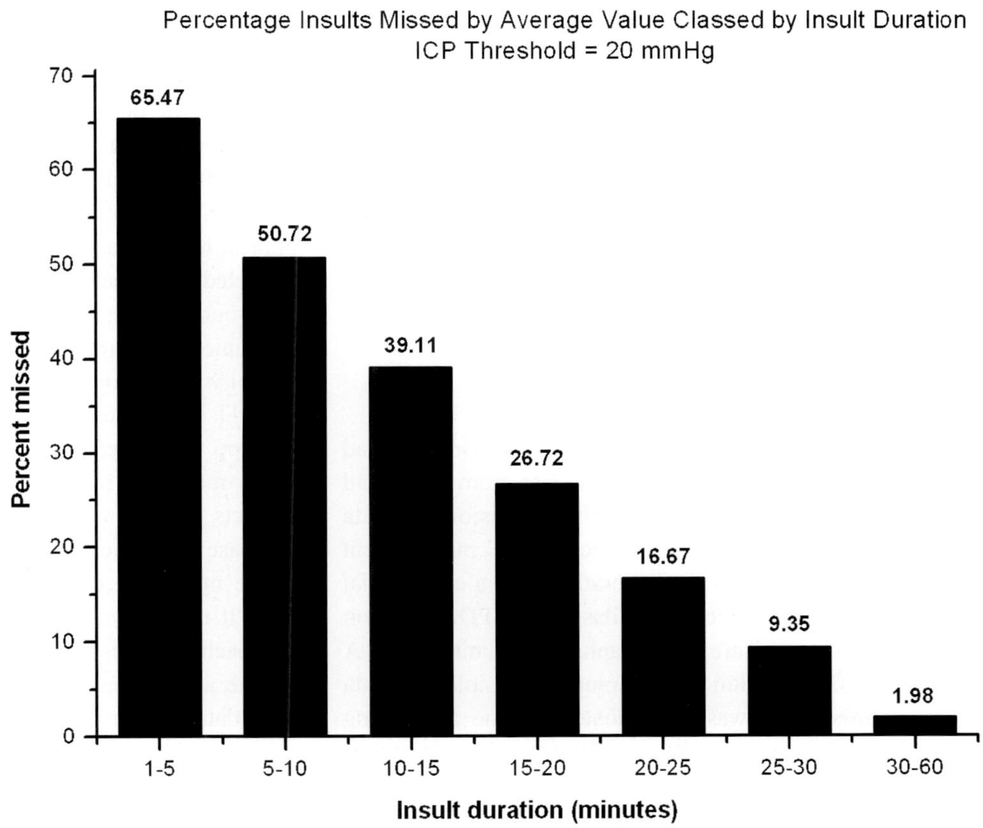

Fig. 3 Analysis of "missed" insults when using hourly averaged readings of ICP

collection of physiological, demographic and clinical information. The aim was to recruit and collect data from ten patients from each centre. Centres were coordinated, by a national validation centre where data validation staff were employed. The structure of the group is shown in Fig. 1 which depicts the way in which individual centres communicated with the national validation centres. There is both vertical and horizontal integration and individual centres are actively encouraged to communicate, participate and share information and resources with other centres.

Over the period between 1998 and 2006 data capture systems were deployed in high dependency and intensive care units in 21 participating neuroscience centres. The demographic information included age, cause, GCS, CT scan, pupil reaction and times of transfer; physiological data included minute by minute recording including ICP, CPP, MAP SaO2.

A web site (http://www.brainit.org) was designed and this provided several different functions:

1. Upload of data from centres
2. Requests for data validation and outcome assessment
3. Distribution information and results
4. Discussion forum for different groups (steering committee, data validators etc)

A data validation system was proposed so that data could be checked for accuracy. Firstly data from the acute phase was transmitted to Glasgow and inconsistencies identified. A request to correct any missing or out of range values was made. Then a request for validation of a 20% sample would be made to the data validator. Six months later a request for outcome assessment would be made, recorded and the validated data entered into the database.

Rules for publication of data were defined simply as publication can only be from validated data and manuscripts must be circulated to the Steering Group for comment prior to submission.

Results

At the end of the Framework V project, BrainIT had successfully collected and validated data from 200 head injured patients. This required the provision of data collection tools for the demographic, clinical management and physiological data. In collaboration with a commercial company (KelvinConnect Ltd, Glasgow) a PDA solution was produced to capture demographic and clinical data. A variety of different solutions to capturing physiological data were required as it was dependent upon the monitoring equipment used in each centre. Three main approaches were used, first a bedside computer running the Edinburgh Browser software [4], second a commercial product (ICUPilot, CMA Microdialysis AB) or thirdly an in-house solution that particular centres had already developed for other research works.

To date a total of 349 patients (277 male and 72 female) have now been transmitted to the coordinating centre in Glasgow (Fig. 2). The age of these patients ranged from 1 to 87 years (median 31). One hundred and forty-five had been involved in a traffic accident and 32 were pedestrians; 78 had suffered a fall and 24 were assaulted. A large amount of physiological data was collected (e.g. BP 2,531 days, ICP 2,212 days in total).

This dataset provides the opportunity to perform unique analyses and the statistical features of blood pressure have been described [5]. Further work related to the diurnal variations in ICP, optimal sampling rate determination and a comparison of summary measures of secondary insults are also underway. For example, Fig. 3 shows the percentage of insults that are missed when end hour averages are compared with minute-by-minute readings. Using a threshold of 20 mmHg the duration of missed insults was calculated for distinct 1-h periods. The percentage of ICP insults missed compared to the end hour value was highest with shorter duration insults although significant numbers of insults greater than 10 min in duration were still missed. The group has also been able to undertake a survey of traumatic brain injury management [2] and the evaluation of a new pressure monitoring device [1].

Discussion

This challenging collaboration has successfully developed a clinical research network focussed on improving the treatment of head injured patients. It has brought together a large number of centres capable and willing to undertake clinical research across the network. It has also successfully collected a vast quantity of high quality data that provides a rich source for analysis and hypothesis testing.

Clinical research networks such as BrainIT have the capacity to undertake research in a new way. The ethos of BrainIT is to foster an open collaboration, to data share and develop computer based data collection standards. The more members that join the BrainIT group, the more projects are formed. This will generate more data for the database and therefore generate more hypotheses. This will create more project ideas that will in turn generate more data. It is this cyclical process with an open collaborative approach which is at the heart of the BrainIT concept

The advantages are the ability to: standardise methods, store data in a compatible format and, because of the size of the network, studies can be done much quicker. There are benefits in bringing together different healthcare professionals who have a combined interest in a particular area to produce a larger intellectual mass. These combine to produce benefits for the patient, the primary goal, but there are also advantages for the academic researchers and industry, both pharmaceutical and device manufactures.

The main disadvantages are that considerable time and effort are required to set up and maintain the network. There are resource implications in running the network by supporting centres and researchers and although considerable effort is made towards fully funding centre activities, during periods of low funding, centres may be required to absorb costs. On-going maintenance is required and individuals or groups may lose focus and/or interest particularly if regular meetings or projects of interest are not provided. Individual differences may surface so that the group is not always working together as a cohesive unit. If these barriers to progress are identified and the group are aware of them then strategies can be put in place to minimise their potential for creating a problem.

A good research network has the ability to answer important clinical questions that are beyond the capability of singles centres. They require a common enthusiasm and but the output from the group can be greater than the sum of the constituent parts.

Acknowledgements We would like to thank all the staff at each of the participating centres for the time and help in collecting the data This work was supported by EEC project: QLRI-2003-01160

Investigators and Participating Centres Barcelona Spain, Prof J Sahuquillo; Cambridge UK., Prof JD Pickard; Edinburgh UK, Prof R Minns, Prof I Whittle, Prof R Minns; Glasgow UK, Mr L Dunn; Gothenburg Sweden, Dr B Rydenhag; Heidelberg, Germany, Dr K Kiening; Iasi Romania, Dr S Iencean; Kaunas Lithuania, Prof D Pavalkis; Leipzig Germany, Prof J Meixensberger; Leuven Belgium, Prof J Goffin; Mannheim Germany, Prof P Vajkoczy; Milano Italy, Prof N Stocchetti; Monza Italy, Dr G Citerio; Newcastle upon Tyne UK, Dr IR Chambers; Novara Italy, Prof F Della Corte; Southampton UK, Dr J Hell; Uppsala Sweden, Prof P Enblad; Torino Italy, Dr L Mascia; Vilnius Lithuania, Prof E Jarzemaskas; Zurich Switzerland, Prof R Stocker.

BrainIT Steering Group members IR Chambers, Middlesbrough; G Citerio, Monza; P Enblad Uppsala; BA Gregson, Newcastle upon Tyne; T Howells, Uppsala; Karl Kiening, J Mattern, Heidelberg; P Nilsson Uppsala; I Piper, Glasgow; A Ragauskas, Kaunas and J Sahuquillo, Barcelona.

Conflict of interest statement We declare that we have no conflict of interest.

References

1. Citerio G, Piper I, Cormio M, Galli D, Cazzaniga S, Enblad P, Nilsson P, Contant C, Chambers I (2004) Bench test assessment of the new Raumedic Neurovent-P ICP sensor: a technical report by the BrainIT group. Acta Neurochir 146(11):1221–1226
2. Enblad P, Nilsson P, Chambers I Citerio G, Fiddes H, Howells T, Kiening K, Ragauskas A, Sahuquillo J, Yau YH, Contant C, Piper I (2004) R3-Survey of traumatic brain injury management in European Brain IT centres year 2001. Intensive Care Med 30 (6):1058–1065
3. EEC Project QLGI-CT-2000-00454
4. Howells T (1994) Edinburgh monitor and browser©; Computer Programme
5. Mitchell P, Gregson BA, Piper I, Citerio G, Mendelow AD, Chambers IR (2007) Blood pressure in head-injured patients. J Neurol Neurosurg Psychiatry 78:399–402
6. Piper I, Citerio G, Chambers I Contant C, Enblad P, Fiddes H, Howells T, Kiening K, Nilsson P, Yau YH (2003) The BrainIT group: concept and core dataset definition. Acta Neurochir 145 (8):615–629
7. Zanier ER, Ortolano F, Ghisoni L, Colombo A, Losappio S, Stocchetti N (2007) Intracranial pressure monitoring in intensive care: clinical advantages of a computerized system over manual recording. 1: Crit Care 11(1):R7

Acta Neurochir Suppl (2008) 102: 229–233

Pilot application of fractal characterisation and its response to change on physiological wave forms

Martin Shaw · Ian Piper

Abstract

Background Over the last two decades the advances in analysis techniques for physiological time series data have been moving from the classical statistics to a more nonlinear or chaos based approach to looking at patterns in the variability of the time series. From this work it can be shown that physiological time series exhibit complex multi-fractal properties. So by designing a classification based on this nonlinear and chaotic nature you can detect changes and alterations in the underlying physiological processes.

Methods Applying a proven relationship between the wavelet modulus maxima representation and the Hölder exponent we could assess the multi fractal nature of the of the signal detection underlying changes in the physiology. Using two distinct techniques one global and the other localised in time, classification of two distinct the time series was carried out firstly via the analysis of the distribution of the Hölder exponents over all scales of the signal and secondly via a moving window application of the mean Hölder function.

Findings The distribution methodology did not return significant results though this is probably more to do with the signal than the technique. The trending approach shows a predictive nature with slope being linked to increased instability in the signal content.

Conclusions Overall this study has shown the applicability of the techniques which definitely warrant further refinement and study.

Ian Piper On behalf of the BrainIT group (http://www.brainit.org).

M. Shaw (✉) · I. Piper
Clinical Physics, Institute of Neurological Sciences,
Southern General Hospital,
1345 Govan Road,
Glasgow G514TF, United Kingdom
e-mail: martin.shaw@nhs.net

Keywords Wavelets · Fractals · Mathematical modelling · Wave form analysis · ICP

Introduction

Over the last two decades analysis techniques are moving from classical stochastic process analysis using basic statistics to more nonlinear systems or chaos theoretical based approaches which are looking at patterns in the variability of the time series.

It has been shown that physiological time series exhibit complex multi-fractal properties [1]. So by designing a classification and analysis based on this nonlinear and chaotic nature of the time series we should be able to detect changes and alterations in the underlying physiological processes. To this end two different analysis scenarios were tested. Firstly looking at the overall wave form and characterising its fractal nature and secondly looking at the trending of a window of its fractal nature with time.

Materials

We studied a number of randomly selected ICP waveforms from the BrainIT dataset (http://www.BrainIT.org) and used the R [4] statistical package for algorithm implementation and analysis.

Methods

The main idea used in the initial analysis and fractal characterisation of the wave form leverages the relationship between the mathematical properties of a wavelet transform and a signals' localised fractal nature.

Wavelets [3] can be thought of as a time–frequency analysis technique analogous to a Fourier transform but without the inherent frequency related problems of the later approach. This is mitigated by the fact the convolution is carried out with a scaling factor and a translational factor applied to the convolving function. Like Fourier transforms they can either be continuous or discrete the former implying that all scales are calculated the later only integer scales. The actual transform is carried out via Eq. 1:

$$\mathrm{Wf}(s,b)=\frac{1}{s}\int_{-\infty}^{\infty} f(x)\psi\left(\frac{x-b}{s}\right)\mathrm{d}x. \tag{1}$$

This is a scalable convolution with the ψ function being called a mother wavelet function and the scaled "s" and translated "b" function is known as a daughter function. In all of the later analytic tests a Mexican hat wavelet mother wavelet (Eq. 2) was used:

$$\psi(p)=\left(1-p^2\right)e^{-p/2}, \tag{2}$$

which is the second derivative of the Gaussian function. Figure 1 below is an example analysis using the above mother function on an ICP wave form.

Mandelbrot [2] defines a fractal as self similar signals repeating at different scales within the same signal. He also defines the Hurst exponent as a way of characterisation of the scaling properties of the signal that then can be thought of as an overview of the whole signal. However as we are interested in the signal on a more localised level, there is a related value known as the Hölder exponent [5]. This essentially characterises the singularities of a signal at a single scale level. Where a singularity is defined to be a discontinuity in a signal where the differential of the signal is not continuous at that point. It can be shown that the Taylor expansion [6] of the wavelet transform that Eq. 3 holds over the set of all singularities of the time series:

$$\mathrm{Wf}(s,x_0)\cong|s|^{h(x_0)} \tag{3}$$

where x_0 is the set of all singularities, $h(x_0)$ is defined to be the Hölder exponent of the singularity at x_0 however calculation of "h" by this method is not efficient but by using an optimised partitioning function and counting argument [6] the "local" Hölder exponent can be calculated.

Firstly the wavelet modulus maxima transform (WMMT) is performed and then the ridges are extracted from this view [5]. This ridge representation of the signal is

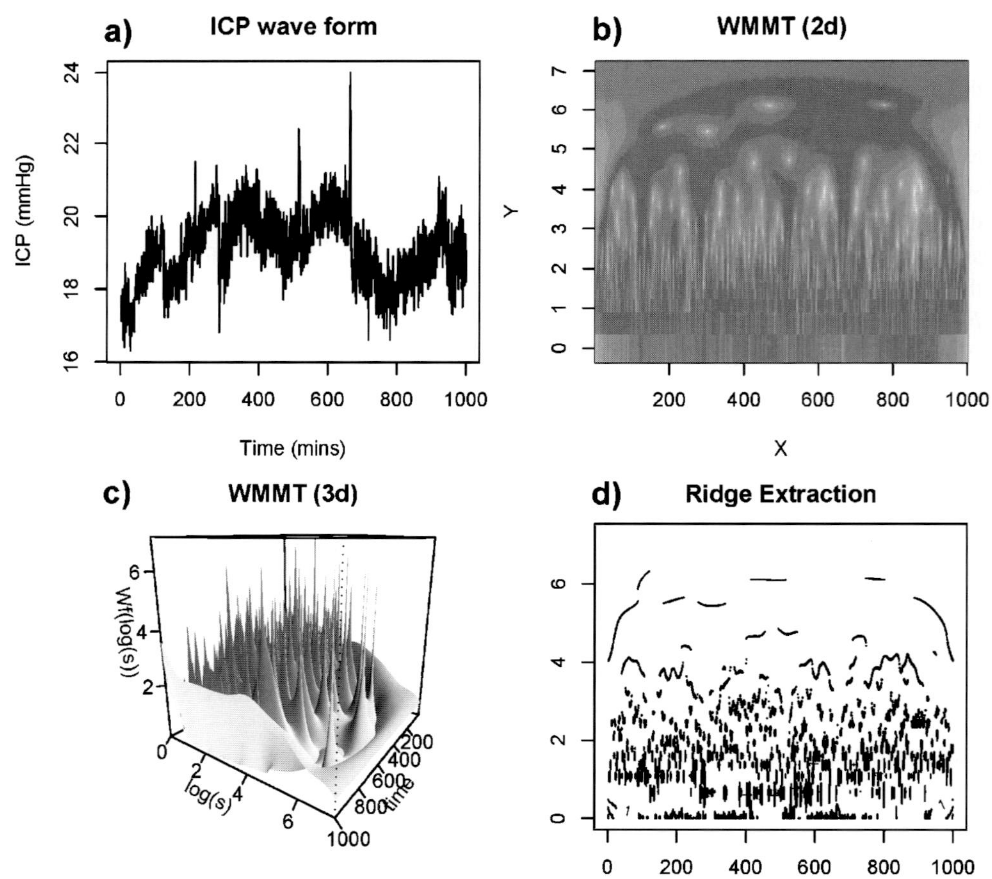

Fig. 1 **a** Original ICP wave form, **b** a 2d wavelet modulus maxima transformation of **a**, **c** a 3d representation of **b**, **d** ridge extraction from **b**

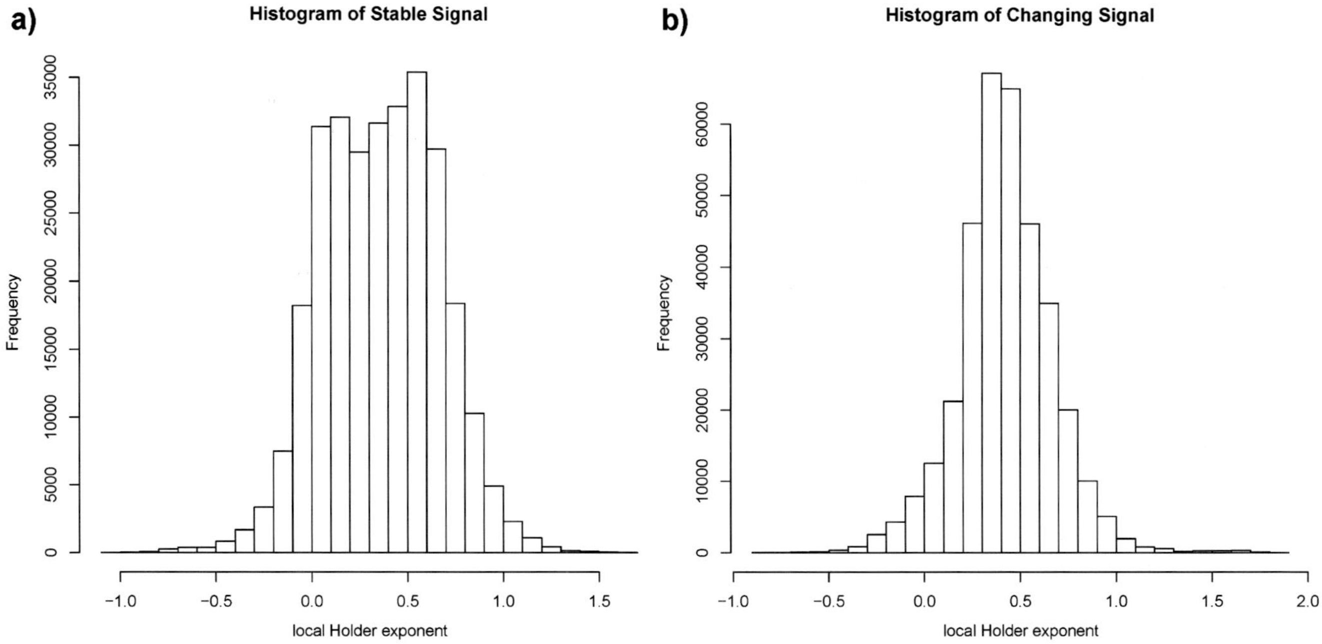

Fig. 2 Comparison of stable (**a**) and changing (**b**) local Hölder exponent distributions over all scales

all that is required to calculate the local Hölder exponent; this should make the analysis more efficient as no redundant information needs to be processed. This form is in fact a representation of all the singularities in the signal. (Fig. 1b,d). Then by using a partitioning function 4 over the set of all singularities $\Omega(s)$ at a given scale "s":

$$Z(s,q) = \sum_{\Omega(s)} (\mathrm{Wf}(\omega_i(s)))^q \tag{4}$$

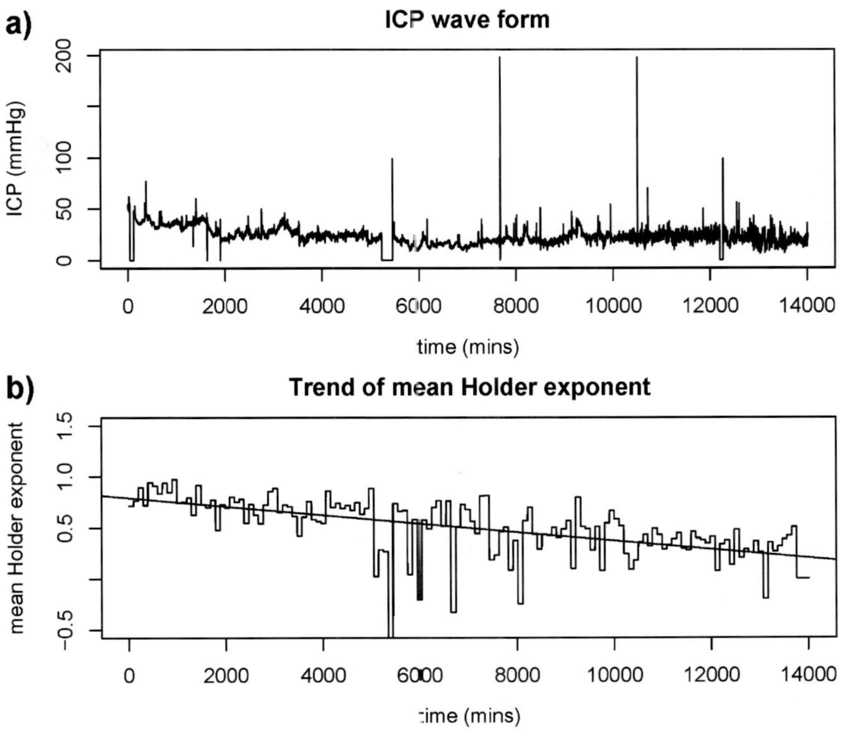

Fig. 3 ICP signal (**a**) and corresponding mean Hölder exponent trend (**b**)

we can define the mean wavelet transform value of all singularities at that scale s to be (Eq. 5):

$$M(s) = \sqrt{\frac{Z(s,2)}{Z(s,0)}}. \tag{5}$$

Finally, the mean Hölder ($\bar{h}$) is then calculated by solving for the slope of the straight line function 6:

$$\log(M(s)) = \bar{h}\log(s) + C. \tag{6}$$

It can then be shown [6] that the local Hölder exponent at scale s is calculated by Eq. 7:

$$\hat{h}_{s_{\mathrm{lo}}}^{s_{\mathrm{SL}}} \cong \frac{\log(\mathrm{Wf}(s_{\mathrm{lo}})) - \log(M(s))}{\log(s_{\mathrm{lo}}) - \log(s_{\mathrm{SL}})} \tag{7}$$

where s_{lo} is the minimum scale used and s_{SL} is the sample length.

This paradigm can be understood in terms that the signal is caused by the physiological system and this signal contains fractal content or information which is directly related to the underlying physiology. This implies that changes to physiology will directly alter the signal and hence the information contained within. By using the above analysis techniques we gain an overview of this fractal information. To test these ideas, two general applications of the above method are presented; Firstly a global signal overview by looking at local Hölder distributions and secondly a more localised view of the data looking at the trending of the mean Hölder exponent.

This first methodology can be thought of as an overview of the signals local Hölder exponent as a global look at the nature of the singularities of the time series and it should present a snap shot of the underlying physiology across the time range something akin to a fingerprint of the signal. ICP wave form data from a random patient was sampled from the BrainIT database and was split into a number of equal lengths. A “stable” segment where the signal remains relatively steady and the amplitude of the number of singularities remain low and a “changing” section, where the signal becomes more unstable and the amount and amplitude of the signal singularities increases. The local Hölder exponent calculated for each of these sections and the distribution of the local Hölder was then plotted and analysed over all scales (Fig. 2).

The second analysis technique to be applied looks at the change or trend in the mean Hölder calculation over the full time course of a signal. It could be thought of as a moving average approach to looking at the Hölder exponents of a signal. This should represent the time course of changes in physiology of the patient.

Again six ICP wave form signals were randomly sampled from the BrainIT dataset and these were then cleaned and a moving window approach was created to allow the repeated application of the mean Hölder function along the time course of the original signal. The window size used for this analysis was 200 min with a 100-min overlap. The choice of this window size was arbitrarily defined by the signal length from previous testing. Once this approach was applied to the signals a linear regression was applied to give an over all trend for the calculated mean Hölder exponents (Fig. 3).

Results

For the first analysis looking at the distributions, as can be seen from (Fig. 2) there is not much if any separation in the distributions and this is representative of all samples tested. In the analysis on trending of the mean Hölder it can be seen to be inversely proportional to signal fractal information content. However quantification of this is difficult as we would need an independent measure of the stability to statistically compare it with the trend gradient. That said over all the samples tested the gradient does show a predictive ability (Fig. 3).

Discussion

In the first analysis the lack of separation is not totally unexpected as the differences in the ICP signals tested between stable and changing sections are not greatly different mathematically, minimally deterministic and based on more random processes and so less likely to have much fractal “information” content. If this is the case then this technique should be then aimed at more deterministic signals such as looking at B wave activity in the ICP trend for example.

In the second analysis the predictive nature of this technique is a promising start though it will require further study to firstly statistically prove and secondly to find the optimal size for the moving window and still provide enough signal to accurately represent it with the mean Hölder exponent.

Overall this pilot study has shown the applicability of the techniques and as such it has only scratched the surface of how these approaches could be applied and of the implications of the links between the physiological systems and the fractal information content of the signals they produce.

Conflict of interest statement We declare that we have no conflict of interest.

References

1. Bassigthwaighte JB, Liebowitch LS, West BJ (1994) Fractal physiology. Oxford University Press, Oxford

2. Mandelbrot BB (1983) The fractal geometry of nature. Freeman, New York
3. Muzy JF, Bacry E, Arneodo A (1991) Wavelets and multifractal formalism for singular signals: application to turbulence data. Phys Rev Lett 67(25):3515–3518 (December 16)
4. R Development Core Team (2007) R: a language and environment for statistical computing. R Foundation for Statistical Computing, Vienna, Austria, ISBN 3-900051-07-0 (URL http://www.R-project.org)
5. Scafettaa N, Griffin L, West BJ (2003) Hölder exponent spectra for human gait. Physica A 328:561–583
6. Struzik ZR (2000) Determining local singularity strengths and their spectra with the wavelet transform. Fractals 8(2):163–179

PART 5:
Neuroimaging

Acta Neurochir Suppl (2008) 102: 237–240

The predictive value of ICP as compared to magnetic resonance imaging in comatose patients after head injury

R. Firsching · F.-W. Roehl · D.-H. Woischneck · N. John · M. Skalej

Summary
Introduction While highly increased intracranial pressure (ICP) is of high predictive value indicating a fatal outcome, the predictive value of moderately increased ICP early after head injury remains uncertain. We compared the predictive value of ICP to the predictive value of magnetic resonance imaging (MRI) early after head injury.
Methods 55 patients with a Glasgow Coma Scale (GCS) of less than 8, for more than 24 hours after head injury were investigated. Outcome was classified according to the Glasgow Outcome Scale (GOS). All patients received registration of ICP upon arrival at the hospital and an initial cranial computerized tomography scan. An MRI study was subsequently performed within 10 days of admission. The highest mean ICP registered within one hour in the first day of admission and the location of lesions as identified by MRI were related with outcome.
Results ICP was neither related with mortality nor with GOS of survivors. The location of lesions as depicted by MRI proved to be statistically significantly related with the GOS ($p<0.001$). Age proved to be clearly and significantly related with outcome ($p=0.019$).
Conclusions Our current MRI findings suggest that the location of the initial brain injury lesion correlates with outcome at 6 months. No such correlation could be identified for intracranial pressure on the first day after head injury ($p=0.766$).

R. Firsching (✉) · N. John
Klinik für Neurochirurgie, Universitaetsklinikum,
Leipziger Strasse 44,
39120 Magdeburg, Germany
e-mail: raimund.firsching@med.ovgu.de

N. John
e-mail: NICI-JEE@gmx.de

D.-H. Woischneck
Klinik für Neurochirurgie, ULM, Universitaetsklinikum,
Ludwig-Heilmeier Str.2,
89312 Guenzburg, Germany
e-mail: DIETER-HEINRICH.
WOISCHNECK@UNIKLINIKUM-ULM.de

M. Skalej
Institut fuer Neuroradiologie, Universitaetsklinikum,
Leipziger Strasse 44,
39120 Magdeburg, Germany
e-mail: MARTIN.SKALEJ@MED.OVGU.DE

F.-W. Roehl
Medizinische Fakultaet Magdeburg, Institut fuer Biometrie und
Medizinische Informatik,
Leipziger Strasse 44,
39120 Magdeburg, Germany
e-mail: FRIEDRICH-WILHELM.ROEHL@med.ovgu.de

Keywords Head injury · Intracranial pressure · Magnetic resonance imaging · Outcome

Introduction

The role of intracranial pressure (ICP) after head injury remains a topic of continued interest. In the older literature, the correlation of ICP and cerebral perfusion pressure (CPP) has been reported to be inferior to or at most similar to the correlations with clinical signs [5]. With magnetic resonance imaging (MRI) emerging as a new option of possible predictive value [2], we present a comparison of the prognostic value of ICP recording and MRI after head injury.

Patients and methods

55 comatose patients received ICP monitoring upon admittance after a head injury. Coma persisted for a minimum of 24 hours. ICP was measured intraventricularly

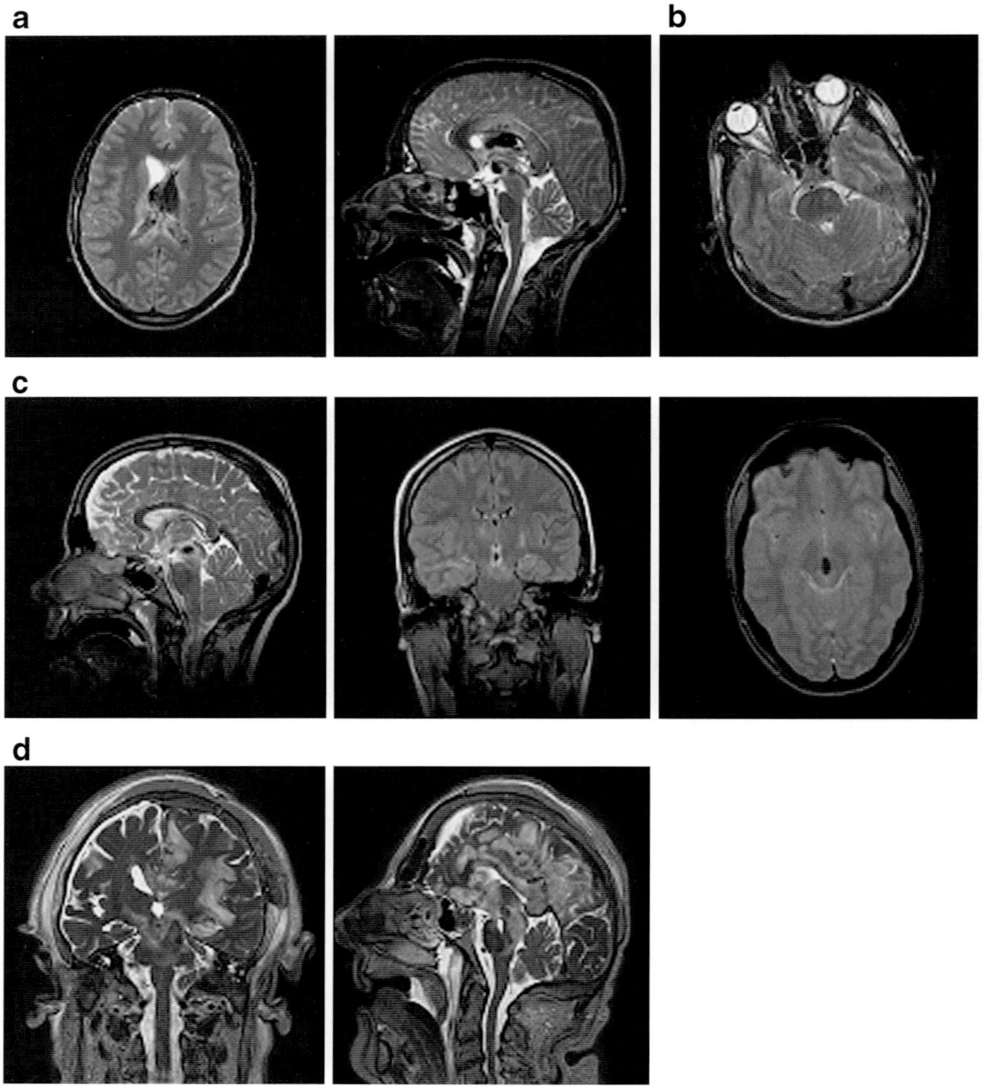

Fig. 1 MRI of patients with **a**) Grade I: supratentorial contusions only **b**) Grade II: unilateral lesion of the mesencephalon **c**) Grade III: bilateral lesion of the mesencephalon **d**) Grade IV: bilateral lesion of the pons

in 49 patients and epidurally in 6 patients. On the first day, high ICP levels are rare. The highest mean ICP over a stretch of 60 minutes within the first 24 hours was taken into account for further analysis. Sedation was reduced to the minimum necessary to ensure adequate ventilation [2]. No measures were taken to treat ICP. MRI was obtained within 10 days of the injury (median 3 days) and evaluated by the neuroradiologist, who was blinded to the clinical findings. The ICP values (Table 3, in mmHg) and the location of lesions on MRI (MRT Grade) were correlated with mortality or outcome of survivors as specified by the Glasgow outcome score (GOS) at 6 months after the injury. The location of lesions on MRI was classified in 4 grades as specified earlier [2]. The classification reflected the correlation of the location of lesions on MRI with outcome as identified with statistical means in an earlier study (see Fig. 1) [2]:

Grade I lesions of the hemispheres only,

Grade II unilateral lesion of the brain stem with or without a grade I lesion,

Table 1 P-values of the correlations between the variables of interest

Variable	GOS	ICP	MRI	AGE
ICP	0.766	-	-	-
MRI	<0.001	0.711	-	-
Age	0.019	0.005	0.897	-
Gender	0.211	0.752	0.317	0.366

Table 2 Contingency table with number of patients (n=55) for combination between MRI Grade and Glasgow Outcome Score

	Glasgow Outcome Score				
MRT Grade	1	2	3	4	5
I	7	0	5	11	5
II	0	2	2	3	0
III	3	9	3	0	0
IV	4	0	1	0	0

Table 3 Raw data of 55 comatose patients who received ICP monitoring upon admittance after a head injury

ID	Gender	Age	GOS	ICP	MRT Grade	Day of MRT
1	f	22	4	30	1	4
2	m	83	1	31	4	2
3	m	34	3	18	3	2
4	m	27	4	15	2	4
5	m	46	5	12	1	3
6	m	55	1	32	1	7
7	m	48	1	38	1	2
8	f	10	3	18	4	2
9	m	23	3	15	1	7
10	m	17	5	15	1	2
11	m	46	2	30	3	6
12	m	76	1	25	1	1
13	m	46	1	40	4	8
14	m	77	4	23	1	2
15	m	15	2	26	3	3
16	f	62	2	18	3	7
17	f	80	4	10	1	3
18	m	28	4	22	1	2
19	m	49	2	25	3	8
20	m	79	1	20	1	6
21	f	28	2	28	2	4
22	m	68	4	15	1	1
23	m	18	2	38	3	4
24	f	52	2	20	3	3
25	m	38	5	23	1	7
101	f	19	3	32	3	1
102	m	10	5	32	1	0
103	m	52	2	17	3	9
104	m	35	4	5	2	0
105	m	39	3	42	1	3
106	f	21	3	40	3	1
107	m	20	5	35	1	5
108	f	16	3	28	1	1
109	m	38	3	33	1	7
110	m	21	4	53	1	6
111	m	53	2	68	2	2
112	m	38	3	20	2	2
113	m	4	4	76	1	6
114	m	30	2	54	3	0
115	m	29	3	13	1	?
116	m	19	4	34	1	4
117	m	46	4	14	1	4
118	m	21	4	30	1	0
119	f	15	4	50	1	1
120	f	16	4	44	2	2
121	m	46	3	40	2	3
122	f	60	2	33	3	0
123	m	80	1	15	1	3
124	m	21	1	60	3	0
125	m	18	1	44	4	0
126	f	40	1	22	4	0
127	m	58	1	38	3	3
128	m	68	1	7	1	1
129	m	56	1	30	1	8
130	m	12	1	46	3	10

Grade III bilateral lesion of the mesencephalon with or without a grade II lesion,

Grade IV bilateral lesion of the pons with or without a grade III lesion.

The univariate correlations between the categorial variables (GOS, MRI, gender) were analysed using Fisher's exact test. An one-way ANOVA was performed with the categorical variables (MRI, GOS, gender) as main factor and the continuous variables (ICP and age, respectively) as dependent variables. A logistic regression with age, ICP and their cross effect respectively age and MRI with interaction was used to identify their effect on the mortality. Finally we computed sensitivity and specificity from contingency tables to look for cut-points in MRI Grade and ICP to identify the prognostic value using a ROC analysis. For this procedure the GOS was divided in two groups (1+2) and (3–5) respectively.

All statistical decisions were 2-tailed with a significance level of $\alpha=5\%$ without α-adjustment. The critical probability p supported the interpretation. The results should be interpreted in an exploratory manner.

Results

Table 1 shows the p-values for the univariate correlation between the variables of interest. MRI ($p<0.001$, Fisher's exact test) and age ($p=0.019$, ANOVA) have a clear influence on the outcome. No such correlation could be identified for intracranial pressure on the first day after head injury ($p=0.660$, ANOVA).

Frequencies of the contingency table (Table 2) show this cross effect between MRI findings and outcome. The correlation between age and ICP is also significant (p=

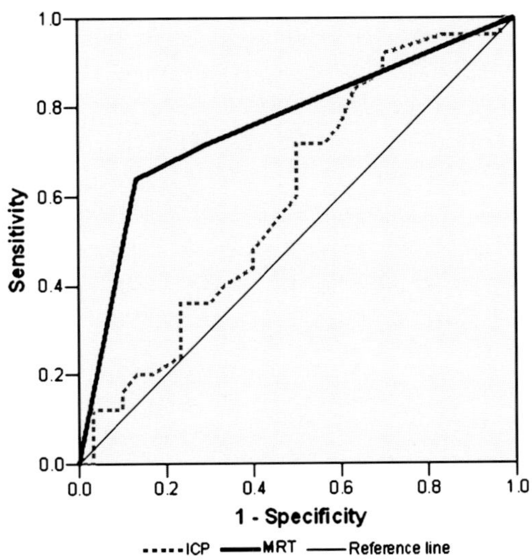

Fig. 2 Prognosis of mortality for different cut-points of ICP (- - - -) and respectively MRI (——)

0.005, Pearson correlation) but no correlation was found between ICP and outcome (p=0.766, ANOVA). Table 3 shows the raw data of the 55 analysed comatose patients.

In logistic regression to modelling the mortality (GOS= 1) with age, MRI and their cross effects, both had a p-value (p=0.008 respectively p=0.024) below the significance level but in a model with age (p=0.028) and ICP (p= 0.094), ICP did not reach this level.

These results were underlined by the ROC analysis. The ROC function of ICP to GOS has no recognizable cut-point (see Fig. 2) and the area under the curve (AUC) with a value of 0.615 is smaller than the AUC of 0.758 for the ROC function of MRI to GOS (see Fig. 2).

Discussion

It was not the purpose of this analysis to question the overall benefit of ICP monitoring for targeted management of increased ICP after head injury [1]. The early warning function and the predictive value of the registration of ICP, however, remains uncertain.

All patients in this series received ICP monitoring upon admittance, but some patients emerged from coma shortly after 24 hours had elapsed with subsequent termination of ICP monitoring. The time frame of 24 hours of ICP monitoring was relevant to all patients analysed.

Age statistically appeared to be the most important predictor of outcome. This is consistent with older reports in the literature [3].

MRI also shows a statistically strong correlation with outcome, but 7 deaths in patients with a slight head injury diminished this correlation. This may well be explained with the multiple injuries of these 7 patients, who were older than the average patient of this series. Death occurred within 6 months of the injury in 2 cases not clearly related with the head injury.

Peak ICP values within the first 24 hours after head injury did not correlate with outcome. Given the importance of secondary events, it would be important to measure ICP over longer periods of time following injury in future studies. Over 20 years ago, Miller et al. [4] reported that exceedingly high ICPs correlate with mortality. Efforts such as our current study continue to further advance our understanding of the relationship between ICP and outcome. From this study we concluded that MRI may be useful adjunct for predicting outcome following TBI. This may be particularly relevant to relatives of the comatose patient.

Conflict of interest statement We declare that we have no conflict of interest.

References

1. Fernandez R, Firsching R, Lobato R, Mathiesen T, Pickard J, Servadei F, Tomei G, Brock M, Cohadon F, Rosenorn J (1997) Guidelines for treatment of hed injury in adults. Zentralbl Neurochir 58:72–74
2. Firsching R, Woischneck D, Klein S, Reissberg S (2001) Classification of severe head injury based on magnetic resonance imaging. Acta Neurochir 143:263–271
3. Frowein R, Firsching R (1990) Classification of head injury. In: Vinken B (ed) Handbook of clinical neurology, vol 13 (57). Elsevier, North Holland Publ. Co., Amsterdam, pp 101–122
4. Miller JD (1986) Prediction of outcome after head injury. A critical review. In: Vigouroux RP (ed) Extracerebral collections Advances in Neurotraumatology, vol I. Springer, Vienna, New York
5. Miller JD, Becker D, Ward J, Sullivan H, Adams W, Rosner M (1977) Significance of intracranial hypertension in severe head injury. J Neurosurg 47:501–516

Acta Neurochir Suppl (2008) 102: 241–245

Metabolic disturbance without brain ischemia in traumatic brain injury: A positron emission tomography study

Nobuyuki Kawai · Takehiro Nakamura · Takashi Tamiya · Seigo Nagao

Abstract

Background Cerebral ischemia is believed to be an important mechanism of secondary neuronal injury in traumatic brain injury (TBI).

Methods In this study, we performed $^{15}O_2$ positron emission tomography (PET) studies to measure the cerebral blood flow (CBF) and oxygen metabolism ($CMRO_2$) in the pericontusional region in a total of 15 patients (11 males, 4 females, aged 15–81 years) who sustained TBI with contusional hematoma. PET studies were performed a mean of 13.5±9.1 days (range 2–33 days) after TBI occurred.

Findings The areas of pericontusional tissues located 10 mm away from the cerebral contusion exhibited mildly decreased CBF (89%) and severely suppressed $CMRO_2$ (67%) when comparison was made with the remote cerebral cortex. Severely suppressed oxygen metabolism in the pericontusional tissue was observed not only in the acute stage, but also in the subacute stage after TBI, whereas blood flow was slightly recovered in the subacute stage. We also compared the PET findings obtained in the acute or subacute stage after TBI and structural abnormalities on late-stage MRI in 5 patients. The area of flow defect on the CBF-PET image developed into irreversible tissue damage (necrosis) in the chronic stage. The area of hypoperfusion surrounding the lesion partly resulted in tissue necrosis: however, a large part of the hypoperfused tissue survived in the chronic stage. Again, a significant portion of oxygen hypometabolism surrounding the lesion did not develop into tissue necrosis.

Conclusions We conclude that impaired cerebral blood flow and metabolism in the pericontusional region is observed even in the subacute stage after TBI and is unlikely to cause severe further neuronal damage.

Keywords Brain contusion · Cerebral blood flow · Cerebral metabolic rate of oxygen · Positron emission tomography

Introduction

In traumatic brain injury (TBI) there is increasing evidence for the existence of a 'traumatic penumbra' - tissue that is most at risk of secondary ischemic neuronal injury and that will be most affected by changes in physiology or therapeutic interventions. Conventional management of TBI is based upon the premise that ischemia is a critical pathophysiological mechanism of secondary injury, and treatment of patients has focused on the maintenance of systemic oxygenation and cerebral perfusion pressure as a means of minimizing such injury. Recent positron emission tomography (PET) studies have demonstrated regional ischemia early after TBI, and shown that increases in ischemic tissue volume correlate with a poor Glasgow Outcome Score 6 months after the injury [2]. In contradistinction to these studies, two previous PET studies performed in TBI patients both early and late after the injury failed to show an elevated incidence of brain ischemia [3, 4]. A recent combined microdialysis and PET

N. Kawai (✉) · T. Nakamura · T. Tamiya · S. Nagao
Department of Neurological Surgery,
Kagawa University School of Medicine,
1750-1 Miki-cho, Kita-gun,
Kagawa 761-0793, Japan
e-mail: nobu@med.kagawa-u.ac.jp

T. Nakamura
Department of Neurobiology,
Kagawa University School of Medicine,
1750-1 Miki-cho, Kita-gun,
Kagawa 761-0793, Japan

study also demonstrated that the incidence of brain ischemia even in severe TBI patients is very low [9].

In this article, we describe the results of PET studies in the cerebral hemodynamics and the oxygen metabolism in perilesional and remote areas of brain contusion from the acute stage to the subacute stage in patients with TBI. We also discuss the ultimate fate of pericontusional areas which showed impaired cerebral blood flow and oxygen metabolism by comparing initial PET images and late-stage MR images.

Materials and methods

Fifteen patients (11 males and 4 females, mean age: 53.9± 18.0 years, mean GCS on admission: 10.2±3.1) with brain contusion were included in this study. Four patients underwent surgery for intracranial hemorrhage. Six patients were treated with brain hypothermia or intensive normothermia. All patients underwent ^{15}O-gas PET to measure the regional cerebral blood flow (rCBF), cerebral metabolic rate of oxygen (rCMRO$_2$) and oxygen extraction fraction (rOEF) at the acute or subacute stage after TBI (2–33 days, mean 13.5±9.1 days). PET data obtained from 13 control subjects (7 males and 6 females, mean age: 60.8± 11.7 years) were used as controls. Patients who required sedatives or anesthetics were not enrolled in this study, because these drugs significantly influence the cerebral blood flow and metabolism. Regions of interest (ROIs) were defined on the co-registered initial CT images superimposed on the PET images. The contusion was determined visually as the area showing an abnormal CT appearance and outlined on the CT and PET co-registered image at the level of the contusion center. The pericontusional and remote cortices were set on the cortex 10 mm and 30 mm, respectively, apart from the contour of the brain damage. Follow-up late-stage MRI studies were performed in 5 patients 2–14 months after the injury. The area of structural abnormality was identified and outlined on the FLAIR or T$_2$-weighted MR images, and then projected onto co-registered PET images of CBF and CMRO$_2$.

^{15}O gas-PET scanning

Before emission scanning, a 5-minute transmission scan using a ^{68}Ge rod source was obtained to correct tissue attenuation. Intermittent arterial blood sampling and radioactivity measurements were performed throughout PET scanning using a catheter implanted in the brachial artery to obtain the arterial input function. One-minute inhalation of C^{15}O (2 GB/min) followed by a 3-minute static scanning and 3 blood samplings were obtained to measure the cerebral blood volume (CBV). The C^{15}O$_2$ slow bolus inhalation method was used to measure the cerebral blood flow (CBF). A 10-minute dynamic scanning (30 s × 1; 15 s × 10; 30 s × 10; 60 s × 2) was performed following 2 min of slow bolus inhalation of C^{15}O$_2$ gas (1.5 GB/min). The arterial input function was determined by frequent arterial blood sampling. Finally, after a 7-minute inhalation of ^{15}O$_2$ (0.5 GB/min), the steady-state O$_2$ image was scanned and 3 arterial blood samplings were obtained for 5 min to determine the oxygen extraction fraction (OEF) and the cerebral metabolic rate for oxygen (CMRO$_2$).

Results

1) PET values in the pericontusional area

Regional hyperemia (rCBF: 55–60 ml/100 g/min) was observed in three patients in perilesional or adjacent areas of the brain contusion. The rCBF in the pericontusional areas was significantly lower than that of the remote areas (39.2±4.4 vs. 44.2±3.7 ml/100 g/min, $P<0.05$) (Fig. 1a). The rCMRO$_2$ in the pericontusional areas was also significantly lower than that of the remote areas (1.67± 0.63 vs. 2.49±0.70 ml/100 g/min, $P<0.01$) (Fig. 1b). The rOEF in the pericontusional areas was 0.28±0.08 and this value was significantly lower compared with that in the remote areas (0.36±0.09) (Fig. 1c). The relative reduction of the rCMRO$_2$ (33%) and the rOEF (25%) in the pericontusional areas compared with the remote areas was much greater than the reduction of the rCBF (12%). The PET values obtained in the remote areas were not significantly different from the values obtained in the controls (CBF: 45.6±2.8 ml/100 g/min, CMRO$_2$: 2.74± 0.43 ml/100 g/min, OEF: 0.34±0.05).

2) PET values in the remote areas: relationship with injury severity and prognosis

When we compared the PET values in the remote areas between patients with severe TBI (initial GCS≤8) and non-severe TBI (initial GCS≥9), there were no significant differences in the rCBF (45.8±4.1 vs. 43.4 ml/100 g/min), rCMRO$_2$ (2.46±0.68 vs. 2.51±0.74 ml/100 g/min) and rOEF (0.36±0.06 vs. 0.37±0.10) (Fig. 2a). Again, when we compared the PET values in the remote areas between patients with favorable outcome (GR/MD) and unfavorable outcome (SD/VS), there were no significant differences in the rCBF (43.5±3.6 vs. 45.3±3.9 ml/100 g/min), rCMRO$_2$ (2.54±0.76 vs. 2.42±0.64 ml/100 g/min) and rOEF (0.37± 0.09 vs. 0.35±0.09) (Fig. 2b).

3) PET findings and ultimate fate of contusional brain injury

We compared the PET findings obtained in the acute or subacute stage after TBI and structural abnormalities on late-stage MRI in 5 patients (Fig. 3). The area of flow defect on the CBF-PET image developed into irreversible

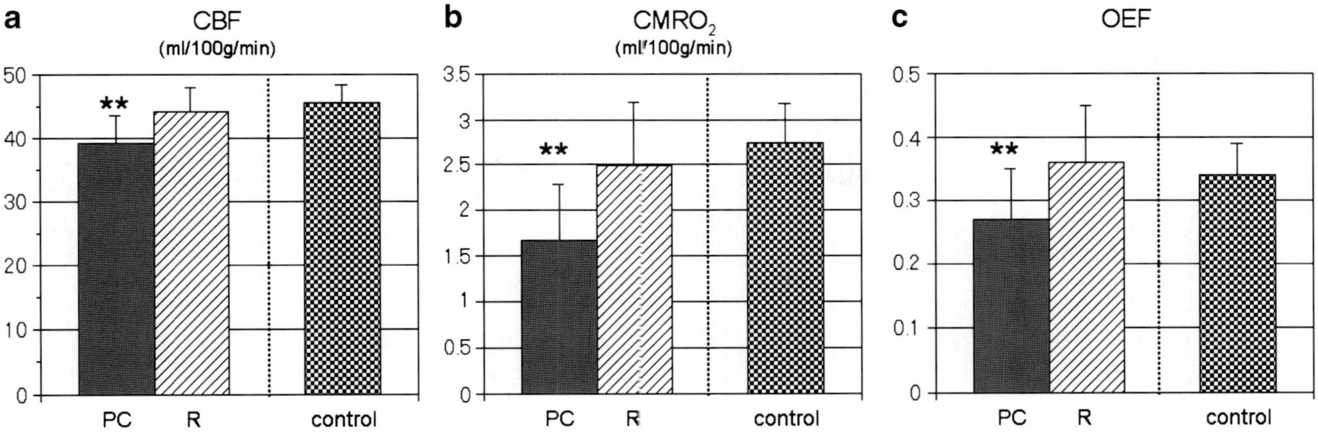

Fig. 1 Comparison of rCBF (**a**), rCMRO$_2$ (**b**) and rOEF (**c**) in the pericontusional (PC) and ipsilateral remote (R) areas of contusional brain injury. Their values were also compared with control subjects (Control). The rCBF, rCMRO$_2$ and rOEF in the pericontusional areas were significantly lower than those in the remote areas (**, $P<0.01$)

tissue damage (necrosis) in the chronic stage. The area of hypoperfusion surrounding the lesion partly resulted in tissue necrosis: however a large part of the hypoperfused tissue survived in the chronic stage. Again, a significant portion of oxygen hypometabolism surrounding the lesion did not develop tissue necrosis. The result was more evident when the PET examination was performed in the subacute stage after TBI.

Discussion

We investigated the cerebral hemodynamics and oxygen metabolism in focal brain lesions and remote areas of brain contusion and contusion with traumatic intracerebral hematoma. Of course, some of the patients might have had pathophysiological factors of diffuse brain injury. The PET values obtained in the remote areas (normal findings on CT

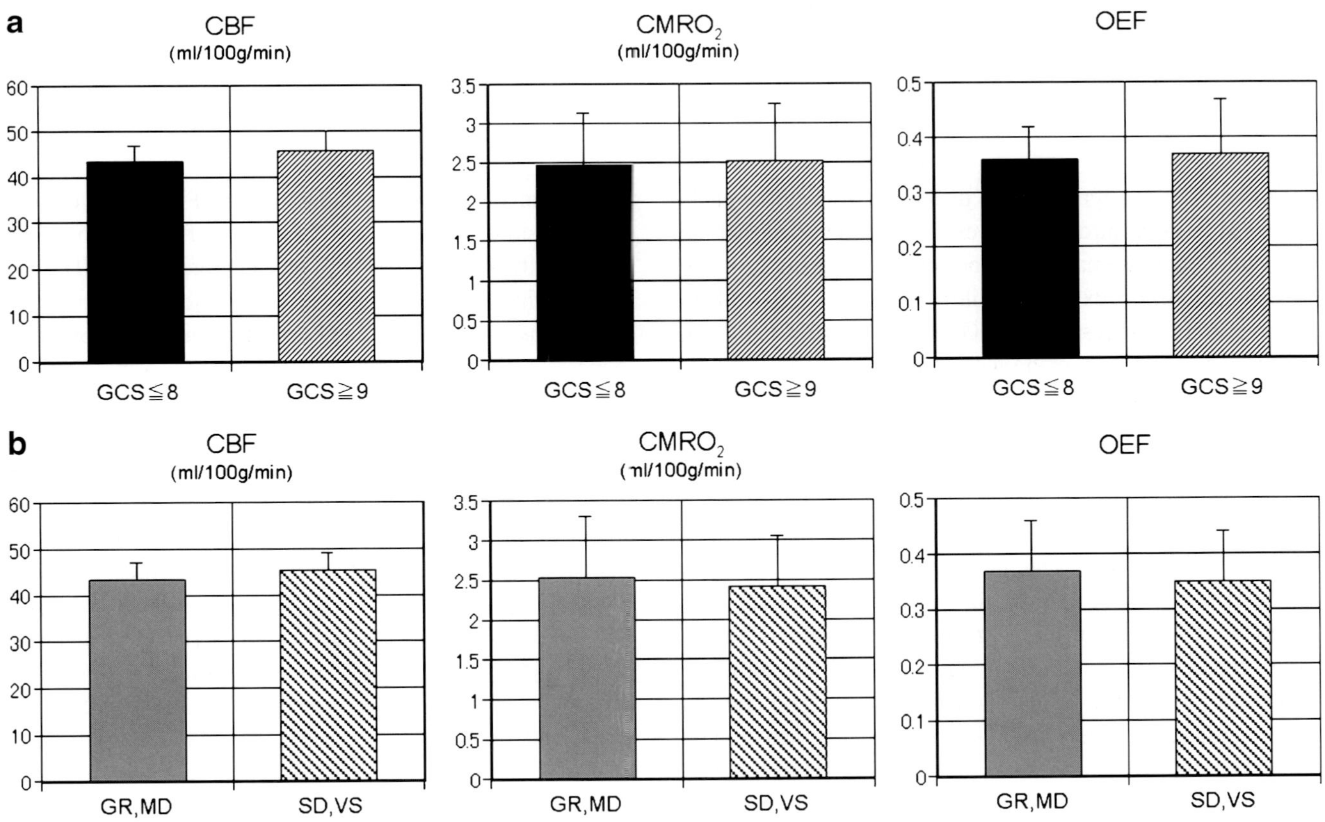

Fig. 2 Comparison of PET values in the remote areas between the patients with severe TBI (initial GCS≤8) and non-severe TBI (initial GCS≥9) (A), and between patients with favorable outcome (GR/MD) and unfavorable outcome (SD/VS) (B). PET values were not correlated with initial injury severity and neurological outcome

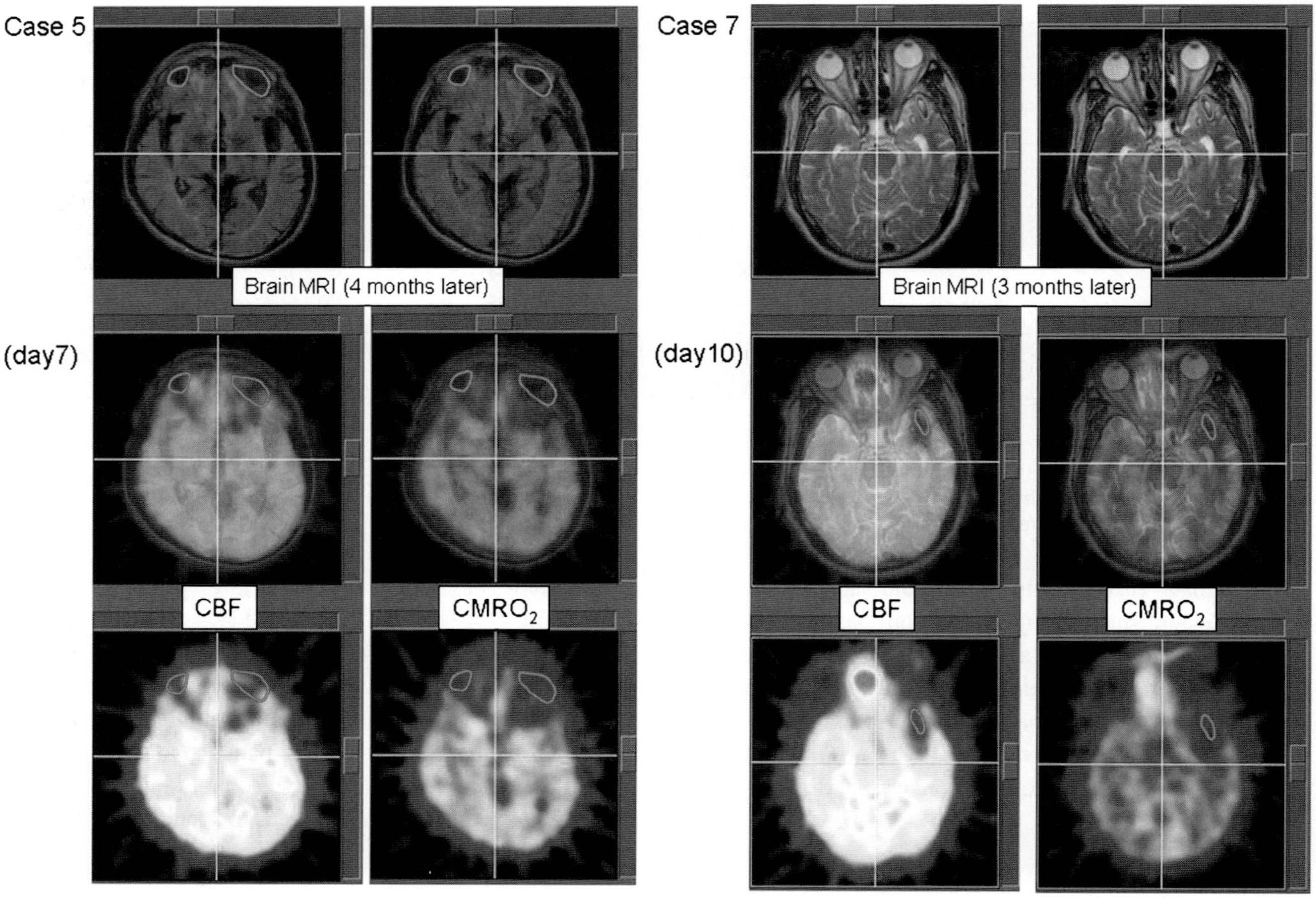

Fig. 3 Comparison of the PET findings obtained in the acute (**a**) or subacute (**b**) stage after TBI and structural abnormalities on late-stage MRI. The area of hypoperfusion and hypometabolism surrounding the lesion partly resulted in tissue necrosis: however, some part of the hypoperfused and hypometabolized tissue survived in the chronic stage

scan) were not significantly different from the values observed in the controls. The present study revealed that the areas of pericontusional tissues located 10 mm away from the cerebral contusion exhibited mildly decreased CBF (88%) with severely suppressed $CMRO_2$ (67%) and OEF (75%), when comparison was made with those in the remote areas. Severely suppressed oxygen metabolism in the pericontusional tissue was observed not only in the acute stage, but also in the subacute stage after TBI, whereas blood flow was slightly recovered in the subacute stage.

Several lines of evidence from clinical brain injury research suggest that ischemia occurs frequently in TBI. These data are most evident for the initial 12 hours after injury in which brain-imaging studies as well as whole brain oxygenation monitoring have shown that ischemia occurs in 30% of the population [1]. In contradistinction to these studies, two previous PET studies performed in TBI patients both early and late after the injury failed to show an elevated incidence of brain ischemia [3, 4]. A recent combined microdialysis (MD) and PET study at a mean time of 36 hours after injury also demonstrated that the incidence of brain ischemia in TBI patients is very low [9].

In focal brain injury, studies of rCBF have revealed a pericontusional zone of low rCBF, which often corresponds to hypodensity on CT scans [7]. A recent intracerebral MD study revealed that the lactate/pyruvate (L/P) ratio in the pericontusional zone was elevated 24 hours after injury and continued to increase during the following 10 hours [5]. A trend toward normalization of the L/P ratio was observed during the next day [5]. The L/P ratio is considered to be a marker of the relative redox state of the tissue, with increases in the L/P ratio indicating cerebral ischemia. Our PET study is restricted to the period of observation from post-injury Day 2 and cannot address the incidence of ultra-early brain ischemia. The mean $CMRO_2$ value observed in the pericontusional area was 1.67 ml/100 g/min (67% of the remote area) in our patients. A previous primate ischemia study demonstrated that $CMRO_2$ was the best predictor of reversible or irreversible brain damage and the critical metabolic threshold level appeared to be a reduction of oxygen metabolism to between 61% and 69% of the corresponding contralateral region [6]. Since $CMRO_2$ is a measure of mitochondrial oxidative function, reduced $CMRO_2$ in the pericontusional area represents mitochondrial dysfunction without apparent local ischemia.

We compared the PET findings obtained in the acute or subacute stage after TBI and structural abnormalities on late-stage MRI in 5 patients. Classic PET studies in stroke patients determined thresholds for irreversible tissue damage for CBF and $CMRO_2$ as 12 ml/100 g/min and 1.4 ml/100 g/min respectively, with a penumbral CBF threshold of around 20 ml/100 g/min [8]. These thresholds, in particular the at-risk CBF threshold of 20 ml/100 g/min, have been used in a number of studies of TBI to define viable and non-viable tissue. Our study demonstrated that the area of flow defect or severe hypoperfusion (usually below 10–15 ml/100 g/min) in the brain contusion developed irreversible tissue damage (necrosis) in the chronic stage. Although the area of hypoperfusion surrounding the lesion partly resulted in tissue necrosis, a large part of the hypoperfused tissue survived in the chronic stage. Again, a significant portion of oxygen hypometabolism surrounding the lesion did not develop into tissue necrosis. This was less evident when the PET study was performed in the acute stage and more evident when the PET examination was performed in the subacute stage after TBI. This result suggests that the area of disturbed oxygen metabolism around the brain contusion was spreading to the surrounding tissue: however, the area of secondary disturbed oxygen metabolism did not develop into tissue necrosis.

Conflict of interest statement We declare that we have no conflict of interest.

References

1. Bouma GJ, Muizelaar JP, Choi SC, Newlon PG, Young HF (1991) Cerebral circulation and metabolism after severe traumatic brain injury: the elusive role of ischemia. J Neurosurg 75:685–693
2. Coles JP, Fryer TD, Smielewski P, Chatfield DA, Steiner LA, Johnston AJ, Downey SPMJ, Williams GB, Aigbirhio F, Hutchinson PJ, Rice K, Carpenter TA, Clark JC, Picakard JD, Menon DK (2004) Incidence and mechanism of cerebral ischemia in early clinical head injury. J Cereb Blood Flow Metab 24:202–211
3. Diringer MN, Videen TO, Yundt K, Zazulia AR, Aiyagari V, Dacey RG Jr, Grubb RL, Powers WJ (2002) Regional cerebrovascular and metabolic effects of hyperventilation after severe traumatic brain injury. J Neurosurg 96:103–108
4. Diringer MN, Yundt K, Videen TO, Adams RE, Zazulia AR, Deibert E, Aiyagari V, Dacey RG Jr, Grubb RL Jr, Powers WJ (2000) No reduction in cerebral metabolism as a result of early moderate hyperventilation following severe traumatic brain injury. J Neurosurg 92:7–13
5. Engström M, Polito A, Reinstrup P, Romner B, Ryding E, Ungerstedt U, Nordström C-H (2005) Intracerebral microdialysis in severe brain trauma: the importance of catheter location. J Neurosurg 102:460–469
6. Frykholm P, Andersson JLR, Valtysson J, Silander HC, Hillered L, Persson L, Olsson Y, Yu WR, Westerberg G, Watanabe Y, Långström B, Enblad P (2000) A metabolic threshold of irreversible ischemia demonstrated by PET in a middle cerebral artery occlusion-reperfusion primate model. Acta Neurol Scand 102:18–26
7. Furuya Y, Hlatky R, Valadka AB, Diaz P, Robertson CS (2003) Comparison of cerebral blood flow in computed tomographic hypodense areas of the brain in head-injured patients. Neurosurgery 52:340–346
8. Powers WJ, Grubb Jr RL, Darriet D, Raichle ME (1985) Cerebral blood flow and cerebral metabolic rate of oxygen requirements for cerebral function and viability in humans. J Cereb Blood Flow Metab 5:600–608
9. Vespa P, Bergsneider M, Hattori N, Wu H-M, Huang S-C, Martin NA, Glenn TC, McArthur DL, Hovda DA (2005) Metabolic crisis without brain ischemia is common after traumatic brain injury: a combined microdialysis and positron emission tomography study. J Cereb Blood Flow Metab 25:763–774

Acta Neurochir Suppl (2008) 102: 247–251

Concordant biology underlies discordant imaging findings: diffusivity behaves differently in grey and white matter post acute neurotrauma

Virginia F. J. Newcombe · Guy B. Williams · Jurgens Nortje · Peter G. Bradley · Doris A. Chatfield · Joanne G. Outtrim · Sally G. Harding · Jonathan P. Coles · Bala Maiya · Jonathan H. Gillard · Peter J. Hutchinson · John D. Pickard · T. Adrian Carpenter · David K. Menon

Abstract
Background Cerebral edema is a common sequelum post traumatic brain injury (TBI). Quantification of the apparent diffusion coefficient (ADC) using diffusion tensor imaging (DTI) may help to characterize the pathophysiology of brain swelling.
Methods Twenty-two patients with moderate-to-severe TBI underwent magnetic resonance (MR) imaging, including DTI, within five days of injury. The mean ADCs in whole brain white matter, whole brain grey matter and entire brain were calculated and compared to twenty-five controls.
Findings A significant decrease in the grey matter ADC ($p<0.001$), significant increase in the white matter ADC ($p<0.001$) and no significant change in the whole brain ADC ($p=0.771$) was observed. No significant correlation was found between DTI parameters in any of the three regions of interest (ROI) and GCS, time to scan, intracranial pressure (ICP) before and during the time of the scan, cerebral perfusion pressure at time of scan, or Glasgow Outcome Score (GCS). Conclusions; The decrease in ADC seen in the grey matter is consistent with cytotoxic edema. The increase in ADC in the white matter indicates damage that has led to an overall less restricted diffusion. This study assists in the interpretation of the ADC by showing that the acute changes are different in the whole brain white and grey matter ROIs post TBI.

Keywords Neurotrauma · Acute brain injury · Diffusion tensor imaging · Magnetic resonance imaging · Cerebral edema

V. F. J. Newcombe (✉) · J. Nortje · P. G. Bradley · D. A. Chatfield · J. G. Outtrim · J. P. Coles · B. Maiya · D. K. Menon
University Division of Anaesthesia, Cambridge University, Box 93, Addenbrooke's Hospital, Hills Road, Cambridge CB2 2QQ, UK
e-mail: vfjn2@wbic.cam.ac.uk

V. F. J. Newcombe · G. B. Williams · J. Nortje · P. G. Bradley · D. A. Chatfield · J. G. Outtrim · S. G. Harding · J. P. Coles · B. Maiya · J. H. Gillard · J. D. Pickard · T. A. Carpenter · D. K. Menon
Wolfson Brain Imaging Centre, Cambridge University, Addenbrooke's Hospital, Cambridge, UK

P. J. Hutchinson · J. D. Pickard
Academic Neurosurgery Unit, Department of Clinical Neurosciences, Addenbrooke's Hospital, Cambridge, UK

J. H. Gillard
Department of Radiology, University of Cambridge, Addenbrooke's Hospital, Cambridge, UK

Introduction

Cerebral edema is common post traumatic brain injury (TBI) and is associated with poorer outcome [14]. Its sequelae may include raised intracranial pressure (ICP), brain herniation with consequent compression of vital structures, and impaired cerebral perfusion and ischaemia. However, its mechanism in diffuse injury after acute neurotrauma is not well understood.

By characterizing the natural displacements of water molecules, diffusion weighted imaging (DWI) may provide insight into the microscopic architecture of the brain in both normal and pathological states. The diffusion tensor can be used to describe the magnitude of water diffusion (apparent diffusion coefficient, ADC), how directional the diffusion is (anisotropy) and the orientation of that direction (eigenvectors/eigenvalues). The latter measures are often used to describe the substantial directional differences in water diffusivity in highly organised white matter. However, as the diffusion in grey matter is usually isotropic in nature, the ADC is a better measure to investigate grey matter. The ADC, often referred to as mean diffusivity, characterizes the overall presence of obstacles to diffusion, which includes cellular and extracellular structures. This makes it a useful tool to describe how brain tissue may change in an edematous state. Previous studies have associated restricted diffusion with cytotoxic edema, and increased diffusion with vasogenic [14]. However, the exact causes of these imaging changes are not well understood [10, 11].

It has been well described that the diffusivity of water is higher in grey than white matter in healthy volunteers, due to their structural differences. Any changes seen post TBI are, therefore, likely to be different in these two tissue types. However, to date, we have not found any studies that contrast changes in pure white and grey matter regions of interest (ROIs). The aim of this analysis was to assess whether the diffusion changes seen in cerebral edema secondary to diffuse traumatic brain injury using DTI are indeed different in grey and white matter, and to see if this informs about the formation of edema in the acute phase of injury.

Methods

Imaging protocol

22 patients (13 male, 9 female) underwent imaging using a 3 Tesla Bruker MedSPEC S300 scanner (Bruker Medical, Ettlington, Germany) within five days of injury. To ensure that our analysis was not dominated by pathophysiology in the vicinity of focal lesions, only patients with a Marshall grade of 3 or less (diffuse injury) were included. The imaging protocol included a 3D T1-weighted structural sequence (fast spoiled gradient echo (SPGR), repetition time (TR) 19•18 ms, echo time (TE) 5.0 ms) as well as spin echo planar diffusion weighted imaging (12 non-collinear directions, 5 b values equally spaced from 329 to 1590 s/mm^2, 4 b=0 images). The diffusion weighted parameters were: 20×20 cm field of view, 100×100 matrix size, 27 axial slices, 5 mm slice thickness, TR=6000 ms, TE=100 ms, diffusion sensitizing duration=23•5 ms (δ); with separation (leading edge to leading edge)=60 ms (Δ). Twenty-five age matched controls (healthy volunteers, 17 male, 8 female) underwent an identical imaging protocol.

Image analysis

ADC maps were created using in-house software based on the method proposed by Basser and Pierperoli [4]. Statistical parametric mapping software (SPM5) was used to segment the SPGR image into grey and white matter regions of interest (ROIs), which were then transformed into binary masks (Fig. 1) [21]. The grey and white matter regions were added together to create a whole brain ROI. To help aid the accuracy of co-registration the skull and extracranial soft tissue were stripped from the SPGR images using the Brain Extraction Tool [20]. The stripped SPGR was then coregistered to the b=0 diffusion weighted image (T2) using the vtkCISG non-linear normalised mutual information algorithm [24]. The coregsistration transformation matrix was applied to the three ROIs and the mean ADC was calculated in each one.

Statistics

All statistical analyses used SPSS (SPSS 14•0, Chicago, IL, USA). Following assessment for normality, non-parametric data comparisons were performed on the diffusion data using Mann-Whitney U (MWU) for unpaired tests. Student's t-test (two-tailed) was used to perform parametric comparisons. The relationships between clinical parameters including ICP (during and before the scan), cerebral perfusion pressure during the scan, mean arterial pressure (MAP) during the scan, Glasgow Coma Score (GCS) and Glasgow Outcome (GOS), and ADC were assessed using Spearman's rank correlation. Results with p-values<0•05 were considered to be statistically significant.

Results

Patient characteristics

The mean (± standard deviation, SD) age for the 22 patients (35±13 years) was not significantly different to that of controls (34±14 years) (p=0.65). The median GCS of the patient group at admission was 6 (range 6 to 11). Subsequent deterioration meant that all patients required sedation and mechanical ventilation for control of intracranial hypertension by the time they were recruited to the study. The median (interquartile range, IQR) Injury Severity Score was 21 (16 to 32) and APACHE II 19 (15.5 to 23). The median time to scan was 28 (range 7 to 109) hours. The mean CPP during the scan was 70±11 mmHg. The mean ICP during the scan was 19.6±4.4 mmHg and in the twelve hours prior the scan 17.8±4.7 mmHg. The median Glasgow Outcome Score (GOS) was 3 (range 1 to 5).

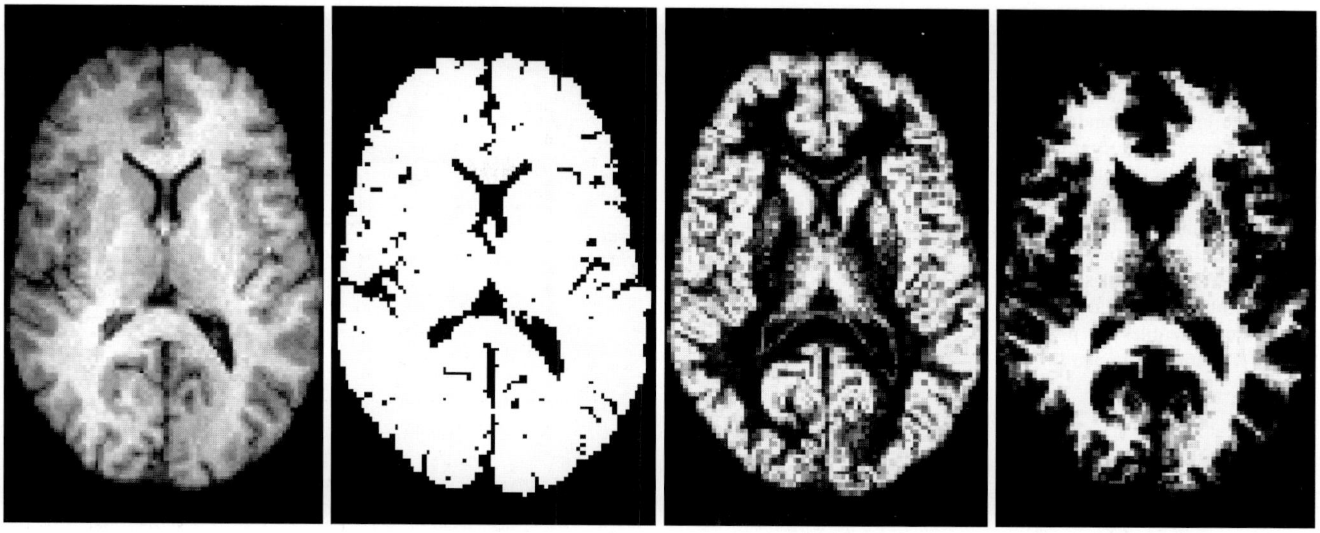

Fig. 1 An example of the probability maps for the grey and white matter regions of interest (ROI) created by segmenting the SPGR in a healthy volunteer. These were converted into binary maps and added together to create a whole brain ROI

Imaging results

There was no significant difference in the ADC in the whole brain ROI (median (interquartile range) ($\times 10^{-3}$ mm^2/s)) between cases and controls (0.74 (0.72 to 0.75) *vs* 0.74 (0.73 to 0.75); p=0.771) (Fig. 2). The cases showed a significant decrease in the grey matter ADC compared to controls (0.77 (0.75 to 0.79) *vs* 0.80 (0.79 to 0.82); $p<0.001$). In contrast there was a significant increase in the ADC of patients in the white matter ROI (0.67 (0.65 to 0.70) *vs* 0.63 (0.62 to 0.65); $p<0.001$). No significant correlation was found between any of the three ROIs and; GCS, injury-scan interval, mean intracranial pressure (ICP) at time of scan or the twelve hours before the scan, cerebral perfusion pressure at time of scan and GOS.

Discussion

This study demonstrates that the changes in MR imaging characteristics, as described using diffusion tensor imaging, are different in grey and white matter after moderate-to-severe acute brain injury. It is likely that similar pathological mechanisms in regions with large structural differences may underlie the apparently discordant results of changes in diffusivity post acute neurotrauma.

There have been several animal studies using DWI following brain injury. Barzo and colleagues, using an impact acceleration model in rats, reported a hyperacute, transient increase in ADC that occurred in the first 60 minutes followed by a reduction in ADC that lasted a for at least 14 days [3]. A controlled cortical impact model in rats looking at the hippocampus also found an increase in ADC. However, in this model it lasted for 72 hours [18]. In contrast, other studies and/or model types have found decreases in ADC in the first 24 to 72 hours [1, 12, 18, 22] with the longer time course investigations reporting a subsequent increase in diffusivity [1, 18, 22]. The variations seen in these studies are likely to reflect not only the specific injury model used, but also the ROIs selected and the time of imaging post injury. Despite their differences, all of these animal models created lateralized/contusional lesions and therefore may not reliably reflect the pathophysiology that occurs in human shearing injuries.

There are few human studies investigating acute neurotrauma (within one week of injury) changes in ADC in non-focal injuries. Maramarou et al. imaged patients with moderate-to-severe TBI between 40 and 220 hours post injury, and found a decrease in ADC that corresponded to an increase in brain water volume in a hemispheric (mixed grey and white matter) region of interest (ROI) [15, 16]. The authors interpreted these findings as showing that cytotoxic edema was the predominant cause of increase in brain volume [15, 16]. While at least one study has found a decease in white matter ADC in the splenium [7], most other studies have found ADC to increase in normal appearing white matter ROIs which included the corpus callosum and internal capsule [6–8]. None of these studies looked at pure grey matter ROIs.

Exactly what the diffusion tensor describes in terms of microstructure, and thus the mechanism of how it changes in pathological states, is not clearly understood. Cytotoxic edema may lead to restricted diffusion in the grey matter by restricting the relatively free diffusion of extracellular water, with a concurrent relative increase in the less easily diffusible intracellular component.

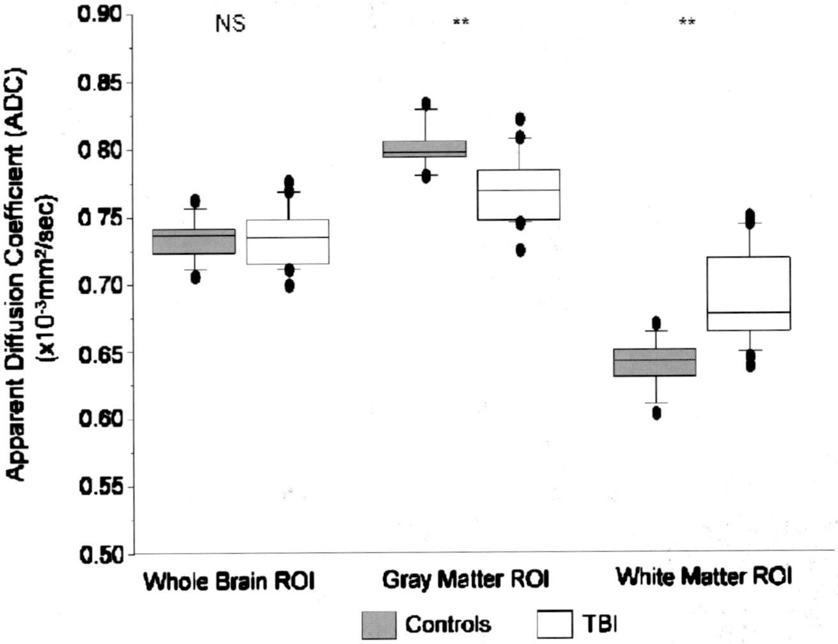

Fig. 2 These graphs illustrate the ADCs found in the whole brain, grey and white matter ROIs both in controls and those post brain injury. The central lines in the boxes denote the median values, the upper and lower edges the 75th and 25th percentiles, the error bars the 90th and 10th percentiles and the closed circles the data outside these percentiles. ** $p < 0{\cdot}001$; NS, non-significant

Changes in the cellular membrane, including regulation of water channels like aquaporin-4 as has been seen in animal studies of TBI, may also help to explain the changes observed [2, 9, 13, 23].

In contrast, white matter is a highly directionally organised structure. The myelin sheaths, microtubules, neurofilaments, and the structuring of the axons into bundles are all thought to contribute to its anisotropic nature. Histological studies have shown that these structural elements may all be damaged in a cascade of molecular events after acute brain injury within the axons [5, 19]. We have previously investigated whole brain white matter in acute brain injury and found a decrease in fractional anisotropy (FA) [17]. This was secondary to an increase in the radial diffusion, but was not accompanied by a change in diffusion on the longitudinal axis. These findings support the hypothesis that the increased diffusivity seen at this stage is consistent with edema, most likely cytotoxic, as a predominant cause. If this is indeed the case, then a concordant microstructural change (i.e. local cellular edema) could explain the discordant imaging findings in grey and white matter.

The lack of significant correlations between various clinical parameters and the ADCs in either the grey or white matter ROIs may be due to the large ROIs analysed. A more targeted analysis with smaller ROIs may make the results more sensitive and specific, as may larger patient numbers. A further limitation is the large variation of timing of scans post injury which ranged from 7 hours to 4.5 days, as changes in ADC secondary to edema acutely are likely to be dynamic during this period.

Conclusions

The changes in diffusivity seen in the acute phase of traumatic brain injury are consistent with cytotoxic edema. The differences seen in the grey and white matter ROIs emphasise the need to separate these structures when investigating pathological changes. Analysis of diffusivity in this way may offer a way to observe cerebral edema over time, thus giving more insight into the disease process. This technique offers the potential to be a biomarker for edema progression and resolution in acute brain injury.

Acknowledgements This research was supported within the framework of an MRC Program Grant (Acute brain injury: heterogeneity of mechanisms, therapeutic targets and outcome effects [G9439390 ID 65883]). VFJ Newcombe was supported by the Gates Cambridge Trust and an Overseas Research Studentship. J Nortje was supported by a British Journal of Anaesthesia/Royal College of Anaesthetics Research Fellowship. JP Coles was supported by an Academy of Medical Sciences/Health Foundation Clinician Scientist Fellowship. PJ Hutchinson was supported by an Academy of Medical Sciences/ Health Foundation Senior Surgical Scientist Fellowship.

Conflict of interest statement We declare that we have no conflict of interest.

References

1. Albensi BC, Knoblach SM, Chew BG, O'Reilly MP, Faden AI, Pekar JJ (2000) Diffusion and high resolution MRI of traumatic brain injury in rats: time course and correlation with histology. Exp Neurol 162:61–72

2. Amorini AM, Dunbar JG, Marmarou A (2003) Modulation of aquaporin-4 water transport in a model of TBI. Acta Neurochir Suppl 86:261–263
3. Barzo P, Marmarou A, Fatouros P, Hayasaki K, Corwin F (1997) Contribution of vasogenic and cellular edema to traumatic brain swelling measured by diffusion-weighted imaging. J Neurosurg 87:900–907
4. Basser PJ, Pierpaoli C (1996) Microstructural and physiological features of tissues elucidated by quantitative-diffusion-tensor MRI. J Magn Reson B 111:209–219
5. Buki A, Povlishock JT (2006) All roads lead to disconnection?–Traumatic axonal injury revisited. Acta Neurochir (Wien) 148:181–193 discussion 193–184
6. Goetz P, Blamire A, Rajagopalan B, Cadoux-Hudson T, Young D, Styles P (2004) Increase in apparent diffusion coefficient in normal appearing white matter following human traumatic brain injury correlates with injury severity. J Neurotrauma 21:645–654
7. Huisman TA, Sorensen AG, Hergan K, Gonzalez RG, Schaefer PW (2003) Diffusion-weighted imaging for the evaluation of diffuse axonal injury in closed head injury. J Comput Assist Tomogr 27:5–11
8. Inglese M, Makani S, Johnson G, Cohen BA, Silver JA, Gonen O, Grossman RI (2005) Diffuse axonal injury in mild traumatic brain injury: a diffusion tensor imaging study. J Neurosurg 103: 298–303
9. Ke C, Poon WS, Ng HK, Lai FM, Tang NL, Pang JC (2002) Impact of experimental acute hyponatremia on severe traumatic brain injury in rats: influences on injuries, permeability of blood-brain barrier, ultrastructural features, and aquaporin-4 expression. Exp Neurol 178:194–206
10. Le Bihan D (2003) Looking into the functional architecture of the brain with diffusion MRI. Nat Rev Neurosci 4:469–480
11. Le Bihan D (2007) The 'wet mind': water and functional neuroimaging. Phys Med Biol 52:R57–R90
12. Mac Donald CL, Dikranian K, Song SK, Bayly PV, Holtzman DM, Brody DL (2007) Detection of traumatic axonal injury with diffusion tensor imaging in a mouse model of traumatic brain injury. Exp Neurol 205:116–131
13. Manley GT, Fujimura M, Ma T, Noshita N, Filiz F, Bollen AW, Chan P, Verkman AS (2000) Aquaporin-4 deletion in mice reduces brain edema after acute water intoxication and ischemic stroke. Nat Med 6:159–163
14. Marmarou A (2007) A review of progress in understanding the pathophysiology and treatment of brain edema. Neurosurg Focus 22:E1
15. Marmarou A, Signoretti S, Aygok G, Fatouros P, Portella G (2006) Traumatic brain edema in diffuse and focal injury: cellular or vasogenic? Acta Neurochir Suppl 96:24–29
16. Marmarou A, Signoretti S, Fatouros PP, Portella G, Aygok GA, Bullock MR (2006) Predominance of cellular edema in traumatic brain swelling in patients with severe head injuries. J Neurosurg 104:720–730
17. Newcombe VF, Williams GB, Nortje J, Bradley PG, Harding SG, Smielewski P, Coles JP, Maiya B, Gillard JH, Hutchinson PJ, Pickard JD, Carpenter TA, Menon DK (2007) Analysis of acute traumatic axonal injury using diffusion tensor imaging. Br J Neurosurg 21:340–348
18. Obenaus A, Robbins M, Blanco G, Galloway NR, Snissarenko E, Gillard E, Lee S, Curras-Collazo M (2007) Multi-modal magnetic resonance imaging alterations in two rat models of mild neurotrauma. J Neurotrauma 24:1147–1160
19. Povlishock JT, Katz DI (2005) Update of neuropathology and neurological recovery after traumatic brain injury. J Head Trauma Rehabil 20:76–94
20. Smith SM (2002) Fast robust automated brain extraction. Hum Brain Mapp 17:143–155
21. SPM5 http://www.fil.ion.ucl.ac.uk/spm/
22. Van Putten HP, Bouwhuis MG, Muizelaar JP, Lyeth BG, Berman RF (2005) Diffusion-weighted imaging of edema following traumatic brain injury in rats: effects of secondary hypoxia. J Neurotrauma 22:857–872
23. Vizuete ML, Venero JL, Vargas C, Ilundain AA, Echevarria M, Machado A, Cano J (1999) Differential upregulation of aquaporin-4 mRNA expression in reactive astrocytes after brain injury: potential role in brain edema. Neurobiol Dis 6:245–258
24. vtkCISG Version www.image-registration.com

Acta Neurochir Suppl (2008) 102: 253–257

Qualitative aspects of cranial CT perfusion scanning in a mixed neurosurgical patient collective

S. Wolf · N. Kuckertz · M. Bauer · L. Schürer · C. Lumenta

Abstract

Background In patients with ischemic stroke, computer tomography (CT) perfusion imaging provides rapid information on the penumbra adjacent to the infarct core. For neurosurgical patients with acute brain injury, the value of CT perfusion is undecided up to now. We present our experience in a series of 78 examinations in 35 patients with acute intracranial pathology.

Methods CT perfusion was performed with a Siemens Emotion Duo® CT scanner using a single slice at the level of the upper basal ganglia. Color maps of time to peak (TTP), cerebral blood flow (CBF) and cerebral blood volume (CBV) were analyzed according to qualitative criteria. Quantitative evaluation with self-defined regions of interest was not performed due to repeatability problems and inconsistent data.

Findings TTP showed an interhemispheric difference in 45% and regional prolongation in 16% of the scans. Global TTP was prolonged in 60%, while global CBF was reduced in 43%. Two patients showed hyperemia. A CBF/CBV mismatch, indicating non-infarcted penumbra at risk, was seen in 67%. Six patients with aneurysmal SAH showed reduced CBF, and consecutive angiography confirmed vasospasm in every case.

Conclusions CT perfusion scanning gives valuable information at a low risk and with negligible additional time after a routine cranial CT. In our opinion, this modality may have considerable impact on the clinical management of severely brain injured patients in future.

Keywords CT perfusion · TTP · CBF/CBV mismatch · Vasospasm

S. Wolf (✉) · L. Schürer · C. Lumenta
Department of Neurosurgery, Academic Hospital Munich-Bogenhausen, Technical University of Munich, Englschalkinger Straße 77, 81925 München, Germany
e-mail: Stefan.Wolf@operamail.com

N. Kuckertz · M. Bauer
Department of Radiology, Academic Hospital Munich-Bogenhausen, Technical University of Munich, Englschalkinger Straße 77, 81925 München, Germany

Introduction

While the theoretical concept was already developed in 1980 [1], computer tomography perfusion (CT perfusion) became technically feasible in the last decade with the advent of CT scanner technology fast enough to perform single examination slices with a frequency of one per second. Using contrast agent, information on cerebral blood flow is acquired with repeated images at the same level [13]. Within minutes after a conventional cranial CT, CT perfusion is able to reliably assess still viable tissue in patients with acute ischemic stroke [15]. A reduction in cerebral blood flow (CBF) with preserved or even elevated cerebral blood volume (CBV), called CBF/CBV mismatch, is indicative for penumbra, while infarcted areas show a matched decrease in both parameters [12]. Neuroradiologic imaging with diffusion-weighted MRI or CT perfusion provides a better understanding of the pathophysiology of acute stroke and is considered gold-standard in contemporary decision making, when considering thrombolysis treatment options [11].

For neurosurgical patients with acute brain injury, the value of CT perfusion scanning is undecided up to now. We present our experience with this emerging modality in a mixed neurosurgical patient collective.

Methods

78 perfusion CT examinations from 35 patients treated in our neurosurgical intensive care unit were analyzed retrospectively. All but three patients were comatose or required sedation to tolerate mechanical ventilation or for control of intracranial pressure. Most frequent diagnosis was severe aneurysmal subarachnoid hemorrhage in 24 patients. CT perfusion examination was performed on clinical demand on the discretion of the attending neurosurgeon. Elevated intracranial pressure was treated in every patient prior to CT perfusion scanning to exclude a possible confounding pressure related impact on cerebral blood flow.

After a conventional native CT examination, CT perfusion was performed with a Siemens Emotion Duo® CT scanner (Siemens AG, Erlangen, Germany) using a single slice usually at level of the upper basal ganglia to visualize parts of the ACA, MCA and PCA vascular territories. 50 ml of iodinated contrast agent (Isovist® 370 mg/ml) was injected with a flow of 8 ml/sec in the proximal port of a central line with a luminal width of 14 G or 16 G. Color maps of time to peak (TTP), cerebral blood flow (CBF) and cerebral blood volume (CBV) were analyzed according to their qualitative appearance, as well as presence or absence of a CBF/CBV mismatch. Every examination was reviewed by a dedicated neuroradiologist and a neurosurgeon specialized in neurosurgical critical care. In case of disagreement on the interpretation, consensus was found via discussion.

For quantitative assessment, regions of interest were marked on the scans and numeric evaluation performed as proposed by the manufacturer of the CT scanner. DICOM copies of the scans were used to assess the repeatability of quantitative values for CBF, CBV and TTP. However, using this approach, we had large discrepancies in values gained between different observers as well serially for the same person. Therefore, a more formal statistical evaluation of quantitative CT perfusion measurement values was not performed in the current study.

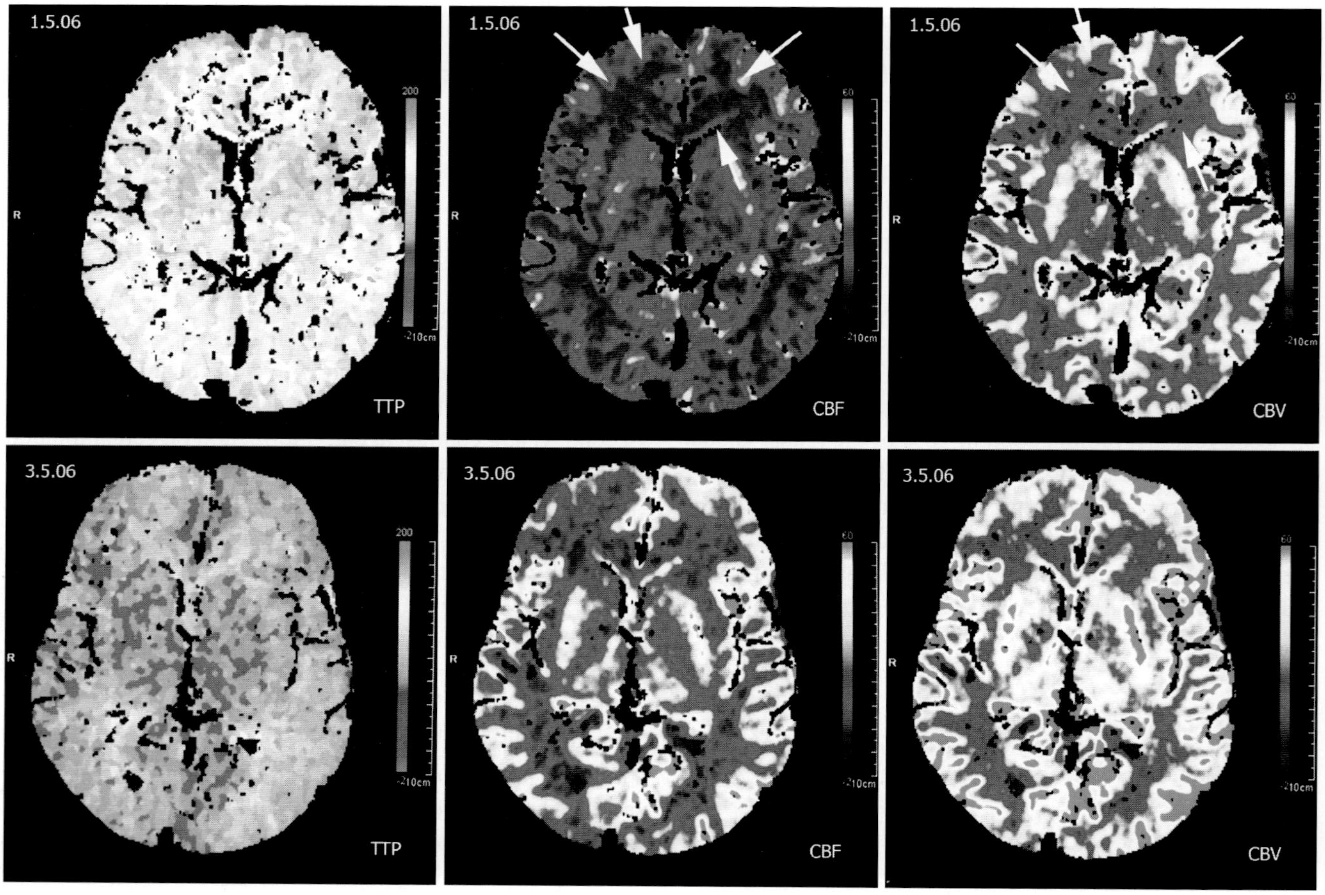

Fig. 1 Time course of vasospasm after aneurysmal hemorrhage. Upper row: Prolonged TTP and CBF/CBV mismatch in the whole brain, particularly in the frontal white matter (arrows). Lower row: Improvement in CBF/CBV mismatch and decrease of TTP after two days of repeated endovascular treatment, with residual interhemispheric difference (right side better)

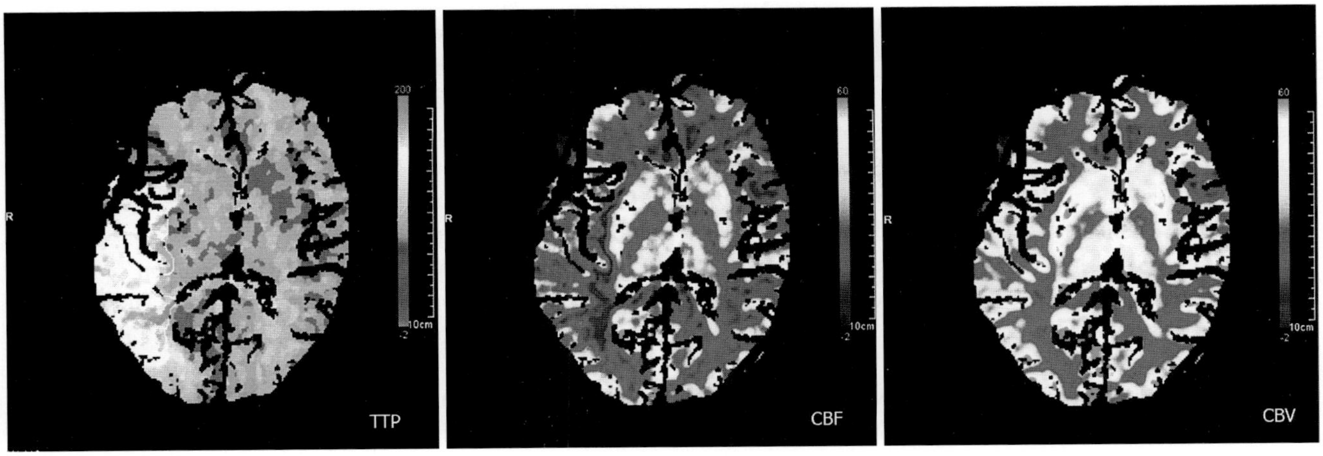

Fig. 2 TTP elevation in the distal MCA after skull base meningioma surgery

Results

In 15 patients one PCT was performed, two in six patients, three in five patients and four in nine patients. In 40 examinations (52%), a hypodense lesion reflecting infarction or the rim of a contusion or intracerebral hemorrhage was noticed on the native slice. Concerning information additionally available from CT perfusion, TTP showed an interhemispheric difference in 35 (45%) and a prolongation in a circumscribed vascular territory, either a branch of MCA or ACA, in 12 scans (16%). Global TTP was increased in 47 (60%), while global CBF was reduced in 34 (43%) of the examinations. Two patients showed hyperemia. A CBF/CBV mismatch, indicating non-infarcted penumbra at risk, was seen in 52 (67%) of the scans. In follow-up examinations, 19 (44%) were rated as improved, five (12%) as unchanged and 19 (44%) showed deterioration.

Following CT perfusion, new neuromonitoring probes for regional CBF and/or brain tissue oxygenation were implanted in four patients and in two other patients the existing probes were relocated to the vascular territory indicated at risk. Together, CT perfusion changed the prior neuromonitoring strategy in 8% of cases. Six patients with aneurysmal SAH who showed reduced CBF with CBF/CBV mismatch proceeded to angiography, where vasospasm was confirmed in every case.

Illustrated findings

Case 1 This patient progressively lost consciousness after aneurysmal SAH and so far uneventful recovery after clipping of a left-sided MCA aneurysm. CT perfusion revealed prolonged TTP and severe CBF/CBV mismatch, predominantly in the frontal white matter (Fig. 1, upper row). Vasospasm was confirmed with immediate angiography. After two days of repeated endovascular treatment with intraarterial nimodipine, TTP showed marked improvement and the CBF/CBV mismatch regressed. The

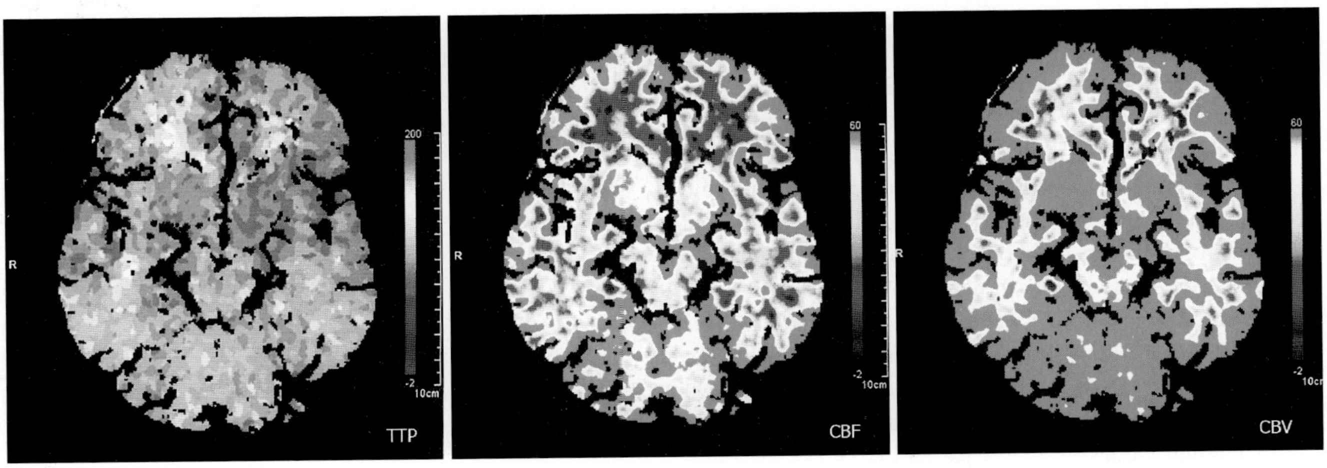

Fig. 3 Hyperemia after aneurysmal SAH

hemispheric difference in TTP was still a pathologic finding (Fig. 1, lower row).

Case 2 This patient underwent surgery for an extensive skull-base meningioma. Intraoperatively, the MCA had to be reconstructed due to tumor involvement. Postoperative examination revealed a weakness of the left arm. CT perfusion showed prolongation of TTP in a distal branch of the MCA territory (Fig. 2). CBF appeared slightly reduced, but without mismatch to CBV.

Case 3 This female patient became drowsy and obtunded on the tenth day after aneurysmal SAH. Transcranial doppler sonography showed marked elevation of flow velocities suggesting vasospasm; however, CT perfusion revealed hyperemia in CBV with adequate CBF and normal TTP. No evidence for vasospasm was seen (Fig. 3). Therefore, repeat angiography was postponed. The patient's clinical status improved markedly after treatment of her elevated blood pressure and further recovery was uneventful.

Discussion

In neurosurgical patients, the primary objective for CT perfusion imaging resembles the clinical situation after ischemic stroke: to facilitate decisions on the therapy for brain tissue still vital, but pending at risk for infarction. This perfusion-oriented therapeutic paradigm is the concept underlying contemporary neuromonitoring with focal brain tissue oxygenation or regional CBF probes [10, 17]. After traumatic brain injury, early CT perfusion may add prognostic information with higher sensitivity and specificity for contusion detection than native scans [20]. In several small series it was used to identify patients with vasospasm after aneurysmal SAH and to guide medical and endovascular therapy [5, 16, 19].

Our results support these findings. TTP prolongation or CBF/CBV mismatch both give additional information compared to conventional brain scans and sole ICP monitoring in the ICU. In 8% of the present series we used the information of CT perfusion to place or replace a regional monitoring probe. If CT perfusion suggested vasospasm after subarachnoid hemorrhage, angiography confirmed the diagnosis in all patients.

The information from CT perfusion discussed so far is qualitative. Two strategies for quantitative CT perfusion evaluation exist. Both involve the drawing of a region of interest (ROI) on the native scans and calculation of the corresponding values of TTP, CBF and CBV.

A simple approach mirrors the ROI at the midline and calculates the radio of the values from both sides [18]. This relies on the presence of uninjured brain to serve as a comparison and tries to avoid the inaccuracies of absolute value calculation described below. However, in contrast to stroke perfusion imaging, in the neurosurgical setting there may be no unaffected hemisphere and global cerebral blood flow impairment may be present, rendering ratio calculation not sufficient for quantitative analysis.

Several algorithms exist for the calculation of absolute values from CT perfusion studies [7]. The Siemens Emotion Duo® software uses contrast agent flow in the superior sagittal sinus for value calculation, without necessitating definition of an arterial structure for contrast inflow. Using this approach, we were unable to get even remotely close to clinically acceptable limits of agreement for either intra- or interobserver repeatability [3]. We consider this a problem of the algorithm and software used for evaluation and not a limitation of CT perfusion technology per se. While the venous outflow in the sagittal sinus is considered crucial for CBF and CBV calculation [8, 14], methods including arterial inflow may be superior [2, 15]. Using these, CT perfusion is able to accurately assess brain perfusion, using Xenon-CT and PET for validation [4, 9]. The next step in research is to correlate calculated CT perfusion parameters with the measurement results from focal monitoring probes, thus providing further insight in both technologies [6].

Conclusion

CT perfusion scanning is easy to implement in the clinical setting and gives valuable information at low risk for the patient with negligible additional time after a routine cranial CT. The pace of technological development is fast and tries to increase the spatial resolution with multi slice scanners as well as reliably providing absolute values for assessment of the cerebral circulation. In our opinion, CT perfusion may have considerable impact on the clinical management of severely brain injured patients in future, either in complementing advanced neuromonitoring or even for defining indications for decompressive surgery.

Conflict of interest statement We declare that we have no conflict of interest.

References

1. Axel L (1980) Cerebral blood flow determination by rapid-sequence computed tomography: theoretical analysis. Radiology 137:679–686
2. Bisdas S, Konstantinou GN, Gurung J et al (2007) Effect of the arterial input function on the measured perfusion values and infarct volumetric in acute cerebral ischemia evaluated by perfusion computed tomography. Invest Radiol 42:147–156

3. Critchley LA, Critchley JA (1999) A meta-analysis of studies using bias and precision statistics to compare cardiac output measurement techniques. J Clin Monit Comput 15:85–91
4. Furukawa M, Kashiwagi S, Matsunaga N et al (2002) Evaluation of cerebral perfusion parameters measured by perfusion CT in chronic cerebral ischemia: comparison with xenon CT. J Comput Assist Tomogr 26:272–278
5. Harrigan MR, Magnano CR, Guterman LR et al (2005) Computed tomographic perfusion in the management of aneurysmal subarachnoid hemorrhage: new application of an existent technique. Neurosurgery 56:304–317
6. Hemphill JC III, Smith WS, Sonne DC et al (2005) Relationship between brain tissue oxygen tension and CT perfusion: feasibility and initial results. AJNR Am J Neuroradiol 26:1095–1100
7. Hoeffner EG, Case I, Jain R et al (2004) Cerebral perfusion CT: technique and clinical applications. Radiology 231:632–644
8. Kealey SM, Loving VA, Delong DM et al (2004) User-defined vascular input function curves: influence on mean perfusion parameter values and signal-to-noise ratio. Radiology 231:587–593
9. Kudo K, Terae S, Katoh C et al (2003) Quantitative cerebral blood flow measurement with dynamic perfusion CT using the vascular-pixel elimination method: comparison with H2(15)O positron emission tomography. AJNR Am J Neuroradiol 24:419–426
10. Meixensberger J, Jaeger M, Vath A et al (2003) Brain tissue oxygen guided treatment supplementing ICP/CPP therapy after traumatic brain injury. J Neurol Neurosurg Psychiatry 74:760–764
11. Muir KW, Buchan A, von Kummer R et al (2006) Imaging of acute stroke. Lancet Neurol 5:755–768
12. Murphy BD, Fox AJ, Lee DH et al (2006) Identification of penumbra and infarct in acute ischemic stroke using computed tomography perfusion-derived blood flow and blood volume measurements. Stroke 37:1771–1777
13. Nabavi DG, Cenic A, Craen RA et al (1999) CT assessment of cerebral perfusion: experimental validation and initial clinical experience. Radiology 213:141–149
14. Sanelli PC, Lev MH, Eastwood JD et al (2004) The effect of varying user-selected input parameters on quantitative values in CT perfusion maps. Acad Radiol 11:1085–1092
15. Schaefer PW, Roccatagliata L, Ledezma C et al (2006) First-pass quantitative CT perfusion identifies thresholds for salvageable penumbra in acute stroke patients treated with intra-arterial therapy. AJNR Am J Neuroradiol 27:20–25
16. Sviri GE, Britz GW, Lewis DH et al (2006) Dynamic perfusion computed tomography in the diagnosis of cerebral vasospasm. Neurosurgery 59:319–325
17. Vajkoczy P, Horn P, Thome C et al (2003) Regional cerebral blood flow monitoring in the diagnosis of delayed ischemia following aneurysmal subarachnoid hemorrhage. J Neurosurg 98:1227–1234
18. van der Schaaf I, Wermer MJ, van der Graaf Y et al (2006) Prognostic value of cerebral perfusion-computed tomography in the acute stage after subarachnoid hemorrhage for the development of delayed cerebral ischemia. Stroke 37:409–413
19. Wintermark M, Ko NU, Smith WS et al (2006) Vasospasm after subarachnoid hemorrhage: utility of perfusion CT and CT angiography on diagnosis and management. AJNR Am J Neuroradiol 27:26–34
20. Wintermark M, van Melle G, Schnyder P et al (2004) Admission perfusion CT: prognostic value in patients with severe head trauma. Radiology 232:211–220

Acta Neurochir Suppl (2008) 102: 259–262

Study of perfusion in and around cerebral contusions by means of computed tomography

Bart Depreitere · Richard Aviv · Sean Symons · Michael Schwartz · Walter Coudyzer · Guy Wilms · Guy Marchal

Abstract

Background Several authors have found low absolute values of cerebral blood flow (CBF) in both contusion core and pericontusional parenchyma of head-injured patients by means of Xenon Computed Tomography (CT). Perfusion CT has become available as a new and validated tool for studying CBF in patients. The aim of the present study was to assess the relation between volume expansion of contusions and pericontusional CBF measured by perfusion CT.

Methods Eight head-injured patients with a contusion on the admission CT head scan underwent a perfusion CT scan within 48 hours post trauma. The patients received standard head injury management. The eventual maximum contusion volume was assessed on the follow up plain CT scans.

Findings Expansion of the contusion was observed in 6 patients. Reduced CBF was found in all contusions with absolute CBF values below 10 ml/100 g/min in the CT hyperdense/mixed density areas and below 20 ml/100 g/min in the surrounding hypodense areas. Penumbra areas, when defined by a mean transit time > 150% and cerebral blood volume > 2 ml/100 g, were limited to thin concentric rims surrounding the ischemic cores. We could not find a pattern of CBF that predicted contusion expansion.

Conclusions Based on the present preliminary data there is no indication that contusion expansion can be predicted on the basis of pericontusional CBF data.

Keywords Cerebral contusion · Cerebral blood flow · Penumbra · Perfusion computed tomography

Introduction

Cerebral contusions often expand in the first days following the trauma. In a study by Oertel et al., in which computed tomography (CT)-scans of the first 24 hours post trauma were analyzed, 51% of the intraparenchymal hemorrhagic lesions had grown during this time interval [7]. The pathophysiology behind this volume expansion is not

B. Depreitere (✉)
Department of Neurosurgery, University Hospitals Leuven,
Herestraat 49,
3000 Leuven, Belgium
e-mail: bart.depreitere@uzleuven.be

R. Aviv · S. Symons
Division of Neuroradiology,
Sunnybrook Health Sciences Centre,
2075 Bayview Avenue,
Toronto, ON M4N 3M5, Canada

R. Aviv
e-mail: richard.aviv@sunnybrook.ca

S. Symons
e-mail: sean.symons@sunnybrook.ca

M. Schwartz
Division of Neurosurgery, Sunnybrook Health Sciences Centre,
2075 Bayview Avenue,
Toronto, ON M4N 3M5, Canada
e-mail: m.schwartz@utoronto.ca

W. Coudyzer · G. Wilms · G. Marchal
Department of Radiology, University Hospitals Leuven,
Herestraat 49,
3000 Leuven, Belgium

W. Coudyzer
e-mail: walter.coudyzer@uzleuven.be

G. Wilms
e-mail: guido.wilms@uzleuven.be

G. Marchal
e-mail: guy.marchal@uzleuven.be

completely understood yet. It is clear, however, that the deleterious events that are induced by the mechanical distortion, are not confined to the contusion core. Excitatory amino acids and inflammatory cytokines diffuse from the contusion core into the pericontusional area, already vulnerable by the sustained – sublethal - mechanical distortion [5, 9, 10]. As a consequence, the pathological cascades going on in the core also set off in the pericontusional area. One pathological phenomenon is that the microvasculature in the pericontusional area is compromised. Ultrastructural studies have revealed two direct causes of arteriolar occlusion: 1/ expression of ICAM-1 (intercellular adhesion molecule) leading to the endothelial adherence of leucocytes and thrombocytes; and 2/ enormous swelling of the astrocytic endfeet (podocytes) by K^+ uptake [3, 8]. This vascular occlusion may lead to ischemia.

Two studies that have used the Xenon-CT technique to assess regional cerebral blood flow (CBF) in and around contusions have demonstrated that contusions behave as ischemic lesions with a necrotic core – with CBF values below the threshold for viability - and a penumbra zone in which the CBF increases with increasing distance to the core. In the study by von Oettingen et al. [11] a series of 17 patients with cerebral contusions underwent Xenon-CT at a mean time interval of two days after the trauma. A close correlation was found between the ischemic CBF values and the volume of atrophy on the follow up CT-scans, the mean CBF in the contusions being 5.9 ml/100 g/min. The CBF studies further indicated pericontusional zones of low CBF at risk for secondary ischemic insults. Schröder et al. [8] performed CBF studies by Xenon-CT in 11 head injured patients with contusions. The Xenon-CT study immediately followed the diagnostic CT-scan. They found a mean CBF value of 17.5 ml/100 g/min for edematous pericontusional areas and mean CBF of 39.9 ml/100 g/min for non-edematous pericontusional tissue.

While CT-scanning has become the mainstay of acute radiological diagnosis, perfusion CT allows for a rapid qualitative and quantitative evaluation of cerebral perfusion by generating maps of CBF, cerebral blood volume (CBV) and mean transit time (MTT). After a contrast bolus is delivered and a time-density curve is derived for each pixel, CBF is calculated based on the concept of conservation of flow. Quantitative CBF values generated by perfusion CT have been validated by comparison with Xenon-CT [12]. The broad availability of CT-scanners and their low cost compared to other techniques make perfusion CT an elegant technique, particularly in those situations where patients already undergo CT for primary diagnostic reasons. The aim of the present study was to assess the relation between volume expansion of contusions and the CBF in the pericontusional area measured by perfusion CT.

Material and methods

Four head-injured patients who were admitted to Sunnybrook Health Sciences Centre (Toronto, Canada) between February and June 2006 and who demonstrated a cerebral contusion on their admission CT head scan underwent a perfusion CT scan following the admission CT scan and at approximately 24 hours post trauma. The CT protocol was performed on a 4- and 64-slice CT GE (Lightspeed and VCT, GE Milwaukee, Wis, USA). Perfusion CT imaging was collected in two phases with the following parameters: 80 kVp, 190 mA, 3–5 s delay, injection of 0.5 ml/kg (30–50 ml) Iohexol (300 mg/ml) at a rate of 4 ml/sec. The first phase consisted of a 45 s cine scan at 1 rotation/second and was followed by a second phase with one rotations/second every 15 seconds for an additional 75 seconds. Perfusion CT studies covered a 20 mm-40 mm slab with 4–8 5 mm sections. These were centered to cover the contusion seen on the plain CT.

In four patients with a cerebral contusion admitted to the University Hospitals Leuven (Leuven, Belgium) between October 2006 and February 2007, a single perfusion CT scan was performed within 48 hours post trauma. The latter was performed on a 16-slice Philips CT (Philips Brilliance, Eindhoven, The Netherlands) using equivalent parameters to the ones described in the Toronto protocol.

Patient ages ranged between 19 and 87 years. Admission Glasgow Coma Scores ranged between 3 and 15. None of the patients had coagulation deficits. All patients received standard head injury management which consisted of neurological observation in 5 patients and of intracranial pressure (ICP) monitoring with management of raised ICP by sedation, optimization of cerebral perfusion pressure and ventricular drainage in 3 patients. All patients underwent a CT head scan 24 hours post trauma. Later CT head scans were performed on clinical indication. All patients underwent at least one later follow up CT head scan in the week following the trauma. For volume measurements, the contusions were defined as the hyperdense or mixed density lesions excluding the hypodense rim surrounding them. The maximum contusion volume was defined as the largest volume of the contusion on the follow up CT head scans performed in the first week post trauma. While standard head injury management also includes surgical removal of mass lesions when indicated, none of the patients in the present study required a decompressive surgical intervention, except for one patient in whom a contusionectomy was performed at day 8. In this patient the contusion volume on the CT scan immediately prior to the surgery was considered the maximum contusion volume. This is a pilot study in preparation of a larger study on perfusion in cerebral contusions. The study was approved by the local

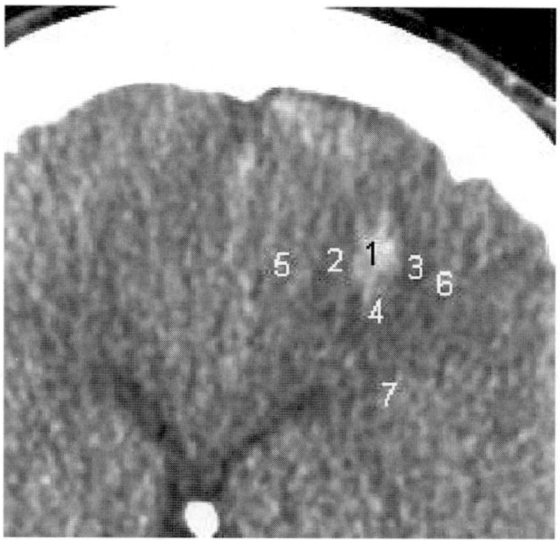

Fig. 1 Left frontal hyperdense contusion with hypodense pericontusional area. CBF values were: 1/6.26 ml/100 g/min; 2/2.16 ml/100 g/min; 3/9.30 ml/100 g/min; 4/20.62 ml/100 g/min; 5/78.71 ml/100 g/min; 6/53.55 ml/100 g/min; and 7/103.04 ml/100 g/min

ethics boards in both centres. All data were analyzed using Philips Extended Brilliance Workspace software.

Results

A total number of 23 contusions were found (1–8 per patient). Twelve contusions were situated in the frontal lobes, 8 in the temporal lobes, 2 in the rolandic area and 1 in the occipital lobes. Timing of the perfusion CT scan ranged between 18 and 48 hours post trauma in Leuven. The Toronto patients had the first perfusion CT scan between 1.5 and 13 hours post trauma and the second between 24 and 37 hours post trauma.

Exploration of CBF in and around the contusions showed that CBF values were lowest in the hyperdense/mixed density zones, with absolute values below 10 ml/100 g/min. In the hypodense pericontusional zones, absolute CBF values generally varied between 5 and 20 ml/100 g/min. The normal appearing parenchyma immediately surrounding this showed a mixture of normal and increased CBF values (Fig. 1). When penumbra zones were visualized using the method developed by Wintermark et al. [13] for ischemic stroke patients (ischemic core defined by CBV < 2 ml/100 g and penumbra defined by MTT > 150% of normal and CBV > 2 ml/100 g), penumbras appeared to be limited to very thin rims (Fig. 2). Comparison of the CBF data in the 4 patients that underwent a perfusion CT scan on admission and about 24 hours later demonstrated that the area of CBF reduction closely followed the contusion and pericontusional hypodense area and expanded where the latter expanded. In six patients a marked expansion of the contusion was seen on the follow up CT scans. We could not find any pattern in the CBF data that allowed for prediction of contusion expansion, either in the hyperdense/mixed density core of the contusion, the hypodense pericontusional area, or in the surrounding, normal appearing parenchyma.

Discussion

No real consensus exists on the management of cerebral contusions. Some authors have promoted a more or less aggressive attitude with early resection of contusions in order to protect the pericontusional tissue [8]. Others, however, emphasize the heterogenity of contusions con-

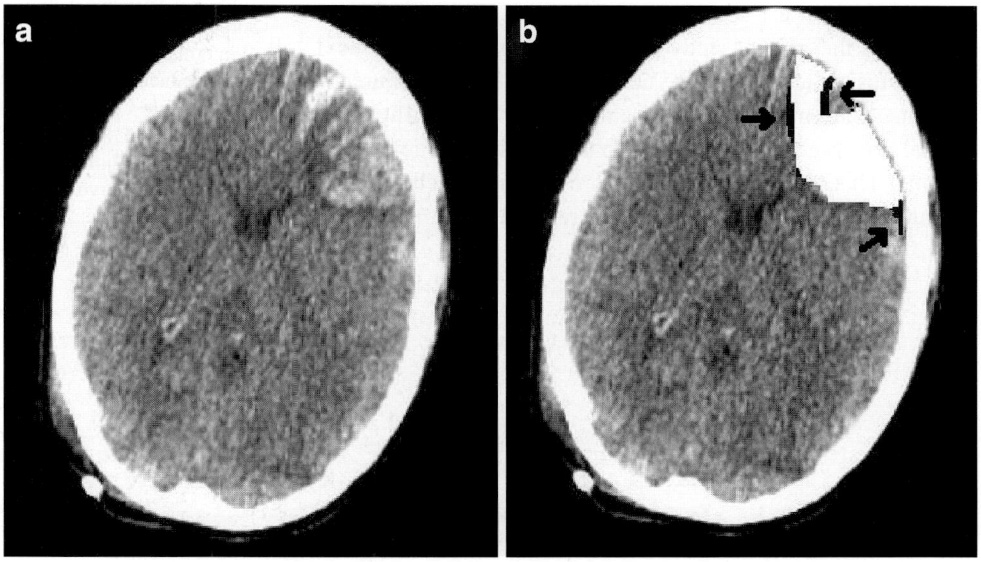

Fig. 2 **a**. Left frontal contusion. **b**. Visualization of ischemic core (white) and penumbra (black+ indicated by arrows) as defined by Wintermark et al. [13] (see text)

taining portions of brain that may be viable [6]. The only consensus seems to be that large contusions with substantial mass effect should be removed surgically [1, 2, 4]. By the time a contusion has expanded it may already have had a detrimental effect. A reliable predictor of contusion expansion would therefore help in the selection of patients that could benefit from early contusionectomy. Based on our preliminary data, there is no indication that contusion expansion can be predicted based on pericontusional CBF. The present study confirms that absolute CBF is reduced inside the contusions and – somewhat less dramatically – in the hypodense pericontusional areas, which is in keeping with the Xenon-CT studies by Schröder et al. and von Oettingen et al. [8, 11]. However, when applying the method by Wintermark et al. [13] to differentiate penumbra from ischemic core (see results section), both contusion and most of the hypodense pericontusional area were included in the ischemic core and the penumbra was limited to a thin rim surrounding the core in all cases. This is different from the situation in ischemic stroke, where large penumbra zones are usually seen. Of course, thresholds for tissue viability are not necessarily the same in traumatic brain injury and ischemic stroke.

In all cases of contusion expansion in the present study, tissue that appeared normal on the initial CT scans evolved to hypodensity on the later scans. CBF values in this tissue, measured before the expansion, were not reduced. More research is definitely required in order to better understand the pathophysiology of expansion of cerebral contusions. This research should not just focus on CBF, but also on $CMRO_2$ and possible mismatches between both. Although the above observation was true in all expanding contusions, the low number of subjects is a limitation of the present study. The ideal timing for perfusion studies after the head injury cannot be deduced, since only 4 patients underwent a repeat perfusion CT scan.

Finally, the present study has convincingly shown to us that perfusion CT can easily be performed in head-injured patients. Current multi-slice scanners allow for sufficient spatial coverage to study CBF in and around large focal lesions. Therefore, perfusion CT is a practical method for obtaining a snapshot of quantitative regional CBF.

Conflict of interest statement We declare that we have no conflict of interest.

References

1. Bullock R, Golek J, Blake G (1989) Traumatic intracerebral hematoma – which patients should undergo surgical evacuation? CT scan features and ICP monitoring as a basis for decision making. Surg Neurol 32:181–187
2. Bullock MR, Chesnut R, Ghajar J et al (2006) Surgical management of traumatic parenchymal lesions. Neurosurg 58(Suppl 3):25–52
3. Busse R, Stoltenburg-Didinger G, von Eltz S et al (2002) Micro-circulatory disturbances after experimental traumatic brain injury – an immunohistochemical and ultrastructural study. Presented at the 7th Annual Conference of the Euroacademia Multidisciplinaria Neurotraumatologica, Newcastle (UK), June 26–29
4. Gallbraith S, Teasdale G (1981) Predicting the need for operation in the patient with an occult traumatic intracranial hematoma. J Neurosurg 55:75–81
5. Holmin S, Höjeberg B (2004) In situ detection of intracerebral cytokine expression after human brain contusion. Neurosci letters 369:108–114
6. McLaughlin MR, Marion DW (1996) Cerebral blood flow and vasoresponsivity within and around cerebral contusions. J Neurosurg 85:871–876
7. Oertel M, Kelly DF, McArthur D et al (2002) Progressive hemorrhage after head trauma: predictors and consequences of evolving injury. J Neurosurg 96:109–116
8. Schröder ML, Muizelaar JP, Bullock MR et al (1995) Focal ischemia due to traumatic contusions documented by stable xenon-CT and ultrastructural studies. J Neurosurg 82:966–971
9. Stoffel M, Eriskat J, Plesnila M et al (1997) The penumbra zone of a traumatic cortical lesion: a microdialysis study of excitatory amino acid release. Acta Neurochir (Wien) 70(Suppl):91–93
10. Tanaka H, Katayama Y, Kawamata T et al (1994) Excitatory amino acid release from contused brain tissue into surrounding brain areas. Acta Neurochir (Wien) 60(Suppl):524–527
11. von Oettingen G, Bergholt B, Gyldensted C et al (2002) Blood flow and ischemia within traumatic cerebral contusions. Neurosurgery 50:781–790
12. Wintermark M, Thiran JP, Maeder P et al (2001) Simultaneous measurement of regional cerebral blood flow by perfusion CT and stable xenon CT: a validation study. AJNR 22:905–914
13. Wintermark M, Flanders AE, Velthuis B et al (2006) Perfusion-CT assessment of infarct core and penumbra. Receiver operating characteristic curve analysis in 130 patients suspected of acute hemispheric stroke. Stroke 37:979–985

Acta Neurochir Suppl (2008) 102: 263–270

Magnetic resonance measurement of blood and CSF flow rates with phase contrast - normal values, repeatability and CO_2 reactivity

Stefan K. Piechnik · Paul E. Summers · Peter Jezzard · James V. Byrne

Summary
Background Similarity in flow pulsatility has been proposed as a basis for semi-automated segmentation of vessel lumens for MR-based flow measurement, but re-examinations of salient aspects of the methodology have not been widely reported.
Methods 12 normal control subjects underwent repeated (3*Baseline+1*5%CO_2) phase contrast measurements of CSF flow through the cerebral aqueduct and foramen magnum, and CBF through the 6 large cranial vessels at the level of the 1st vertebra. Average flows were calculated for regions temporally correlated ($0.3 \leq$ Rthreshold ≤ 0.95) to user defined seed points and their 3x3 neighbours.
Results Arterial CBF averaged 710ml/min, with low variability (±4%/17%, intra-individual/group CV respectively) and was the only flow to respond significantly to 5%/mmHg CO_2. Venous outflow was much smaller (298ml/min ±10%/72%), possibly due to the weak venous pulse and variable venous anatomy. Average CSF flows exceeded the classical 0.4ml/min CSF production rate and were highly variable – aqueduct: 0.6ml/min (±50%/93%), foramen magnum: -2.7ml/min (±158%/226%).
Conclusions This preliminary analysis identified procedural steps that can improve the accuracy and repeatability of MR flow measurements, but the process remains user-dependent for the weakly pulsatile foramen magnum CSF and venous flows where variability remains a significant confound even to relatively large perturbations such as CO_2 administration

Keywords Magnetic resonance imaging · Flow measurement · Phase contrast · Cerebral blood flow

S. K. Piechnik (✉) · P. Jezzard
Department of Clinical Neurology,
Centre for Functional Magnetic Resonance Imaging of the Brain (FMRIB), University of Oxford, John Radcliffe Hospital, Headington,
Oxford OX3 9DU, UK
e-mail: stefanp@fmrib.ox.ac.uk

P. Jezzard
e-mail: peterj@fmrib.ox.ac.uk

P. E. Summers · J. V. Byrne
Department of Neuroradiology, Nuffield Department of Surgery, University of Oxford,
Level 01, West Wing, The John Radcliffe Hospital, Headley Way, Oxford, OX3 9DU, UK

P. E. Summers
e-mail: paul.summers@unimore.it

J. V. Byrne
e-mail: james.byrne@nds.ox.ac.uk

P. E. Summers
Department of Biomedical Sciences,
University of Modena and Reggio Emilia,
Via G. Campi 287,
41100 Modena, Italy

Introduction

Magnetic Resonance Imaging (MRI) has long-established its place in the diagnosis of neurological disorders through conventional static images of cerebral anatomy, and increasingly through functional measures reflecting perfusion and responses to neuronal activation. In this latter category is the measurement of tissue movement including the dynamics of blood flow, which can be observed through velocity-induced phase differences (phase contrast PC) encoded by incorporating additional magnetic field gradients into the imaging sequence [4]. PC MRI was originally developed for cardiac applications, but measurements are possible in smaller vessels and other fluid conduits in the body. In the neurosurgical domain, CSF stroke volume through the cerebral aqueduct has been used to predict hydrocephalus outcome [10]. Measurements of the pulsatile arterial inflow, venous outflow and the CSF oscillations at the level of the Foramen Magnum (FM) are of further interest as they form the elements of a recently proposed method for non-invasive prediction of the intracranial pressure (ICP) based on modelling of the dynamic volume balance in the intracranial cavity [3].

A consideration affecting the readiness of PC-MRI for routine use is the need for reproducible methods for identifying the region of flow, be it a blood vessel or a CSF-filled canal. This is particularly an issue in-vivo where independent measurements are not generally feasible. Simple methods such as signal intensity thresholds, and user defined boundaries are prone to significant inter- and intra-operator variability, with consequent adverse effects on reproducibility. The similarity in the temporal characteristics of the flow waveform between voxels within each vessel has been proposed as a means to reduce the user-dependence of boundary definition. Typically however, such a technique still requires a user-defined starting point and cut-off threshold for operation. We have found few literature reports on in-vivo methodological assessment of the importance of these user choices. Moreover, there is a need to verify the limited experimental data against the normal values, accuracy, repeatability and physiological sensitivity of flow measurements obtained by PC-MRI for describing the intracranial blood and CSF flows. In this study, therefore, we present a preliminary examination of the impact of the choices underlying pulsatility-based segmentation on the net flows in a small group of normal subjects under repeated measurements in normo- and hypercapnic conditions.

Methods

12 normal control subjects (10 male, 2 female; aged, 32± 10 years) underwent MR imaging in a 1.5 T MRI Scanner (Signa Excite, GE Medical Systems, WI, USA). All subjects provided written informed consent for their participation in this study, which was approved by the appropriate regional ethical committee for medical research. The examination lasted approximately 1 hour. The scanning parameters and positioning for the primary investigation were based on the recommendations prescribed for non-invasive ICP measurement [1, 3]. The protocol involved planning scans followed by 2D gradient-echo PC flow measurements of 1) CSF flow through the cerebral aqueduct (venc=9 cm/s, range 6–11 cm/s); 2) FM (venc=7 cm/s); and 3) blood flow through the 6 large cranial vessels at the level of 1st vertebra (venc=70 cm/s). PC images were reconstructed for 32 time points in a single heart cycle at the resolution of 0.55×0.55 mm and 6 mm slice thickness. The acquisitions were retrospectively triggered using a finger pulse oximeter, and had repetition and echo times in the ranges TR=13.3–16.4 ms, and TE=6.3–7.4 ms respectively (specific value fixed depending on the value of venc) and a flip angle $\alpha=15°$. The flow measurements were initiated after the subjects had acclimatized to the scanner environment for roughly 15 minutes, during which time conventional MRI and MRA sequences of the head were performed. These included sagittal T1-weighted MRI, time-of-flight and phase contrast MRA, as well as required planning scans which were performed once at the outset of measurements for each subject and subsequently used to define the imaging plane for each scan perpendicular to the expected flow pathway(s). In each subject we performed 4 cycles of three measurements (aqueduct, FM and blood flow) with each cycle requiring roughly 9 minutes. The first three cycles were performed under normoxia (baseline) and the final cycle during administration of a 5% CO_2 mixture in air (hypercapnia) using a simple non-rebreathing circuit The order of the three measurements was permuted for each cycle, with the initial order having been randomly chosen for each subject. If phase wrapping was noted in any of the measurements as part of the first cycle of measurement for a subject, the velocity encoding range for the corresponding subsequent measurements on that subject was increased according to the estimated peak velocity.

Signal magnitude and phase subtraction images were generated for each PC scan, and all images were exported in DICOM format for off-line analysis. The PC signal magnitude images were segmented to exclude low signal regions such as bone and air cavities and phase maps were then phase unwrapped [8]. For each subject, the internal carotid arteries, internal jugular veins and vertebral arteries were examined bilaterally (total 6 vessels) to describe the intracranial blood supply while the aqueduct and FM were extracted to characterize CSF flow. The region of interest (ROI) for each structure was segmented based on a threshold of the correlation coefficient (R) describing temporal similarity of the velocity waveform in neighbouring voxels to a seed point (SPt) manually placed near the vessel centre in fashion similar to that proposed by Alperin [2]. Region-growing from the SPt incorporated into the ROI those contiguous points with R above a preset value. The resulting area was used to calculate average velocity waveforms across its constituent voxels, the volumetric flow and their respective mean values. For background correction, the temporal and spatial mean velocity was determined for each image from a large ROI located in the midbrain or spinal cord that contained no visible vessels or CSF spaces. The values were subtracted from each image and the flow values re-calculated.

Two tests to measure the robustness of the segmentation process were undertaken. In the first, the ROIs were defined from at least two SPts placed manually near the centre of the vessel using correlation thresholds ranging from R=0.3 to R=0.95. In the second, the impact of small variations in seed placement was examined by recalculating the flow parameters using each member of the 3×3 neighbourhood of the original SPts. Software for SPt placement and flow calculations was written in-house using IDL (Interactive Data Language, ver. 6.1, Research Systems Inc., CO, USA).

Results

Out of a total of 48 imaging data sets (each with three scans), one normocapnia dataset was lost due to a reconstruction error and two were incomplete - one hypercapnia measurement due to excessive heart rate variability which prevented correct acquisition, and one normocapnia measurement. The duration of a single measurement depended strongly on the heart-rate and its variability, but there was no significant difference between the average duration of 2.7±0.9 minutes under normocapnia and 2.8±1.6 min under hypercapnia (p=0.4).

An example of the dependencies of the flow velocity estimates in the studied anatomic structures on R-threshold and seed point for a single subject is presented in Fig. 1. Decreasing the correlation threshold from 0.95 to 0.3 resulted in a steady decrease in the flow velocity in arteries and cerebral aqueduct. This resembles the pattern expected of laminar flow with a central maximum velocity diminishing toward the lumen edge. In contrast, the velocities in the FM and veins showed more complicated patterns of dependence on R and placement of seed points. Differences in velocity waveforms within the veins, coupled with the muted venous pulsation led to progressive inclusion of different sub-parts of the vein as R decreased from high to intermediate values, and then inclusion of boundary voxels and eventual leakage into neighbouring structures as the R

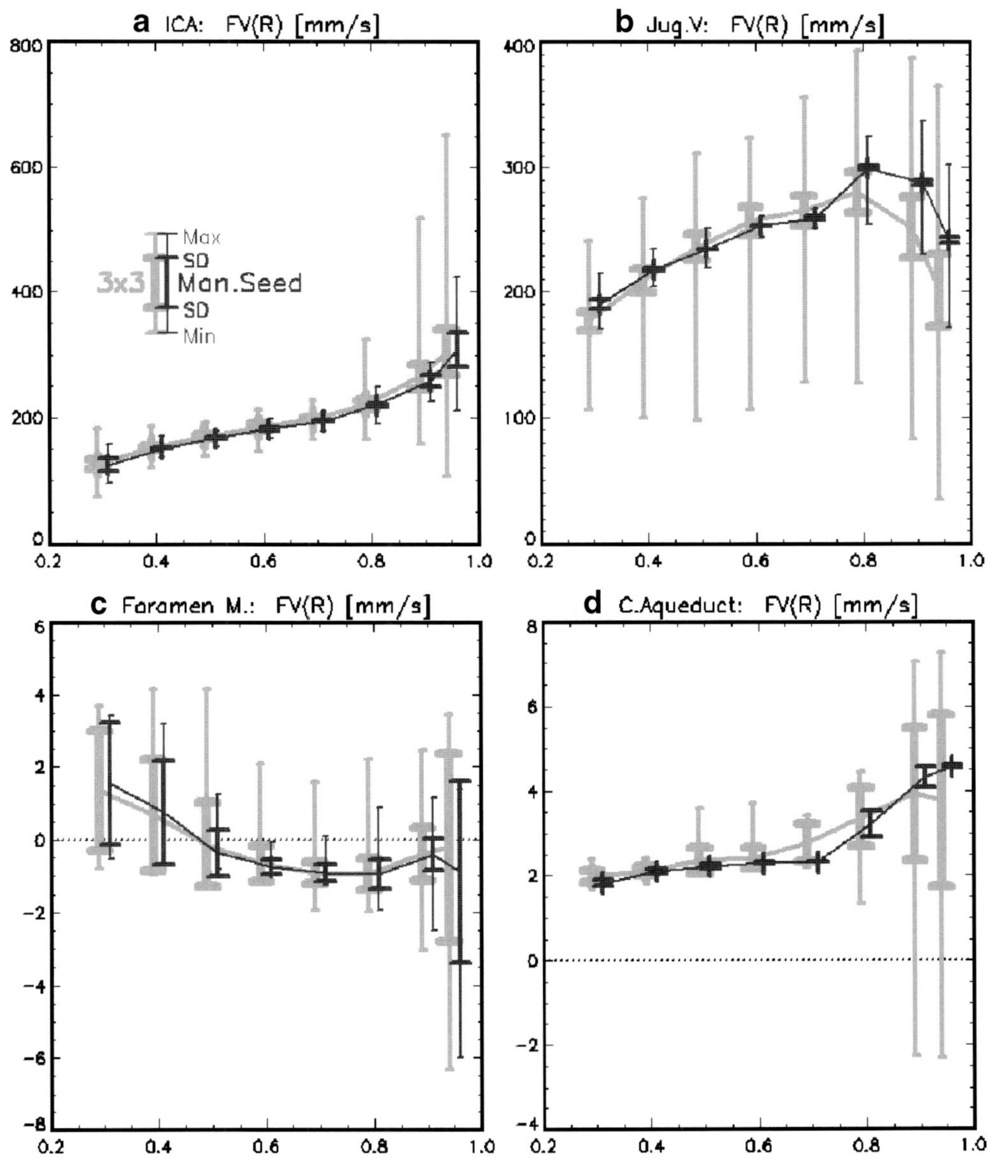

Fig. 1 Example (case 1, normocapnia) dependencies of calculated flow velocity on the correlation threshold R (x-axis) used to define flow lumen area for **a**) a single internal carotid artery (ICA) showing a decrease in mean velocity with decreasing R-value consistent with laminar flow decreasing from a central maximum, and sensitivity to the placement of seed points only for the highest values of R. This pattern is reproduced in neither **b**) a jugular vein nor **c**) the foramen magnum. These structures are also characterised by relatively large variability due to manual placement of seed points (Man. Seed), further exacerbated for seed points in their closest (3× 3) neighbourhood. **d**) Although the sensitivity to the seed placement in the cerebral aqueduct is pronounced at high R-values the flow estimates are relatively consistent at intermediate R-values indicating that the segmented regions converge

decreases further. On the basis of velocity, the transition to including non-flowing adjacent tissue was not apparent for the arteries, veins or aqueduct. But for the FM, contamination by adjoining veins was seen and likely explains the substantial increase in mean velocity at low R thresholds.

The volumetric flow was calculated as the product of cross-sectional mean velocity and area at each time point and integrated over all the cardiac cycle. The resulting values are shown across the range of correlation thresholds and seed points in Fig. 2 for the same subject as Fig. 1. Over a range of low to intermediate correlation thresholds, the total arterial and aqueductal flows were only weakly dependent on the correlation coefficient. In both these cases, the impact of the positioning of the seed point diminished with decreasing R threshold. A similar appearance holds for the FM at correlation coefficients above 0.5, but the variance increased consistently as the R threshold decreased. The venous flow, presented in Fig. 2b, did not show a plateau region but increased in close to linear fashion with decreasing R-threshold. The calculation of the venous flow depended strongly on the selection of the seed point as indicated by relatively large whiskers. Even departures by 1 pixel from the manually positioned seed points seemed to have significant potential to change the estimates of flow.

The above observations indicate that the value of the correlation threshold was a major predicting factor affecting the final flow estimates. On the basis of maximizing intra-individual consistency of flow values, R=0.6, appears to present a reasonable default value for arterial, aqueduct and foramen flow measurement, for which representative numeric assessments and measures of their variation are reported in Table 1. The table also shows the substantial effect of background correction on the CSF flow estimates, bringing them significantly closer to physiologically expected values, while having very little effect on the blood flow values. A

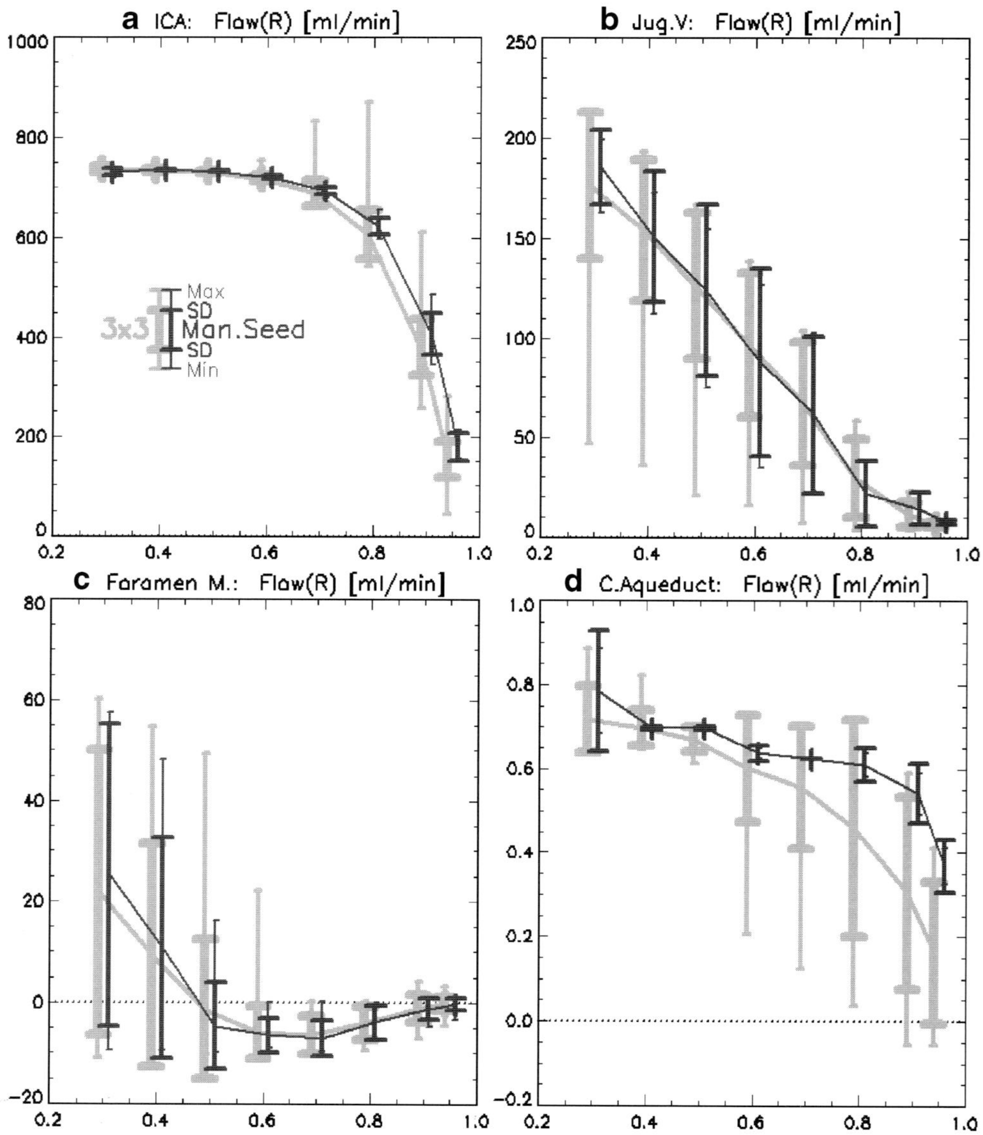

Fig. 2 Dependence of the volumetric flow on the correlation threshold R and the position of seed points for the tested anatomic structures in the same case as in Fig. 1. **a**) For both manually placed seeds (Man. Seed) and their neighbours (3×3) the flow in single carotid artery (ICA) reaches a relatively stable value for R<0.7. **b&c**) this pattern is not reproduced in the jugular vein, and exists only over a limited range for the foramen magnum. Both depend heavily on the selection of thresholds or small deviation in positioning of the seed points. **d**) The estimate of flow in cerebral aqueduct reached a narrow plateau across mid-range R, but suboptimal choice of seed points has large effect on the values

consequence, however, of reducing the mean values of CSF flow was that the variance relative to the mean tended to increase. Background correction had the further effect of reducing the representative sensitivity of flow estimates to the choice of R value around R=0.6 (determined by the ratio of fractional change in flow estimate to the mean flow for a change of 0.1 in correlation coefficient threshold).

Overall, there were no significant differences in the mean values obtained for the user-selected seed points and those of their 3×3 neighbourhood. The alternative seed points (Figs. 1 and 2) did, however, result in a wider range of values, indicating that outliers due to misplacement of seed points could be several times larger than intra- or inter-individual variance and up to ten times the average values (Table 1, (Fmax-Fmin)/F).

The end-tidal CO_2 increased from 5.6±0.3% at the baseline to 6.6±0.3% during hypercapnia, corresponding to a change of roughly 7 mmHg partial pressure. The effects of hypercapnia are presented in Fig. 3 by grey data points and summarized in the last two columns of Table 1 for R=0.6. Only the arterial flow response was statistically significant. The average change in veins was similar, but overshadowed by the high variability of vein measurements, which was 4 times larger than for the arteries. The relative changes in CSF flow were unexpectedly large at up to 70% of baseline values, depending on the method of calculation. [Note the differences in values and sign between the appearance of plots in Fig. 3 (reactivity of averages) and the values in Table 1 (calculated as average of individual relative changes)]. However, none of these changes was statistically significant due to the effect of the large variability in flow measurements in these structures (Table 1).

Discussion

We found MR measurement of arterial cerebral inflow to be particularly robust, with intra-individual repeatability of measurements within 5% and total flow estimates of about 700 ml/min. There was also very little impact of introduced jitter in the positioning of manual seed points, and the estimates were stable for correlation thresholds below 0.7. These good metrological properties can be attributed to the combination of a relatively large flow velocity giving rise to a high signal-to-noise ratio, and a distinctive pulsatility pattern throughout the large vessel, making it compatible with the chosen method of data analysis. On average, the identified venous outflow accounted only for about 50% of arterial inflow and was highly dependent on both the placement of the seed points and the choice of the arbitrary threshold for the waveform shape correlation. Although the jugular veins are also characterised by relatively fast flow velocity, the flow waveform is flattened, and variable across the vessel cross-section. These factors contribute to measurement variability through incomplete identification of the complete venous lumen at high R-thresholds, and leakage out of the vessel at low thresholds. The non-uniform flow profiles may be a consequence of using a single imaging plane to capture all six main cerebral vessels. While in this preliminary report, we have not specifically addressed PC image acquisition and its impact, it must be noted that priority was given to achieving perpendicular orientation for the internal carotid arteries and, where possible, the vertebral arteries. In consequence, the veins were often acquired obliquely, which may exacerbate errors in estimating flow. Known variability and asymmetry in venous outflow accounts for the fact that bilateral jugular veins could be identified in fewer than half of our subjects, with disregarded smaller veins likely to explain the remaining underestimation in venous flow. This might be resolved by application of additional measurements focused on reducing venous variability.

The cerebral aqueduct was the smallest structure investigated and the impact of the selection of the seed point was large for high correction thresholds, but diminished rapidly as R was reduced. This can be expected for such a small

Table 1 Estimated flow (F), its variability and the changes due to CO_2 calculated at the correlation threshold R=0.6

		mean flow (F) [ml/min]		CV		$\frac{F_{max}-F_{min}}{F}$	Sensitivity	$\frac{F_{co2}-F_{rest}}{F_{rest}}$	
Background corrected		Man SPt	3×3 nbhd	Group	Individual	Worst case	$(\Delta F/\Delta R)$/(F/R)	Man SPt	3×3 nbhd
Blood	Arteries	715.9	705.3	19%	4%	72%	−0.45	35±15%	36±16%
	Jug.V.	298.	278.4	67%	13%	170%	−1.00	29±55%	38±69%
CSF	F. Magn.	−2.7	−2.5	226%	158%	1268%	0.22	66±335%	−39±355%
	Aqueduct	0.6	0.6	93%	50%	155%	−0.07	−16±49%	−14±53%
Without background correction									
Blood	Arteries	710.1	698.6	16%	4%	72%	−0.45	35±15%	36±16%
	Jug.V.	298.3	277.4	68%	12%	214%	−0.98	36±63%	47±79%
CSF	F. Magn.	22.4	21.1	33%	17%	279%	−1.88	32±68%	36±60%
	Aqueduct	1.7	1.6	46%	19%	156%	−0.57	−7±24%	−4±23%

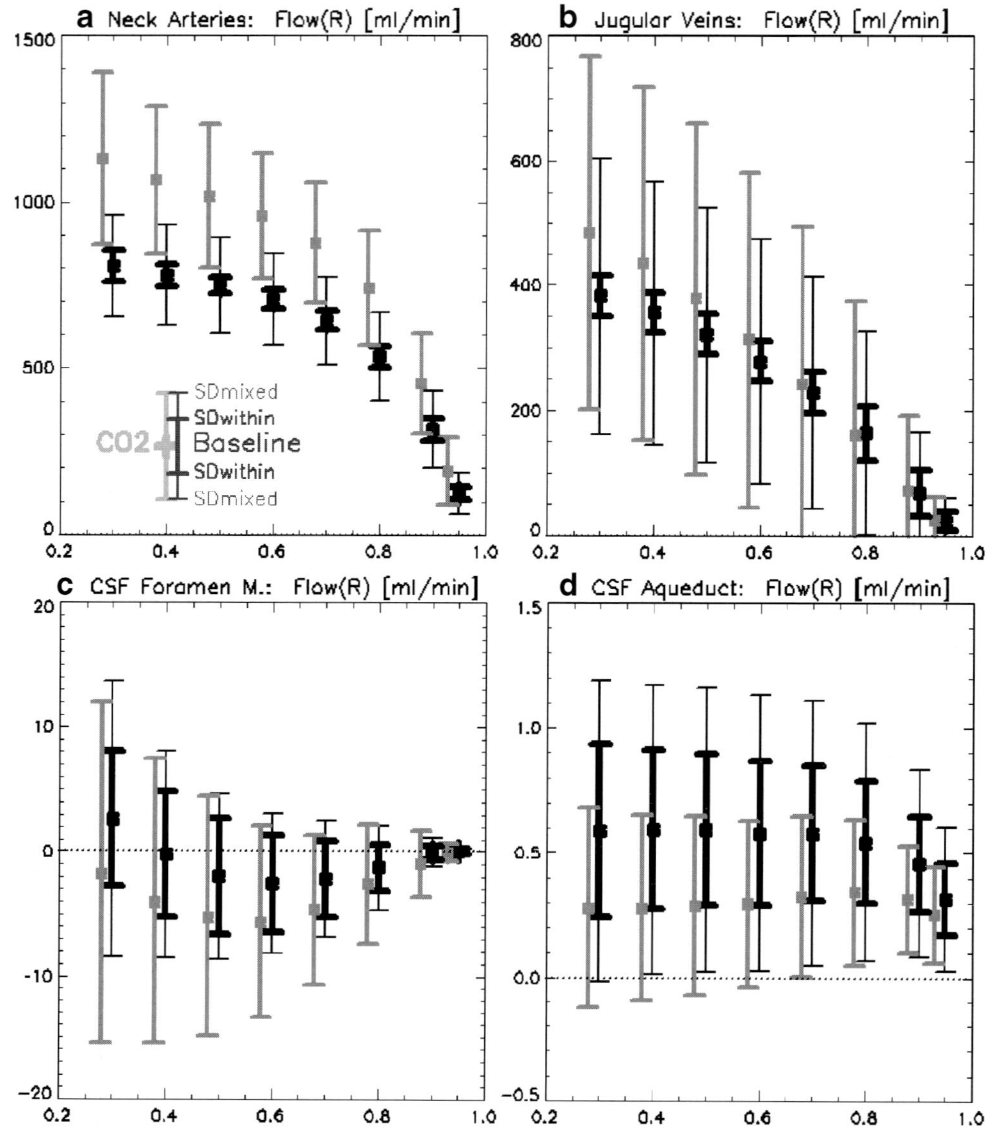

Fig. 3 Estimates of volumetric flow across the repeated experiments at the baseline (black points, whiskers: thick=intraindividual SD, thin=group SD) and under hypercapnia (grey points, whiskers=group SD) show the same traits identified in Fig. 2. **a**) The arterial flow is estimated most reliably. **b**) Mean venous blood flow is about 50% less then arterial input (note Y-axis range) and there is no clear evidence of an asymptotic value. **c**) CSF flow through the FM has complex shape and large variability measures, but the mean is fairly stable for intermediate R-values. **d**) CSF flow in aqueduct is relatively independent of R below a value of 0.8, but its mean is larger than the estimates of bulk production of CSF in ventricles. Hypercapnia induced changes are only statistically significant for the internal carotid artery

structure, where displacement of the seed by a single pixel can correspond to the difference between mid-lumen and lumen edge which have slightly different pulsation properties and are distinctly segmented at high R thresholds. Even at low thresholds, the clear pulsation isolates the few voxels with flow from the background and a stable flow estimate is obtained. As the ratio between aqueduct size and the image resolution is only about four, flow overestimation of up to 100% due to partial volume effects is possible [12]. This could account for the bulk inflow rate of CSF to the 4th ventricle in our study being higher than expected from 0.4 ml/min CSF secretion rate. While, background correction reduced our net flow rates (0.6±0.6 vs 1.7±0.8 ml/min) improving the agreement with established CSF secretion rates, both were within 2 SD of the literature values for CSF production, and within the range of previous MR reports [5, 6, 10]. It must be recognized however, that reported values for CSF flow by MR differ substantially, with Huang et al. [6], reporting a mean physiological rate of CSF production in the range of 0.4 ml/min while Luetmer et al. [10] and Florez et al. [5], using other segmentation methods, report average values of 1.5 and 3.8 ml/min respectively. From the reports it is difficult to identify the basis for these differences, which are comparable to the effect of the application of background correction to our data.

A range of intermediate R values give rise to consistent mean flow values for the foramen magnum, but the measurement variability is pronounced both for repeat measurements and for the choice of seed points. Inspection of individual ROIs overlaid over the FM revealed that for very low values of R, the segmentation tends to spill over into neighbouring venous structures, giving rise to the rapid change in the FM flow and inversion of the flow direction. Even without leakage of the segmentation into closely adjacent vessels, their proximity may contaminate each FM flow measurement due to overlap of the point spread function of their voxels [9]. The impact of background correction is particularly profound in the case of the FM where the average flow was returned to physiological values close to zero. This has the effect of greatly exacerbating the variability of the results when expressed as CVs and, in turn, the CO_2 reactivity which is determined here as a fraction of the mean normocapnic flow.

As well as the absolute accuracy of the flow estimates, their sensitivity to physiological manoeuvres, such as hypercapnic challenge, is often useful in clinical decision-making. In this study, the $EtCO_2$ increased by 18% under CO_2 challenge, eliciting a 35% increase in blood flow rates. Assuming a baseline $EtCO_2$ of 40 mmHg at typical atmospheric pressure, this corresponds to an arterial flow reactivity of 5%/mmHg, well within the range of current PET estimates [7]. The change in arterial flow was statistically significant, but venous flow changes of similar magnitude were not significant due to the their greater variability. The changes in CSF fluxes due to CO_2 were similarly not statistically significant on an inter-individual basis. However, the differences between average group estimates of CSF flow under baseline and hypercapnia presented in Fig. 3c and d appear as consistent across the range of correlation thresholds as for the blood flow and may be interpreted as yet another indicator of contamination with blood partial volume in CSF structures or reference regions.

Our underlying purpose in this preliminary analysis was to define a robust analysis process for further investigations of blood and CSF flows. Choosing of the optimal threshold segmentation is essential for such applications. Based on minimizing the inter- and intra-individual variability of flow measurements, we found an R-value of 0.6 to be generically applicable to the pulsatile flows in arteries, aqueduct and foramen magnum. This threshold avoids both the variability in seed point selection effects in the aqueduct and the leakage of the FM segmentation into adjoining vessels. This agrees well with the findings of Alperin et al. who based their assessment on the pattern of change in the cross-sectional area with the value of R [2]. For the veins, however, a single seed point and threshold are inadequate in terms of yielding reproducible results. Multiple seed points together with operator judgement of adequacy of the segmentation [2] could be adopted to improve the vein and FM segmentation reproducibility, but it necessitates further user intervention, and is likely to vary between operators. While the analytic methods can be yet further improved [2, 6], in view of the presented underlying variability of repeated measurements it seems necessary to further investigate the basis of any final estimates.

Background correction made a significant difference to both aqueduct and FM flow rates, which is attractive on the basis of improving agreement with expected values of CSF production. A risk, however, is that background correction is applied in a subjective manner to achieve this end, and the values become the result of operator expectations rather than their physiological basis. Moreover, as different strategies for estimating background velocity values are likely to yield different results, a consistent approach is needed. Direct calculation of the Maxwell terms for MR PC acquisition with phantom verification would provide a truly objective strategy to background correction, but are difficult to incorporate into routine clinical practice without the assistance of the scanner manufacturer. Our background correction, based on nearby regions of interest that should have no net displacement over the cardiac cycle is commonplace, but may not be optimal. In particular, as in-vivo phase contrast measurements are sensitive to all sources of flow, any preferential direction of cerebral blood in unresolved vessels (of which there are several scales, see for example [11]) may lead to bias in the background correction.

Summary MR PC imaging can be achieved using sequences available as standard features on commercial MR scanners but image post-processing requires close attention. PC offers a reliable estimate of bulk flow in large arterial vessels, but the accuracy and repeatability of the measurements in CSF and venous flow compartments is much lower and strongly depends on the application of background correction. Using correlation to the pulsation waveform, segmentation and flow values depend on subjective manual placement of seed points and operator chosen threshold values. Arbitrary choices of these parameters, or inconsistent application of background correction can significantly confound interpretation of the results and overshadow even relatively large responses such as to CO_2 administration. The impact of these factors has been shown for average flow values, but the sensitivity of secondary parameters such as amplitudes and waveform-shape and their implication on subsequent estimations of ICP or brain compliance has yet to be determined.

Acknowledgements SKP and PJ are supported by UK Medical Research Council grants. This work is also funded by NIHR Biomedical Research Centre Programme. We thank Dr. R. Wise (CUBRIC, Univ. Cardiff, UK) for the help with the design of the CO_2 part of the experiment.

Conflict of interest statement We declare that we have no conflict of interest.

References

1. Alperin N (2006) Protocol For MRI Investigations of Intracranial Dynamics (MRIIID). Personal communication, Oxford
2. Alperin N, Lee SH (2003) PUBS: pulsatility-based segmentation of lumens conducting non-steady flow. Magn Reson Med 49:934–944
3. Alperin NJ, Lee SH, Loth F, Raksin PB, Lichtor T (2000) MR-Intracranial pressure (ICP): a method to measure intracranial

elastance and pressure noninvasively by means of MR imaging: baboon and human study. Radiology 217:877–885
4. Firmin DN, Nayler GL, Kilner PJ, Longmore DB (1990) The application of phase shifts in NMR for flow measurement. Magn Reson Med 14:230–241
5. Florez YN, Moratal D, Forner J, Marti-Bonmati L, Arana E, Guajardo-Hernandez U, Millet-Roig J (2006) Semiautomatic analysis of phase contrast magnetic resonance imaging of cerebrospinal fluid flow through the aqueduct of Sylvius. Magma 19:78–87
6. Huang TY, Chung HW, Chen MY, Giiang LH, Chin SC, Lee CS, Chen CY, Liu YJ (2004) Supratentorial cerebrospinal fluid production rate in healthy adults: quantification with two-dimensional cine phase-contrast MR imaging with high temporal and spatial resolution. Radiology 233:603–608
7. Ito H, Kanno I, Ibaraki M, Hatazawa J, Miura S (2003) Changes in human cerebral blood flow and cerebral blood volume during hypercapnia and hypocapnia measured by positron emission tomography. J Cereb Blood Flow Metab 23:665–670
8. Jenkinson M (2003) Fast, automated, N-dimensional phase-unwrapping algorithm. Magn Reson Med 49:193–197
9. Lagerstrand KM, Vikhoff-Baaz B, Starck G, Forssell-Aronsson E (2006) Quantitative phase-contrast flow MRI measurements in the presence of a second vessel closely positioned to the examined vessel. J Magn Reson Imaging 23:156–162
10. Luetmer PH, Huston J, Friedman JA, Dixon GR, Petersen RC, Jack CR, McClelland RL, Ebersold MJ (2002) Measurement of cerebrospinal fluid flow at the cerebral aqueduct by use of phase-contrast magnetic resonance imaging: technique validation and utility in diagnosing idiopathic normal pressure hydrocephalus. Neurosurgery 50:534–543 discussion 543–534
11. Paxinos G, Huang XF (1995) Atlas of the human brainstem. Academic Press San Diego, London
12. Tang C, Blatter DD, Parker DL (1993) Accuracy of phase-contrast flow measurements in the presence of partial-volume effects. J Magn Reson Imaging 3:377–385

PART 6:
ICP: Brain complicance, biophysics, and biomechanics

Acta Neurochir Suppl (2008) 102: 273–277

Motor trephine syndrome: A mechanistic hypothesis

Shirley I. Stiver · Max Wintermark · Geoffrey T. Manley

Abstract

Background In our neurotrauma practice, "motor trephine syndrome" was defined as a contralateral monoparesis that developed as a delayed and reversible complication in patients treated with decompressive hemicraniectomy for traumatic brain injury (TBI). The goal of this study was to define causal factors associated with this syndrome.

Methods We retrospectively reviewed clinical records and imaging studies of all patients undergoing decompressive hemicraniectomy followed by cranioplasty repair in our comprehensive database of TBI patients. Detailed analysis of motor function from the time of injury to 6 months following cranioplasty repair identified three patterns of motor recovery.

Results Blossoming of contusions, CSF circulation dysfunction, and longer times to cranioplasty repair were strongly associated with "motor trephine syndrome". We hypothesize that "motor trephine syndrome" arises from decompensated CSF flow with transgression of CSF fluid and edema into brain parenchyma, together with associated decrements in cerebral blood flow.

Conclusion Prior contusion injury, decreased skull resistance with large hemispheric decompressions, and longer intervals to cranioplasty repair facilitate transparenchymal flow of CSF and edema. "Motor trephine syndrome" is rapidly reversible following cranioplasty repair. CSF and edema fluid changes within the parenchyma and CBF normalize, coincident with improvements in the patient's motor function, upon replacement of the bone.

Keywords Craniotomy · Craniocerebral trauma · Head injuries closed · Paresis · Cerebral spinal fluid · Cerebrovascular circulation · Cranioplasty · Trephining

Introduction

Creation of holes in the cranium or "trephination" is an ancient procedure dating back to the Neolithic ages 10,000 years ago [4]. At that time, trephination may have been performed as a tribal ritual with survivors deemed to have special powers or as a treatment to allow evil demons, thought responsible for headaches, convulsions and mental disorders, to escape from the skull. In 1939 Grant et al. [6] first described "Syndrome of the Trephined" as a constellation of neurological, cognitive and psychological symptoms that developed following craniectomy. Common symptoms of the syndrome include headache, fatigue, and mood disturbance, which evolve in a delayed and insidious fashion. Case examples and small series have noted motor weakness following trephination [5, 6, 10–12]. The pathophysiology underlying "Syndrome of the Trephined" is poorly understood. Cogent arguments have debated the

S. I. Stiver · G. T. Manley
Department of Neurological Surgery,
University of California San Francisco,
San Francisco, CA, USA

G. T. Manley
e-mail: manleyg@neurosurg.ucsf.edu

M. Wintermark
Department of Radiology, University of California San Francisco,
Box 0628, San Francisco, CA, USA
e-mail: max.wintermark@radiology.ucsf.edu

S. I. Stiver (✉) · G. T. Manley
Department of Neurosurgery,
University of California San Francisco,
San Francisco General Hospital,
1001 Potrero Avenue, Room 101,
San Francisco, CA 94110-0899, USA
e-mail: sstiver@neurosurg.ucsf.edu

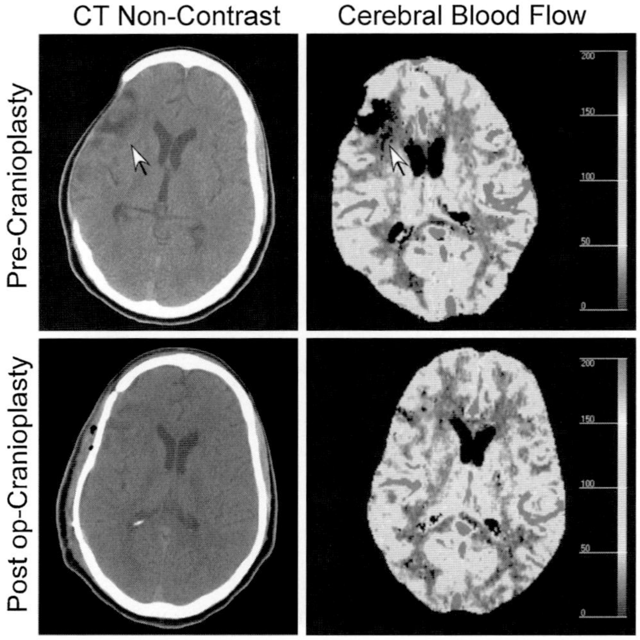

Fig. 1 Case illustration pre- and post-cranioplasty imaging. **a**) Non-contrast CT, immediately prior to cranioplasty, showed areas of hypoattenuation (arrow), consistent with transgression of CSF and edema into brain parenchyma. **b**) CT perfusion CBF map evidenced decreased CBF (arrow) in, and adjacent to, the areas of hypoattenuation. **c**) Five days following cranioplasty repair, the area of hypoattenuation has nearly completely resolved on the non-constrast CT scan. **d**) Coincident with resolution of the area of hypoattenuation, CT perfusion imaging showed improvement in the CBF map. Resolution of the CSF and CBF flow disturbances was coincident with marked recovery of the patient's grip strength over the 5 day period following cranioplasty repair

role of direct atmospheric pressure on the unprotected brain, cerebral blood flow (CBF) disturbances, and cerebral spinal fluid (CSF) circulation abnormalities [1–3, 5, 9, 12–14]. In our neurotrauma practice, we observed the onset of delayed motor weakness in a number of patients treated with decompressive hemicraniectomy for severe traumatic brain injury (TBI). The goal of this study was to define causal factors underlying the onset of delayed motor deficit, which we termed "motor trephine syndrome", following decompressive hemicraniectomy for TBI.

Case illustration

A 33 year-old male fell from a moving vehicle while intoxicated. His Glasgow Coma Scale score was 13 in the field and 8 (E1V2M5) in the emergency department. He localized symmetrically with all four extremities. CT scan of the head demonstrated an acute right sided subdural hematoma with midline shift and obliteration of the basal cisterns. A large frontotemporal-parietal-occipital decompressive hemicraniectomy was performed during which the dura was opened to the bone edges and the bone (~10× 15 cm) was left out and stored in the tissue bank. CT scan within 24 hours following surgery showed blossoming of a right frontal contusion. Post-operative left sided weakness, arm greater than leg, slowly improved and recovered completely by 3 months. At 6 months he began to complain of weakness in his dominant left hand. One month later, at time of admission for cranioplasty repair, his left grip strength was MRC grade 4/5, biceps 4+/5, and he had a mild left pronator drift. Non-contrast CT and CT perfusion imaging studies were performed before and after cranioplasty (Fig. 1). Within 5 days following cranioplasty repair, the pronator drift had resolved and grip strength had improved markedly to 4+/5, with full, 5/5 motor recovery noted at his 1 month follow-up.

Methods

During the 5 year period, July 2001– July 2006, patients undergoing cranioplasty repair were retrospectively identified from our comprehensive database of traumatic brain injury (TBI) patients treated with decompressive hemicraniectomy. We examined inpatient, clinic, and rehabilitation medical records and recorded patient demographics, injury specifics, and neurological examinations with specific attention to the motor testing throughout the hospitalization, rehabilitation, and follow up periods. Serial CT imaging studies were evaluated for the presence of epidural/subdural and intraparenchymal mass lesions, contusions, midline shift, cisternal compression, hygromas, hydrocephalus,

Table 1 Study groups

Group	Number of patients (%)	Characterstics
Motor Trephine Sydrome	10 (26%)	Full motor recovery following decompressive hemicraniectomy; onset of delayed contralateral upper extremity weakness, distal >> proximal, beginning a median of 4.5 months later; slowly progressive; prompt improvement in strength following cranioplasty repair
Uneventful recovery	20 (53%)	Recovery of full motor function without relapse
Stable deficit	8 (21%)	Significant motor weakness post-decompressive hemicraniectomy that remained stable and did not improve with cranioplasty repair

ischemic infarction, edematous change, and encephalomalacia. Rotterdam Score on head CT was also assigned [8]. Statistical comparisons employed t-tests, Fisher exact, and Mann Whitney rank sum analyses and were considered significant for two-sided probabilities less than 0.05.

Results

Over the five-year study period, motor recovery in 38 patients who had undergone decompressive hemicraniectomy and cranioplasty repair for TBI was categorized into 3 groups (Table 1). Thirty patients recovered full motor function following their injury and decompressive hemicraniectomy. Ten patients (26%) developed "motor trephine syndrome", manifested by delayed onset of weakness, contralateral to the site of hemicraniectomy. Patients with "motor trephine syndrome" are summarized in Table 2. By contrast, twenty patients (53%) recovered full motor strength without relapse in their motor function. Eight patients (21%) suffered dense hemiparesis following injury and decompressive hemicraniectomy, which did not fully recover.

Patients with "motor trephine syndrome" complained of new onset weakness in the distal contralateral upper extremity, beginning a median of 4.5 months following their decompressive hemicraniectomy. Typical problems comprised difficulties with writing, fastening buttons, and dropping objects. The hand was most impaired, and the weakness was slowly progressive. In severe cases, the grip did not close, and metacarpal extensors demonstrated only a flicker of extension. The lower extremity was comparatively spared. Sensory and language function was not impaired. The delayed motor weakness was reversible. All patients with "motor trephine syndrome" experienced improvement in motor strength within the first 72 hours following cranioplasty repair and full recovery generally within one month afterwards.

Comparison of the 10 patients (26%) who developed "motor trephine syndrome" following decompressive hemicraniectomy with the 20 patients (53%) who recovered without a relapse in their motor strength demonstrated no difference between the two groups in the incidence of motor deficit from the initial injury itself. However, following decompressive hemicraniectomy, motor weakness was significantly more common in the group that developed "motor trephine syndrome" ($p=0.05$). We attributed this increased incidence of motor weakness to blossoming of contusions following decompression. On post-operative CT imaging, contusions were present in 80% of patients who developed "motor trephine syndrome", as compared to 45% of patients who recovered without delayed deficit ($p=0.1$).

Table 2 Demographics, injury characteristics, and motor examinations from time of injury to cranioplasty repair for patients with Motor Trephine Syndrome[†]

Patient	Age	Sex	Mechanism of Injury	GCS Emergency Department	Injury motor deficit at presentation	Motor deficit post -decompressive hemicraniectomy (DHC)	Time to full motor recovery (days)	Time from DHC to onset of delayed motor deficit (days)	Time from DHC to cranioplasty (months)	Cranioplasty pre-operative motor deficit	Cranioplasty follow-up: Time to full motor recovery (months)
1	24	M	PVA	7 E1V2M4	No	Yes, mild	105	180	7	Grip and finger extensors 4/5, mild pronator drift	1
2	43	M	MVA	7 E1V1M5	Yes	No	8	120	8	Biceps 4+/5, Grip 4/5	1
3	33	F	PVA	6T E1VTM4	No	Yes, mild	60	120	4.5 (became infected and removed) Repeat 18 mo	Grip 3/5, finger extensors 1–2/5	1
4	49	F	PVA	14 E4V4M6	No	Yes, moderate	40	210	13	Grip 4+/5	6
5	21	M	Assault	12 E3V4M5	No	No	2	150	6	Grip 3/5	6
6	17	F	MVA	4 E1V1M2	No	No	14	90	6	Grip 4+/5	1
7	42	M	Assault	13 E4V4M5	No	Yes, severe	60	120	7	Biceps, triceps, grip 3/5	6
8	14	M	Assault	3 E1V1M1	No (GCS 3)	Yes, moderate	90	180	6	Grip 4+/5	1
9	47	M	Found down	7 E2V1M4	No	No	22	120	8	Biceps 4/5, Grip 3/5	0.5
10	31	M	Fall from standing	14 E4V5M6	Yes	Yes, moderate	40	180	5.5	Biceps 4−/5, triceps and grip 3/5	2

DHC - decompressive hemicraniectomy; MVA -motor vehicle accident; PVA pedestrian versus auto accident

[†] Motor grading scale: 5 -full resistance, 4+ -moderate resistance, 4 -some resistance, 3 -antigravity, 2 -movement not against gravity, 1 -flicker

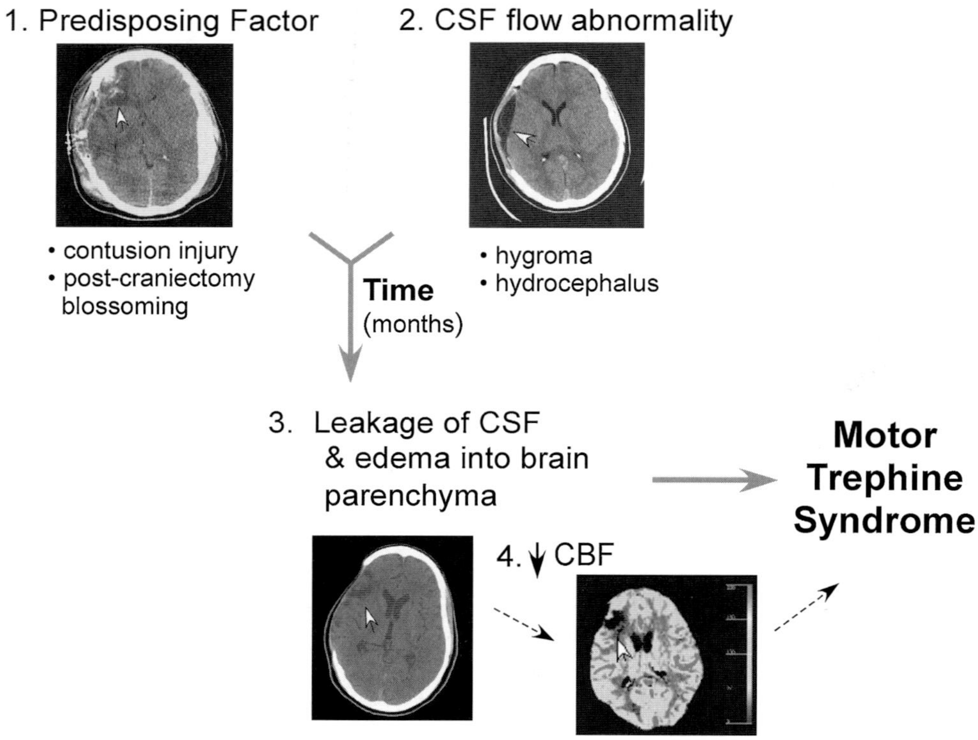

Fig. 2 Mechanistic schematic for "motor trephine syndrome". The presence of prior intraparenchymal contusions as well as CSF flow abnormalities are requisite factors - neither alone is sufficient to induce the syndrome. 1. Contusions may be caused by the injury itself or blossoming following decompression (arrow). 2. CSF hygromas (arrow) are present on early post-operative CT imaging in 90% of patients who develop "motor trephine syndrome". Over time, dysfunction of CSF flow persists and reserve mechanisms decompensate. Long intervals to cranioplasty repair increase the likelihood of developing the syndrome. 3. CSF and edema transgress into brain parenchyma underlying the skull defect, manifested by areas of hypoattenuation on CT imaging (arrow). Previous intraparenchymal contusion injury, decreased resistance to fluid accumulation in the absence of bone enclosure, and long intervals to cranioplasty facilitate transgression of CSF and edema into brain parenchyma underlying the craniectomy defect. 4. CBF flow by CT perfusion imaging is diminished in areas of hypoattenuation (arrow) corresponding to CSF and edema leakage. Transgression of CSF and edema, acting through or in concert with, diminished CBF leads to "motor trephine syndrome"

CSF flow disturbances were strikingly more common in patients who developed "motor trephine syndrome". Within the first month of decompressive hemicraniectomy, CT imaging demonstrated extra-axial hygromas in 90% of patients who later developed "motor trephine syndrome", statistically increased over a 45% incidence of hygromas in patients who recovered without a relapse in motor strength ($p=0.02$). CSF circulation abnormalities persisted. CT imaging prior to cranioplasty repair evidenced that hygromas (83%) and hydrocephalus (83%) were common in patients who developed "motor trephine syndrome". By contrast, hydrocephalus was present in 33% and hygromas in 45% of patients who recovered motor function uneventfully ($p=0.1$ and 0.3, respectively). The combination of contusions, particularly temporal contusions, during the acute hospitalization together with hydrocephalus at the time of cranioplasty repair was statistically more common in patients who developed "motor trephine syndrome" ($p=0.01$). Furthermore, the time interval between decompressive hemicraniectomy and cranioplasty repair was significantly longer in patients who developed motor trephine syndrome (median 6.5 versus 4.5 mo, $p=0.002$).

CSF circulation dysfunction in patients who developed "motor trephine syndrome" was marked by distinctive areas of hypoattenuation, compatible with efflux of CSF and edema into brain parenchyma. These hypoattenuations were differentiated from resolving contusions, ischemic lesions, or encephalomalacia based on their imaging characteristics and distribution. Large areas of CSF and edema transgression into brain parenchyma were observed in 67% of patients with motor trephine syndrome, but were absent in patients without motor relapse ($p=0.01$). CT perfusion imaging performed in two patients with "motor trephine syndrome" demonstrated diminished cerebral blood flow in and adjacent to these areas of hypoattentuation. Most interestingly, hypoattenuated areas of edema and CSF transgression resolved and CBF improved within days of cranioplasty repair. Serial CT and CT perfusion imaging studies during this timeframe demonstrated sequential improvements in CSF and edematous change within the parenchyma, as well as the cerebral blood flow, that were coincident with progressive improvement in motor function following replacement of the bone flap.

Discussion

In our neurotrauma population, a high incidence of contusion blossoming following decompression and long intervals to cranioplasty repair may have given us the opportunity to recognize "motor trephine syndrome" as a reversible complication of decompressive hemicraniectomy for TBI patients. We termed this delayed onset of contralateral monoparesis complicating decompressive hemicraniectomy as "motor trephine syndrome" in reference to Grant et al's [6] description of "Syndrome of the Trephined," first reported in 1939.

A mechanistic hypothesis for the pathophysiology underlying "motor trephine syndrome" is depicted in Fig. 2. Our results suggest that "motor trephine syndrome" is primarily a disorder of CSF flow acting in concert with a number of predisposing factors. We hypothesize that contusion injury, either at the time of impact or from blossoming with decompression, acts synergistically with impaired CSF flow to induce "motor trephine syndrome". Following decompressive hemicraniectomy, early postoperative CT imaging evidenced contusions in 80% and CSF hygromas in 90% of patients who developed "motor trephine syndrome". CSF flow continues to be impaired, and long intervals to cranioplasty repair predispose to development of "motor trephine syndrome". Over time, reserve pathways of CSF flow decompensate, and CSF and edema transgress into brain parenchyma underlying the skull defect, as manifested by areas of hypoattenuation on CT scan. This process is akin to transependymal flow and periventricular hypoattenuations that hallmark CSF impairment in late stage hydrocephalus [7] Contusion damage to brain parenchyma and large, hemispheric bony decompressions diminish resistance to egress of CSF flow and edema into brain parenchyma, when other routes of CSF diversion have fatigued. Prompt resolution of CSF and edema hypoattenuations, together with related CBF abnormalities, occurred with concurrent dramatic improvements in motor strength upon cranioplasty repair of skull defects in patients with "motor trephine syndrome".

Conclusions

Delayed, motor weakness or "motor trephine syndrome" is a reversible complication of decompressive hemicraniectomy in TBI patients. Motor strength, particularly grip and metacarpal extensor strength, should be monitored closely until the skull defect has been repaired. Patients with contusions, CSF flow impairment, and long intervals to cranioplasty repair are at higher risk of developing "motor trephine syndrome". Transparenchymal flow of CSF and edema into brain parenchyma and associated changes in CBF may be important mechanistic events in development of "motor trephine syndrome". "Motor trephine syndrome" resolves promptly with cranioplasty repair and early bone replacement may prevent development of this delayed complication of decompressive hemicraniectomy.

Conflict of interest statement We declare that we have no conflict of interest.

References

1. Czosnyka M, Copeman J, Czosnyka Z, McConnell R, Dickinson C, Pickard JD (2000) Post-traumatic hydrocephalus: influence of craniectomy on the CSF circulation. J Neurol Neurosurg Psychiatry 68:246–248
2. Dujovny M, Agner C, Aviles A (1999) Syndrome of the trephined: theory and facts. Crit Rev Neurosurg 9:271–278
3. Dujovny M, Fernandez P, Alperin N, Betz W, Misra M, Mafee M (1997) Post-cranioplasty cerebrospinal fluid hydrodynamic changes: magnetic resonance imaging quantitative analysis. Neurol Res 19:311–316
4. Finger S (1994) Origins of neuroscience. Oxford University Press,, New York, NY
5. Fodstad H, Love JA, Ekstedt J, Friden H, Liliequist B (1984) Effect of cranioplasty on cerebrospinal fluid hydrodynamics in patients with the syndrome of the trephined. Acta Neurochir (Wien) 70:21–30
6. Grant F, Norcross N (1939) Repair of cranial defects by cranioplasty. Ann Surg 110:488–512
7. James AE Jr., Strecker EP, Sperber E, Flor WJ, Merz T, Burns B (1974) An alternative pathway of cerebrospinal fluid absorption in communicating hydrocephalus. Transependymal movement. Radiology 111:143–146
8. Maas AI, Hukkelhoven CW, Marshall LF, Steyerberg EW (2005) Prediction of outcome in traumatic brain injury with computed tomographic characteristics: a comparison between the computed tomographic classification and combinations of computed tomographic predictors. Neurosurgery 57:1173–1182
9. Richaud J, Boetto S, Guell A, Lazorthes Y (1985) Effects of cranioplasty on neurological function and cerebral blood flow. Neurochirurgie 31:183–188
10. Schiffer J, Gur R, Nisim U, Pollak L (1997) Symptomatic patients after craniectomy. Surg Neurol 47:231–237
11. Segal DH, Oppenheim JS, Murovic JA (1994) Neurological recovery after cranioplasty. Neurosurgery 34:729–731
12. Stula D (1985) Intracranial pressure measurement in large skull defects. Neurochirurgia (Stuttg) 28:164–169
13. Suzuki N, Suzuki S, Iwabuchi T (1993) Neurological improvement after cranioplasty. Analysis by dynamic CT scan. Acta Neurochir (Wien) 122:49–53
14. Winkler PA, Stummer W, Linke R, Krishnan KG, Tatsch K (2000) Influence of cranioplasty on postural blood flow regulation, cerebrovascular reserve capacity, and cerebral glucose metabolism. J Neurosurg 93:53–61

Acta Neurochir Suppl (2008) 102: 279–282

Biomechanical modeling of decompressive craniectomy in traumatic brain injury

Chun Ping Gao · Beng Ti Ang

Summary

Background Decompressive craniectomy is the final phase in the graded scheme of critical care management of refractory raised intracranial pressure following severe traumatic brain injury. We aim to define the optimal size for decompressive craniectomy so that a good balance is achieved between reduction of raised ICP and the extent of trans-calvarial herniation. Provision of such quantitative data will also allow for improved data comparison in clinical trials addressing the surgical management of severe head injury.

Methods In this study, we utilize a finite element mesh model and focus on the effect of size of both unilateral and bifrontal decompressive craniectomy on intracranial pressure and brain herniation.

Findings The finite element mesh model is able to effect modeling of brain deformation and intracranial pressure changes following both unilateral fronto-parietal-temporal and bifrontal decompressive craniectomy.

Conclusions Finite element mesh modeling in the scenario of reafractory raised intracranial pressure following severe head injury may be able to guide the optimal conduct of decompressive surgery so as to effect a reduction in intracranial pressure whilst minimizing trans-calvarial brain herniation.

Keywords Decompressive craniectomy · Intracranial pressure · Finite element · Herniation

C. P. Gao · B. T. Ang (✉)
Department of Neurosurgery, National Neuroscience Institute,
11 Jalan Tan Tock Seng,
Tan Tock Seng, Singapore 308433
e-mail: bengti.ang@gmail.com

Introduction

Raised intracranial pressure (ICP) is a fundamental pathophysiologic process following traumatic brain injury. The latter is a major public health issue as it affects young persons in the prime of their lives. Herniation of brain tissue may lead to extensive secondary injury culminating in cellular damage and death which translates clinically into a poor or less desirable outcome. An attempt at improving outcome of head injury survivors will have an enormous impact on society. The lack of meaningful quantitative analyses of brain swelling and ICP elevation in this setting hinders the development of improved protocols and techniques for decompressive craniectomy. This preliminary study aims to create a finite element mesh model to study the efficacy of both unilateral and bifrontal decompressive craniectomy. Computational analysis is carried out using a generic model and with patient data. Emphasis is given to the effect of craniectomy size on the extent of ICP change and brain cortical deformation.

Materials and methods

Biomechanical modeling of the brain tissue

The brain parenchyma was modeled as a biphasic poroelastic medium with soft tissue characteristics [4]. The volume occupied by the solid component corresponds to neural and glial elements, while the voids correspond to the extracellular space. The mechanics of a poroelastic material is described by using Biot's theory. It is based on several principles including constitutive equations of solid phase, Darcy's law governing the diffusion of fluid through the pores of tissue, equilibrium conditions of a stress field and

Table 1 Material properties used for the simulation studies

Symbol	Value	Units	Symbol	Value	Units
E, white and gray	350	Pa	K_1, white*	9.2×10^{-9}	m^4 /Ns
V	0.35	no unit	K_2, white *	4.6×10^{-9}	m^4 /Ns
g_k^p	0.285	Pa	K_3, white *	2.3×10^{-9}	m^4 /Ns
τ_1	3.1	s	K_1, gray*	45.9×10^{-9}	m^4 /Ns
τ_2	27	s	K_2, gray *	22.9×10^{-9}	m^4 /Ns
τ_3	410	s	K_3, gray *	11.5×10^{-9}	m^4 /Ns

* Three different values to simulate tissue swelling due to elevated intracranial pressure

the conservation of mass of the two phases. It can be generalized into equations corresponding to a set of elasticity (Eq. 1) and diffusion (Eq. 2) with a pressure-coupling term p.

$$\mu\frac{\partial^2 u_i}{\partial x_j x_j} + (\mu+\lambda)\frac{\partial e}{\partial x_i} - \frac{\partial p}{\partial x_i} = 0 \tag{1}$$

$$K_{ij}\frac{\partial^2 p}{\partial x_j \partial x_j} - \frac{\partial e}{\partial t} = 0 \tag{2}$$

where μ is Lame's shear modulus, λ is Lame's elastic modulus; u_i is displacement of the solid phase; x_i is space coordinates; K_{ij} is hydraulic permeability; e is Cauchy's strain, p is fluid pressure; t is time.

To account for the intrinsic viscoelastic behavior of the solid phase, a time-dependant relaxation function is also incorporated into the biphasic model [1, 3]. The constitutive relation of the viscoelastic solid matrix is represented in the form of a Prony series as Eq. 3.

$$\mu(t) = \mu_0\left[1 - \sum_{k=1}^{n} g_k\left(1 - e^{-\frac{t}{\tau_k}}\right)\right] \tag{3}$$

where $\mu(t)$ is the time-dependent shear relaxation modulus that characterizes the material response, μ_0 is the instantaneous shear modulus in the undeformed state, τ_k is characteristic time, g_k are relaxation coefficients, and n is the order of the relaxation function. A varying capillary permeability of the brain tissue due to the intracranial pressure elevation was used to simulate brain swelling [2, 8]. All these aforementioned parameters were taken from available literature [1, 2, 7] and are listed in Table 1.

FEM modeling and simulation

First, a simplistic spherical model was constructed in an analogous fashion to the human brain to provide an insight into our modeling approach. It contained five radial layers representing the skull, dura mater, gray matter, white matter, and ventricles. Discretizing the computational domain into a mesh of elements, the finite element method (FEM) was used to solve the underlying partial differential equations governing its physical behaviour. The more complicated patient-specific brain FEM models were constructed based on clinical data. Computed Tomography (CT) images of patients following decompressive craniectomy were obtained to derive anatomical information. These images were processed using a mesh generation scheme developed in a previous study [3]. The three dimensional (3-D) model reconstruction and FEM mesh generated are shown in Fig. 1. Hexahedral elements were chosen to model the brain for its suitability in soft tissue modelling [5].

For simplification in this preliminary study on decompressive craniectomy, the nominal intracranial pressure used herein was assumed to be equal to the interstitial pressure at the boundary surface [6]. For the simplistic model, an initial pressure of 5000 Pa (about 38 mmHg) was created to the model surface to simulate a pre-surgical ICP, while for the patient model clinical measurements from ICP recordings were used. The part of the surface representing the craniectomy defect was set to be displacement-free, while other parts of the model surface were assumed to be fixed in all degrees of freedom (Dirichlet boundary condition) simulating the brain cortex covered by the skull

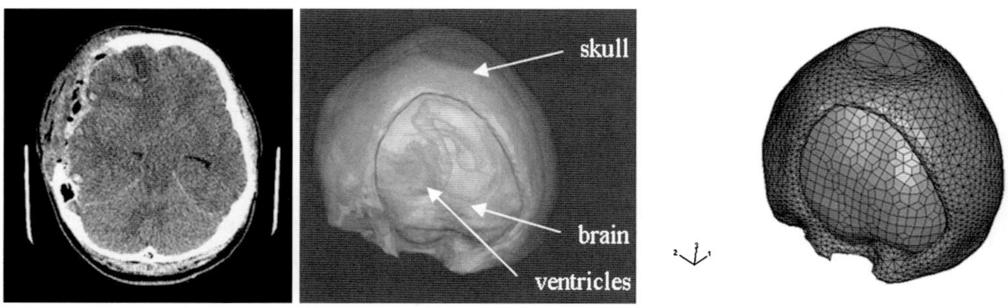

Fig. 1 (left) An axial CT section of a patient post-decompressive cranioctomy; (middle) 3D reconstruction of the patient with craniectomy; (right) the corresponding finite element mesh

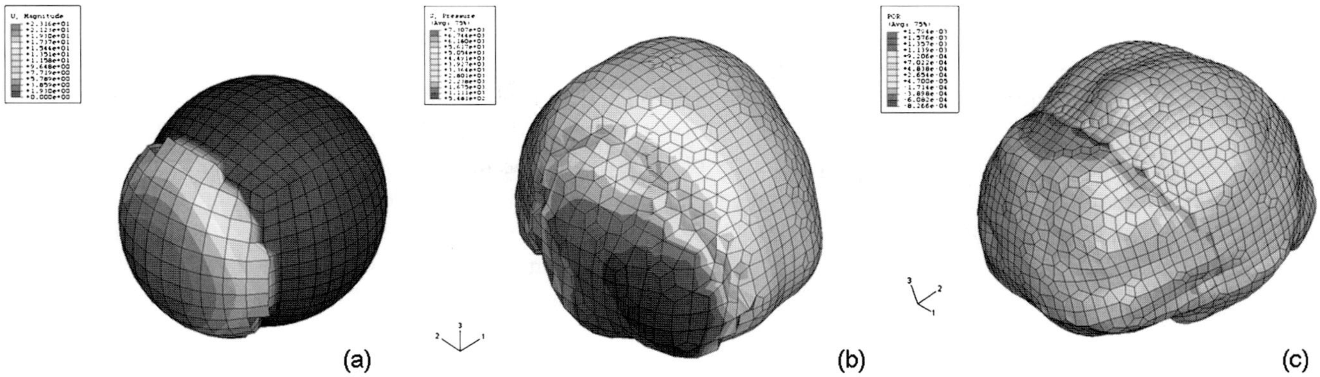

Fig. 2 Simulation results on brain models after craniectomy: (**a**) brain deformation field of simplistic model; (**b**) mean stress distribution on patient model with unilateral bone flap removal; (**c**) interstitial pressure distribution on patient model with bifrontal bone flap removal

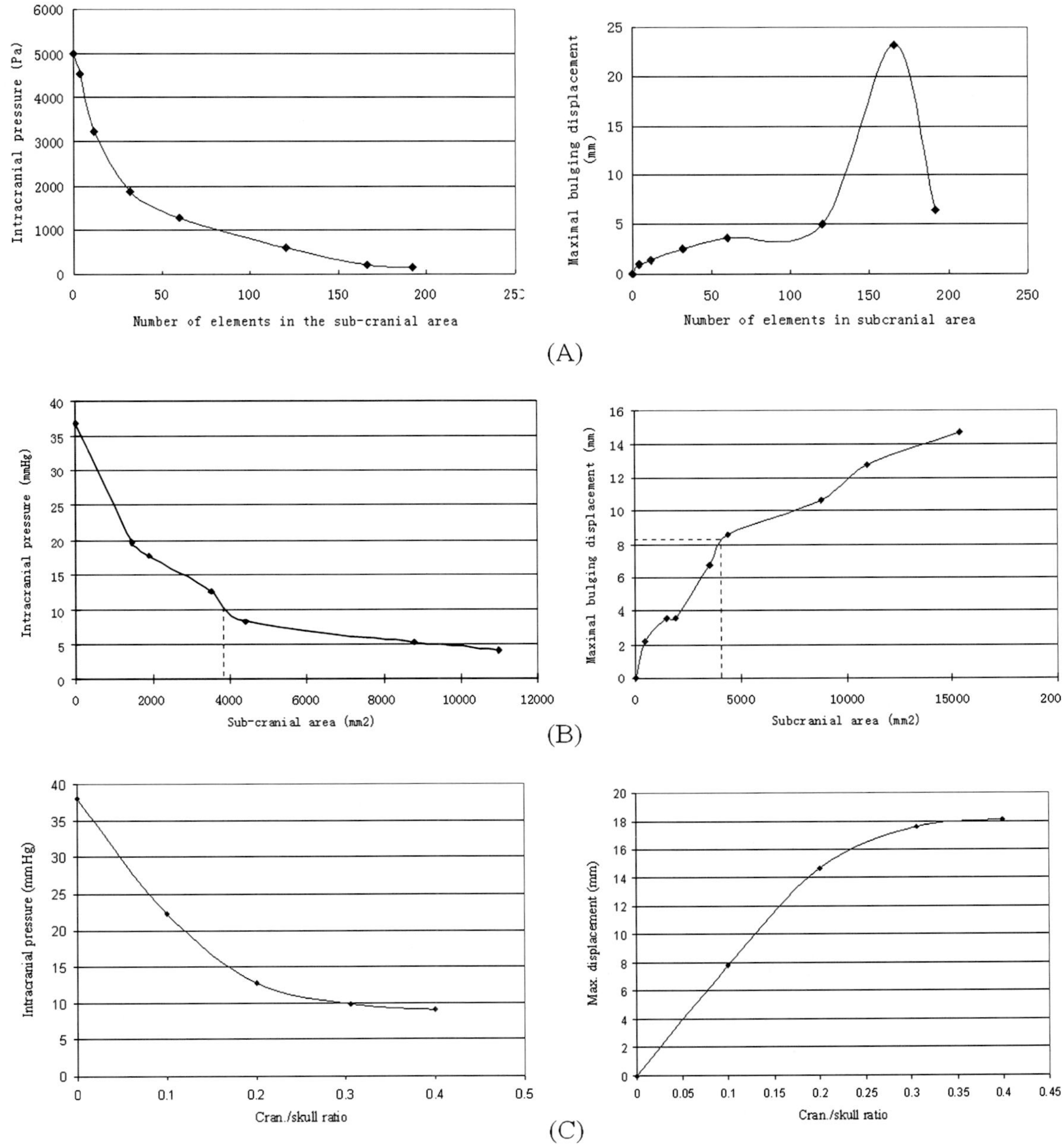

Fig. 3 (left) ICP changes with craniectomy size; (right) Maximal bulging displacement change with craniectomy size: (**a**) simplistic spherical model; (**b**) with unilateral bone flap removal; (**c**) with bifrontal bone flap removal

and dura. The *sub-craniectomy area* was parameterized into a circular-shaped moving boundary condition with a varying radius. Both unilateral bone flap and bilateral bone flap situations were taken into account. A series of simulations with various craniectomy sizes or numbers of "free elements" were performed using a commercial nonlinear FEM package ABAQUS (ABAQUS, Inc., Pawtucket, RI).

Results

Brain cortical deformation and pressure distributions following a decompressive procedure were observed in the simulation results (Fig. 2a-c). The typical bulging deformation under the bone flap was reproduced by the computational analysis. The predicted maximal displacement and nominal pressure changes of patient models were compared with those measured from medical images and pressure monitoring and found to be consistent.

Figure 3 shows the simulation results of the effect of the craniectomy size on brain deformation and intracranial pressure changes. The initial elevated ICP generally decreases while the craniectomy size increases. The ICP change rate also decreases until the pressure reaches a certain point from which it keeps relatively static. The maximal bulging displacement of the sub-craniectomy area, which can be assumed to be related to tissue herniation, becomes larger when the craniectomy size increases. However, this displacement starts to decrease drastically when it reaches a transitional point where the craniectomy size is considerable large (Fig. 3a).

For the case of unilateral decompressive craniectomy, the pre-surgical raised ICP decreased after bone flap removal, especially in the centre of the craniectomy which eventually falls to a normal ICP level of 4.1 mmHg for the maximal-sized craniectomy. The larger the craniectomy size, the more the pressure decreases. For instance, when a desired 10 mmHg pressure drop is required, the minimal craniectomy size would be about 3780 mm^2 in area, corresponding to a predicted minimal cortical bulging displacement of 8.30 mm in the centre. Similar trends of ICP change and cortical deformation were observed in the case of bifrontal bone flap removal. For bifrontal decompressive craniectomy, a crani./skull ratio is defined which denotes the ratio of the mid-sagittal width of the craniectomy (ie. from the basal edge of the craniectomy defect in the midline to the posterior/ superior edge of the craniectomy in the midline) to the distance from the nasion to the inion. This could constitute a clinically relevant index which allows one to plan the extent of craniectomy.

Discussion

The simulation results were generally consistent with obtained clinical data. From both the theoretical simplistic model and clinical case simulation, a clear relationship between ICP/cortical deformation and craniotomy size was seen. This suggests that there exists an optimal size of a craniectomy which could strike a balance between ICP lowering and the extent of tissue deformation.

The quantitative analysis shows that intracranial pressure is related to craniectomy size in a nonlinear fashion, where the rate of ICP change decreases with progressive increase in the craniectomy size.

It is worth noting that the maximum stress regions of the herniated brain tissue is found to be around the craniectomy edges, which suggests more attention should be given to this region in order to reduce the risk of tissue injury at the craniectomy edges. This preliminary model will be improved upon so as to incorporate more detailed neuroanatomy together with fluid-structure interactions.

Conflict of interest statement We declare that we have no conflict of interest.

References

1. Cheng S, Bilston LE (2007) Unconfined compression of white matter. J Biomech 40:117–124
2. Dumpuri P, Thompson RC, Dawant BM, Cao A, Miga MI (2007) An atlas-based method to compensate for brain shift: preliminary results. Med Image Anal 11:128–145
3. Gao C, Tay FEH, Nowinski WL (2004) Constructing a detailed finite element brain model for neurosurgery simulation based on a brain atlas. Acta Press, Anaheim, CA, United States Marbella, Spain, pp 937–942
4. Kyriacou SK, Mohamed A, Miller K, Neff S (2002) Brain mechanics for neurosurgery: modeling issues. Biomech Model Mechanobiol 1:151–164
5. Mendis KK, Stalnaker RL, Advani SH (1995) A constitutive relationship for large deformation finite element modeling of brain tissue. J Biomech Eng 117:279–285
6. Miga MI, Paulsen KD, Hoopes PJ, Kennedy FE, Hartov A, Roberts DW (2000) In vivo modeling of interstitial pressure in the brain under surgical load using finite elements. J Biomech Eng 122:354–363
7. Miga MI, Paulsen KD, Lemery JM, Eisner SD, Hartov A, Kennedy FE, Roberts DW (1999) Model-updated image guidance: initial clinical experiences with gravity-induced brain deformation. IEEE Trans Med Imaging 18:866–874
8. Reulen HJ, Graham R, Spatz M, Klatzo I (1977) Role of pressure gradients and bulk flow in dynamics of vasogenic brain edema. J Neurosurg 46:24–35

Acta Neurochir Suppl (2008) 102: 283–285

Coupling of sagittal sinus pressure and cerebrospinal fluid pressure in idiopathic intracranial hypertension - a preliminary report

J. D. Pickard · Z. Czosnyka · M. Czosnyka · B. Owler · J. N. Higgins

Abstract

Background Narrowing of the cranial dural venous sinuses has been implicated as contributing to elevated intracranial pressure in idiopathic intracranial hypertension [IIH]. Such narrowing may be either a fixed stenosis or secondary to raised ICP. We have investigated whether narrowing of the venous sinuses may reflect direct coupling between cerebrospinal fluid pressure and sagittal sinus pressure.

Methods Nine patients with the clinical features of IIH [8F, 1M; mean age 41 (range 22–55)] were studied as part of their standard clinical investigations by simultaneous lumbar CSF infusion study and direct retrograde cerebral venography whereby a catheter is placed within the sagittal sinus under fluoroscopic guidance.

Findings In all cases, both CSF pressure (Pcsf) and sagittal sinus pressure (Pss) were elevated with Pcsf slightly exceeding Pss (27.0 +/− 2.3 mm Hg. 25.2 +/− 7.5 mm Hg; difference P=0.026; correlation R=0.97, P=0.0032). There was a gradient of pressure along the sagittal and transverse sinuses. CSF infusion provoked rises in both Pcsf and Pss (R=0.97, P<0.0007). During drainage of CSF after the test (8 cases), Pcsf decreased to values lower than Pss (−3.26 +/− 3.9 mm Hg; P=0.0097). There was excellent correlation between slow waves of Pcsf and Pss (mean R=0.9) and between baseline pulse amplitudes of both pressures (R= 0.91; P=0.03).

Conclusions in the 9 patients studied with IIH, Pcsf and Pss were coupled both statically (mean values) and dynamically (vasogenic components). During drainage, both pressures decreased until probably central venous pressure was reached and then Pcsf decreased further while Pss remained constant. This suggests that, in many cases of IIH, there is functional obstruction of venous outflow through the dural sinuses. Raised Pcsf partly obstructs venous sinus outflow, thereby increasing Pss which, in turn, leads to a further rise in Pcsf, et sequor. This vicious cycle can be interrupted by draining CSF.

Keywords Idiopathic intracranial hypertension · Pseudotumor cerebri · Cerebrospinal fluid · Intracranial pressure

J. D. Pickard (✉) · Z. Czosnyka · M. Czosnyka · B. Owler
Academic Neurosurgery Division, University of Cambridge,
P. O. Box 167, Addenbrookes Hospital,
Cambridge CB20QQ, UK
e-mail: JDPsecretary@medschl.cam.ac.uk

J. N. Higgins
Department of Radiology,
Addenbrookes Hospital,
Cambridge CB2 0QQ, UK

B. Owler
Department of Neurosurgery, Institute of Clinical Neuroscience,
Royal Prince Alfred Hospital,
Missenden Rd,
Camperdown, NSW 2050, Australia

Introduction

There is convincing evidence that impairment of CSF absorption in the presence of patent subarachnoid spaces is a major contributor to raised CSF pressure in idiopathic intracranial hypertension (IIH; pseudotumor cerebri, benign intracranial hypertension) [6, 7, 8, 12, 13]. The block to CSF absorption may be at the level of the arachnoid granulations or within the cranial dural venous system or beyond.

The determination of CSF outflow resistance (Rcsf) relies on Davson's Eq. (2):

$$Pcsf = (If \times Rcsf) + Pss$$

where If is the CSF formation rate and Pss is the sagittal sinus pressure.

As Davson emphasized [3], it is important to confirm that Pss is independent of the pressure in the subarachnoid space. Cerebral venous thrombosis has long been known to be a cause of pseudotumor cerebri but recent studies have implicated narrowing of the cranial venous sinuses as contributing to elevated ICP in 'idiopathic' cases [9, 10]. Such narrowing may be either a fixed stenosis or secondary to raised ICP [5, 11]. Hence, Pss may not always be expected to remain constant during CSF infusion studies performed to measure Rcsf. There may also be a gradient of pressure along the sagittal and transverse sinuses so that Pss is not a single number in such cases. Endovascular stenting of the transverse sinus is a successful treatment in some patients [4].

We have investigated whether narrowing of the venous sinuses in IIH may reflect direct coupling between Pcsf and Pss.

Material and methods

Nine patients [8 female, 1 male; mean age 41 (range 22–55)] were studied who had the clinical features of IIH conforming to the modified Dandy criteria [signs and symptoms of raised ICP, no localizing neurological signs, normal neuroimaging apart from MR venography, raised CSF pressure and normal CSF constituents]. As part of their standard clinical investigations, they had simultaneous lumbar CSF infusion studies [1] and direct retrograde cerebral venography (DRCV) whereby a catheter is placed within the sagittal sinus under fluoroscopic guidance [9, 10].

Results

In all cases, both CSF pressure (Pcsf) and sagittal sinus pressure (Pss) were elevated with Pcsf slightly exceeding Pss (27.0 +/− 2.3 mm Hg. 25.2 +/− 7.5 mm Hg; difference P=0.026; correlation R=0.97, P=0.0032). There was a gradient of pressure along the sagittal and transverse sinuses. CSF infusion (1 ml/min) provoked rises in both Pcsf and Pss (R=0.97, P<0.0007) - see Fig. 1. During drainage of CSF after the test (8 cases), Pcsf decreased to values lower than Pss (−3.26 +/− 3.9 mm Hg; P=0.0097). There was excellent correlation between slow waves of Pcsf and Pss (mean R=0.9) and between baseline pulse amplitudes of both pressures (R=0.91; P=0.03).

Discussion

These results confirm that a significant component of the inability of the craniospinal axis of many patients with IIH to cope with increased CSF flow reflects a functional

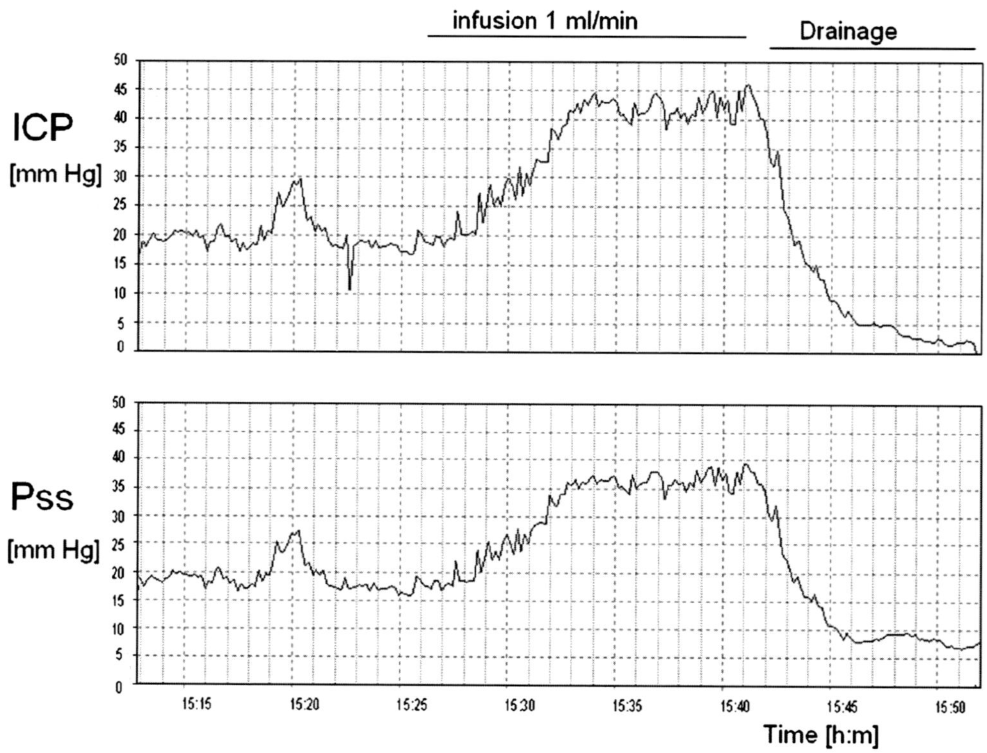

Fig. 1 Example of mean CSF pressure (ICP) and sagittal sinus pressure (Pss) recorded during infusion test and subsequent drainage of CSF from lumbar space

obstruction of venous outflow through the dural sinuses, typically in the distal transverse sinuses [5, 9, 10, 11]. Pcsf and Pss were coupled in these nine cases of IIH both statically (mean values) and dynamically (vasogenic components). During drainage, both pressures decreased until probably central venous pressure was reached whereafter Pcsf decreased further while Pss remained constant. Hence, in IIH with secondary venous sinus narrowing, the normal mechanism whereby increased Pcsf leads to increased CSF drainage and restoration of normal CSF pressure is compromised, leading to a further rise in Pcsf, et sequor [8, 13]. The prolonged benefit in some IIH patients produced by a single lumbar puncture may be understood if removal of CSF reduces secondary venous compression and improves CSF drainage for a period much longer than that required to replace the volume of CSF removed. Permanent decoupling of cerebral venous pressure by stenting an obstructive lesion may be therapeutic, regardless of the initial aetiology [4, 8, 13].

The results of CSF infusion studies in IIH are difficult to interpret without knowledge of the response of Pss. In many IIH patients, Pss is not a single number - there is a gradient along the sagittal and tranverse sinuses. The dilated subarachnoid spaces, optic nerve and root sheaths seen in IIH attest to the impairment of CSF absorption. Our results confirm earlier studies [6, 12] that proposed that there are at least two interdependent mechanisms underlying raised Pcsf in IIH: a rise in Pss and an increase in CSF outflow resistance. Davson's equation, assuming constant Pss, may not provide an adequate description of CSF dynamics under these circumstances.

Conflict of interest statement We declare that we have no conflict of interest.

References

1. Czosnyka M, Whitehouse H, Smielewski P, Simac S, Pickard JD (1996) Testing of cerebrospinal compensatory reserve in shunted and non-shunted patients: a guide to interpretation based on an observational study. J Neurol Neurosurg Psychiat 60:549–568
2. Davson H, Hollingsworth JR, Segal MD (1970) The mechanism of drainage of the cerebrospinal fluid. Brain 96:329–336
3. Davson H, Welch K, Segal MB (1987) Physiology and PAthophysiology of the Cerebrospinal Fluid. Churchill Livingstone, UK
4. Higgins JN, Owler B, Cousins C, Pickard JD (2002) Venous sinus stenting for refractory intracranial hypertension. Lancet 359:228–230
5. Higgins JN, Pickard JD (2004) Lateral sinus stenoses in idiopathic intracranial hypertension resolving after CSF diversion. Neurology 62:1907–1908
6. Janny P, Chazel J, Colnet G, Irthum B, Georget AM (1981) Benign intracranial hypertension and disorders of CSF absorption. Surg Neurol. 15:168–174
7. Johnston I (1973) The reduced CSF absorption syndrome: a reappraisal of benign intracranial hypertension and related conditions. Lancet 2:418–420
8. Johnston IH, Owler BK, Pickard JD (2007) The Pseudotumor Cerebri Syndrome. Cambridge University Press, UK
9. Karahalios DG, Rekate HL, Khayata MH et al (1996) Elevayed intracranial venous pressure as a universal mechanism in pseudotumor cerebri of varying aetiologies. Neurology 46:198–202
10. King JO, Mitchell PJ, Thomson KR et al (1995) Cerebral venography and manometry in idiopathic intracranial hypertension. Neurology 45:2224–2228
11. King JO, Mitchell PJ, Thomson KR et al (2002) Manometry combined with cervical puncture in idiopathic intracranial hypertension. Neurology 58:26–30
12. Malm J, Kristensen B, Markgren P et al CSF hydrodynamics in idiopathic intracranial hypertension: a long-term study. Neurology 42:851–858
13. Owler BK, Perker G, Halmagyi GM, Johnston IH, Besser M, Pickard JD, Higgins JN (2005) Cerebral venous outflow obstruction and pseudotumor syndrome. Adv Tech Stand Neurosurg 30:108–178

Acta Neurochir Suppl (2008) 102: 287–289

The measurement of brain tissue stiffness in-vivo

I. R. Chambers · D. Martin · A. Clark · A. Nicklin · A. D. Mendelow · P. Mitchell

Summary

Background There is considerable interest in surgical decompression as a management strategy (RescueICP) for intractable intracranial hypertension. After such an operation measurements of intracranial pressure (ICP) and thus cerebral perfusion pressure (CPP) become less meaningful. Measurements of the biomechanical properties of the brain may be one measure capable of detecting changing status of such patients. However these properties of the brain are neither documented or well understood. We have developed an indentation probe capable of making measurements of human brain stiffness.

Method The device consists of an indenting tip of depth 2 mm and diameter 12 mm surrounded by an annular body of 20 mm diameter. Measurements are made by two load cells, connected through interface electronics to a laptop computer.

Findings Laboratory measurements show the probe to provide accurate and repeatable measurements over a range of zero to 10N. Inter-operator variability from six health-care professionals had a coefficient of variance of 8.75%. Measurements obtained during surgery from a patient undergoing tumour resection were towards the lower end of the device's measurable range.

Conclusions We have determined that this indentation device has a linear response and that the inter- and intra-operator variability is low. Although the device is still in an early stage of development, preliminary results during intracranial surgery demonstrate that this device is capable of measuring *in-vivo* tissue stiffness. Further work is required to derive a quantitative "stiffness index" from the two load curves. In addition a standard operation method is required so that consistent and repeatable measurements are made. The device may be of value in assessing patients after decompressive craniectomy.

Keywords Head injury · Tissue stiffness · Surgical decompression · Measurement

D. Martin · A. Clark · A. Nicklin
Regional Medical Physics Department,
Newcastle General Hospital,
Newcastle Upon Tyne, UK

A. D. Mendelow · P. Mitchell
Department of Neurosurgery, Newcastle General Hospital,
Newcastle Upon Tyne, UK

I. R. Chambers (✉)
The James Cook University Hospital,
Marton Road,
Middlesbrough TS4 3BW, UK
e-mail: iain.chambers@stees.nhs.uk

Introduction

Intracranial hypertension remains a significant cause of mortality in head injured patients. Medical management often fails to control the vasodilatory cascade of increasing oedema that raises the intracranial pressure (ICP), reducing cerebral perfusion pressure (CPP) and blood flow. There is widespread interest in surgical decompressive craniectomy as a management strategy in which segments of the skull are removed and the brain is allowed to expand uninhibited [4, 6].

Little is known regarding the evolving physical processes after surgical decompression. Intracranial pressure and CPP become less meaningful but it may still be beneficial to be able to monitor changes to the brain. Whilst the mechanical properties of tissues such as the skin, lungs, vasculature and musculo-skeletal system have been the focus of a significant amount of research [2, 3, 5], in contrast, tissues which do not bear mechanical loads (such as kidney, liver, and brain) have not been so well characterised. Because of this, a device that could make repeated measurements of brain tissue stiffness may be able

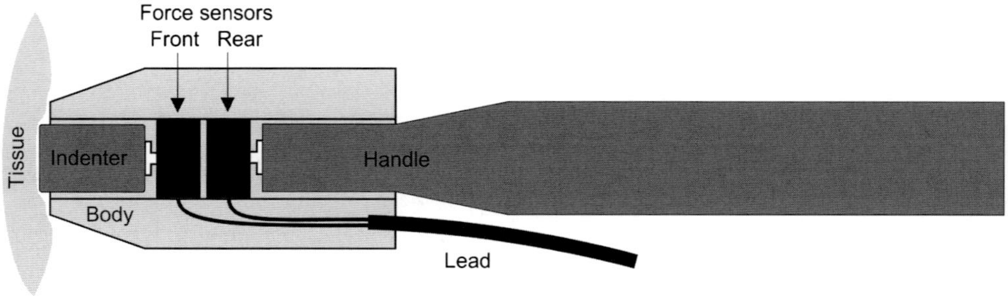

Fig. 1 Design schematic of the indentation probe

to provide relevant information on how the brain responds after surgery and thereby inform clinical management.

The primary objective of this work was to develop an instrument capable of measuring the tissue stiffness of human brain.

Materials and methods

There are two main approaches that can be used to measure tissue stiffness; elastography/vibrography, and indentation. Although vibrography reports promising results, in order to produce an instrument capable of repeated use at the bedside, an indentation design was chosen. The indentation probe is most similar to that described by Arakoski [1]. The probe consists of an indenting tip with a depth of 2 mm and diameter 12 mm surrounded by an annular ring (diameter 20 mm) connected to the probe body. The two load cells separately measure the total applied load and the indenter load and are connected to electronic amplifiers. The tissue indentation device contains two load cells (Honeywell FSG series, www.honeywell.com) capable of measuring applied force up to 1500 g. The load cells are arranged in a back-to-back fashion, with the 'front' load cell attached to the indenting tip, and the 'rear' load cell attached to the handle (see Fig. 1). Arranging the two load cells in this way removes the need to control or standardise the depth, force and rate of probe application. The relationship between the outputs from the two load cells can be used to characterise the stiffness of the material. The handle is a polyacetal H rod by which the device is held when in use. Force applied is transmitted via the handle and body to the rear load cell. The indenting tip is a polyacetal H insert 2 mm in depth and 10 mm in diameter. It transmits the force applied to its face to the front load cell (Fig. 1).

The analogue signals from the load cells are converted to digital values by a modified analogue-digital convertor (Pico ADC-11, www.picotech.com). The modification was required to provide access to the internal 5 volt USB power rail and 2.5 volt ADC reference signal. The advantage of this modification was that it was a cheap, ready-made and stand alone means of acquiring power for the load cells within the probe.

The amplified digital output was displayed on a desktop computer screen (PicoScope and PicoLog Recorder, www.picotech.com) during bench work. For clinical data collection, custom software was developed using Borland Delphi 7 which runs under Windows 2000/XP. The software controls the timing and acquisition of the two force signals from the analogue interface and allows the display and recording of these values. The probe conforms to IEC 601.01, the relevant safety standard for medical equipment.

The system has been assessed with laboratory studies and a pilot clinical study. The compact handheld design, attached to a laptop trolley, makes it suitable for use in the operating room or at the bedside. Approval from the local Research Ethics Committee was obtained prior to commencing the study.

The probe was initially calibrated, by recording the voltage output that corresponded with a given force applied using weights of varying magnitude (2–100 g). The reading taken with a constant weight was repeated 60 times. The mean and standard deviation from 60 separate measurements were calculated as an assessment of the normality of the distributions.

To assess inter- and intra- operator variability, foam samples of differing densities (40 kg/m^3, 65 kg/m^3, 70 kg/m^3,

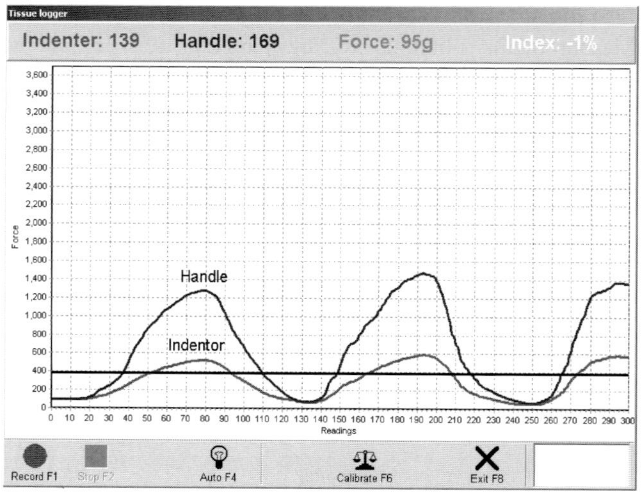

Fig. 2 Software programme displaying measurements from the surface of human brain

90 kg/m^3) were assessed. Measurements were taken by 6 operators on the same day. The force applied was such that the voltage output from the front load cell was constant. Similarly, the same piece of foam was measured in the same fashion by one operator, on 6 separate occasions, taking 6 readings each time. In addition the rate at which the probe was applied was varied. In-vivo measurements were made by a neurosurgeon during a craniotomy for tumour resection. The probe was placed, by hand, on the cortical surface tissue prior to removal of the tumour and the applied force varied so that the tissue was lightly deformed.

Results

Laboratory testing showed good probe linearity over a range from 0 to 10N ($r>0.999$). Intra-operator variability was assessed from 6 readings taken on successive days and inter-operator variability by readings taken by 6 operators. The coefficients of variance were 7.57% and 8.75% respectively. The device was applied to sections of foam of differing density (40, 65 and 70 kg/m^3); the values obtained showed clear differences between each of the foam types. Three clinical measurements have been taken and indicate that values of brain tissue stiffness appear to be towards the low end of the device's measurable range (Fig. 2).

Conclusions

Results from laboratory studies have determined that this indentation device has a linear response and that the inter- and intra- operator variability is low. To our knowledge this is the first time that brain tissue stiffness has been measured in-vivo. Although the device is still in an early stage of development, the preliminary results from intra-operative readings demonstrate that this device is capable of measuring *in-vivo* tissue stiffness. The values obtained are at the lower end of the measurement range of the device and further work is required to derive a quantitative "stiffness index" from the two load curves. In addition a standard method of use is required so that consistent and repeatable measurements are made. These are promising results and the device may be of value in assessing patients after decompressive craniectomy by providing relevant information on how the brain responds after surgery.

Conflict of interest statement We declare that we have no conflict of interest.

References

1. Arokoski JPA, Surakka J, Ojala T, Kolari P, Jurvelin JS (2005) Feasibility of the use of a novel soft tissue stiffness meter. Physiol Meas 26:215–228
2. Fung YC (1981) Biomechanics. mechanical properties of living tissues. Springer, New York
3. Gao P, Yuan Z (1999) Development of a micromechanical probe-measuring instrument for surface properties characterization. Meas Sci Technol 10:N105–108
4. RescueICP trial. www.rescueICP.com
5. Schmid-Schonbein GW, Woo SL-Y, Zweifach BW (1986) Frontiers in Biomechanics. Springer, New York
6. Taylor A, Butt W, Rosenfeld J, Shann F, Ditchfield M, Lewis E, Klug G, Wallace D, Henning R, Tibballs J (2001) A randomized trial of very early decompressive craniectomy in children with traumatic brain injury and sustained intracranial hypertension. Childs Nervous System 17(3):154–62

PART 7:
Stroke, subarachnoid hemorrhage, and intracerebral hematoma

Acta Neurochir Suppl (2008) 102: 293–297

Changes in brain biochemistry and oxygenation in the zone surrounding primary intracerebral hemorrhage

Ernest Wang · Chi Long Ho · Kah Keow Lee · Ivan Ng · Beng Ti Ang

Abstract
Background While the management of primary intracerebral hemorrhage (ICH) remains controversial, there remains a subset of patients that undergo clot evacuation. This study aims to characterize brain physiology and biochemistry after surgery for this condition.
Methods Thirty-six consecutive patients requiring ventilation for primary ICH had intracranial pressure (ICP), tissue oxygenation (PbO_2) and cerebral microdialysis (CMD) monitoring. 28 patients with a Glasgow Outcome Score (GOS) of 1–3 formed group 1 while 5 patients with a GOS of 4–5 formed group 2. The control group consisted of 3 patients managed conservatively without surgery.
Findings The mean PbO_2 (24.5±20.8 mmHg) was higher in the patients in group 1 (poor outcome) compared with those in the control group (13.6±9.0 mmHg) ($p<0.001$). Compared to patients in group 2, the patients in group 1 also had a higher PbO_2 ($p=0.02$) together with worse levels of lactate/pyruvate (L/P) ratio and glycerol ($p<0.001$). In all 3 groups, ICP reduction to < 20 mmHg was achieved together with a return to of pressure reactivity (PRx) to <0.3.
Conclusions In spontaneous ICH, derangements in the peri-lesional tissue demonstrated by local techniques of PbO_2 monitoring and CMD are not seen in global indices such as the PRx.

Keywords Intracerebral hemorrhage · Cerebral microdialysis · Brain tissue oxygenation · Ischemic penumbra

E. Wang · C. L. Ho · K. K. Lee · I. Ng · B. T. Ang (✉)
Department of Neurosurgery, National Neuroscience Institute,
11, Jalan Tan Tock Seng,
Singapore, Singapore 308433
e-mail: bengti.ang@gmail.com

Introduction

While primary intracerebral hemorrhage accounts for 15–20% of all strokes [7], its consequences are most devastating, contributing to greater than 50% of deaths that occur [19]. No effective or definitive treatment exists and there remains a lack of data demonstrating the efficacy of surgery [4, 8]. Yet surgical clot evacuation is still practiced as there could possibly be a subgroup of patients that would benefit. Understanding the pathophysiological consequences [11] of an intracerebral hematoma is therefore important in our attempt to limit secondary damage. Focal injury follows mechanical pressure from the hematoma, the theoretical possibility of ischemic injury in the peri-hematoma zone and the direct neuropathic effects of blood itself. Simultaneous global injury may also occur as a result of raised ICP, cerebral dysautoregulation, hemodynamic alterations and metabolic changes. Consequently, monitoring physiologic parameters [12] such as mean arterial blood pressure (MAP), ICP, PbO_2 and brain temperature together with biochemical parameters through cerebral microdialysis (CMD) may provide important new insights. Both global and local indices that reflect autoregulatory and metabolic alterations may then be discerned. We utilize multi-modality monitoring techniques in this study to characterize brain physiology and neurochemistry in spontaneous supratentorial ICH after surgical clot evacuation.

Methods

Patient selection

From January 2003 to January 2006, 36 consecutive patients with spontaneous ICH were included into this prospective observational study. Institutional ethics committee approval had been obtained prior to commencement of the study. All patients recruited had an initial computerized tomographic (CT) scan demonstrating a supratentorial ICH with its epicentre over the putamen, a clinical deterioration to a Glasgow Coma Scale (GCS) score of 8 or less together with radiological evidence of mass effect (midline shift of > 5 mm or effacement of the basal cisterns). Patients with a devastating brain injury (not expected to survive beyond 24 hours), fixed and dilated pupils or who had bleeding diathesis were excluded from the study.

Treatment protocol

Early surgical evacuation of hematoma via a standard frontal parietal-temporal craniotomy was offered to all patients. The hematoma was removed under magnification and the bone flap deliberately left out in anticipation of post-operative brain swelling. Intraparenchymal neuromonitoring probes were then placed within 1 cm of the hematoma cavity with their positions verified with a post-operative CT scan. ICP was continuously monitored using a fiberoptic intraparenchymal device (Codman and Shurtleff, Raynham, MA, USA) with brain temperature and brain tissue oxygenation measured with LICOX polarographic Clark-type microcatheters (Integra Neuroscience, Plainsboro, NJ, USA). Microdialysis catheters (CMA 70; microdialysis, Solna, Sweden) were also inserted via the same bolt system.

Clinical management was in accordance to established neurocritical care guidelines. Inspired oxygen fraction (FiO2) was adjusted to achieve an arterial oxygenation saturation of > 95% with haemoglobin levels maintained above 10 g/dl. ICP control measures included nursing in the 30 degrees head up position, sedation with propofol (2–10 mg/kg/hour), boluses of 20% mannitol (2 ml/kg up to a plasma osmolarity of 320 mosmol/L) according to ICP spikes and mild hypocapnia by regulating pCO_2 at 30–35 mmHg.

Monitoring

PbO_2 monitoring was commenced 3 hours after insertion of the LICOX probe, thus giving values from the latter adequate time to stabilise. Cerebral microdialysates were collected every hour and analysed for glucose, lactate, pyruvate and glutamate. The pressure reactivity index, PRx, was calculated as a moving correlation coefficient between the last 30 consecutive samples of values for ICP and arterial blood pressure (ABP) averaged for a period of 10 seconds.

Outcome groups

The patients were divided into 2 outcome groups. Twenty-eight patients with a poor outcome, defined as a Glasgow Outcome Score (GOS) of 1–3, were placed in group 1. Group 2 consisted of 5 patients with a good outcome (GOS of 4–5). Patients whose relatives did not consent to surgical evacuation had insertion of intraparenchymal probes alone for neuromonitoring. They were treated with medical therapy alone and formed the control group consisting of 3 patients.

Statistical analysis

The demographic and clinical data collected that were continuous variables are reported as mean ± standard deviation or as median with interquartile range (IQR) when non-normally distributed. The two sample paired t-test was performed if the normality and equality of variances assumption was satisfied; otherwise the Mann Whitney U test was used. The mean values of MAP, ICP, PRx, PbO_2 and dialysates of CMD were studied over a 72-hour period. Analysis of the differences in the means of the variables in the 3 groups was carried out using ANOVA and subsequent post-hoc comparisons using Bonferroni's t-test. A probability value (p value) of less than 0.05 was considered the criterion of significance for all comparisons. Statistical analyses were performed using the commercially available statistical program SPSS version 12.0 (SPSS Inc., Chicago, Illinois, USA).

Results

Patient characteristics

The ages of the 36 patients evaluated ranged from 31 to 81 years with a mean age of 54.8±12.3 years. There were 24 male (67%) and 12 female (33%) patients. Median GCS scores on admission was 6 [interquartile range (IQR) 3 to 13]. The average volume of hematoma was 75.4±28.3 milliliters (IQR 26 to 160) on admission. Data from the first 72 hours of monitoring were analyzed for the purposes of this study. There were no significant differences between the 3 groups with respect to age, ethnicity, gender, volume of hematoma, hemoglobin levels and GCS score on admission.

Outcome

Six months after the initial hemorrhage, 10 patients (28%) had died with 26 (72%) patients surviving. Amongst those that died, 9 (25%) were from group 1 and 1 (3%) was from the control group. In those that survived, 8 (7 from group 1 and 1 from the control group) remained vegetative at 6 months while 12 (all from group 1) were severely disabled. Four patients had a moderate disability (3 from group 2 and 1 from the control group) and only 2 patients had a good recovery (both from group 2).

Brain physiology and biochemistry

The mean PbO_2 (24.5±20.8 mmHg) was higher in the patients with a poor outcome (group 1) compared with those in the control group (13.6±9.0 mmHg) ($p<0.001$). When compared to the PbO_2 in patients belonging to group 2 with a good outcome (17.7±5.6 mmHg), the PbO_2 in patients in group 1 was also significantly higher ($p=0.02$). Patients in group 1 also had the highest number of epochs of ischemia ($PbO_2<10$ mmHg). In all 3 groups, ICP reduction to < 20 mmHg and maintenance of cerebral perfusion pressure (CPP) of > 70 mmHg was achieved together with a return of the PRx to <0.3. Patients in group 1 with a poor outcome also had significantly higher lactate, lactate/pyruvate (L/P) ratio and glycerol when compared to patients in group 2 with a good outcome ($p<0.001$) (Table 1). In addition, group 1 patients had higher glucose and glutamate levels ($p=0.04$) (Table 1). Except for glycerol levels, there was no difference in microdialysate levels of glucose, glutamate, lactate and L/P between group 1 and the control group. In patients with a good outcome, there was an improvement in PbO_2 with time. In patients with a good outcome, the mean hourly PbO_2 values were in the ischemic range of < 15 mmHg after surgery but this improved to normal levels of > 20 mm Hg 24 hours after surgery (Fig. 1). This was mirrored by a gradual worsening of PbO_2 with time in the control group (Fig. 1). However, the mean hourly PbO_2 in the poor outcome group (group 1) was consistently highest of all 3 groups (Fig. 1). Lactate and the L/P ratio remained higher in the poor outcome patients compared to those with a good outcome and this similar trend in time was also seen in the glycerol levels. The temporal profile of PRx demonstrated a return of autoregulation even in the poor outcome group but there was a surge of dysautoregulation in the good outcome group after 48 hours (Fig. 2).

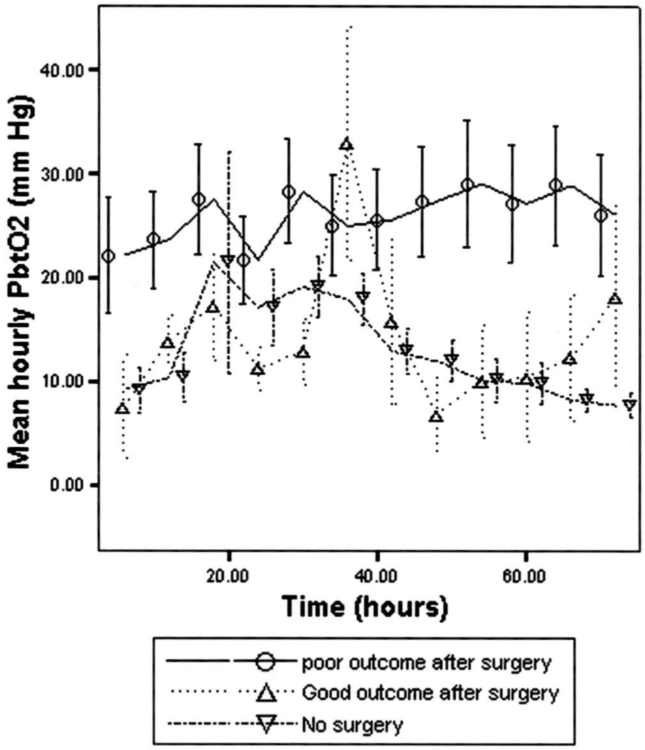

Fig. 1 Brain tissue oxygenation in all 3 groups. The temporal profile of mean hourly PbO2 in patients with a good outcome demonstrated an improvement with time, peaking 40 hours after surgery. The converse was true for in the control group. Patients with a poor outcome had seemingly 'normal' hourly PbO2 values that did not change with time

Discussion

Brain injury from spontaneous primary ICH results from the rupture of small arterioles or penetrating arteries. They typically occur in certain locations like the putamen, thalamus, pons and cerebellum and understandably have far reaching consequences. Many studies, including meta-analyses, have failed to show the benefit of surgical clot evacuation for spontaneous ICH [9, 18]. The recently published multi-center STICH (Surgical Trial in Intracerebral Hemorrhage) trial [11] showed no overall benefit from early surgery when compared with initial conservative treatment. Yet, when acute neurological deterioration from mass effect occurs, surgical evacuation is often considered. Small studies of less invasive forms of surgery like endoscopy and stereotactic aspiration [3, 5] have suggested that these may hold promise in improving outcome and reducing morbidity. It therefore follows that a clearer understanding of the underlying pathophysiological processes that occur in spontaneous ICH may help to define future optimal therapy. In limiting this study to putaminal hemorrhages, we identify neuromonitoring characteristics of these patients who achieved a favorable outcome after surgery.

Table 1 Brain biochemistry in both outcome groups

	Gp 1	Gp 2	P - value
Lactate (mmol/L)	4.0	1.6	<0.001
L/P (mmol/L)	78.0	28.1	<0.001
Glycerol (μmol/L)	78.0	15.3	<0.001
Glutamate (μmol/L)	11.3	5.1	0.04
Glucose (mmol/L)	1.1	2.2	0.04

A large intraparenchymal clot causes a rise in ICP with displacement of surrounding structures from cerebral edema and mass effect. Not only does the hematoma cause local injury to the surrounding tissue through multiple metabolic

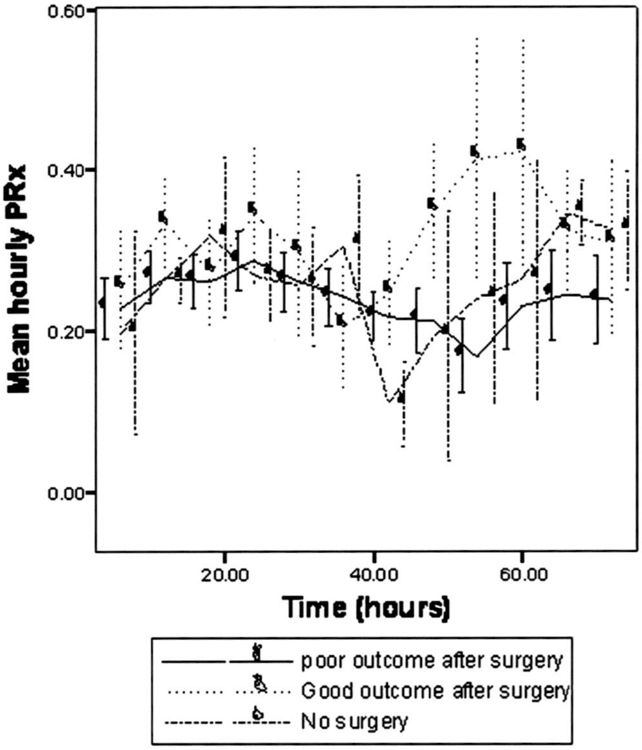

Fig. 2 Temporal profile of PRx in all 3 groups. There was a return in mean hourly PRx values to < +0.3 after 40 hours in all 3 groups. However, patients with a good outcome (group 2) experienced a large rise in PRx after 40 hours corresponding to a rise in PbO2

and biochemical processes [1, 10], the rest of the brain parenchyma may be susceptible to systemic insults and secondary injury. ICP monitoring, CPP maintenance and other standard neurointensive care measures in modern neurointensive care units provide continuous monitoring [12] of global insults to the brain. The PRx [6] is an index derived from such global monitoring techniques which capitalizes on the relationship between MAP and ICP. It quantifies cerebrovascular reactivity and has been strongly correlated with outcome in head trauma. Following surgical decompression in acute brain injury, the PRx [22] has been demonstrated to return to normal values with time. In this study, there was a similar normalization of ICP and PRx in all 3 groups where maintenance of CPP to > 70 mmHg was also achieved. Despite achieving these conventional targets, 28 (78%) patients had a poor outcome. This illustrates the challenge and limitations of utilizing global indices alone in targeted therapy.

The value of other modalities of monitoring to guide treatment therefore becomes of interest. The utility of CMD in measuring brain biochemistry [17, 21] has been demonstrated mainly in traumatic brain injury and sub-arachnoid hemorrhage. The dialysates measured reflect energy failure and membrane degradation in the immediate area around the probe [10]. Post-operative CT scans in our study confirm the localization of microdialysis probes inserted within 1 cm of the peri-hematoma zone in all cases. This could be achieved as the probes were inserted under direct vision during open surgery. The metabolic profile of the peri-lesional tissue in spontaneous ICH could lead to answers to questions regarding its viability and vulnerability to further insults. The Swedish group have recently shown that the area close to an evacuated ICH has a biochemical profile [16] similar to that in the penumbra surrounding focal traumatic brain contusions and this may have clinical implications influencing surgical decision making. Our findings complement this but with an added control group of patients with a similar severity of injury that were managed only with ICP/PbO_2/CMD monitoring alone. Ironically, PbO_2 was highest in the poor outcome group. It reached a mean value of 24.5±20.8 mmHg over the first 72 hours, suggesting that the peri-lesional tissue is no longer hypoxic. However, this group had the highest number of epochs of abnormal PbO_2 over the monitoring period. Significantly, compared to the group with a good outcome, this poor outcome group of patients had a worse biochemical profile indicating a greater severity of insult in the area surrounding the hematoma. The elevated L/P ratio, lactate, glutamate and glycerol levels, together with the low extracellular glucose levels, were no different from that in the control group of patients. It is noteworthy that the control group had the highest lactate values reflective of worst local injury as the other two groups were treated with surgical evacuation and decompression.

Cerebral hypoxia as an important predictor of poor outcome [2, 14, 20] is somewhat contradicted by the highest PbO_2 levels in these patients. The value of following temporal changes and examining trends [2, 22] become appreciated. After 72 hours, PbO_2 in patients with a poor outcome showed no improvement with time but the converse was true in those that did well after 6 months. These results are echoed in a separate local study on the changes in brain biochemistry following acute head injury where survivors with a good outcome demonstrated a change in PbO_2 from ischemic to normal values. The seemingly ‘normal’ PbO_2 in patients with a poor outcome could possibly be due to a ‘diffusion’ impairment, cellular dysfunction leading to an inability to utilize oxygen or a hyperaemic flow itself secondary to the acute brain injury. However, cerebral blood flow (CBF) studies [13] and other metabolic measurements such as that in PET scans would be necessary to validate this. The changes in cerebral hemodynamics and brain oxygenation that occur after surgery for spontaneous putaminal hemorrhages have also been described [15] where a fall in ICP following surgery was accompanied with improvements in PbO_2, CPP and CBF.

Despite a return to an autoregulatory state, the trends in PRx in all 3 groups of this study were unremarkable.

Interestingly, a sudden increase in PRx in patients with a good outcome after 40 hours was preceded by a surge in PbO_2. This may represent a period of dysautoregulation in this group of patients with a good outcome which reverted after 72 hours.

To the best of our knowledge, this is the only study examining the viability of peri-lesional tissue in spontaneous ICH where a control group is employed. The differences in brain physiology and biochemistry in the immediate tissue adjacent to the clot cavity is demonstrated in different clinical outcome groups. This suggests that a zone of vulnerable tissue does exist in spontaneous ICH. While global indices including ICP monitoring likely remain important, neuromonitoring of local tissue changes can give additional relevant information. The provision of such local indices of monitoring gives the clinician the ability to make clinical decisions that might potentially improve outcome.

Conclusion

These results suggest that monitoring of PbO_2 and CMD dialysates in spontaneous supratentorial ICH can demonstrate local derangements in the peri-hematoma cavity that are not reflected in global indices such as the PRx. Multimodality monitoring of both local and global indices simultaneously may become important tools in the management of spontaneous ICH.

Conflict of interest statement We declare that we have no conflict of interest.

References

1. Ackerman RH, Alpert NM, Correia JA et al (1979) Correlations of PET scans with TCT scans and clinical course. Acta Neurol Scand 60(suppl 72):230–231
2. Ang BT, Wong J, Lee KK, Wang E, Ng I (2007) Temporal changes in cerebral tissue oxygenation with cerebrovascular pressure reactivity in severe traumatic brain injury. J Neurol Neurosurg Psychiatry 78(3):298–302
3. Barrett RJ, Hussain R, Coplin WM, Berry S, Keyl PM, Hanley DF, Johnson RR, Carhuapoma JR (2005) Frameless stereotactic aspiration and thrombolysis of spontaneous intracerebral hemorrhage. Neurocrit Care 3(3):237–245
4. Batjer HH, Reisch JS, Allen BC, Plaizier LJ, Su CJ (1990) Failure of surgery to improve outcome in hypertensive putaminal hemorrhage. A prospective randomized trial. Arch Neurol 47 (10):1103–1106
5. Cho DY, Chen CC, Chang CS, Lee WY, Tso M (2006) Endoscopic surgery for spontaneous basal ganglia hemorrhage: comparing endoscopic surgery, stereotactic aspiration, and craniotomy in noncomatose patients. Surg Neurol 65(6):547–555
6. Czosnyka M, Smielewski P, Kirkpatrick P, Laing RJ, Menon D, Pickard JD (1997) Continuous assessment of the cerebral vasomotor reactivity in head injury. Neurosurgery 41:11–19
7. Daverat P, Castel JP, Dartigues JF, Orgogozo JM (1991) Death and functional outcome after spontaneous intracerebral hemorrhage: A prospective study of 166 cases using multivariate analysis. Stroke 22:1–6
8. Fernandes HM, Mendelow AD (1999) Spontaneous intracerebral hemorrhage: a surgical dilemma. Br J Neurosurg 13(4):389–394
9. Hankey GJ (1997) Surgery for primary intracerebral hemorrhage: Is it safe and effective? A systematic review of case series and randomized trials. Stroke 28:2126–2132
10. Hutchinson PJ, O'Connell MT, Al-Rawi PG, Maskell LB, Kett-White R, Gupta AK, Richards HK et al (2000) Clinical cerebral microdialysis: a methodological study. J Neurosurg 93:37–43
11. Mendelow AD, Gregson BA, Fernandes HM, Murray GD, Teasdale GM, Hope DT, Karimi A, Shaw MD, Barer DH (2005) Early surgery versus initial conservative treatment in patients with spontaneous supratentorial intracerebral haematomas in the International Surgical Trial in Intracerebral Hemorrhage (STICH): a randomised trial. Lancet 365(9457):387–397
12. Mulvey JM, Dorsch NW, Mudaliar Y, Lang EW (2004) Multimodality monitoring in severe traumatic brain injury: the role of brain tissue oxygenation monitoring. Neurocrit Care 1(3):391–402
13. Murphy BD, Fox AJ, Lee DH, Sahlas DJ, Black SE, Hogan MJ, Coutts SB et al (2006) Identification of penumbra and infarct in acute ischemic stroke using computed tomography perfusion-derived blood flow and blood volume measurements. Stroke 37 (7):1771–1777
14. Ng I, Lee KK, Wong J (2005) Brain tissue oxygenation monitoring in acute brain injury. Acta Neurochir Suppl 95:447–51 Review
15. Ng I, Yap E, Lim J (2005) Changes in cerebral hemodynamics and cerebral oxygenation during surgical evacuation for hypertensive intracerebral putaminal hemorrhage. Acta Neurochir Suppl 95:97–101
16. Nilsson OG, Polito A, Saveland H, Ungerstedt U, Nordstrom CH (2006) Are primary supratentorial intracerebral hemorrhages surrounded by a biochemical penumbra? A microdialysis study. Neurosurgery 59(3):521–528
17. Persson L, Hillered L (1992) Chemical monitoring for neurosurgical intensive care patients using intracerebral microdialysis. J Neurosurge 76:72–80
18. Prasad K, Shrivastava A (2004) Surgery for primary supratentorial intracerebral hemorrhage (Cochrane Review), in The Cochrane Library, Chichester, England, John Wiley & Sons
19. Sacco RL, Mayer SA (1994) Epidemiology of intracerebral hemorrhage. In: Feldmann E (ed) Intracerebral hemorrhage. Futura Publishing Company, New York
20. Sarrafzadeh AS, Kiening KL, Callsen TA, Unterberg AW (2003) Metabolic changes during impending and manifest cerebral hypoxia in traumatic brain injury. Br J Neurosurg 17:340–346
21. Sarrafzadeh AS, Sakowitz OW, Callsen TA, Lanksch WR, Unterberg AW (2000) Bedside microdialysis for early detection of cerebral hypoxia in traumatic brain injury. Neurosurg Focus 9:1–6
22. Wang EC, Ang BT, Wong J, Lim J, Ng I (2006) Characterization of cerebrovascular reactivity after craniectomy for acute brain injury. Br J Neurosurg 20:24–30

Acta Neurochir Suppl (2008) 102: 299–303

Prediction of early mortality in primary intracerebral hemorrhage in an asian population

Beng Ti Ang · Siew Pang Chan · Kah Keow Lee · Ivan Ng

Abstract
Background Primary intracerebral hemorrhage accounts for the relative minority of all strokes and yet is more fatal and disabling. Various prognostic models for mortality and functional outcome following primary intracerebral hemorrhage have been proposed, however there is little data which focuses on a multi-racial population profile characteristic of communities in South-East Asia. A reliable grading scale for this condition will allow for accurate risk stratification, treatment selection, resource allocation and possibly also aid in the definition of common enrollment criteria for clinical trials.
Methods This study investigates an Asian population of primary intracerebral hemorrhage patients and defines using a variety of data mining techniques the clinical variables that significantly impact on early mortality. The models produced are then compared to ascertain which one optimally predicts this outcome.
Findings Past history of stroke, known atrial fibrillation, use of warfarin, glucose level, presenting Glasgow Coma Scale (GCS) and pupil abnormality, post-resuscitation GCS and pupil abnormality, initial international normalized ratio (INR) and prothrombin time (PT) results, vomiting, seizure, total volume of clot, ventricular extension and hydrocephalus were significantly associated with early mortality. Logit with backward elimination showed that only age, presenting GCS, 1st INR result and total volume of clot were significantly associated with mortality in the final multivariate model. The use of the other data mining techniques yielded comparable results.
Conclusions The predictors for early mortality and poor outcome in primary intracerebral hemorrhage are similar in Asian and Western populations.

Keywords Intracerebral hemorrhage · Outcome · Mortality · Prediction

B. T. Ang (✉) · K. K. Lee · I. Ng
Department of Neurosurgery, National Neuroscience Institute, Singapore,
11, Jalan Tan Tock Seng,
Singapore 308433, Singapore
e-mail: bengti.ang@gmail.com

S. P. Chan
Department of Clinical Epidemiology, Tan Tock Seng Hospital, Singapore, Singapore

Introduction

The ability to precisely predict the outcome following primary intracerebral hemorrhage (ICH) would facilitate clinical management. One would then be able to offer accurate prognosis and perhaps even begin to stratify the ICH patient population into subgroups that may benefit from different therapeutic regimes. Although various scoring scales in ICH have been developed, there is scant information in the field with regards to a multi-racial society which constitutes the community in many parts of South-East Asia. This study thus aims to develop an ICH grading scale specific to our local population; in addition, we evaluate a variety of data mining techniques to investigate their relative prediction efficiencies.

Materials and methods

Study population

We studied 634 patients that were consecutively admitted to our institution for primary ICH from July 2004 to December 2006. Demographic characteristics, admission clinical data and imaging data were collated. The various parameters/clinical variables considered for analysis are listed in Table 1. Approval of this study was obtained from the ethics

Table 1 Study sample variables

	30-Day Mortality Died (n=179)	Alive (n=455)
Age (years) *		
Mean	67.7	63.5
Range	17–95	18–109
Gender		
Male	100 (27.0%)	271 (73.0%)
Female	79 (30.0%)	184 (60.0%)
History of stroke *		
Yes	59 (35.5%)	107 (64.5%)
No	119 (25.5%)	348 (74.5%)
Known hypertension		
Yes	125 (27.2%)	334 (72.8%)
No	54 (31.0%)	120 (69.0%)
Known atrial fibrillation *		
Yes	19 (42.2%)	26 (57.8%)
No	160 (27.2%)	429 (72.8%)
Warfarin use *		
Yes	26 (57.8%)	19 (42.2%)
No	153 (26.0%)	436 (74.0%)
Antiplatelet use		
Yes	31 (30.4%)	71 (69.6%)
No	136 (26.4%)	379 (73.6%)
Glucose (mmol/L) *		
Mean	9.0	8.5
Range	4.9–19.7	3.7–23.4
Presenting GCS *		
≤8	103 (70.6%)	43 (29.4%)
>8	73 (15.2%)	407 (84.8%)
Post resuscitation GCS *		
≤8	99 (68.3%)	46 (31.7%)
>8	34 (9.5%)	322 (90.5%)
Presenting pupil abnormality *		
Yes	37 (88.1%)	5 (11.9%)
No	97 (20.1%)	386 (79.9%)
Post resuscitation pupil abnormality *		
Yes	42 (91.3%)	4 (8.7%)
No	82 (20.5%)	319 (79.5%)
1st PT result (sec) *	17.0	14.0
Mean		
Range	11.7–59.4	10.2–67.6
1st PTT result (sec)		
Mean	32.3	30.1
Range	20.9–115.0	3.0–101.5
1st INR result *		
Mean	1.4	1.1
Range	0.9–4.9	0.6–13.0
Vomiting *		
Yes	47 (36.7%)	81 (63.3%)
No	132 (26.1%)	373 (73.9%)
Seizure *		
Yes	5 (27.8%)	13 (72.2%)
No	127 (28.1%)	325 (71.9%)
Total volume of clot /ml *		
Mean	56.7	16.7
Range	0.3–224.5	0.0–176.9
Ventricular extension *		
Yes	111 (44.4%)	139 (55.6%)
No	61 (16.4%)	312 (83.6%)
Hydrocephalus *		
Yes	62 (45.3%)	75 (54.7%)
No	108 (22.3%)	376 (77.7%)

(* Significant at P≤0.05 level of significance)

Table 2 Comparison between logit and Bayesian logit

Covariates	Logit OR (95% C.I.)	Bayesian Logit OR (95% C.I.)
Age (years)	1.03 (1.01–1.05) *	1.00 (0.99–1.01)
Presenting GCS		
≤8	Reference	Reference
>8	0.11 (0.06–0.22) *	0.67 (0.08–0.22) *
1st INR result	1.44 (1.09–1.90) *	1.06 (0.98–1.08)
Total volume of clot /ml	1.03 (1.02–1.04) *	1.00 (0.99–1.01)
Hydrocephalus		
Yes	Reference	Reference
No	0.92 (0.48–1.74)	0.99 (0.52–1.68)

(* Significant at P≤0.05 level of significance)

Fig. 1 Receiver operating characteristic curves for the various models. ▶ (Logit: multivariate logistic regression; Bayesian Logit: Bayesian multivariate logistic regression; GAM: generalized additive model; CART: classification tree; LOTUS: logistic trees with unbiased selection)

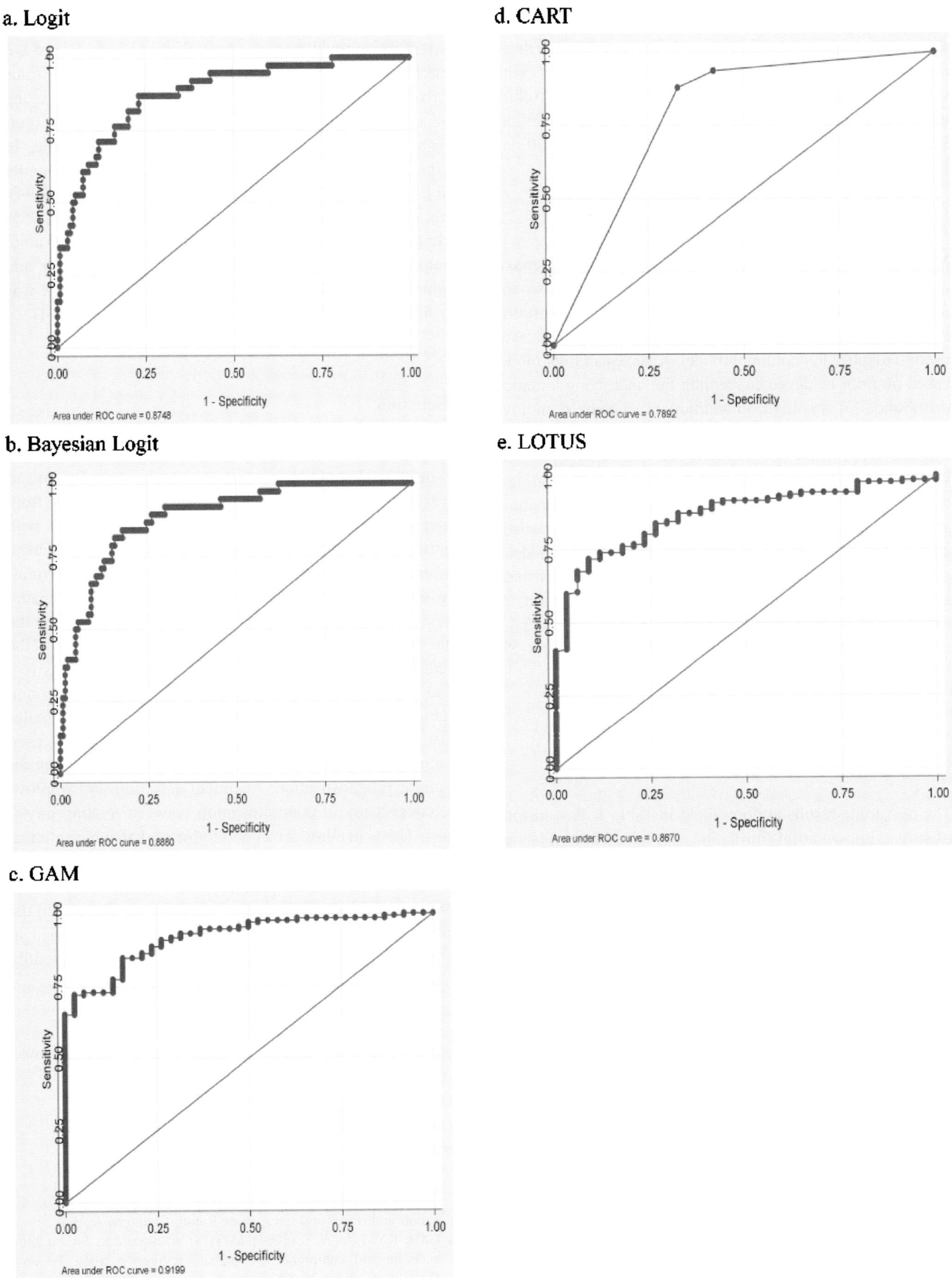
a. Logit
Sensitivity
1 - Specificity
Area under ROC curve = 0.8748
b. Bayesian Logit
Sensitivity
1 - Specificity
Area under ROC curve = 0.8880
c. GAM
Sensitivity
1 - Specificity
Area under ROC curve = 0.9199
d. CART
Sensitivity
1 - Specificity
Area under ROC curve = 0.7892
e. LOTUS
Sensitivity
1 - Specificity
Area under ROC curve = 0.8670

committee of our institution. ICH was defined as a sudden focal neurological event with confirmation of hemorrhage on computed tomography of the brain. Patients with secondary causes of intracerebral hemorrhage were excluded. ICH volume was calculated using the ABC/2 method. Early mortality was defined as mortality within the 30 day period following admission.

Statistical methods

Multivariate logistic regression (logit) [6] based on maximum likelihood was employed to ascertain how the identified clinical variables were associated with mortality at 30 days (1: dead; 0: alive). Backward elimination was used to identify the optimum model. A Bayesian logit model based on prior evidence concerning the underlying binomial distribution [3] was built. In addition, a generalized additive model (GAM) [10], Classification Tree (CART) [1] and Logistic Trees with Unbiased Selection (LOTUS) [2] models were also applied. To validate the models, the original sample was randomly divided into two sub-samples (training set: 60%, test set: 40%). Receiver operating characteristic curves (ROC) [5, 7] were generated to compare the models' predictive accuracy with observations in the test set. Univariate analyses were performed with chi-square tests or Mann-Whitney tests [9]. Analysed with Stata 9.0, R 2.4.1 and LOTUS 2.3, all statistical tests were conducted at 5% level of significance.

Results

The univariate results are presented in Table 1. Past history of stroke, known atrial fibrillation, use of warfarin, glucose level, presenting Glasgow Coma Scale (GCS) and pupil abnormality, post-resuscitation GCS and pupil abnormality, initial international normalized ratio (INR) and prothrombin time (PT) results, vomiting, seizure, total volume of clot, ventricular extension and hydrocephalus were significantly associated with early mortality. However, logit with backward elimination showed that only age, presenting GCS, 1st INR result and total volume of clot were significantly associated with mortality in the final multivariate model (Table 2) The model is found to be satisfactory by means of the Hosmer-Lemeshow test (p: 0.27). Based on a previous study [8], the Bayesian model utilized a prior beta distribution, namely Beta[1.0656, 1.8944] for handling the unknown parameter of the underlying binomial distribution. Its close-form estimated odds ratios (ORs) are quite different from those of the logit model (Table 2). Only presenting GCS was statistically significant. The following models do not generate ORs for analysis. Instead, they generate probabilities for predicting patient mortality: for CART, the optimal tree was first obtained with a least CP criterion with 1 standard error, after pruning with 10 folds of cross validation. The same result was obtained when the deviance was used as the criterion. Presenting GCS (cut-off≤8) was identified as the only significant variable. In terms of prediction using the test set (Fig. 1), the area under the ROC (AUC) for Logit was 0.87 (95% C.I.: 0.81–0.94). This was lower than that of the Bayesian model (0.88; 95% C.I.: 0.83–0.94) and GAM (0.92; 95% C.I.: 0.88–0.96), but comparable to LOTUS (0.87; 95% C.I.: 0.81–0.92) and higher than CART (0.79; 95% C.I.: 0.73–0.85). There was no significant difference between the models' AUC [4].

Discussion

Using a sample of significant size, we were able to derive a set of clinical variables predictive of early mortality for our local population. Our study is consistent with data from other groups and is able to predict early mortality with equivalent performance. It reinforces how specific clinical and radiological factors play an important role in determining outcome. In particular, the presence of coagulopathy does not bode well for the ICH patient. The use of the data mining techniques does not appreciably increase the predictive accuracy. In other neurosurgical conditions (eg. head injury, aneurysmal subarachnoid hemorrhage and arteriovenous malformations), well-defined grading scales exist. ICH however suffers from a lack of a widely used grading system. A consistent grading system will allow for accurate prognostication, treatment selection and improve the consistency of communication between treating physicians. This will allow for a good framework in which clinical decisions are made. With increasing computer sophistication, the use of advanced statistical modeling methods does not necessarily mandate excessive time commitment. In the near future, grading scales will inevitably be modified to include proteomic and genetic biomarkers to hopefully improve predictive accuracy.

Conflict of interest statement We declare that we have no conflict of interest.

References

1. Breiman L, Friedman JH, Olshen RA, Stone CJ (1998) Classification and regression trees. Boca Raton, Chapman & Hall
2. Chan KY, Low WY (2004) LOTUS: an algorithm for building accurate and comprehensive logistic regression trees. J Comp Grap Stats 13(4):826–852

3. Das S, Dey DK (2006) On Bayesian analysis of generalised liner models using the Jacobian technique. JASA 60(3):264–268
4. Hanley JA, McNeil BJ (1983) A method of comparing the areas under receiver operating characteristic curves derived from the same case. Radiology 148:839–843
5. Hanley JA, McNeil BJ (1982) The meaning and use of the area under a receiver operating characteristic (ROC) curve. Radiology 143:29–36
6. Hosmer DW, Lemeshow S (2000) Applied logistic regression. Wiley, New York
7. Metz CE (1978) Basic principles of ROC analysis. Semin Nucl Med 8:283–298
8. Nilsson OG, Lindgren A, Brandt L, Saveland H (2002) Prediction of death in patients with primary intracerebral haemorrhage: a propsectiv study of a defined population. J Neurosurg 97:531–536
9. Siegel S, Castellan NJ (2000) Nonparametric statistics for the behavioural science. McGraw-Hill, Singapore
10. Wood SN (2006) Generalized additive model: an introduction with R. Boca Raton, CRC Press

Acta Neurochir Suppl (2008) 102: 305–306

Ischemic events after carotid interventions in relationship to baseline cerebrovascular reactivity

George K.-C. Wong · Stephanie C.-P. Ng ·
Matthew T.-V. Chan · David T.-F. Sun ·
Winnie W.-M. Lam · Joseph M.-K. Lam · Wai S. Poon

Abstract

Background We aimed to investigate whether baseline cerebrovascular reactivity could predict subsequent ischemic event after intervention and identify the patient group for more aggressive medical and interventional management paradigms.

Methods Patients with more than 70% cervical carotid stenosis (from ultrasonography) were reviewed. Patients, who had baseline cerebrovascular reactivity test before intervention and had either carotid endarterectomy (CEA) or carotid angioplasty and stenting (CAS) performed, were recruited for analysis. Transcranial Doppler ultrasonography was used to examine the reactivity of the middle cerebral artery in response to 5% carbon dioxide in oxygen. The mean follow up period was 66 months.

Findings Twenty-six patients had symptomatic carotid stenosis and ten patients had asymptomatic carotid stenosis. There were four subsequent ischemic events during follow up. None of the nine patients with impaired baseline ipsilateral cerebrovascular reactivity had subsequent ischemic event.

Conclusions In this current study, impaired baseline cerebrovascular reactivity did not predict the subsequent stroke risk after carotid intervention. Cerebrovascular reactivity testing may not serve as an indicator for aggressive medical and surgical treatments.

Keywords Carotid stenosis · Cerebrovascular reactivity · Stroke · Carotid endarterectomy

Introduction

The benefits of carotid endarterectomy (CEA) for symptomatic severe cervical carotid stenosis have been clearly established from the results of 3 major prospective randomized trials: the NASCET, the European Carotid Surgery Trial

G. K.-C. Wong · S. C.-P. Ng · D. T.-F. Sun · J. M.-K. Lam ·
W. S. Poon
Division of Neurosurgery, The Chinese University of Hong Kong,
Hong Kong, China

M. T.-V. Chan
Department of Anesthesiology and Intensive Care,
The Chinese University of Hong Kong,
Hong Kong, China

W. W.-M. Lam
Department of Diagnostic Radiology and Organ Imaging,
The Chinese University of Hong Kong,
Hong Kong, China

W. S. Poon (✉)
Division of Neurosurgery, Department of Surgery,
Prince of Wales Hospital,
Shatin, New Territories,
Hong Kong, SAR, China
e-mail: wpoon@surgery.cuhk.edu.hk

G. K.-C. Wong · S. C.-P. Ng · D. T.-F. Sun · J. M.-K. Lam ·
W. S. Poon
Department of Surgery, 4th Floor, Prince of Wales Hospital,
Shatin, NT,
Hong Kong, SAR, China

M. T.-V. Chan
Department of Anaesthesiology and Intensive Care, 5th Floor,
Prince of Wales Hospital,
Shatin, NT,
Hong Kong, SAR, China

W. W.-M. Lam
Department of Diagnostic Radiology and Organ Imaging,
Ground Floor, Prince of Wales Hospital,
Shatin, NT,
Hong Kong, SAR, China

(ECST), and the Veterans Affair Cooperative Study Program [1–5]. In patients at high surgical risk, endovascular treatment with angioplasty and stenting was utilized with similar benefits. Among symptomatic patients with TIAs or minor strokes and high-grade carotid stenosis, the trials showed an absolute risk reduction with carotid endarterectomy, in the magnitude of 17% stroke risk reduction in 2 years. Nevertheless, the figures implied that there still exists a recurrent stroke risk in this group of patients (9% in 2 years).

We aimed to explore whether impaired baseline cerebrovascular reactivity would also predict the subsequent stroke risk after carotid intervention and identify the patient group for more aggressive medical and interventional management paradigms.

Patients and method

Patients with more than 70% cervical carotid stenosis (from ultrasonography) were reviewed. Patients, who had baseline cerebrovascular reactivity test before intervention and had either carotid endarterectomy (CEA) or carotid angioplasty and stenting (CAS) performed, were recruited for analysis. Transcranial Doppler ultrasonography via transtemporal route was used to examine the reactivity of the middle cerebral artery (MCA) to 5% carbon dioxide in oxygen. Normal cerebrovascular reactivity was defined by >1% increase per unit change in end-tidal CO2 in ipsilateral middle cerebral artery velocity in response to 5% carbon dioxide in oxygen. Statistical analysis was carried out with SPSS Version 14.0. Data were given as means and standard deviations (SD). X^2 test or Fisher exact test was applied to univariate dichotomous variables. A 2-tailed value of $P< 0.05$ was considered statistically significant.

Results

The male to female ratio was 7:2. Age was 66.3 +/−9.0 (mean+/−SD) years, ranging from 41 years to 88 years. 58% of patients had hypertension. 25% of patients had diabetes. 25% of patients had hyperlipidemia. 47% of patients had cardiac disease including ischemic heart disease or atrial fibrillation. 11% patients had history of radiotherapy treatment for nasopharyngiocarcinoma. 22% of patients had history of smoking. 42% of patients had associated contralateral carotid stenosis.

Thirty six patients had either carotid endarterectomy (CEA) or carotid angioplasty and stenting (CAS) performed and were recruited for analysis. Twenty nine patients had carotid endartectomy done and seven patients had carotid angioplasty and stenting done. Follow up period was 66.3 +/−21.6 (mean +/−SD) months, ranged from 1 to 96 months. With the exception of one early mortality case from laryngectomy one month after carotid endarterectomy, all patients had a follow up period more than 24 months. None of the nine patients with impaired baseline ipsilateral cerebrovascular reactivity had a subsequent ischemic event. There was no statistical relationship between impaired baseline cerebraovascular reactivity and subsequent ischemic stroke event after carotid intervention (P=0.55).

Discussion

Identification of the high risk group for subsequent stroke after intervention for carotid stenosis may help to select target for future trials of more aggressive medical treatment or even simultaneous interventional treatment for any concomitant intracranial stenoses. We thus explored the possibility of baseline cerebrovascular reactivity may be a predictive factor. In this study, impaired ipsilateral cerebrovascular reactivity before carotid intervention was not indicative of subsequent ischemic event. Further stroke risk seems to have different mechanisms other than hemodynamic compromise in this group of patients. Small vessel disease and emboli may be some of the possible mechanisms involved for subsequent ischemic event after carotid interventions (CEA and CAS).

Conclusion

Impaired baseline cerebrovascular reactivity did not predict the subsequent stroke risk after carotid intervention. Cerebrovascular reactivity testing may not serve as an indicator for aggressive medical and surgical treatments.

Conflict of interest statement We declare that we have no conflict of interest.

Reference

1. Barnett HJ, Taylor DW, Eliasziw M, Fox AJ, Ferguson CG, Gaynes RB, Rankin RN, Clagett GP, Hachinski VC, Sackett DL, Thorpe KE, Meldrum HE, Spence JD (1998) Benefit of carotid endarterectomy in patients with symptomatic moderate or severe stenosis: North American Symptomatic Carotid Endarterectomy Trial Collaborators. N Engl J Med 339:1415–1425
2. Mayberg MR, Wilson SE, Yatsu F, Weiss DG, Messina L, Hershey LA, Colling C, Eskridge J, Deykin D, Winn HR (1991) Carotid endarterectomy and prevention of cerebral ischemia in symptomatic carotid stenosis: Veterans Affairs Cooperative Studies Program 309 Trialist Group. JAMA 266:3289–3294
3. MRC European Carotid Surgery Trial (1991) Interim results for symptomatic patient with severe or with mild carotid stenosis: European Carotid Surgery Trialists' Collaborative Group. Lancet 337:1235–1243
4. No authors listed. Beneficial effect of carotid endarterectomy in symptomatic patients with high-grade carotid stenosis: North American Symptomatic Carotid Endarterectomy Trial Collaborators. N Engl J Med 1991;325:445–453
5. No author listed. Randomized trial of endarterectomy for recently symptomatic carotid stenosis: final results of the MRC European Carotid Surgery Trial (ECST). Lancet 1998;351:1379–1387

Acta Neurochir Suppl (2008) 102: 307–310

Effect of increased intracranial pressure on cerebral vasospasm in SAH

Krissanee Karnchanapandh

Abstract

Introduction Increased ICP is common and might precipitate cerebral vasospasm (VSP)-induced ischemic events in aneurysmal SAH (ASAH).Our objective was to determine if there is an association between increased ICP and transcranial colour coded Doppler-angiographic VSP (TCCD-A VSP) in relation to delayed neurological deficit (DND) and poor outcome.

Methods A retrospective study was undertaken in 30 patients who were status-post clipping of a ruptured anterior circulation aneurysm causing subarachnoid hemorrhage. ICP monitoring, post-operative MCA/A1 flow velocity measurements, and post-operative angiography were evaluated, with clinical follow up for at least 3 months. Statistical testing was done using a Chi- square analysis between ICP>=30 mmHg, CPP<=70 mmHg, TCCD-A VSP and DND and poor outcome.

Findings Mean ICP was 34.2 mmHg and mean CPP was 71.3 mmHg. ICP>=30 mmHg occurred in 56.7%. of patients and CPP<=70 mmHg occurred in 36.7%. TCCD-A VSP, DND and poor outcome occurred in 56.7%, 60% and 33% of patients respectively. Overall ICP>=30 mmHg was significantly related to TCCD-A VSP (p value=0.046) but not with DND or poor outcome. CPP<=70 mmHg had no significant relationship with TCCD-A VSP, DND or poor outcome. TCCD-A VSP was significantly related to DND (p value=0.00018) but not to poor outcome. The combination of TCCD-A VSP and ICP>=30 mmHg was significantly related to poor outcome (p value=0.015), but the combination of TCCD-A VSP with CPP<=70 mmHg was not.

Conclusions Increased ICP, not decreased CPP, was related to VSP. The combination of TCCD-A VSP and increased ICP was predictive of poor outcome. Management of acute ASAH should include reduction of increased ICP especially when there is concomitant TCCD-A VSP.

Keywords ICP · CPP · vasospasm · SAH

K. Karnchanapandh (✉)
Department of Neurosurgery, Rajavithi Hospital,
Department of Medical Services, Ministry of Public Health,
College of Medicine, Rangsit University,
2 Payathai Road,
Bangkok 10400, Thailand
e-mail: patarapol@hotmail.com

Introduction

Excluding aneurysmal rupture and rebleeding, cerebral vasospasm (VSP) is a leading cause of unfavorable outcome in aneurysmal SAH (ASAH). [5, 11, 12, 15] Vasospasm does not always result in development of a neurological deficit. Factors other than VSP itself such as increased ICP or decreased CPP might also precipitate an ischemic event. The objective of this study was to determine if there is an association between increased ICP, decreased CPP, and VSP in relation to delayed neurological deficit (DND) and poor outcome.

Methods

A retrospective study was performed involving 30 patients with SAH status-post clipping of a ruptured anterior circulation aneurysm at Rajavithi Hospital during 1995–1997. Hunt and Hess [8] scores ranged from 2 to 4. All patients underwent aneurysm clipping within 7 days of rupture, post-operative ICP monitoring, post- operative transcranial Doppler ultrasound MCA (middle cerebral

Table 1 ICP, CPP, TCCD-A VSP and Fisher's grouping

Fisher's grouping	n	mean ICP	ICP>=30	mean CPP	CPP<=70	TCCD-A VSP
F2	4	28	2 (50%)	71	1 (25%)	1 (25%)
F3	18	39.3	12 (66.7%)	69.2	7 (38.8%)	12 (66.7%)
F4	8	27.6	3 (37.5%)	87.3	3 (37.5%)	4 (50%)

artery)/A1 (anterior cerebral artery) flow velocity measurements, post-operative 4 vessel cerebral angiographay, and post-operative CT scanning. Clinical follow up was for at least 3 months. Continuous ICP monitoring was performed using a subdural fiberoptic monitor (Camino) inserted during dural closure after aneurysm clipping. Significantly increased ICP was defined as an ICP>=30 mmHg for more than 15 minutes. Measures used to reduce elevated ICP included spinal drainage, ventricular drainage, mannitol/glycerosteril infusion, and right/left wide craniectomy decompression. MCA/A1 flow velocity (MCA/A1 FV) measurement was performed bilaterally intermittently using transcranial colour coded Doppler ultrasonography (TCCD) with a 2 MHz handheld probe (Acuson) via the transtemporal window ipsilateral to the craniotomy. VSP was defined by TCCD if MCA/A1 mean FV>=120 cm/s [1] and 4 vessel cerebral angiography demonstrated severe narrowing of the insonated artery. Management for vasospasm included hypervolemic-hypertensive-hemodilution and nimodipine infusion. Delayed neurological deficit (DND) was defined as a new neurological deficit occurring after admission. An irreversible neurological deficit (IRND) and ultimate clinical outcome was determined 3 months after aneurysm rupture. Outcome was assessed using the National Institutes of Health Stroke Scale (NIHSS) [2] with a good outcome defined as no neurological deficit (NIHSS =0), a fair outcome defined as a minor IRND resulting in an NIHSS=1–7, and poor outcome defined as a major IRND resulting in an NIHSS>7 or death. Chi- square testing was performed to assess the relationship between ICP>=30 mmHg, CPP<=70 mmHg, TCCD-A VSP, DND, IRND and poor outcome. Statistical significance was defined as a p value<=0.05.

Results

Ages ranged from 32–75 with a mean age of 49.6 years. 14 cases (46.6%) involved anterior communicating aneurysms, 8 cases (26.6%) involved MCA aneurysms, 6 cases (20%) were of posterior communicating artery aneurysms, and 2 cases (6.7%) involved ventral internal carotid artery aneurysms. Two patients (6.7%) had a Hunt and Hess [8] (H/H) score of 2, 19 patients (63.3%) were H/H 3, and 9 patients (30%) were H/H 4. According to Fisher's grouping (F) [4], 4 cases (13.3%) were F 2, 18 cases (60%) were F 3, 8 cases (26.6%) were F 4. Mean interval time from most recent hemorrhage to clipping was 3.8 days. Of 30 total patients, 30% had spinal drainage, 33% had ventricular drainage, 50% developed cerebral edema and none had recurrent intracranial hemorrhage on post-operative CT-scan.

ICP, CPP and TCCD-A VSP/DND, IRND and outcome

Mean duration of ICP monitoring was 8 days. Mean ICP and mean CPP were 34.2 and 71.3 mmHg respectively. 17 patients (56.7%) developed ICP>=30 mmHg and 10 (33.3%) had ICP>=40 mmHg. 11 patients (36.7%) had CPP<=70 mmHg with 5 (16.7%) having CPP<=60 mmHg. TCCD-A VSP was able to be performed in 17 patients (56.7%). The incidence of TCCD-A VSP in patients with ICP>=30, ICP>=40, CPP<=70 and CPP<=60 mmHg was 41.2%, 60%, 54.5% and 40% respectively. ICP, CPP and TCCD-A VSP according to Fisher's Grouping [4] are summarized in Table 1. 18 patients (60%) developed DND. Nine (30%) had IRND. The incidence of DND relative to Fisher group was F 2 (0%), F 3 (72.3%,n=13), and F 4

Table 2 ICP, CPP,TCCD-A VSP and DND, IRND, poor outcome

ICP,CPP,TCCD-A VSP	n	DND	IRND	Poor outcome
ICP>=30	17	10 (58%)	6 (35%)	7 (41%)
CPP<=70	11	7 (63%)	3 (27%)	3 (27%)
ICP>=30+CPP<=70	7	4 (57%)	2 (28%)	2 (28%)
TCCD-A VSP	17	15 (88%)	7 (41%)	8 (47%)
ICP>=30+TCCD-A VSP	7	7 (100%)	4 (57%)	5 (71%)
CPP<=70+TCCD-A VSP	6	6 (100%)	2 (33%)	2 (33%)
ICP>=30+CPP<=70+TCCD-A VSP	3	3 (100%)	1 (33%)	1 (33%)

(62.5%, n=5). The incidence of IRND relative to Fisher group was F 2 (0%), F3 (38.8%,n=7), and F4 (25%,n=2). Ninteen patients (63.3%) achieved a good outcome and poor outcome occurred in 10 patients (33%). Poor outcome relative to Fisher grouping was F2 (25%, n=1), F3 (44.4%, n=8), and F4 (12.5%, n=1). DND, IRND and poor outcome according to various combinations of ICP, CPP and TCCD-A VSP are summarized in Table 2.

Statistical analysis

ICP>=30 mmHg was significantly related with TCCD-A VSP (p value=0.046) but not with DND (p=0.59), IRND (p=0.38), or poor outcome (p=0.26). CPP<=70 mmHg had no significant relationship with TCCD-A VSP, DND, IRND, or poor outcome (p values=0.579, 0.533, 0.571, 0.431). TCCD-A VSP was strongly related to DND (p value=0.00018) but not with IRND (p=0.12), or poor outcome (p=0.06). The presence of both TCCD-A VSP and ICP>=30 mmHg together was related to poor outcome (p value=0.015) but the presence of both TCCD-A VSP and CPP<=70 mmHg together was not (p value=0.69).

Discussion

In the study, TCCD-A VSP and increased ICP were both most common in patients with Fisher Group 3 subarachnoid hemorrhage on CT [4]. Increased ICP (as defined in this study) was statistically significantly related to TCCD-A VSP. An association between increased ICP and VSP in acute aneurysmal SAH has been considered previously in other studies [3, 7, 11, 14, 16, 17]. Increased ICP might result from TCCD-VSP when arterial narrowing results in ischemic cerebral edema [16] or might represent the increased cerebral blood volume associated with distal dilatation of intraparenchymal vessels accompanying the proximal VSP.[6] Hayashi et al. found that the initial increased ICP was follow by depression of ICP and that the initial increase was concomitant with the beginning of arterial spasm while the secondary rise of ICP thereafter was likely due to ischemic cerebral edema. [7] Thus, the pathophysiology of increased ICP in conjunction with TCCD-A VSP in SAH may consist of multiple mechanisms at different time points in the disease course. In this study, DND was strongly associated with TCCD-A VSP but neither TCCD-A VSP nor increased ICP and/or decreased CPP were significantly associated with poor outcome. Jakuboski et al. were unable to demonstrate any relationship among ICP, CPP or MAP and neurological outcome in ASAH. [9] It is possible that ICP>=30 mmHg is not high enough and CPP<=70 mmHg is not low enough to effect outcome. Siessjo and Zwetnow demonstrated that increased ICP did not disturb oxidative metabolism of brain tissue until CPP decreased to about 30 mmHg. [13] In our study, the combination of TCCD-A VSP and increased ICP was significantly related to poor outcome while the combination of TCCD-A VSP and decreased CPP was not. Thus, the mechanism by which increased ICP interacts with TCCD-A VSP and outcome may not be through decreased CPP. Farrar studied the role of ICP in cerebral arterial spasm by using isolated rabbit carotid and femoral arteries perfused at a constant pressure to examine the variation of flow with transmural pressure and demonstrated that in the presence of active vasoconstriction any increase in extraluminal pressure or ICP resulted in increase in arterial resistance. This suggested that chronic arterial spasm is the result of mild constriction amplified by the simultaneous occurrence of increased ICP. [3] This might explain a positive correlation between increased ICP and TCCD-A VSP and a positive correlation between TCCD-A VSP with simultaneous increased ICP and poor outcome in our study.

Conclusion

Increased ICP, but not decreased CPP (as defined in this study), is related to post-SAH vasospasm. While DND is related principally to VSP, poor outcome is associated with the presence of both TCCD-A VSP and increased ICP occurring in combination. This suggests that the management of ASAH should include reduction of elevated ICP, especially when there is concomitant TCCD-A VSP.

Conflict of interest statement We declare that we have no conflict of interest.

References

1. Aaslid R, Huber P, Nornes H (1984) Evaluation of cerebrovascular spasm with transcranial Doppler ultrasound. J Neurosurg 60:37–41
2. Brott T, Adams HP, Olinger CP, Master JR, Barsan WG, Biller J, Spilker J, Holleran R, Eberle R, Hertzberg V (1989) Measurements of acute cerebral Infarction; A clinical examination scale. Stroke 20:864–870
3. Farrar JK (1975) Chronic cerebral arterial spasm. J Neurosurg 43:408–417
4. Fisher M, Kistler JP, Davis JM (1980) Relation of cerebral vasospasm to subarachnoid hemorrhage visualized by computerized tomographic scanning. Neurosurgery 6:1–9
5. Fogelholm R, Hernesniemi J, Vapalahti M (1993) Impact of early surgery on outcome after aneurysmal subarachnoid hemorrhage: a population based study. Stroke 24:1649–1654
6. Grubb RL, Raichle ME, Eichling JO, Gado MH (1977) Effects of subarachnoid hemorrhage on cerebral blood volume, blood flow and oxygen utilization in humans. J Neurosurg 46:446–453

7. Hayashi M, Murokawa S, Fujii H, Kitano T, Kobayashi H, Munemoto S, Yamamoto S (1978) Intracranial pressure in patients with diffuse cerebral arterial spasm following ruptured intracranial aneurysms. Acta Neurochir 44:81–195
8. Hunt WE, Hess RM (1968) Surgical risk as related to time of intervention in the repair of intracranial aneurysms. J Neurosurg 28:14–20
9. Jakuboski J, Bell BA, Symon L et al (1982) A primate model of subarachnoide hemorrhage: change in regional cerebral blood flow, autoregulation, carbondioxide reactivity and conduction time. Stroke 13:601–611
10. Kassell NF, Peerless SJ, Reilly PL (1976) Intracranial pressure, aneurysms and subarachnoid hemorrhage. In: Beck JWF, Bosch DA, Brock M (eds) Intracranial pressure III. Springer Verlag, Berlin, Heidelberg, New York, pp 147–151
11. Kassell NF, Torner JC, Jane JA, Haley EC, Ddams HP (1990) The international cooperative study on timing of aneurysm surgery. Part 2: Surgical results. J Neurosurg 73:37–47
12. Longstreth WT, Nelson LM, Koepsell TD, Van Belle G (1993) Clinical course of spontaneous subarachnoid hemorrhage: a population - based study in King County, Washington. Neurology 43:712–718
13. Siessjo BK, Zwetnow NN (1970) Effect of increased cerebrospinal fluid pressure upon adenine nucleotides and upon lactate and pyruvate in rat brain tissue. Acta Neurol Scandinav 46:187–202
14. Takagi H, Yada K, Ohwada T, Saito T (1976) An analysis of operative time course of ICP in 35 cases with intracranial aneurysms. In: Beck JWF, Bosch DA, Brock M (eds) Intracranial pressure III. Springer Verlag, Berlin, Heidelberg, New York, pp 152–156
15. Tettenborn D, Dyeka J (1990) Prevention and treatment of delayed ischemic function in patient with aneurysmal subarachnoid hemorrhage. Stroke 21:85–89
16. Vapalahti M, Nieminen V, Kari K (1976) Incidence of hydrocephalus and increased ventricular fluid pressure in patient with ruptured supratentorial aneurysms. In: Beck JWF, Bosch DA, Brock M (eds) Intracranial pressure III. Springer Verlag, Berlin, Heidelberg, New York, pp 135–138
17. Volby B, Enevoldsen EM (1980) Intraventricular pressure, CSF lactate and vasospasm in ruptured intracranial aneurysm. In: Shulman K, Marmarou A, Miller JD, Becker DP, Hockwald GM, Brock M (eds) Intracranial pressure IV. Springer Verlag, Berlin, Heidelberg, New York, pp 211–214

Acta Neurochir Suppl (2008) 102: 311–316

Cerebral blood flow thresholds predicting new hypoattenuation areas due to macrovascular ischemia during the acute phase of severe and complicated aneurysmal subarachnoid hemorrhage. A preliminary study

A. Chieregato · A. Tanfani · A. Noto · S. Fronza · F. Cocciolo · E. Fainardi

Summary
Background Focal ischemia may affect patients with aneurysmal subarachnoid hemorrhage (SAH), and the potential evolution of cerebral infarction may greatly influence the patients' outcome. The aim of the study was to assess the values of regional cortical cerebral blood flow (rCBF) thresholds predictive for ischemia during the acute phase of SAH.
Methods In 34 patients affected by poor grade or complicated SAH, 52 pairs of Xenon-CT (Xe-CT) studies of regional CBF were analyzed, in which the follow-up Xe-CT study was obtained no later than 72 hours after the baseline study. Corresponding cortical ROIs were placed in the perimeter of the cortex on both the Xe-CT studies. A blinded, experienced neuroradiologist classified for each ROI, the development of a new hypoattenuation at the unenhanced CT images included in the follow-up Xe-CT, while another independent investigator collected rCBF levels of the ROI in the baseline Xe-CT study.
Findings New hypoattenuation developed in 3.94% of the ROIs in the paired follow-up Xe-CT studies, and these evolving ROIs were associated with a lower rCBF in baseline Xe-CT. However, the positive predictive value of rCBF levels for the development of new hypoattenuation was only moderately predictive (28.3%) for very low physiological values (5 ml/100gr/min).
Conclusions The results suggest that there is no absolute rCBF threshold of ischemia in severe and complicated SAH patients and that the rCBF values are only moderately predictive at levels lower than previously described.

Keywords Cerebral blood flow · Intracranial pressure · Ischemia · Physiological thresholds · Subarachnoid hemorrhage

A. Chieregato · A. Tanfani · S. Fronza · F. Cocciolo
UO Anestesia e Rianimazione, Dipartimento di Emergenza,
AUSL-Cesena, Ospedale M. Bufalini,
Viale Ghirotti, 286,
47023 Cesena, Italy

A. Noto
Unità Cardiovascolare e Chirurgia Toracica, Ospedale G. Martino,
Università di Messina,
Via Consolare Valeria,
98122 Messina, Italy

E. Fainardi
UO Neuroradiologia, Dipartimento di Neuroscienze,
Azienda Ospedaliera Universitaria, Arcispedale S. Anna,
Corso Giovecca 203,
44100 Ferrara, Italy

A. Chieregato (✉)
UO Anestesia e Rianimazione,
AUSL-Cesena, Ospedale M. Bufalini,
Viale Ghirotti, 286,
47023 Cesena, Italy
e-mail: achiere@ausl-cesena.emr.it

Introduction

The principal aims in the management of patients with aneurysmal subarachnoid hemorrhage (SAH) are the prevention of rebleeding by clipping or coiling of the aneurysm and the avoidance of secondary ischemic complications. In the physiopathology of large vessel hypoperfusion, a reduction in regional cerebral blood flow (rCBF) clearly has a relevant role. This concept has been thoroughly explored in experimental and clinical studies of acute ischemic stroke where physiological thresholds of CBF for the development of irreversible tissue damage [6], and for the penumbra, have been defined. However, the applicability of thresholds for tissue damage derived from stroke studies to patients with SAH is questionable. In fact, although macrovascular occlusion may result in a topographically well-defined ischemic core surrounded by an area of penumbra in both SAH and ischemic stroke, the exact timing of hypoperfusion development may be difficult to establish in SAH due to the already compromised clinical conditions of these patients and the more progressive evolution of ischemic abnormalities, at least in the case of vasospasm. In addition, ischemia related to SAH can be promoted by different mechanisms, which may sometimes simultaneously coexist. Furthermore, the changes in metabolism following SAH have not yet been elucidated. Cerebral metabolism is commonly reduced after SAH [1], even in good grade patients, most likely as a result of the effects of initial bleeding. Consequently, as in SAH low CBF may be associated to preserved flow–metabolism coupling, CBF thresholds for tissue survival might be expected to be lower than in stroke. The effects of deep sedation on CBF in patients with poor grade or complicated SAH with elevation of intracranial pressure could contribute to a further reduction in ischemic thresholds, as in Traumatic Brain Injury (TBI) [3, 5]. Notwithstanding these limitations, SAH and acute stroke could share some of the mechanisms leading to the formation of ischemia in major vascular territories. In this context, Xenon-CT (Xe-CT) has proven to be a powerful imaging modality to quantitatively identify cerebral hemodynamic disturbances occurring in patients with SAH [9] providing evidence that thresholds for irreversible ischemia and penumbra can be employed in SAH setting. However, the ability of Xe-CT technology to detect new ischemic lesions during SAH monitoring is, in severely compromised patients, hampered by the fact that a reliable neurological examination cannot be performed, as well as by the confounding effects of sedation which make the clinical suspicion of major changes in rCBF less probable. The aim of this study was to determine, in the acute phase after SAH, physiological thresholds for the development of cortical tissue damage occurring in territories of major cerebral vessels according to a methodology previously proposed in stroke [6] and TBI [3], by using positron emission tomography (PET). To achieve these objectives, in patients with poor grade or complicated SAH during the acute phase, the occurrence of any new region of hypoattenuation, considered suggestive of the development of a new ischemic area, was evaluated in pairs of two consecutive Xe-CT studies, separated by a time interval of no more than 72 hours. The analysis consisted in the calculation of the probability of cortical rCBF to predict the evolution in hypoattenuation of the corresponding cortical area at the unenhanced CT images included in the follow-up Xe-CT.

Material and methods

Patient inclusion and general management

From a case material of 362 Xenon-CT (Xe-CT) studies performed in 159 patients with SAH from January 2000 to February 2007, we selected patients having at least 2 studies with a time delay of no longer than 72 hours between them. Of these, the initial scan was considered as the baseline Xe-CT study, whereas the subsequent scan was classified as the follow-up Xe-CT study. The time window interval was selected as being adequate to allow a critical perfusion level to induce a consistent water accumulation due to cytotoxic edema valuable by means of the unenhanced CT images included in the follow-up Xe-CT. Each study was analyzed for artifacts and for the overlap in the positioning, angle of the CT, scans and parameters used. According to this criteria, the final case mix for the analysis was consistent with 90 Xe-CT studies, 52 pairs of Xe-CT studies from 34 patients with SAH. For the patients who were critical at entry, the management involved sedation and ventilation, the placement of an intraventricular catheter to measure ICP and to treat acute hydrocephalus. Shortly afterwards, patients underwent angiography to detect aneurysms and to evaluate whether embolization or surgery was appropriate. A staircase treatment protocol was applied to maintain ICP below 20–25 mmHg and a cerebral perfusion pressure (CPP) around 70 mmHg as previously described [2]. The protocol widely applied deep sedation with benzodiazepine, plus fentanyl, and, if needed, metabolic suppression with propofol or barbiturate was applied to control elevated ICP or to protect the brain from risk factors for ischemia.

Xe-CT measurement and CBF analysis

Xe-CT studies were planned to help clinical management in the following cases to verify: 1) the presence of reduced rCBF levels, potentially due to vasospasm, or increased rCBF values due to hyperemia when a routinely measured transcranial doppler showed elevated or a progressive increase in flow velocity values after repeated measurements. In the case of suspected vasospasm, Xe-CT examination was followed by diagnostic angiography; 2) the time course of areas with critical rCBF values, when a previous Xe-CT revealed an unexplained rCBF asymmetry between the two middle cerebral artery territories; 3) the effect of sedation on CBF levels or the presence of hyperemia as a potential cause of elevated ICP, in the setting of refractory ICP elevation; 4) the adequacy of CPP; 5) the viability of previously identified low density areas; 6) the rCBF in vascular territories when the surgeon or the neuroradiologist suggested a possible vascular occlusion.

CBF studies were conducted using a CT scanner equipped for Xe-CT CBF imaging (Xe-CT system-2^{TM}, Diversified Diagnostic Products, Inc., Houston, TX). According to Pindzola and Yonas [8], wash–in protocol was performed using four contiguous 10-mm axial sections separated by 20-mm intervals, located on the cerebral hemispheres, with the head aligned along the orbitomeatal plane. Two baseline scans separated by a time interval of 30 seconds were obtained at each of the four chosen levels. After a delay of 33 seconds and during approximately 4.5 minutes, six additional Xe-CT scans were obtained at each level while the patient inhaled a mixture of 28% Xenon, 40% Oxygen and 32% room air. The comparability of the values obtained by our Xe-CT apparatus (from site to site and from scanner manufacturer to manufacturer) was made possible by testing with phantom values. In each of the four selected Xe-CT levels, rCBF was by dedicated software (Xe-CT System Version 1.0 w ©, 1998, Diversified Diagnostic Products, Inc, Houston, TX), expressed in ml/100 gr/min. In each of the four slices, multiple circular regions of interest (ROIs), with size comprised between 250 and 300 mm^2, were placed along the whole perimeter of the cortex of both cerebral hemispheres, excluding the sub-tentorial regions and the skull base. In each Xe-CT study, the ROI was located symmetrically in both cerebral hemispheres without any overlap and the number of cortical ROIs obtained per patient was, on average, 63.2 (SD 4.6). In both baseline and follow-up Xe-CT images, on each of the 4 slices, cortical ROIs were placed in the same corresponding order and matched by one of the authors (AT). The ROIs affected by artifacts potentially influencing rCBF measurements, were excluded from the analysis in both baseline Xe-CT and follow-up Xe-CT studies. The artifacts considered were: altered confidence values on Xe-CT software analysis, the presence of ICP catheter or air within a cortical ROI, the proximity of a cortical ROI with the bone (especially in the basal frontal lobe ROIs), or metal devices positioned inside (clip or coiling) or outside (on the scalp) the brain. The ROIs including hyperintensities or mixed densities and those which were already completely hypoattenuated in the baseline Xe-CT were also excluded. The ROIs initially analyzed were 5694, but after the aforementioned criteria of exclusion, the ROIs eligible for the study became 4886, consistent with 2795 pairs of ROIs. Parenchymal hypoattenuation on the CT scans associated to corresponding Xe-CT maps was defined as a visually well-recognized cerebral region of abnormally increased radiolucency relative to other parts of the same structure, or to its contralateral counterpart by an experienced neuroradiologist (EF). In each ROI examined, the appearance in the follow-up study of a new low density area, or an expansion of a low density area previously identified in the baseline study, was considered as a new hypoattenuation. As a consequence, ROIs characterized by the presence of a new hypoattenuation were categorized as *evolving*, whereas those in which any new hypoattenuation occurred were judged as *non-evolving*. The location, the shape of hypoattenuation at CT, as well as surgical and neuroadiological and clinical information, were used to distinguish between new hypoattenuation due to macro-vessels hypoperfusion and those related to post-surgical or perilesional focal edema. Based on multiple clinical data, the hypoattenuation due to impairment of large vessels was classified as due to macrovessel occlusion or stenosis (caused by inadvertent post-clipping or post-coiling vascular occlusion or caused by vascular compression during focal herniation of the brain) or due to vasospasm. In each ROI analyzed, rCBF levels were measured in both baseline and follow-up Xe-CT studies and were matched in a dedicated data base with corresponding data concerning ROI evolution (the development or not of new hypoattenuated lesions). For the calculation of the prediction probability for a new hypoattenuation, the match was settled between the rCBF in the baseline Xe-CT and the outcome measure in term of parenchimal hypoattenuation measured in the unenhanced CT images included in the follow-up Xe-CT.

Statistical analysis

Descriptive statistics was based on the comparison between evolving and non evolving ROIs. In this way, rCBF, area in mm^2, and Hounsfield units measured for each ROI analyzed in baseline and follow-up Xe-CT studies were compared within the same study, between two different groups of ROIs (evolving versus non evolving), and between two paired studies, within the same ROI. According to the data distribution, Mann-Whitney test (rCBF data) and the pooled t-test (area and Hounsfield units data) were used to compare data within the same Xe-CT study. The variables values of each paired ROI were compared between baseline and follow-up XeCT study with a Wilcoxon Signed Rank test (CBF data) or a paired t-test (area and Hounsfield unit data) (DataDesk 6.1, Ithaca, New York USA). The predictive ability of CBF was analyzed using the complete set rCBF levels obtained from all evolving and non evolving ROIs in the follow-up Xe-CT (MedCalc 9.3.0, Frank Shoonjans, Belgium). The ability of a proposed threshold to predict eventual tissue evolution in the acute phase after SAH was given by the positive predictive value (PPV; the probability that a ROI with a definite rCBF levels in baseline Xe-CT will demonstrate it was a truly evolving ROI, with the formation of a hypoattenuated lesion, in the follow-up Xe-CT) and negative predictive value (NPV; the probability that a ROI having a certain rCBF levels will become a truly non evolving ROI). To determine these values, the prior likelihood of a rCBF-ROIs being evolving or non evolving also had to be accounted for. To obtain an estimate of this probability, we determined the proportion of evolving and non evolving ROIs in the four follow-up Xe-CT slices for all patients. Using Bayes' theorem, we then determined the PPV and NPV of every rCBF measured value.

Results

Patient characteristics

Mean age of the 34 patients included was 48.5 + 11.2 years, and 19 (55.9%) patients were female. According to the Hunt & Hess scale, patients with grade V and IV were 10 (29.4%), those with grade III were 15 (44.1%), whereas those with grade II and I were 9 (26.5%). The Fisher score was IV in 3 patients (8.8%), III in 23 patients (67.6%), II in 9 patients (23.6%). Twenty seven (79.4%) patients were discharged alive. The most common location of the aneurysm was the anterior communicating cerebral artery (16 patients, 47.1%), followed by the middle cerebral artery (8 patients, 23.5%), carotid artery (4 patients, 11.8%), posterior communicating cerebral artery (4 patients, 11.8%), anterior cerebral artery (1 patient, 2.9%) and corioid artery (1 patient, 2.9%). Twenty three (67.6%) patients underwent clipping, 9 patients (26.5%) coiling while, in 2 patients (5.9%) the aneurysm was not treated due to the extreme degree of the patient severity.

Characteristics of the Xe-CT studies

At least one paired study was analyzed per patient (22 patients, 64.7%), but multiple measurements were taken in 12 patients (two paired Xe-CT studies were obtained in 8 patients, 23.5%; three studies in 3 patients, 8.8%, and five studies in 1 patient, 2.9%). The median time which elapsed between bleeding and the Xe-CT study was, respectively, 93 hours (interquartile range, IQR 103) for baseline Xe-CT studies and 146.5 hours (IQR 103) for the follow up Xe-CT study. The median time interval between the two Xe-CT studies was 48 hours (IQR 27).

Evolving and non evolving ROIs

In the follow-up Xe-CT study, 110/2795 (3.94%) paired ROIs were found to develop a new hypoattenuation (evolving ROI) due to vascular causes, while in further 21 paired studies of ROIs the new hypoattenuation was due to edema which were not primarily due to macrovascular ischemia and so were not considered. The ROIs due to macrovessels hypoperfusion was located predominantly in the cortical mantle of the middle cerebral artery (n 50, 54.6%), then of the anterior cerebral artery (n 40, 40.9%) and of the posterior cerebral artery (n 5, 4.5%). The causes of the hypoattenuation were vasospasm in 68 ROIs (61.8%) and vascular occlusion or stenosis in 42 (38.2%). Mean rCBF levels in the baseline Xe-CT study were lower in evolving than in non evolving ROIs ($p \leq 0.0001$) (Table 1). A reduction of the mean rCBF level was observed from the baseline to the follow-up Xe-CT in evolving ROIs ($p=0.0002$), but not for non evolving ROIs ($p=0.3592$) (Table 1). An appreciable reduction in mean Hounsfield units values was found from the baseline to the follow-up Xe-CT studies in evolving ROI ($p \leq 0.0001$), while the mean Hounsfield units values were lower in the evolving ROIs in respect to non evolving ROIs in the follow up Xe-CT ($p \leq 0.0001$) (Table 1).

Determination of probabilities and examination of rCBF thresholds

Evolving ROIs comprised 3.94% of the entire ROI population, giving estimates of prior probability of evolving and non evolving outcome of 0.0394 and 0.9606. Figure 1 shows the PPV and NPV of evolving outcomes plotted against cortical rCBF levels in baseline Xe-CT studies. While the NPV was very high for all values of rCBF variables (as would be expected since evolving ROIs constitute only a very small proportion of the whole cortical ROIs), the PPV increase only at extremely low values of the physiological range of rCBF. Thus, only at these very low values of rCBF (5 ml/100 gr/min) is there some probability (PPV=28.3%) of a ROI becoming an evolving ROI with a new hypoattenuation. Looking to the slop of PPV (Fig. 1), it can be observed that below 20 ml/100 gr/min, PPV progressively increases. However, due to its very low PPV (6.6%), the threshold of 20 ml/100 gr/min seems to be of questionable utility in clinical judgment.

Discussion

The present study suggests that in deeply sedated patients with poor grade or complicated SAH, when a perfusion study is planned for various clinical reasons, and not specifically to detect ischemia due to macrovessel hypoperfusion, the prevalence of new cortical hypoattenuation is rare. Our results also seem to indicate that no absolute rCBF threshold of ischemia can been found in severe and complicated SAH patients and that the rCBF values moderately predict new ischemic cortical lesions. The levels of rCBF associated with more probability to new hypoattenuation seems to be lower than previously suggested [9].

There are several reasons which can explain the moderate prediction of rCBF for ischemia, as defined by

Table 1 Regional cerebral blood flow (rCBF) levels, Hounsfield Units and areas (mm^2) in evolving and non-evolving ROIs

ROIs	n (%)	Baseline Xe-CT	follow up Xe-CT	p*	Baseline Xe-CT	follow up Xe-CT	p*	Baseline Xe-CT	follow up Xe-CT	p*
		rCBF (ml/100 gr/min)			Hounsfield			mm^2		
		Median (IQR)			Mean (SD)			Mean (SD)		
Non evolving	2685 (96.06)	33 (23.8)	32.5 (24.2)	0.3592	31.6 (4.4)	31.4 (4.3)	≤0.0001	286.7 (6.0)	286.8 (5.8)	0.5962
Evolving	110 (3.94)	23.3 (21.9)	9.5 (26.2)	0.0002	32.5 (4.5)	29.4 (5.5)	≤0.0001	285.5 (6.0)	285.2 (5.8)	0.2954
p^		≤0.0001	≤0.0001		0.0269	≤0.0001		0.0343	0.2804	

Data from paired baseline and follow-up Xe-CT studies are shown.
* statistical comparison between paired baseline and follow-up Xe-CT ROIs variables
^ comparison between non-evolving and evolving ROIs variables.

standard CT criteria. First, an efficient threshold value for irreversible neuronal damage in SAH patients studied with Xe-CT was described in 1990 by Yonas and coworkers [9] who selected patients with clinical vasospasm. Therefore, the a priori probability of developing a new ischemic lesion, as well as the disease prevalence, was more consistent in comparison to our study. It is intuitive to consider that the rarer the disease is, that we want to diagnose, the less likely we are to reach a correct diagnosis. The low prevalence detected for new ischaemic lesions in our investigation may also be the consequence of the different clinical indications used to schedule Xe-CT studies which were performed either to improve the diagnosis when risk of ischemia was considered high, or even just simply to monitor patient conditions. In the latter case, the rate of new hypoattenuated lesions observed in follow up studies was potentially low, thus reducing the prevalence of evolving ROIs and, consequently, the PPV. Second, although CT is currently considered to be a sensitive and practical imaging method in detecting ischemic lesions, it is widely accepted that its ability to demonstrate the presence of ischemic penumbra is limited and less consistent compared to functional imaging [7]. Therefore, since, in this study the definition of evolving ROI was based on CT criteria represented by the occurrence of a new hypoattenuation in follow up studies, our probability of losing the effect of more subtle degrees of ischemia was high. In order to overcome this shortcoming future studies matching quantitative measurement of CBF and cytotoxic edema by means of diffusion Magnetic Resonance Imaging would be useful. Third, the deep sedation applied may alter the ratio between CBF and ischemia by affecting $CMRO_2$. The case mix selected in the present study was probably consistent with the most severe patients in whom the clinical history was complicated and a specialist medical treatment of intracranial hypertension was performed [2]. Consequently, the thresholds for ischemia could be lowered by sedation which is known to reduce $CMRO_2$ [5]. This could explain why the thresholds rCBF levels here found (lower than 5 ml/100 gr/min) could

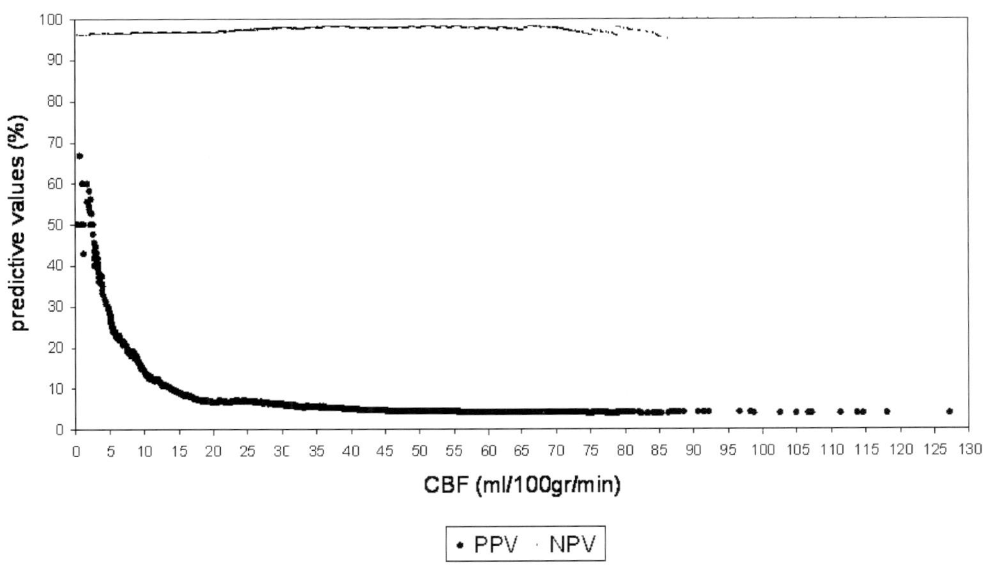

Fig. 1 Positive predictive value (continuous line) and negative predictive value (dashed line) for cerebral blood flow (CBF) levels in the baseline Xe-CT studies

be well below that of 15 ml/100 gr/min suggested by Yonas [9] who probably dealt with patients less profoundly sedated. Furthermore, the spontaneous reduction in $CMRO_2$ which seems to occur in SAH [1] could represent a further explanation for our finding that only extremely low rCBF values may be of some relevance. These converging findings suggest that hypometabolism might be one of the reasons explaining the low prevalence of new ischemic lesions at follow-up Xe-CT. Fourth, the time windows selected between the baseline and follow-up studies may be either too wide or too narrow. In fact, a critical elusive decline in rCBF might occur, after having already obtained the baseline Xe-CT study, and modify the tissue density in the remaining time elapsing before the follow-up Xe-CT study. Conversely more subtle, prolongated or phasic impairment of rCBF could lead to a reduction in CT densities which become appreciable after the 72 hours selected in this study. However, these are limitations due to the appropiateness of the timing chosen which affect every kind of functional imaging.

In conclusion, the current study suggests that in severe and complicated SAH, like in severe TBI [3], there is no absolute rCBF threshold that provides accurate prediction of the evolution of cortical tissue in ischemia. The CBF measurement alone seems not enough to predict the fate of brain tissue appropriately, at least when the risk factors are not clearly defined a priori to plan a Xe-CT study. Considering the snapshot nature of Xe-CT imaging, an accurate case mix and timing selection are most likely needed [4, 9]. Marginally, our study seems to support the concept that in poor grade or complicated SAH patients, the range of rCBF thresholds of ischemia might be consistently lower than previously described, probably also as a consequence of the deep sedation applied to control intracranial hypertension or to reduce metabolic requirements.

Conflict of interest statement We declare that we have no conflict of interest.

References

1. Carpenter DA, Grubb RL, Tempel LW, Powers WJ (1991) Cerebral oxygen metabolism after aneurysmal subarachnoid hemorrhage. J Cereb Blood Flow Metab 11:837–844
2. Chieregato A, Sabia G, Tanfani A, Compagnone C, Tagliaferri F, Targa L (2006) Xenon-CT and transcranial Doppler in poor-grade or complicated aneurysmatic subarachnoid hemorrhage patients undergoing aggressive management of intracranial hypertension. Intensive Care Med 32:1143–50
3. Cunningham AS, Salvador R, Coles JP, Chatfield DA, Bradley PG, Johnston AJ, Steiner LA, Fryer TD, Aigbirhio FI, Smielewski P, Williams GB, Carpenter TA, Gillard JH, Pickard JD, Menon DK (2005) Physiological thresholds for irreversible tissue damage in contusional regions following traumatic brain injury. Brain, 128:1931–1942
4. Horn P, Vajkoczy P, Bauhuf C, Munch E, Poeckler-Schoeniger C, Schmiedek P (2001) Quantitative regional cerebral blood flow measurement techniques improve noninvasive detection of cerebrovascular vasospasm after aneurysmal subarachnoid hemorrhage. Cerebrovasc Dis 12:197–202
5. Johnston AJ, Steiner LA, Chatfield DA, Coleman MR, Coles JP, Al-Rawi PG, Menon DK, Gupta AK (2003) Effects of propofol on cerebral oxygenation and metabolism after head injury. Br J Anaesth 91:781–786
6. Marchal G, Benali K, Iglesias S, Viader F, Derlon J-M, Baron J-C (1999) Voxel-based mapping of irreversible ischaemic damage with PET in acute stroke. Brain; 122:2387–400
7. Muir KW, Buchan A, von Kummer R, Rother J, Baron J-C (2006) Imaging of acute stroke. Lancet Neurol 5:755–768
8. Pindzola RR, Yonas H (1998) The xenon-enhanced computed tomography cerebral blood flow method. Neurosurgery 43:1488–1492
9. Yonas H, Sekhar L, Johnson DW, Gur D (1989) Determination of irreversible ischemia by xenon-enhanced computed tomographic monitoring of cerebral blood flow in patients with symptomatic vasospasm. Neurosurgery 24:368–372

Acta Neurochir Suppl (2008) 102: 317–320

Hyperbaric oxygen preconditioning activates ribosomal protein S6 kinases and reduces brain swelling after intracerebral hemorrhage

Z. Qin · Y. Hua · W. Liu · R. Silbergleit · Y. He · R. F. Keep · J. T. Hoff · G. Xi

Abstract
Background New protein synthesis is key to ischemic tolerance induced by preconditioning and ribosomal protein S6 kinases (p70 S6 K) are important enzymes in protein synthesis. Hyperbaric oxygen preconditioning (HBOP) reduces ischemic brain damage. This study investigated if HBOP can activate p70 S6 K and increase new protein synthesis and if HBOP induces brain tolerance against brain swelling after intracerebral hemorrhage (ICH).
Methods There were two parts of the studies. 1) Rats received five consecutive sessions of HBOP. Twenty-four hours after HBOP, the rats had an ICH and were sacrificed one or three days later for brain edema measurement. 2) Rats received five sessions of HBOP or control pretreatment and were sacrificed for Western blot analysis and immunohistochemistry of activated p70 S6 K and heme oxygenase-1 (HO-1).
Findings Five sessions of HBOP significantly reduced brain edema in the ipsilateral basal ganglia after ICH Western blot analysis showed that HBOP activated p70 S6 K and increased HO-1 levels in the basal ganglia. Strong activated p70 S6 K immunoreactivity was also found in the basal ganglia.
Conclusions Our results suggest activation of p70 S6 K may have a role in heat shock protein synthesis after HBOP and may contribute to HBOP-induced brain protection.

Keywords Hyperbaric oxygen · Preconditioning · Cerebral hemorrhage · Heme oxygenase-1

This study was supported by grants NS-017760, NS-039866 and NS-047245 from the National Institutes of Health (NIH) and the Scientist Development Grant 0435354Z from American Heart Association (AHA). The content is solely the responsibility of the authors and does not necessarily represent the official views of the NIH and AHA.

Z. Qin · Y. Hua · W. Liu · Y. He · R. F. Keep · J. T. Hoff · G. Xi (✉)
Department of Neurosurgery, Room 5018, BSRB,
University of Michigan,
Ann Arbor, MI 48109-2200, USA
e-mail: guohuaxi@umich.edu

R. Silbergleit
Department of Emergency Medicine, University of Michigan,
TC B1354/0303, 1500 East Medical Center Drive,
Ann Arbor, MI 48109-0303, USA

Introduction

Intracerebral hemorrhage (ICH) is a common and often fatal subtype of stroke and produces severe neurologic deficits in survivors. At present, there is no effective treatment that attenuates brain edema and improves long-term outcome in ICH, but recent evidence indicates that hemoglobin, hemoglobin breakdown products and an enzyme, heme oxygenase (HO)-1, that breaks down heme (it also called heat shock protein 32), maybe involved in ICH-induced brain injury [12].

Brain tolerance can be achieved by many preconditioning stimuli including a brief period of global or focal ischemia, hypoxia, hyperthermia, cytokines and intracerebral low dose thrombin infusion [4, 15]. Recent studies demonstrate hyperbaric oxygen preconditioning (HBOP) can also reduce ischemic brain damage [8]. However, it is not clear whether or not HBOP can reduce brain injury after ICH.

Although the precise mechanisms involved in preconditioning-induced brain tolerance have not been fully determined, new protein synthesis is required in delayed neuroprotection induced by preconditioning [2]. p70 S6 K activation is important for protein synthesis. The major target of the activated kinase is the 40S ribosomal protein

S6, a major regulator of protein synthesis [1]. In heart, activation of ribosomal protein S6 kinases (p70 S6 K) is essential for preconditioning-induced protection [5, 6].

This study investigated whether or not HBOP can activate p70 S6 K, increase HO-1 synthesis, and induce brain tolerance against brain swelling after ICH.

Materials and methods

Experimental groups

There were two parts to the study. In the first part, male Sprague-Dawley rats received 5 consecutive sessions of HBOP (3 ATA, 100% oxygen, one hour daily for 5 days). Control animals received normobaric room air. Twenty-four hours after HBOP, the rats received an intracaudate injection of autologous whole blood (100-μl) and were sacrificed one or three days later for brain edema measurement. In the second part, rats received either 5 sessions of HBOP or control pretreatment and were sacrificed for Western blot analysis and immunohistochemistry of activated p70 S6 K and HO-1 24 h after the final HBOP.

Animal preparation and intracerebral infusion

The animal study protocol was approved by the University of Michigan Committee on the Use and Care of Animals. Male Sprague-Dawley rats (Charles River Laboratories, Portage, MI) weighing 300 to 350 g were used in this study. Rats were allowed free access to food and water before and after experiment. Animals were anesthetized with intraperitoneal injection of pentobarbital (45 mg/kg). ICH was induced by intracaudate injection of autologous whole blood as previously described [13].

HBO administration

Animals in HBO treatment groups were placed in a small rodent HBO chamber (Marine Dynamics Corp. Long Beach, CA). HBO-treated animals were pressurized over 15 minutes to a plateau pressure of 3 ATA (Atmosphere Absolute) with 100% oxygen supplied continuously and maintained for 60 minutes. Decompression was then carried out over 25–30 minutes. Control animals were also transferred into the HBO chamber but received normobaric room air.

Brain water content measurement

Animals were reanesthetized (pentobarbital 60 mg/kg, i.p.) and decapitated 24 hours after ICH to measure brain water content using wet/dry weight method as described previously [13]. Tissue water content (%) was then calculated as: 100×(Wet Weight-Dry Weight)/Wet Weight.

Western blot analysis

Western blot analysis was performed as described previously [14]. The primary antibodies were polyclonal rabbit anti-p70 S6 K (1:1000 dilution, Cell Signaling Technology) and polyclonal rabbit anti-HO 1 (1:2000 dilution, Cell Signaling Technology). The second antibody was peroxidase-conjungated goat anti-rabbit (Bio-Rad, Hercules, CA). The antigen-antibody complexes were demonstrated with a chemiluminescence system (Amersham Pharmacia) and exposed to photosensitive film (X-OMAT; Kodak, Rochester, NY). Relative densities of the band were analyzed using NIH Image software (version 1.62).

Immunohistochemistry

Animals were reanesthetized and perfused intracardially with 4% paraformaldehyde in 0.1M PBS (pH 7.4). Immunohistochemistry was performed using the avidin-biotin complex technique as previously described [11]. The primary antibodies were polyclonal rabbit anti-p70 S6 K and polyclonal rabbit anti-HO 1 (1:400 dilution, Cell Signaling Technology). The second antibody was goat anti-rabbit immunoglobulin G (1:500 dilution, Vector Laboratories). Normal rabbit immunoglobulin G (Vector Laboratories) was used as a negative control.

Statistical analysis

Data are expressed as mean ± standard deviation (S.D.). Statistical significance was analyzed by two-tailed Student t test. Probability value of $p<0.05$ was considered statistically significant.

Results

All physiological variables, including mean blood pressure, blood gases, hematocrit, and blood glucose level were controlled within normal ranges.

The effects of HBOP on ICH-induced brain swelling were examined. Five sessions of HBOP reduced brain water content in the ipsilateral basal ganglia (79.5±0.9% vs. 81.2±1.1% in the control group, n=9–10, $p<0.01$) one day after ICH. HBOP also attenuated perihematomal edema (81.3±1.4% vs. 82.8±1.1% in controls,n=6, $p<0.05$) at 72 hours after ICH. HBOP did not affect brain water content in the contralateral hemisphere or cerebellum.

Western blot analysis showed an increase of activated p70 S6 K protein levels in the basal ganglia after five sessions of HBOP (e.g. 85 kDa subunit: 3814±1311 vs. 1273±526 pixels in controls, $p<0.05$, Fig. 1a). Immunohistochemistry also demonstrated a few activated p70 S6 K positive cells in the basal ganglia and p70 S6 K immunoreactivity was increased after HBOP (Fig. 1b). To examine whether or not HBOP increased the synthesis of new proteins, we measured brain HO-1 levels. HBOP increased HO-1 protein levels (2590±724 vs. 834±549 pixels in controls, $p<0.05$, Fig. 2).

Discussion

HBOP-induced ischemic tolerance was firstly reported in gerbils [10]. It has been reported that HBOP rather than simple hyperbaricity induces ischemic tolerance [3]. HBOP-induced ischemic tolerance may be strain dependent since HBOP can induce ischemic tolerance in SV129 mice but not in C57BL/6 mice [8]. In the present study, we demonstrate HBOP-induced tolerance against hemorrhagic brain damage.

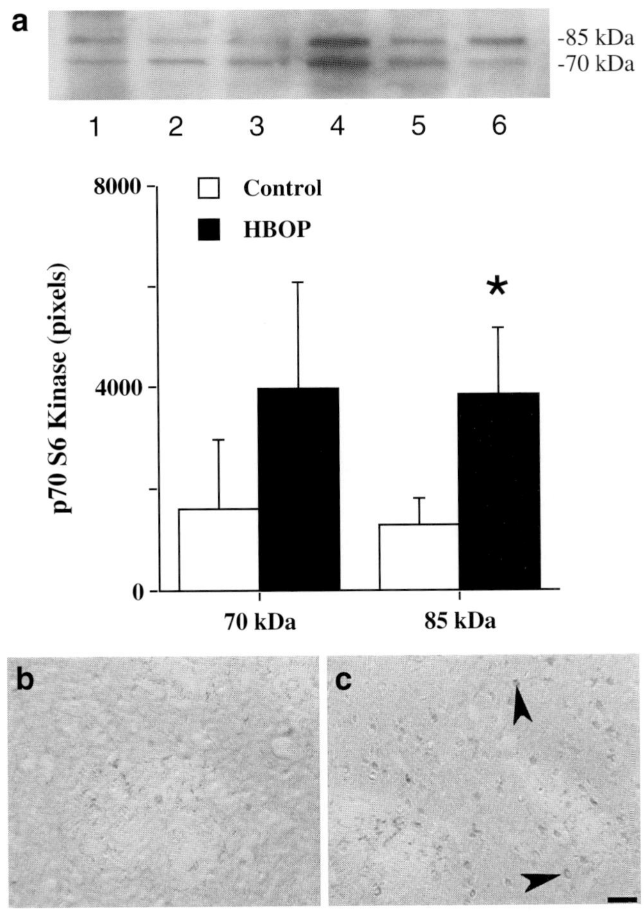

Fig. 1 Activated p70 S6 K levels in the brain after HBOP. **(a)** Western blots showing an increase of activated p70 S6 K in the basal ganglia (control, lanes 1–3;HBOP, lanes 4–6). Values are mean ± S.D., *$p<0.05$ vs. control. **(b & c)** Immunoreactivity of activated p70 S6 K in the basal ganglia 24 hours after HBOP (B: control, C: HBOP). Arrow heads indicate positive cells. Scale bar=50 μm

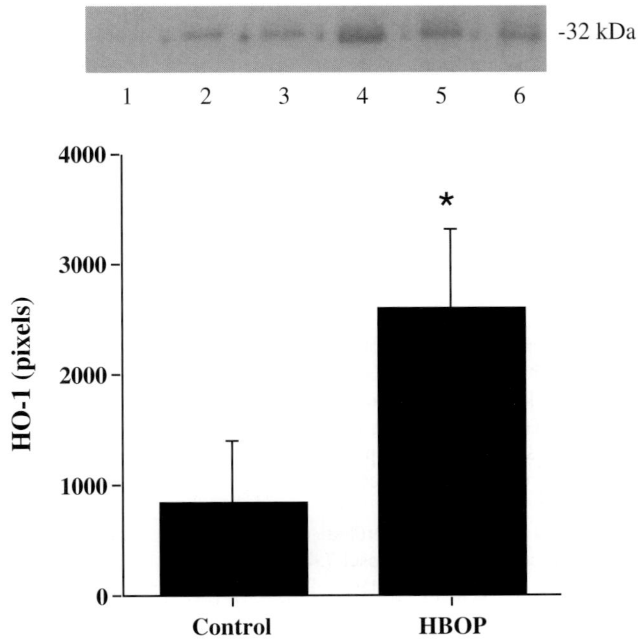

Fig. 2 Western blot analysis showing HO-1 levels in the basal ganglia. Lanes 1–3: control; Lanes 4–6: HBOP. Values are mean ± S.D., *$p<0.05$ vs. control

Ischemic preconditioning studies have been an excellent method for elucidating neuroprotective mechanisms, but the potential medical importance of this phenomenon has been questioned. A major advantage of HBOP compared to other preconditioning agents is that it is noninvasive and currently available for use in patients. Neurosurgical procedures may result in edema, hemorrhage and/or ischemia. Therefore, it is important to develop a safe way to precondition brain before surgery. HBOP may be such a pretreatment.

Protein synthesis is a major component of preconditioning in the brain and the requirement for synthesis underlies the delay that is required between the preconditioning stimulus and the injury-inducing event for protection to occur [2]. Here we showed that p70 S6 K, which may have a key role in protein synthesis, are activated by HBOP. The major target of the activated p70 S6 K is the 40S ribosomal protein S6, a major regulator of protein synthesis [1]. In heart, there is evidence that activation of p70 S6 K is crucial for preconditioning [5, 6].

HO-1 degrades heme into iron, carbon monoxide and biliverdin. Upregulation of brain HO-1 levels by HBOP may be neuroprotective. Studies suggest that HO-1 is involved in cytoprotection from different kinds of injuries [9], although HO-1 upregulation may also induce brain injury by increasing cellular iron content [7].

In summary, HBOP activates p70 S6 K, causes HO-1 upregulation, and induces brain tolerance against brain swelling after ICH. Pretreatment with hyperbaric oxygen might be beneficial for patients undergoing neurosurgical operations.

Conflict of interest statement We declare that we have no conflict of interest.

References

1. Berven LA, Crouch MF (2000) Cellular function of p70S6 K: a role in regulating cell motility. Immunol Cell Biol 78:447–451
2. Dirnagl U, Simon RP, Hallenbeck JM (2003) Ischemic tolerance and endogenous neuroprotection. Trends Neurosci 26:248–254
3. Dong H, Xiong L, Zhu Z, Chen S, Hou L, Sakabe T (2002) Preconditioning with hyperbaric oxygen and hyperoxia induces tolerance against spinal cord ischemia in rabbits. Anesthesiology 96:907–912
4. Gidday JM (2006) Cerebral preconditioning and ischaemic tolerance. Nat Rev Neurosci 7:437–448
5. Hausenloy DJ, Mocanu MM, Yellon DM (2004) Cross-talk between the survival kinases during early reperfusion: its contribution to ischemic preconditioning. Cardiovasc Res 63: 305–312
6. Hausenloy DJ, Tsang A, Mocanu MM, Yellon DM (2005) Ischemic preconditioning protects by activating prosurvival kinases at reperfusion. Am J Physiol Heart Circ Physiol 288:H971–976
7. Huang F, Xi G, Keep RF, Hua Y, Nemoianu A, Hoff JT (2002) Brain edema after experimental intracerebral hemorrhage: role of hemoglobin degradation products. J Neurosurg 96:287–293
8. Prass K, Wiegand F, Schumann P, Ahrens M, Kapinya K, Harms C, Liao W, Trendelenburg G, Gertz K, Moskowitz MA, Knapp F, Victorov IV, Megow D, Dirnagl U (2000) Hyperbaric oxygenation induced tolerance against focal cerebral ischemia in mice is strain dependent. Brain Res 871:146–150
9. Sharp FR, Massa SM, Swanson RA (1999) Heat-shock protein protection. TINS 22:97–99
10. Wada K, Ito M, Miyazawa T, Katoh H, Nawashiro H, Shima K, Chigasaki H (1996) Repeated hyperbaric oxygen induces ischemic tolerance in gerbil hippocampus. Brain Res 740:15–20
11. Xi G, Hua Y, Keep RF, Duong HK, Hoff JT (2001) Activation of p44/42 mitogen activated protein kinases in thrombin-induced brain tolerance. Brain Res 895:153–159
12. Xi G, Keep R, Hoff J (2006) Mechanisms of brain injury after intracerebral hemorrhage. Lancet Neurol 5:53–63
13. Xi G, Keep RF, Hoff JT (1998) Erythrocytes and delayed brain edema formation following intracerebral hemorrhage in rats. J Neurosurg 89:991–996
14. Xi G, Keep RF, Hua Y, Xiang JM, Hoff JT (1999) Attenuation of thrombin-induced brain edema by cerebral thrombin preconditioning. Stroke 30:1247–1255
15. Xi G, Reiser G, Keep RF (2003) The role of thrombin and thrombin receptors in ischemic, hemorrhagic and traumatic brain injury: deleterious or protective? J Neurochem 84:3–9

Acta Neurochir Suppl (2008) 102: 321–325

Stroke with subarachnoid hemorrhage: assessment of cerebrovascular pressure regulation and simulated cerebrovascular resistance

Michael L. Daley · Nithya Narayanan · Charles W. Leffler · Per Kristian Eide

Abstract
Background Monitoring methods designed to assess cerebrovascular regulation and increased cerebrovascular resistance (CVR) of patients with subarachnoid hemorrhage (SAH) would facilitate therapeutic intervention and potentially reduce secondary complications. The aim of this study was to assess changes of cerebrovascular regulation and CVR by evaluating changes of cerebrovascular pressure transmission in patients with SAH.
Methods Admission Hunt-Hess grades, Fisher scores, Glasgow Outcome Scores (GOS) at 6 months, and pressure recordings were obtained from 20 patients. Biomechanical models of cerebrovascular pressure transmission were constructed over one-minute intervals for the initial and final two hours of post-hemorrhage monitoring.
Findings Classified according to the GOS score at 6 months, eight patients died (GOS 1), five were severely disabled (GOS 3), and seven patients were moderately disabled (GOS 4). During the initial monitoring period 100%, 80%, and 28.6% of groups with GOS 1, 3, and 4 demonstrated impairment of cerebrovascular regulation, whereas, in the final monitoring period 100%, 100%, and 14.3% respectively demonstrated impairment. Between monitoring periods, simulated CVR (sCVR) significantly increased ($p<0.001$) for patients with GOS 1 and 3 and decreased for those with GOS 4 with mean resistance for the latter group significantly lower ($p<0.001$) than other means.
Conclusions Loss of cerebrovascular regulation and increased sCVR were observed in SAH patients with poor outcome.

Keywords Subarachnoid hemorrhage · Cerebrovascular pressure transmission · Cerebrovascular pressure regulation · Cerebrovascular resistance

M. L. Daley (✉) · N. Narayanan
Department of Electrical and Computer Engineering,
The University of Memphis,
Engineering Science Building, Rm. 208B,
Memphis, TN 38152-3180, USA
e-mail: mdaley@memphis.edu

C. W. Leffler
Department of Physiology,
The University of Tennessee Health Science Center,
301 Nash Building,
Memphis, TN 38163, USA

P. K. Eide
Department of Neurosurgery, The National Hospital,
Sognvannsvn 20,
Oslo, Norway

Introduction

Studies of patients with subarachnoid hemorrhage indicate that cerebral hemodynamics are commonly disturbed in the acute stage [6, 7, 9]. CBF may be reduced in patients without vasospasm [6, 9], and parenchymal arterial vessels demonstrate an impaired autoregulatory response [12]. Previously we have shown that changes in the highest modal frequency (HMF) of cerebrovascular transmission, the dynamic relationship between arterial blood pressure (ABP) and intracranial pressure (ICP), relative to changes of cerebral perfusion pressure (CPP) are a means to assess cerebrovascular regulation [3, 4]. More recently we have developed a two-step modeling technique by incorporating the derived modal frequencies of the numerical model and the actual blood pressure recording with a physiologically-based biomechanical model to produce a simulated ICP recording which matches the actual ICP recording. Using this modeling procedure we have validated that laboratory measures of changes of arteriolar resistance and compliance during induced hypercapnia are replicated by the physiologically-based biomechanical model [5]. The aim of this retrospective study was to assess changes of cerebrovascular regulation and cerebrovascular resistance by evaluating changes of cerebrovascular pressure transmission in patients with SAH using the two-step modeling method.

Materials and methods

Standard ICP and ABP monitoring is part of the management strategy of adult stroke patients with SAH at The National Hospital, Oslo, Norway with the aim of controlling ICP. The attributes of the patients grouped according to the Glasgow Outcome Score (GOS) at 6 months and the admission Hunt-Hess (HH) grades and Fisher scores (FS), age, and sex are given in Table 1. Pressure recordings were labeled such that the patients could not be identified by the Memphis group. The Institutional Review Board at The University of Memphis approved the data analysis protocol.

Pressure recordings were digitized at 100 Hz. Recording periods were not uniform; mean monitoring time (+/−S.D.) was 33.6 (+/−30.4) hrs. Some patients were monitored periodically over a course of several days while others were only monitored once. Mean values of HMF and simulated CVR (sCVR) were determined for the first two hours, middle two hours, and last two hours of each monitoring period.

The first step of the two step modeling method uses system identification modeling; the details of which have been previously reported [3–5]. Briefly, system identification modeling method produces a black box model with a simulated output that matches the actual output by minimization of the least squares difference between the actual and simulated digitized output files [8]. Of critical importance in the use of system identification modeling is the selection of the generalized description of the dynamic process to be modeled [8]. We chose the physiologically-based biomechanical model proposed by Czosynka and his colleagues [2] to provide the basis of the required generalized dynamic differential equation of cerebrovascular pressure transmission. The autoregressive moving average system identification technique is applied to each 6000 paired samples of digitized pressure recordings of 1 min to obtain the modal frequencies of cerebrovascular pressure transmission. For the brief period on which each numerical model is based, perturbations of ABP and ICP are assumed to be small and cerebrovascular pressure transmission is considered to be linear and time-invariant.

In the second step of the modeling method, the mean (±S.D.) values of the modal frequencies derived by the first step numerical model and the corresponding 1 min interval of ABP recording are used to guide the manipulation of the parameters of the physiologically-based biomechanical model such that the simulated 1 min recording of ICP matches the actual ICP recording over the interval. Of note is that three modifications of the biomechanical model proposed by Czosynka and colleagues [2] were implemented. First, the description of the resistance of the arterial-arterioles and capillaries was slightly modified by placing the element representing the arterial-arteriolar bed at the mid-point, as it has been modeled by others previously [1]. Secondly, in addition to the representation of the bridging veins with variable resistance, cerebral veins within the parenchyma were also considered to vary. Thirdly, the representation of venous sinus pressure was replaced by an external venous resistance component and connected to a central venous pressure generator as previously modeled [11]. A MATLAB algorithm was constructed to simulate ICP for each ABP recording and the corresponding numerically derived modal frequencies for each modeling interval. The algorithm was designed to manipulate the resistance and compliance elements of the physiologically-based biomechanical model such that the minimum square error for the 6000 sample interval between: 1) the actual and simulated ICP recordings; 2) the numerical model value of HMF and the biomechanical value of HMF; and 3) the numerical model value of dampening factor (DF) of the lower modal frequencies and the DF of the biomechanical model were minimized. Use of DF to represent the dominant two modal frequencies simplified the computation of matching the numerically derived and simulated lower modal frequencies by eliminating the need for complex arithmetic for the condition of an under-damped system. Values of each element of the biomechanical model were manipulated. Initially a normal CBF was assumed to be 600 ml/min [10, 11] which corresponds to a CBF value of 40 ml/100 gm/min for an

Table 1 Attributes of patients grouped according to 6-month GOS: sex, age, and fisher score and hunt-hess grade

GOS	N	Sex (F(M))	Age (+/−SD) yrs	FS[1] (+/−SD)	HH[1] (+/−SD)
1	8	5(3)	51.6 (10.8)	3.2 (0.3)	4.5 (1.1)
3	5	2(3)	61.8 (10.4)	3.4 (0.2)	3.4 (1.1)
4	7	4(3)	53.4 (7.1)	2.6 (0.6)[1]	3.6 (1.4)

[1] Assumes continuous scale of category driven integer scales

assumed brain weight of 1.5 kg. Both compliance and resistance elements were allowed to range over values consistent with those reported in previously published modeling studies of intracranial pressure dynamics [2, 11]. Compliance of the arterial/arteriolar bed was allowed to range from 0.01 to 10 ml/mmHg while venous and intracranial compliance were varied between 0.01 to 2 ml/mmHg [2, 11]. Venous resistance varied from 0.025 to 0.5 mmHg−100gm-min/ml and resistance of the arterial-arteriolar bed varied from 0.5 to 6.25 mmHg−100gm-min/ml [11]. A maximum mean squared error of 0.6 and the minimum correlation value of 0.9 between both actual and simulated ICP recording was used to initialize the square error values. The parameters of the biomechanical model were estimated by a recursive algorithm to obtain a best fit set which produced a minimum least square error and maximum correlation value for each segment of the pressure recording. Because simulated arterial-arteriolar capillary resistance was much greater than simulated cerebral venous resistance, we used the value of arterial-arteriolar capillary resistance to provide an assessment of simulated cerebrovascular resistance (sCVR).

All mean values are reported ± standard deviation. In all cases, the degree of significance between two mean values was determined by using the paired t-test. The correlation between highest modal frequency (HMF) and CPP was used to assess cerebrovascular regulation for the initial and final two-hours of monitoring [3, 4]. An inverse relationship between HMF and CPP is an indication of intact regulation; whereas, a direct relationship indicates impaired regulation. For each GOS-group the percentage of patients with impaired regulation during the initial two hours and final two hours was computed.

Results

Mean values of HMF and sCVR at the beginning, middle, and final hour of each monitoring session are plotted (see Fig. 1). GOS 1 and GOS 3 groups demonstrated trends of a decrease in HMF and an increase of sCVR with time; whereas, patients in the GOS 4 group demonstrated opposite trends. Grand mean regression line slopes (+/−SD) of HMF/day and sCVR/day for GOS 4 patients were 2.95 (+/−0.56) Hz/day and −0.45 (+/−0.07) mmHg −100 gm-min/ml/day and significantly ($p<0.001$) different compared to the corresponding mean values of both GOS 1 and GOS 3 groups at −0.58 (+/−0.17) Hz/day and 0.39 (+/−0.16) mmHg−100 gm-min/ml/day and −2.23 Hz/day and 0.41 (+/−0.06) mmHg−100 gm-min/ml/day, respectively. The relationship between HMF and CPP was used to assess cerebrovascular regulation [3, 5]. An inverse relationship between HMF and CPP is an indication of intact regulation; whereas, a direct relationship indicates impaired regulation [3, 5]. Mean values of initial two hours and final two hours monitored post hemorrhage, CPP, sCVR, and the percentage of the number of patients within each GOS group determined to have impaired regulation for the initial two hours and final two hours of monitoring post hemorrhage were computed (see Table 2). During the initial two hours of monitoring 100%, 80%, and 28.6% of patients with 6-month GOS of 1, 3, and 4 respectively demonstrated impaired regulation; whereas, in the final hours 100%, 100%, and 14.3% respectively demonstrated impairment. Between the initial two hours and final two hours of monitoring, sCVR significantly increased ($p<0.001$) for patient with GOS 1 and 3 and decreased for those with GOS 4. The mean value of sCVR during the final monitoring hour was significantly lower ($p<0.001$) for patients with GOS 4. This value reflected a simulated CBF of 26.4 ml/100 gm-min; whereas the corresponding mean values of sCVR for the GOS 1 and GOS 3 groups reflected ischemic simulated CBF values of 9.9 and 9.5 ml/100 gm-min respectively.

Discussion

The new findings of this retrospective study are that loss of cerebrovascular regulation occurs early following SAH and model-derived cerebrovascular resistance progressively increases for patients with poor outcome after SAH. In contrast, intact cerebrovascular regulation with a progressive decrease of model-derived cerebrovascular resistance was observed in those with more favorable outcome. These findings are also consistent with an earlier study that found assessment of loss of cerebrovascular pressure regulation within 1–6 days post-hemorrhage is predictive of poor outcome [7].

Caution should be made when relating these data to outcome since the numbers of patients in each group were low and only two two-hour periods, the first two hours and final two hours of monitoring, were used to evaluate cerebrovascular pressure regulation. Neither angiographic images nor evaluations for symptomatic vasospasm nor measurement of CBF were available. Delayed vasospasm of conducting vessels may have contributed to the loss of cerebrovascular regulation. However, because of the early loss of cerebrovascular regulation and significant increase of the model derived resistance sCVR it is likely that increasing more generalized vasoconstriction of the microvasculature is involved.

In conclusion, in the early stages of Aneurysmal SAH loss, of cerebrovascular regulation and increasing values of model-derived cerebrovascular resistance were observed in patients with poor outcome.

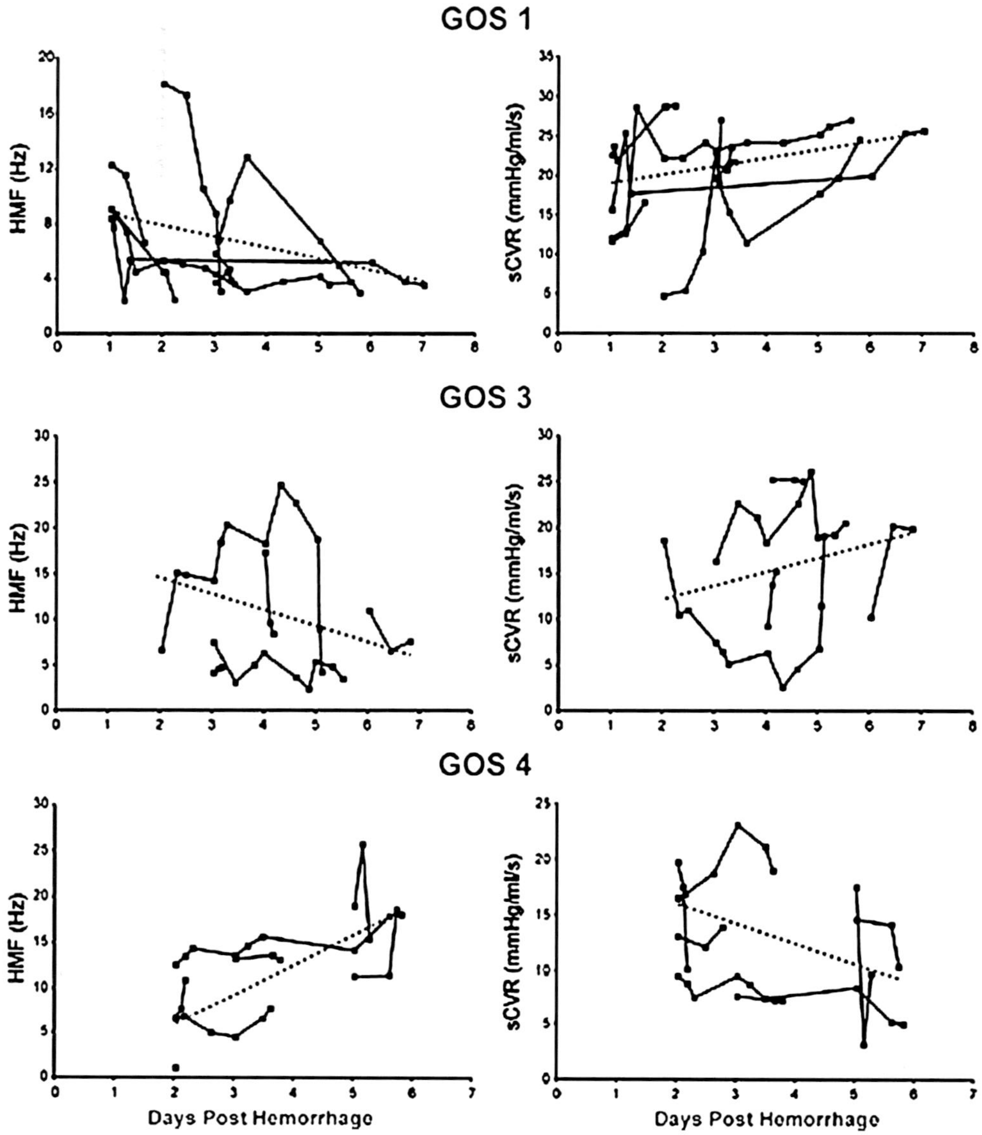

Fig. 1 Trends in HMF and sCVR for GOS 1, 3, and 4 Patients vs. Days from Occurrence of Subarachnoid Hemorrhage. Each symbol represents a mean value of 120 one-minute values plotted with standard error bars within each symbol. Thin solid line connects values at the initial, middle, and final two hour intervals of each contiguous monitoring period. Top. GOS 1: As HMF decreased ($p<0.001$) with time sCVR increased ($p<0.05$). Middle. GOS 3: As HMF decreased ($p<0.001$) with time sCVR increased ($p<0.025$). Bottom. GOS 4: As HMF increased ($p<0.001$) with time sCVR decreased ($p<0.01$). Dashed line through each graph represents regression line of all values

Table 2 Mean(± S.D) of initial and final monitoring time post hemorrhage, CPP, sCVR and percent impaired regulation at each monitoring period

GOS	N	Initial Time (+/−SD) day	CPP (+/−SD) mmHg	Percent Impaired Reg.	sCVR (+/−SD) mmHg-min−100 g/ml	Final Time (+/−SD) day	CPP (+/−SD) mmHg	Percent Impaired Reg.	sCVR (+/−SD) mmHg-min−100 g/ml
1	8	1.7 (0.8)[1,2]	74.8 (9.7)	100 8/8	4.3 (2.2)[3]	3.8 (1.5)	68.1 (9.0)	100 (8/8)	6.9 (0.8)[3]
3	5	3.6 (1.5)[1]	72.3 (9.8)	80.0 (4/5)	3.0 (0.20)[4]	4.8 (1.3)	69.5 (10.8)	100 (5/5)	7.3 (0.9)[4]
4	7	3.0 (1.4)[2]	77.4 (8.3)	28.6 (2/7)	3.9 (1.5)	3.6 (1.4)	79.1 (8.0)	14.3 (1/7)	3.0 (1.6)[5]

[1,2] Significant difference of GOS 1 initial time from GOS 3 ($p<0.01$)[1] and GOS 4 ($p<0.05$)[2]

[3] Significant difference between initial and final mean values of SCVR for GOS 1 ($p<0.01$)

[4] Significant difference between initial and final mean values of SCVR for GOS 3 ($p<0.001$)

[5] Significant difference of mean SCVR GOS 4 from means of GOS 1 and 3 at $p<0.001$

Conflict of interest statement We declare that we have no conflict of interest.

References

1. Agarwal GC, Berman BM, Stark L (1969) A lumped parameter model of the cerebrospinal fluid system. IEEE Trans Biomed Eng 16:45–53
2. Czosnyka M, Piechnik P, Richards HK, Kirpatrick P, Smielewski P, Pickard J (1997) Contribution of mathematical modeling to the interpretation of bedside tests of cerebrovascular autoregulation. J Neurol Neurosurg Psychiatry 63:721–731
3. Daley ML, Leffler CW, Czosnyka M, Pickard JD (2006) Intracranial pressure monitoring: modeling cerebrovascular pressure transmission. Acta Neurochir Suppl 96:103–107
4. Daley ML, Pourcyrous M, Timmons SD, Leffler CW (2005) Assessment of cerebrovascular autoregulation: changes of the highest modal frequency of cerebrovascular pressure transmission with cerebral perfusion pressure. Stroke 35:1952–1956
5. Daley ML, Pourcyrous M, Timmons SD, Leffler CW (2007) Mode Changes of cerebrovascular pressure transmission induced by cerebral vasodilation. J Neurotrauma 24:559–566
6. Frykholm P, Andersson JL, Langstrom B, Enblad P (2004) Haemodynamic and metabolic disturbances in the acute stage of subarachnoid hemorrhage demonstrated by PET. Acta Neurol Scand 109:25–32
7. Lang EW, Diehl RR, Mehdorn M (2001) Cerebral autoregulation testing after aneurismal subarachnoid hemorrhage: the phase relationhip between arterial blood pressure and cerebral blood flow velocity. Crit Care Med 29:158–163
8. Ljung L (1987) System identification: theory for the user. Prentice Hall, Upper Saddle River, NJ, pp 1–12
9. Miranda P, Lagares A, Alen J, Perez-Nunzes A, Arrese I, Lobato RD (2006) Early transcranial Doppler after subarachnoid hemorrhage: clinical and radiological correlations. Surg Neurol 65:247–252
10. Ramsay SC, Murphy K, Shea SA, Friston KJ, Lammertsma AA, Clark JC, Adams L, Guz A, Frackowiak RSJ (1993) Changes in global cerebral blood flow in humans: effect on regional cerebral blood flow during a neural activiation task. J Physiol 471:521–534
11. Ursino M (1988) A mathematical study of human intracranial hydrodynamics Part I-The cerebrospinal fluid pulse pressure. Ann Biomed Eng 16:379–401
12. Yundt KD, Grubb RL, Diringer MN, Powers WJ (1998) Autoregulatory vasodilation of parenchymal vessels is impaired during cerebral vasospasm. J Cereb Blood Flow Metab 18:419–424

Acta Neurochir Suppl (2008) 102: 327–330

Effects of melatonin in early brain injury following subarachnoid hemorrhage

Robert E. Ayer · Takashi Sugawara · John H. Zhang

Abstract

Background Aneurysmal subarachnoid hemorrhage (SAH) is a devastating disease that is associated with significant morbidity and mortality. There is substantial evidence to suggest that oxidative stress is significant in the development of acute brain injury following SAH. Melatonin is a strong antioxidant that has low toxicity and easily passes through the BBB. Previous studies have shown that melatonin provides neuroprotection in other models of CNS injury.

Methods This experiment evaluates melatonin as a neuroprotectant against early brain injury following SAH. The endovascular perforation model of SAH was performed in male Sprague Dawley rats followed by the administration of melatonin two hours after the insult. Mortality and brain water content were assessed 24 after SAH.

Findings A significant reduction in 24 h mortality was seen following treatment with 150 mg/kg of melatonin. Brain water content was evaluated in the high dose treatment group to see if a reduction in brain edema was associated with reduced mortality. High dose melatonin tended to reduce brain water content following SAH.

Conclusions Large doses of melatonin significantly reduced mortality and brain water content in rats following SAH.

R. E. Ayer · T. Sugawara · J. H. Zhang
Department of Physiology and Pharmacology,
Loma Linda University Medical Center,
Loma Linda, CA, USA

J. H. Zhang
Department of Neurosurgery,
Loma Linda University Medical Center,
11234 Anderson Street, Room 2562B,
Loma Linda, CA, USA

J. H. Zhang (✉)
Department of Anesthesiology,
Loma Linda University Medical Center,
Loma Linda, CA, USA
e-mail: johnzhang3910@yahoo.com

Keywords Subarachnoid hemorrhage · Melatonin · Early brain injury · Mortality

Introduction

Spontaneous subarachnoid haemorrhage (SAH) represents 5–7% of all strokes and carries significant morbidity and mortality [19]. Risk of mortality approaches 50% within 30 days [22, 23]. Early brain injury immediately following SAH has been linked to poor outcomes [4, 25]. Many studies have provided evidence that oxidative stress plays a significant role in the pathophysiology of early brain injury following SAH [10, 16]. Melatonin is a pharmacological agent that is recognized as a powerful anti-oxidant that easily passes across the blood-brain barrier (BBB) [3, 6, 7]. It has been shown to be neuroprotective in several models of CNS injury that include: traumatic brain injury, ischemic stroke, and spinal cord injury [1, 11–13, 17, 20, 28]. This study hypothesizes that the administration of melatonin following SAH will provide neuroprotection and improve outcomes.

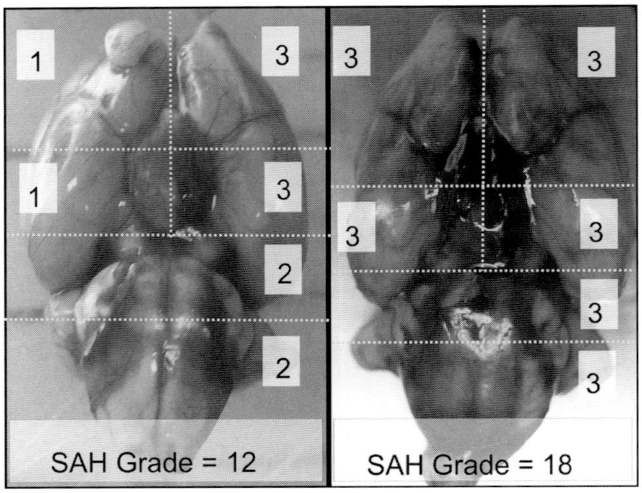

Fig. 1 Demonstration of the SAH grading system by Sugawara et al. Each section is scored 0–3: 0 = no blood; 1 = minimal blood; 2 = thick clot present with visible vessels; 3 = loss of visualization of the Circle of Willis in that sector. The left image is a sample that was excluded for small blood volume. The right image represents the most severe SAH in the grading system

Materials and methods

Male Sprague Dawley rats weighing 300 to 380 grams were used for this study. Four treatment groups were initially evaluated: [1] SHAM surgery given vehicle (10% ethanol); [2] SAH given vehicle; [3] SAH plus 15 mg/kg melatonin; [4] SAH plus 150 mg/kg melatonin. Rats were anesthetized with i.p. Ketamine (100 mg/kg) and Xylazine (10 mg/kg), intubated, and mechanically ventilated throughout surgery. Core body temperature was kept at 37±0.5°C. Rats underwent endovascular perforation of the internal carotid artery bifurcation as previously described [24]. In summary, a sharpened 4–0 nylon suture was thread through the ICA to puncture the cerebral vasculature at the ICA bifurcation. Due to the considerable variability in the hemorrhage volume produced by this model, the amount of blood in the basal cisterns was quantified according to the scale developed by Sugawara et al [27]. Animals received a score

Table 1 Distribution of SAH severity

Mortality	n	Average ±S.E.
SAH+Vehicle	27	16.333±0.207
SAH+15 mg/kg	13	16.462±0.291
SAH+150 mg/kg	10	16.800±0.327
Brain Water Content	n	Average ±S.E.
SAH+Vehicle	7	15.857±0.261
SAH+150 mg/kg	6	16.500±0.428

Distribution of SAH Grades. All groups that were compared had no significant differences in the volumes of subarachnoid blood. All animals with SAH grades less than 15 were excluded from analysis.

Table 2 Brain water content in animals 24 h after SAH

Experimental Group	n	Right Hemisphere	Left Hemisphere	Cerebellum	Brain Stem
SHAM+Vehicle	10	0.789±0.000442	0.792±0.00339	0.790±0.00110	0.736±0.00563
SAH+Vehicle	7	0.795±0.00253*	0.804±0.00551*	0.806±0.00723*	0.745±0.00371
SAH+150 mg/kg	6	0.793±0.00158	0.797±0.00313	0.801±0.00563	0.735±0.00875

High dose melatonin prevents significant increases in brain water content at 24 h.
*Indicates significant increase in brain water content vs sham ($p<0.05$ ANOVA)

for SAH severity that ranged from no blood (0), to severe hemorrhage (18) in the basal cisterns. Only those animals with hemorrhages ≥ 15 were evaluated. Melatonin, N-Acetyl-5-methoxytryptamine, (Sigma Aldrich) was dissolved in 10% ethanol for intraperitoneal (i.p.) injection. Dosages of 15 mg/kg and 150 mg/kg were given two hours after SAH. Survival was assessed continuously for 8 hours and again at 24 hours after SAH. Animals were then sacrificed and brain edema was evaluated at 24 h by measuring brain water content. Briefly, brains were sectioned, immediately weighed, dried for 48 h, and weighed again to calculate brain water content: [(wet weight – dry weight)/wet weight] x 100.

Results

A total of one hundred and sixty seven rats were used in this experiment. One hundred and forty five underwent endovascular SAH, 16 underwent SHAM surgery, and 6 animals had cardiopulmonary arrest after the induction of anesthesia leading to their exclusion from analysis. Fifty one percent (75/145) of rats that underwent endovascular puncture received an SAH score of fifteen or higher according to the Sugawara grading scale (Fig. 1). Animals with SAH grades less than 15 were excluded because there is no mortality in the model with these smaller bleeds. All groups evaluated at 24 h had similar SAH grade distributions (Table 1). In animals that survived the initial two hours and received treatment, the 24 h survival rate was significantly improved by 150 mg/kg of melatonin (Grehan Breslow Survival analysis $p<0.05$) versus vehicle or 15 mg/kg treatment. The brain water content of the 150 mg/kg group was measured to determine if reduced brain edema was associated with the decreased mortality. High dose melatonin prevented significant increases in brain water content (Table 2).

Discussion

Treatment with high dose melatonin significantly reduced mortality, which was associated with a reduction in brain water content at 24 h. Brain edema is identified on CT following SAH in 6–8% of cases and is an independent predictor of poor outcome and mortality [8]. Both cytogenic [5, 21, 26], and vasogenic [9] edema have been described in the acute period following SAH. Edema formation following SAH is initiated by the global cerebral ischemia that immediately follows aneurysm rupture [13, 15, 17, 28]. Melatonin treatment has been linked to reductions in cerebral edema following global ischemia. It has been shown that melatonin treatment reduced edema in the cerebellum as a result of the down regulation of vascular endothelial growth factor (VEGF) and astrocytic aquaporin 4 (AQP4) protein expression [15]. VEGF gene and protein expression are increased under hypoxic conditions and are linked to the breakdown of the BBB following SAH [14, 18]. The perivascular astrocyte protein, AQP4, a water transport molecule, has been implicated in cerebral edema formation, and these proteins are also elevated in SAH [2, 29]. Melatonin before and after middle cerebral artery occlusion (MCAO) resulted in reductions in cerebral edema that were greatest in areas of increased astrocyte density [17, 28]; improvements in BBB integrity and cerebral edema were also found with melatonin treatment following cold induced infarction [13]. Together, these studies demonstrate a role for melatonin in the reduction of cerebral edema following ischemic insults. This current study suggests that melatonin may also be effective at reducing brain edema following SAH.

Acknowledgements This study was partially supported by grants from NIH NS45694, HD43120, and NS43338 to JHZ.

Conflict of interest statement We declare that we have no conflict of interest.

References

1. Aydin MV, Caner H, Sen O, Ozen O, Atalay B, Cekinmez M, Altinors N (2005) Effect of melatonin on cerebral vasospasm following experimental subarachnoid hemorrhage. Neurol Res 27:77–82
2. Badaut J, Brunet JF, Grollimund L, Hamou MF, Magistretti PJ, Villemure JG, Regli L (2003) Aquaporin 1 and aquaporin 4 expression in human brain after subarachnoid hemorrhage and in peritumoral tissue. Acta Neurochir Suppl 86:495–498
3. Barchas J, DFSS (1967) Acute pharmacology of melatonin. Nature 214:919–920
4. Broderick JP, Brott TG, Duldner JE, Tomsick T, Leach A (1994) Initial and recurrent bleeding are the major causes of death following subarachnoid hemorrhage. Stroke 25:1342–1347
5. Busch E, Beaulieu C, de Crespigny A, Moseley ME (1998) Diffusion MR imaging during acute subarachnoid hemorrhage in rats. Stroke 29:2155–2161
6. Cheung RT (2003) The utility of melatonin in reducing cerebral damage resulting from ischemia and reperfusion. J Pineal Res 34:153–160
7. Cheung RT, Tipoe GL, Tam S, Ma ES, Zou LY, Chan PS (2006) Preclinical evaluation of pharmacokinetics and safety of melatonin in propylene glycol for intravenous administration. J Pineal Res 41:337–343
8. Claassen J, Carhuapoma JR, Kreiter KT, Du EY, Connolly ES, Mayer SA (2002) Global cerebral edema after subarachnoid hemorrhage: frequency, predictors, and impact on outcome. Stroke 33:1225–1232
9. Doczi T (1985) The pathogenetic and prognostic significance of blood-brain barrier damage at the acute stage of aneurysmal subarachnoid haemorrhage. Clinical and experimental studies. Acta Neurochir (Wien) 77:110–132
10. Gaetani P, Pasqualin A, Baena R, Borasio E, Marzatico F (1998) Oxidative stress in the human brain after subarachnoid hemorrhage. J Neurosurg 89:748–754

11. Genovese T, Mazzon E, Crisafulli C, Esposito E, Di Paola R, Muia C, Di Bella P, Bramanti P, Cuzzocrea S (2007) Effects of combination of melatonin and dexamethasone on secondary injury in an experimental mice model of spinal cord trauma. J Pineal Res 43:140–153
12. Genovese T, Mazzon E, Muia C, Bramanti P, De Sarro A, Cuzzocrea S (2005) Attenuation in the evolution of experimental spinal cord trauma by treatment with melatonin. J Pineal Res 38:198–208
13. Gorgulu A, Palaoglu S, Ismailoglu O, Tuncel M, Surucu MT, Erbil M, Kilinc K (2001) Effect of melatonin on cerebral edema in rats. Neurosurgery 49:1434–1441
14. Josko J, Gwozdz B, Hendryk S, Jedrzejowska-Szypulka H, Slowinski J, Jochem J (2001) Expression of vascular endothelial growth factor (VEGF) in rat brain after subarachnoid haemorrhage and endothelin receptor blockage with BQ-123. Folia Neuropathol 39:243–251
15. Kaur C, Sivakumar V, Zhang Y, Ling EA (2006) Hypoxia-induced astrocytic reaction and increased vascular permeability in the rat cerebellum. Glia 54:826–839
16. Kaynar MY, Tanriverdi T, Kemerdere R, Atukeren P, Gumustas K (2005) Cerebrospinal fluid superoxide dismutase and serum malondialdehyde levels in patients with aneurysmal subarachnoid hemorrhage: preliminary results. Neurol Res 27:562–567
17. Kondoh T, Uneyama H, Nishino H, Torii K (2002) Melatonin reduces cerebral edema formation caused by transient forebrain ischemia in rats. Life Sci 72:583–590
18. Mayhan WG (1999) VEGF increases permeability of the blood-brain barrier via a nitric oxide synthase/cGMP-dependent pathway. Am J Physiol 276:C1148–C1153
19. McCormick WF, Nofzinger JD (1965) Saccular Intracranial Aneurisms: An Autopsy Study. J Neurosurg 22:155–159
20. Mesenge C, Margaill I, Verrecchia C, Allix M, Boulu RG, Plotkine M (1998) Protective effect of melatonin in a model of traumatic brain injury in mice. J Pineal Res 25:41–46
21. Orakcioglu B, Fiebach JB, Steiner T, Kollmar R, Juttler E, Becker K, Schwab S, Heiland S, Meyding-Lamade UK, Schellinger PD (2005) Evolution of early perihemorrhagic changes–ischemia vs. edema: an MRI study in rats. Exp Neurol 193:369–376
22. Schievink WI (1997) Intracranial aneurysms. N Engl J Med 336:28–40
23. Schievink WI, Riedinger M, Jhutty TK, Simon P (2004) Racial disparities in subarachnoid hemorrhage mortality: Los Angeles County, California, 1985–1998. Neuroepidemiology 23:299–305
24. Schwartz AY, Masago A, Sehba FA, Bederson JB (2000) Experimental models of subarachnoid hemorrhage in the rat: a refinement of the endovascular filament model. J Neurosci Methods 96:161–167
25. Sehba FA, Bederson JB (2006) Mechanisms of acute brain injury after subarachnoid hemorrhage. Neurol Res 28:381–398
26. Sibon I, Menegon P, Rouanet F, Dousset V, Orgogozo JM (2004) MRI of acute brainstem ischaemia: cytotoxic versus vasogenic oedema? Eur J Neurol 11:497–498
27. Sugawara T, Ayer RE, Jadhav V, Zhang JH (2008) A New Grading System Evaluating Bleeding Scale in Filament Perforation Subarachnoid Hemorrhage Rat Model. J Neurosci Methods 167(2):327–334
28. Torii K, Uneyama H, Nishino H, Kondoh T (2004) Melatonin suppresses cerebral edema caused by middle cerebral artery occlusion/reperfusion in rats assessed by magnetic resonance imaging. J Pineal Res 36:18–24
29. Yatsushige H, Ostrowski RP, Tsubokawa T, Colohan A, Zhang JH (2007) Role of c-Jun N-terminal kinase in early brain injury after subarachnoid hemorrhage. J Neurosci Res 85:1436–1448

Acta Neurochir Suppl (2008) 102: 331–333

Decompressive craniectomy for hemispheric infarction: predictive factors for six month rehabilitation outcome

G. K. Wong · J. Kung · S. C. Ng · X. L. Zhu · W. S. Poon

Abstract

Background Decompressive craniectomy after hemispheric infarction has been shown to reduce mortality and functional outcome in selected patients. However, the optimal timing for surgery and patient most likely to benefit from this procedures was not known. We aimed to determine possible factors predictive of outcome following decompressive craniectomy for ischemic infarction from review of oneurological outcome in our patients at six months.

Methods We retrospectively reviewed 21 patients who underwent decompressive craniectomy for hemispheric infarction over a three year period in a regional neurosurgical center in Hong Kong. All patients were recruited subsequently for active in-patient rehabilitation, when suitable.

Findings The median age was 53 and the male to female ration was 1:3. Four patients (19%) achieved independent activity of daily living at six months after rehabilitation. Neither early surgery, within 24–48 hours after admission, nor side of infarction correlated with six month neurological outcome. All four patients with favourable neurological outcome at six month demonstrated favourable clinical improvement even at one month.

Conclusions Early decompressive hemicraniectomy is not predictive of neurological outcome, determined by Glasgow outcome score, at six months (P=1.00, NS).

G. K. Wong · J. Kung · S. C. Ng · X. L. Zhu · W. S. Poon
Division of Neurosurgery, Prince of Wales Hospital,
The Chinese University of Hong Kong,
Hong Kong, China

G. K. Wong (✉) · J. Kung · S. C. Ng · X. L. Zhu · W. S. Poon
Division of Neurosurgery, Department of Surgery,
Prince of Wales Hospital,
4/F, Prince of Wales Hospital, Shatin, NT,
Hong Kong, China
e-mail: georgewong@surgery.cuhk.edu.hk

Keywords Cerebral infarction · Decompressive craniectomy · Outcome · Rehabilitation

Introduction

The high mortality that follows a large cerebral infarction is in part due to cerebral oedema. Oedema causes mass effect with raised intracranial pressure and herniation. Medical therapies are used to reduce intracranial pressure but outcome is poor in spite of treatment. Decompressive surgery is a well-described neurosurgical procedure that attempts to relieve high intracranial pressure. Case series in recent years have shown that decompressive craniectomy in younger patients with hemispheric infarction improves both survival rates and functional outcome. [1, 3, 6, 8, 10, 12, 13] Recent pooled analysis of three randomized controlled trials [11] has confirmed the efficacy of decompressive craniectomy in reducing case fatality and disability. [7] Although the pooled analysis showed that decompressive surgery group had a significant reduction in mortality and bedbound survivors, the optimal timing of surgery and predictive factors for favourable neurological recovery remained uncertain. We aimed to review our patient data with respect to neurological outcome at six months for possible predictive factors.

Patients and methods

We retrospectively reviewed all patients who underwent decompressive craniectomy for hemispheric infarction between January 2003 and June 2006 in a regional neurosurgical center in Hong Kong. All patients were recruited subsequently for active in-patient rehabilitation, if suitable. Rehabilitation programme included ten sessions per week with a physiotherapist, occupational therapist and speech therapist as appropriate. There were 21 patients in total and all were retrieved for analysis. Demographic data as age, sex, hypertension, diabetes mellitus, hyperlipidemia, atrial fibrillation, valvular and ischemic heart disease were recorded. Side of infarction were also recorded. All operative records were reviewed for the operative procedures and time of surgery in relationship to time of admission.

Our procedure for the decompressive craniectomy was described as below. Patient was put under general anaesthesia with endotracheal intubation. The patient was put in supine oblique position with a shoulder roll ipsilateral to the side of infarction. A large question mark flap incorporating to the frontal, temporal and parietal areas were fashioned up to midline and posterior to the parietal eminence. Maximal bone flap was fashioned with the aids of craniotome and high speed drill. The bone flap was stored frozen in the tissue bank. Stellate durotomies with inlay dural substitutes were made. A ventricular catheter would be inserted frontally. Wound was then closed after haemostasis. The patient would be kept sedated, intubated, ventilated and transferred to intensive care unit. The patient would be closed monitored for intracranial pressure and perfusion pressure. Sedation and osmotherapy would be stopped once the oedema subsided. In this respect, we had a standardised surgical paradigm for patients in this cohort.

Favourable neurological outcome was defined by independency in daily activity of living according to the Glasgow Outcome Scale (Good recovery and moderate disability) [5]. Outcome data was retrieved at time points of one month, three months, and six months after surgery. Six month neurological outcome was assumed to be representative of the long-term neurological outcome. Categorical data were analyzed by Fisher's Exact Test and continuous data was analysed by unpaired t-test. Statistical significance was defined by a two-tailed probability value of less than 0.05. Data were analysed with statistics software SPSS Version 14.0.

Results

The age (mean+/-SD) was 53.0+/-9.3 years and the male to female ratio was 1:3. All patients had hemispheric infarction of the anterior circulation. No intravenous or intra-arterial thrombolysis was attempted in this group of patients. No patient had haemorrhagic transformation. Six had internal carotid artery territory infarct and 15 had middle cerebral artery territory infarct. Ten (48%) patients had hypertension; three (14%) patients had diabetes mellitus; two (10%) patients had hyperlipidemia; six (29%) patients had atrial fibrillation; five (24%) patients had ischemic or valvular heart disease. Four patients (19%) achieved independent activity of daily living at six months after rehabilitation and four patients (19%) was dead at six months. There was no statistical significant difference in the distribution of the above factors in relationship to six-month neurological outcome.

Eight had left-sided infarct and four had early surgery. Neither early surgery within 24–48 hours after admission nor side of infarction correlated with six month neurological outcome, $p=1.00$. All four patients with favourable neurological outcome at six month were indicated by their clinical improvement at one month. Two patients were able to return to work. None of the patients with unfavourable outcome at one month were able to regain independency of daily activity of living at six months.

Discussion

Hemispheric cerebral infarction is commonly associated with variable degree of cerebral oedema. In severe case, this may lead to uncal or transtentorial herniation. Serious oedema formation usually manifests between the second to the fifth day after cerebral infarction. [4] Patients with hemispheric infarction and massive cerebral oedema have a poor prognosis. Mortality as high as 80% was reported in case series of maximal medical treatment. [2] Decompressive craniectomy has been suggested as a way to salvage this group of patients. [1, 3, 6, 8, 10–13] However, many of the patients who survive are left dependent and severely disabled. Clinicians often need to select patients, who will benefit from subsequent rehabilitation and regain functional independency, based on their clinical judgement. Though our case series was uncontrolled, and bias in patient selection would be actually a reflection of actual patient selection for hemicraniectomy in clinical practice.

Recent case series have shown that decompressive craniectomy in younger patients with malignant middle cerebral artery (MCA) territory infarction improves both survival rates and functional outcome. [1, 3, 6, 8, 10, 12, 13] Rabinstein AA et al have performed a systematic analysis on factors predicting prognosis after decompressive craniectomy for hemispheric infarction. [9] In their analysis, side of infarction and timing of surgery did not predict outcome. In their multivariate analysis, older age independently predicted poor recovery.

We were able to confirm the finding that side of infarction and timing of surgery did not predict eventual neurological outcome at six months. In our cohort, there was no difference in age between the patients with favourable and unfavourable neurological recovery, with a mean age of 53 years in both groups. The overall young age of the patients in this study may explain the lack of effect of age on neurological outcome. We noted that at one month after surgery, the patients who could achieve independency at daily living would be the group who had favourable neurological outcome eventually at six months. This may serve as an early indicator for patient and family counselling as well as to optimize timing and utilization of rehabilitation resources.

Conclusion

Timing of decompressive hemicraniectomy is not predictive of neurological outcome in terms of Glasgow outcome score, p=1.00. One month neurological outcome is indicative of neurological outcome at six months.

Conflict of interest statement We declare that we have no conflict of interest.

References

1. Foerch C, Lang JM, Krause J, Raabe A, Sitzer M, Seifert V, Steinmetz H, Kessler KR (2004) Functional impairment, disability and quality of life outcome after decompressive hemicraniectomy in malignant middle cerebral artery infarction. J Neurosurg 101:248–254
2. Hacke W, Schwab S, Horn M, Spraner M, De Georgeia M, von Kummer R (1996) Malignant middle cerebral artery territory infarction: clinical course and prognostic signs. Arch Neurol 53:309–315
3. Harscher S, Reichart R, Terborg C, Hagemann G, Kalff R, Witte OW (2006) Outcome after decompressive craniectomy in patients with severe ischemic stroke. Acta Neurochir (Wien) 148:31–37
4. Hofmeijer J, Amelink GJ, Algra A, van Gijn J, Macleod MR, Kappelle LJ, van der Worp HB, the HAMLET investigators (2006) Hemicraniectomy after middle cerebral artery infarction with life-threatening edema trial (HAMLET). Protocol for a randomised controlled trial of decompressive surgery in space-occupying hemispheric infarction. Trial 7:29
5. Jennett B, Bond M (1975) Assessment of outcome after severe brain damage. A practical scale. Lancet 1:480–484
6. Lanzino DJ, Lanzino G (2000) Decompressive craniectomy for space-occupying supratentorial infarction: rational, indications, and outcome. Neurosurg Focus 8(5):Article 3
7. Morley NCD, Berge E, Cruz-Flores S, Whittle IR (2002) Surgical decompression for cerebral oedema in acute ischaemic stroke. Cochrane Database of Systemic Reviews, Issue 3. Art. No.: CD003435
8. Pillai A, Menon SK, Kumar S, Rajeev K, Kumar A, Panikar D (2007) Decompressive hemicraniectomy in malignant middle cerebral artery infarction: an analysis of long-term outcome and factors in patient selection. J Neurosurg 106:59–65
9. Rabinstein AA, Mueller-Kronast N, Maramattom BV, Zazulia AR, Bamlet WR, Diringer MN, Wijdicks EF (2006) Factors predicting prognosis after decompressive craniectomy for hemispheric infarction. Neurology 67:891–893
10. Robertson SC, Lennarson P, Hasan DM, Traynelis VC (2004) Clinical course and surgical management of massive cerebral infarction. Neurosurgery 55:55–62
11. Vahedi K, Hofmeijer J, Juettler E, Vicaut E, George B, Algra A, Amelink GJ, Schmiedeck P, Schwab S, Rothwell PM, Bousser MG, van der Worp HB, Hacke W, for the DECIMAL, DESTINY and HAMLET investigators (2007) Early decompressive surgery in malignant infarction of the middle cerebral artery: a pooled analysis of three randomised controlled trials. Lancet Neurol 6:215–222
12. Walz B, Zimmermann C, Bottger S, Haberl RL (2002) Prognosis of patients after hemicraniectomy in malignant middle cerebral artery infarction. J Neurol 249:1183–1190
13. Yao Y, Liu W, Yang X, Hu W, Li G (2005) Is decompressive craniectomy for malignant middle cerebral artery territory infarction of any benefit for elderly patients? Surgical Neurology 64:165–169

Acta Neurochir Suppl (2008) 102: 335–338

Effects of temperature changes on cerebral biochemistry in spontaneous intracerebral hematoma

Ernest Wang · Chi Long Ho · Kah Keow Lee · Ivan Ng · Beng Ti Ang

Abstract
Background Fever worsens outcome in acute brain injury, presumably by accelerating secondary damage. Improved understanding of the pathophysiological processes that occur in spontaneous intracerebral hemorrhage (ICH) may help to determine if controlled normothermia might be of clinical benefit.
Methods In this prospective observational study over a period of 18 months at the National Neuroscience Institute, Singapore, we examined the effects of temperature changes on brain biochemistry and tissue oxygenation in 25 consecutive patients with spontaneous primary putaminal hemorrhage. The patients were divided into 3 groups according to the mean brain temperature over a 72-hour monitoring period following surgery and standard medical measures to control post-operative brain swelling and secondary injury.
Findings Patients that become spontaneously hypothermic with a mean brain temperature of less than 36 degrees centigrade (°C) had greater impairment in brain biochemistry as reflected by the worst brain lactate/pyruvate (L/P) ratio, glutamate and glucose dialysates. Brain tissue oxygenation, on the other hand, was highest and within normal limits in these spontaneously hypothermic patients. The hyperthemic group had similar L/P ratio, glycerol and glutamate levels when compared to the normothermic group. The glucose levels were found to be significantly different in all 3 groups.
Conclusions Extremes of temperature in spontaneous ICH, in particular - spontaneous hypothermia with a mean brain temperature of less than 36°C, are associated with a poor outcome. Cerebral microdialysis can be used to detect these detrimental changes that occur.

E. Wang · C. L. Ho · K. K. Lee · I. Ng · B. T. Ang (✉)
Department of Neurosurgery, National Neuroscience Institute,
11, Jalan Tan Tock Seng,
Singapore 308433, Singapore
e-mail: bengti.ang@gmail.com

Keywords Intracerebral hemorrhage · Brain temperature · Microdialysis · Brain tissue oxygenation

Introduction

Pyrexia exacerbates neurological damage in acute brain injury and contributes to deleterious secondary effects [9]. Fever is associated with worse outcome [1, 3], especially if it occurs early after the initial primary injury. This has lead many to advocate for aggressive attempts to maintain normothermia after acute brain injury. However, most of these studies have focused on traumatic brain injury (TBI) and ischemic stroke [9]. Given the paucity of data on the effects of temperature changes on neuronal activity in hemorrhagic stroke, we utilized multimodality monitoring techniques to characterize the impact of temperature on brain biochemistry and tissue oxygenation in spontaneous ICH.

Methods

Patient selection and treatment protocol

Over a period of 18 months, we prospectively studied 25 consecutive patients with spontaneous ICH. They were treated in accordance with standard neurointensive care protocols and consent procedures were approved by the institutional ethics committee. Patients were included in the study if an initial computerized tomographic (CT) scan showed a supratentorial putaminal ICH causing a clinical deterioration to a Glasgow Coma Scale (GCS) score of ≤ 8 or if radiological evidence of mass effect (midline shift of >5 mm or effacement of the basal cisterns) was evident. They were excluded if the pupils were fixed and dilated or if there was a suspected underlying abnormality (a vascular malformation, aneurysm or bleeding diathesis).

A unilateral frontal parietal-temporal decompressive craniectomy was employed to evacuate the hematoma in all cases. Three intraparenchymal neuromonitoring probes were inserted under direct vision and positioned 1cm from the hematoma cavity. The perihematoma location of the probes was verified with a post-operative CT scan. Brain temperature and brain tissue oxygenation (PbO_2) were measured with LICOX polarographic Clark-type probes (Integra Neuroscience, Plainsboro, NJ, USA). Microdialysis catheters (CMA 70; microdialysis, Solna, Sweden) were used to measure brain glucose, lactate, pyruvate and glutamate. Intracranial pressure (ICP) was continuously monitored using a fiberoptic intraparenchymal probe (Codman and Shurtleff, Raynham, MA, USA). Core body temperature was measured with a rectal thermistor probe. Inspired oxygen fraction (FiO_2) was adjusted to achieve an arterial oxygen saturation of >95% with hemoglobin levels maintained above 10 g/dl. ICP control measures included nursing in the 30 degrees head up position, sedation with propofol (2–10 mg/kg/hour), boluses of 20% mannitol (2 ml/kg up to a plasma osmolarity of 320 mosmol/L) according to ICP spikes and mild hypocapnia by regulating pCO_2 at 30–35 mmHg. No barbiturates were administered. Fever was defined clinically as having occurred if the core body temperature was > 38.5°C. Febrile episodes were treated with 1g of oral acetoaminophen (up to a maximum of 4g per day) and if left unabated for more than an hour, a surface cooling blanket was employed together with the application of a Blanketrol II (Cincinnati Sub-zero products Inc, Cincinnati, Ohio, USA) intermittently to the body and thighs. There was no active induction of hypothermia through invasive systemic means. Intravenous antibiotics were started on clinical grounds and given appropriate to the corresponding cultures. To achieve stabilization and accurate PbO_2 measurements, monitoring was commenced 3 hours after insertion of the LICOX probes. Analysis of cerebral microdialysates was performed on hourly collections. The pressure reactivity index, PRx, was calculated as a moving correlation coefficient between the last 30 consecutive samples of values for ICP and arterial blood pressure (ABP) averaged for a period of 10 seconds.

Temperature groups

The patients were divided into 3 groups according to the mean brain temperature over the first 72-hour period after surgery. There were 6 patients in the hyperthermic group 1 (brain temperature > 39°C), 10 patients in the spontaneous hypothermic group 2 (brain temperature < 36°C) and 9 patients in the normothermic temperature group 3 (brain temperature 36–39°C). Their ages ranged from 31 to 81 years with a mean age of 55 (± 13.6) years. There were 18 male (72%) and 7 female (28%) patients. The median GCS score on admission was 8 [interquartile range (IQR) 3 to 13]. The average volume of hematoma was 80.8± 30.3 milliliters (IQR 26 to 160). Data from the first 72 hours of monitoring were analyzed for the purposes of this study. The outcome at 6 months was determined by the Glasgow Outcome Score (GOS).

Statistical analysis

Data that were continuous variables are reported as mean ± standard deviation while non-normally distributed data are reported as median with IQR. When the normality and equality of variances assumption was satisfied, the two sample paired t-test was performed. The Mann Whitney U test was otherwise used. The mean values of MAP, ICP, PRx, PbO_2 and cerebral microdialysates were studied over a 72-hour monitoring period. Analysis of the differences in the means of the variables in the 3 groups was carried out using ANOVA and subsequent post-hoc comparisons using Bonferroni's t-test. All comparisons were considered significant when the probability (p value) of < 0.05 was achieved. Statistical analyses were performed using the commercially available statistical program SPSS version 12.0 (SPSS Inc., Chicago, Illinois, USA).

Results

The outcome of patients in the 3 temperature groups is summarized in Table 1. Altogether, 8 (32%) patients died and 17 (68%) survived. Only 4 (16%) patients had a good outcome (GOS 4–5) while 21 (84%) patients had a poor outcome (GOS 1–3). All 10 (100%) patients in the spontaneous hypothermic group had a poor outcome while 5 out of 6 (83%) patients in the hyperthermic group had a poor outcome. In the normothermic group, 3 out of 9 (33%) patients had a poor outcome while 6 (67%) patients had a good outcome. With regards to age, GCS on admission and volume of hematoma, no differences were seen when comparing all 3 groups. The mean temperature difference between rectal temperature and PbO_2 was 1.3°C. While ICP reduction to < 20 mm Hg was achieved in all 3 groups, the normothermic group had the lowest mean ICP of 11.5± 6.4 mm Hg ($p<0.001$). The hyperthermic and hypothermic groups had a mean ICP of 19.4±11.8 mm Hg and 18.2± 8.4 mm Hg respectively ($p=0.07$). The maintenance of a

Table 1 Outcome of the 3 temperature groups over a 72 hour monitoring period

Groups	Brain temperature (°C)	Poor outcome (GOS 1–3)	Favorable outcome (GOS 4,5)	Total
Hyperthermia (Gp 1)	>=39	5 (83%)	1 (17%)	6
Spontaneous hypothermia (Gp 2)	<36	10 (100%)	0	10
Normothermia (Gp 3)	36–38.9	6 (67%)	3 **(33%)**	9

cerebral perfusion pressure (CPP) of > 70 mm Hg with a return to cerebral autoregulation (PRx<0.3) was achieved in all 3 groups. The mean PbO_2 was highest in the spontaneous hypothermic group ($p<0.001$) when compared to the other 2 groups (see Table 2). The hypothermic group also had the worst lactate, L/P ratio, glutamate and glucose levels ($p<0.001$) (Table 2). The hyperthemic group had similar L/P ratio, glycerol and glutamate levels when compared to the normothermic group ($p=0.1$). There were no differences in glycerol levels demonstrated amongst the 3 temperature groups while glucose was significantly different across all 3 groups ($p=0.001$).

Discussion

Elevated brain temperatures worsen neurological injury through various mechanisms [11]. Hyperpyrexia is known to increase extracellular levels of excitatory amino acids and oxygen free radicals, accelerate enzymatic inhibition of protein kinase activity [6], impede energy metabolism recovery, worsen cytoskeletal proteolysis [17] and aggravate blood-brain barrier breakdown. That fever is a common occurrence in neurosurgical patients [14] and a prognostic indicator of an unfavourable outcome has many clinical implications. The experimental body of evidence in favour of treating fever is substantial [2, 6] while the ensuing clinical evidence is conflicting. Clifton et al [8] in a large multicenter trial of 392 patients did not demonstrate any improvement in outcome from induced hypothermia (to 33°C) despite earlier encouraging results [7]. Gupta et al suggested that decreasing the brain temperature below 35°C may impair brain tissue oxygenation [10]. While much of this evidence pertains to TBI and ischemic stroke [5], Schwartz et al [20] did demonstrate worse outcomes with elevated rectal temperatures of > 37.5°C in a study of 196 patients with spontaneous ICH. Nonetheless, more work is necessary to examine the effect of temperature in hemorrhagic stroke.

Table 2 Brain biochemistry in all 3 temperature groups

	Gp 1 (Hyperthermia)	Gp 2 (Hypothermia)	Gp 3 (normal)
PbO_2 (mm Hg)	16.0±8.5	30.8±24.3	12.2±9.5
Lactate (mmol/L)	3.4±2.0	4.7±2.9	2.6±2.9
L/P (mmol/L)	54.9±38.3	91.7±111.0	45.9±3.2
Glutamate	6.4±12.	17.8±32.9	8.7±16.4
Glucose	2.0±1.1	0.9±0.9	1.4±1.3

In our study, fever was defined as a brain temperature of >39°C as the latter has been demonstrated to usually exceed rectal temperature by 0.5–1.5°C [19]. This value was chosen for the following reason: the conventional definition of fever being a core body temperature of > 38.5°C, and the average difference between core body temperature and brain temperature known to be about 0.5 degrees [15], a patient fulfilling the former requirement would have a corresponding brain temperature of >39°C. Sustained fever [13] and fever that develops soon after the onset of stroke are strongly associated with poor outcome [4, 11]. A mean brain temperature monitored continuously over the first 72 hours following surgery therefore becomes relevant as it represents an important monitoring period during which potential therapeutic measures can be instituted.

In a study of 18 acutely brain injured patients, Stocchetti et al [21] did not show any significant neurochemical alterations, such as in L/P ratio and glutamate levels, during either effervescence or defervescence. It was postulated that this was because cerebral blood flow and oxygenation were able to meet up with metabolic demands. Moreover, their analysis was restricted to episodes cumulating to less than 8% of the data collected in contrast to a 72-hour continuous monitoring period in this study. Our results demonstrate otherwise in this cohort of spontaneous ICH patients who underwent surgical hematoma evacuation, as we found that the neurochemical changes reflective of the worst neuronal injury were most pronounced in the group of patients who were spontaneous hypothermic. This is not surprising as the worst outcome occurred in this group of patients as well. Of note, others have found that hypothermia on admission was associated with a greater severity of injury and worse outcomes than was normothermia on admission [8]. This suggests that spontaneous hypothermia may be a consequence of more severe brain injury.

Mean PbO_2 in the spontaneous hypothermic group was actually highest and within normal limits (30.8±24.3 mm Hg). This is concordant with our results in a separate local study on the changes in brain biochemistry following acute head injury. Patients with the worst outcome had significantly better PbO_2 values compared to survivors with a good outcome (however, those with a good outcome demonstrated a significant improvement of PbO_2 from ischemic to normal values, a change that did not occur in those with an unfavourable outcome). In this study, we postulate that normal values in spontaneously hypothermic patients could possibly be due to a 'diffusion' impairment [16]: cellular dysfunction leading to an inability to utilize oxygen or a hyperaemic flow itself secondary to the acute brain injury.

Cerebral microdialysates were similar between the hyperthermic and normothermic groups despite a difference in outcome. This suggests that a reversibility of brain injury could possibly exist in those that can potentially survive and improve to a good outcome [18]. Glucose, being a very sensitive marker of neuronal metabolism [12] was significantly different across all 3 temperature groups while glycerol, a less sensitive marker of membrane damage, was not different.

To date and to the best of our knowledge, there is no study characterizing the effect of temperature regulation on brain biochemistry in spontaneous ICH alone [20]. Overall analysis of our data shows a difference in brain neurochemistry with respect to temperature regulation in spontaneous ICH. In the treatment of hemorrhagic stroke where different therapeutic measures remain controversial, a deeper understanding of the pathophysiological differences that exists in spontaneous ICH may help clarify optimal treatment.

Conclusion

In spontaneous ICH, extremes of temperature worsen neurological injury and are associated with worse outcomes. Such injury is reflected in local changes in brain neurochemistry adjacent to the hematoma cavity. Patients with spontaneous hypothermia with a brain temperature of $< 36°C$ have the most devastating injury while hyperthermic patients with a brain temperature of $> 39°C$ also have a similar outcome. However, further studies are necessary to determine if active systemic treatment of temperature with controlled cooling techniques would lead to better outcomes.

Conflict of interest statement We declare that we have no conflict of interest.

References

1. Azzimondi G, Bassein L, Nonino F, Florani L, Vignatelli L, Re G, D'Alessandro R (1995) Fever in acute stroke worsens prognosis: A prospective study. Stroke 26:2040–2043
2. Baena RC, Busto R, Dietrich WD, Globus MY, Ginsberg MD (1997) Hyperthermia delayed by 24 hours aggravates neuronal damage in rat hippocampus following global ischemia. Neurology 48:768–773
3. Busto R, Dietrich WD, Globus MY, Valdes I, Scheinberg P, Ginsberg MD (1987) Small differences in intraischemic brain temperature critically determine the extent of ischemic neuronal injury. J Cereb Blood Flow Metab 7:729–738
4. Castillo J, Davalos A, Marrugat J, Noya M (1998) Timing for fever-related damage in acute ischemic stroke. Stroke 29:2455–2460
5. Castillo J, Martinez F, Leira R et al (1994) Mortality and morbidity of acute cerebral infarction related to emperature and basal analytical parameters. Cerebrovasc Dis 4:56–71
6. Churn SB, Taft WC, Billingsley MD, Blair RE, DeLorenzo RJ (1990) Temperature modulation of ischemic neuronal death and inhibition of calcium/calmodulin-dependent protein kinase II in gerbils. Stroke 21:1715–1721
7. Clifton GL, Allen S, Barrodale P et al (1993) A phase II study of moderate hypothermia in severe brain injury. J Neurosurg 79:354–362
8. Clifton GL, Miller ER, Choi SC et al (2001) Lack of effect of induction of hypothermia after acute brain injury. N Engl J Med 344:556–563
9. Ginsberg MD, Busto R (1998) Combating hyperthermia in acute stroke: A significant clinical concern. Stroke 29:529–534
10. Gupta AK, Al-Rawi PG, Hutchinson PJ, Kirkpatrick PJ (2002) Effect of hypothermia on brain tissue oxygenation in patients with severe head injury. Br J Anaesth 88(2):188–192
11. Hindfelt B (1976) The prognostic significance of subfertility and fever in ischemic cerebral infarction. Acta Neurol Scand 53:72–79
12. Hutchinson PJ, O'Connell MT, Al-Rawi PG, Maskell LB, Kett-White R, Gupta AK, Richards HK et al (2000) Clinical cerebral microdialysis: a methodological study. J Neurosurg 93:37–43
13. Jones PA, Andrews PJ, Midgley S et al (1994) Measuring the burden of secondary insults in head-injured patients during intensive care. J Neurosurg Anesthesiol 6:4–14
14. Kilpatric MM, Lowry DW, Firlik AD, Yonas H, Marion DW (2000) Hyperthermia in the neurosurgical intensive care unit. Neurosurgery 47(4):850–856
15. Marion DW (2004) Controlled normothermia in neurologic intensive care. Crit Care Med 32(No. 2 Suppl):S43–S45
16. Menzel M, Doppenberg EMR, Zauner A, Soukup J, Reinert MM, Bullock MR (1999) Increased inspired oxygen concentration as a factor in improved brain tissue oxygenation and tissue lactate levels after severe human head injury. J Neurosurg 91:1–10
17. Morimoto T, Ginsberg MD, Dietrich WD, Zhao W (1997) Hyperthermia enhances spectrin breakdown in transient cerebral ischemia. Brain Res 746:43–51
18. Ng I, Lee KK, Wong J (2005) Brain tissue oxygenation monitoring in acute brain injury. Acta Neurochir Suppl 95:447–451 Review
19. Rumana CS, Gopinath SP, Uzura M et al (1998) Brain temperature exceeds systemic temperature in head-injured patients. Crit Care Med 26:562–567
20. Schwarz S, Hafner K, Aschoff A et al (2000) Incidence and prognostic significance of fever following intracerebral hemorrhage. Neurology 54:354–361
21. Stocchetti N, Protti A, Lattuda M, Magnoni S, Longhi L, Ghisoni L, Egidi M, Zanier ER (2005) J Neurol Neurosurg Psych 76:1135–1139

Acta Neurochir Suppl (2008) 102: 339–343

Increased levels of CSF heart-type fatty acid-binding protein and tau protein after aneurysmal subarachnoid hemorrhage

E. R. Zanier · L. Longhi · M. Fiorini · L. Cracco · A. Bersano · T. Zoerle · V. Branca · S. Monaco · N. Stocchetti

Summary

Background Heart-type Fatty Acid-Binding Protein (H-FABP) and tau protein (τ) have been shown to be novel biomarkers associated with brain injury and, therefore, they could represent a useful diagnostic tool in patients with subarachnoid hemorrhage (SAH). The goal of this study was to measure H-FABP and τ in cerebrospinal fluid (CSF) following SAH to test the hypothesis that a relationship exists between SAH severity and H-FABP/τ values.

Methods Twenty-seven consecutive SAH patients admitted to our ICU were studied. Serial CSF samples were obtained in every patient starting on the day of SAH and daily for up to 2 weeks post-SAH. H-FABP/τ levels were measured by enzyme-linked immunosorbent assay.

Results Patients with severe SAH showed significantly higher peak levels of H-FABP and τ compared to mild-SAH patients (FABP: $p=0.02$; τ: $p=0.002$). In addition the peak concentrations of H-FABP and τ in CSF from SAH patients correlated significantly with Glasgow Coma Scale motor score (H-FABP: Spearman $r=-0.52$, $p=0.006$; τ: Spearman $r=-0.63$, $p=0.0004$). Based on outcome at discharge from the hospital, patients were categorized into survivors and non-survivors. Peak concentrations of both proteins in the non-survivors group were significantly higher than in the survivors.

Conclusions H-FABP and τ CSF levels are proportional to SAH severity and may be novel biomarkers that can be used to predict the severity of outcome following clinical SAH.

Keywords Aneurysmal subarachnoid hemorrhage · Biomarkers · Heart-type fatty acid-binding protein · Tau protein

E. R. Zanier · L. Longhi · T. Zoerle · N. Stocchetti
Neurosurgical Intensive Care Unit,
Department of Anesthesia and Critical Care Medicine,
Fondazione IRCCS Ospedale Maggiore Policlinico,
Mangiagalli e Regina Elena,
University of Milano,
Milano, Italy

A. Bersano
Department of Neurology,
Fondazione IRCCS Ospedale Maggiore Policlinico,
Mangiagalli e Regina Elena, University of Milano,
Milano, Italy

V. Branca
Department of Neuroradiology,
Fondazione IRCCS Ospedale Maggiore Policlinico,
Mangiagalli e Regina Elena,
Milano, Italy

M. Fiorini · L. Cracco · S. Monaco
Department of Neurological and Visual Sciences,
University of Verona,
Ospedale GB Rossi,
Verona, Italy

E. R. Zanier (✉)
Neurosurgical Intensive Care Unit,
Department of Anesthesia and Critical Care Medicine,
Fondazione IRCCS Ospedale Maggiore Policlinico,
Mangiagalli e Regina Elena, Via Sforza 35,
Milano, Italy
e-mail: ezanier@policlinico.mi.it

Introduction

Aneurysmal subarachnoid hemorrhage (SAH) is a severe condition due to the rupture of an intracranial aneurysm in the subarachnoid space. SAH is an important cause of premature death and disability worldwide, affecting approximately 10/100,000 hospitalized patients every year

[1, 2]. Studies on mortality and morbidity of SAH in Italy report a mortality rate of 25% within 6 months from the first bleeding and severe neurological disability in 15% of the survivors [2].

The severity of initial brain damage is one of the most important factors associated with the outcome following SAH [3]. Global brain ischemia associated with acute increases in intracranial pressure (ICP) and/or focal brain ischemia due to tissue compression are two major mechanisms of initial brain damage following SAH [3]. The identification of biomarkers, known to be released during brain ischemia following SAH could therefore represent an aid for the clinician.

Recently, Heart-type Fatty Acid-Binding Protein (H-FABP), a small cytoplasmic lipid-binding protein, with a molecular weight of 15 kDa, was found to be increased in patients with mild traumatic brain injury [4] and those with ischemic and hemorrhagic stroke [5]. Following acute ischemic stroke, an acute increase in plasma H-FABP was significantly associated with the severity of the neurological deficit and the functional outcome [6]. H-FABP in the brain is predominately localized in the neuronal cell body, and following cellular damage, is rapidly released into the blood. CSF levels of H-FABP following SAH are unknown and we wondered whether this protein might therefore have the potential to be a rapid diagnostic marker of injury severity following human SAH.

Tau protein (τ), a phosphorylated microtubule-associated protein, has been recognized as a nonspecific marker of axonal injury [7]. Following brain damage τ is released into the extracellular space and may be increased in the cerebrospinal fluid (CSF). In a study conducted on 17 SAH patients, it was reported that the increased CSF τ observed following SAH was correlated with both injury severity and clinical outcome [8]. In the present study we assessed CSF levels of both H-FABP and τ in SAH patients to determine whether a direct relationship exists between expression of these proteins and SAH severity.

Materials and methods

Patient population The study was approved by the Local Research Ethics Committee of the Ospedale Maggiore Policlinico, Milano. Written informed consent was obtained from the patient or, in the comatose patients, from the next of kin. Twenty seven consecutive patients admitted to our neuro-critical Intensive Care Unit (ICU) with a diagnosis of SAH were enrolled.

Clinical Management The severity of SAH, on admission and before surgery/endovascular treatment, was recorded according to the World Federation of Neurological Surgeons (WFNS) grading scale [9].

Admission CT scan was classified using the Fisher scale [10]. Management goals included the early clipping/coiling of the aneurysm (in our Center, 77% of ruptured intracranial aneurysms are secured within the first 24 hours) and, where indicated, surgical evacuation of the intracranial hematoma. Hydrocephalus was drained using an intraventricular catheter and intracranial pressure (ICP) was monitored in all patients with the goal to maintain ICP levels below 20 mmHg, and cerebral perfusion pressure (CPP) above 70 mmHg. In addition a jugular bulb catheter was inserted in all comatose patients to allow determination of jugular saturation of oxygen (SjO_2) and calculation of arterio-jugular difference of oxygen content ($AjDO_2$). Outcome criteria included survival recorded at discharge from the hospital.

CSF sampling CSF samples (5 ml) were collected beginning on day 1 and thereafter twice daily (8 A.M./ 8 P.M.). CSF samples were treated with an anticoagulant, centrifuged and then frozen and kept at - 80 degrees Celsius. Samples were analyzed using specific ELISA kits for H-FABP and τ as described [11].

Groups From the analysis of serial CSF samples, the maximum (peak) values were determined in each patient for both proteins. To explore the relationship between CSF H-FABP/τ and injury severity, patients were grouped and dichotomized according to the WFNS score of mild SAH (WFNS 1–3) and severe SAH (WFNS 4–5). However, since GCS can vary from 12 to 3 amongst patients with WFNS 4 and 5, we investigated the relationship between the motor component of GCS (mGCS) and CSF levels of H-FABP and τ. To explore the relationship between CSF H-FABP/τ and outcome, patients were also categorized into survivors and non-survivors assessed at hospital discharge.

Statistical analysis ICP data are presented as mean ± standard error of the mean. H-FABP and τ (whose distribution was not normal) are presented as median values (+ range). Mann Whitney test was used to assess the relationship between peak H-FABP/τ values and WFNS, and outcome at discharge. Correlation and linear Spearman regression analyses were used to determine the relationship between CSF biomarkers and injury severity (assessed with the mGCS). A "p" value of < 0.05 was accepted as significant.

Results

Patients The clinical features of the patients are presented in Table 1. Median age was 56 (range 30–78) and 23 patients (85%) were female. Seventeen patients (63%) were in poor clinical conditions on admission (WFNS 4–5).

Anterior circulation aneurysms were present in 18 (69%) patients. On admission CT, 2 patients were Fisher scale grade 3 (localized clots and/or vertical layers of blood 1 mm or greater in thickness), and 25 were grade 4 (diffuse or no subarachnoid blood with intracerebral or intraventricular clots). Aneurysms were treated with an endovascular approach in 20 patients (74%) and with surgical clipping in 4 patients. In 3 patients with irreversible brain damage (GCS 3 and absence of upper brainstem reflexes after hydrocephalus drainage), aneurysm closure was not performed. Intracranial pressure was monitored in 23 patients (85%).

CSF H-FABP/τ concentrations following SAH All but 2 patients had at least one pathologically elevated concentration of H-FABP CSF values (median 6607 pg/ml; range 603–100000 pg/ml) compared to published reference population (CSF H-FABP of 1729 pg/ml: [12]). Similarly, all but 3 patients had at least one pathologically elevated concentration of CSF τ (median 4322 pg/ml; range 146–17222 pg/ml) compared to published reference population (CSF τ normal values up to 672 pg/ml: [7]).

Relationship between H-FABP/τ values and degree of brain damage following SAH Patients with severe SAH (WFNS 4–5) showed significantly higher peak H-FABP (median 14388 pg/ml; range 603–100000 pg/ml) and τ (median 7194 pg/ml; range 710–17222 pg/ml) levels compared to mild-SAH (H-FABP median 3622 pg/ml; range 1800–8873 pg/ml; Mann Whitney test $p=0.02$. τ median 1572 pg/ml; range 146–5627 pg/ml; Mann Whitney test $p=0.002$). Additionally, we investigated the relationship between mGCS and CSF biomarkers. The peak concentrations of H-FABP and τ in SAH CSF correlated significantly with mGCS (H-FABP: Spearman $r=-0.52$, $p=0.006$; τ: Spearman $r=-0.63$, $p=0.0004$) (Fig. 1).

Table 1 Characteristics of patients with subarachnoid haemorrhage

Patient Code n°	Age	Sex	WFNS*	Fisher	Aneurysm location	Aneurism treatment	Timing (hours)	mean ICP mmHg	max ICP mmHg
1	56	F	5	4	MCA	Coil	<24	34	66
2	51	F	5	4	PCA	Coil	<24	77	110
3	48	F	5	4	ICA	NT	-	68	91
4	65	M	5	4	VA	NT	-	44	117
5	39	F	5	4	ACoA	Coil	24–72	56	63
6	68	F	5	4	Unknown*	NT	-	50	51
7	73	F	5	4	ACoA	Coil	24–72	10	80
8	68	F	4	4	ACoA	Coil	<24	12	39
9	58	F	4	4	MCA	Clip	<24	14	24
10	78	F	4	4	Peric.	Clip	<24	9	20
11	38	F	1	3	PCA	Coil	<24	26	36
12	56	F	5	4	PICA	Coil	<24	16	32
13	47	F	1	4	ACoA	Coil	<24	-	-
14	54	F	4	4	Basilar A	Coil	<24	13	28
15	62	F	5	4	MCA	Clip	<24	12	18
16	73	F	1	4	ACoA	Coil	<24	-	-
17	37	F	1	3	PCoA	Coil	<24	-	-
18	63	F	2	4	Basilar A	Coil	24–72	9	18
19	63	F	1	4	SCA	Coil	<24	-	-
20	30	F	1	4	ICA	Coil	24–72	18	26
21	41	M	2	4	PICA	Coil	<24	11	25
22	67	F	4	4	MCA	Coil	<24	11	22
23	45	F	4	4	ACA	Coil	>72	14	16
24	68	F	1	4	PCA	Coil	<24	10	25
25	63	M	5	4	ACoA	Clip	<24	22	76
26	49	F	3	4	ACA	Coil	<24	12	25
27	42	M	5	4	ACA	Coil	<24	14	30

Legend: WFNS, World Federation of Neurological Surgeons grading scale. MCA = Middle cerebral artery, AcoA = Anterior communicating artery, PcoA = Posterior communicating artery, PCA = Posterior cerebral artery, ACA = Anterior cerebral artery, PICA = Posterior inferior cerebellar artery, SCA = Superior cerebellar artery, Peric. = Pericallosal artery, VA = Vertebral artery, ICA = Internal carotid artery. * No aneurysm found or patient died before ancillary investigations could be performed. Clip = surgical treatment, Coil = endovascular treatment, NT = no treatment, Timing = timing of treatment from SAH, ICP = intracranial pressure.

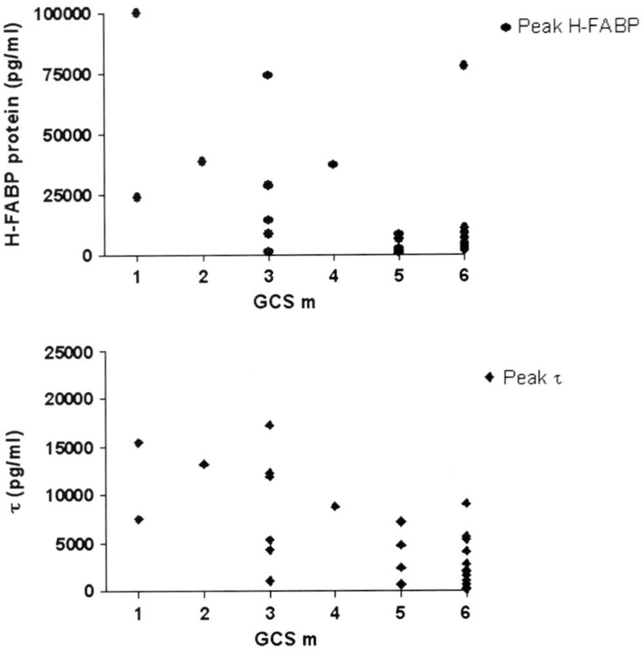

Fig. 1 Relationship between CSF mGCS and peak H-FABP (upper panel) and τ (lower panel) levels. H-FABP: Spearman r=−0.52, p=0.006; τ Spearman r=−0.63, p=0.0004

CSF H-FABP and τ concentration in Survivors versus Non-Survivors Based on outcome at hospital discharge, patients were categorized into survivors and non-survivors (Table 2). The peak concentration of H-FABP (28793 pg/ml with range of 1800–100000 pg/ml) in the non-survivors group was significantly higher than that recorded from survivors: 3925 (603–37127) pg/ml (Mann Whitney: p=0.005). Similarly the peak concentration of τ was 9066 (719–17222) pg/ml in the non-survivors group, it was significantly higher than the peak concentration in survivors: 2260 (146–12296) pg/ml (Mann Whitney: p=0.008).

Influence of Surgery In case of surgical clipping of the aneurysm (4 patients) CSF sampling occurred after surgery. To address the influence of surgery on CSF H-FABP and τ levels, we repeated all analyses after exclusion of these 4 patients. The relationship between peak concentrations of H-FABP and τ with WFNS, mGCS and outcome at discharge remained significant, suggesting that CSF protein increases are associated with SAH and not with surgical manipulation.

Discussion

In this study we observed that, following SAH, (i) a significant increase of CSF concentrations of H-FABP and τ occurs and (ii) a direct relationship exists between the degree of brain damage, evaluated with WFNS scale, mGCS and outcome at discharge, and CSF levels of H-FABP and τ. Several studies have reported that alterations in CSF H-FABP [5, 12, 13] and τ [7, 8, 14, 15] levels are associated with cellular brain damage in patients with acute neurological disorders and degenerative disease. SAH is characterized by a loss of structural integrity of glial and neuronal cells and release of cell-specific proteins into the CSF. The majority of patients in our study showed increased levels of H-FABP (92%) and τ (89%) compared to normal references. Perhaps more importantly, the degree of H-FABP and τ increase reflected the clinical condition of the patient. In addition, we observed higher levels of these proteins in non-survivors when compared to survivors. Among non-survivors, the cause of death was acute brain injury in 5 cases (median mGCS=3; range 1–3), and vasospasm associated delayed cerebral ischemia (DCI) in the other 4 patients (median mGCS=6; range 3–6). Even though the relationship between DCI and CSF H-FABP/τ levels was not the aim of our study, high levels of these proteins in this sub-group of patients suggest that the relationship of H-FABP/τ with intracranial secondary insults should be investigated.

The latency between time of SAH and H-FABP and τ elevation is an important variable that was not determined in the present study, future directions for our work will be

Table 2 Survivors versus non-survivors

Characteristic	All	Survivors	Non-survivors	Significance
H-FABP (pg/ml)	6607 (603–100000)	3925 (603–37127)	28793 (1800–100000)	0.005
tau (pg/ml)	4322 (146–17222)	2260 (146–12296)	9066 (719–17222)	0.008
Age, years	56 (30–78)	56 (30–73)	63 (39–78)	ns
Sex (F:M)	23: 4	16: 2	7: 2	ns
WFNS	4 (1–5)	3.5 (1–5)	5 (1–5)	0.038
mGCS	5.5 (1–6)	6 (3–6)	3 (1–6)	ns (p=0.06)
Fisher	4 (3–4)	4 (3–4)	4 (4–4)	ns
Number	27	18	9	

Legend: The median (+ range) is shown. Ns = not significant; WFNS = World Federation of Neurological Surgeons; mGCS = Glasgow Coma Scale motor score.

to address the temporal relationship of FABP/τ changes and the occurrence of secondary ischemia. Furthermore, we limited our analysis to the CSF, particularly since τ is a neuronal protein that cannot be detected in plasma. Conversely, H-FABP is not brain-specific and can increase in plasma following extracranial injuries [16, 17]. It would therefore be important to investigate the relationship between CSF and plasma H-FABP levels following SAH to discriminate between H-FABP increases due to brain damage and plasma CSF contamination due to the initial bleeding. Moreover in this pilot study we only investigated the relationship between H-FABP/τ changes and the outcome at time of discharge. Additional work is needed to establish whether high H-FABP/τ values are related or not to functional outcome 6 month after SAH, and to address which role, if any, these proteins might play as predictive factors for late morbidity and mortality following SAH. Despite these limitations, our findings provide new evidence that H-FABP and τ in the CSF increase substantially in patients following SAH and that a direct relationship between injury severity and H-FABP and τ levels exist.

Conflict of interest statement We declare that we have no conflict of interest.

References

1. Cahill J, Calvert JW, Zhang JH (2006) Mechanisms of early brain injury after subarachnoid hemorrhage. J Cereb Blood Flow Metab 26:1341–1353
2. Citerio G, Gaini SM, Tomei G, Stocchetti N (2007) Management of 350 aneurysmal subarachnoid hemorrhages in 22 Italian neurosurgical centers. Intensive Care Med 33:1580–1586
3. Macdonald RL, Pluta RM, Zhang JH (2007) Cerebral vasospasm after subarachnoid hemorrhage: the emerging revolution. Nat Clin Pract Neurol 3:256–263
4. Pelsers MM, Hanhoff T, Van d, V, Arts B, Peters M, Ponds R, Honig A, Rudzinski W, Spener F, de Kruijk JR, Twijnstra A, Hermens WT, Menheere PP, Glatz JF (2004) Brain- and heart-type fatty acid-binding proteins in the brain: tissue distribution and clinical utility. Clin Chem 50:1568–1575
5. Zimmermann-Ivol CG, Burkhard PR, Floch-Rohr J, Allard L, Hochstrasser DF, Sanchez JC (2004) Fatty acid binding protein as a serum marker for the early diagnosis of stroke: a pilot study. Mol Cell Proteomics 3:66–72
6. Wunderlich MT, Hanhoff T, Goertler M, Spener F, Glatz JF, Wallesch CW, Pelsers MM (2005) Release of brain-type and heart-type fatty acid-binding proteins in serum after acute ischaemic stroke. J Neurol 252:718–724
7. Ost M, Nylen K, Csajbok L, Ohrfelt AO, Tullberg M, Wikkelso C, Nellgard P, Rosengren L, Blennow K, Nellgard B (2006) Initial CSF total tau correlates with 1-year outcome in patients with traumatic brain injury. Neurology 67:1600–1604
8. Kay A, Petzold A, Kerr M, Keir G, Thompson E, Nicoll J (2003) Temporal alterations in cerebrospinal fluid amyloid beta-protein and apolipoprotein E after subarachnoid hemorrhage. Stroke 34: e240–e243
9. Rosen DS, Macdonald RL (2005) Subarachnoid hemorrhage grading scales: a systematic review. Neurocrit Care 2:110–118
10. Fisher CM, Kistler JP, Davis JM (1980) Relation of cerebral vasospasm to subarachnoid hemorrhage visualized by computerized tomographic scanning. Neurosurgery 6:1–9
11. Bersano A, Fiorini M, Allaria S, Zanusso G, Fasoli E, Gelati M, Monaco H, Squintani G, Monaco S, Nobile-Orazio E (2006) Detection of CSF 14-3-3 protein in Guillain-Barre syndrome. Neurology 67:2211–2216
12. Steinacker P, Mollenhauer B, Bibl M, Cepek L, Esselmann H, Brechlin P, Lewczuk P, Poser S, Kretzschmar HA, Wiltfang J, Trenkwalder C, Otto M (2004) Heart fatty acid binding protein as a potential diagnostic marker for neurodegenerative diseases. Neurosci Lett 370:36–39
13. Guillaume E, Zimmermann C, Burkhard PR, Hochstrasser DF, Sanchez JC (2003) A potential cerebrospinal fluid and plasmatic marker for the diagnosis of Creutzfeldt-Jakob disease. Proteomics 3:1495–1499
14. Strand T, Alling C, Karlsson B, Karlsson I, Winblad B (1984) Brain and plasma proteins in spinal fluid as markers for brain damage and severity of stroke. Stroke 15:138–144
15. Hesse C, Rosengren L, Vanmechelen E, Vanderstichele H, Jensen C, Davidsson P, Blennow K (2000) Cerebrospinal fluid markers for Alzheimer's disease evaluated after acute ischemic stroke. J Alzheimers Dis 2:199–206
16. Puls M, Dellas C, Lankeit M, Olschewski M, Binder L, Geibel A, Reiner C, Schafer K, Hasenfuss G, Konstantinides S (2007) Heart-type fatty acid-binding protein permits early risk stratification of pulmonary embolism. Eur Heart J 28:224–229
17. O'Donoghue M, de Lemos JA, Morrow DA, Murphy SA, Buros JL, Cannon CP, Sabatine MS (2006) Prognostic utility of heart-type fatty acid binding protein in patients with acute coronary syndromes. Circulation 114:550–557

PART 8:
Experimental studies and models

Acta Neurochir Suppl (2008) 102: 347–351

DNA vaccination against neurite growth inhibitors to enhance functional recovery following traumatic brain injury

Yi Zhang · Beng Ti Ang · Zhi Cheng Xiao · Ivan Ng

Abstract

Background Myelin-associated proteins contribute to failure of axon regeneration in the injured central nervous system of the adult.

Methods In this study, we employed a recombinant DNA vaccine encoding the myelin-derived inhibitors NogoA, myelin-associated glycoprotein (MAG) and tenascin-R (TnR), so as to effect the production of antibodies against these myelin-related antigens, in a rodent head injury model and ascertained its potential for promoting axonal plasticity and functional recovery. Adult rats underwent lateral fluid percussion at the left sensorimotor cortex (SMC) and treatment with the DNA vaccine before or after injury. Behavioral tests and neuroanatomical tract tracing was carried out.

Findings The vaccinated rats showed improved cortico-rubral plasticity and functional recovery compared to control groups.

Conclusions This suggests that a DNA vaccination approach may provide a promising strategy for promoting repair after traumatic brain injury.

Keywords Myelin-associated inhibitors · DNA vaccine · Head injury · Axonal regeneration

Y. Zhang · B. T. Ang (✉) · I. Ng
Department of Neurosurgery, National Neuroscience Institute,
11 Jalan Tan Tock Seng,
Tan Tock Seng, Singapore 308433
e-mail: bengti.ang@gmail.com

Z. C. Xiao
Institute of Molecular and Cell Biology,
61 Biopolis Drive,
Proteos, Singapore 138673

Introduction

Trauma remains a major contributor to morbidity and mortality in societies worldwide. A large number of victims have long-term disabilities due to the lack of effective therapy to affect central nervous system repair, and this results in a significant economic burden to families, caregivers and society [1]. Axonal plasticity leading to functional recovery in the immature central nervous system (CNS) following brain injury has been demonstrated [2]. Following unilateral cortical damage, an increased crossed projection of intact corticospinal tract axons occurs to the contralateral red nucleus at the level of the midbrain [3]. Functional and anatomical repair is very limited in the adult brain compared with that in the immature CNS, primarily due to axonal growth inhibitors associated with myelin [4] such as NogoA [5], myelin-associated glycoprotein (MAG) [6], oligodendrocyte myelin glycoprotein (OMgp) [7] and TnR [8]. Many studies have identified that neutralization of inhibitors at the time of or after brain injury result in functional regeneration [9, 10]. These studies highlight the importance of these inhibitors as critical targets for effecting repair of the injured CNS. However, blockade of all myelin proteins may be of limited utility as certain components have important biological functions [11, 12]. To address this, a recombinant DNA vaccine, encoding fragments which compose the immunoglobulin domains 1–5 of rat MAG, the epidermal growth factor-like repeats of human TnR and the N-terminal and 66 amino acid extracellular loop of human NogoA, has been generated. Our previous study has shown that the DNA vaccine could produce antibodies to neutralize these inhibitors and promote long-distance axonal regeneration and functional recovery of locomotor functions after spinal cord injury [13]. The present study aims to investigate whether neutralization of these inhibitors allows compensatory structural plasticity and restoration of function in response to TBI upon prophylactic and therapeutic treatment with the DNA vaccine.

Materials and methods

Construction of the DNA vaccine

This has been described in detail previously [13].

Animals

Female Sprague-Dawley rats were supplied by the Animal Breeding Center of the National University of Singapore. Animals were housed three to a cage and maintained in a temperature and humidity controlled room on a 12-h light: 12-h dark cycle. All testing was performed during the light portion of the cycle. The animals were divided into the following groups: (A) No lesion or sham lesion and no treatment (n=10); (B) Lesion and no treatment (n=19); (C) The *prophylactic* group: DNA vaccine (pcDNA-NGIs) was given until autoantibodies were induced and traumatic brain injury (TBI) was performed as soon as autoantibodies were detectable (n=19); (D) the control group: The control vaccine (empty vector; pcDNA3.1) was given via the same way as that for the DNA vaccine and TBI was performed at the same time point as that in the prophylactic group (n= 19); (E) The *therapeutic* group: TBI was performed and pcDNA-NGIs was given immediately until autoantibodies were induced (n=19).

DNA vaccine administration

100 μg of pcDNA-NGIs was intramuscularly injected into the tibial muscle once a week until specific autoantibodies were detectable. pcDNA3.1 (control empty vector) was administered in the same manner.

Traumatic brain injury

The procedure of lateral fluid percussion (LFP) head injury in the rat is a common model that has been well described [14].

Neurobehavioral testing

All animals were trained for 3 weeks and baseline measurements were recorded. Neurological scoring was performed daily in all animals. Neurological function was evaluated by a trained observer using an ordinal scale. Animals were scored up to eight weeks following brain injury.

Skilled forelimb reaching test

Forelimb coordination and fine digit motor control was measured using the skilled forelimb reaching test [10]. All animals were trained to reach through bars to grasp food pellets. The paw not being evaluated was constrained. Animals were presented with a total of 20 sugar pellets in succession. A score of 1 was given for each pellet the animals obtained by using the forelimb to reach and grasp the pellet with its digits and carry the pellet to its mouth.

Beam walk

The beam walk measures the animals' complex neuromotor function [15]. The apparatus was a painted wooden beam 90 cm long, 4 cm wide, and 1.5 cm thick elevated 38 cm above a foam bed. A 100-W incandescent light bulb and white noise generator (110 dB) was located at the starting point, and a darkened box (the goal) was located at the opposite end of the beam. The noise was terminated after the rat crossed the beam and entered the box. Animals readily learned to cross the beam and enter the goal box to escape the noise and light. Before LFP brain injury, each rat was trained to cross the beam in 10 s or less on three consecutive trials. The maximum trial length was 60 s. Baseline performance was reestablished on the day of brain injury. The rats were then retested following injury. The mean beam crossing latency of three trials on each test day was used as the measure of performance.

Neuroanatomical tract tracing

After 8 weeks of behavioral testing, eight animals of each group were anesthetized and placed in a stereotactic frame. The bone overlying the motor cortex contralateral to the traumatic lesion was removed, and a 10% biotinylated dextran amine (BDA: Molecular Probes: 10% wt/vol solution in sterile saline) solution in 0.01 M phosphate buffer was pressure injected (1 ul per injection at a rate of 0.2 ul per 1 min) using a Hamilton syringe at four sites within the motor cortex. Animals were returned to standard housing condition for 2 weeks and then sacrificed.

Immunohistological assessment

The brains were removed following transcardiac perfusion with 4% Paraformaldehyde. After overnight post-fixation in 4% Paraformaldehyde followed by 30% sucrose, 30 µl serial sections were cut using a cryostat and collected for semi-free floating processing with avidin-biotin-peroxidase complex for visualization of BDA.

Quantification of cortico-rubral fibers

Axons crossing at the level of the red nucleus were counted under 20× magnification; the midline was identified by using the aqueduct and the paired red nuclei as landmarks. For each case, we averaged the number of the BDA-labeled axons that crossed to the denervated side in the three standard coronal sections that included the red nucleus; Results are from at least eight randomly selected animals from each group.

Statistics

For behavioral data, a one-way analysis of variance (ANOVA) with a Tukey post hoc test was used to compare the mean values across all groups. For anatomical data, ANOVA was used to compare the mean fiber at the level of the red nucleus between groups. For all analyses, $p<0.05$ was considered significant.

Results

Functional recovery

With regards to the skilled forelimb researching task (Fig. 1): at baseline testing before TBI, all animals achieved the same success score in grasping pellets with no significant difference between groups ($p>0.05$). At 1 week after TBI, all injured groups showed a severe deficit in pellet grasping (mean of four pellets), with no significant differences between groups ($p>0.05$). From 4 weeks after TBI, rats in the prophylactic group had significantly greater mean pellet-reaching success scores compared with other groups ($p<0.05$). There was no difference between the therapeutic group and the lesion only/control group until week 8 ($p>0.05$). At week 8 after TBI, animals in the therapeutic group (mean of twelve pellets) also had significantly greater mean pellet-reaching success than the lesion only/control group ($p<0.05$). There were no significant differences between the lesion only and control group at any time point ($p>0.05$).

Beam walk performance is shown in Fig. 2. The mean crossing latency for all groups prior to surgery was 6.5± 0.25 s. Statistical analysis demonstrated no difference between groups in the first three weeks after TBI ($p>0.05$). Crossing latencies for animals in the prophylactic group were significantly decreased after TBI compared to other groups on test weeks 4–8 ($p<0.05$). In contrast, beam crossing latencies for animals in the therapeutic group did not differ significantly from lesion only/control groups till 8 weeks ($p>0.05$). Latencies for animals in the control group did not differ significantly from the lesion only group ($p>0.05$).

Corticorubral plasticity

To relate behavioral changes to the development of new cortico-efferent pathways, the corticorubral projection from the spared hemisphere was examined with the anterograde tracer, BDA (Fig. 3). In the intact CNS, the corticorubral pathway is primarily an ipsilateral projection with a minor contralateral component. In all groups, the region of the red nucleus on the BDA-injection side showed more dense BDA-positive fibers than on the contralateral side. The number of crossed corticorubral fibers was less in the lesion only/control group compared with the immunized groups () ($p<0.05$). There was no significant difference between the

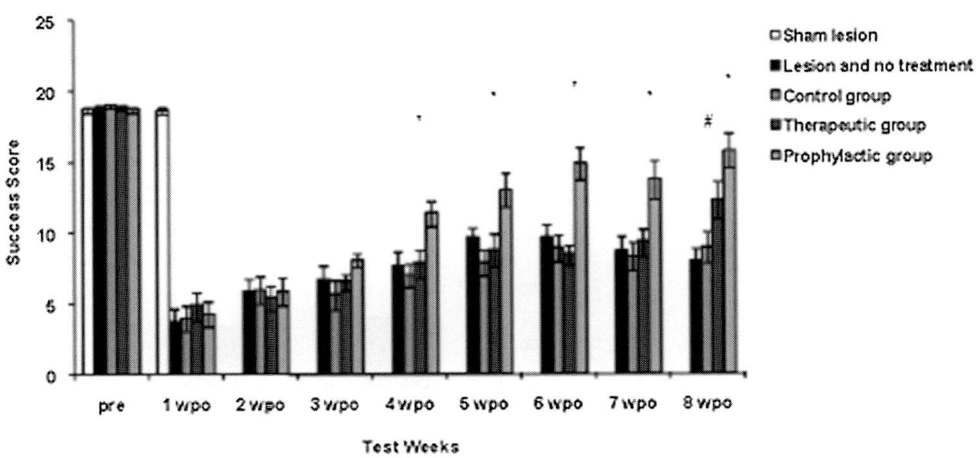

Fig. 1 Forelimb reaching performance from 1 week post-injury (wpo) to 8 wpo after TBI shows functional improvement in the DNA vaccine treatment group. Data are presented as mean±SD. *, $p<0.05$, the prophylactic group compared with other groups. #, $p<0.05$, the therapeutic group compared with the lesion with no treatment group and control group

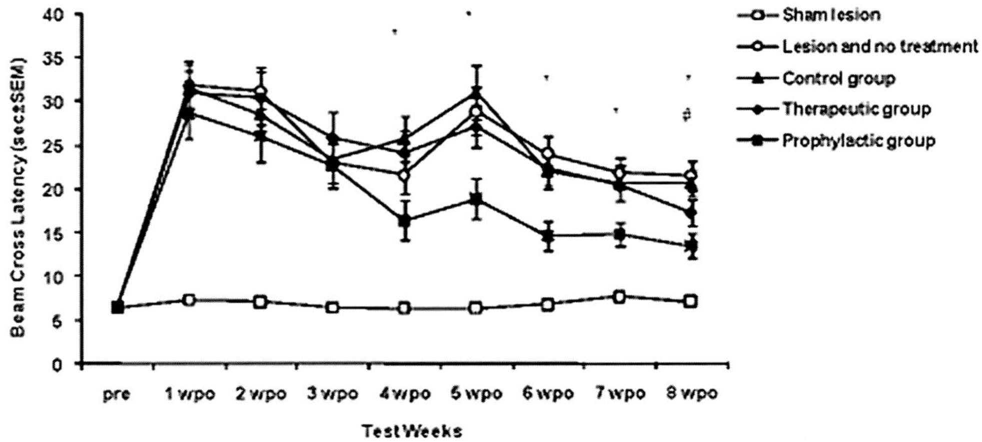

Fig. 2 Beam walk performance from 1 week po to 8 wpo after TBI shows functional improvement in the DNA vaccine treatment group. Data are presented as mean±SD. *, p<0.05, the prophylactic group compared with other groups. #, p<0.05, the therapeutic group compared with the lesion and no treatment group and control group

lesion and the control group (p>0.05). DNA vaccination had a greater effect on increasing the number of crossed corticorubral axons in rats in the prophylactic group as compared with the therapeutic group (p<0.05).

Discussion

Our results show that treatment with a DNA vaccine against myelin-associated inhibitors before and after TBI in adult

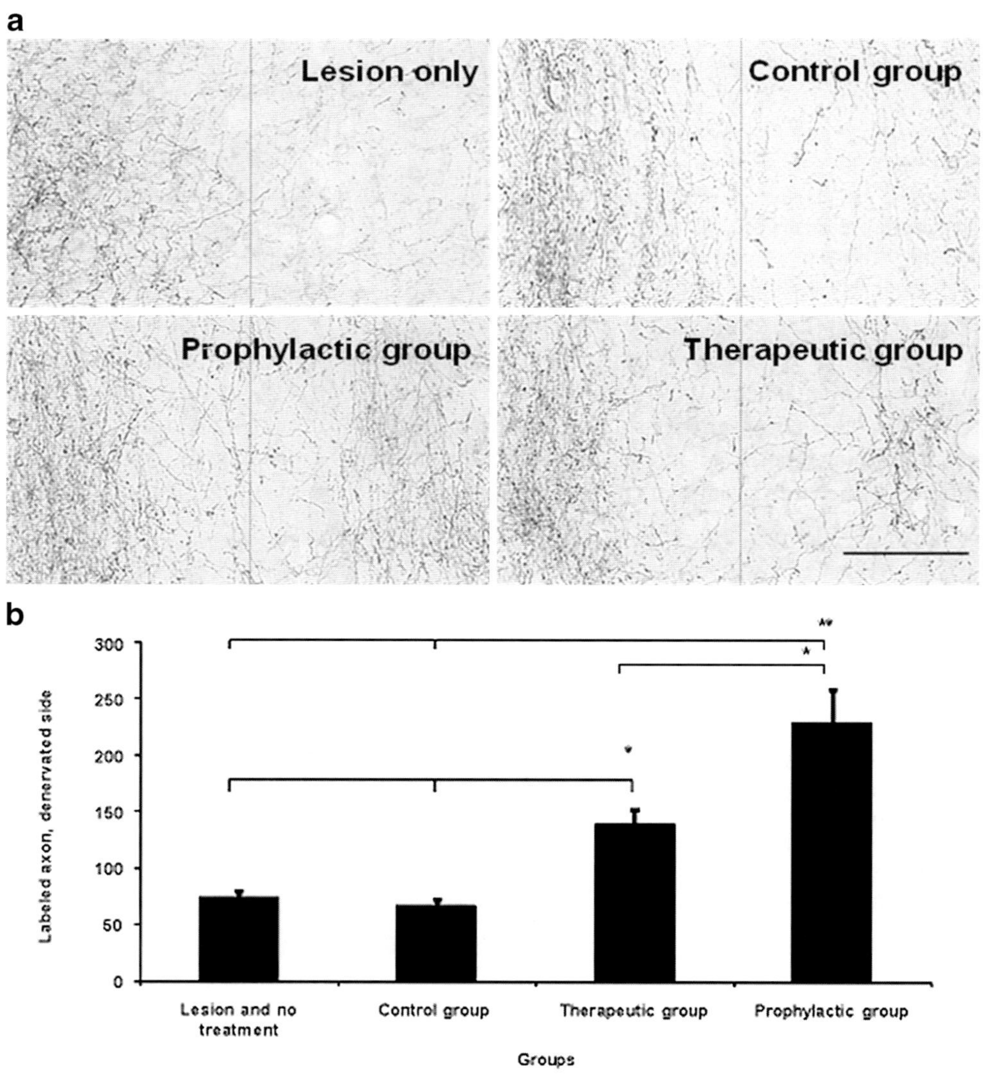

Fig. 3 Axon crossings at the level of the red nucleus. **a**: labeled axons in the vicinity of the red nucleus are shown on the normal side (Left) and on the denervated (Right) side. **b**: Quantitation of axons crossing from the intact side of the brain to the denervated red nucleus. *, p<0.05, **, p<0.001. (ANOVA). (Bar in d=50 μm)

rats promotes recovery of function,as measured by a forelimb task and beam walking, and also promotes improved structural axonal plasticity. DNA vaccination induces cell-mediated immunity [16, 17]. Intra-muscular DNA vaccination results in the rapid movement of foreign DNA out of local sites and an immuno-stimulatory response sustained at distant sites [18]. Activated T cells are able to cross the blood brain barrier (BBB) and enter the CNS parenchyma [19]. The safety concern with regards to induction of an auto-immune demyelination in view of the fact that the vaccine contained myelin antigens has been addressed in our previous study [13]. In the present study, the improvement for the therapeutic group compared to control occurred later than in the prophylactic group. Previous reports have also described differences in time of functional recovery following treatment with antibodies directed against NogoA at the point of CNS injury compared to a delayed treatment [10]. Our results suggest that DNA vaccination could be a possible strategy to enhance CNS repair after head injury.

Conflict of interest statement We declare that we have no conflict of interest.

References

1. Marshall LF (2000) Epidemiology and cost of central nervous system injury. Clin Neurosurg 46:105–112
2. Barth TM, Stanfield BB (1990) The recovery of forelimb-placing behavior in rats with neonatal unilateral cortical damage involves the remaining hemisphere. J Neurosci 10:3449–3459
3. Murakami F, Higashi S (1988) Presence of crossed corticorubral fibers and increase of crossed projections after unilateral lesions of the cerebral cortex of the kitten: a demonstration using anterograde transport of Phaseolus vulgaris leucoagglutin. Brain Res 447:98–108
4. Schwab ME (2004) Nogo and axon regeneration. Curr Opin Neurobiol 14(1):118–124
5. Chen MS, Huber AB, Van Der Haar ME, Frank M, Schnell L, Spillmann AA, Christ F, Schwab ME (2000) Nogo-A is a myelin-associated neurite outgrowth inhibitor and an antigen for monoclonal antibody IN-1. Nature 403:434–439
6. Mukhopadhyay G, Doherty P, Walsh FS, Crocker PR, Filbin MT (1994) A novel role for myelin-associated glycoprotein as an inhibitor of axonal regeneration. Neuron 13:757–767
7. Kottis V, Thibault P, Mikol D, Xiao ZC, Zhang R, Dergham P, Braun PE (2002) Oligodendrocyte-myelin glycoprotein (OMgp) is an inhibitor of neurite outgrowth. J Neurochem 82:1566–1569
8. Xiao ZC, Taylor J, Montag D, Rougon G, Schachner M (1996) Distinct effects of recombinant tenascin-R domains in neuronal cell functions and identification of the domain interacting with the neuronal recognition molecule F3/11. Eur J Neurosci 8(4): 766–782
9. Emerick AJ, Neafsey EJ, Schwab ME, Kartje GL (2003) Functional reorganization of the motor cortex in adult rats after cortical lesion and treatment with monoclonal antibody IN-1. J Neurosci 23(12):4826–4830
10. Seymour AB, Andrews EM, Tsai SY, Markus TM, Bollnow MR, Brenneman MM, O'Brien TE, Castro AJ, Schwab ME, Kartje GL (2005) Delayed treatment with monoclonal antibody IN-1 1 week after stroke results in recovery of function and corticorubral plasticity in adult rats. J Cereb Blood Flow Metab 25(10): 1366–1375
11. Götz B, Scholze A, Clement A, Joester A, Schütte K, Wigger F, Frank R, Spiess E, Ekblom P, Faissner A (1996) Tenascin-C contains distinct adhesive, anti-adhesive, and neurite outgrowth promoting sites for neurons. J Cell Biol 132(4):681–699
12. Xiao ZC, Revest JM, Laeng P, Rougon G, Schachner M, Montag D (1998) Defasciculation of neurites is mediated by tenascin-R and its neuronal receptor F3/11. J Neurosci Res 52 (4):390–404
13. Xu G, Nie DY, Chen JT, Wang CY, Yu FG, Sun L, Luo XG, Ahmed S, David S, Xiao ZC (2004) Recombinant DNA vaccine encoding multiple domains related to inhibition of neurite outgrowth: a potential strategy for axonal regeneration. J Neurochem 91(4):1018–1023
14. McIntosh TK, Vink R, Noble L, Yamakami I, Fernyak S, Soares H, Faden AL (1989) Traumatic brain injury in the rat: characterization of a lateral fluid-percussion model. Neuroscience 28(1):233–244
15. Feeney DM, Gonzalez A, Law WA (1982) Amphetamine, haloperidol, and experience interact to affect rate of recovery after motor cortex injury. Science 217:855–857
16. McDonnell WM, Askari FK (1996) DNA vaccines. N Engl J Med 334(1):42–45
17. Bhardwaj N (2001) Processing and presentation of antigens by dendritic cells: implications for vaccines. Trends Mol Med 7 (9):388–394 Review
18. Dupuis M, Denis-Mize K, Woo C, Goldbeck C, Selby MJ, Chen M, Otten GR, Ulmer JB, Donnelly JJ, Ott G, McDonald DM (2000) Distribution of DNA vaccines determines their immunogenicity after intramuscular injection in mice. J Immunol 165(5):2850–2858
19. Flugel A, Matsumuro K, Neumann H, Klinkert WE, Birnbacher R, Lassmann H, Otten U, Wekerle H (2001) Anti-inflammatory activity of nerve growth factor in experimental autoimmune encephalomyelitis: inhibition of monocyte transendothelial migration. Eur J Immunol 31(1):11–22

Acta Neurochir Suppl (2008) 102: 353–356

Ischemic blood-brain barrier and amyloid in white matter as etiological factors in leukoaraiosis

Ryszard Pluta · Sławomir Januszewski · Marzena Ułamek

Abstract
Background Pathology of white matter, which is observed in ischemic brain, indicates that similar processes contribute to Alzheimer's disease development. These injuries have been seen in the subcortical and periventricular regions. Periventricular white matter changes in ischemic and Alzheimer's disease brain, referred to as leukoaraiosis, are responsible for changes in memory, cognition and behavior. It is not clear whether the blood-brain barrier in ischemic periventricular white matter is altered in aged animals.
Methods We studied blood-brain barrier changes with amyloid precursor protein staining around blood-brain barrier vessels. Rats were made ischemic by *cardiac arrest*. Blood-brain barrier insufficiency, accumulation of amyloid precursor protein and platelets around blood-brain barrier vessels were investigated in ischemic periventricular white matter up to 1-year survival.
Findings Ischemic periventricualr white matter demonstrated enduring blood-brain barrier changes. Toxic fragments of amyloid precursor protein deposits were associated with the blood-brain barrier vessels. Moreover our investigation revealed platelet aggregates in- and outside blood-brain barrier vessels. Toxic parts of amyloid precursor protein and platelet aggregates correlated very well with blood-brain barrier permeability.
Conclusions Progressive injury of the ischemic periventricular white matter may be caused not only by a degeneration of neurons destroyed during ischemia but also by damage in blood-brain barrier. Chronic ischemic blood-brain barrier insufficiency with accumulation of toxic components of amyloid precursor protein in the periventricular white matter perivascular space, may gradually over a lifetime, progress to leukoaraiosis and finally to severe dementia.

Keywords Leukoaraiosis · Blood-brain barrier · Axonal leakage · Amyloid precursor protein

R. Pluta (✉) · S. Januszewski · M. Ułamek
Laboratory of Ischemic and Neurodegenerative Brain Research, Department of Neurodegenerative Disorders, Medical Research Centre, Polish Academy of Sciences, Pawinskiego 5 Str, 02–106 Warsaw, Poland
e-mail: pluta@cmdik.pan.pl

Introduction

Alzheimer's disease is progressive disorder with unknown etiological mechanisms. Recent clinical and experimental studies propose connection of ischemic processes with Alzheimer's dementia [6, 8]. Slowly developing ischemic- and Alzheimer's-type dementia probably has connection with white matter changes, too. Several reports have been suggested that axonopathy and axonal transport deficits may also play a causative role during progression of Alzheimer's disease [2, 4, 11]. Here the overlap of ischemia with Alzheimer's disease will be a central issue, as will the white matter changes frequently seen in dementia. Periventricular white matter injury occurs in both neuropathological processes and is termed leukoaraiosis. In this paper we will try understand the role of ischemic blood-brain barrier permeability for platelets and toxic fragments of amyloid precursor protein in leukoaraiosis process.

Materials and methods

In female Wistar rats (n=24, 3 months old, 150–180 g body weight) blood-brain barrier, amyloid precursor protein and platelets pathology were investigated following 10 min of ischemic brain injury due to *cardiac arrest* [7]. Experiments were performed with rats aged 2 and 7 days and 6 and 12 month. Sham-operated control animals (n=16) were sacrificed at the same time points. Animals were perfusion fixed for light and electron microscopic analysis.

Horseradish peroxidase served as an indicator of blood-brain barrier permeability. It was injected *i.v.* and was allowed to circulate for 30 min. Left hemisphere was cut at 50–80 μm slices in the coronal plane with a vibratome and incubated in a solution of 3,3′-diaminobenzidine tetrahydrochloride and examined by light microscopy [7]. Sections from the right hemisphere were selected for electron microscopy studies [7]. Antibodies recognizing different parts of amyloid precursor protein were used to stain paraffin sections [7].

Results

Until one year after ischemic brain injury periventricular white matter regions contained single and scattered areas of horseradish peroxidase extravasations (Fig. 1a) (Table 1). Horseradish peroxidase extravasations involved arterioles, capillaries, venules and veins. Extravasated horseradish peroxidase appeared to be restricted to branches and bifurcations of leaking blood-brain barrier vessels. In summary ischemic periventricular white matter presented random blood-brain barrier permeability (Fig. 1a).

After short-term survival following experimental *cardiac arrest* N- and C-terminal of amyloid precursor protein and β-amyloid peptide immunoreactivity were found around the blood-brain barrier vasculature (Table 1). Following long-term survival, staining only for the neurotoxic β-amyloid peptide and C-terminal of amyloid precursor protein was found (Fig. 1c) (Table 1). Multiple and abundant β-amyloid peptide and C-terminal of amyloid precursor protein accumulations embraced or adjoined the blood-brain barrier vessels (Fig. 1c). The staining size was different and irregular in shape. Blood-brain barrier vessel lumens and their inner and outer side of walls were also stained. The halo of β-amyloid peptide and C-terminal of amyloid precursor protein immunoreactivity in the perivascular space of blood-brain barrier vessels suggests that both proteins can easy cross walls of blood-brain barrier vasculature. In general, perivascular deposits of different parts of amyloid precursor protein took the same forms as extravasated horseradish peroxidase.

Investigation of leaky sites of blood-brain barrier vessel demonstrated single or aggregating platelets sticking and adhering to the blood-brain barrier vessel walls at all time points [7, 8]. Intravascular aggregates of platelets predominated in blood-brain barrier branches and bifurcations,

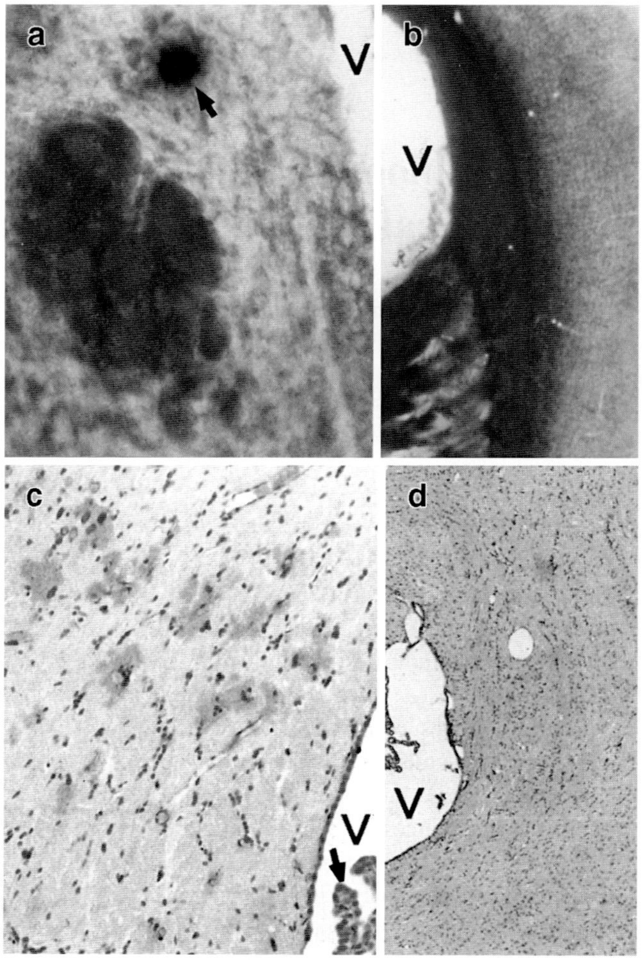

Fig. 1 **a**. Vibratome section reacted for horseradish peroxidase, extravasated horseradish peroxidase (arrow) is noted in periventricular space. 10-min brain ischemia, 1 year survival. (× 60). **b**. No staining for horseradish peroxidase in vibratome section is seen in periventricular space. Sham-operated rat, 1 year survival. (× 40). **c**. Numerous perivascular, extracellular and intracellular C-terminal of amyloid precursor protein deposits in periventricular space (brown color). C-terminal of amyloid precursor protein-positive staining in plexus chorioideus (arrow). 10-min brain ischemia, 1 year survival. (× 100). **d**. C-terminal of amyloid precursor protein staining is not identified in perivascular, extracellular and intracellular space in periventricular area. Sham-operated rat, 1 year survival. (× 60). Ventricles of brains (v) are shown

which correlated very well with blood-brain barrier permeability. Many vessels were plugged by platelets completely and completely blocked blood flow. Moreover single and aggregating platelets were noted on the abluminal side of blood-brain barrier vessels [7, 8]. Platelets pathology was single, scattered, and random. These kind of alterations occurred in arterioles, capillaries, venules and veins independent of survival time. Toxic components of amyloid precursor protein and platelet aggregates correlated well with blood-brain barrier permeability.

Control animals showed no horseradish peroxidase leakage (Fig. 1b) (Table 1) and no luminal or abluminal

Table 1 Immunoreactivity for N-terminal of amyloid precursor protein (NAPP), β-amyloid peptide (βA) and C-terminal of amyloid precursor protein (CAPP) and horseradish peroxidase (HRP) in the periventricular white matter perivascular space following 10-min brain ischemia

Group	NAPP	βA	CAPP	HRP
Controls	–	–	±	–
Short-term survival				
2 days	++	++	++	+
7 days	++	++	++	+
Long-term survival				
6 months	–	+++	+++	+
12 months	–	+/++	+/++	+

The staining intensity was categorized into five grades as follows: - no staining; ± staining for cytoplasm of single cells; + a single and diffuse areas; ++ a few and diffuse areas; +++ many strong and diffuse areas

platelet aggregation around blood-brain barrier vessels [7, 8]. In control brains staining for different fragments of amyloid precursor protein was not observed (Fig. 1d) except for weak immunoreactivity for the C-terminal of amyloid precursor protein in the cytoplasm of single cells (Table 1).

Discussion

We identified a novel lesion in periventricular white matter myelinated fibers caused by blood-brain barrier permeability for toxic components of amyloid precursor protein in ischemic brain. Therefore it is possible that the focal abnormal levels of β-amyloid peptide and C-terminal of amyloid precursor protein may locally destroy axon membranes [7, 10, 13] and induce local axonal leakage e.g. for amyloid precursor protein (Fig. 2). Moreover, if an axon is destroyed at certain sites [7], amyloid precursor protein may leak out and, *via* proteolysis of amyloid precursor protein, β-amyloid peptide may be produced in a *vicious cycle* [1, 3] (Fig. 2). Both circulatory and intraaxonal components, including amyloid precursor protein, may leak out of ischemic blood-brain barrier and ischemic axons and produce leukoaraiosis, in same way as in Alzheimer's brain (Fig. 2). This process may be involved in the formation of amyloid plaques that may also play a causative role in the cognitive deficits in Alzheimer's disease [6, 8] (Fig. 2).

In the present study we found that swollen axons are accompanied by ischemic blood-brain barrier changes and accumulation of the C-terminal of amyloid precursor protein and β-amyloid peptide [7]. The occurrence of astrocytic and axonal swelling in perivascular space one year following brain ischemia [7] represents vasogenic edema. On the other hand the available evidence indicates that impaired axonal transport is a key process inducing axon swelling in an experimental model of Alzheimer's disease [11]. Axonal swelling phenotypes that resemble the dystrophic neurites related to senile plaques have been described not only in Alzheimer's disease mouse models but also in the brains of aged and Alzheimer's disease patients. It is possible that the swollen axons [7] with local membrane damages [10, 13] may in turn affect axonal transport and make axonal leakage worse. Amyloid precursor protein is a functional membrane protein in axons that is produced in the cell body and transported to the distal part of axons [5]. If amyloid precursor protein leakage occurs at the site of axonal leakage, then β-amyloid

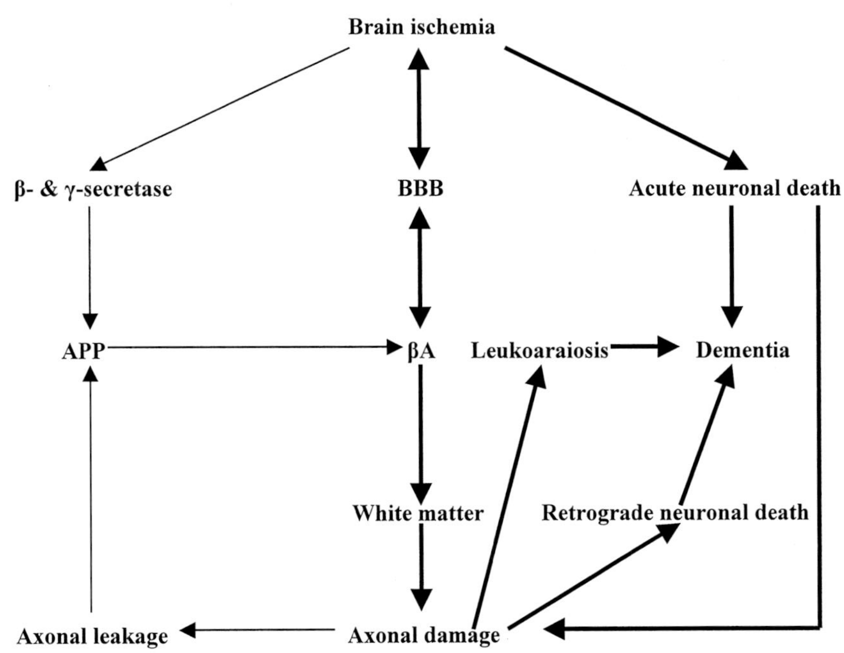

Fig. 2 Schematic diagram of the ischemic pathological processes in leukoaraiosis development. BBB-Blood-brain barrier; APP-amyloid precursor protein; βA-β-amyloid peptide

peptide production and deposition may occur in the extraaxonal space by proteolysis of amyloid precursor protein [1, 3] (Fig. 2). Accumulation of β-amyloid peptide and other products of amyloid precursor protein cleavage, at the sites of axonal injury after traumatic experimental brain lesion, suggest that this may also be the case in Alzheimer's disease [1, 3, 11]. The uncontrolled leakage of amyloid precursor protein and other proteins may decrease the availability of these components in the distal part of axons, resulting in dystrophic neurites and diminished neurotransmitters at the axon terminals. Axonal leakage may be an additional factor in the development of leukoaraiosis and retrograde secondary neuronal death in ischemic injury and dementia (Fig. 2).

If ischemic blood-brain barrier with different fragments of amyloid precursor protein and axonal leakage appear together around the neurites as we observed in our experimental model, this may explain the formation of amyloid plaques with dystrophic neurites typical of Alzheimer's disease. It is understandable that axonal leakage is due to direct and indirect ischemic membrane damage of axons. Any factor that affects membrane stability and results in severe blood-brain barrier [7, 12] and axonal leakage, may precipitate Alzheimer's disease-like neuropathological change. Our findings indicate that blood-brain barrier and axonal leakage may be a key pathological change that would help to explain mechanisms of leukoaraiosis. These changes may possibly lead to the development of amyloid plaques and neurofibrillary tangles and may contribute to Alzheimer's dementia (Fig. 2). Finally we can observe age-related loss of myelinated fibers and neurons [9]. The development of substances stabilizing axonal transport might represent a novel therapeutic target to prevent cognitive decline in neurodegenerative disorders such as ischemia and Alzheimer's disease.

Acknowledgements This study was supported in part by founds from: Medical Research Centre, Polish Ministry of Science and Higher Education (2007–2010-Cost/253/2006) and European Union (Cost Action B-30).

Conflict of interest statement We declare that we have no conflict of interest.

References

1. Chen XH, Siman R, Iwata A, Meaney DF, Trojanowski JQ, Smith DH (2004) Long-term accumulation of amyloid-beta, beta-secretase, presenilin-1, and caspase-3 in damaged axons following brain trauma. Am J Pathol 165:357–371
2. Dai J, Buijs RM, Kamphorst W, Swaab DF (2002) Impaired axonal transport of cortical neurons in Alzheimer's disease is associated with neuropathological changes. Brain Res 948:138–144
3. Dietrich WD, Kraydieh S, Prado R, Stagliano NE (1998) White matter alterations following thromboembolic stroke: a β-amyloid precursor protein immunocytochemical study in rats. Acta Neuropathol (Wien) 95:524–531
4. Gotz J, Itter LM, Kins S (2006) Do axonal defects in tau and amyloid precursor protein transgenic animals model axonopathy in Alzheimer's disease? J Neurochem 98:993–1006
5. Kamal A, Almenar-Queralt A, LeBlanc JF, Roberts EA, Goldstein LS (2001) Kinesin-mediated axonal transport of a membrane compartment containing beta-secretase and presenilin-1 requires amyloid precursor protein. Nature 414:643–648
6. Pluta R (2006) Ischemia-reperfusion factors in sporadic Alzheimer's disease. In: Welsh, EM (eds) New research on Alzheimer's disease. Nova Science Publishers, Inc, Hauppauge, pp 183–234
7. Pluta R, Ułamek M, Januszewski S (2006) Micro-blood-brain barrier openings and cytotoxic fragments of amyloid precursor protein accumulation in white matter after ischemic brain injury in long-lived rats. Acta Neurochir (Wien) (Suppl) 96: 267–271
8. Pluta R (2007) Role of ischemic blood-brain barrier on amyloid plaques development in Alzheimer's disease brain. Curr Neurovasc Res 4:121–129
9. Saver JL (2006) Time is brain - quantified. Stroke 37:263–266
10. Silberberg DH, Manning MC, Schreiber AD (1984) Tissue culture demyelination by normal human serum. Ann Neurol 15: 575–580
11. Stokin GB, Lillo C, Falzone TL, Brusch RG, Rockenstein E, Mount SL, Raman R, Davies P, Masliah E, Williams DS, Goldstein LS (2005) Axonopathy and transport deficits early in the pathogenesis of Alzheimer's disease. Science 307:1282–1288
12. Ueno M, Tomimoto H, Akiguchi I, Wakita H, Sakamoto H (2002) Blood-brain barrier disruption in white matter lesions in a rat model of chronic cerebral hypoperfusion. J Cereb Blood Flow Metab 22:97–104
13. Wagner KR, Dean C, Beiler S, Bryan DW, Packard BA, Smulian AG, Linke MJ, de Courten-Myers GM (2005) Plasma infusions into porcine cerebral white matter induce early edema, oxidative stress, pro-inflammatory cytokine gene expression and DNA fragmentation: implications for white matter injury with increased blood-brain barrier permeability. Curr Neurovasc Res 2:149–155

Acta Neurochir Suppl (2008) 102: 357–361

Matrix metalloproteinase inhibition attenuates brain edema after surgical brain injury

Vikram Jadhav · Mitsuo Yamaguchi · Andre Obenaus · John H. Zhang

Abstract

Background Neurosurgical operations can result in inevitable brain injury due to the procedure itself. This surgical brain injury (SBI) can cause post-operative complications such as brain edema following blood-brain barrier (BBB) disruption leading to neurological deficits.

Methods We tested whether inhibition of matrix metalloproteinases (MMPs) 9 and 2 provided neuroprotection against SBI. A rodent SBI model, which involves a partial frontal lobe resection, was used to evaluate two treatment regimens of MMP inhibitor-1 (inhibitor of MMP-9 and MMP-2); a single dose (5 mg/kg, pretreatment) and daily dose treatment (5 mg/kg x 3, pre- and post-treatment). Post-operative assessment at different time periods included brain water content (brain edema), immunohistochemical analysis, zymography for MMP enzymatic activity, and neurological assessment.

Findings The results indicate that SBI caused localized edema around the site of surgical resection with concomitant increase in MMP-9 and MMP-2 activity. Both treatment regimens with MMP inhibitor-1 decreased brain edema and attenuated the rise in MMP-9 and MMP-2 activity. An increased expression of MMP-9 was also seen in the neurons and neutrophils in the affected brain tissue at the periphery of surgical resection.

Conclusions The study suggests a potential role for MMP inhibition as preoperative therapy before neurosurgical procedures.

Keywords Surgical brain injury · MMP-9 · MMP-2 · Brain edema

V. Jadhav · M. Yamaguchi · J. H. Zhang
Department of Physiology and Pharmacology,
Loma Linda University,
RH 228, 11041 Campus Street,
Loma Linda, CA, USA

J. H. Zhang (✉)
Division of Neurosurgery,
Loma Linda University Medical Center,
11234 Anderson Street, Room 2562B,
Loma Linda, CA 92354, USA
e-mail: johnzhang3910@yahoo.com

J. H. Zhang
Department of Anesthesiology,
Loma Linda University School of Medicine,
Loma Linda, CA, USA

A. Obenaus
Department of Radiation Medicine, Loma Linda University,
Loma Linda, CA 92354, USA

Introduction

Neurosurgical procedures can result in inevitable brain injury by different means including direct trauma, retractor stretch, hemorrhage, and electrocautery causing postoperative complications such as brain edema, ischemia, BBB disruption and neuronal cell death leading to neurological deficits and adverse outcomes [2, 6, 7, 11, 12, 19, 22, 23]. Our previous in-vivo study showed that brain edema following disruption of blood-brain barrier (BBB) played an important role in pathophysiology of this surgical brain injury (SBI) [13, 14, 20]. Thus, therapies aimed at reducing the BBB disruption are likely to attenuate post-operative brain edema after SBI and decrease the neurological deficits.

Matrix metalloproteinases (MMPs) are proteases which cause disruption of BBB [3, 10, 21] and are well implicated in different types of ischemic, hemorrhagic and neurodegenerative brain pathologies in clinical as well as experimental animal studies [1, 5, 8, 18, 25]. Furthermore, MMPs such as MMP-9 and MMP-2 play a critical role in cerebral edema formation and increasing permeability of BBB after different brain injuries [3, 8, 10, 21]. In this study we tested MMP inhibitor-1, an inhibitor of MMP-9 and MMP-2, for neuroprotective effect in the rodent model of SBI [13, 14, 20]. We hypothesized that MMP-9 and MMP-2 inhibition will decrease brain edema and improve neurological deficits after SBI.

Materials and methods

Surgical brain injury modeling and treatment groups

This protocol was evaluated and approved by the Institutional Animal Care and Use Committee at Loma Linda University, Loma Linda, CA. The rodent model of SBI was used as described before [13, 14, 20]. In brief, following anesthesia with ketamine (100 mg/kg) plus xylazine (10 mg/kg) i.p, a square cranial window was drilled such that the left lower corner of the square was at the bregma. The dura was incised and reflected to expose the underlying right frontal lobe. Using a flat blade (6 mm × 1.5 mm), two incisions were made along sagittal and coronal planes leading away from the bregma to sever an area of brain 2 mm lateral of sagittal and 1 mm proximal of coronal planes. The depth of the incisions extended to the base of the skull. The weight of the sectioned brain was not significantly different between the various groups. Hemostasis was achieved by intraoperative packing and saline irrigation. Sham surgery included only craniotomy and replacement of the bone flap without any dural incisions. A total of 67 Sprague-Dawley male rats (300–350 g) were divided randomly into four groups; (1) SBI with vehicle treatment (0.1% DMSO), (2) SBI with single pretreatment of MMP inhibitor-1 (an inhibitor of MMP-9 and MMP-2 administered i.p. 5 mg/kg 60 mins before surgery, Calbiochem, CA, U.S.A.), (3) SBI treated daily (total three doses of MMP Inhibitor-1, 5 mg/kg each at 60 mins before surgery, and at 24 and 48 hrs post-surgery) and (4) sham surgical group.

Brain water content

The animals were sacrificed under deep anesthesia at different time points and the brains were divided into frontal ipsilateral, frontal contralateral, parietal ipsilateral, parietal contralateral, brain stem and cerebellum on ice. These parts were weighed immediately (wet weight) and weighed again after drying in oven at 105°C for 48 hours (dry weight) [26]. The percent of water content was calculated as [(wet weight – dry weight)/ wet weight] x 100%.

Immunohistochemistry

Immunofluorescence studies were performed on ten micrometer-thick frozen brain sections (Leica cryostat CM 3050 S) using standard protocols [15] with minor modifications. Sections were incubated with primary polyclonal antibodies (1:50 dilution in buffer solution) which included MMP-9, MMP-2, myeloperoxidase (MPO) (Santa Cruz Biotechnology, CA, U.S.A.) and NeuN (Chemicon International, Temecula, CA) at 4°C for 24 hours, followed by appropriate fluorescent secondary antibodies (FITC, TRITC, 1:50). Digitalized microphotographs of immunofluorescent sections were obtained using Olympus BX51 fluorescent microscope and analyzed by MagnaFire SP 2.1B software.

Matrix metalloproteinase zymography

MMP-9 and MMP-2 gelatinase activities were determined by gel zymography as described previously [25]. Briefly, aforementioned brain parts (similar to those used for brain water content) were homogenized in lysis buffer including protease inhibitors on ice. After centrifugation, the supernatant was collected and the total protein concentrations were determined using the Bradford assay (Bio-Rad, Hercules, CA, U.S.A.). Equal amounts (50 μg) of total protein extracts were prepared and separated by 10% Tris-glycine gel with 0.1% gelatin as substrate (Bio-Rad). After separation, the gel was renatured and incubated with developing buffer (at 37°C for 24 hours). After developing, the gel was stained with 0.5% Coomassie Blue R-250 for 30 minutes and then destained. Gelatinolytic activity was determined as clear zones or bands at the appropriate molecular weights and the results were quantified using Image J software from http://www.nih.gov (http://rsb.info.nih.gov/ij/). Mouse MMP-9 and human MMP-2 (Chemicon, Temecula, CA, U.S.A.) were used as standards.

Neurological evaluation

Neurological status of the animals was evaluated at different time points using a 27 point scoring system modified from Garcia et al [9].

Data analysis

Data are expressed as mean ± S.E.M. Statistical significance was verified by one-way ANOVA for multiple comparisons and student's 't'test wherever applicable. Probability value of $p<0.05$ was considered statistically significant.

Results

Brain edema after SBI

Our pilot experiments using 10 animals and previously published reports showed that brain water content is maximal 24–72 hours after SBI, however, only in the ipsilateral frontal lobe i.e brain tissue surrounding the surgical resection site [13, 14, 20]. In this study, there was increased brain water content (83.58±0.5%) in the ipsilateral frontal lobe after SBI as compared to sham-operated animals (79.20±0.09%). Brain water content in rest of the brain areas was not statistically different between the SBI and sham groups. This animal model had zero mortality in any of the groups.

MMP enzymatic activity in brain after SBI

Both MMP-9 and MMP-2 gelatinase activities determined by gel zymography [25] were increased only in the ipsilateral frontal lobe (data not shown) and over days 1–14 as compared to pre-surgery levels (Table 1). MMP-9 enzymatic activity, however, was greatly increased over days 1–3 after surgical-injury ($P<0.05$ vs preoperative values, n=4 in each grp).

MMP Expression in Different Cell Types

Fluorescent double-immunostaining was done on brain tissue bordering SBI (bregma +2.20 mm) to examine the cells expressing MMP-9 and MMP-2 at 48 hours post-surgery. MMP-9 immunoreactivities were increasingly expressed on the neurons (labeled with NeuN) and neutrophils (labeled with MPO) after SBI (Fig. 1). MMP-2 immunoreactivities, however, were not seen in any cell types (data not shown). In negative controls, immunoreactivities of MMP-9, NeuN, and MPO were not observed after omitting the respective primary antibodies (data not shown).

Table 1 Time course of MMP-9 and MMP-2 activity

	MMP-9 (Expressed in arbitrary density units)	MMP-2
Day 0	0	0
Day 1	9209±2746*	851±265*
Day 3	8793±4307*	3443±785*
Day 7	2792±830*	2576±510*
Day 14	1763±446*	2547±427*

depicts the MMP-9 and MMP-2 enzymatic activity in ipsilateral frontal lobe tissue at different time points after SBI. Both MMP-9 and MMP-2 activities are significantly higher than pre-operative levels ($P<0.05$ vs preoperative values, n=4 in each grp); however, MMP-9 activity is greatly increased.

Protective effects of MMP inhibitor-1

Both treatment regimens (single and daily treatment) with MMP Inhibitor-1 effectively attenuated the MMP-9 and MMP-2 gelatinase activities in ipsilateral frontal lobe 72 hrs post-surgical injury (Table 2). Both the treatment regimens also significantly attenuated the brain edema in frontal ipsilateral brain as compared to vehicle treated group 72 hrs post-surgical injury (Table 2), however they did not improve the neurological score over 24–72 hrs.

Discussion

This study shows that inhibition of MMP-9 and MMP-2 using MMP Inhibitor-1 attenuates brain edema after SBI. Localized brain edema was seen in the brain tissue bordering the surgical injury. Increased MMP-9 and MMP-2 enzyme activities were also observed in the same brain tissue at time points corresponding to maximal brain edema. MMP-9 activity was greatly increased, especially over 1–3 days. On the other hand, MMP-2 enzyme activity, though significant, was only mildly increased after surgical injury (Table 1). Immunohistochemical evidence further suggested that the MMP-9, but not MMP-2 expression, increases in neurons and recruited neutrophils in the affected brain tissue (Fig. 1). Neutrophils have been implicated as major cellular source of MMP-9 after brain injury [16, 17]. MMP-9 promotes BBB breakdown secondary to microvascular lamina proteolysis leading to edema formation after brain injuries [8, 21]. Thus, increased MMP activity especially that of MMP-9 could produce BBB disruption leading to brain edema after SBI.

Accordingly, we tested MMP Inhibitor-1 which inhibited MMP-9 and MMP-2 enzymatic activity in the affected brain. Treatment with two different regimens (single dose and daily treatment) of MMP inhibitor-1 resulted in reduced brain edema. Although, the present data suggests a prominent role of MMP-9 in SBI induced brain edema, the exact contributions of MMP-9 and MMP-2 will have to be addressed in future studies using selective inhibitors. It would be interesting to study if MMP-9 disrupts tight junction proteins after SBI leading to enhanced BBB permeability as shown recently in a rat model of focal ischemia [24].

All animals subjected to SBI showed neurological deficits, however the MMP inhibitor-1 treatment regimens did not improve the neurological scores. The subtle neurobehavioral deficits resulting from frontal lobe injury will require specific and sensitive neurobehavioral tests [4, 9] to detect outcomes.

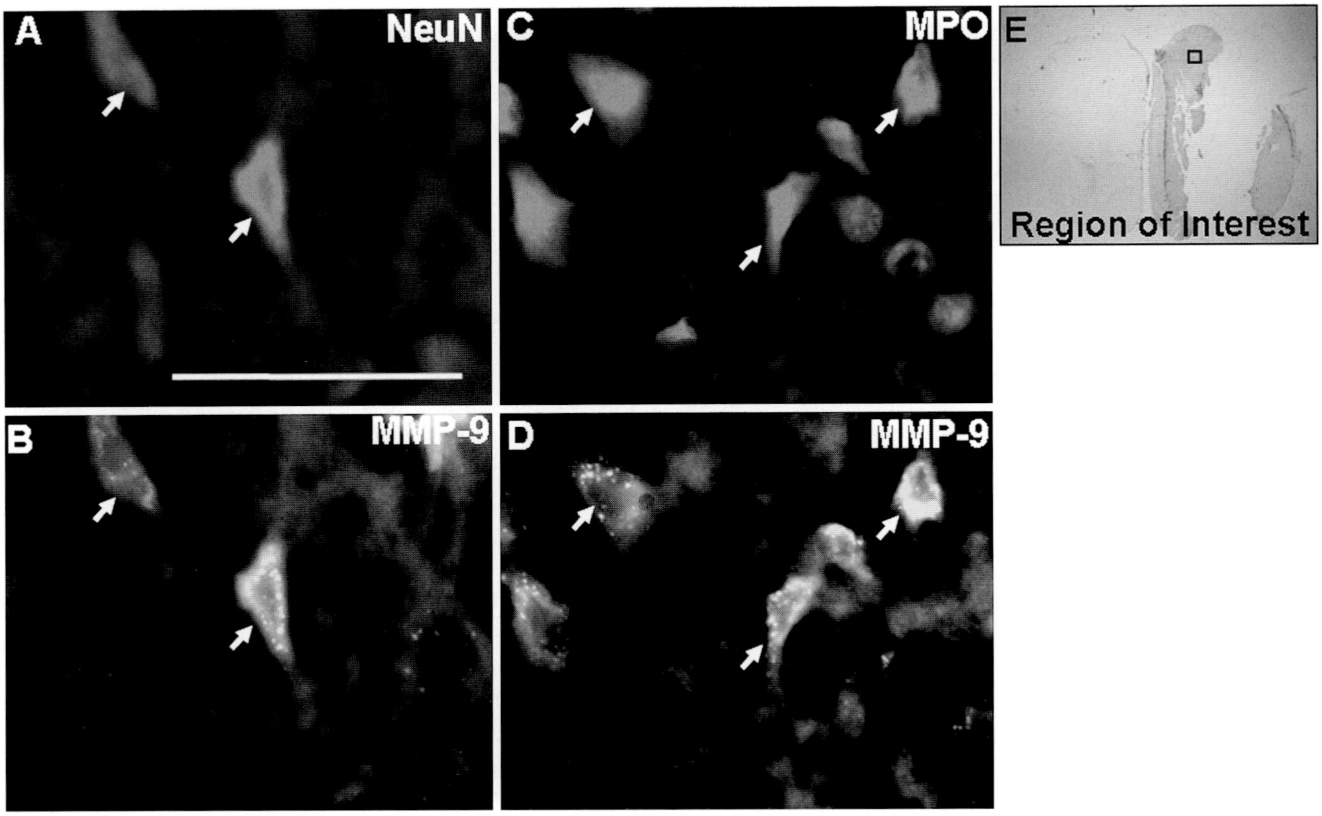

Fig. 1 Co-localization of MMP-9 with different cell markers. MMP-9 immunoreactivities were co-localized with NeuN (neuronal marker) and myeloperoxidase (MPO, neutrophilic marker) immunoreactivities. All individual pictographs are appropriately labeled and representative of 4 experiments each. The arrows show some of the cells that colocalize the markers for panels A and B and for panels C and D respectively. The IgG-immunostained brain section (panel E, bregma +2.20 mm) indicates the region of interest in the brain tissue adjoining the surgical-injury. Scale bar for panels A-D represents 50 μm

These experiments are planned for future studies. The study suggests that inhibition of activity of type IV collagenases such as MMP-2 and MMP-9 activity reduces brain edema. This effect of MMP Inhibitor-1 is mechanistically distinct from presently used clinical modalities such as diuretics, steroids and osmotic agents. MMP inhibition may have potential therapeutic utility in clinical neurosurgical practice for prevention of post-operative edema.

Table 2 Effect of MMP inhibitor-1

MMP Levels (Arbitrary Density Units)			
	Vehicle Treated	Single Treatment	Daily Treatment
MMP-9	8739±4307	2662±675*	1707±655*
MMP-2	3433±785	973±348*	740±550*
Neurological Score (Max=21 points)			
Day 1	20.5±0.7	24.5±1.0	22.8±1.4
Day 2	23.0±0.4	25.7±0.7	25.5±1.2
Day 3	25.0±0.3	25.3±0.9	25. 8±0.6
Brain Edema (% Brain Water Content)			
Sham			
78.8±0.03	82.2±0.26*	80.9±0.05*#	80.2±0.12*#

depicts the effects of MMP Inhibitor-1 with a single and daily dose treatment regimen. Both single and daily dose regimens inhibit the MMP-9 and MMP-2 enzymatic activity compared to the vehicle treated at 72 hrs post-surgery, $P<0.05$ for * vs vehicle-treated group, n=4. Both treatment regimens also attenuate the brain edema indicated by decreased brain water content at 72 hrs after surgical-injury. $P<0.05$ for * vs sham and # vs vehicle treated respectively. The neurological scores were significantly decreased in all animals with SBI modeling as compared to sham-operated animals, however, there was no significant difference between the individual groups.

Acknowledgements This study was partially supported by grants from NIH NS45694, NS53407, and NS43338 to JHZ.

Conflict of interest statement We declare that we have no conflict of interest.

References

1. Adair JC, Charlie J, Dencoff JE, Kaye JA, Quinn JF, Camicioli RM, Stetler-Stevenson WG, Rosenberg GA (2004) Measurement of gelatinase B (MMP-9) in the cerebrospinal fluid of patients with vascular dementia and Alzheimer disease. Stroke 35(6): e159–62
2. Andrews RJ, Muto RP (1992) Retraction brain ischaemia: cerebral blood flow, evoked potentials, hypotension and hyperventilation in a new animal model. Neurol Res 14:12–18
3. Aoki T, Sumii T, Mori T, Wang X, Lo EH (2002) Blood-brain barrier disruption and matrix metalloproteinase-9 expression during reperfusion injury: mechanical versus embolic focal ischemia in spontaneously hypertensive rats. Stroke 33(11):2711–2717
4. Basso DM, Beattie MS, Bresnahan JC (1996) Graded histological and locomotor outcomes after spinal cord contusion using the NYU weight-drop device versus transection. Exp Neurol 139: 244–256
5. Castellanos M, Leira R, Serena J, Pumar JM, Lizasoain I, Castillo J, Davalos A (2003) Plasma metalloproteinase-9 concentration predicts hemorrhagic transformation in acute ischemic stroke. Stroke 34(1):40–46
6. Deletis V, Sala F (2001) The role of intraoperative neurophysiology in the protection or documentation of surgically-induced injury to the spinal cord. Ann N Y Acad Sci 939:137–144
7. Fasano VA, Penna G (1992) [Postoperative complications in neurosurgery]. Minerva Anestesiol 58:15–21
8. Fujimura M, Gasche Y, Morita-Fujimura Y, Massengale J, Kawase M, Chan PH (1999) Early appearance of activated matrix metalloproteinase-9 and blood-brain barrier disruption in mice after focal cerebral ischemia and reperfusion. Brain Res 842(1):92–100
9. Garcia JH, Wagner S, Liu KF, Hu XJ (1995) Neurological deficit and extent of neuronal necrosis attributable to middle cerebral artery occlusion in rats. Statistical validation. Stroke 26:627–634
10. Gasche Y, Copin JC, Sugawara T, Fujimura M, Chan PH (2001) Matrix metalloproteinase inhibition prevents oxidative stress-associated blood-brain barrier disruption after transient focal cerebral ischemia. J Cereb Blood Flow Metab 21(12):1393–400
11. Hellwig D, Bertalanffy H, Bauer BL, Tirakotai W (2003) Pontine hemorrhage. J Neurosurg 99:796–797
12. Hernesniemi J, Leivo S (1996) Management outcome in third ventricular colloid cysts in a defined population: a series of 40 patients treated mainly by transcallosal microsurgery. Surg Neurol 45:2–14
13. Jadhav V, Matchett G, Hsu FP, Zhang JH (2007) Inhibition of Src tyrosine kinase and effect on outcomes in a new in vivo model of surgically induced brain injury. J Neurosurg 106(4):680–686
14. Jadhav V, Solaroglu I, Obenaus A, Zhang JH (2007) Neuroprotection against surgically induced brain injury. Surg Neurol 67 (1):15–20 discussion 20. Review
15. Jadhav V, Jabre A, Lin SZ, Lee TJ (2004) EP1- and EP3-receptors mediate prostaglandin E2-induced constriction of porcine large cerebral arteries. J Cereb Blood Flow Metab 24:1305–1316
16. Justicia C, Panes J, Sole S, Cervera A, Deulofeu R, Chamorro A, Planas AM (2003) Neutrophil infiltration increases matrix metalloproteinase-9 in the ischemic brain after occlusion/reperfusion of the middle cerebral artery in rats. J Cereb Blood Flow Metab 23 (12):1430–1440
17. Kaczorowski SL, Schiding JK, Toth CA, Kochanek PM (1995) Effect of soluble complement receptor-1 on neutrophil accumulation after traumatic brain injury in rats. J Cereb Blood Flow Metab 15(5):860–864
18. Lee SR, Tsuji K, Lee SR, Lo EH (2004) Role of matrix metalloproteinases in delayed neuronal damage after transient global cerebral ischemia. J Neurosci 24(3):671–678
19. Manninen PH, Raman SK, Boyle K, el-Beheiry H (1999) Early postoperative complications following neurosurgical procedures. Can J Anaesth 46:7–14
20. Matchett G, Hahn J, Obenaus A, Zhang J (2006) Surgically induced brain injury in rats: The effect of erythropoietin. J Neurosci Methods 158(2):234–241
21. Rosenberg GA, Estrada EY, Dencoff JE (1998) Matrix metalloproteinases and TIMPs are associated with blood-brain barrier opening after reperfusion in rat brain. Stroke 29(10):2189–2195
22. Solaroglu I, Beskonakli E, Kaptanoglu E, Okutan O, Ak F, Taskin Y (2004) Transcortical-transventricular approach in colloid cysts of the third ventricle: surgical experience with 26 cases. Neurosurg Rev 27:89–92
23. Tommasino C (1992) [Postoperative cerebral edema. Physiopathology of the edema and medical therapy]. Minerva Anestesiol 58:35–42
24. Yang Y, Estrada EY, Thompson JF, Liu W, Rosenberg GA (2007) Matrix metalloproteinase-mediated disruption of tight junction proteins in cerebral vessels is reversed by synthetic matrix metalloproteinase inhibitor in focal ischemia in rat. J Cereb Blood Flow Metab 27(4):697–709
25. Wang X, Jung J, Asahi M, Chwang W, Russo L, Moskowitz MA, Dixon CE, Fini ME, Lo EH (2000) Effects of matrix metalloproteinase-9 gene knock-out on morphological and motor outcomes after traumatic brain injury. J Neurosci 20(18):7037–7042
26. Xi G, Hua Y, Keep RF, Younger JG, Hoff JT (2002) Brain edema after intracerebral hemorrhage: the effects of systemic complement depletion. Acta Neurochir Suppl 81:253–256

Acta Neurochir Suppl (2008) 102: 363–366

Thrombin enhances glioma growth

Ya Hua · Lingling Tang · Richard F. Keep · Julian T. Hoff · Jason Heth · Guohua Xi · Karin M. Muraszko

Abstract
Background Our previous studies have demonstrated that argatroban, a specific thrombin inhibitor, reduces brain edema and neurological deficits in rat glioma models. The present study investigated whether or not thrombin enhances glioma growth in vivo and in vitro.
Methods There were two parts in this study. In the first part, rat C6 glioma cells were treated with or without thrombin. These cells were then injected into the right caudate of adult male Fischer 344 rats. Rats underwent behavioral testing prior to sacrifice 12 days later for tumor mass measurement. In the second part, C6 cells were incubated in serum-free medium for 24 hours and then treated with thrombin with or without argatroban, a thrombin inhibitor. DNA synthesis was examined using a 5-bromo-2′-deoxyuridine (BrdU) ELISA kit. Cell proliferation was determined using a 3-(4,5-dimethylthiazol-2-yl)-2,5-diphenyl-tetrazolium bromide (MTT) assay.
Findings Treatment of C6 cells with thrombin prior to intracerebral implantation resulted in a larger tumor mass and worse neurological deficits at day12. In vitro, thrombin increased DNA synthesis in C6 glioma cells, and this effect was blocked by argatroban. MTT assay showed that thrombin significantly increased glioma cell proliferation in vitro.
Conclusions In summary, thrombin enhances C6 glioma growth in vivo and cell proliferation in vitro suggesting that thrombin may be a target of glioma therapy.

Keywords Thrombin · Gliomas · Argatroban · Proliferation

This study was supported by grants NS-017760 and NS-039866 from the National Institutes of Health.

Y. Hua (✉) · L. Tang · R. F. Keep · J. T. Hoff · J. Heth · G. Xi · K. M. Muraszko
Department of Neurosurgery, University of Michigan,
Room 5018, BSRB,
Ann Arbor, MI 48109-2200, USA
e-mail: yahua@umich.edu

Introduction

Brain tumors kill thousands of people every year. There are, as yet, no effective treatments for malignant gliomas. Investigations indicate that thrombin may be a key factor contributing to tumor proliferation, invasion, angiogenesis and metastasis [10, 13–15]. Our recent studies have also found that thrombin has a key role in the growth of gliomas [7]. Thrombin activity is increased in C6 cell gliomas and intracerebral infusion of argatroban, a specific thrombin inhibitor, reduces brain edema and neurological deficits in a rat glioma model [5, 7]. Systemic use of argatroban also reduces tumor mass, attenuates glioma-induced neurological deficits and prolongs survival time [8].

In the present study, we investigated whether or not thrombin enhances glioma growth in vitro and in vivo.

Materials and methods

Experimental groups

The study comprised in vivo and in vitro experiments. In the first part, rat C6 glioma cells (6×10^5 in 10 μl saline) were treated with or without rat thrombin (0.5 U/ ml) for one hour at room temperature and infused into the right caudate of adult male Fischer 344 rats. Rats underwent behavioral testing at days 3, 6, 9 and 12 and were euthanized at day 12 for tumor mass measurement. In the second part of the study, C6 cells were cultured in 96-well plates. The cells were incubated in serum-free medium for 24 hours and then treated with human thrombin (0.5 U/ml) with argatroban (10 μM) or with vehicle. Cell proliferation was determined 72 hours later by 3-(4,5-dimethylthiazol-2-yl)-2,5-diphenyl-tetrazolium bromide (MTT) assay. DNA synthesis was measured 24 hours later using a 5-bromo-2′-deoxyuridine (BrdU) ELISA kit (Roche Molecular Biochemical).

Cell culture

C6 glioma cells (passage number 36 to 42) were obtained from American Type Culture Collection (Manassas, VA). They were grown at 37°C in air with 5% CO_2 in Ham's F-10 medium with 2.5% fetal bovine serum and 15% horse serum [7].

Experimental brain tumor

The University of Michigan Committee on the Use and Care of animals approved the protocols for these animal studies. The animals were anesthetized with pentobarbital (40 mg/kg, ip). Core temperature was maintained at 37°C with use of a feedback-controlled heating pad. The rats were positioned in a stereotaxic frame (Kopf Instrument), and a cranial burr hole (1 mm) was drilled on the right coronal suture 3.5 mm lateral to the midline. Approximately 6×10^5 C6 cells pre-treated with rat thrombin (Sigma) or vehicle for one hour and suspended in10 μl saline were infused into the right caudate nucleus (coordinates: 0.2 mm anterior, 5.5 mm ventral and 3.5 mm lateral to the bregma) at a rate of 1 μl/min using a 10-μl Hamilton syringe.

Tumor mass

We used the weight difference between the ipsilateral (tumor side) and contralateral hemisphere to estimate the tumor mass [5].

Behavioral tests

Two tests were used to assess tumor-induced behavioral deficits. A vibrissae-stimulated forelimb placing test (10 trials per side for each rat) was used to examine sensorimotor/ proprioceptive capacity. A corner turn test was also used as a measure of sensorimotor function. In that test, the rat was allowed to proceed into a corner, the angle of which was 30 degrees. To exit the corner, the rat could turn either to the left or right, and this was recorded. This was repeated 10 to 15 times, with at least 30 seconds between trials, and the percentage of right turns was calculated. All behavior was scored by experimenters who were blind to both neurological and treatment conditions. These tests are highly correlated with extent of striatal injury without being influenced by repeated testing [5, 6].

Cell proliferation measurement in vitro

Cells were plated in 96-well plates at a cell density of 10,000 cells/well. Cells were incubated in serum-free medium for 24 hours and treated with 0.5 U/ml of human thrombin with or without 10 μM argatroban for 72 hours. A 3-(4,5-dimethylthiazol-2-yl)-2,5-diphenyl-tetrazolium bromide (MTT; Sigma) assay was performed 72 hours later. Metabolically active cells can change yellow tetrazolium salt MTT to indigo-blue formazan. For the assay, MTT (5 mg/ml, 25 μl per 200 μl medium) was filtered, added to the wells and incubated at 37°C for 4 hours. Medium was removed gently using a 21-gauge needle and 200 μl dimethyl sulfoxide (DMSO) was added. After 10 minutes, the optical density was measured on a micro plate reader at a wavelength of 540/630 nm. All in vitro studies were repeated three times.

DNA synthesis assay

C6 glioma cells were cultured in 96-well plates at a cell density of 20,000 cells/well. Cells were incubated in serum-free medium for 24 hours and then treated with 0.5 U/ml of thrombin with or without argatroban (10 μM) for another 24 hours. DNA synthesis was determined using a BrdU ELISA kit (Roche Molecular Biochemical). The optical density was measured at a wavelength of 370/490 nm. Studies were repeated three times.

Statistical analysis

All data in this study are presented as mean ± standard deviation. Data were analyzed with Student's t test or Wilcoxon test. Significance levels were set at $p<0.05$.

Results

In vivo, intracerebral implantation of C6 glioma cells treated with thrombin resulted in a larger tumor mass (98±45 vs. 41±36 mg in controls, $p<0.05$, Fig. 1A) and also induced worse neurological deficits. Both forelimb placing (day 12: 21±8% vs. 60±8% in controls, $p<0.01$) and corner turn (day 9: 75±11% vs. 44±4% in controls, $p<0.05$) scores were worse in animals injected with thrombin-treated cells (Fig. 1B and C).

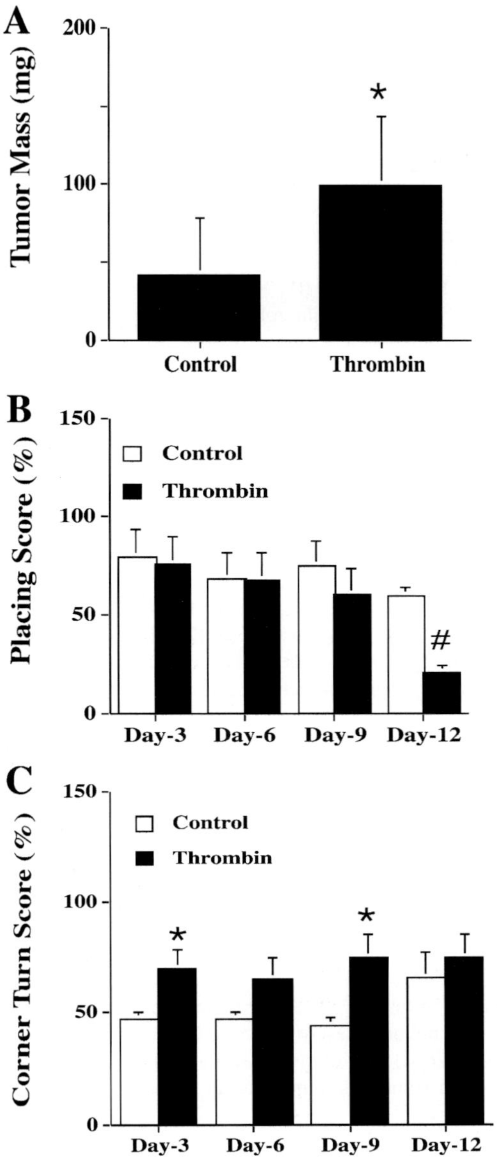

Fig. 1 (**A**) Tumor mass (the weight difference between the ipsilateral and the contralateral hemisphere) in rats 12 days after C6 cell implantation. The C6 cells were pre-treated with or without thrombin (0.5 U/ml). (**B**) Forelimb placing scores of the rats after C6 cell implantation. (**C**) Corner turn scores of the rats after C6 cell implantation. Values are mean ± SD, *p<0.05, #p<0.01 vs. the control group

In vitro, thrombin treatment increased DNA synthesis in C6 glioma cells (p<0.05, Fig. 2A), an effect blocked by the thrombin inhibitor, argatroban (optical density of BrdU: 1.21±0.08 vs. 1.46±0.07 in thrombin only group, p<0.01, Fig. 2A). An MTT assay also showed that thrombin increased glioma cell proliferation (optical density of MTT: 1.02±0.14 vs. 0.59±0.10 in controls, p<0.01, Fig. 2B). Argatroban again blocked thrombin-induced glioma cell proliferation (p<0.01, Fig. 2B).

Discussion

Our recent studies have demonstrated that thrombin and thrombin receptors are present in gliomas and that argatroban, a thrombin inhibitor, can reduce tumor mass and prolong survival time in a rat glioma model. In this study, we found that thrombin enhances glioma growth in vitro and in vivo. The results suggest a crucial role of thrombin in glioma pathogenesis.

Multiple sources of thrombin, including glioma- and plasma-derived thrombin, may contribute to glioma growth. Prothrombin mRNA and thrombin immunoreactivity have been detected in cultured glioma cells and in both rat and human dissected gliomas [7]. Thrombin can also be produced immediately in the brain and/or in the glioma resulting from an influx of prothrombin from blood following blood-brain barrier breakdown or a leaky blood-tumor barrier. In addition, normal parenchyma cells can produce thrombin [17].

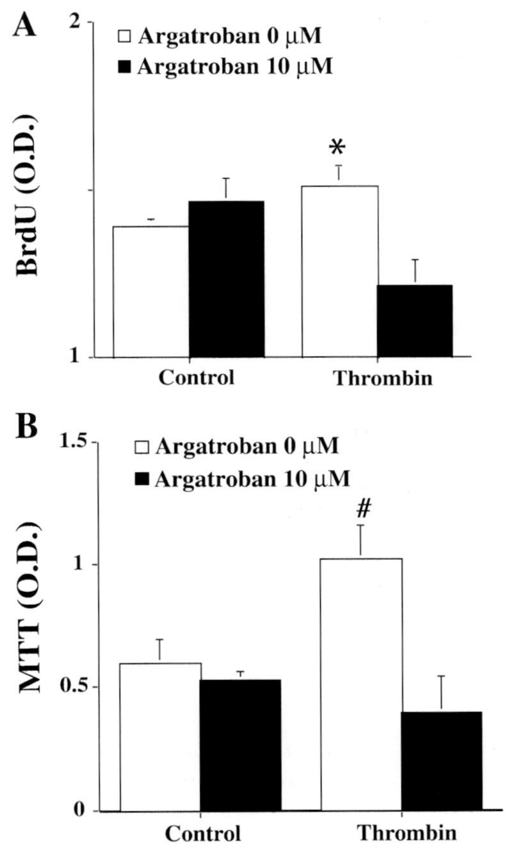

Fig. 2 C6 glioma cells were treated with control or thrombin (0.5 U/ml) with or without argatroban (10 µM) for 24 hours. (**A**) DNA synthesis was determined using a BrdU ELISA kit. (**B**) C6 cell proliferation was examined by MTT assay. O.D.: optical density. Values are mean ± SD, *p<0.05 vs. control p<0.01 vs. argatroban 10 µM. #p<0.01 vs. control and argatroban 10 µM

Thrombin stimulation may enhance glioma development via an increase in glioma cell proliferation. Our in vitro studies have found that thrombin can increase DNA synthesis and proliferation in C6 glioma cells. Studies from other research groups have also shown that thrombin can regulate cell proliferation. Neveu et al have reported that thrombin enhances the synthesis and secretion of nerve growth factor [12] and stimulates proliferation in glial cells [1]. Others have demonstrated that thrombin may act as a direct growth factor for other types of tumor cells [2, 11] including human glioma cells [16].

Thrombin mediates its widespread actions on cells through the activation of G-protein-coupled protease activated receptors (PARs) [3]. Recent studies indicate that PARs mediate some of the pathological effects of thrombin. For example, PAR-1 activation by thrombin receptor activating peptide results in angiogenesis [9, 10] and activation of PAR-1 is also associated with tumor cell invasion [4]. Whether or not thrombin-mediated glioma growth is through thrombin receptor activation needs further investigation.

In summary, thrombin stimulates glioma growth in vitro and in vivo, suggesting that thrombin may be a therapeutic target of gliomas.

Conflict of interest statement We declare that we have no conflict of interest.

References

1. Cavanaugh KP, Gurwitz D, Cunningham DD, Bradshaw RA (1990) Reciprocal modulation of astrocyte stellation by thrombin and protease nexin-1. J Neurochem 54:1735–1743
2. Chinni C, de Niese MR, Tew DJ, Jenkins AL, Bottomley SP, Mackie EJ (1999) Thrombin, a survival factor for cultured myoblasts. J Biol Chem 274:9169–9174
3. Coughlin SR (1999) How the protease thrombin talks to cells. Proc Natl Acad Sci U S A 96:11023–11027
4. Even-Ram SC, Maoz M, Pokroy E, Reich R, Katz BZ, Gutwein P, Altevogt P, Bar-Shavit R (2001) Tumor cell invasion is promoted by activation of protease activated receptor-1 in cooperation with the alpha vbeta 5 integrin. J Biol Chem 276:10952–10962
5. Hua Y, Keep RF, Schallert T, Hoff JT, Xi G (2003) A thrombin inhibitor reduces brain edema, glioma mass and neurological deficits in a rat glioma model. Acta Neurochir Suppl 86:503–506
6. Hua Y, Schallert T, Keep RF, Wu J, Hoff JT, Xi G (2002) Behavioral tests after intracerebral hemorrhage in the rat. Stroke 33:2478–2484
7. Hua Y, Tang L, Keep RF, Schallert T, Fewel M, Muraszko K, Hoff JT, Xi G (2005) The role of thrombin in gliomas. J Thromb Haemost 3:1–7
8. Hua Y, Tang LL, Fewel ME, Keep RF, Schallert T, Muraszko KM, Hoff JT, Xi GH (2005) Systemic use of argatroban reduces tumor mass, attenuates neurological deficits and prolongs survival time in rat glioma models. Acta Neurochir Suppl 95:403–406
9. Maragoudakis ME, Tsopanoglou NE (2000) On the mechanism(s) of thrombin induced angiogenesis. Adv Exp Med Biol 476:47–55
10. Maragoudakis ME, Tsopanoglou NE, Andriopoulou P, Maragoudakis MM (2000) Effects of thrombin/thrombosis in angiogenesis and tumour progression. Matrix Biology 19:345–351
11. Medrano EE, Cafferata EG, Larcher F (1987) Role of thrombin in the proliferative response of T-47D mammary tumor cells. Mitogenic action and pleiotropic modifications induced together with epidermal growth factor and insulin. Exp Cell Res 172:354–364
12. Neveu I, Jehan F, Jandrot-Perrus M, Wion D, Brachet P (1993) Enhancement of the synthesis and secretion of nerve growth factor in primary cultures of glial cells by proteases: a possible involvement of thrombin. J Neurochem 60:858–867
13. Nierodzik ML, Chen K, Takeshita K, Li JJ, Huang YQ, Feng XS, D'Andrea MR, Andrade-Gordon P, Karpatkin S (1998) Protease-activated receptor 1 (PAR-1) is required and rate-limiting for thrombin-enhanced experimental pulmonary metastasis. Blood 92:3694–3700
14. Nierodzik ML, Kajumo F, Karpatkin S (1992) Effect of thrombin treatment of tumor cells on adhesion of tumor cells to platelets in vitro and tumor metastasis in vivo. Cancer Res 52:3267–3272
15. Nierodzik ML, Plotkin A, Kajumo F, Karpatkin S (1991) Thrombin stimulates tumor-platelet adhesion in vitro and metastasis in vivo. J Clin Invest 87:229–236
16. Ogiichi T, Hirashima Y, Nakamura S, Endo S, Kurimoto M, Takaku A (2000) Tissue factor and cancer procoagulant expressed by glioma cells participate in their thrombin-mediated proliferation. J Neuro-Oncol 46:1–9
17. Xi G, Reiser G, Keep RF (2003) The role of thrombin and thrombin receptors in ischemic, hemorrhagic and traumatic brain injury: deleterious or protective? J Neurochem 84:3–9

Acta Neurochir Suppl (2008) 102: 367–371

The antioxidant effects of melatonin in surgical brain injury in rats

Steve Lee · Vikram Jadhav · Robert Ayer · Hugo Rojas · Amy Hyong · Tim Lekic · Gary Stier · Robert Martin · John H. Zhang

Abstract

Background Surgical brain injury (SBI) to normal brain tissue can occur as inevitable sequelae of neurosurgical operations. SBI can contribute to post-operative complications such as brain edema following blood-brain barrier (BBB) disruption leading to neurological deficits. Melatonin is a commonly used drug with known antioxidant properties and neuroprotective effects in experimental animal studies (Chen et al., J Pineal Res 41:175–182, 2006; Chen et al., J Pineal Res 40(3):242–250, 2006; Cheung, J Pineal Res 34:153–160, 2003; Lee et al., J Pineal Res 42(3):297–309, 2007; Reiter et al., Exp Biol Med (Maywood) 230(2):104–117, 2005).

Methods We tested different concentrations of melatonin (5 mg/kg, 15 mg/kg and 150 mg/kg) administered 1 hour before surgery for neuroprotection against SBI using a rodent model. Post-operative assessment included brain water content (brain edema), lipid peroxidation assays (oxidative stress), and neurological assessment.

Findings The results showed a trend in decreasing brain edema with lower doses of melatonin (5 mg/kg and 15 mg/kg), however, high concentration of melatonin (150 mg/kg) significantly increased brain edema compared to all other groups. This deleterious effect of high-dose melatonin was also observed in lipid-peroxidation assay wherein lower-dose melatonin (15 mg/kg) attenuated oxidative stress, but high-dose melatonin (150 mg/kg) increased oxidative stress as compared to vehicle-treated group. Furthermore, high-dose melatonin also worsened neurological outcomes compared to other groups whereas; the low-dose melatonin group (15 mg/kg) showed some improved neurological parameters.

Conclusions The study suggests that low-dose melatonin may provide neuroprotective effects against SBI. Further studies are needed to confirm this. More importantly, the findings of the study stress the need to carefully reassess safety issues with high doses of melatonin, which is considered to be a practically non-toxic drug.

V. Jadhav · R. Ayer · H. Rojas · A. Hyong · T. Lekic · J. H. Zhang
Department of Physiology and Pharmacology,
Loma Linda University School of Medicine,
Loma Linda, CA, USA

V. Jadhav
e-mail: drjadhav@gmail.com

R. Ayer
e-mail: rayer77@yahoo.com

H. Rojas
e-mail: hrojas06m@llu.edu

A. Hyong
e-mail: amyhyong@gmail.com

T. Lekic
e-mail: tlekic07m@llu.edu

J. H. Zhang
Department of Neurosurgery,
Loma Linda University School of Medicine,
Loma Linda, CA, USA

S. Lee · G. Stier · R. Martin · J. H. Zhang
Department of Anesthesiology,
Loma Linda University School of Medicine,
Loma Linda, CA, USA

S. Lee
e-mail: steveparklee@yahoo.com

G. Stier
e-mail: gstier@llu.edu

R. Martin
e-mail: rdmartin@llu.edu

J. H. Zhang (✉)
Division of Neurosurgery, Loma Linda University Medical Center,
11234 Anderson Street, Room 2562B,
Loma Linda, CA 92354, USA
e-mail: johnzhang3910@yahoo.com

Keywords Surgical brain-injury · Melatonin · Brain-edema · Neuroprotection

Introduction

5-methoxy-N-acetyltryptamine, commonly known as melatonin is a well known dietary supplement used in the USA for over a decade [13]. Melatonin is a naturally secreted hormone from the pineal gland with important function in the regulation of circadian and seasonal rhythms. However, it has also been reported to have powerful antioxidant properties and acts as a direct free radical scavenger [22, 23]. Melatonin also activates antioxidant enzymes such as superoxide dismutase, catalase, and gluthathione peroxidase [25]. Recent experimental studies have shown that melatonin provides neuroprotection against cerebral ischemia and decreases hemorrhagic transformation seen after stroke [2, 3, 6, 18, 24]. Melatonin was observed to decrease oxidative stress, prevent the disruption of blood brain barrier (BBB) integrity, and attenuate resulting brain edema after stroke. Melatonin is also thought to be neuroprotective in other forms of brain injury induced by excitotoxicity, trauma and neurodegenerative diseases such as Alzheimer's disease and Parkinson's disease [1, 4, 19].

In our previous studies, we have seen that BBB disruption leading to brain edema is critical in pathophysiology of surgical brain injury (SBI) after neurosurgical procedures [15, 16, 20]. Postoperative complications such as brain edema are commonly encountered after neurosurgical procedures leading to neurological deficits [9, 21, 26]. Our studies have shown that there is also increased oxidative stress in the brain tissue affected by SBI [20]. Free radical toxicity and oxidative stress are pivotal in disruption of BBB and formation of brain edema [8, 12]. Melatonin is known to freely cross the BBB and is widely distributed in all tissues allowing it to act on all cells [7]. We hypothesized that melatonin will decrease oxidative stress and attenuate post operative complications such as brain edema after neurosurgical procedures and provide neuroprotection against SBI.

Materials and methods

Surgical brain injury modeling

This protocol was evaluated and approved by the Institutional Animal Care and Use Committee at Loma Linda University, Loma Linda, CA. The rodent model of SBI was used as described before [15, 16, 20]. In brief, following anesthesia with ketamine (100 mg/kg) plus xylazine (10 mg/kg) i.p, a square cranial window was drilled such that the left lower corner of the square was at the bregma. The dura was incised and reflected to expose the underlying right frontal lobe. A sharp flat steel blade with the cutting surface measuring 6 mm × 1.5 mm, was used to make two incisions along sagittal and coronal planes leading away from the bregma to resect an area of brain 2 mm lateral in the sagittal plane and 1 mm ventral in the coronal plane. The depth of the incisions extended to the base of the skull. The weight of the sectioned brain was not significantly different between the various groups. Hemostasis was achieved by intraoperative packing and saline irrigation. After confirmation of hemostasis, the drilled bone piece was placed back in position and the skin sutured with 3-0 Ethicon on reverse cutting needle. Vital parameters including temperature (maintained between 35–37°C) and respiration were monitored intraoperatively and postoperatively. All animals received post-operative fluids. Sham surgery included only craniotomy and replacement of the bone flap without any dural incisions.

Treatment groups

Melatonin (5-methoxy-N-acetyltryptamine) was purchased from Sigma Aldrich (St. Louis, United States). The compound was dissolved in 0.9% NS with 10% ethanol to create solutions for dosages for treatment groups receiving 5 mg/kg, 15 mg/kg and 150 mg/kg intraperitoneally 1 hr before the surgery (pretreatment). All other group received equal volume of vehicle.

Brain water content

The animals were sacrificed under deep anesthesia at 24 hrs and the brains were divided into frontal ipsilateral (right), frontal contralateral (left), parietal ipsilateral, parietal contralateral, brain stem and cerebellum and placed on ice. These parts were weighed immediately (wet weight) and weighed again after drying in oven at 105°C for 48 hours (dry weight) [15, 16, 20]. The percent of water content was calculated as [(wet weight — dry weight)/wet weight] × 100 % as reported previously [27].

Lipid peroxidation assay

The animals were anesthetized, perfused with ice cold PBS, and brain samples were collected at 24 hours after surgical injury. The level of lipid peroxidation products (malondialdehyde [MDA]) in the ipsilateral (right) frontal lobe were measured using a LPO-586 kit (OxisResearch; Portland, United States) as previously described [20]. The brain tissue from right frontal lobe (ipsilateral) was homogenized in 20 mmol/L phosphate buffer (pH 7.4) with 0.5 M butylated hydroxytoluene (Sigma Aldrich Corp, MO, USA) in acetonitrile. The homogenates were centrifuged at 20,800 g for 10 min at 4°C and the supernatants were collected (Centrifuge 5417C/R from Eppendorf AG, 22331 Hamburg, Germany). Protein concentration was measured by DC protein assay (Bio-Rad) and the samples were reacted with a chromogenic reagent at 45°C for 60 min. After incubation, the samples were centrifuged at 20,800 g for 10 min at 4°C and supernatants were measured at 586 nm. The level of MDA was calculated as picomoles per milligram protein according to the standard curve.

Neurological evaluation

Neurological status of the animals was evaluated at 24 hrs using a 18 point scoring system modified from Garcia et al. [10]. Briefly, the scoring system consisted of six tests with scores of 1–3 for each test. These six tests included: (i) spontaneous activity; (ii) symmetry in the movement of four limbs; (iii) forepaw outstretching; (iv) climbing; (v) body proprioception; and (vi) response to vibrissae touch. The score given to each rat at the completion of the evaluation was the summation of all six individual test scores. The minimum neurological score was 3 and the maximum was 18. Vibrissae stimulation (paw placement test) as described previously [14]. Briefly, right and left whiskers were stimulated separately and the ability of left paw to be placed on table was noted. Similarly, right paw was tested, however the contralateral side (left paw) was affected and only those results were reported. Beam walking tests were also performed using 2.4 cms diameter× 100 cms length beam and 5 cms diameter × 60 cms length beam. Time spent on beam, distance traveled and speed were calculated.

Data analysis

Data are expressed as mean ± S.E.M. Statistical significance was verified by one-way ANOVA for multiple comparisons. Probability value of $p<0.05$ was considered statistically significant.

Results

Brain Edema after SBI

There was increased brain water content in the right and left frontal lobes in all rats subjected to SBI. The lower doses of melatonin showed a trend towards lowering brain edema, though not statistically significant (p = 0.13 for 15 mg/kg, right frontal lobe). On the other hand, high-dose melatonin (150 mg/kg) significantly increased brain edema in both right and left frontal lobes compared to all other groups (p< 0.05, Table 1). Brain water content in rest of the brain areas was not different between the SBI and sham groups. This animal model had zero mortality in any of the groups.

Melatonin effects on oxidative stress

Lipid peroxidation assay (malondialdehyde levels) showed that low-dose melatonin (15 mg/kg) significantly attenuated the oxidative stress after SBI compared to vehicle group. This, however, was reversed with the high-dose melatonin (150 mg/kg) which showed significantly higher oxidative stress as compared to all other groups (Table 2).

Neurological evaluation

Numerous neurological tests were performed to evaluate the sensorimotor deficits 24 hours after surgery (Table 3). A modification of scoring system by Garcia et al. [10] using an 18 point maximum scoring showed that all rats subjected to SBI had worsened scores. However, the rats treated with high-dose melatonin (150 mg/kg) had further worsening of neurological deficits compared to vehicle control as well as low dose melatonin groups (5 mg/kg and 15 mg/kg) (Table 3A). The vibrissae stimulation studies reported here depict response of left paw (contralateral side) after stimulation of left and right whiskers separately. All animals subjected to SBI showed significantly lower scores, compared to controls after the following stimulation paradigms a) left whisker left paw and b) right whisker

Table 1 Effects of melatonin on brain edema

	Sham (8)	Vehicle (8)	5 mg/kg Melatonin (8)	15 mg/kg Melatonin (7)	150 mg/kg Melatonin (6)
Right Frontal Lobe	79.20±0.09	81.44±0.25*#	81.38±0.23*#	80.97±0.22*#	82.20±0.24*
Left Frontal Lobe	78.59±0.18	79.87±0.22*	79.66±0.05*#	79.61±0.11*#	80.27±0.31*

The table shows brain water content in right (ipsilateral) and left (contralateral) frontal lobes 24 hrs after surgical injury. Both right and left frontal lobes show increased brain water content compared to respective sham group, however, the right lobe has much higher water content (almost 2%). Low-dose melatonin (5 mg/kg and 15 mg/kg) show a trend towards lowering brain water content, though not statistically significant (p=0.13 with 15 mg/kg dose, right frontal lobe). The high-dose melatonin (150 mg/kg), however, significantly increases brain water content compared to other groups. ($P<0.05$ is considered significant,* and # indicates v/s sham and 150 mg/kg melatonin respectively). n numbers of animals per group are indicated in parentheses.

Table 2 Antioxidant effects of melatonin (lipid peroxidation assay)

	Sham (4)	Vehicle (6)	15 mg/kg Melatonin (8)	150 mg/kg Melatonin (6)
Lipid Peroxidation pmole/mg protein	2.21±1.52	15.25±1.43*	8.28±1.415*#	19.78±0.95*# $

The table shows malondialdehyde levels for lipid peroxidation assay from right (ipsilateral) frontal lobe 24 hrs after surgical injury. All rats subjected to SBI showed increased oxidative stress compared to sham group. Low-dose melatonin (15 mg/kg) significantly attenuated the oxidative stress compared to vehicle; however, high-dose melatonin (150 mg/kg) significantly increased oxidative stress compared to all other groups. ($P<0.05$ is considered significant, * # and $ indicates v/s sham and v/s vehicle and v/s 15 mg/kg melatonin respectively). n numbers of animals per group are indicated in parentheses.

left paw (Table 3B) as outlined in methods. The low dose melatonin group (15 mg/kg) significantly improved these scores compared to the vehicle-treated group, which were significantly reversed by the high-dose melatonin (150 mg/kg) treatment (Table 3B). With the exception of 'time spent on beam' parameter for 5 cm diameter beam, all vehicle treated rats showed significant deficits compared to sham animals on both 2.4 cm and 5 cm diameter beams for all parameters tested (Table 3C). Both doses of melatonin (15 mg/kg and 150 mg/kg) significantly lowered the time spent on 2.4 cm diameter beam, to almost sham group levels.

Discussion

The present study showed that low-dose melatonin showed a trend towards lowering brain edema after surgical injury caused during neurosurgical procedures. On the other hand, high-dose melatonin (150 mg/kg) notably increased brain edema following SBI. Furthermore, high-dose melatonin also worsened the neurological outcomes in rats after SBI. Presently, melatonin is freely available as a supplement mainly used as an aid for better sleep. It is generally thought to be a very safe drug with practically no toxicity [5, 6]. However, few studies have examined the effects of high-dose melatonin. One study which used high concentration of melatonin (100 mg/kg), comparable to the dose used in the present study, reported a lack of antioxidant effects of melatonin at that particular dose [11]. In the present study we also observed that, high-dose melatonin (150 mg/kg) did not lower oxidative stress, and on the contrary, it significantly worsened the oxidative stress following SBI.

The low-dose melatonin (15 mg/kg) treatment decreased oxidative stress. Though, not statistically significant, there was a trend towards decreased brain edema with this dose as well. Melatonin at this dose improved some neurological parameters as seen in beam-walking tests. The low dose

Table 3 Neurological evaluation tests

	Sham	Vehicle	5 mg/kg Melatonin	15 mg/kg Melatonin	150 mg/kg Melatonin
(A) Neurological Tests	(n=11)	(n=15)	(n=8)	(n=13)	(n=14)
Scores (Max=18)	18.0±0.0	14.7±0.9*#	15.7±1.4#	15.7±0.6*#	12.0±0.4*
(B) Paw Placement Test	(n=11)	(n=15)	(n=8)	(n=13)	(n=14)
Lt Whisker Lt Paw	10.0±0.0	1.6±0.8*$	2.2±1.2*	5.0±1.3*	1.5±0.9*$
Rt Whisker Lt Paw	10.0±0.0	1.6±0.7*$	2.5±1.2*	5.1±1.2*	2.1±0.9*$
(C) Walking Beam Tests					
2.4 cm Diameter Beam	(n=11)	(n=8)		(n=6)	(n=14)
Time (secs)	22.7±3.7	73.7±1.8*	–	23.0±6.1&	28.4±4.4&
Distance (cms)	15.3±4.3	4.0±2.4*	–	5.8±2.2*	1.6±1.2*
Speed (cms/sec)	4.9±1.1	1.8±0.9*	–	0.8±0.4*	0.8±0.6*
5 cm Diameter Beam	(n=4)	(n=8)		(n=6)	(n=14)
Time (secs)	60.0±0.0	52.5±4.9	–	50.3±4.0	42.8±5.2
Distance (cms)	125.1±14.0	18.1±16.7*	–	6.6±6.6*	1.0±0.5*
Speed (cms/sec)	15.7±5.3	2.0±1.5*	–	0.5±0.5*	0.7±0.4*

(A) An 18 point neurological scoring system showed that all rats subjected to SBI had significant neurological deficits; however, the groups treated with high-dose melatonin had significantly more neurological deficits as compared to all other groups. * and # indicates v/s sham and 150 mg/kg melatonin respectively. (B) Paw placement test showed worsening scores in all groups subjected to SBI. The 15 mg/kg dose significantly improved the scores compared to vehicle group which were reversed with 150 mg/kg dose. * and $ indicates v/s sham and 15 mg/kg melatonin respectively. (C) The beam walking tests on both 2.4 cms and 5 cms diameter beams indicated that vehicle-treated animals showed worsening parameters such as increased time taken, decreased distance traversed and lesser speed on the beams as compared to sham group. Both melatonin groups showed significant improvement in time taken on 2.4 cms beam with no significant changes in other parameters. * and & indicates v/s sham and vehicle respectively. $P<0.05$ is considered significant. n numbers of animals per group are indicated in parentheses melatonin group (15 mg/kg) performed significantly better on 2.4 cms diameter beam but not on the 5 cms diameter beam as compared to the vehicle-treated group. The narrower beam is more likely to unmask the subtle neuroprotective benefits provided by this dose of melatonin for SBI. Whether these neuroprotective effects are through other mechanisms which secondarily affect brain edema needs to be investigated. Studies using other models have suggested that melatonin has anti-apoptotic, anti-inflammatory, anti-excitotoxicity actions in addition to activating neuroprotective signaling pathways [17, 28]. Future studies on melatonin in SBI can be directed towards studying these neuroprotective mechanisms.

Most importantly, this study shows that high-dose melatonin may be deleterious by paradoxically increasing oxidative stress, worsening brain edema and adversely affecting neurological outcomes. Further studies with high-dose melatonin are warranted given the easy availability and inexpensive nature of the drug [5, 6].

Acknowledgements This study was partially supported by grants from NIH NS45694, NS53407, and NS43338 to JHZ.

Conflict of interest statement We declare that we have no conflict of interest.

References

1. Alvira D, Tajes M, Verdaguer E, Acuna-Castroviejo D, Folch J, Camins A, Pallas M (2006) Inhibition of the cdk5/p25 fragment formation may explain the antiapoptotic effects of melatonin in an experimental model of Parkinson's disease. J Pineal Res 40 (3):251–258
2. Chen H, Chen T, Lee M, Chen S, Hsu Y et al (2006) Melatonin decreases neurovascular oxidative/nitrosative damage and protects against early increases in the blood-brain barrier permeability after transient focal cerebral ischemia in mice. J Pineal Res 41:175–182
3. Chen T, Lee M, Chen H, Kuo Y, Lin S et al (2006) Melatonin attenuates the postischemic increase in blood-brain barrier permeability and decreases hemorrhagic transformation of tissue-plasminogen activator therapy following ischemic stroke in mice. J Pineal Res 40(3):242–250
4. Cheng Y, Feng Z, Zhang QZ, Zhang JT (2006) Beneficial effects of melatonin in experimental models of Alzheimer disease. Acta Pharmacol Sin 27(2):129–139 Review
5. Cheung RT, Tipoe GL, Tam S, Ma ES, Zou LY, Chan PS (2006) Preclinical evaluation of pharmacokinetics and safety of melatonin in propylene glycol for intravenous administration. J Pineal Res 41:337–343
6. Cheung RT (2003) The utility of melatonin in reducing cerebral damage resulting from ischemia and reperfusion. J Pineal Res 34:153–160
7. Costa EJ, Lopes RH, Lamy-Freund MT (1995) Permeability of pure lipid bilayers to melatonin. J Pineal Res 19:123–126
8. Del Zoppo GJ (2006) Stroke and neurovascular protection. N Engl J Med 354:553–555
9. Fasano VA, Penna G (1992) Postoperative complications in neurosurgery. Minerva Anestesiol 58:15–21
10. Garcia JH, Wagner S, Liu KF, Hu XJ (1995) Neurological deficit and extent of neuronal necrosis attributable to middle cerebral artery occlusion in rats. Statistical validation. Stroke 26:627–634
11. Gul S, Celik SE, Kalayci M, Tasyurekli M, Cokar N, Bilge T (2005) Dose-dependent neuroprotective effects of melatonin on experimental spinal cord injury in rats. Surg Neurol. 64:355–361
12. Gursory-Ozdemir Y, Can A, Dalkara T (2004) Reperfusion-induced oxidative/nitrative injury to neurovascular unit after focal cerebral ischemia. Stroke 35:1449–1453
13. Harderland R, Pandi-Perumal S, Cardinali D (2006) Melatonin. Int J Biochem Cell Biol 28:313–316
14. Hua Y, Schallert T, Keep RF, Wu J, Hoff JT, Xi G (2002) Behavioral tests after intracerebral hemorrhage in the rat. Stroke 33(10):2478–2484
15. Jadhav V, Matchett G, Hsu FP, Zhang JH (2007) Inhibition of Src tyrosine kinase and effect on outcomes in a new in vivo model of surgically induced brain injury. J Neurosurg 106(4):680–686
16. Jadhav V, Solaroglu I, Obenaus A, Zhang JH (2007) Neuroprotection against surgically induced brain injury. Surg Neurol 67 (1):15–20 discussion 20. Review
17. Kilic U, Kilic E, Reiter RJ, Bassetti CL, Hermann DM (2005) Signal transduction pathways involved in melatonin-induced neuroprotection after focal cerebral ischemia in mice. J Pineal Res 38:67–71
18. Lee MY, Kuan YH, Chen HY, Chen TY, Chen ST, Huang CC, Yang IP, Hsu YS, Wu TS, Lee EJ (2007) Intravenous administration of melatonin reduces the intracerebral cellular inflammatory response following transient focal cerebral ischemia in rats. J Pineal Res 42(3):297–309
19. Lee SH, Chun W, Kong PJ, Han JA, Cho BP, Kwon OY, Lee HJ, Kim SS (2006) Sustained activation of Akt by melatonin contributes to the protection against kainic acid-induced neuronal death in hippocampus. J Pineal Res 40(1):79–85
20. Lo W, Bravo T, Jadhav V, Titova E, Zhang JH, Tang J (2007) NADPH oxidase inhibition improves neurological outcomes in surgically-induced brain injury. Neurosci Lett 414(3):228–232
21. Manninen PH, Raman SK, Boyle K, el-Beheiry H (1999) Early postoperative complications following neurosurgical procedures. Can J Anaesth 46:7–14
22. Reiter RJ, Tan DX, Terron MP, Flores LJ, Czarnocki Z (2007) Melatonin and its metabolites: new findings regarding their production and their radical scavenging actions. Acta Biochim Pol 54(1):1–9
23. Reiter RJ, Acuna-Castroviejo D, Tan DX, Burkhardt S (2001) Free radical-mediated molecular damage. Mechanisms for the protective actions of melatonin in the central nervous system. Ann N Y Acad Sci 939:200–215 Review
24. Reiter RJ, Tan DX, Leon J, Kilic U, Kilic E (2005) When melatonin gets on your nerves: its beneficial actions in experimental models of stroke. Exp Biol Med (Maywood) 230(2):104–117 Review
25. Tomas-Zapico C, Coto-Montes A (2005) A proposed mechanism to explain the stimulatory effect of melatonin on antioxidative enzymes. J Pineal Res 39:99–104
26. Tommasino C (1992) Postoperative cerebral edema. Physiopathology of the edema and medical therapy. Minerva Anestesiol 58:35–42
27. Xi G, Hua Y, Keep RF, Younger JG, Hoff JT (2002) Brain edema after intracerebral hemorrhage: the effects of systemic complement depletion. Acta Neurochir Suppl 81:253–256
28. Yon JH, Carter LB, Reiter RJ, Jevtovic-Todorovic V (2006) Melatonin reduces the severity of anesthesia-induced apoptotic neurodegeneration in the developing rat brain. Neurobiol Dis 21:522–530

Acta Neurochir Suppl (2008) 102: 373–376

Dymanics of matrix-metalloproteinase 9 after brain trauma – results of a pilot study

Martin Kolar · Jan Pachl · Helena Tomasova · Pavel Haninec

Abstract

Background Secondary brain injury contributes to poor outcome for patients sustaining brain trauma. Matrix metalloproteinase-9 (MMP-9) is a potential marker, as well as effector of secondary brain injury. This enzyme degrades components of extracellular matrix, and thus it can contribute to blood-brain barrier disruption.

Methods We studied dynamics of MMP-9 in jugular venous blood of 15 patients sustaining either an isolated head injury or a head injury as a part of major trauma, and requiring intensive care (Glasgow Coma Scale ≤8 at the time of admission). Blood samples were taken at the 1st, 3rd and 5th day, levels of MMP-9 in plasma were assessed using ELISA. Outcome quality was assessed at the time of discharge from our hospital.

Findings Our results show an increase of MMP-9 levels on the 1st day after the brain trauma, followed by a drop on the 3rd day and a rise on day 5. This biphasic time-course was observed in all patients, but no statistically significant differences between each group (major trauma vs. isolated brain trauma, good outcome vs. poor outcome) were found.

Conclusions Initially increased MMP-9 levels in the 1st posttraumatic day is probably related to transient blood-brain barrier dysruption. The decrease of MMP-9 levels observed on the 3rd day can be explained by restoration of blood-brain barrier integrity and its reduced permeability. The second rise of MMP-9 levels observed in the 5th day probably indicates a developing secondary brain injury during which MMP-9 is produced in the brain as a part of an inflammatory response. Results of our study suggest that MMP-9 could play an important role in pathogenesis of secondary brain injury.

Keywords Brain trauma · Secondary brain injury · Blood-brain barrier · Martix metalloproteinase 9

M. Kolar (✉)
Department of Anesthesiology and Critical Care Medicine, Charles University in Prague, Third Faculty of Medicine, Ruska 87,
100 34 Prague, 10, Czech Republic
e-mail: martinuv.mail@seznam.cz

J. Pachl
Department of Anesthesiology and Critical Care Medicine, Teaching Hospital Kralovske Vinohrady,
Srobarova 50,
100 34 Prague, 10, Czech Republic
e-mail: pachl@fnkv.cz

H. Tomasova
Department of Biochemistry, Teaching Hospital Motol,
Prague, Czech Republic

P. Haninec
Department of Neurosurgery,
Teaching Hospital Kralovske Vinohrady,
Prague, Czech Republic

Introduction

Traumatic brain injury is a leading cause of death and disability in patients younger than 40 years of age. Incidence of brain trauma in European countries is about 235 / 100,000 and mortality is 15 / 100,000 per year (16). In European Union countries, costs of a hospital stay of a patient with a severe head trauma ranged from 5,500 to 9,000 € (7) (approx. 7,600 to 12,500 USD). Prognosis of the patient is worsened by secondary brain injury which develops as a result of pathophysiologic reactions caused by the primary insult. Aggressive managment in the intensive care can prevent or reduce the consequences of the secondary brain injury.

Post-traumatic secondary brain injury results from increased intracranial pressure and negatively affected brain tissue perfusion. An increased permeability of the blood-brain barrier is a part of the complex mechanisms responsible for intracranial hypertension. Mediators acting in the blood-brain barrier disruption are, among others, **matrix metalloproteinases (MMPs)**. These zinc-dependent endopeptidases degrade both fibrillar and amorphous components of extracellular matrix (13, 19). In physiological conditions, MMPs play an important role in ontogenesis, e.g. in tissue differentiation and angiogenesis, wound healing, bone remodelling or implantation (19). In disease, a role for MMPs has been described in multiple sclerosis (11), vascular dementia (1), growth of tumor and metastases (10) or airway remodelling in pulmonary asthma, idiopatic pulmonary fibrosis or chronic obstructive pulmonary disease (5).

Extensive experiments in animal models demonstrated the importance of **matrix metalloproteinase-9** (gelatinase B) for brain injury development in various pathologic conditions (3, 12, 18, 20). Upregulation of MMP-9 mRNA on the 1st and 4th posttraumatic day was observed in experimental brain trauma (18). Increased activity of MMP-9 in mice with brain contusion was observed not only in the injured hemisphere, but also in the contralateral hemisphere (20). Blood-brain barrier disruption and increased MMPs levels have been observed after focal ischaemia (3, 14) as well as in diffuse brain injury caused by inflammation (12). The extent of injury after focal ischaemia (3, 4) as well as after brain contusion (20) was significantly smaller in MMP-9 knock-out mice. Administration of broad-spectrum MMPs inhibitors reduced injury in a murine model of focal brain ischaemia (3). Also, MMP-9 levels in cerebrospinal fluid and the extent of cortical lesion were reduced in animals with pneumococcal meningitis treated by an MMPs inhibitor (12).

Despite extensive animal model research, a limited number of human studies have monitored the levels of MMP-9. A prior study, performed in intensive care patients, showed increased MMP-9 levels in the first day following brain trauma (15). Our study further investigates the temporal profile of MMP expression with patients studied at multiple time points to 5 days following traumatic brain injury.

Aim of this study was to describe the dynamics of MMP-9 plasma levels in human traumatic brain injury during the first 5 days following injury.

Material and methods Study performed in Dept. of Anesthesiology and Critical Care Medicine of Teaching Hospital Kralovske Vinohrady, Prague, Czech Republic, included patients who sustained either an isolated head injury or a head injury as a part of major trauma, and required intensive care (Glasgow Coma Scale ≤8 at the time of admission). All patients were treated according to Brain Trauma Foundation guidelines. A catheter into the dominant jugular bulb was inserted within the first 24 hours.

Blood samples were taken from jugular bulb on the 1st, 3rd and 5th day after the injury (day of the injury = day 0). Noncoagulant blood was centrifuged for 5 minutes at 6,000 rpm and the plasma was stored at – 30°C. Levels of matrix metalloproteinase 9 were assessed using ELISA kits by Amersham Bioscience (sensitive for both activated enzymes and their proenzymes). Quality of outcome was assessed at the time of discharge from our hospital using Glasgow Outcome Score. The patients were not observed after discharge from the hospital. Data were statistically evaluated by Mann-Whitney and Wilcoxon's tests.

Results

Characteristics of the experimental group

Fifteen patients (14 men and 1 woman, age 20–87 years, median 35 years) with a brain trauma (12 brain contusions, 8 epidural and/or subdural hematomas, 4 traumatic subarachnoidal bleedings) admitted between November 2001 and July 2002 were enrolled into the study. Eight patients sustained an isolated brain injury and 7 sustained a polytrauma (7 long bones or pelvis fractures and 4 lung contusions). The length of hospital stay was 1 to 36 days (median 16 days). Good outcome (GOS 4 and 5) at the time of discharge from our hospital was observed in 8 patients and poor outcome (GOS 1-3) in 7 patients (including 3 deaths).

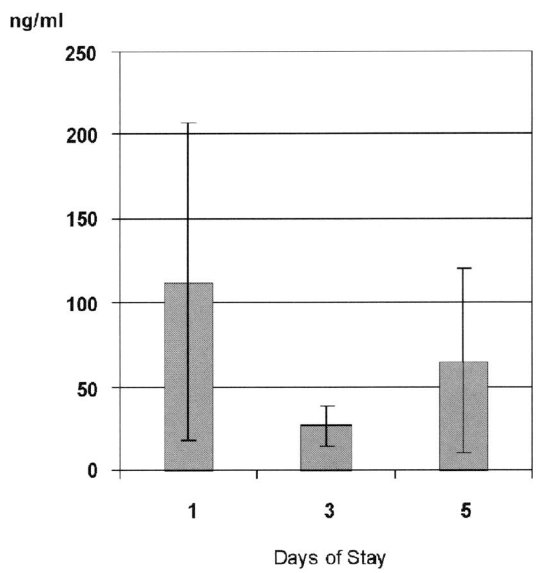

Fig. 1 Levels of MMP-9 in jugular blood on the 1st, 3rd and 5th day after severe brain trauma (Glasgow Coma Scale ≤8 at the time of admission). Average ± standard deviation shown

Changes in MMP-9 levels

Obtained data show elevated MMP-9 levels in jugular blood during the first day (111,9±94,5 ng/mL, average ± standard deviation), followed by a decrease on the 3rd day (26,6±12,1 ng/mL) and an increase on the 5th day (65,3± 55,0 ng/mL). (Fig. 1) This dynamics was observed in all patients (statistically not significant due to a wide range of values and a small number of patients). Higher levels of MMP-9 on the 1st day were observed in polytraumatised patients compared to patients with isolated craniotrauma.

Although the data show higher MMP-9 levels in patients with poor outcome (Glasgow Outcome Score 1-3), this in not statistically significant due to a small number of patients (data not shown). Higher levels of MMP-9 in jugular venous blood on the first day after brain trauma were observed in patients with poor outcome (p=0,1).

Discussion Our results show an increase of MMP-9 levels on the 1st day after the brain trauma, followed by a drop on the 3rd day and a rise on day 5. These findings agree with results of a study performed in neurocritical care patients in which an increase of MMP-9 jugular blood levels occurred during the first 24 hours of traumatic brain injury (15). High variability of MMP-9 levels obtained in our study may be related to the timing of the first blood sampling. This was always performed on the first posttraumatic day, but not exactly 24 hours after the injury.

The decrease of MMP-9 levels observed on the 3rd day can be explained by restoration of blood-brain barrier integrity at this time point. Animal model studies prove that rapid and transient changes in blood-brain barrier permeability can be observed immediately after traumatic brain injury. Rapid changes of blood-brain barrier permeability occur immediately following the brain injury as well as a rapid reversal of the increased permeability outside the injured area within first 60 minutes after experimental traumatic brain injury (8). In a study by Tanno (17) the greatest blood-brain barrier permeability to horseradish peroxidase was observed during the first hour and to IgG in the first 24 hours following traumatic brain injury. Magnetic resonance imaging studies have shown that maximal blood-brain barrier permeability occurs within 30 minutes after an experimental craniotrauma (6).

The second rise of MMP-9 levels observed in the 5th day may indicate developing secondary brain injury, during which MMP-9 is produced as a part of an inflammatory response. This corresponds with previous findings (9) showing a biphasic course of post-traumatic brain oedema, with the first peak during the first 24 hours followed by a second peak on days 3 to 5. Polymorphonuclear cells, which can be a source of MMP-9, infiltrate the brain tissue during this time frame.

Conclusion

In our 15 patients with traumatic brain injury, elevated levels of MMP-9 were observed in the early posttraumatic period. The levels were elevated within the first 24 hours, decreased on the 3rd day, and then increased again on the 5th day following traumatic brain injury. The increase of MMP-9 levels in the 5th day may be related to inflammatory mechanisms of secondary brain injury. Matrix metalloproteinase 9 can contribute to the development of a secondary brain injury by increasing the blood-brain barrier permeability and enabling influx of inflammatory mediators. Current studies with exact timing of blood sampling and prolonged monitoring periods are ongoing in order to determine the dynamics of matrix metalloproteinases action after brain trauma.

Acknowledgement This study is supported by Grant Agency of Charles University project no. 112507. The author would like to thank Dr. Iva Rohousova (Charles University) and Dr. Jan Kolar (Czech Academy of Sciences) for helpful assistance.

Conflict of interest statement We declare that we have no conflict of interest.

References

1. Adair JC, Charlie J, Dencoff JE, Kaye JA, Quinn JF, Camicioli RM, Stetler-Stevenson WG, Rosenberg GA (2004) Measurement of gelatinase B (MMP-9) in the cerebrospinal fluid of patients with vascular dementia and alzheimer disease. Stroke 35:e159–e162
2. Aoki T, Sumii T, Mori T, Wang X, Lo EH (2002) Blood-brain barrier disruption and matrix metalloproteinase-9 expression during reperfusion injury: mechanical versus embolic focal ischemia in spontaneously hypertensive rats. Stroke 33:2711–2717
3. Asahi M, Asahi K, Jung JC, del Zoppo GJ, Fini ME, Lo EH (2000) Role for matrix metalloproteinase 9 after focal cerebral ischemia: effects of gene knock-out and enzyme inhibition with BB-94. J Cereb Blood Flow Metab 20:1681–1689
4. Asahi M, Wang X, Mori T, Sumii T, Jung J, Moskowitz MA Fini ME, Lo EH (2001) Effects of matrix metalloproteinase 9 gene knock-out on the proteolysis of blood-brain barrier and white matter components after cerebral ischemia. J Neurosci 21:7724–7732
5 Atkinson JJ, Senior RM (2003) Matrix Metalloproteinase-9 in Lung Remodelling. Am J Respir Cell Mol Biol 28:12–24
6. Barzó P, Marmarou A, Fatouros P, Corwin F, Dunbar J (1996) Magnetic resonance imaging-monitored acute blood-brain barrier changes in experimental traumatic brain injury. J Neurosurg 85:1113–21
7. Berg J (2004) Economic evidence in trauma: a review. Eur J Health Econom Suppl 1:S84–91
8. Fukuda K, Tanno H, Okimura Y, Nakamura M, Yamaura A (1995) The blood-brain barrier disruption to circulating proteins in the early period after fluid percussion brain injury in rats. J Neurotrauma 12:315–324
9. Holmin S, Soderlund J, Biberfeld P, Mathiesen T (1998) Intracerebral inflammation after human brain contusion. Neurosurgery 42:291–298

10. John A, Tuszynski G (2001) The Role of Matrix Metalloproteinases in Tumor Angiogenesis and Tumor Metastasis. Pathol Oncol Res 7:14–23
11. Leppert D, Ford J, Stabler G, Grygar C, Lienert C, Huber S, Miller KM, Hauser SL, Kappos L (1998) Martix metalloproteinase-9 (gelatinase B) is selectively elevated in CSF during relapses and stable phases of multiple sclerosis. Brain 121:2327–34
12. Lieb SL, Leppert D, Clements J, Tauber MG (2000) Matrix metalloproteinases contribute to brain damage in experimental pneumococcal meningitis. Infect Immun 68:615–620
13. Lo HE, Wang X, Cuzner ML (2002) Extracellular proteolysis in brain injury and inflammation: role for plasminogen activators and matrix metalloproteinases. J Neurosci Res 69:1–9
14. Planas AM, Solé S, Justicia C, Farré ER (2000) Estimation of Gelatinase Content in Rat Brain: Effect of Local Ischemia. Biochem Biophys Res Commun 278:803–807
15. Suehiro E, Fujisawa H, Akimura T, Ishihara H, Kajiwara K, Kato S, Fuki M, Yamashita S, Maekawa T Suzuki M (2004) Increased matrix metalloproteinase-9 in blood in association with activation of interleukin-6 after traumatic brain injury: influence of hypothermic therapy. J Neurotrauma 21:1706–11
16. Tagliaferri F, Compagnone C, Korsic M, Servadei F, Kraus J (2006) A systematic review of brain injury epidemiology in Europe. Acta Neurochir 148:255–268
17. Tanno H, Nockels RP, Pitts LH, Noble LJ (1992) Breakdown of the blood-brain barrier after fluid percussive brain injury in the rat. Part 1: Distribution and time course of protein extravasation. J Neurotrauma 9:21–32
18. von Gertten C, Holmin S, Mathiesen T, Nordquist ACS (2003) Increase in matrix metalloproteinase-9 and tissue inhibitor of matrix metalloproteinase-1 mRNA after cerebral contusion and depolarisation. J Neurosci Res 73:803–810
19. Vu TH, Werb Z (2000) Matrix metalloproteinases: effectors of development and normal physiology. Genes Develop 14:2123–33
20. Wang X, Jung J, Asahi M, Chwang W, Russo L, Moskowitz MA, Dixon CE, Fini ME, Lo EH (2000) Effects of matrix metalloproteinase-9 gene knock-out on morphological and motor outcomes after traumatic brain injury. J Neurosci 20:7037–7042

Acta Neurochir Suppl (2008) 102: 377–379

Experimental subarachnoid hemorrhage in the rat: influences of nimodipine

Sebastian Thomas · Birte Herrmann · Madjid Samii · Thomas Brinker

Abstract

Background Traumatic subarachnoid hemorrhage is a common finding following traumatic brain injury. Clinical studies revealed a positive influence of Nimodipine. However, till now no experimental studies have been performed. The aim of the present study was to determine the influence of early Nimodipine administration on outcome and histological findings in the rat:

Methods Diffuse brain injury was produced in Sprague-Dawley rats using a brass weight falling from a predetermined height. Traumatic subarachnoidal hemorrhage was produced by administration of heparin before the injury. A total number of 52 animals were divided in 4 groups.

Findings Mortality increased following administration of heparin. Mortality was not reduced following administration of Nimodipine. The histological investigation revealed less cell loss in animals with administration of Nimodipine as well as increased GFAP immunoexpression.

Conclusions Administration of Heparin results in a marked traumatic subrachnoidal hemorrhage following diffuse traumatic brain injury. Administration of Nimodipine does not reduce overall mortality. However, histological investigations indicate a positive effect of Nimodipine on cell loss.

S. Thomas (✉)
Department of Neurosurgery, Klinikum Saarbrücken,
Winterberg 1,
66119 Saarbrücken, Germany
e-mail: thomas_seb@hotmail.com

B. Herrmann
Kinderkrankenhaus auf der Bult,
Hannover, Germany

M. Samii · T. Brinker
International Neuroscience Institute,
Hannover, Germany

Keywords Traumatic subarachnoidal hemorrhage · Rat · Nimodipine · Animal model

Introduction

Following traumatic brain injury traumatic subarachnoid hemorrhage (t-SAH) is a common finding. In a study by Eisenberg et al. based on analysis of computed tomography scans of 753 patients t-SAH was found in 39% cases [1]. In a Japanese series by Kobayashi et al. (1988) t-SAH was detected in 23% of 414 severely head injured patients [6]. A European multicenter study known as the Head Injury Trial No.2 (HIT II) indicated t-SAH in 33% of 819 patients [5]. A clinical study of 750 unselected patients from the European Brain Injury Consortium Survey of Head Injuries confirm the adverse prognostic significance of t-SAH [2].

Although the importance of t-SAH on outcome following trauma has been described in several clinical studies there is a lack of animal studies.

In a previous study from our laboratory a t-SAH model in the rat was established for the first time. Utilizing this model we found a significant increase of ICP due to the actual bleeding with an overall mortality of 41.6% [9, 10].

The aim of the present study was to determine the influence of Nimodipine administration on outcome and histological findings in the rat following t-SAH.

Materials and methods

As approved by the Animal Research Committee of the Medical School of Hannover, 52 male Sprague-Dawley rats (350–400g) were used for this study. They were initially anesthetized with isoflurane and then intubated and artificially ventilated with a gas mixture of N_2O (70%) and O_2 (30%) with isoflurane. Body temperature was maintained between 37° and 38° C with a thermostatically controlled heating pad. Polyethylene catheters (PE-50) were placed into the femoral artery and connected to a Statham-transducer. The impact acceleration model modified by Thomas et al. [3, 9] was used to produce trauma. Following a midline scalp incision the skin and periosteum were reflected and the skull carefully dried. A stainless steel disc serving as a helmet was mounted on the skull using cyanoacrylate. A brass weight (500 g) fell free onto from a predetermined height of 1 m on the helmet guided by a 2m Plexiglas tube. The animal was placed in prone position on a foam (20×20×50 cm; spring constant 30/50) held within a wooden frame. The lower end of the tube was placed directly above the helmet. Immediately following the initial impact the box was slid horizontally to prevent a second impact. Five minutes before trauma a high-dose heparin (10 I.E./g) was injected i.a.. Five minutes following trauma the equivalent dose of protamine (10 I.E./g) was administered and anesthesia reinstated.

2 cc Nimodipine was administered s.c. for consecutive 7 days.

Animals were divided in four groups: Group 1: Heparin/Nimodipine (N=17); Group 2: Heparin/no Nimodipine (N=15); Group 3: no Heparin/Nimodipine (N=10); Group 4: no Heparin/no Nimodipine (N=10).

All brains were processed for hematoxylin-eosin staining using 20 μm coronal sections. Surviving animals were perfused transcardially 8 days after injury and the brains were removed and post-fixed in 4%-formalin. Sections were studied under light-microscopy for cell loss and hemorrhages.

All measurements were stored and analyzed using commercially available software (Dasy Lab V.3, Datalog, Mönchengladbach, Germany). Statistical analysis was performed using a z-test. Data are presented as mean ± SD.

Results

The mortality rates for different groups were as follows: Group 1: 59%; group 2: 60%; group 3: 20%; group 4: 10%. Statistical analysis revealed a significant difference between heparinized (Group 1+Group 2) and non-heparinized (Group 3+Group 4) animals. The highest mortality rate was on day 1 in all groups. In addition heparinized animals exhibited a significant increase of mortality on day 3. (Fig. 1).

The histological investigation indicated in group 2 (Heparin/no Nimodipine) multiple necroses in cortical and hippocampal areas. There were a few intact pyramidal cells. In 50% of all animals in that group there were extravasated erythrocytes in cortical and hippocampal areas and necrosis. In group 1 (Heparin/Nimodipine) there was less blood vessel dilatation compared to group 2. There was less cortical necrosis (40%) and more intact pyramidal cells found. Cortical bleeding was detected in 43% of those

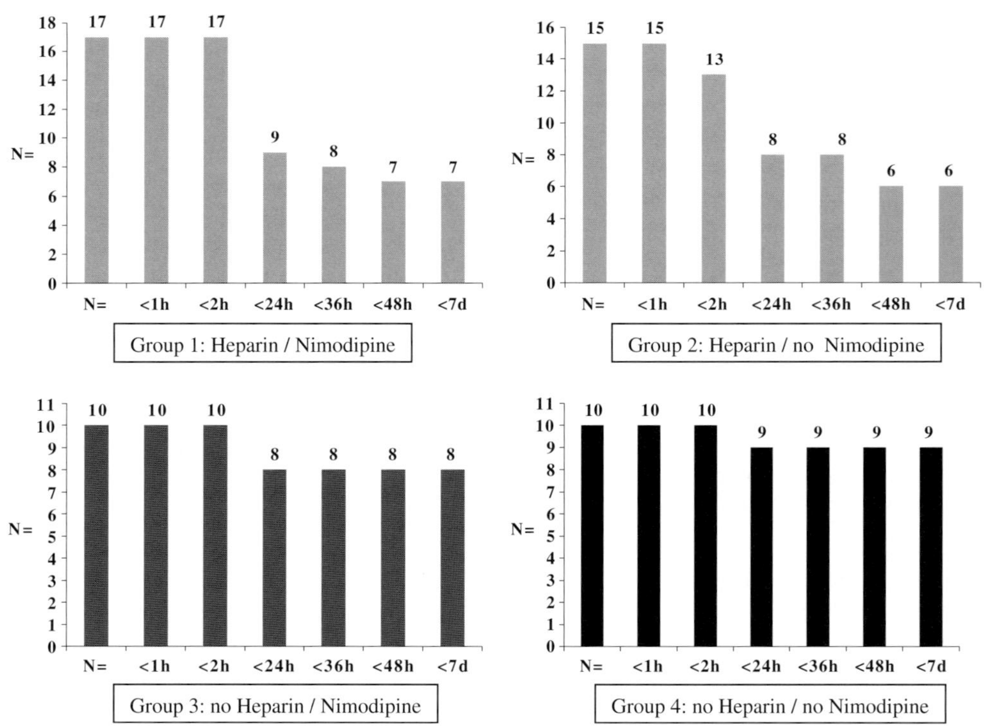

Fig. 1 Time course of the survival rate within the four different study groups

animals. In group 4 (no Heparin/no Nimodipine) there was no cortical bleeding. Compared to the heparinized animals there were less cortical necrosis and less necrosis in the hipocampal CA3 layer. However, necrosis was detected in all CA1 layers. In group 3 (no Heparin/Nimodipine) animals there was no cortical bleeding. There were almost no cortical necrosis. Necrosis in CA3 was only detected in one brain. Necrosis in CA1 layer was detected in 40% of the animals.

Discussion

The present study represents for the first time the utilization of the previously established t-SAH model to investigate the influence of Nimodipine administration following t-SAH. The results indicate no benefit of Nimodipine on mortality. However, the histological investigations indicated less cell loss in the hippocampal areas. Further investigations regarding the histological changes are under way.

Within the last few years, several clinical studies have been undertaken to determine the influence of Nimodipine in patients with traumatic brain injury and particular emphasis on t-SAH. The first published studies indicated promising results [4, 5]. However, till now those initial positive data have not been verified. In a study on 750 patients from the European Brain Injury Consortium Survey of Head Injuries, the adverse prognostic significance of t-SAH was confirmed. It was concluded that death among patients with t-SAH is related to the initial mechanical damage, rather than to the effects of delayed vasospasm and secondary ischemic brain damage [2]. Those results were in contradiction to data from Lee et al. indicating that hemodynamically significant vasospasm, which typically occurs in the setting of SAH, is strongly associated with poor outcome [7].

The clinical studies reveal quite contradictory results and conclusions regarding pathophysiological mechanisms and pharmaceutical influence. Therefore, more experimental studies are required.

Conflict of interest statement We declare that we have no conflict of interest.

References

1. Eisenberg HM, Gary HE, Aldrich EF et al (1990) Initial CT findings in 753 patients with severe head injury. A report from the NIH Traumatic Coma Data Bank. J Neurosurg 73:688–698
2. European study group on Nimodipine in severe head injury (1994) A multicenter trial of the efficacy of Nimodipine on outcome after severe head injury. J Neurosurg 80:797–804
3. Marmarou A, Foda MAA, Van den Brink W, Campbell J, Kita H, Demetriadou K (1994) A new model of diffuse brain injury in rats. Part I: Pathophysiology and biomechanics. J Neurosurg 80:291–300
4. Kakarieka A, Braakman R, Schakel EH (1994) Clinical significance of the finding of subarachnoidal blood on CT scan after head injury. Acta Neurochir 129:1–5
5. Kakarieka A (1997) Traumatic subarachnoidal haemorrhage. Springer, Berlin, Heidelberg, New York, p 21
6. Kobayashi S, Nakazawa S, Hiroyuki Y et al (1988) Traumatic subarachnoid hemorrhage in acute severe head injury. No To Shinkei 40:1131–1135
7. Lee JH, Martin NA, Alsina G, McArthur DL, Zaucha K, Hovda DA, Becker DP (1997) Hemodynamically significant cerebral vasospasm and outcome after head injury: A prospective study. J Neurosurg 87:221–233
8. Servadei F, Murray GD, Teasdale GM, Dearden M et al (2002) Traumatic subarachnoid hemorrhage: Demographic and clinical study of 750 patients from the European Brain Injury Consortium survey of Head Injuries. Neurosurgery 50:261–269
9. Thomas S, Tabibinia F, Herrmann B, Schuhmann MU, Brinker T, Samii M (1999) Traumatic subarachnoidal hemorrhage: Introduction of a new experimental model in the rat. J Neurotrauma 16:959
10. Thomas S, Tabibnia F, Schuhmann MU, Brinker T, Samii M (2000) ICP and MABP following traumatic subarachnoidal hemorrhage in the rat. Acta Neurochir (Suppl) 76:203–205

Acta Neurochir Suppl (2008) 102: 381–384

Neuroprotective effect of C1-inhibitor following traumatic brain injury in mice

L. Longhi · C. Perego · E. R. Zanier · F. Ortolano · P. Bianchi · N. Stocchetti · M. G. De Simoni

Abstract
Background The goal of the study was to evaluate the effects of C1-inhibitor (C1-INH), an endogenous glycoprotein endowed with multiple anti-inflammatory actions, on cognitive and histological outcome following controlled cortical impact (CCI) brain injury.
Methods Male C57Bl/6 mice (n=48) were subjected to CCI brain injury. After brain injury, animals randomly received an intravenous infusion of either C1-INH (15 U either at 10 minutes or 1 hour postinjury) or saline (equal volume, 150 μl at 10 min postinjury). Uninjured control mice received identical surgery and saline injection without brain injury. Cognitive function was evaluated at 4 weeks postinjury using the Morris Water Maze. Mice were subsequently sacrificed, the brains were frozen and serial sections were cut. Traumatic brain lesion was assessed by dividing the area of the ipsilateral hemisphere for the area of the contralateral one at the level of the injured area of the brain.
Findings Brain-injured mice receiving C1-INH at 10 min postinjury showed attenuated cognitive dysfunction compared to brain-injured mice receiving saline ($p<0.01$). These mice also showed a significantly reduced traumatic brain lesion compared to mice receiving saline ($p<0.01$). Mice receiving C1-INH at 1 hour post injury did not show a significant improvement in either cognitive or histological outcome.
Conclusions Our results suggest that administration of C1-INH at 10 minutes postinjury attenuates cognitive deficits and histological damage associated with traumatic brain injury.

Keywords Traumatic brain injury · Inflammation · Complement · Neuroprotection

L. Longhi · E. R. Zanier · F. Ortolano · P. Bianchi · N. Stocchetti
Department of Anesthesia and Critical Care Medicine, Ospedale Maggiore Policlinico, Neurosurgical Intensive Care Unit, Via Sforza n 35, 20122 Milano, Italy

L. Longhi · C. Perego · E. R. Zanier · F. Ortolano · M. G. De Simoni (✉)
Laboratory of Inflammation and Nervous System Diseases, Mario Negri Institute, Via La Masa n 19, 20156 Milano, Italy
e-mail: desimoni@marionegri.it

Introduction

Each year traumatic brain injury (TBI) affects 2,000,000–3,000,000 people in the USA [1] and almost 1,000,000 people in the European Union [2]. Traumatic brain injury is associated with primary injury (due to the biomechanical effects of the impact) and with secondary molecular and cellular events that are initiated minutes after the injury and may last for months. These secondary processes activated by trauma interact in a complex network leading to cell death or recovery [3]. Since the major part of traumatic brain damage is known to evolve over days following the impact, an opportunity exists for pharmacological therapeutic intervention to attenuate secondary pathological cascades. The complement system is a major component of innate immunity that includes plasma and membrane proteins that mediate pathways of enzymatic reactions. Complement activation results in deposition of complement components on pathologic targets and liberation of proinflammatory molecules. Excessive complement activation is believed to be deleterious to host-tissue [4, 5]. Intracerebral complement activation has been documented both in humans and in animal models following TBI [6–8], where it has been suggested to mediate post-traumatic cell death by increasing post-traumatic inflammatory response, apoptosis and direct cell lysis [9]. The C1-inhibitor (C1-INH) is an endogenous serine-protease inhibitor of the complement and contact-kinin pathways [10] that has been previously approved for clinical use as treatment in cases of hereditary angioedema [10]. In models of ischemia-reperfusion, C1-INH attenuated acute neurobehavioral deficits, ischemic volume at 48 hours post-ischemia, and neurodegeneration at 4 days post-ischemia [11, 12]. Potential mechanisms associated with C1-INH neuroprotection include 1) reduced leukocyte infiltration at 24 hours post-ischemia, 2) reduced microglial activation/ macrophage infiltration up to 48 hours, 3) reduced gene expression of intracellular adhesion molecule (ICAM)-1, p-selectin, and pro-inflammatory cytokines such as tumor necrosis factor (TNF)-alpha (α), Interleukin (IL)-18 and procaspase 3, and 4) increased gene expression of the anti-inflammatory cytokine IL-10 and of markers of neuronal integrity such as heavy-chain neurofilament protein (NFH) [12, 13]. Therefore, complement inhibitors such as the glycoprotein C1-INH might act on multiple targets in the complex cascades of post-traumatic secondary brain injury. In the present study we tested the hypothesis that plasma-derived C1-INH (Berinert CSL-Behring, Marburg, Germany) might reduce the cognitive dysfunction and histological damage associated with controlled cortical impact (CCI) brain injury in mice.

Material and methods

Experimental brain injury

All procedures were conducted conform the institutional guidelines that are in compliance with national (D.L. n.116, G.U. suppl. 40, 18 February 1992) and international laws and policies (EEC Council Directive 86/609, OJL 358,1; Dec.12,1987; NIH Guide for the Care and Use of Laboratory Animals, U.S. National Research Council 1996).

Male C57Bl/6 (20–25 g, n=36) were anesthetized with intraperitoneal (i.p.) administration of sodium pentobarbital, (65 mg/kg) and placed in a stereotactic frame. An eye lubricant ointment (Xantervit ophthalmic ointment, Sifi, CT, Italy) was applied to protect corneal membranes during surgery. Mice were subjected to a craniectomy followed by CCI brain injury. Our model of injury uses a 3 mm rigid impactor driven by a pneumatic piston, rigidly mounted at an angle of 20° from the vertical plane and applied perpendicularly to the exposed *dura mater* over the left parieto-temporal cortex, between bregma and lambda at a velocity of 5 meter/second and depth of 1 mm [14]. At 10 minutes postinjury, animals randomly received an intravenous infusion of either C1-INH (15 U, n=12) or saline (equal volume, 150 μl, n=12). An additional group of mice received the C1-INH at 1 hour postinjury (n=12). A fourth group of mice received identical anesthesia, surgery without CCI and saline infusion to serve as uninjured controls (sham, n=12).

Behavior

Cognitive function was evaluated at 4 weeks postinjury using the Morris Water Maze (MWM). Our MWM is a white circular pool (1 meter diameter) that is filled with water (18–20°C) made opaque by adding non-toxic water-soluble white paint. The postinjury visuo-spatial learning task requires that animals learn how to locate a submerged platform placed 0.5 cm under the surface of the water using external visual cues. The essential feature of this task is that the mice learn how to escape from the cold water onto the platform after being randomly placed into the pool at one of four sites. Latencies to reach and climb onto the platform are recorded for each trial, with a maximum of 60 seconds per trial. The learning task consists of 8 trials/day for three consecutive days for a total of 24 trials. The cognitive performance of all animals was obtained by averaging the latencies of 24 trials over three days as previously described [14].

Histology

All animals were sacrificed one month postinjury for histological analysis. Under deep anesthesia (Equitensin i.p., 120 μl/mouse), they were transcardially perfused with 20 ml of phosphate buffer saline (PBS) 0.1 M, pH 7.4, followed by 50 ml of chilled paraformaldehyde (4%) in PBS, and then the brains carefully removed from the skull. The brains were postfixed overnight at 4°C, then transferred to 30% sucrose in 0.1 M phosphate buffer for 24 hr until equilibration. Subsequently the brains were frozen and 40-micron-thick serial sections cut using a cryostat from bregma + 1 mm to bregma – 3.25 mm. Finally the section were stained using Neutral Red (Neutral Red Gurr Certistain, BDH, Poole, UK) [11]. The ipsilateral and contralateral hemispheres were manually outlined. Subsequently the injured area was calculated by dividing the ipsilateral hemisphere for the contralateral hemisphere at the coordinates mentioned above.

Statistical analysis

Learning latencies and traumatic injured area are presented as mean ± standard error of the mean. The comparison between groups was performed by 2 way analysis of variance (ANOVA) followed by Bonferroni post-hoc test. A "p" value < 0.05 was considered significant.

Results

Cognitive function

At 4 weeks following brain injury all the animals were able to swim in the MWM without notable impairment in swimming ability and were able to learn the visuo-spatial task, as reflected by decreasing latencies to find the platform over the three-day period. The learning ability (latency) of brain-injured animals was significantly worse than the performance of sham-injured animals ($p<0.01$). By 4 weeks post-treatment, the learning ability of the mice receiving C1-INH at 10 minutes postinjury was significantly better than that of vehicle-treated, brain-injured mice ($p<0.01$). Mice receiving C1-INH at 1 hour postinjury showed a learning ability that was intermediate between that of the mice treated at 10 minutes postinjury and the mice receiving saline (Fig. 1).

Neuroprotection

At 1 month postinjury we observed a trauma-induced area of tissue loss (involving the ipsilateral cortex and hippocampus) extending from bregma + 0.6 to bregma − 3.25 mm. Mice receiving C1-INH at 10 min postinjury showed a significantly reduced traumatic cerebral lesion compared to mice receiving saline ($p<0.01$). Brain-injured mice receiving C1-INH at 1 hour postinjury showed a similar traumatic cerebral lesion compared to brain-injured mice receiving saline (Fig. 2).

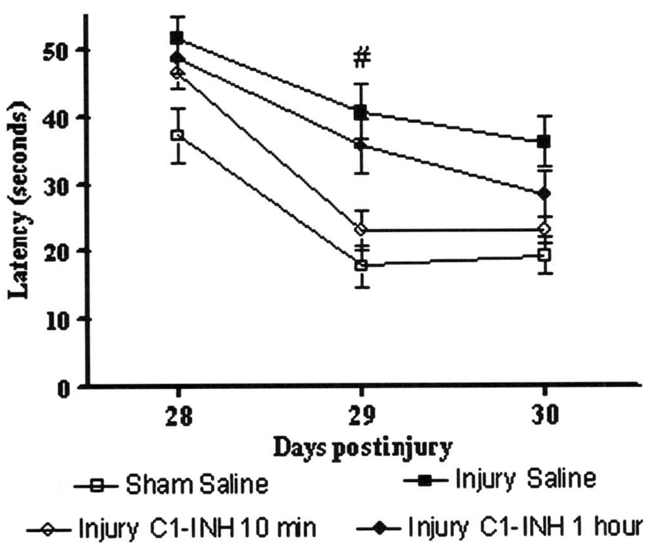

Fig. 1 Mice receiving C1-INH at 10 min postinjury showed a better cognitive performance compared to mice receiving saline. Two way ANOVA: time effect: $p<0.001$; group effect: $p<0.001$. Bonferroni: # $p<0.01$ Injury C1-INH 10 min vs. Injury Saline

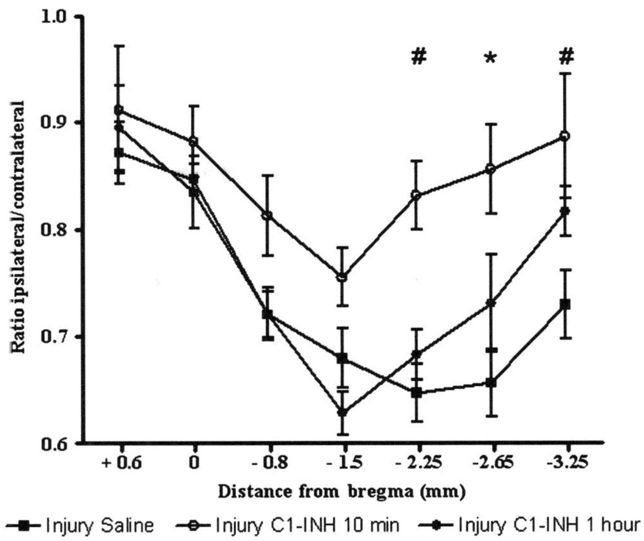

Fig. 2 Mice receiving C1-INH at 10 min postinjury showed a significantly reduced traumatic brain lesion compared to mice receiving saline. Two way ANOVA: progression of the lesion: $p<0.001$; group effect: $p<0.001$. Bonferroni: # $p<0.01$ Injury C1-INH 10 min vs. Injury Saline; * $p<0.001$ Injury C1-INH 10 min vs. Injury Saline

Discussion

In the present study we observed that administration of the endogenous glycoprotein C1-INH at 10 minutes following CCI brain injury significantly attenuated cognitive dysfunction and reduced histological damage at 1 month postinjury in mice. C1-INH is a multifaceted anti-inflammatory compound, one of its best known actions being inhibition of the complement cascade.

Following TBI, intracerebral synthesis of complement factors occurs together with passive leakage across the open BBB as shown by accumulation of complement factors into the injured brain [9]. Complement activation occurs via 3 pathways: the classical, the alternative and the lectin that converge to the common protein C3 and result in the generation of C3a and C5a (anaphylotoxins), C4b and C3b (opsonins) and the membrane attack complex (MAC) [4]. Following acute brain damage, the anaphylotoxins C3a and C5a might increase blood brain barrier (BBB) permeability leading to leukocytes recruitment into the injured brain and subsequent free radicals production [9]. The fragment C5a might also induce apoptosis in neuronal and glial cells by binding to the C5a receptor (R) [9]. Finally, the membrane attack complex (MAC) might induce cell lysis [9]. The relevance of the different complement pathways in the pathogenesis of TBI has been studied by others investigators using genetically engineered mice and pharmacological strategies aimed at blocking complement activation. It has been shown that antagonism of C3 using transgenic mice overexpressing the endogenous C3 convertase inhibitor [15] or pharmacological compounds [16, 17] resulted in attenuated functional deficits and histological damage following TBI, suggesting that complement inhibition at the point where all activation pathways converge is a promising strategy. In addition, the selective antagonism of the alternative pathway using knockout mice for factor B and a monoclonal anti-factor B antibody also resulted in functional improvement and attenuation of post-traumatic histological damage, suggesting that the alternative pathway may exert a major role in the post-traumatic complement-mediated neuroinflammatory response [18, 19]. However more recently, You et al. subjected C1q (-/-), C3 (-/-), C4 (-/-) and wild type (WT) mice to CCI brain injury and observed that only the C4 (-/-) mice showed reduced neurological motor deficits and contusion volume when compared to WT mice, suggesting that C4 (a mediator of classic and lectin pathways) mediates post-traumatic cell death and dysfunction via mechanisms that are independent from C3 activation [20]. Since recent studies have used different TBI models and differing outcome measures at different time points, the most optimal therapeutic strategy of complement modulation following TBI has yet to be determined. Although C1-INH is a major inhibitor of C1q, the first step of activation of the classical pathway, this compound is also an inhibitor of the lectin and alternative pathways [10, 21]. In a model of ischemia and reperfusion, the neuroprotective effect of C1-INH occurrred also in C1q (-/-) mice, suggesting that inhibition involved the classical complement pathway is not required for C1-INH neuroprotective action [12]. Despite the fact that the mechanisms underlying C1-INH mediated neuroprotection following TBI remain unknown, they are likely to be associated with modulation of the inflammatory and apoptotic cascades.

Conflict of interest statement We declare that we have no conflict of interest.

Reference

1. Anonymous (1999) Consensus conference. Rehabilitation of persons with traumatic brain injury. NIH Consensus Development Panel on Rehabilitation of Persons With Traumatic Brain Injury [see comments]. JAMA 282:974–983
2. Tagliaferri F, Compagnone C, Korsic M, Servadei F, Kraus J (2006) A systematic review of brain injury epidemiology in Europe. Acta Neurochir (Wien) 148:255–268
3. McIntosh TK, Saatman KE, Raghupathi R, Graham DI, Smith DH, Lee VM, Trojanowski JQ (1998) The Dorothy Russell Memorial Lecture. The molecular and cellular sequelae of experimental traumatic brain injury: pathogenetic mechanisms. Neuropathol Appl Neurobiol 24:251–267
4. Walport MJ (2001) Complement. First of two parts. N Engl J Med 344:1058–1066
5. Walport MJ (2001) Complement. Second of two parts. N Engl J Med 344:1140–1144
6. Bellander BM, von Holst H, Fredman P, Svensson M (1996) Activation of the complement cascade and increase of clusterin in the brain following a cortical contusion in the adult rat. J Neurosurg 85:468–475
7. Bellander BM, Singhrao SK, Ohlsson M, Mattsson P, Svensson M (2001) Complement activation in the human brain after traumatic head injury. J Neurotrauma 18:1295–1311
8. Tornqvist E, Liu L, Aldskogius H, Holst HV, Svensson M (1996) Complement and clusterin in the injured nervous system. Neurobiol Aging 17:695–705
9. Stahel PF, Morganti-Kossmann MC, Kossmann T (1998) The role of the complement system in traumatic brain injury. Brain Res Brain Res Rev 27:243–256
10. Caliezi C, Wuillemin WA, Zeerleder S, Redondo M, Eisele B, Hack CE (2000) C1-Esterase inhibitor: an anti-inflammatory agent and its potential use in the treatment of diseases other than hereditary angioedema. Pharmacol Rev 52:91–112
11. De Simoni MG, Storini C, Barba M, Catapano L, Arabia AM, Rossi E, Bergamaschini L (2003) Neuroprotection by complement (C1) inhibitor in mouse transient brain ischemia. J Cereb Blood Flow Metab 23:232–239
12. De Simoni MG, Rossi E, Storini C, Pizzimenti S, Echart C, Bergamaschini L (2004) The powerful neuroprotective action of C1-inhibitor on brain ischemia-reperfusion injury does not require C1q. Am J Pathol 164:1857–1863
13. Storini C, Rossi E, Marrella V, Distaso M, Veerhuis R, Vergani C, Bergamaschini L, De Simoni MG (2005) C1-inhibitor protects against brain ischemia-reperfusion injury via inhibition of cell recruitment and inflammation. Neurobiol Dis 19:10–17
14. Longhi L, Watson DJ, Saatman KE, Thompson HJ, Zhang C, Fujimoto S, Royo N, Castelbuono D, Raghupathi R, Trojanowski JQ, Lee VM, Wolfe JH, Stocchetti N, McIntosh TK (2004) Ex vivo gene therapy using targeted engraftment of NGF-expressing human NT2N neurons attenuates cognitive deficits following traumatic brain injury in mice. J Neurotrauma 21:1723–1736
15. Rancan M, Morganti-Kossmann MC, Barnum SR, Saft S, Schmidt OI, Ertel W, Stahel PF (2003) Central nervous system-targeted complement inhibition mediates neuroprotection after closed head injury in transgenic mice. J Cereb Blood Flow Metab 23:1070–1074
16. Leinhase I, Schmidt OI, Thurman JM, Hossini AM, Rozanski M, Taha ME, Scheffler A, John T, Smith WR, Holers VM, Stahel PF (2006) Pharmacological complement inhibition at the C3 convertase level promotes neuronal survival, neuroprotective intracerebral gene expression, and neurological outcome after traumatic brain injury. Exp Neurol 199:454–464
17. Kaczorowski SL, Schiding JK, Toth CA, Kochanek PM (1995) Effect of soluble complement receptor-1 on neutrophil accumulation after traumatic brain injury in rats. J Cereb Blood Flow Metab 15:860–864
18. Leinhase I, Holers VM, Thurman JM, Harhausen D, Schmidt OI, Pietzcker M, Taha ME, Rittirsch D, Huber-Lang M, Smith WR, Ward PA, Stahel PF (2006) Reduced neuronal cell death after experimental brain injury in mice lacking a functional alternative pathway of complement activation. BMC Neurosci 7:55
19. Leinhase I, Rozanski M, Harhausen D, Thurman JM, Schmidt OI, Hossini AM, Taha ME, Rittirsch D, Ward PA, Holers VM, Ertel W, Stahel PF (2007) Inhibition of the alternative complement activation pathway in traumatic brain injury by a monoclonal anti-factor B antibody: a randomized placebo-controlled study in mice. J Neuroinflammation 4:13
20. You Z, Yang J, Takahashi K, Yager PH, Kim HH, Qin T, Stahl GL, Ezekowitz RA, Carroll MC, Whalen MJ (2007) Reduced tissue damage and improved recovery of motor function after traumatic brain injury in mice deficient in complement component C4. J Cereb Blood Flow Metab
21. Jiang H, Wagner E, Zhang H, Frank MM (2001) Complement 1 inhibitor is a regulator of the alternative complement pathway. J Exp Med 194:1609–1616

Acta Neurochir Suppl (2008) 102: 385–388

Up-regulation of L type amino acid transporter 1 after spinal cord injury in rats

Terushige Toyooka · Hiroshi Nawashiro · Nariyoshi Shinomiya · Akiko Yano · Hidetoshi Ooigawa · Atsushi Ohsumi · Yoichi Uozumi · Youichi Yanagawa · Hirotaka Matsuo · Katsuji Shima

Abstract
Background L-type amino acid transporter 1 (LAT1) is proposed to be a major nutrient transporter at the blood brain barrier. LAT1 requires the heavy chain of 4F2 cell surface antigen (4F2hc) for functional expression.
Methods We investigated the expression of this heterodimeric transporter after traumatic spinal cord injury in rat by using immunohistochemical and western blot analyses.
Findings LAT1 immunoreactivities were up-regulated in the capillary endothelia in close to the injury epicenter 24 hours after injury. It reached a peak at 48 hours after injury, and thereafter decreased. 4F2hc was abundant and unchanged all through the time course after SCI. Western blot analysis under reductive and non-reductive conditions showed that LAT1 and 4F2hc were conjugated as a heterodimeric transporter and the functional regulation was dependent on the light chain, LAT1.
Conclusions We suggest that LAT1 may be transiently up-regulated as part of the tissue-repair process after traumatic contusion injury in the spinal cord.

T. Toyooka (✉) · H. Nawashiro · A. Yano · H. Ooigawa · A. Ohsumi · Y. Uozumi · K. Shima
Department of Neurosurgery, National Defense Medical College, 3-2 Namiki, Tokorozawa, Saitama 359-8513, Japan
e-mail: teru-toy@mse.biglobe.ne.jp

N. Shinomiya · H. Matsuo
Department of Integrative Physiology and Bio-Nano, National Defense Medical College, 3-2 Namiki, Tokorozawa, Saitama 359-8513, Japan

Y. Yanagawa
Department of Emergency Medicine, National Defense Medical College, 3-2 Namiki, Tokorozawa, Saitama 359-8513, Japan

Keywords L type amino acid transporter 1 · LAT1 · 4F2 heavy chain · Spinal cord injury

Introduction

L type amino acid transporter is a heterodimeric system which comprises the combination of a low molecular weight protein (light chain) and a high molecular weight glycoprotein (heavy chain) exert its functions as a highly selective transporter [4]. L-type amino acid transporter 1 (LAT1), one of the light chains, is proposed to be a major nutrient transporter at the blood brain barrier because it has been shown to be expressed in the brain capillary endothelial cells [7]. LAT1 requires an additional single membrane spanning protein, the heavy chain of 4F2 cell surface antigen (4F2hc) for the functional expression as an amino acid transporter [4, 6]. LAT1 is thought to be up-regulated to support the high protein synthesis for cell growth because of the high level of its expression in tumor cell lines [3]. LAT1 has been also proved to be over-expressed in glioma cells, and linked to the gliomatous proliferation [8]. Recent experimental study related to exhaustive analyses about gene expression after traumatic brain injury disclosed that the mRNA of LAT1 was up-regulated 24 hours after traumatic severe brain injury in rat [5]. In the present study, we elucidated, for the first time, the temporal and spatial expression level of this transporter after traumatic contusion injury in the rat spinal cord.

Materials and methods

Animal experimental procedures

Adult female Sprague-Dawley rats 10-weeks old and weighing 220 to 280 g were used. The rats were anesthetized by intraperitoneal injection of pentobarbital sodium (50 mg/Kg) and inhalation of a 30:70 mixture of oxygen and nitrous oxide gas through a facemask. The rectal temperature was measured with a rectal probe and maintained constant at a level of 37.0–37.5°C using a heating pad. The lamina of the tenth thoracic vertebrae were removed and a New York University (NYU) weight-drop device [2] used to produce a severe spinal cord contusion injury with a 10 gram weight dropped from a height of 25 mm. After injury, the rats were placed on a heating pad to maintain a temperature of approximately 37°C until they could move independently. All animal procedures were approved by the National Defense Medical College.

Tissue preparation

For Western blot analysis, a 10-mm spinal cord segment containing the injury epicenter was dissected using intraperitoneal anesthesia with pentobarbital sodium (50 mg/Kg) at 24, 48 and 72 hours and 5, 7 days after injury (n=4 per time point). The cord segment was immediately submerged in ice-cold artificial CSF (0.6 mmol/L NaH2PO4/2H2O, 3.35 mmol/L dextrose, 21 mmol/L sodium bicarbonate, 2.5 mmol/L calcium chloride, and 1 mmol/L magnesium chloride) and the dura surrounding the excised cord were removed. The tissue specimens were frozen and kept at −80°C. Sham-control animals were subjected to a laminectomy but did not receive a weight-drop injury and were sacrificed 48 hours after injury (n=4). For immunohistochemical analysis, the rats were perfused transcardially with normal saline followed by 4% buffered paraformaldehyde at the same time points after injury (n=3 per time point). Longitudinal 10-mm spinal cord segment centering on the injury epicenter was removed and embedded in paraffin after fixation in 4% buffered paraformaldehyde, followed by 0.1 mmol/L phosphate-buffered saline (pH 7.4) for 3–4 hours at 4°C. Sagittal and axial sections (5-μm thick) were prepared.

Immunohistochemistry

Immunohistochemistry was performed by the dextran polymer peroxidase complex method using an ENVISION+ Kit (DAKO, Kyoto, Japan). After deparaffinization and hydration, the sections in 10 mmol/L sodium citrate (pH 6.0) were boiled for 5 minutes at 95°C. Endogenous peroxidase was blocked with 3% hydrogen peroxidase. The sections were incubated with phosphate-buffered saline containing 10% normal goat serum at room temperature to eliminate any nonspecific binding and were incubated overnight at 4°C with primary antibodies against LAT1 and 4F2hc (Transgenic, Kumamoto, Japan). The sections were incubated with secondary antibody linked with dextran polymer peroxidase complex for 30 minutes at room temperature. Peroxidase was demonstrated with DAB. For an evaluation of morphological changes, adjacent sections were counterstained with hematoxylin. For negative

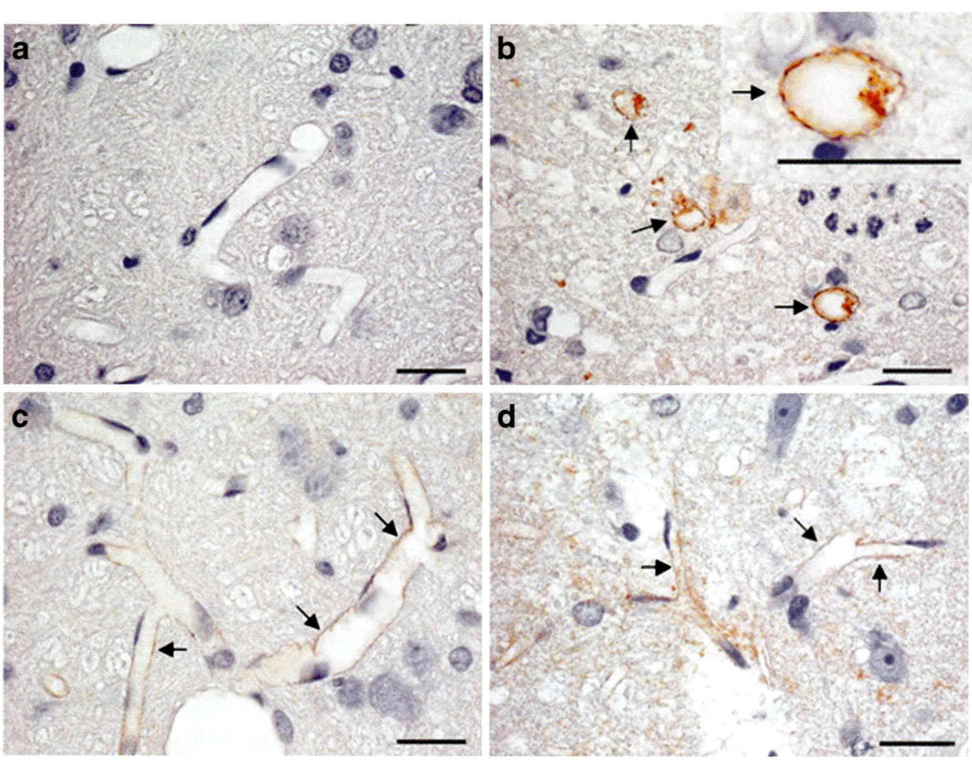

Fig. 1 Photomicrographs show the spatial distribution of immunoreactivity for LAT1 and 4F2hc in the spinal cord after traumatic contusion injury. Sections were prepared from the sham control (**a**, **c**) and the rats at 48 hours after injury (**b**, **d**). No significant immunostaining of LAT1 was seen in the sham operated control (**a**). After injury, remarkable immunoreactivity for LAT1 was detected on capillary endothelia in the region adjacent to the injury epicenter (**b**, arrows). Immunoreactivity for 4F2hc was ubiquitously detected on the capillary endothelia in the sham operated control (**c**, arrows), and was not changed very much in anywhere all through the time course after injury (D, arrows). All bars=20 μm

controls, normal rabbit IgG was used instead of the primary antibodies.

Western blot analysis

The frozen tissue specimens were homogenized with ice-cold histolysis buffer. The protein content was determined using a detergent compatible protein assay reagent kit (Bio-Rad, Hercules, CA, U.S.A.). Equal amounts of protein (50 μg) were loaded in each lane on 12% tris polyacrylamide gel (sodium dodecyl sulfate-polyacrylamide gel electrophoresis with loading buffer (100 mM Tris-hydrochloride, pH 6.8; 4% sodium dodecyl sulfate, 20% glycerol, 0.2% bromophenol blue, 200 mM dithiothreitol). The protein was loaded without dithiothreitol when the electrophoresis was performed under non-reductive conditions. The protein was transferred to polyvinylidene difluoride membranes (GE healthcare, Buckinghamshire, United Kingdom) at 250 mA for 3 hours. The membranes were then incubated overnight at 4°C with primary antibody against monoclonal LAT1, 4F2hc (Transgenic) and GAPDH (Santa Cruz, CA, U.S.A.) antibodies. The membranes were incubated for 1 hour at room temperature with horseradish peroxidase conjugated secondary antibodies (Santa Cruz). The blots were developed with ECL detection reagent (GE healthcare).

Statistical analysis

All data are expressed as the means±SD. The densities of the immunoblotting band were quantified by using an ImageJ analyzer, version 1.38× (National Institute of Health, Bethesda, MD, U.S.A.). Statistical significance for Western blots was determined using an analysis of ANOVA followed by a post hoc Bonferroni/Dunn test. A p value of <0.05 was considered statistically significant.

Results

Immunohistochemical results of LAT1 and 4F2hc are summarized in Fig. 1. Strong immunoreactivity for LAT1 was detected on capillary endothelia in the region adjacent to the injury epicenter from 24 hours after injury, although no significant immunostaining was seen in the sham operated controls (Fig. 1a,b). Expression was strongest from 48 to 72 hours after injury, and thereafter decreased gradually. On the contrary, immunoreactivity for 4F2hc was ubiquitously detected on the capillary endothelia in the sham operated control, and was unchanged all through the time course after injury (Fig. 1c,d).

The results of Western blot analysis were consistent with those of immunohistochemistry (Fig. 2). LAT1 immunoreactivity significantly increased 24 hours after injury

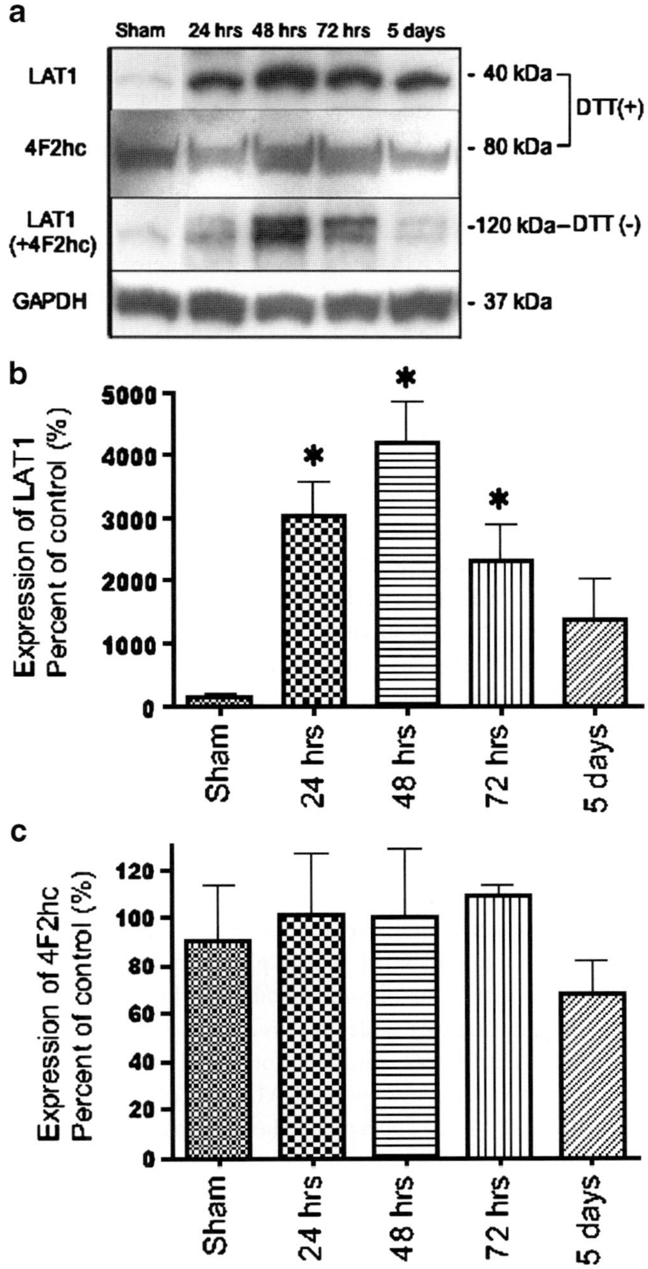

Fig. 2 Time cource of LAT1 and 4F2hc expression in the contused spinal cord. Picture of representative Western blot analysis showed the time course of immunoreactivity for LAT1, 4F2hc and GAPDH under reductive condition (dithiothreitol +), and LAT1 under non-reductive condition (dithiothreitol -) (**a**). The bar graph shows quantitative data of the optical density in LAT1 (**b**) and 4F2hc (**c**) with equal amounts of isolated protein from the 10-mm length spinal cord containing the injury epicenter 24, 48 and 72 hours and 5 days after injury (n=4 per time point). The values of LAT1 and 4F2hc were corrected relative to that of GAPDH and are represented as percentage of the sham control (mean ± SD). *P < compared with the sham-operated rats (n=4 per time point)

compared with sham-control rats, and reached a peak 48 hours after injury, and thereafter gradually decreased. 4F2hc immunoreactivity did not represent significant change all through the time course after injury. Under non-reductive conditions, the heterodimeric bands (120 kDa) representing LAT1 (40 kDa) combined with 4F2hc (80 kDa) appeared to show the same pattern as LAT1 alone.

Discussion

We indicated that LAT1 immunoreactivity increased during a few days after traumatic spinal cord injury. By immunohistochemical findings, it was confirmed that this occurred in the capillary endothelia adjacent to the injury epicenter. LAT1 has been identified to be up-regulated by an activation of lymphocytes, a hormonal stimulation [1, 3], and strong LAT1 expression has been detected in some tumor cell lines including glioma [4, 10]. We have previously reported that high LAT1 expression correlated with poor survival of patients with high grade glioma [8]. These data indicate that LAT1 may be highly regulated to supply cells with more essential amino acids to support cell growth and proliferation. It is noteworthy that up-regulation of LAT1 was observed to also occur in traumatic spinal cord injury. In addition, this up-regulation after trauma was seen in the endothelium of the vasculature in the region of the epicenter of the injury. This up-regulation may occur as part of the tissue repair processes after injury.

In contrast, 4F2hc was abundant in normal spinal cord and its immunoreactivity was unchanged all through the time course after SCI. 4F2hc has been reported to be expressed ubiquitously [9], and to play a functional role in transferring LAT1 to plasma membrane. Interestingly, in Western blot analysis under non-reductive conditions and probing with primary antibody for LAT1, heterodimeric bands (120 kDa) representing LAT1 (40 kDa) combined with 4F2hc (80 kDa) appeared as the same pattern of LAT1. We confirmed in the traumatic SCI that LAT1 and 4F2hc were conjugated as a heterodimeric transporter.

Conflict of interest statement We declare that we have no conflict of interest.

References

1. Gaugitsch HW, Prieschl EE, Kaithoff F, Huber NE, Baumruker T (1992) A novel transiently expressed, integral membrane protein linked to cell activation. J Biol Chem 267:11267–11273
2. Gruner JA (1992) A monitored contusion model of spinal cord injury in the rat. J Neurotrauma 9:123–128
3. Kanai Y (1997) Family of neutral and acidic amino acid transporters: molecular biology, physiology and medical implications. Curr Opin Cell Biol 9:565–572
4. Kanai Y, Segawa H, Miyamoto K, Uchino H, Takeda E, Endou H (1998) Expression cloning and characterization of a transporter for large neutral amino acids activated by the heavy chain of 4F2 antigen (CD98). J Biol Chem 273:23629–23632
5. Li HH, Lee SM, Cai Y, Sutton RL, Hovda DA (2004) Differential gene expression in hippocampus following experimental brain trauma reveals distinct features of moderate and severe injuries. J Neurotrauma 21:1141–1153
6. Mastroberardino L, Spindler B, Pfeiffer R, Skelly PJ, Loffing J, Shoemaker CB, Verrey F (1998) Amino-acid transport by heterodimers of 4F2hc/CD98 and members of a permease family. Nature 395:288–291
7. Matsuo H, Tsukada S, Nakata T, Chairoungdua A, Kim DK, Cha SH, Inatomi J, Yorifuji H, Fukuda J, Endou H, Kanai Y (2000) Expression of a system L neutral amino acid transporter at the blood-brain barrier. Neuroreport 11:3507–3511
8. Nawashiro H, Otani N, Shinomiya N, Fukui S, Ooigawa H, Shima K, Matsuo H, Kanai Y, Endou H (2006) L-type amino acid transporter 1 as a potential molecular target in human astrocytic tumors. Int J Cancer 119:484–492
9. Parmacek MS, Karpinski BA, Gottesdiener KM, Thompson CB, Leiden JM (1989) Structure, expression and regulation of the murine 4F2 heavy chain. Nucleic Acids Res 17:1915–1931
10. Sang J, Wang S, Panzia MA, Bassily NH, Thompson NL (1996) TA1, a highly conserved oncofetal complementary DNA from rat hepatoma, encodes an integral membrane protein associated with liver development, carcinogenesis, and cell activation. Cancer Res 55:1152–1159

Acta Neurochir Suppl (2008) 102: 389–390

Cortical expression of prolactin (PRL), growth hormone (GH) and adrenocorticotrophic hormone (ACTH) is not increased in experimental traumatic brain injury

J. C. Goodman · L. Cherian · C. S. Robertson

Abstract

Background Cerebral cortical expression of the pituitary hormones prolactin (PRL) and growth hormone (GH) have reported in ischemic damage. Both hormones may be involved in vascular tone regulation and angiogenesis, and growth hormone is thought to be neuroprotective while prolactin stimulates astrogliosis.

Methods We examined expression of prolactin, growth hormone and adrenocorticotrophic hormone (ACTH) using tissue microarray technology in the controlled cortical impact model of traumatic brain injury (TBI).

Findings No increased expression of these hormones was seen.

Conclusions Unlike ischemia, traumatic brain injury does not result in up-regulation of the pituitary hormones PRL and GH in cerebral cortex.

J. C. Goodman · L. Cherian · C. S. Robertson
Department of Neurosurgery, MS:BCM650,
Baylor College of Medicine,
One Baylor Plaza,
Houston, TX 77030, USA

L. Cherian
e-mail: lcherian@bcm.tmc.edu

C. S. Robertson
e-mail: claudiar@bcm.tmc.edu

J. C. Goodman
Department of Neurology, Baylor College of Medicine,
Houston, TX 77030, USA

J. C. Goodman (✉)
Department of Pathology, MS BCM 315,
Baylor College of Medicine,
One Baylor Plaza,
Houston, TX 77030, USA
e-mail: jgoodman@bcm.tmc.edu

Keywords Traumatic brain injury · Prolactin · Growth hormone · ACTH · Tissue microarray · Vasoinhibin

Introduction

Recent reports suggest that neocortical expression of the pituitary hormones prolactin and growth hormone is increased following ischemic brain damage, and that growth hormone may be neuroprotective while prolactin may stimulate astrogliosis [4, 5]. Both hormones are also now recognized as a new class of compounds known as "vasoinhibins" that suppress vasodilatation and angiogensis while accelerating vascular regression via endothelial apoptosis [2, 3]. We used tissue microarrays of rat neocortex to assess expression levels and cellular localization of prolactin, growth hormone and ACTH following controlled cortical impact to see if these findings also occur in traumatic brain injury.

Based on previous reports of participation of neocortical extra-pituitary GH and PRL is ischemia models, we hypothesized that neocortical expression of pituitary hormones would be increased near a neocortical contusion induced by controlled cortical impact.

Of particular interest are the vasoinhibin effects of PRL and GH since defects of vascular reactivity are common in TBI. In the CCI model, regional cerebral blood flow is reduced immediately following injury. This CBF reduction is associated with reduced nitric oxide production and pharmacological maneuvers that restore nitric oxide levels also restore CBF to physiological ranges and result in reduction in cortical contusion volumes [1]. If PRL and GH are upregulated in CCI, their suppression of vasodilatation could contribute to the post-traumatic reduction in CBF. A portion of the action of vasoinhibins is interference with nitric oxide induced vasodilatation due to inhibition of endothelial nitric oxide synthase.

Tissue microarrays permit rapid high throughput analysis of protein expression in tissue cores from a large number of experimental animals. Since tissue cores preserve cellular morphology, both cytological localization and relative expression levels can be examined [6].

Materials and methods

Forty-eight rats were subjected to moderate controlled cortical impact and the brains were obtained 14 days after injury. From the paraffin blocks containing coronal sections with the cortical contusions and underlying hippocampus, a tissue microarray block was prepared containing tissue cores from the injured and uninjured cerebral cortex. The TBI microarray was created using a ATA-27 Automated Tissue Microarray System (Beecher Instruments, Silver Spring, MD) in the M.D. Anderson Cancer Center Tissue Microarray Core courtesy of Dr. Gregory N. Fuller.

With this tissue microarray, we performed immunohistochemical staining for prolactin, growth hormone and ACTH simultaneously on 96 tissue cores. Two cores were obtained from injured and uninjured cerebral cortex and hippocampus in each animal. Semi-quantitative analysis of intensity of tissue staining and cellular localization was performed.

Results

There was extremely low expression of PRL in endothelial cells and none in glia or neurons. No neocortical GH or ACTH immunoreactivity was detected in either injured or un-injured neocortex. Antigen preservation was confirmed by robust GFAP, neurofilament, CD34 and vimentin staining. Rat pituitary gland served as a positive control and demonstrated strong PRL, GH and ACTH immunoreactivity.

There was no increase or change in localization of prolactin immunoreactivity. No growth hormone or ACTH was seen in injured neocortex following controlled cortical impact.

Discussion

Unlike findings reported in ischemic injury, we did not find elevated cerebral cortical prolactin or growth hormone following traumatic brain injury. In addition, we did not see any ACTH expression. Ischemia is an important mechanism of secondary injury in TBI but in this particular experimental setting does not appear to activate cortical pituitary hormone expression. It is important to recognize that the reports of neocortical pituitary hormone activity in ischemic injury were done in younger animals and at earlier time points than our study; additional studies may be appropriate to see if growth hormone and prolactin expression in the cortex may be relevant to traumatic brain injury in younger individuals or a more acute setting.

The vasoinhibins are recently described endogenous regulators of angiogenesis and vascular tone. Immediately following TBI there is dysregulation of vascular tone leading to decreased CBF. This reduction in blood flow can be sufficiently severe to reach ischemic levels and may contribute to secondary injury following TBI. Experimentally, this phenomenon is associated with reduced regional nitric oxide production and restoration of nitric oxide levels leads to normalization of blood flow and reduced cortical contusion volume. It is possible that vasoinhibin activation which directly inhibits endothelial nitric oxide synthase activity may contribute to this reduction in nitric oxide production. Additional studies of prolactin expression at early time points may be informative regarding this possibility.

Acknowledgement This work was performed in compliance of the Baylor College of Medicine Animal Use and Care Regulations. Funding from NIH NINDS PO-1 NS38660–01 is gratefully acknowledged.

Conflict of interest statement We declare that we have no conflict of interest.

References

1. Cherian L, Chacko G, Goodman C, Robertson CS (2003) Neuroprotective effects of L-arginine administration after cortical impact injury in rats: dose response and time window. J Pharmacol Exp Ther 304:(2)617–623
2. Clapp C et al (2006) Vasoinhibins: endogenous regulators of angiogenesis and vascular function. Trends Endocrinol Metab 17 (8):301–307
3. Clapp C, Gonzalez C, Macotela Y, Aranda J, Rivera JC, Garcia C, Guzman J, Zamorano M, Vega C, Martin C, Jeziorski MC, de la Escalera GM (2006) Vasoinhibins: a family of N-terminal prolactin fragments that inhibit angiogenesis and vascular function. Front Horm Res 35:64–73
4. Moderscheim TA et al (2007) Prolactin is involved in glial responses following a focal injury to the juvenile rat brain. Neuroscience 145(3):963–973
5. Scheepens A et al (2001) Growth hormone as a neuronal rescue factor during recovery from CNS injury. Neuroscience 104(3):677–687
6. Wang H, Wang H, Zhang W, Fuller GN (2002) Tissue microarrays: applications in neuropathology research, diagnosis, and education. Brain Pathol 12(1):95–107

Acta Neurochir Suppl (2008) 102: 391–394

Simvastatin attenuates cerebral vasospasm and improves outcomes by upregulation of PI3K/Akt pathway in a rat model of subarachnoid hemorrhage

Takashi Sugawara · Vikram Jadhav · Robert Ayer · John Zhang

Abstract
Background Cerebral vasospasm is a common sequelae of subarachonoid hemorrhage (SAH), however, the mechanism of cerebral vasospasm is still unclear. Recently, statins have been shown to have efficacy in ameliorating cerebral vasospasm. The present study investigates whether simvastatin attenuates cerebral vasospasm after subarachnoid hemorrhage (SAH) via upregulation of the PI3K/Akt pathway
Methods 47 adult male Sprague-Dawley rats were divided into 6 groups: sham-operated, SAH treated with vehicle, SAH treated with low dose simvastatin (1 mg/kg), high dose simvastatin (20 mg/kg), SAH treated with simvastatin plus the PI3K inhibitor (wortmannin), and sham-operated plus wortmannin. Simvastatin was administered intraperitoneally 30 minutes after SAH created by the standard endovascular perforation model. Histological parameters of the ipsilateral internal carotid artery (ICA - diameter, perimeter, and wall thickness) and neurological score were assessed at 24 hours.
Findings Mortality was reduced to zero in both the treated groups as compared to 20% in the vehicle-treated and 36% in the simvastatin plus wortmannin-treated groups. The decrease in ICA diameter and perimeter observed in vehicle-treated group (203.2±10.3 μm, 652.7±29.0 μm) as compared to sham (259.7±10.6, 865.4±39.5) were significantly attenuated by high-dose simvastatin (267.4±8.0, 882.4±30.0). The increase in wall thickness (vehicle 29.50±2.42 μm v/s sham 9.52±0.56 μm) was significantly attenuated by both high and low dose simvastatin (11.87±1.56, 19.75±1.40). These effects of simvastatin were blocked with the addition of wortmannin (162.7±20.6, 528.9±65.9, 29.19±1.97). High dose simvastatin improved the neurological deficits after SAH, but this was also blocked by wortmannin.
Conclusions The beneficial effects of high dose simvastatin in ameliorating cerebral vasospasm are likely mediated by upregulation of the PI3K/Akt pathway.

Keywords Subarachnoid hemorrhage · Cerebral vasospasm · Statin · PI3K

T. Sugawara · V. Jadhav · R. Ayer · J. Zhang (✉)
Department of Physiology, Loma Linda University,
11041 Campus Street, RH214,
Loma Linda, CA 92350, USA
e-mail: johnzhang3910@yahoo.com

T. Sugawara
e-mail: sugataka@nifty.com

V. Jadhav
e-mail: drjadhav@gmail.com

R. Ayer
e-mail: rayer77@yahoo.com

J. Zhang
Department of Neurosurgery,
Loma Linda University Medical Center,
11234 Anderson Street, Room 2562B,
Loma Linda, CA 92354, USA

J. Zhang
Department of Anesthesiology, Loma Linda University,
11234 Anderson Street, Room 2562B,
Loma Linda, CA 92354, USA

Introduction

Cerebral vasospasm is a common sequelae of subarachonoid hemorrhage (SAH) which may lead to secondary ischemia [3]. The mechanism of cerebral vasospasm is still unclear. Recently, statins have been shown to have efficacy in ameliorating cerebral vasospasm in experimental and clinical studies [5, 6, 9]. In this study, we investigated the involvement of one of the putative mechanisms of statin mediation of cerebral vasospasm, the PI3K/Akt/eNOS pathway, using a rat SAH endovascular perforation model.

Materials and methods

Experimental animals and groups

All experiments were approved by the Institutional Animal Care and Use Committee of Loma Linda University. Forty seven adult male Sprague-Dawley rats (Harlan, Indianapolis, IN) weighing between 250 and 350 g were divided randomly into 6 weight-matched groups: sham-operated (sham n=6), SAH treated with 10% ethanol in normal saline as vehicle (SAH n=10), SAH treated with low dose simvastatin (Calbiochem, CA, USA) (1 mg/kg) (S(1) n=7) or high dose simvastatin (20 mg/kg) (S(20) n=6), SAH treated with simvastatin plus a PI3K inhibitor (wortmannin) (w+S(20) n=14), and sham-operated plus a PI3K inhibitor (sham+w n=4).

Induction of SAH

General anesthesia was induced with ketamine (100 mg/kg i.p.) and xylazine hydrochloride (10 mg/kg i.p.) followed by atropine (0.1 mg/kg s.c.). After intubation, the animals were ventilated with an animal ventilator (Harvard Apparatus). A heating pad and a heating lamp were used to maintain rectal temperature at 36.0±0.5°C.

SAH was induced by endovascular perforation of the internal carotid artery (ICA) bifurcation with a sharpened 4—0 nylon suture as described previously, with slight modifications [1]. Shortly, after exposing the left common carotid artery (CCA), external carotid artery (ECA), and ICA, the ECA was ligated and fashioned into a stump. The suture was advanced into the ICA from the ECA stump through the common carotid bifurcation. The suture was further advanced into the intracranial ICA until resistance was felt (~15 to 18 mm from the common carotid bifurcation) and then pushed 3 mm further to perforate the ICA wall. Then the suture was withdrawn into the ECA and the ICA was reperfused. Operative procedures were the same for the sham group, except that the suture was removed once resistance was felt without puncture. The rats were allowed to recover after the incision was closed and they were housed individually until euthanization. All animals had free access to food and water.

Drug administration

Thirty minutes after the procedure, the treated groups received an intraperitoneal injection of simvastatin administered at a dose of 1 or 20 mg/kg. Sham, SAH, and sham+w groups received vehicle (1.5 ml of 10% ethanol in normal saline). The W+S(20) and sham+w groups received an intravenous injection of wortmannin at a dose of 15 μg/kg 15 minutes before endovascular perforation.

Neurological scoring

Neurological scores were evaluated at 24 hours in a blinded fashion with a modification of the scoring system reported by Garcia et al. [4]. An 18-point scoring system was used to evaluate the sensorimotor deficits.

Assessment of cerebral vasospasm

Histological parameters of the ipsilateral ICA such as diameter, perimeter and wall thickness were used for assessment of cerebral vasospasm 24 hours after surgery, as shown before [2]. 6 animals from each group excepting the sham+w group (n=4) were used for assessment. Animals presenting with mild SAH were not included for assessment of vasospasm and neurological scoring in order to exclude unaffected vessels. SAH grades (mild, moderate and severe) were categorized based on an 18 point scale corresponding to the amount of subarachnoid blood clots, and this correlated with the degree of vasospasm and neurological score (data not shown). At 24 hours after surgery, under deep anethesia the rats were transcardially perfused with 0.1 mol/L PBS (pH 7.4) followed by fixation with 10% formaline as described previously with a pressure of 60–80 mmHg for 15 minutes [7]. The whole brains were quickly exenterated and postfixed in 10% formaline followed by 30% sucrose (w/v) for 3 days. Ten-micron-thick coronal sections cut by a cryostat (Leica Microsystems LM3050S) were mounted on poly-L-lysine- coated slides (Richard Allen, Kalamazoo, MI). These sections were stained with hematoxylin and eosin (H&E). We measured the major and minor axis, perimeter and wall thickness of ipsilateral intracranial ICA using Image J freeware from the National Institutes of Health, MD. The diameter of the intracranial ICA was defined by taking the average of major and minor axes.

Statistical analysis

Statistical differences between the various groups were assessed using a 1-way ANOVA with Holm-Sidak posthoc analysis. Data are expressed as mean±SEM. A value of $P < 0.05$ was considered statistically significant

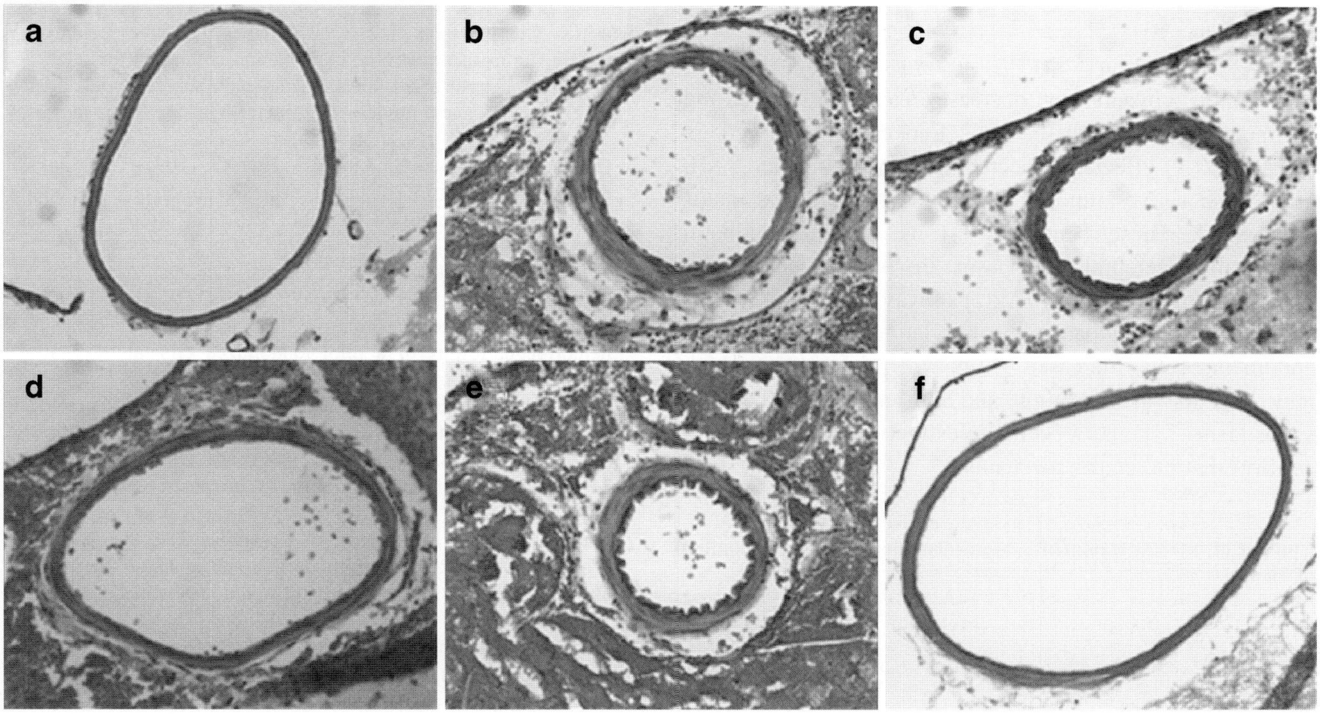

Fig. 1 Cerebral vasospasm. A-F shows the representative cross-sections of ipsilateral IC from each group, sham, SAH treated with vehicle, SAH treated with low dose simvastatin, SAH treated with high dose simvastatin, SAH treated with high dose simvastatin plus wortmannin, and sham treated with wortmannin respectively

Results

Mortality was reduced to zero in both the treated groups as compared to 20% in the vehicle-treated and 36% in the simvastatin plus wortmannin-treated groups. There was no mortality in the group treated with wortmannin alone. The decrease in ICA diameter and perimeter observed in the vehicle-treated group (203.2±10.3 μm, 652.7±29.0 μm) as compared to sham (259.7±10.6, 865.4±39.5) were significantly attenuated by high-dose simvastatin (267.4±8.0, 882.4±30.0). The increase in wall thickness (vehicle 29.50±2.42 μm v/s sham 9.52±0.56 μm) was significantly attenuated by both high and low dose simvastatin (11.87±1.56, 19.75±1.40). These effects of simvastatin were blocked with the addition of wortmannin (162.7±20.6, 528.9±65.9, 29.19±1.97). High dose simvastatin improved the neurological deficits observed after SAH and this was also blocked with the addition of wortmannin. The group treated with wortmannin alone did not have significantly different morphological parameters and neurological scores as compared to the sham group (Fig. 1 and Table 1).

Table 1 Cerebral vasospasm.and neurological score

–	Cerebral Vasospasm			Neurological Score
	Diameter (μm)	Perimeter (μm)	Thickness (μm)	
Sham	259.7±10.6	865.4±39.5	9.52±0.56	18.00±0.00
SAH	203.2±10.3 *	652.7±29.0 **	29.50±2.42 **	14.33±1.43 *
S(1)	186.3±22.9 **	610.6±69.0 **	19.75±1.56 **	15.50±1.14
S(20)	267.4±8.0 #	882.4±30.0 ##	11.87±1.40 ##	17.66±0.21 #
W+S(20)	162.7±20.6 **	528.9±65.9 **	29.19±1.97 **	12.33±1.17 **
Sham+W	261.9±29.1 #	855.1±86.3 #	11.76±0.86 ##	18.00±0.00 #

Table shows morphological parameters i.e ICAl diameter, perimeter, and wall thickness from each group. High dose simvastatin prevented the decrease in vessel diameter and perimeter and the increase in vessel wall thickening and also improved neurological deficits after SAH. All these effects of high dose simvastatin were obliterated in the presence of wortmannin. Sham: sham, SAH: SAH treated with vehicle, S(1): SAH treated with low dose simvastatin, S(20): SAH treated with high dose simvastatin, W+S(20): SAH treated with high dose simvastatin plus wortmannin, and Sham+W: sham treated with wortmannin respectively. *: vs sham $p<0.05$, **: vs sham $p<0.01$, #: vs SAH $p<0.05$, ##: vs SAH $p<0.01$

Discussion

Simvastatin prevented the decrease in vessel diameter and perimeter and the vessel wall thickening observed in arteries (ICA) affected by cerebral vasospasm after SAH. High dose simvastatin also improved neurological status after SAH. These beneficial effects of simvastatin in ameliorating cerebral vasospasm are likely mediated by upregulation of the PI3K/Akt pathway. This pathway is suggested to upregulate phosphorylation of eNOS in the endothelial cells. Increased eNOS will produce NO which may lead to vasorelaxation and thus ameliorate cerebral vasospasm [8]. However, investigation is needed to confirm this molecular mechanism.

Conflict of interest statement We declare that we have no conflict of interest.

References

1. Bederson JB, Germano IM, Guarino L (1995) Cortical blood flow and cerebral perfusion pressure in a new noncraniotomy model of subarachnoid hemorrhage in the rat. Stroke 26:1086–1091
2. Cahill J, Calvert JW, Solaroglu I, Zhang JH (2006) Vasospasm and p53-induced apoptosis in an experimental model of subarachnoid hemorrhage. Stroke 37:1868–1874
3. de Oliveira JG, Beck J, Ulrich C, Rathert J, Raabe A, Seifert V (2007) Comparison between clipping and coiling on the incidence of cerebral vasospasm after aneurysmal subarachnoid hemorrhage: a systematic review and meta-analysis. Neurosurg Rev 30:22–30
4. Garcia JH, Wagner S, Liu KF, Hu XJ (1995) Neurological deficit and extent of neuronal necrosis attributable to middle cerebral artery occlusion in rats. Statistical validation. Stroke 26:627–634
5. Lynch JR, Wang H, McGirt MJ, Floyd J, Friedman AH, Coon AL, Blessing R, Alexander MJ, Graffagnino C, Warner DS (2005) Simvastatin reduces vasospasm after aneurysmal subarachnoid hemorrhage: results of a pilot randomized clinical trial. Stroke 36:2024–2026
6. McGirt MJ, Pradilla G, Legnani FG, Thai QA, Recinos PF, Tamargo RJ, Clatterbuck RE (2006) Systemic administration of simvastatin after the onset of experimental subarachnoid hemorrhage attenuates cerebral vasospasm. Neurosurgery 58:945–951
7. Parra A, McGirt MJ, Sheng H, Laskowitz DT, Pearlstein RD, Warner DS (2002) Mouse model of subarachnoid hemorrhage associated cerebral vasospasm: methodological analysis. Neurol Res 24:510–516
8. Santhanam AV, Smith LA, Akiyama M, Rosales AG, Bailey KR, Katusic ZS (2005) Role of endothelial NO synthase phosphorylation in cerebrovascular protective effect of recombinant erythropoietin during subarachnoid hemorrhage-induced cerebral vasospasm. Stroke 36:2731–2737
9. Tseng MY, Czosnyka M, Richards H, Pickard JD, Kirkpatrick PJ (2005) Effects of acute treatment with pravastatin on cerebral vasospasm, autoregulation, and delayed ischemic deficits after aneurysmal subarachnoid hemorrhage: a phase II randomized placebo-controlled trial. Stroke 36:1627–1632

Acta Neurochir Suppl (2008) 102: 395–399

HIF-1 alpha inhibition ameliorates neonatal brain damage after hypoxic-ischemic injury

Wanqiu Chen · Vikram Jadhav · Jiping Tang · John H. Zhang

Abstract

Background Hypoxia-inducible-factor-1alpha (HIF-1α) has been considered as a regulator of both prosurvival and prodeath pathways in the nervous system. This study was designed to elucidate the role of HIF-1α in neonatal hypoxia-ischemia (HI) brain injury.

Methods 2-methoxyestradiol (2ME2), a HIF-1α inhibitor, was tested at different dosages (1.5, 15 and 150 mg/kg) and a therapeutic window was tested by administrating 2-methoxyestradiol (15 mg/kg) immediately or 3 hours after the induction of a hypoxic ischemic injury. Infarct size using TTC staining and brain edema were measured at 48 hours post hypoxia-ischemia. Blood-brain barrier (BBB) permeability was examined by IgG staining. Vascular endothelial growth factor (VEGF) and HIF-1α expression and distribution were studied by immunohistochemistry and western blotting analysis.

Findings 2ME2 exhibited dose-dependent neuroprotection by decreasing infarct volume and attenuating brain edema. 2ME2 also attenuated BBB disruption, and decreased HIF-1α and vascular endothelial growth factor (VEGF) expression. The neuroprotection, however, was lost when 2ME2 was administered 3 hours after neonatal HI.

Conclusion The study shows that the acute inhibition of HIF-1α is neuroprotective in neonatal hypoxic-ischemic injury by preserving BBB integrity and reducing brain edema.

Keywords Hypoxia inducible factor · Neonatal hypoxia ischemia · Neuroprotection · Vascular endothelial growth factor (VEGF)

Introduction

Perinatal hypoxic-ischemic brain injury is a major cause of morbidity and mortality in infants and children, with an incidence of 2–9 per 1000 babies [11] and it could lead to neurological deficits, cerebral palsy and epilepsy [7]. However, clear evidence-based strategies for treatment are not available now [7].

Hypoxia-inducible factor 1α (HIF-1α) is a heterodimeric transcriptional factor and it regulates genes which are important in tissue survival, such as vascular endothelial growth factor (VEGF) and erythropoietin (EPO). There are conflicting reports regarding the effects of HIF-1α in the brain after ischemic injury [2, 6]. 2-Methoxyestradiol (2ME2) is a naturally occurring derivative of estradiol and an inhibitor of HIF-1α [8]. It has been shown to effectively inhibit tumor growth and angiogenesis in vivo and inhibits several stages of angiogenesis besides endothelial cell proliferation [8].

The present study was designed to clarify the role of HIF-1α in the neonatal brain in response to hypoxia/ischemia. We evaluated the effects of 2ME2 in the established Rice-Vannucci neonatal hypoxia/ischemia rat model and tested the hypothesis that the acute inhibition of HIF-1α provides neuroprotection against neonatal HI.

W. Chen · V. Jadhav · J. H. Zhang (✉)
Department of Physiology and Pharmacology,
Loma Linda University Medical Center,
Risley Hall, Room 214,
Loma Linda, CA 92354, USA
e-mail: johnzhang3910@yahoo.com

J. Tang
Department of Physiology and Pharmacology,
Loma Linda University Medical Center,
Risley Hall, Room 219,
Loma Linda, CA 92354, USA

Materials and methods

Animal modeling

The experimental protocol was approved by the Institutional Committee for Animal Care and Handling. A modified Rice-Vannucci model [9] was adopted as follows: Under anesthesia (3% isoflurane), the right common carotid artery of seven-day-old (P7) Sprague-Dawley rat pup was identified, exposed, and permanently ligated. After recovering for 2 hours, the pups were exposed to hypoxia (8% oxygen balanced with nitrogen, 37°C) for 2 h. After hypoxia, the animals returned to their dams and the ambient temperature was maintained at 37°C for 24 h. Sham animals underwent anesthesia and the common carotid artery was exposed without ligation and hypoxia.

Drug administration

2ME2 (Sigma-Aldrich Corp, MO), an HIF-1α inhibitor, was administered at three dosages of 1.5 mg/kg, 15 mg/kg and 150 mg/kg. The drug was dissolved in DMSO and further diluted in PBS [13] to a final volume of 100 μl, and was administered by intraperitoneal injection immediately after HI. To assess the window period for effective treatment, 2ME2 (15 mg/kg) was also administered at 3 h after HI. The vehicle-treated HI group received DMSO diluted with PBS at the same volume as the treatment group.

Brain water content

At 48 hours after HI, brain tissue was removed for brain water content analysis and divided into the following three parts: right hemisphere, left hemisphere, and cerebellum. The hemispheres were separated by a midline incision and weighed on a high precision balance (Denver Instrument, sensitivity ± 0.001 g) immediately after removal (wet weight) and again after drying in an oven at 105°C for 24 hours as described by others [12]. The percentage of water content was calculated as $[(\text{wet weight} - \text{dry weight})/\text{wet weight}] \times 100\%$.

Histology and immunohistochemistry

At 24 h post-HI, animals were perfused with PBS followed by 4% paraformaldehyde. Paraffin-embedded brains were sectioned into 10-μm-thick slices by cryostat (CM3050S; Leica Microsystems). Immunohistochemistry was performed [3] for HIF-1α and VEGF (Santa Cruz Biotechnology). IgG staining used for detecting BBB breakdown was performed at 24 h after HI and conjugated goat anti-rat IgG-biotin (sc-2041; Santa Cruz Biotechnology) was used [10].

Infarct volume measurement

2,3,5-triphenyltetrazolium chloride monohydrate (TTC) staining was used to measure infarct volume as previously described [14]. Briefly, at 48 h after HI the brains were removed and sectioned into 2 mm slices, then immersed into 2% TTC solution at 37°C for 5 minutes, followed by 10% formaldehyde. The infarct volume was traced and analyzed by Image J software (NIH), version 1.32.

Western blotting

Western Blot analysis was performed as described previously [3]. Animals were euthanized at 24 h after HI (n=8 for each group). The primary antibodies used were rabbit polyclonal anti-HIF-1α and rabbit polyclonal anti-VEGF. Immunoblots were then probed with an ECL kit and visualized with the imagine system (Bio-Rad, Versa Doc, model 4000). The data were analyzed by the software Quantity one 4.6.1 (Bio-Rad).

Statistics

Data were expressed as mean ± SD. Statistical differences between more than two groups were analyzed by using 1-way ANOVA followed by Tukey post-hoc analysis. A P value of <0.05 was considered statistically significant.

Results

HIF-1α inhibition decreased infarct volume

HIF-1α inhibition by administration of 2ME2 immediately following induction of HI reduced infarct volume in a dose-dependent manner. The mean infarct volume was significantly decreased at 15 mg/kg dosage (17.8±2.4%) compared with vehicle-treated HI group (30.4±1.2%), but not at 1.5 mg/kg dosage (30.4±2.4%). At the high dosage of 150 mg/kg there was a maximal decrease in infarct volume (10.3±3.4%); however, it was accompanied with a high mortality (6 of 10 pups died). The other groups had zero mortality. 15 mg/kg was considered appropriate dosage for rest of the experiments and for elucidating molecular mechanisms.

To evaluate the window period of treatment, 2ME2 was administered at 15 mg/kg dosage immediately or at 3 hours after HI and infarct volume was measured at 48 hours after HI. Administration of 2ME2 3 hours after HI did not decrease infarct volume as compared to controls (27.8± 2.9% vs. 30.4±1.2%). Administration of 2ME2 immediately after HI was used for the rest experiments for further exploring the mechanisms.

HIF-1α inhibition decreased BBB disruption and brain edema

Brain edema as indicated by increased brain water content was seen at 48 hours after HI. In the vehicle-treated HI group, the water content in the hemisphere (ipsilateral to the carotid ligation) increased significantly compared with both the normal control groups (vehicle-treated vs. control groups = 90.5±0.47% vs. 87.2±0.19%, $P<0.001$) and sham groups (vehicle-treated vs. sham = 90.5±0.47% vs. 87.0±0.11%, $P<0.001$). In the 2ME2-treated group, with dosage of 15 mg/kg and administered immediately after HI, the mean water content of the ipsilateral hemispheres differed significantly from the vehicle-treated HI group (2ME2-treated vs. vehicle-treated = 88.0±0.28% vs. 90.5± 0.47%, $P<0.001$). The contralateral hemisphere and cerebellum did not show significant changes in brain water content (Fig. 1a). The ratio of water content in the ipsilateral hemisphere compared to the contralateral hemisphere of each group also demonstrated that there is a significant decrease in 2ME2-treated group compared with the vehicle-treated HI group (Fig. 1b).

IgG staining demonstrated BBB leakage in the ipsilateral cortex and hippocampus area in the vehicle-treated HI group. It represented by the brown stains which is IgG passing through the disrupted BBB and penetrating into the brain parenchyma. The 2ME2-treated group demonstrated a smaller IgG-stained region as compared with the vehicle-treated HI group (Fig. 1c).

2ME2 inhibits HIF-1α and VEGF expression

Immunohistochemical analysis of brain sections revealed that both HIF-1α and VEGF were extensively upregulated

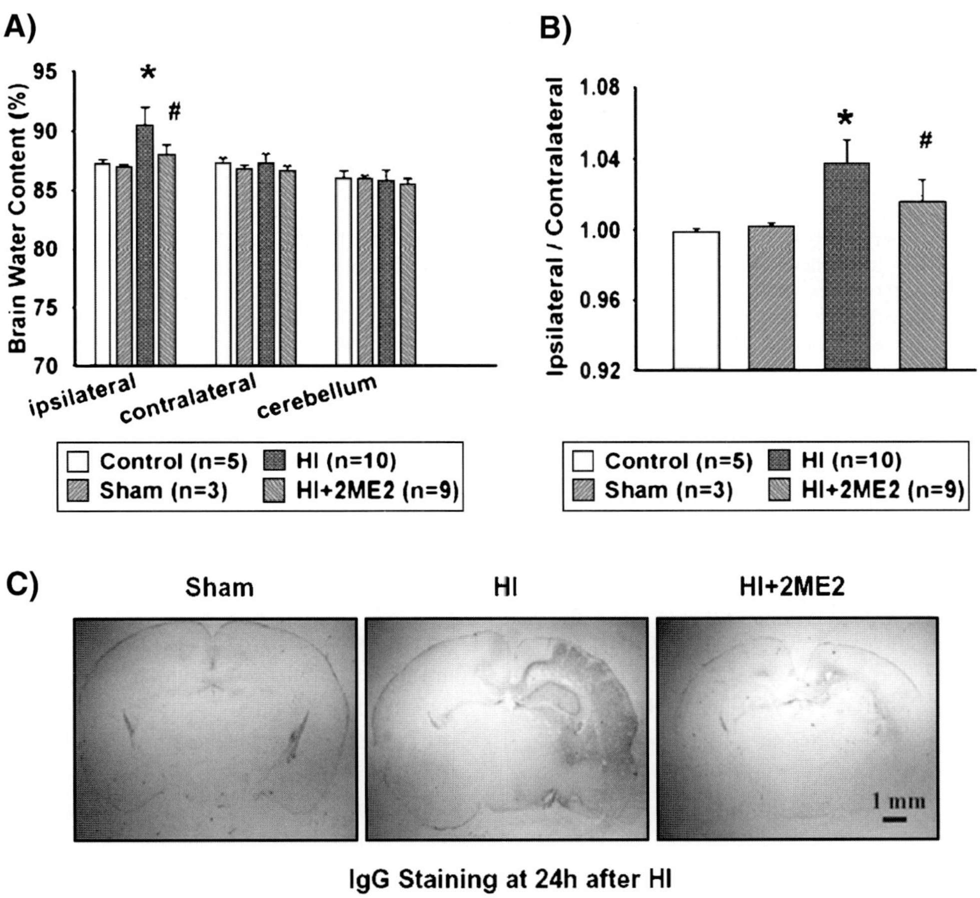

Fig. 1 2ME2 attenuated brain edema after neonatal HI. **(a)** Quantification of brain water content in the cerebellum, ipsilateral and contralateral brain hemisphere 48 h after HI. Compared with the sham and normal control groups, the brain water content was markedly increased in the HI group (*$P<0.001$ vs. control and sham. The control group contained normal pups without any surgery or treatment. Vertical bars indicate standard deviation). 2ME2 treatment significantly decreased ipsilateral hemisphere water content as compared with the vehicle-treated HI group (#$P<0.001$ vs. HI). There was no statistical difference among the groups in contralateral hemisphere and cerebellum water content. **(b)** Brain water content was expressed as the ratio of ipsilateral/contralateral hemisphere water content (*$P<0.001$, versus control and sham; # $P<0.001$, versus HI; vertical bars indicate standard deviation). (C) IgG staining in sections of the rat brain of sham, HI and HI+2ME2 groups, respectively. There is no staining in the sham section. A dense IgG staining (brown stain) was seen in the ipsilateral cortex and hippocampus in the HI group, and it was reduced in HI+2ME2 group

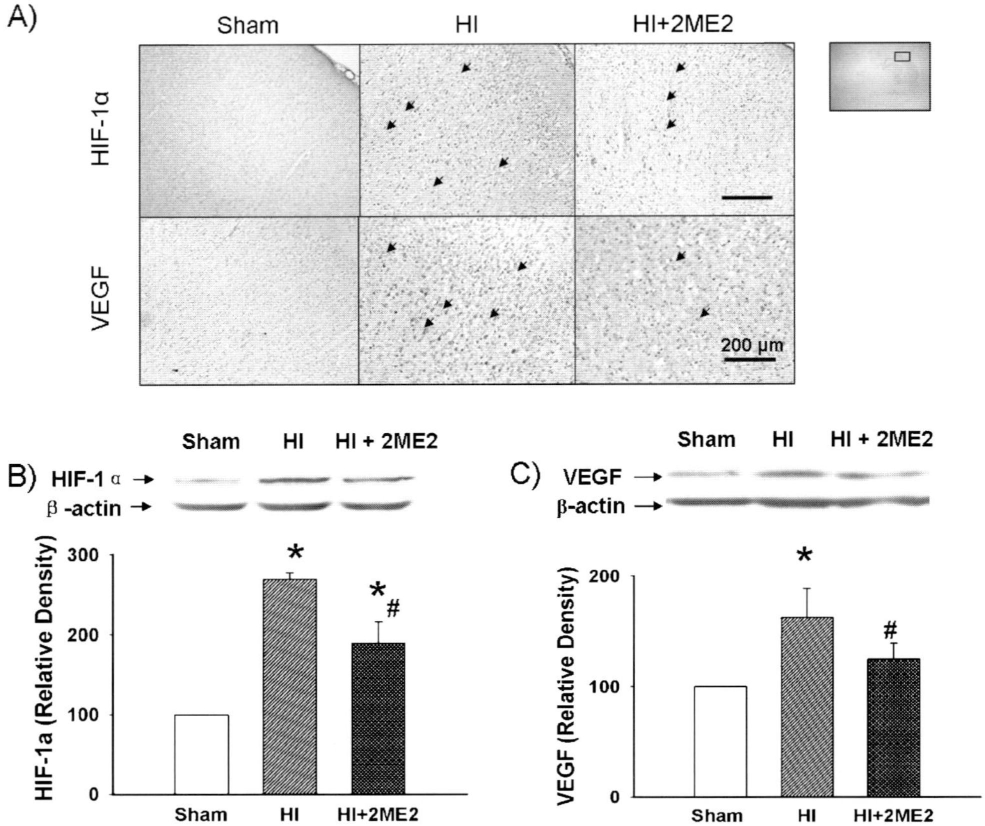

Fig. 2 Immunohistochemistry and Western blot for HIF-1α and VEGF expression after hypoxia-ischemia injury. **(a)** Immunostaining for HIF-1α and VEGF in the cortex of the ipsilateral hemisphere at 24 h after HI. Compared with the vehicle-treated HI group, HIF-1α and VEGF expression was reduced in the 2ME2 group. Arrowheads indicate cells that are positive for HIF-1α and VEGF. Scale bar represents 200 μm. **(b and c)** Representative Western blot analysis showed that HIF-1α and VEGF (with β-actin as a loading control) were expressed in the ipsilateral hemisphere at 24 h after HI respectively. Quantification of the Western blot analysis showed increased HIF-1α and VEGF in the vehicle-treated HI group compared with both the sham group and the 2ME2-treated group (*$P<0.001$, versus sham; #$P<0.05$, versus HI; vertical bars indicate standard deviation)

in the cerebral cortex region of the ipsilateral hemisphere at 24 h after HI, which was seen as brown granular deposits within cells. After 2ME2 treatment, there were fewer cells with condensed staining of HIF-1α and VEGF (Fig. 2a). Western blot analysis showed an upregulation of HIF-1α 24 h after HI by about 2 fold, and it was significantly inhibited by 2ME2 (Fig. 2b, P<0.05, HI+2ME2 vs. HI, ANOVA). Similarly, VEGF was upregulated at 24 h after HI (Fig. 2c, $P<0.001$, HI vs. sham, ANOVA) and it was significantly decreased in the 2ME2-treated group (Fig. 2b, P<0.05, HI+2ME2 vs. HI, ANOVA).

Discussion

In the present study, we evaluated the therapeutic effects of acute inhibition of HIF-1α using 2ME2 after HI brain injury in neonatal rats. The results indicated that HIF-1α inhibition attenuated the disruption of BBB, reduced brain edema and decreased infarct volume.

HIF-1α inhibition using different dosages of 2ME2 was seen to reduce the infarct volume. Thus, pilot experiments indicated that 15 mg/kg was an appropriate dose. Then we evaluated the therapeutic window of 2ME2 treatment. Our previous study showed that HIF-1α protein levels peaked during the initial reperfusion phase (from 0 to 4 h). They then steadily declined as reperfusion persisted (from 4 to 24 h), and remained elevated up to 24 h after the insult [3]. Administration of 2ME2 3 hours after induction of HI was not effective, possibly due to the early surge of HIF-1α levels. HIF-1α is known to regulate pro-apoptotic members such as p53 and BNIP3 in cerebral ischemic injuries [1, 4, 5]. The reduction of infarct volume after 2ME2 treatment may be due to the inhibition of pro-apoptotic pathways. Future studies, however, are required to evaluate if anti-apoptotic mechanisms are involved in the neuroprotective effects of 2ME2 in neonatal HI.

2ME2 may provide neuroprotection by inhibiting HIF-1α and some of its target genes such as VEGF. VEGF can promote angiogenesis and restore cerebral microvascular circulation. However, VEGF is a potent vascular permeability factor which could enhance the BBB leakage especially in the early stages after HI injury [15]. It has been suggested that the inhibition of VEGF at the acute stage of stroke may be beneficial. In this study, reduced IgG staining and brain water content in the 2ME2-treated group indicated a decreased BBB breakdown and less brain edema respectively. VEGF expression was increased after HI and it was attenuated in the 2ME2-treated group, suggesting that the reduced BBB permeability and brain edema may be involved in the inhibition of VEGF in the acute stage of stroke.

HIF-1α has a dual role as a regulator of both prosurvival and prodeath pathways in the nervous system. Some studies have shown that HIF-1α induction by hypoxic preconditioning or HIF-1α inducer such as deferoxamine is required for neuronal protection [2, 15]. Other studies have also shown pro-death responses of HIF-1α through p53 and BNIP3 [1, 5]. In vivo studies using HIF-1α knockout mice showed that the conditional brain depletion of HIF-1α is neuroprotective in acute hypoxia injury [6]. These contradictory effects of HIF-1α can be attributed to the duration and severity of ischemic injury. It has been suggested that mild hypoxia induces adaptive gene expression, whereas under severe or sustained hypoxia HIF-1α can promote apoptotic cell death.

In summary, acute HIF-1α inhibition by 2ME2 has a neuroprotective effect by preserving BBB integrity, ameliorating brain edema and reducing infarct volume after neonatal hypoxia-ischemia injury. This therapeutic effect of HIF-1α inhibition is likely associated with the inhibition of VEGF.

Acknowledgments This study was partially supported by grants from the NIH HD43120, NS43338, and NS54685 to J.H. Zhang.

Conflict of interest statement We declare that we have no conflict of interest.

References

1. Althaus J, Bernaudin M, Petit E, Toutain J, Touzani O, Rami A (2006) Expression of the gene encoding the pro-apoptotic BNIP3 protein and stimulation of hypoxia-inducible factor-1alpha (HIF-1alpha) protein following focal cerebral ischemia in rats. Neurochem Int 48:687–695
2. Bergeron M, Gidday JM, Yu AY, Semenza GL, Ferriero DM, Sharp FR (2000) Role of hypoxia-inducible factor-1 in hypoxia-induced ischemic tolerance in neonatal rat brain. Ann Neurol 48:285–296
3. Calvert JW, Cahill J, Yamaguchi-Okada M, Zhang JH (2006) Oxygen treatment after experimental hypoxia-ischemia in neonatal rats alters the expression of HIF-1alpha and its downstream target genes. J Appl Physiol 101:853–865
4. Carmeliet P, Dor Y, Herbert JM, Fukumura D, Brusselmans K, Dewerchin M, Neeman M, Bono F, Abramovitch R, Maxwell P, Koch CJ, Ratcliffe P, Moons L, Jain RK, Collen D, Keshert E (1998) Role of HIF-1alpha in hypoxia-mediated apoptosis, cell proliferation and tumour angiogenesis. Nature 394:485–490
5. Halterman MW, Miller CC, Federoff HJ (1999) Hypoxia-inducible factor-1alpha mediates hypoxia-induced delayed neuronal death that involves p53. J Neurosci 19:6818–6824
6. Helton R, Cui J, Scheel JR, Ellison JA, Ames C, Gibson C, Blouw B, Ouyang L, Dragatsis I, Zeitlin S, Johnson RS, Lipton SA, Barlow C (2005) Brain-specific knock-out of hypoxia-inducible factor-1alpha reduces rather than increases hypoxic-ischemic damage. J Neurosci 25:4099–4107
7. Nelson KB, Lynch JK (2004) Stroke in newborn infants. Lancet Neurol 3:150–158
8. Pribluda VS, Gubish ER, Lavallee TM, Treston A, Swartz GM, Green SJ (2000) 2-methoxyestradiol: an endogenous antiangiogenic and antiproliferative drug candidate. Cancer Metastasis Rev 19:173–179
9. Rice JE, Vannucci RC, Brierley JB (1981) The influence of immaturity on hypoxic-ischemic brain damage in the rat. Ann Neurol 9:131–141
10. Tsubokawa T, Yamaguchi-Okada M, Calvert JW, Solaroglu I, Shimamura N, Yata K, Zhang JH (2006) Neurovascular and neuronal protection by E64d after focal cerebral ischemia in rats. J Neurosci Res 84(4):832–840 Sep
11. Vannucci RC (1990) Experimental biology of cerebral hypoxia-ischemia: relation to perinatal brain damage. Pediatr Res 27:317–326
12. Xi G, Hua Y, Keep RF, Younger JG, Hoff JT (2002) Brain edema after intracerebral hemorrhage: the effects of systemic complement depletion. Acta Neurochir Suppl 81:253–256
13. Yan J, Chen C, Lei J, Yang L, Wang K, Liu J, Zhou C (2006) 2-methoxyestradiol reduces cerebral vasospasm after 48 hours of experimental subarachnoid hemorrhage in rats. Exp Neurol 202:348–356
14. Yin D, Zhou C, Kusaka I, Calvert JW, Parent AD, Nanda A, Zhang JH (2003) Inhibition of apoptosis by hyperbaric oxygen in a rat focal cerebral ischemic model. J Cereb Blood Flow Metab 23:855–864
15. Zhang ZG, Zhang L, Jiang Q, Zhang R, Davies K, Powers C, Bruggen N, Chopp M (2000) VEGF enhances angiogenesis and promotes blood-brain barrier leakage in the ischemic brain. J Clin Invest 106:829–838

Acta Neurochir Suppl (2008) 102: 401–404

Simvastatin treatment in surgically induced brain injury in rats

Steve Lee · Vikram Jadhav · Tim Lekic · Amy Hyong · Martin Allard · Gary Stier · Robert Martin · John Zhang

Abstract

Background HMG-CoA reductase inhibitors (Statins) have been shown to reduce blood brain barrier (BBB) disruption and improve neurologic outcome in cerebrovascular disorders. Brain injury due to neurosurgical procedures can lead to post-operative complications such as brain edema and altered neurologic function. The objective of this study was to evaluate whether simvastatin reduces brain edema by preventing BBB disruption and improves neurologic status after surgically-induced brain injury (SBI).

Methods Animals were pretreated for seven days with vehicle or simvastatin i.p. daily, after which they underwent SBI. Neurologic evaluation was assessed at 24 hours post-SBI and the animals were sacrificed for brain water content calculation and BBB evaluation.

Findings Brain water content was significantly increased in the right frontal lobe in all SBI groups as compared to the left frontal lobe. There was no significant difference in brain water content in the right frontal lobe between simvastatin and vehicle treated groups. Evans blue testing did not show a significant difference in disruption of the BBB between groups. Neurologic scores were not significantly different.

Conclusions Simvastatin did not reduce brain water content, protect the BBB, or improve neurologic scores after SBI.

Keywords Simvastatin · Surgically induced brain injury · Neuroprotection · Brain edema

S. Lee · M. Allard · G. Stier · R. Martin
Department of Anesthesiology,
Loma Linda University Medical Center,
11234 Anderson Street, Room 2534,
Loma Linda, CA 92354, USA

V. Jadhav · T. Lekic · A. Hyong
Department of Physiology, Loma Linda University,
11234 Anderson Street, Room 2562B,
Loma Linda, CA 92354, USA

J. Zhang (✉)
Division of Neurosurgery,
Loma Linda University Medical Center,
11234 Anderson Street, Room 2562B,
Loma Linda, CA 92354, USA
e-mail: johnzhang3910@yahoo.com

Introduction

Surgical procedures inevitably lead to inflammation from the use of electrocautery, scalpel blades, and retraction. Tissue at the edge of the surgical resection site is especially vulnerable to neuronal cell death [4]. Tissue inflammation leads to cascades which cause edema and consequential volumetric expansion. The problem encountered with cerebral inflammation is the limited intracranial space in which it is confined. Current therapies for surgically induced brain injury, such as steroids and diuretics, are relatively non-specific and focus mostly on reducing the post-operative edema that has already transpired [8]. Therefore, discovering direct anti-inflammatory mediators could result in more effective reduction in brain edema, improve neurologic outcomes, and could ultimately reduce post-operative morbidity and mortality.

Statins, or 3-hydroxy-3-methyl-glutaryl-coenzyme A (HMG-CoA) reductase inhibitors, have recently become the focus of numerous clinical trials. Outside of their cholesterol lowering effects, this class of drugs has been shown to reduce the incidence of stroke independent of cholesterol levels [10, 11]. This finding has led to

Table 1 Brain water content at 24 hours post surgically induced brain injury

Experimental group	n	Right frontal lobe	Left frontal lobe
SBI + Vehicle	7	81.00±0.52%*	79.66±0.18%
SBI + Statin 1 mg/kg	8	81.29±0.34%*	79.88±0.10%
SBI + Statin 10 mg/kg	6	81.96±0.31%*	80.21±0.28%

SBI caused a significant increase in brain water content in the ipsilateral Right Frontal Lobe when compared to the contralateral, nonsurgical Left Frontal Lobe. There was no significant difference between Vehicle and Statin treatment groups

* Indicates significance versus Left Frontal Lobe, $p<0.05$ ANOVA

investigation of the various pleiotropic effects of statin medications. Statins have been found to induce angiogenesis and neurogenesis, improve endothelial function, increase nitric oxide bioavailability, and exert antioxidant and anti-inflammatory properties [1, 2]. Statins have also been found to protect the blood brain barrier (BBB) [7]. Ifergan et al. [5] found simvastatin and lovastatin reduced BBB permeability and restricted leukocyte migration.

Given these qualities, the objective of this study was to evaluate whether simvastatin pretreatment reduces brain edema by preventing BBB disruption and improves neurologic status after surgically-induced brain injury (SBI).

Materials and methods

36 male Sprague Dawley rats weighing between 275 and 325 g were divided randomly into four groups: sham surgery (n=4), vehicle (n=13), simvastatin 1 mg/kg (n=8), and 10 mg/kg (n=11). As a comparison, 40 mg per day is the maximum dose given in humans. Animals were pretreated with vehicle or simvastatin intraperitoneal injections daily over 7 days after which they were subjected to SBI. The rodent SBI model was used as previously described [6]. Briefly, following anesthesia with ketamine (100 mg/kg) and xylazine (10 mg/kg) i.p, a 3 mm by 3 mm cranial window was drilled 1 mm anterior and 2 mm lateral to the coronal and sagittal sutures respectively. The dura was incised and reflected to expose the underlying right frontal lobe. Using a flat blade (6 mm × 1.5 mm), two incisions were made along the sagittal and coronal planes leading away from the bregma and extending to the edge of the craniotomy window. The depth of the incisions extended to the base of the skull. The weight of the sectioned brain was not significantly different between groups. Hemostasis was achieved by temporary intraoperative packing with sterile gauze and saline irrigation. Electrocautery was utilized for 2 seconds along the medial coronal and posterior sagittal borders at an RF power level consistent with the coagulation setting used in the operating room. Sham surgery included only craniotomy and replacement of the bone flap without any dural incisions. Core temperature was monitored using a rectal probe and was maintained at 37±0.5°C.

At 24 hours post-SBI, animals underwent behavior tests based on the modified Garcia score [3] and vibrissae-stimulated forelimb placement test [9] 24 hours post surgery, after which they were sacrificed for brain water content and BBB analysis using Evans Blue. The brains were sectioned into left frontal, right frontal, left parietal, right parietal, brain stem and cerebellum sections on ice. Each section was weighed immediately (wet weight) and weighed again after drying in an oven at 105°C for 48 hours (dry weight) [12]. The percent brain water content was calculated as $[(\text{wet weight} - \text{dry weight})/\text{wet weight}] \times 100\%$.

To quantitatively assess BBB compromise, Evans Blue was administered through a left femoral cannula 24 hours post-SBI. Animals were then sacrificed and brains were sectioned as described above. The right frontal lobe was homogenized and processed for Evans Blue dye extravasation under a spectrophotometer. Significance was set at $p< 0.05$. This protocol was evaluated and approved by the Institutional Animal Care and Use Committee at Loma Linda University, Loma Linda, CA.

Results

Brain water content

Surgically induced brain injury caused brain water content to significantly increase in the right frontal lobe in all three SBI groups (Vehicle: 81.64±0.52%; Simvastatin: 1 mg/kg: 81.29±0.34%; 10 mg/kg: 81.96±0.31%; Table 1) compared to the left frontal lobe (Vehicle: 79.66±0.18%; Simvastatin: 1 mg/kg: 79.88±0.10%; 10 mg/kg: 80.2±0.28%). The simvastatin treated groups did not show a significant

Table 2 Blood brain barrier analysis of the Right Frontal Lobe using Evans Blue

Experimental group	n	ug/mL Evans Blue/g brain
Sham Surgery	4	0.748±0.402
SBI + Vehicle	6	3.877±0.402*
SBI + Statin 10 mg/kg	5	3.317±0.815*

SBI caused a significant increase in Evans Blue extravasation when compared to Sham surgery. There was no significant difference between Vehicle and Statin treated groups in protecting the BBB

* Indicates significance versus Sham Surgery, $p<0.05$ ANOVA

difference in brain water content in the right frontal lobe when compared to vehicle.

BBB evaluation using Evans Blue

SBI caused a significant amount of Evans blue dye leakage when compared to sham surgery (Sham: 0.77±0.39 ug/mL/g; Vehicle 3.88±0.40 ug/mL/g; Table 2); however, simvastatin treatment did not show a significant difference in disruption of the BBB when compared to vehicle (Simvastatin 10 mg/kg: 3.32±0.81 ug/mL/g).

Neurologic testing

There was no significant difference in the modified Garcia score (maximum score = 18) between simvastatin and vehicle treated groups (Sham: 18±0; Vehicle: 15.44±0.60; Simvastatin: 1 mg/kg: 16.00±0.63; 10 mg/kg: 15.37±0.50). SBI caused a significant reduction in left forelimb placement testing (maximum score = 10) when compared to sham surgery. There was no significant change in vibrissae-stimulated forelimb placement scores between treatment and vehicle treated groups.

Discussion

Despite its antioxidant and anti-inflammatory properties, simvastatin was not able to demonstrate a decrease in brain edema in the surgically induced brain injury model. In addition, based on the Evans Blue analysis, it appears that simvastatin was not able to prevent BBB compromise as had been previously reported. Despite the multiple pleiotropic effects of statins, it is difficult to distinguish which pathway, whether anti-inflammatory, antioxidant, or BBB proponent contributes greatest in reducing post-operative edema.

This surgically induced brain injury model attempts to replicate the instrumentation, resection, and electrocautery employed during a tumor resection or other neurosurgical procedure. These procedural manipulations entail removal of brain tissue, compromise of the BBB, and stimulation of inflammatory pathways that together lead to the formation of brain edema. The goal was to design a model with consistent tissue loss that leads to the reproducible inflammatory effects involved in these types of surgeries. Brain edema in the rat model peaks in the first 24 to 48 hrs and begins to decrease by 72 hours. Based on clinical experience, this rat model produces an accelerated time course as compared to humans. We acknowledge that there are differences between the brain of a rat and a human including geometry and distributions of neurons. However, numerous studies have used the rat model in evaluating inflammatory pathways.

In studying the behavioral effects of a partial frontal lobe resection, the modified Garcia score and vibrissae-stimulated forelimb placement test were employed to uncover any subtle changes in behavior caused by the surgery and ensuing edema. Given the lack of efficacy of simvastatin in reducing brain edema and BBB compromise in comparison to controls, it was not surprising to see the lack of effect of the drug in producing behavioral changes. Perhaps the short time frame of 24 hours may have contributed for the lack of difference in behavioral testing. Longer time frames could possibly produce differences, but was beyond the scope of this study.

Steroids, diuretics, and other interventions attempting to reduce and prevent the morbidity associated with brain edema, do not specifically attack the inflammatory cascade. Despite the lack of effect of simvastatin in this surgically induced brain injury model, more studies are required to elucidate the effects of other drugs that may directly address inflammatory mediators that lead to brain edema. Discovering more effective medications could potentially allow for surgeries in more anatomically confined regions and ultimately reduce post-operative complications.

Acknowledgements Funding provided by NIH Grant NS 53407 and NS 43338 and support from the Department of Anesthesiology at Loma Linda University.

Conflict of interest statement We declare that we have no conflict of interest.

References

1. Chen J, Zhang C, Jiang H, Li Y, Zhang L, Robin A (2005) Atorvastatin induction of VEGF and BDNF promotes brain plasticity after stroke in mice. J Cereb Blood Flow & Metab 25:281–290
2. Endres M, Laufs U, Liao J, Moskowitz M (2004) Targeting eNOS for stroke protection. Trends Neurosci 27(5):283–289
3. Garcia JH, Wagner S, Liu KF, Hu XJ (1995) Neurological deficit and extent of neuronal necrosis attributable to middle cerebral artery occlusion in rats. Statistical validation. Stroke 26:627–634
4. Herman M, Pozzi-Mucelli RS, Skrap M (1996) CT and MRI findings after stereotactic resection of brain lesions. Eur J Radiol 23(3):228–234
5. Ifergan I, Wosik K, Cayrol R, Kebir H, Auger C, Bernard M, Bouthillier A, Moumdjian R, Duquette P, Prat A (2006) Statins reduce human blood-brain barrier permeability and restrict leukocyte migration: relevance to multiple sclerosis. Ann Neurol 60(1):45–55
6. Jadhav V, Matchett G, Hsu FP, Zhang JH (2007) Inhibition of Src tyrosine kinase and effect on outcomes in a new in vivo model of surgically induced brain injury. J Neurosurg 106(4):680–686
7. Kalayci R, Kaya M, Elmas I, Arican N, Ahishali B, Uzun H, Bilgic B, Kucuk M, Kudat H (2005) Effects of atorvastatin on blood-brain barrier permeability during L-NAME hypertension followed by angiotensin-II in rats. Brain Res 1042(2):184–193

8. Marmarou A, Signoretti S, Fatouros P, Portella G, Aygok G, Bullock MR (2006) Predominance of cellular edema in traumatic brain swelling in patients with severe head injuries. J Neurosurg 104:720–730
9. Napieralski JA, Banks RJA, Chesselet MF (1998) Motor and somatosensory deficits following uni- and bilateral lesions of the cortex induced by aspiration or thermocoagulation in the adult rat. Exp Neurol 154:80–88
10. Stepien K, Tomaszewski M, Czuczwar S (2005) Neuroprotective properties of statins. Pharmacol Res 57:561–569
11. Vaughan CJ (2003) Prevention of stroke and dementia with statins. Effects beyond lipid lowering. Am J Cardiol 91 (Suppl):23B–29B
12. Xi G, Hua Y, Keep RF, Younger JG, Hoff JT (2002) Brain edema after intracerebral hemorrhage: the effects of systemic complement depletion. Acta Neurochir Suppl 81:253–256

Acta Neurochir Suppl (2008) 102: 405–408

3% Hypertonic saline following subarachnoid hemorrhage in rats

Steve Lee · Gary Stier · Suzzanne Marcantonio · Tim Lekic · Martin Allard · Robert Martin · John Zhang

Abstract
Background Hypertonic saline (HTS) has been proposed as a treatment after aneurysmal subarachnoid hemorrhage (SAH) to minimize ischemic brain injury due to its osmotic and rheologic properties. Although the benefits of 7.2% HTS use in brain injury have been studied, there is a paucity of data on the use of 3%HTS.
Methods We investigated whether 3%HTS can reduce brain water content and improve neurologic function after SAH in the rodent model compared to 0.9% saline solution (NS). Neurologic testing was conducted at 24 hours post-SAH prior to sacrificing animals for brain water content evaluation.
Findings There was significant potentiation of brain water content in the right hemisphere between 3%HTS and NS groups. The modified Garcia score was not significantly different between the two groups; however, the vibrissae-stimulated forelimb placement test showed significantly lower scores in the HTS group. 3%HTS does not decrease brain edema or improve neurologic deficits as compared to NS. In fact, our study showed 3%HTS potentiated brain edema and worsened neurologic deficits in the rat SAH model.
Conclusions Given the potential adverse effects of HTS therapies, including hyperchloremic acidosis, and the lack of benefit found in our study, more investigation is required to evaluate the clinical use of 3%HTS in the setting of SAH.

Keywords Hypertonic saline · Subarachnoid hemorrhage · Brain edema · Intracranial pressure

S. Lee · G. Stier · S. Marcantonio · M. Allard · R. Martin · J. Zhang
Department of Anesthesiology, Loma Linda University Medical Center, 11234 Anderson Street, Room 2534, Loma Linda, CA 92354, USA

T. Lekic · J. Zhang
Department of Physiology, Loma Linda University, Loma Linda, CA 92354, USA

J. Zhang (✉)
Division of Neurosurgery, Loma Linda University Medical Center, 11234 Anderson Street, Room 2562B, Loma Linda, CA 92354, USA
e-mail: johnzhang3910@yahoo.com

Introduction

Aneurysmal subarachnoid hemorrhage (SAH) is a devastating disease with high morbidity and mortality. As many as 40% of patients die within the first few weeks and 50% of patients die in the first 6 months. Of survivors, more than 33% develop major neurologic deficits. [7, 12, 13] These outcomes may develop as a result of acute brain damage caused by global cerebral ischemia in the context of initial aneurysmal rupture or focal ischemia in the setting of vasospasm later on. In the acute post aneurysmal SAH period, intracranial pressure (ICP) increases due to fluid expansion in the cranium from extravasation of blood [17], as well as cerebral edema secondary to the disruption of the blood brain barrier (BBB). [3, 14] These effects lead to a precipitous drop in cerebral blood flow (CBF) as increasing ICP overwhelms the systemic blood pressure, ultimately leading to brain ischemia [1].

Table 1 Brain water content in animals at 24 hours after SAH

Experimental Group	n	Right Hemisphere	Left Hemisphere
SAH+3% HTS	7	79.9±0.3% *, #	80.3±0.3%#
SAH+0.9% NS	9	79.2±0.3%	80.0±0.4%#
Sham+3% HTS	9	78.9±0.1%	78.9±0.1%
Sham+0.9% NS	8	78.8±0.1%	79.0±0.2%

*Indicates significance versus SAH+0.9% NS, $p<0.05$ ANOVA
Indicated significance versus Sham, $p<0.05$ ANOVA

Much focus has been placed on minimizing the complications associated with SAH, including vasospasm, rerupture of the aneurysm, and cerebral edema. Of interest, recent research has proposed the use of hyperosmolar agents to reduce brain edema. The rationale being that a reduction in edema leads to a decrease in ICP, thereby restoring CBF and preventing secondary ischemia due to low global cerebral perfusion. Hyperosmolar agents, which are being increasingly used in patients with traumatic brain injury, could be applied in the setting of SAH based on similar principles for reducing ICP. These agents have been theorized to control ICP by reducing brain water content and cerebral spinal fluid volume [6], as well as improving the rheologic properties of blood through the optimization of blood viscosity and enhancement of oxygen delivery [2, 9, 11].

Previous studies support the benefits of 7.5% HTS and 7.5% HTS with colloid, both nonclinical concentrations, in ameliorating acute sequelae after SAH injury in rats. [1, 19] In patients with SAH and hyponatremia, 3% hypertonic saline is frequently used clinically to correct natremic status. In a retrospective study by Suarez et al. [16], 3% NaCl-acetate was used to correct mild hyponatremia in 29 patients with SAH and was found to have no associated complications with HTS infusion, but also, interestingly, no changes on CBF. Despite its uses in the setting of hyponatremia and the growing literature supporting the theoretical advantages of nonclinical concentrations (> 3% HTS) of hyperosmolar agents in reducing ICP, no studies have addressed the effects of 3% concentration HTS on brain water content in normonatremic patients in comparison to 0.9% saline solution.

The objective of this study was to determine whether 3% hypertonic saline reduces brain water content and improves neurologic function when compared to 0.9% saline solution (NS) in the setting of SAH in rats.

Materials and methods

A total of 52 adult male Sprague Dawley rats weighing between 300 and 350 g were used and after accounting for mortality from severe SAH (35%), 33 rats remained for study analysis. The animals were randomly divided into four groups: SAH+3% HTS (n=7), SAH+0.9% NS (n=9), Sham Surgery+3% HTS (n=9), and Sham Surgery+0.9% NS (n=8). The rats were anesthetized using isoflurane, intubated, and mechanically ventilated throughout surgery and infusion. Core body temperature was monitored and maintained at 37 ± 0.5^0 C. The left femoral vein and artery were cannulated using PE 50 polyethylene tubing for arterial blood gas, hemodynamic monitoring, and infusion access. Animals underwent SAH modeling using the suture puncture method as previously described. [15] Briefly, a sharpened 4.0 nylon suture was threaded through the internal carotid artery (ICA) to the bifurcation of the ICA and middle cerebral artery for puncture and SAH. To simulate the time required for a patient to arrive to the hospital and receive testing, animals remained under anesthesia for one hour post puncture. They were then administered 3% HTS or 0.9% NS with a bolus rate of 4 mL/kg over 20 minutes followed by a maintenance rate of 0.5 mL/kg/hr over 100 minutes, for a total treatment time of 2 hours.

Neurologic function of the animals was evaluated at 24 hours post-SAH using a 27 point scoring system modified from Garcia et al. [4] and the vibrissae forelimb placement test [10] prior to sacrificing animals for brain water content evaluation by the wet-dry method. Briefly, brains were sectioned, immediately weighed, dried for 48 hours, and weighed again to calculate brain water content: [(wet weight - dry weight)/wet weight] × 100%. Significance was set at $p<0.05$. This protocol was evaluated and approved by the Institutional Animal Care and Use Committee at Loma Linda University, Loma Linda, CA.

Table 2 Vibrissae-stimulated forelimb placement test (Maximum score=10)

Experimental Group	Rt Whisker Rt Forelimb	Lt Whisker Rt Forelimb	Lt Whisker Lt Forelimb	Rt Whisker Lt Forelimb
SAH+3% HTS	0.27±0.27 *	1.20±0.49*	3.07±1.98*	3.53±2.09*
SAH+0.9% NS	5.41±1.56	5.33±1.54	8.37±1.10	7.93±1.08
Sham+3% HTS	10.00±0.00	10.00±0.00	10.00±0.00	10.00±0.00
Sham+0.9% NS	10.00±0.00	10.00±0.00	10.00±0.00	10.00±0.00

*Indicates significance versus SAH+0.9% NS, $p<0.05$ ANOVA

Results

Brain edema after SAH

SAH caused a significant increase in brain water content in both hemispheres: Left hemisphere (SAH+HTS: 80.3±0.3% vs. Sham + HTS: 78.9±0.1%; Table 1) and right hemisphere (SAH+HTS: 79.9±0.3% vs. Sham+HTS: 78.9±0.1%), respectively, in the HTS treated animals. In the NS treated animals, only the left hemisphere showed a significant increase in brain water content (SAH + NS: 80.0±0.4% vs. Sham + NS: 79.0±0.2%). Additionally, there was no significant difference in brain water content between HTS versus NS in the left hemisphere following SAH, however there was a significant potentiation of brain water content in the right hemisphere between 3% HTS and NS (SAH + HTS: 79.9±0.3% vs. SAH + NS: 79.2±0.3%).

Neurologic testing

Neurologic scoring using the modified Garcia scale was not significantly different between the HTS group and NS group. However, the vibrissae-stimulated forelimb placement test showed significantly lower scores after SAH in the HTS as compared to the NS group (Table 2).

Discussion

The goal of this research was to study whether 3% hypertonic saline, a clinically used concentration, would show benefit over maintenance fluid, 0.9% saline solution, in reducing brain water content and improving neurologic scores. Previous research had demonstrated decreases in ICP using 7.5% HTS [1, 19], a concentration not typically used clinically. No studies that we are aware of have looked at 3% HTS in normonatremic patients as a novel method of reducing cerebral edema after SAH.

This study demonstrated that 3% HTS did not reduce brain edema in the hemisphere ipsilateral to the vessel puncture and actually potentiated brain water content in the contralateral hemisphere when compared to NS. The modified Garcia score was not able to detect differences between the NS and HTS groups. The motor-sensory vibrissae-stimulated forelimb placement test showed a significant worsening in scores for the HTS group when compared to the NS and sham groups. This behavior did support the brain water content data that 3% HTS appears to be detrimental in SAH rats.

Although side effects are rare, HTS is not a benign substance and several complications have been reported in the literature. Several studies found renal damage and metabolic disturbances such as hyperchloremic acidosis in patients administered with hypertonic saline. [5, 8, 18] In addition, rapid infusion rates of HTS can lead to central pontine myelinolysis and encephalopathy. Given these potential side effects along with the current findings, the clinical use of hypertonic saline solely to reduce brain edema in normonatremic SAH patients cannot be supported. It is acknowledged that there are several limitations to this research, however given the novelty of the study, further research is needed to determine the clinical appropriateness of hypertonic saline use in the setting of aneurysmal rupture subarachnoid hemorrhage.

Acknowledgements Funding provided by NIH Grant NS 53407 and NS 43338 and support from the Department of Anesthesiology at Loma Linda University.

Conflict of interest statement We declare that we have no conflict of interest.

References

1. Bermueller C, Thal SC, Plesnila N, Schmid-Elsaesser R, Kreimeier U, Zausinger S (2006) Hypertonic fluid resuscitation from Subarachnoid hemorrhage in rats: A comparison between small volume resuscitation and mannitol. J Neurol Sci 241:73–82
2. Burke AM, Quest DO, Chien S, Cerri C (1981) The effects of mannitol on blood viscosity. J of Neurosurg 55(4):550–553
3. Doczi T (1985) The pathogenetic and prognostic significance of blood-brain barrier damage at the acute stage of aneurysmal Subarachnoid haemorrhage. Clinical and experimental studies. Acta Neurochir 77(3–4):110–132
4. Garcia JH, Wagner S, Liu KF, Hu XJ (1995) Neurological deficit and extent of neuronal necrosis attributable to middle cerebral artery occlusion in rats. Statistical validation. Stroke 26:627–634
5. Huang PP, Stucky FS, Dimick AR, Treat RC, Bessey PQ, Rue LW (1995) Hypertonic sodium resuscitation is associated with renal failure and death. Ann Surg 221(5):543–54
6. Jafar JJ, Johns LM, Mullan SF (1986) The effect of mannitol on cerebral blood flow. J Neurosurg 64(5):754–759
7. McCormick WF, Nofzinger JD (1965) Saccular Intracranial Aneurysms: An Autopsy Study. J Neurosurg 22:155–159
8. Moon PF, Kramer GC (1995) Hypertonic saline-dextran resuscitation from hemorrhagic shock induces transient mixed acidosis. Crit Care Med 23(2):323–331
9. Muizelaar JP, Lutz HA 3rd, Becker DP (1984) Effect of mannitol on ICP and CBF and correlation with pressure autoregulation in severely head-injured patients. J Neurosurg 61(4):700–706
10. Napieralski JA, Banks RJA, Chesselet MF (1998) Motor and somatosensory deficits following uni- and bilateral lesions of the cortex induced by aspiration or thermocoagulation in the adult rat. Exp Neurol 154:80–88
11. Ogden AT, Mayer SA, Connolly ES Jr. (2005) Hyperosmolar agents in neurosurgical practice: The evolving role of hypertonic saline. Neurosurgery 57(2):207–215
12. Schievink WI (1997) Intracranial aneurysms. N Engl J Med 336 (1):28–40
13. Schievink WI, Riedinger M, Jhutty TK, Simon P (2004) Racial disparities in subarachnoid hemorrhage mortality: Los Angeles County, California, 1985–1998. Neuroepidemiology 23(6):299–305
14. Scholler, K, Trinkl A, Klonotowski M, Thal SC, Plesnila N, Trabold R, Hamann GF, Schmid-Elsaesser R, Zausinger S (2007)

Characterization of microvascular basal lamina damage and blood-brain barrier dysfunction following Subarachnoid hemorrhage in rats. Brain Res 20; 1142: 237–246
15. Schwartz AY, Masago A, Sehba FA, Bederson JB (2000) Experimental models of subarachnoid hemorrhage in the rat: a refinement of the endovascular filament model. J Neurosci Methods 96(2):161–167
16. Suarez JI, Qureshi AI, Parekh PD, Razumovsky A, Tamargo RJ, Bhardwaj A, Ulatowski JA (1999) Administration of hypertonic (3%) sodium chloride/acetate in hyponatremic patients with symptomatic vasospasm following Subarachnoid hemorrhage. J Neurosurg Anesthesiol 11:178–184
17. Suarez JI, Tarr RW, Selma WR (2006) Aneurysmal Subarachnoid Hemorrhage. New Engl J of Med 354(4):387–396 Review
18. Yildizdas D, Altunbasak S, Celik U, Herguner O (2006) Hypertonic saline treatment in children with cerebral edema. Indian Pediatr 43(9):771–779
19. Zausinger S, Thal SC, Kreimeier U, Messmer K, Schmid-Elsaesser R (2004) Hypertonic fluid resuscitation from Subarachnoid hemorrhage in rats. Neurosurgery 55(3):679–686

Acta Neurochir Suppl (2008) 102: 409–413

Effect of traumatic brain injury on cognitive function in mice lacking p55 and p75 tumor necrosis factor receptors

L. Longhi · F. Ortolano · E. R. Zanier · C. Perego · N. Stocchetti · M. G. De Simoni

Abstract

Background Tumor necrosis factor (TNF)-α has been suggested to play both a deleterious and beneficial role in neurobehavioral dysfunction and recovery following traumatic brain injury (TBI). The goal of this study was to evaluate the specific role of tumor necrosis factor (TNF) receptors p55 and p75 in mediating cognitive outcome following controlled cortical impact (CCI) brain injury by comparing post-traumatic cognitive function in mice with genetically engineered deletion of the gene for either p55 (-/-) or p75 (-/-) receptors.

Method Male C57Bl/6 mice (WT, n=29), and mice genetically engineered to delete p55 TNF (p55 (-/-), n=8) or p75 TNF (p75 (-/-), n=23) receptors were used. They were anesthetized with intraperitoneal (i.p.) administration of sodium pentobarbital (65 mg/kg) and subjected to CCI brain injury of moderate severity. Sham-injured control mice were anesthetized and surgically prepared similarly but they received no impact. Assessment of mRNA expression of inflammatory, proapoptotic and antiapoptotic genes was done by real time-polymerase chain reaction (RT-PCR). Cognitive outcome was evaluated at 4 weeks postinjury using the Morris water maze (MWM).

Findings mRNA expression of inflammatory, proapoptotic and antiapoptotic genes prior to TBI did not reveal any baseline difference between p55 and p75 (-/-) mice. WT mice showed greater baseline expression of inflammatory genes. The learning ability of p55 (-/-) brain-injured mice was significantly better than that observed in p75 (-/-) brain-injured mice ($p<0.05$). Cognitive learning in WT control mice fell between the p55 (-/-) and p75 (-/-) mice.

Conclusions These data suggest that TNF-α may both exacerbate cognitive dysfunction via p55 receptor and attenuate it via p75 receptor.

Keywords Traumatic brain injury · Inflammation · TNF-α · Apoptosis

L. Longhi · F. Ortolano · E. R. Zanier · N. Stocchetti
Neurosurgical Intensive Care Unit,
Department of Anesthesia and Critical Care Medicine,
Fondazione IRCCS Ospedale Maggiore Policlinico,
University of Milano,
Mangiagalli e Regina Elena,
Milano, Italy

L. Longhi · F. Ortolano · E. R. Zanier · C. Perego ·
M. G. De Simoni (✉)
Laboratory of Inflammation and Nervous System Diseases,
Mario Negri Institute,
Via La Masa n 19,
20156 Milano, Italy
e-mail: desimoni@marionegri.it

Introduction

Traumatic brain injury (TBI) is, in part, an inflammatory disease. Post-traumatic inflammatory cascades have been documented both in clinical and experimental TBI including the entry of leukocytes into the injured brain [1, 2] that occurs through the open blood-brain barrier (BBB). These cells (neutrophils and activated macrophages) may release reactive oxygen species (ROS). Post-traumatic inflammation is also associated with upregulation of intercellular adhesion molecules, such as (ICAM)-1, that are involved in vascular adhesion, transendothelial migration of leukocytes [3], and the upregulation and release of cytokines, such as tumor necrosis factor-α (TNF-α) [4, 5] and various interleukins such as IL-1, IL-6, IL-10 and IL-12. Since these cytokines have been implicated in post-traumatic neuropathologic damage/recovery [6], initial pharmacological strategies have been developed to interfere with cytokine production and/or activity with the aim of reversing secondary post-traumatic damage [7]. However, the role of some of these inflammatory molecules following TBI remains unclear. Namely, studies using genetically engineered mice lacking TNF-α or its receptors have provided contradictory results, suggesting that this cytokine might be detrimental in the acute phase following TBI while participating in regenerative/healing processes in the chronic phase [8, 9].

TNF-α binds two receptors: TNFR1 or p55 and TNFR2 or p75 [10]. Interestingly, these receptors may mediate opposing functions. This has been recently demonstrated in a murine model of retinal ischemia, where p55 receptor aggravated tissue destruction, while p75 receptor was protective [11]. Similarly, in a kainic acid-induced seizure model p55 exacerbated, while p75 inhibited, seizures [12]. In the present study we tested the hypothesis that TNF-α could exert opposite roles following TBI, via its p55 and p75 receptors, by comparing the cognitive behavioral response to TBI in wild type (WT), p55 (-/-) and p75 (-/-) mice. Baseline cytokine expression was also evaluated for each genotype.

Materials and methods

Experimental brain injury

All procedures conform to the institutional guidelines in compliance with national (D.L. n.116, G.U. suppl. 40, 18 February 1992) and international laws and policies (EEC Council Directive 86/609, OJL 358, 1; Dec.12, 1987; NIH Guide for the Care and Use of Laboratory Animals, U.S. National Research Council 1996).

The C57Bl/6 strain was used to generate homozigous p55 (-/-) and p75 (-/-) mice [13]. We used male C57Bl/6 WT mice (n=29), p55 (-/-) (n=28) and p75 (-/-) (n=23). They were anesthetized with sodium pentobarbital (65 mg/kg, i.p.) and placed in a stereotactic frame. An eye lubricant ointment (Xantervit ophthalmic ointment, Sifi, CT, Italy) was applied to protect corneal membranes during the surgery. The mice were subjected to a craniectomy followed by CCI brain injury. Our model of injury uses a 3 mm rigid impactor driven by a pneumatic piston, rigidly mounted at an angle of 20° from the vertical plane and applied perpendicularly to the exposed dura mater over the left parieto-temporal cortex, between bregma and lambda at a velocity of 5 meter/second and to a depth of 1 mm [14]. A second group of mice received the same anesthesia and surgery without the brain injury to serve as uninjured controls (sham, n=29). A third group of mice (n=6 /genotype) was sacrificed prior to TBI to provide tissue for baseline (pre-injury) gene expression analysis.

RNA isolation, cDNA synthesis and Real-time PCR

Total RNA was isolated from cortices according to a method previously described [15]. Complementary DNA was synthesized starting from 1.5 μg total RNA from each sample and reverse-transcribing it with random hexamer primers using Multi Scribe reverse transcriptase (TaqMan Reverse transcription reagents, Applied Biosystems, Foster City, CA, USA). The following thermal cycling protocol was used: 10 min at 25°C for incubation, 30 min at 48°C for reverse transcription and 5 min at 95°C for inactivation.

Real-time PCR was performed using a GeneAmp 5700 Sequence Detection System (Applied Biosystems Foster City, CA, USA) as shown previously [16]. Fifty nanograms of cDNA and gene specific primers (200 nM final concentration) were added to Master Mix SYBR Green Dye, TaqMan DNA polymerase, dNTPs with dUTP and optimal buffer components (Applied Biosystems, Foster City, CA, USA) and subjected to PCR amplification in a total volume of 25 μl. The PCR protocol included one cycle at 50°C for 2 min, one cycle at 95°C for 10 min, 40 cycles at 95°C for 15 s and 60°C for 1 min. Real-time PCR was conducted in triplicate with each of the RNA samples. The amplified transcripts were quantified using the comparative cycle threshold method (Applied Biosystems users bulletin #2). Primer optimizing procedures and validation experiments were performed to demonstrate that efficiencies of target and reference were equal. Primers were designed using the Primer Express Software (Applied Biosystems Foster City, CA, USA) based on the following GenBank accession numbers: NM031144 (β-actin, the housekeeping gene), NM031512 (IL-1β), NM012967 (ICAM-1), NM 009741(BCL2) NM 009812(CSP8), AY 902320 (MMP9), NM 010548 (IL10), NM 013693 (TNFα), M 64404(IL-1IRa), NM_007527 (BAX). Gene expression profiles are expressed as fold induction compared to control group.

Behavior

Cognitive function (learning) was evaluated at 4 weeks postinjury using the Morris Water Maze (MWM). Animals were not pre-trained before the injury. Our MWM is a white circular pool (1 meter diameter) that is filled with water (18–20°C) made opaque by adding non-toxic water-soluble white paint. The post-injury visuo-spatial learning task requires that animals learn how to locate a submerged platform placed 0.5 cm under the surface of the water using external visual cues. The essential feature of this task is that the mice learn how to escape from the cold water onto the platform after being randomly placed into the pool at one of four sites. Latencies to reach and climb onto the platform are recorded for each trial, with a maximum of 60 seconds per trial. The learning task consists of 8 trials/day for three consecutive days for a total of 24 trials [14].

Statistical analysis

Learning latencies and baseline cytokines expression are presented as mean ± standard error of the mean. The comparison between groups has been performed by 2 way analysis of variance (ANOVA) followed by Bonferroni post-hoc test. A "p" value<0.05 was considered significant.

Results

Real-time PCR

Both p55 and p75 (-/-) mice showed significantly lower baseline mRNA expression of TNF-α compared to WT ($p<0.01$). In addition, p75 (-/-) mice also showed a significantly lower baseline expression of IL-1 and IL-1ra when compared to WT mice ($p<0.05$). A trend towards lower baseline expression of the other inflammatory genes considered (IL-10, ICAM-1, MMP-9) in p55 and p75 (-/-) mice was observed, but it did not reach statistical significance (Fig. 1).

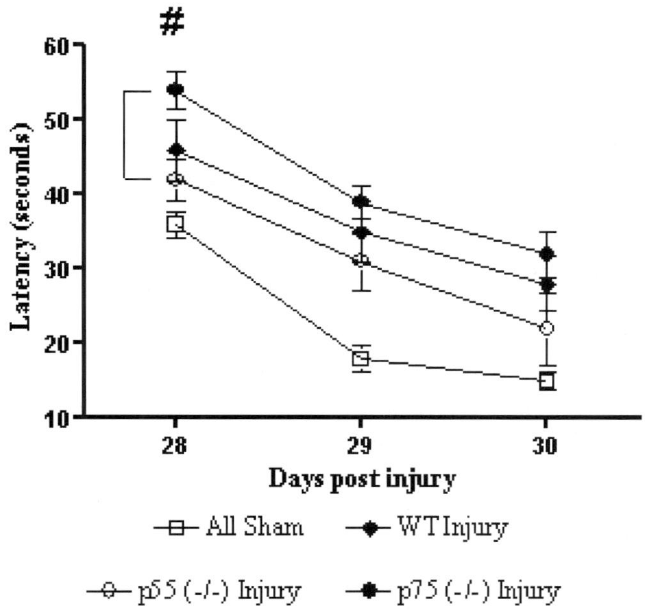

Fig. 2 Cognitive function at 4 weeks postinjury: Since we did not find any difference in the learning ability between sham WT, p55 and p75 (-/-) mice (data not shown), we pooled the data of the all sham animals together. Two way ANOVA: learning over time: $p<0.001$; difference between all groups: $p<0.001$. Bonferroni post-hoc: # $p<0.05$ for p55 (-/-) vs p75 (-/-)

Cognitive function

At 4 weeks following brain injury all the animals were able to swim in the MWM without impairment and were able to

Fig. 1 TNF-α: tumor necrosis factor-α; IL-1 and IL-10: interleukin-1 and interleukin-10; IL-1Ra: interleukin-1 receptor antagonist; ICAM-1: intercellular adhesion molecule-1; MMP-9: matrix metalloproteinase-9; csp-8: caspase-8. #: $p<0.01$ for WT vs both p55 and p75 (-/-). *: $p<0.05$ for WT vs p75 (-/-) mice

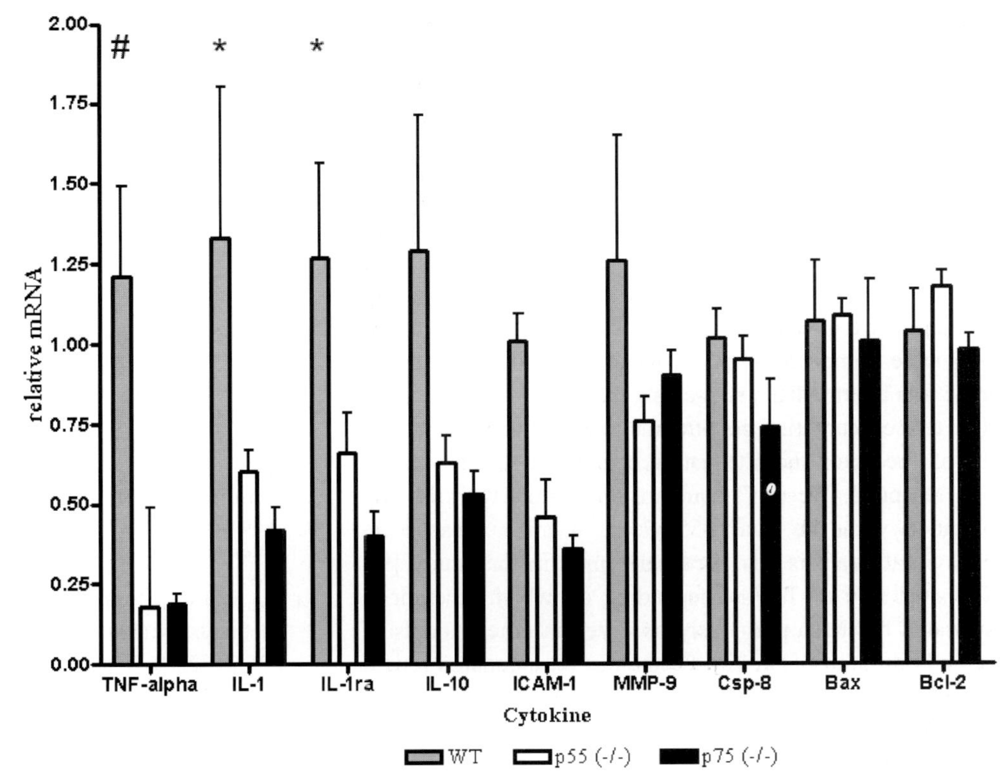

learn the visuo-spatial task, as reflected by decreasing latencies to find the platform over the three-day period. The learning ability of the p55 (-/-) brain-injured mice was significantly better than that of p75 (-/-) brain-injured mice ($p<0.05$) (Fig. 2).

Discussion

In the present study we observed that: 1) p55 (-/-) mice showed attenuated post-traumatic cognitive dysfunction compared to p75 (-/-) mice; 2) p55 and p75 (-/-) mice had reduced TNF-α expression compared to WT mice; 2) p75 (-/-) mice also had reduced gene expression of IL-1 and IL-1ra compared to WT mice and generally both p55 and p75 deficient mice showed a trend towards a lower expression of inflammatory genes.

The role of TNF following TBI is very complex. Early pharmacological studies showed that the antagonism of TNF-α was associated with reduced neurobehavioral sequelae [17, 18]; however studies using genetically engineered mice gave contradictory results [8, 9]. When TNF (-/-) mice were subjected to CCI brain injury of mild severity and compared with their WT littermates, they exhibited attenuated cognitive deficits and motor dysfunction during the first week postinjury. However, WT mice showed a marked recovery over time, while TNF (-/-) mice displayed persistent motor deficits up to 4 weeks postinjury with an accompanying increased injury cavity. These data suggested that while this cytokine may play a deleterious role during the acute postinjury period, it may have a beneficial role in long-term behavioural recovery and tissue repair in the more chronic period after brain injury [8]. Following CCI brain injury, Sullivan et al. showed that knockout mice, lacking both p75 and p55 TNF receptors, showed a greater lesion volume and alteration in BBB permeability in the acute post-traumatic period compared with WT littermates, suggesting that TNF-α might be protective also early after TBI [9]. In our study at 4 weeks postinjury we observed attenuated cognitive deficits in p55 (-/-) mice compared to p75 (-/-) mice, while that of WT mice was intermediate, suggesting that TNF-α might have a dual role exacerbating post-traumatic cognitive sequelae via the p55 receptor and attenuating cognitive sequelae via the p75 receptors. We can't compare our results with the one obtained with the dual p55-p75 (-/-) mice because no behavioral analysis has been performed in that study [9]. The dual role of TNF-α depending on which receptor is activated has been previously observed in other models of acute brain injury [11, 12]. Fontaine et al. compared retinal damage in a model of retinal ischemia followed by reperfusion in p55 and p75 and TNF (-/-) mice and observed that, in comparison with WT, the p55 pathway aggravated tissue destruction, while the p75 pathway was protective via activation of PKB/Akt [11]. Balosso et al. studied seizure susceptibility to kainic acid in WT mice and in mice lacking p55, or p75 or the dual p55-p75 TNF receptors, and observed a reduced duration and number of seizures in p55 (-/-) mice that was associated with a marked upregulation in p75 receptor. Viceversa, mice lacking p75 receptors showed prolonged seizures [12]. Recently, in a model of cryoinjury it has been observed that p55 (-/-) mice showed a reduced number of apoptotic cells at 3 and 7 days postinjury, compared with WT and p75 (-/-) mice, that was associated with reduced inflammatory response and reactive astrocytosis [19]. To understand the molecular mechanisms associated with attenuated functional outcome in our study, we studied whether WT, p55 and p75 (-/-) mice differ in the baseline preinjury expression of genes that are linked to the post-traumatic inflammatory response and apoptotic pathways. Consistently, we observed lower values of baseline expression of the inflammatory genes in p55 and p75 (-/-) mice, suggesting that the deletion of either receptor down-regulates this cytokine pathway. The meaning of the observed difference has to be yet determined. However p55 and p75 (-/-) mice were similar in terms of pre-injury expression of inflammatory and apoptotic genes, suggesting that the differences we observed in post-traumatic behavioral learning between p55(-/-) and p75(-/-) mice might be due to different post-injury responses in the two strains. Post-traumatic molecular changes in p55 and p75 (-/-) mice await further analysis. These data might have implications for the development of novel therapeutic strategies to treat TBI.

Conflict of interest statement We declare that we have no conflict of interest.

References

1. Holmin S, Mathiesen T, Shetye J, Biberfeld P (1995) Intracerebral inflammatory response to experimental brain contusion. Acta Neurochir (Wien) 132:110–119
2. Soares HD, Hicks RR, Smith D, McIntosh TK (1995) Inflammatory leukocytic recruitment and diffuse neuronal degeneration are separate pathological processes resulting from traumatic brain injury. J Neurosci 15:8223–8233
3. Carlos TM, Clark RS, Franicola-Higgins D, Schiding JK, Kochanek PM (1997) Expression of endothelial adhesion molecules and recruitment of neutrophils after traumatic brain injury in rats. J Leukoc Biol 61:279–285
4. Fan L, Young PR, Barone FC, Feuerstein GZ, Smith DH, McIntosh TK (1996) Experimental brain injury induces differential expression of tumor necrosis factor-alpha mRNA in the CNS. Brain Res Mol Brain Res 36:287–291
5. Shohami E, Novikov M, Bass R, Yamin A, Gallily R (1994) Closed head injury triggers early production of TNF alpha and IL-6 by brain tissue. J Cereb Blood Flow Metab 14:615–619

6. Morganti-Kossman MC, Lenzlinger PM, Hans V, Stahel P, Csuka E, Ammann E, Stocker R, Trentz O, Kossmann T (1997) Production of cytokines following brain injury: beneficial and deleterious for the damaged tissue. Mol Psychiatry 2:133–136
7. McIntosh TK, Juhler M, Wieloch T (1998) Novel pharmacologic strategies in the treatment of experimental traumatic brain injury: 1998. J Neurotrauma 15:731–769
8. Scherbel U, Raghupathi R, Nakamura M, Saatman KE, Trojanowski JQ, Neugebauer E, Marino MW, McIntosh TK (1999) Differential acute and chronic responses of tumor necrosis factor-deficient mice to experimental brain injury. Proc Natl Acad Sci U S A 96:8721–8726
9. Sullivan PG, Bruce-Keller AJ, Rabchevsky AG, Christakos S, Clair DK, Mattson MP, Scheff SW (1999) Exacerbation of damage and altered NF-kappaB activation in mice lacking tumor necrosis factor receptors after traumatic brain injury. J Neurosci 19:6248–6256
10. Wajant H, Pfizenmaier K, Scheurich P (2003) Tumor necrosis factor signaling. Cell Death Differ 10:45–65
11. Fontaine V, Mohand-Said S, Hanoteau N, Fuchs C, Pfizenmaier K, Eisel U (2002) Neurodegenerative and neuroprotective effects of tumor Necrosis factor (TNF) in retinal ischemia: opposite roles of TNF receptor 1 and TNF receptor 2. J Neurosci 22:RC216-
12. Balosso S, Ravizza T, Perego C, Peschon J, Campbell IL, De Simoni MG, Vezzani A (2005) Tumor necrosis factor-alpha inhibits seizures in mice via p75 receptors. Ann Neurol 57:804–812
13. Peschon JJ, Torrance DS, Stocking KL, Glaccum MB, Otten C, Willis CR, Charrier K, Morrissey PJ, Ware CB, Mohler KM (1998) TNF receptor-deficient mice reveal divergent roles for p55 and p75 in several models of inflammation. J Immunol 160:943–952
14. Longhi L, Watson DJ, Saatman KE, Thompson HJ, Zhang C, Fujimoto S, Royo N, Castelbuono D, Raghupathi R, Trojanowski JQ, Lee VM, Wolfe JH, Stocchetti N, McIntosh TK (2004) Ex vivo gene therapy using targeted engraftment of NGF-expressing human NT2N neurons attenuates cognitive deficits following traumatic brain injury in mice. J Neurotrauma 21: 1723–1736
15. De Simoni MG, Perego C, Ravizza T, Moneta D, Conti M, Marchesi F, De Luigi A, Garattini S, Vezzani A (2000) Inflammatory cytokines and related genes are induced in the rat hippocampus by limbic status epilepticus. Eur J Neurosci 12:2623–2633
16. Bergamaschini L, Rossi E, Storini C, Pizzimenti S, Distaso M, Perego C, De Luigi A, Vergani C, De Simoni MG (2004) Peripheral treatment with enoxaparin, a low molecular weight heparin, reduces plaques and beta-amyloid accumulation in a mouse model of Alzheimer's disease. J Neurosci 24:4181–4186
17. Shohami E, Bass R, Wallach D, Yamin A, Gallily R (1996) Inhibition of tumor necrosis factor alpha (TNFalpha) activity in rat brain is associated with cerebroprotection after closed head injury. J Cereb Blood Flow Metab 16:378–384
18. Shohami E, Gallily R, Mechoulam R, Bass R, Ben Hur T (1997) Cytokine production in the brain following closed head injury: dexanabinol (HU-211) is a novel TNF-alpha inhibitor and an effective neuroprotectant. J Neuroimmunol 72:169–177
19. Quintana A, Giralt M, Rojas S, Penkowa M, Campbell IL, Hidalgo J, Molinero A (2005) Differential role of tumor necrosis factor receptors in mouse brain inflammatory responses in cryolesion brain injury. J Neurosci Res 82:701–716

Acta Neurochir Suppl (2008) 102: 415–419

Brain contusions induce a strong local overexpression of MMP-9. Results of a pilot study

A. Vilalta · J. Sahuquillo · M. A. Poca · J. De Los Rios · E. Cuadrado · A. Ortega-Aznar · M. Riveiro · J. Montaner

Summary
Background Brain contusions are inflammatory evolutive lesions that induce intracranial pressure increase and edema, contributing to neurological outcome. Matrix metalloproteinases (MMPs) 2 and 9 can degrade the majority of the extracellular matrix components, and are implicated in blood-brain barrier disruption and edema formation. The aim of this study was to investigate MMP-2 and MMP-9 profiles in human brain contusions using zymography.
Methods A prospective study was conducted in 20 traumatic brain injury patients where contusion brain tissue was resected. Brain tissues from lobectomies were used as controls. Brain homogenates were analysed by gelatin zymography and in situ zimography was performed to confirm results, on one control and one brain contusion tissue sample.
Findings MMP-2 and MMP-9 levels were higher in brain contusions when compared to controls. MMP-9 was high during the first 24 hours and at 48 to 96 hours, whereas MMP-2 was slightly high at 24 to 96 hours. In situ zymography confirmed gelatin zymography results. A relation between outcome and MMP-9 levels was found; MMP-9 levels were higher in patients with worst outcome.
Conclusions Our results indicate strong time-dependent gelatinase expression primarily from MMP-9, suggesting that the inflammatory response induced by focal lesions should be considered as a new therapeutic target.

Keywords Neuroinflammation · Matrix metalloproteinases · Traumatic brain injury · Brain contusion · Zymography

A. Vilalta · J. De Los Rios
Neurosurgery and Neurotraumatology Research Unit (UNINN), Vall d'Hebron University Hospital, Universitat Autònoma de Barcelona, Passeig Vall d'Hebron 119-129, 08035 Barcelona, Spain

J. Sahuquillo (✉) · M. A. Poca
Department of Neurosurgery, Vall d'Hebron University Hospital, Universitat Autònoma de Barcelona, Passeig Vall d'Hebron 119-129, 08035 Barcelona, Spain
e-mail: sahuquillo@neurotrauma.net

E. Cuadrado · J. Montaner
Neurovascular Research Laboratory and Neurovascular Unit, Vall d'Hebron University Hospital, Universitat Autònoma de Barcelona, Passeig Vall d'Hebron 119-129, 08035 Barcelona, Spain

A. Ortega-Aznar
Department of Pathology (Neuropathology), Vall d'Hebron University Hospital, Universitat Autònoma de Barcelona, Passeig Vall d'Hebron 119-129, 08035 Barcelona, Spain

M. Riveiro
Neurotraumatology Intensive Care Unit, Vall d'Hebron University Hospital, Universitat Autònoma de Barcelona, Passeig Vall d'Hebron 119-129, 08035 Barcelona, Spain

Introduction

Brain contusions are highly dynamic and evolutive lesions that induce cytotoxic and vasogenic edema, mass effect, neurological deterioration and early or delayed increase in intracranial pressure (ICP). They often increase in size over hours to days after injury due to the evolving events related to the interplay of hemorrhage, vasogenic edema and ischemic necrosis [3]. An increasing body of evidence implicates the inflammatory processes in the evolution and maturation of contusions, in vasogenic edema development, and points to their influence on the neurological outcome [15]. Vasogenic brain edema results from blood-brain barrier (BBB) disruption and massive infiltration of the damaged tissue by leukocytes and extravasation of plasma proteins [17]. Several studies have reported the role of MMPs, a family of neutral proteases that degrade or modify almost all extracellular matrix (ECM) components, in increasing the BBB permeability [4, 9, 12, 13] and in the pathophysiology of traumatic brain injury (TBI). Brain inflammatory cascades are triggered by the production of proinflammatory cytokines, both by intrinsic (microglia, astrocytes and neurons) and extrinsic pathways (infiltrating macrophages, lymphocytes and leukocytes) [18]. Thus, by opening the structurally intact BBB, plasma molecules and inflammatory cells enter the brain neuropil. Because gelatinases (MMP-2 and MMP-9) are the main enzymes that degrade type IV collagen, their role as mediators in BBB opening is very intriguing. Experimental models of TBI suggest a role for MMPs in the pathophysiology of severe TBI [5, 10, 11, 14]. To date, two studies have assessed the changes of MMP levels in the peripheral blood of TBI patients [16, 19] but information regarding MMP levels in human brain contusion tissue is still lacking.

The main goal of our study was to investigate the MMP-2 and MMP-9 profiles in surgically resected human brain contusions, measuring their levels by using both gelatin zymography and *in situ* zymography techniques.

Materials and methods

Prospective study of TBI patients who underwent surgical treatment of brain contusions. Brain samples were obtained from surgically resected areas of contusions. Brain tissue samples obtained from lobectomies performed for extra-axial tumors, that did not present MRI abnormalities in T1 and T2 weighted images, were used as controls. Depending on time from injury to brain resection, four groups were arbitrarily formed: A) <24 hrs, B) 24–48 hrs, C) 48–96 hrs, D) >96 hrs. Clinical outcome was assessed at 6 months by using the Extended Glasgow Outcome Scale (GOSE). Scores obtained were pooled in two groups: Bad outcome: 1–4; Good outcome: 5–8.

Brain homogenates

Resected brain tissue was snap-frozen in liquid nitrogen and stored at −80°C. Brain homogenates were obtained from these tissues by adding lysis buffer and protease inhibitors as described previously [2]. Total protein content of supernatants was determined by bicinchoninic acid (BCA) assay (Pierce, Rockford, IL, USA).

Gelatin zymography

Samples were adjusted to equal protein content (20 μg) and were mixed 1:1 with a Tris-Glycine SDS sample buffer (2x) from Novex®. Proteins were separated by electrophoresis in pre-cast zymogram gels from Novex® at 100 V for two and a half hours at room temperature. A MMP2-9 recombinant protein (Chemicon International, Inc) was loaded in one lane of each gel as a positive control. Following electrophoresis, gels were washed to remove SDS with a zymogram renaturing buffer from Novex® for 45 minutes two times. Zymograms were incubated at 37°C for 18 hours in Novex® developing buffer. Enzymatic activity bands were visualized after staining of the gels in Amido Black 1% for one hour and destaining for a total of 100 minutes, in alternating periods of 20 minutes, in a solution of 30% (v/v) methanol and 10% (v/v) glacial acetic acid. Band intensities were read with a Kodak Gel Logistic 440 imaging system and analysis was performed with a Kodak 1D image analysis software. Band's intensities were compared with the positive control and expressed as ratios.

In situ zymography

Frozen cryostat brain sections (14 μm thick) from one sample from a brain contused specimen and a control patient were used to perform *in situ* zymography as described elsewhere [6]. Nuclei were stained with 4′6-diamidino-2-phenylindole (DAPI) (Vector Laboratories, Inc. Burlingame, CA 94010, USA). Reactivity was analysed with a fluorescence microscope (Olympus IX71, Center Valley, Pennsylvania, USA).

Statistical analysis

The Mann-Whitney Rank Sum Test was used to compare contusion levels for each target protein with controls. Summary variables are expressed as the median and range or mean±SD. Depending on the distribution of the variables either One-way ANOVA (with Bonferroni post-hoc test) or Kruskal-Wallis ANOVA on Ranks were used to compare differences between several groups or phases. Statistical significance was assumed when $P<0.05$.

Results

We studied twenty specimens obtained from resected contusions in TBI patients with a mean age of 44±17 and a mean GCS of 10±4. Median time from TBI to brain contusion extraction was 42 hrs (range:5–205) (Table 1). Six specimens, as described above, were used as controls.

Gelatinase expression

An increase of MMP-2 ($P=0.026$) and MMP-9 ($P=0.012$) was found in brain contusion homogenates when compared

Table 1 Descriptive data from patients and brain specimens

Brain Specimens	GCS	Sex	Age	Hrs from TBI	Contusion phase (n)	GOSE
Brain contusions n=20	10±4	M: 17 F: 3	44±17	42 hrs (range:5–205)	A: 6; B: 6; C:6; D:2	4±2
Brain controls n=6	-	M: 1 F: 5	49±20	-	-	-

Descriptive data from brain control specimens and from contused brain resected at different times from injury. GCS: Glasgow Coma Score; Hrs: hours; Contusion phase: A: ≤24 hrs, B: >24–48 hrs, C: >48–96 hrs; D: >95 hrs (sample size is shown); GOSE: Extended Glasgow Outcome Scale. M:male; F: female

with brain control homogenates. A significant increase in MMP-9 levels was found in the first 24 hours ($P<0.05$) and at 48 to 96 hours ($P<0.05$) after injury (Fig. 1), with a decrease towards normality after 96 hours from TBI. MMP-2 levels were slightly increased from 24 to 96 hours after injury (24–48 hrs: $P=0.065$, 48–96 hrs: $P=0.066$) but this increase was remarkably less pronounced than in MMP-9 (Fig. 1). Patients with bad outcome at six months had significantly higher levels of MMP-9 ($P<0.05$) than patients with good outcome when compared to controls (Fig. 2). The differences in MMP-2 levels in both groups did not reach statistical significance.

Mean intensities, expressed as ratios from arbitrary units, for the tissues assayed with *in situ* zymography were as follows: brain contusion proMMP-9 tissue level was 2.11 and brain control proMMP-9 tissue level was 1.05; brain contusion proMMP-2 tissue level was 1.27 and brain control proMMP-2 tissue level was 0.8 (Fig. 3).

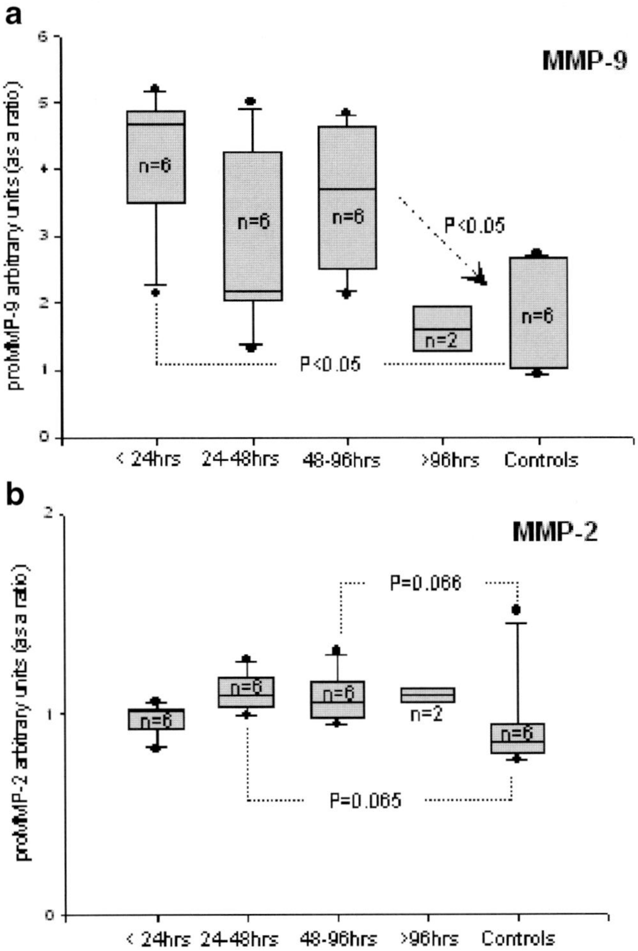

Fig. 1 Box-plot graph summarizing proMMP-9 (**a**) and proMMP-2 (**b**) levels in brain contusion specimens and controls at different times from injury. Only statistically significant differences are shown

In situ gelatinase activity

In situ zymography performed on a single specimen of brain contusion showed a strong gelatinase activity (green fluorescence; grey in the figure) when compared with a control. In this control, the signal obtained was significantly weaker (Fig. 3). High magnification images revealed that in the control tissue, a weak gelatinase signal was detected within the cytoplasm of cells whereas in the contusion tissue a strong reactivity was found primarily in the cell boundaries.

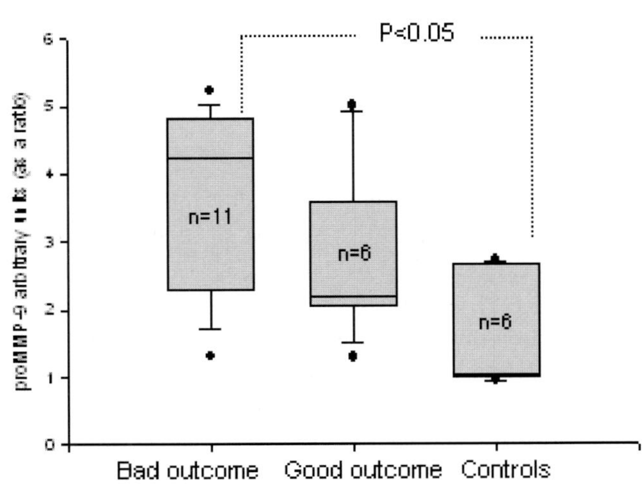

Fig. 2 Box-plot graph showing proMMP-9 levels in patients with bad and good outcome compared with controls. Only statistically significant differences are shown

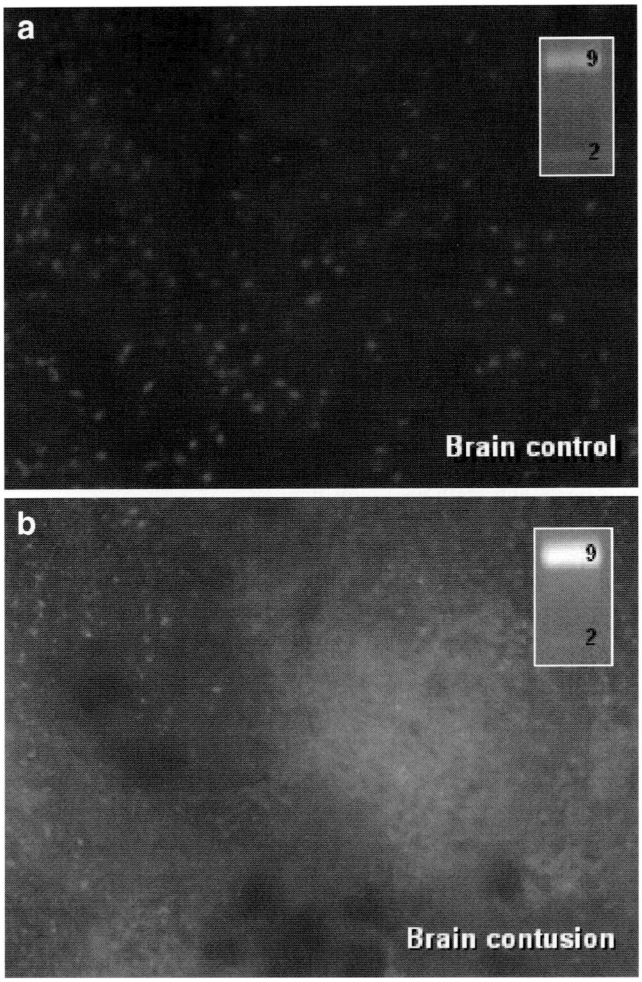

Fig. 3 *In situ* zymography images from a control specimen (**a**) and surgically removed brain contusion tissue (**b**). High gelatinase activity (green fluorescence, grey in the figure) is observed in the brain contusion tissue. In the right-upper side of each image (inset) the gelatin zymograms are shown for each specimen. In zymogram a higher degradation intensity is also observed in tissue from contusion. 9: Band corresponding to proMMP-9 and 2: Band corresponding to proMMP-2

Discussion

The progression of brain damage after trauma involves a complex and multifactorial cascade of secondary events, including excitotoxicity, oxidative stress, inflammation and apoptosis [20]. Emerging evidence suggests that neuro-inflammation plays an important role in the pathophysiology of TBI and MMPs might be implicated in this process. MMPs can degrade several substrates, including the components of the basal lamina of cerebral blood vessels and thus, they could increase BBB permeability [14]. At the macroscopic level, brain edema, whether intra- or extracellular, is still the most important cause of increased ICP after TBI [8]. Several studies suggest that a large amount of edema occurs in the core of brain contusions and that this could contribute to the development of early massive edema [7]. According to the pivotal studies of Katayama's group, the edema found in the core of contusions is not strictly cytotoxic nor vasogenic but passive attraction of water due to the high osmolality that necrosis of neurons, glial cells and vessels exhibit in the contusion core [7]. However, vasogenic edema from the disrupted BBB also facilitates an increase in brain water content.

In our study we found high levels of MMP-9 in brain contusions compared with controls. This was also confirmed by *in situ* zymography, which showed higher gelatinase activity in brain contusion tissue compared with control brain tissue. These findings were in consonance with gelatin zymography results (Fig. 3). MMPs also presented a different temporal profile in brain contusions, MMP-9 was highest over the whole of the period with 2 peaks at the first 24 hours and at 48–96 hours from TBI. Although MMP-2 levels were higher in brain contusions at 24–96 hours from TBI, the small differences between groups and the wide range of MMP-2 detected in controls, preclude the extraction of any conclusion regarding MMP-2 levels, although perhaps with an increased sample size the differences would be greater. Additionally, in a recent study we found high levels of gelatinases in plasma and brain extracellular fluid from patients with TBI [19]. Our findings, if replicated in a larger series, would suggest an early overexpression of MMP-9, creating new targets for a novel approach to the pharmacological treatment of brain contusions. Recent experimental studies have shown that pharmacological blocking of the extracellular regulated kinase can reduce MMP-9 levels attenuating brain edema and tissue damage in mice [10], opening a new whole avenue for pharmacological intervention in TBI patients.

Several studies have also demonstrated a close relationship between different cytokines and the early increase in MMPs. In these studies [1, 16] there was a significant correlation between levels of blood MMP-9 and IL-6, suggesting that in an incompletely understood way IL-6 is associated with the observed increases in MMP-9. In a preliminary study we found an increase in IL-6 and IL-1β levels at 24 to 48 hours from TBI, followed by an increase in MMP-9 levels at 48 to 96 hours from injury in brain contusion homogenates (unpublished results).

Our study was not designed to investigate the outcome in relation to MMP levels. However, we found that patients with a bad oucome had higher MMP-9 levels than patients with a good outcome. Although this should not be disregarded as a spurious finding: adequately powered studies are necessary to verify or refute these results.

MMPs are pathologically overexpressed in TBI but they can also have an important role in repairing brain damage [21]. Further research is needed in order to determine whether an increase of any gelatinase implicated in the neuroinflammation cascade should be blocked or partially blocked, depending on its repairing nature or pathological condition and especially depending on the expression time from TBI.

In conclusion, our preliminary results showed a significant increase in MMP-9 levels in the acute phase of TBI patients with brain contusions. This suggests that these lesions induce a strong local and early infammatory response that may play a very important role in their pathophysiology, in their evolving mass effect and in clinical deterioration of some patients. If this is confirmed in additional studies, the modulation of the inflammatory response induced by focal lesions should be considered a new therapeutic target for improving outcome in these patients.

Acknowledgements This study was supported by Fondo de Investigaciones Sanitarias de la Seguridad Social (FIS) grant number PI051092 to J. Sahuquillo. A Vilalta is the recipient of a pre-doctoral grant from the Institut Fundació de Recerca, Vall d'Hebron University Hospital, Universitat Autònoma de Barcelona, Barcelona, Spain.

Conflict of interest statement We declare that we have no conflict of interest.

References

1. Aibiki M, Maekawa S, Ogura S, Kinoshita Y, Kawai N, Yokono S (1999) Effect of moderate hypothermia on systemic and internal jugular plasma IL-6 levels after traumatic brain injury in humans. J Neurotrauma 16:225–232
2. Asahi M, Asahi K, Jung JC, Zoppo del GJ, Fini ME, Lo EH (2000) Role for matrix metalloproteinase 9 after focal cerebral ischemia: effects of gene knockout and enzyme inhibition with BB-94. J Cereb Blood Flow Metab 20:1681–1689
3. Bullock R, Maxwell WL, Graham DI, Teasdale GM, Adams JH (1991) Glial swelling following human cerebral contusion: an ultrastructural study. J Neurol Neurosurg Psychiatry 54:427–434
4. Gasche Y, Soccal PM, Kanemitsu M, Copin JC (2006) Matrix metalloproteinases and diseases of the central nervous system with a special emphasis on ischemic brain. Front Biosci 11:1289–1301
5. Jaworski DM (2000) Differential regulation of tissue inhibitor of metalloproteinase mRNA expression in response to intracranial injury. Glia 30:199–208
6. Justicia C, Panes J, Sole S, Cervera A, Deulofeu R, Chamorro A, Planas AM (2003) Neutrophil infiltration increases matrix metalloproteinase-9 in the ischemic brain after occlusion/reperfusion of the middle cerebral artery in rats. J Cereb Blood Flow Metab 23:1430–1440
7. Katayama Y, Kawamata T (2003) Edema fluid accumulation within necrotic brain tissue as a cause of the mass effect of cerebral contusion in head trauma patients. Acta Neurochir Suppl 86:323–327
8. Marmarou A, Signoretti S, Fatouros PP, Portella G, Aygok GA, Bullock MR (2006) Predominance of cellular edema in traumatic brain swelling in patients with severe head injuries. J Neurosurg 104:720–730
9. Morganti-Kossmann MC, Rancan M, Stahel PF, Kossmann T (2002) Inflammatory response in acute traumatic brain injury: a double-edged sword. Curr Opin Crit Care 8:101–105
10. Mori T, Wang X, Aoki T, Lo EH (2002) Downregulation of matrix metalloproteinase-9 and attenuation of edema via inhibition of ERK mitogen activated protein kinase in traumatic brain injury. J Neurotrauma 19:1411–1419
11. Morita-Fujimura Y, Fujimura M, Gasche Y, Copin JC, Chan PH (2000) Overexpression of copper and zinc superoxide dismutase in transgenic mice prevents the induction and activation of matrix metalloproteinases after cold injury-induced brain trauma. J Cereb Blood Flow Metab 20:130–138
12. Mun-Bryce S, Rosenberg GA (1998) Gelatinase B modulates selective opening of the blood-brain barrier during inflammation. Am J Physiol 274:1203–1211
13. Mun-Bryce S, Rosenberg GA (1998) Matrix metalloproteinases in cerebrovascular disease. J Cereb Blood Flow Metab 18:1163–1172
14. Rosenberg GA (1995) Matrix metalloproteinases in brain injury. J Neurotrauma 12:833–842
15. Schmidt OI, Heyde CE, Ertel W, Stahel PF (2005) Closed head injury-an inflammatory disease? Brain Res Rev 48:388–399
16. Suehiro E, Fujisawa H, Akimura T, Ishihara H, Kajiwara K, Kato S, Fujii M, Yamashita S, Maekawa T, Suzuki M (2004) Increased matrix metalloproteinase-9 in blood in association with activation of interleukin-6 after traumatic brain injury: influence of hypothermic therapy. J Neurotrauma 21:1706–1711
17. Unterberg AW, Stover J, Kress B, Kiening KL (2004) Edema and brain trauma. Neuroscience 129:1021–1029
18. Vecil GG, Larsen PH, Corley SM, Herx LM, Besson A, Goodyer CG, Yong VW (2000) Interleukin-1 is a key regulator of matrix metalloproteinase-9 expression in human neurons in culture and following mouse brain trauma in vivo. J Neurosci Res 61:212–224
19. Vilalta A, Sahuquillo J, Rosell A, Poca MA, Riveiro M, Montaner J (2008) Moderate and severe traumatic brain injury induce early overexpression of systemic and brain gelatinases. Intensive Care Med 34(8):1384–1392 (Aug)
20. Wang X, Jung J, Asahi M, Chwang W, Russo L, Moskowitz MA, Dixon CE, Fini ME, Lo EH (2000) Effects of matrix metalloproteinase-9 gene knock-out on morphological and motor outcomes after traumatic brain injury. J Neurosci 20:7037–7042
21. Zhao BQ, Tejima E, Lo EH (2007) Neurovascular proteases in brain injury, hemorrhage and remodeling after stroke. Stroke 38:748–752

Acta Neurochir Suppl (2008) 102: 421–424

Shock wave-induced brain injury in rat: Novel traumatic brain injury animal model

Atsuhiro Nakagawa · Miki Fujimura · Kaoruko Kato · Hironobu Okuyama · Tokitada Hashimoto · Kazuyoshi Takayama · Teiji Tominaga

Abstract

Background In blast wave injury and high-energy traumatic brain injury, shock waves (SW) play an important role along with cavitation phenomena. However, due to lack of reliable and reproducible technical approaches, extensive study of this type of injury has not yet been reported. The present study aims to develop reliable SW-induced brain injury model by focusing micro-explosion generated SW in the rat brain.

Methods Adult male rats were exposed to single SW focusing created by detonation of microgram order of silver azide crystals with laser irradiation at a focal point of a truncated ellipsoidal cavity of 20 mm minor diameter and the major to minor diameter ratio of 1.41 after craniotomy. The pressure profile was recorded using polyvinylidene fluoride needle hydrophone. Animals were divided into three groups according to the given overpressure: Group I: Control, Group II: 12.5±2.5 MPa (high pressure), and Group III: 1.0±0.2 MPa (low pressure). Histological changes were evaluated over time by hematoxylin-eosin staining.

Findings Group II SW injuries resulted in contusional hemorrhage in reproducible manner. Group III exposure resulted in spindle-shaped changes of neurons and elongation of nucleus without marked neuronal injury.

Conclusions The use of SW loading by micro-explosion is useful to provide a reliable and reproducible SW-induced brain injury model in rats.

A. Nakagawa (✉) · M. Fujimura · K. Kato · H. Okuyama · T. Tominaga
Department of Neurosurgery,
Tohoku University Graduate School of Medicine,
1-1 Seiryo-machi, Aoba-ku,
Sendai 980-8574, Japan
e-mail: tonkan@zc5.so-net.ne.jp

M. Fujimura
e-mail: fujimur@nsg.med.tohoku.ac.jp

K. Kato
e-mail: Kaoruko@nsg.med.tohoku.ac.jp

H. Okuyama
e-mail: okuyama@nsg.med.tohoku.ac.jp

T. Tominaga
e-mail: Tominaga@nsg.med.tohoku.ac.jp

T. Hashimoto
Interdisciplinary Shock Wave Research Center,
Institute of Fluid Science, Tohoku University,
2-1-1, Katahira, Aoba-ku,
Sendai, Miyagi 980-8577, Japan
e-mail: hasimoto@rainbow.ifs.tohoku.ac.jp

K. Takayama
Nanomedicine,
Tohoku University Bioengineering Research Organization,
2-1-1, Katahira, Aoba-ku,
Sendai, Miyagi 980-8577, Japan
e-mail: k.takayama@mac.com

Keywords Blast wave · Cavitation · Silver azide · Traumatic brain injury

Introduction

Underwater shock waves (SW) created by micro-explosion generate a sudden increase in pressure of readily 100 MPa for a few microseconds and propagate into various media having different acoustic impedances in a well-controlled fashion. Because of the similarity in the acoustic impedance of water and the soft tissues of the body, SW transmit into the soft tissues without causing a substantial loss of energy. SW transmitting through layers of different acoustic impedances dissipates energy which induce mechanical destruction of tissues, depending on their pressure value. In addition, this can trigger simultaneously high-speed micro jet penetration due to the collapse of bubbles upon shock loading. These phenomena are well studied as SW-bubble interaction or bubble dynamic effects [7, 8]. SW are considered to play an important role in blast injury [1, 2], and high energy traumatic brain injury (TBI), [3, 4] presumably as a result of the cavitation phenomena or SW-bubble interaction. However, no extensive studies have been performed regarding the effect of SW on blast generated brain injury due to absence of reliable and reproducible technical approaches.

To address these issues, we have developed reliable experimental setup to evaluate SW-induced brain injury using micro-explosive (silver azide pellet), and have examined pressure-dependent effects of SW on the rat brain.

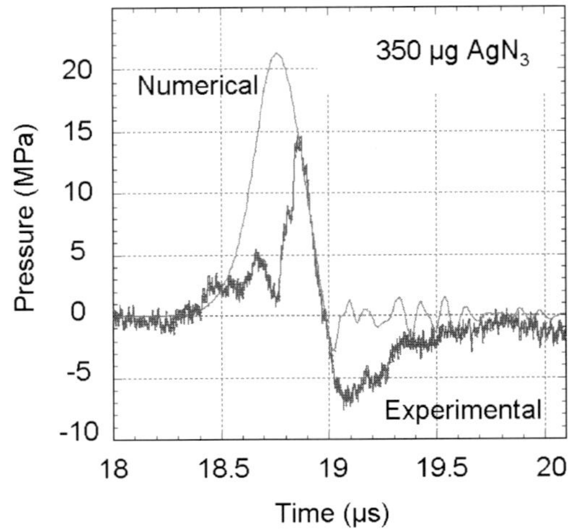

Fig. 1 Typical pressure profile of a shock wave (SW) using silver azide

Materials and methods

SW source: SW were generated by underwater micro-explosion [6–9] of microexplosive (silver azide (AgN_3)) weighing 1–400 µg (in terms of total energy 15 mJ to 6 J) attached on the tip of 0.6 mm diameter quartz optical fiber. The charge was placed at the focal point inside a half truncated ellipsoidal cavity (exit diameter; 20 mm, ratio of major to minor radii: 1.41) and detonated by irradiation of Nd: YAG laser (Neodymium: Yttrium-Aluminum-Garnet laser; Laser Photonics. Ltd. 1064 µm wavelength, 7 ns pulse width, and 25 mJ/pulse) beam through it. The contribution laser energy to SW formation was negligibly small. Upon the detonation, spherical SW thus formed attenuated from the cavity exit and the reflected SW from the curved cavity wall and spontaneously focused at a focal point outside the cavity. To assure the focal pressure during the experiments, we monitored transient pressure at the sidewall of the truncated ellipsoidal cavity using a polyvinylidene fluoride (PVDF) needle hydrophone (serial no. 300/25/289, Imotec Messtechnik, Wärendorf, Germany) with a 0.5-mm-diameter sensing element of 0.34PC/bar, and a rise time of 50 nsec. These values are suited for temporally and spatially accurate pressure measurements. Measured data were stored and displayed on a digital oscilloscope (model DL 716; Yokogawa, Kyoto, Japan). The axial variation of pressures was measured in series of preparatory experiments and the dependence of charge weight on the focus pressure was calibrated. The typical pressure profile of generated shock wave is shown in Fig. 1. Average SW focus pressure was 1.0±0.2 MPa for 1–3 µg, and 12.5±2.5 MPa for 100–350 µg charges, respectively.

SW-induced brain injury rat model Eight-week-old male Sprague-Dawley rats weighing between 250 and 270 g were anesthetized systemically and locally, and a single crainiotomy in the left skull (5–7 mm) was carefully performed under a microscope using a drill without injuring the underlying dura mater. The truncated ellipsoidal SW generator was filled with 5 ml physiological saline (38°C) and an acrylic holder (inner diameter 28 mm, outer diameter 30 mm, height 35 mm) was tightly fixed between the rat's skull with skin flap and the bone window [10]. SW was delivered to left cerebral hemisphere as shown in Fig. 2. Animal procedures were approved by the Institutional Animal Care and Use Committee of the Tohoku University Graduate School of Medicine. The rats were divided into three groups according to the SW exposure patterns. Group I (n=4) served as a control, and did not receive SW exposure. Group II (n=4) received high-overpressure SW (average overpressure: 12.5 MPa). Group III (n=4) received low-overpressure SW (average overpressure: 1.0 MPa). Specimens were obtained at 24 hours after SW exposure and were stained with hematoxylin and eosin and examined with an optical microscope.

Results

After the application of high-overpressure SW (Group II), intracerebral hemorrhage was observed in both the cortical

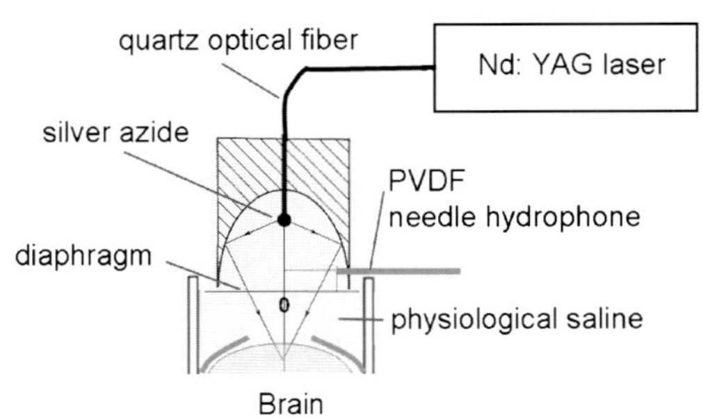

Fig. 2 Schematic diagram of experimental settings

and subcortical regions (Fig. 3a,c). Contusional brain injury was also observed around the hemorrhage. After low-overpressure SW application (Group III), although significant hemorrhage was not evident in the cortical and subcortical regions (Fig. 3b), mild morphological changes such as spindle-shaped changes of neurons and elongation of nucleus toward SW exposure were observed (Fig. 3d). Interestingly, there was no significant histological damage in the theoretical SW focus, which was located at left hippocampus for both in Group II and III animals.

Discussion

Although the relationship between SW and resultant cavitation effects were discussed previously to explain TBI studies [2, 3, 5], no appropriate experimental animal model has yet been developed due to difficulties in generating an accurate and well controlled SW to a targeted area. Among available SW sources, it is not always easy to measure shock focusing pressure profiles. However, the use of micro-explosives with a truncated ellipsoidal cavity enable us to produce a SW well suited for brain injury studies. Minute silver azide pellets in the magnitude of microgram, in particular, can be a reliable energy source for this type of injury. This model is backed up by extensive computational fluid dynamic and SW dynamic studes [7, 11] and we have previously applied it to evaluate liver injury [8].

Present study demonstrates that the application of high-overpressure SW (average overpressure: 12.5 MPa) on rat brain resulted in cerebral contusional hemorrhage in repeatable manner, which is similar to the findings in other organs, such as kidney, liver, and lung, where the injury is considered to be due to small vessel rupture [4]. On the other hand, low-overpressure SW (1 MPa) resulted in mild morphological changes such as spindle-shaped changes of neurons and elongation of nucleus toward SW exposure without marked neuronal injury in repeatable fashion. The underlying mechanism and clinical implication of such morphological changes are not known, although similar findings are reported in the SW-induced liver injury in rabbit [8]. Based on the present study, the threshold of pressure-dependent SW-induced brain injury is speculated to be under 1 MPa, which is much lower than that in other organs such as kidney (3–19 MPa), lung (2–10 MPa) and liver (10 MPa) [4], probably due to the intrinsic vulnerability of neurons as well as the presence of rich fine vessels in the cortical region. Various factors including overpressure, loading duration, impulse, number of exposures and administration rate are considered to be involved in the occurrence of tissue damage [4]. Among them, overpressure would play a dominant role in rupturing vessel, leading to tissue damage [10]. It would also be of interest to determine wheter SW and resulting bubble rupture also contribute to TBI. The combination of these effects may further promote high energy TBI [3, 5], blast injury [1, 2],

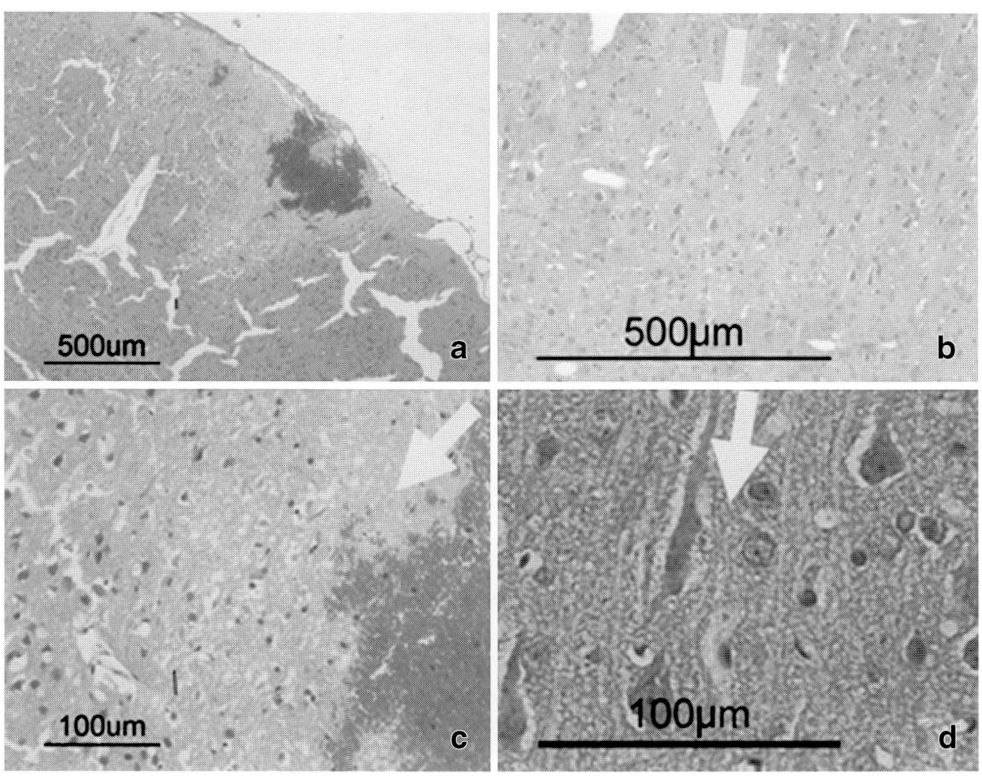

Fig. 3 Histological findings after application of high-overpressure SW (Group II) (**a**, **c**) and low-overpressure SW (Group III) (**b**, **d**) at 24 hours after SW exposure. In Group II, intracerebral hemorrhage was observed both at cortical and subcortical region. Contusional brain injury was evident around the hemorrhage (**a**, **c**). In Group III, spindle-shaped changes of neurons and elongation of nucleus toward SW exposure without any significant hemorrhage was observed (**b**, **d**). Arrow: direction of SW application. Hematoxylin-eosin stain. Original magnification A, B× 100. C, D×400

and gunshot injury. These issues could be clarified in the future study, in which we believe the new method presented here would be useful. This new model of blast injury could also be used to explore new ways of protecting the brain from this type of injury. We have already reported brain protection using materials that contain air, such as cottnoids, expanded polytetrafluoroethylene, and GORE-TEX® dura substitute[11], The development of this new model of blast injury, along with continuing experimental and numerical analysis may contribute to the reduction of SW-induced blast injury in the future.

Acknowledgements This work was supported in part by a Grant-in-Aid for Scientific Research (B) (No. 18390388), (No. 19399372), Grant-in-Aid for Young Scientist (A) (No. 19689028) offered by the Japanese Ministry of Education, Culture, Sports, Science, and Technology. We also acknowledge Takayuki Hirano, Hiroshi Uenohara, Mariko Kambe, and Tomohiro Ohki, Hidenori Ojima, and Tomizo Shiratori for technical assistance.

Conflict of interest statement We declare that we have no conflict of interest.

References

1. Axelsson H, Hjelmqvist H, Medin A, Persson JKE, Suneson A (2000) Physiological changes in pigs exposed to a blast wave from a detonating high-explosive charge. Milit Med 165:119–126
2. Cernak I, Wang Z, Jing J, Bian X, Savic J (2001) Ultrastructural and functional characteristics of blast injury-induced neurotrauma. J Trauma 50:695–706
3. Czurko A, Toth Z, Doczi T, Gallyas F (1997) Intracranial pressure waves generated by high-energy short laser pulses can cause morphological damage in remote areas: comparison of the effects of 2.1-micron Ho:YAG and 1.06-micron Nd:YAG laser irradiations in the rat brain. Lasers Surg Med 21:444–455
4. Delius M (2002) Twenty years of shock wave research at the institute for surgical research. Eur Surg Res 34:30–36
5. Dorheide J, Hoyer HE (1984) Holographic investigation of the impact response of human heads. J Neurosurg 60:718–723
6. Kato K, Fujimura M, Nakagawa A, Saito A, Ohki T, Takayama K, Tominaga T (2007) Pressure-dependent effect of shock waves on rat brain: induction of neuronal apoptosis mediated by a caspase-dependent pathway. J Neurosurg 106:667–676
7. Kleine H, Timofeev E, Takayama K (2005) Laboratory-scale blast wave phenomena - optical diagnostics and applications. Shock Waves 14:343–357
8. Kodama T, Uenohara H, Takayama K (1998) Innovative technology for tissue disruption by explosive-induced shock waves. Ultrasound Med Biol 24:1459–1466
9. Kuwahara M, Kambe K, Kurosu S, Kageyama S, Ioritani N, Orikasa S, Takayama K (1987) Clinical application of extracorporeal shock wave lithotripsy using microexplosions. J Urol 137:837–840
10. Nakagawa A, Kusaka Y, Hirano T, Sato T, Shirane R, Takayama K, Yoshimoto T (2003) Experimental application of shock waves in the vicinity of brain and skull. J. Neurosurg. 99:156–160
11. Saito T, Voinovich PA, Nakagawa A, Hosseini SHR, Takayama K, Hirano T (2004) On the efficiency of Gore-Tex layer for brain protection from shock wave damage in cranioplasty. Rev Sci Instrum 75:4789–4796

Acta Neurochir Suppl (2008) 102: 425–429

Modulation of AQP4 expression by the selective V1a receptor antagonist, SR49059, decreases trauma-induced brain edema

Keisuke Taya · Salih Gulsen · Kenji Okuno · Ruth Prieto · Christina R. Marmarou · Anthony Marmarou

Abstract

Background Currently, there are no pharmacological treatments available for traumatically induced brain edema and the subsequent rise of ICP. Evidence indicates that Aquaporin-4 (AQP4) plays a significant role in the pathophysiology of brain edema. Previously we have reported that SR49059 reduced brain edema secondary to ischemia. We, therefore, examined whether the selective V1a receptor antagonist, SR49059, reduces brain edema by modulating AQP4 expression following cortical contusion injury (CCI).

Methods Traumatic brain injury (TBI) was produced in thirty-two adult male Sprague-Dawley rats by lateral CCI (6.0 m/sec, 3 mm depth). Animals were randomly assigned to vehicle (n=16) or SR49059 treatment (n=16) groups and administered drug (960 μl/hr i.v.) immediately after injury over a 5 hr period. Animals were sacrificed for assessment of brain water content by Wet/Dry method and AQP4 protein expression by immunoblotting expressed as the ratio of AQP4 and Cyclophilin-A densitometries.

Findings Elevated AQP4 expression levels and water content were observed on the right injured side in both the right anterior (RA) and right posterior (RP) section compared to the left non-injured side inclusive of the left anterior (LA) and right anterior (RA) sections. The average AQP4 expression levels in contused areas for animals receiving SR drug treatment (RA: 1.313±0.172, RP: 1.308±0.175) were significantly decreased from vehicle-treated animals (RA: 2.181±0.232, RP: 2.303±0.370, p=0.001, p=0.003). Water content levels on SR treatment (78.89±0.14) was also significantly decreased from vehicle levels (80.38±0.38, p<0.01) in the traumatized hemisphere.

Conclusions SR49059 significantly reduced trauma-induced AQP4 up-regulation in the contused hemisphere. Moreover, brain water content was also significantly reduced paralleling the AQP4 suppression. These data provide further support that vasopressin (AVP) and V1a receptors can control water flux through astrocytic plasma membranes by regulating AQP4 expression. Taken in concert, these results affirm our laboratories contention that AQP4 can be effectively modulated pharmacologically.

Keywords AQP4 · AVP · V1a receptor · SR49059 · Brain edema

Introduction

Brain edema, the infiltration and accumulation of excess fluid in the brain, leads to an increase in brain tissue

K. Taya · S. Gulsen · K. Okuno · R. Prieto · C. R. Marmarou · A. Marmarou (✉)
Department of Neurosurgery,
Virginia Commonwealth University Medical Center,
1101 East Marshall Street, P. O. Box 980508, Richmond, VA 23298-0508, USA
e-mail: marmarou@abic.vcu.edu

S. Gulsen
Baskent Universitesi Tip Fakultesi Hastanesi,
Beyin Cerrahisi Bolumu,
Bahcelievler,
06490 Ankara, Turkey

R. Prieto
Department of Neurosurgery,
Clínico San Carlos University Hospital Planta Sexta - Ala Sur,
Profesor Martin Lagos s/n, Madrid,
28040 Madrid, Spain

K. Okuno
Department of Emergency Medicine,
Jikei University School of Medicine,
3-25-8 Nishi-shinbashi Minato-ku,
Tokyo, Japan 105-8461

volume, a key determinant of morbidity and mortality following TBI [16, 21, 25]. The cellular and molecular mechanism contributing to the development and resolution of TBI-associated brain edema are not well understood and current treatments are unsatisfactory.

Aquaporins (AQPs), a family of water channel proteins ubiquitously distributed throughout the body, comprises at least 12 members in mammals [1, 34] and mediate rapid trans-membrane movement of water. One member of this family, aquaporin-4(AQP4), is abundantly expressed in brain, specifically located in astrocytes and ependymal cells, the cells facing capillaries and pia mater [22]. Recently, many groups have proposed that AQP4 plays a significant role in the pathophysiology of brain edema [1, 2, 25, 33]. Manley et al. demonstrated that AQP4 deletion protected mice from brain swelling in two models of primarily cytotoxic edema: water intoxication and permanent focal cerebral ischemia [20].

It has long been known that the brain contains an intrinsic arginine vasopressin (AVP) fiber system [4, 9, 18]. Many authors have described that centrally released AVP plays an important role in the brain capillary water permeability [12, 26] and ionic homeostasis [6, 8]. Of great theoretical and practical importance, there exists some possibility that of AVP may be mediated by modulation of a specialized water channel, AQP4 [17, 23]. AVP regulates AQP4 expression and translocation in kidney collecting ducts, modulating water re-absorption and maintaining water homeostasis. In brain, in-vitro studies suggest that AQP4-mediated water flux is facilitated by vasopressin V1a receptor agonist [23]. The non-peptide V1 receptor antagonist OPC-21268 significantly reduced brain edema after cold brain injury [3].

Recently, our laboratory has explored the hypothesis that modulation of AQP4 channels may represent a potential avenue for therapeutic intervention. Kleindienst et al. (2006) demonstrated that the selective vasopressin receptor (V1a) antagonist, SR49059, reduced AQP4 expression and brain edema following middle cerebral artery occlusion (MCA-O) [17]. The results of these studies suggest that AQP4 modulation may be linked to AVP release and receptor activation and be part of the mechanisms controlling brain capillary water permeability and ionic homeostasis. Therefore, the objective of this study is to examine whether the selective V1a receptor antagonist, SR49059, reduces brain edema by modulating AQP4 expression following TBI.

Materials and methods

Animals

All experimental procedures involving animals were approved by the Virginia Commonwealth University (VCU) Institutional Animal Care and Use Committee (IACUC) and were conducted in accordance with the recommendations provided in the National Institutes of Health (NIH) guide for the Care and Use of Laboratory Animals. Experiments were carried out on 350 to 430 g adult male Sprague- Dawley rats (Harlan, Indianapolis IN). Rats were housed at 22±1°C with 60% humidity, 12-hour light/12-hour dark cycles, and pellet food and water ad libitum.

Controlled cortical impact injury and surgical procedure

A well-established controlled cortical impact injury models as previously described was used to cause TBI [11]. Rats were initially anesthetized with isoflurane (4.0%), intubated and then artificially ventilated with a gas mixture of nitrous oxide (70%), oxygen (30%) and isoflurane (0.4–2.0%). Rectal temperature was maintained at 37±0.5°C using a heat lamp. Catheters (P.E.50, Becton Dickenson and Company, Sparks, MD) were placed into the femoral artery and femoral vein. Mean arterial blood pressure (MABP), arterial blood gas levels, and brain temperature were monitored and recorded continuously using a data acquisition system (ADInstruments, Colorado Springs, CO). Animals were mounted on stereotaxic frame and secured by two ear bars and an incisor bar. A midline scalp incision was made, and, the skin and periosteum were retracted from the skull surface. A 10-mm-diameter craniotomy was made midway between bregma and lambda on the right side, with the medial edge of the craniotomy 1 mm lateral to midline. Injury was produced using a pneumatic impactor mounted at an angle of 10° from the vertical plane. A single impact at a velocity of 6 m/sec with a deformation depth of 3.0 mm was delivered. After injury, the removed skull section was replaced and sealed with bone wax and the incision was closed with 4-0 silk suture.

Study protocol and drug preparation

The objective of these experiments was to assess the effect of intravenous SR49059 administration on brain swelling and AQP4 expression following CCI. The animals were randomly assigned to vehicle infusion (n=16), or SR49059 infusion (n=16). SR49059 (Sanofi Recherche, Montpellier, France) was dissolved in 1.5% dimethyl sulfoxide (DMSO) as vehicle solution (Sigma-Aldrich, St Louis, MO). SR49059 was injected intravenously immediately after injury and then used in amount of 2.76 mg/kg at the same speed (960 μL/hr) on the basis of information obtained from the literature [29, 30] and our studies [17]. The drug was intravenously administered for 5 hours by using a continuous infusion pump (sp210w syringe pump, KD Scientific, Holliston, MA). Five hours after TBI, animals were sacrificed by an overdose of isoflurane (5%) and decapitated in order to remove the brain. Cerebral tissue was used to measure brain water content by wet /dry weight method and AQP4 expression was assessed by immunoblotting.

Brain edema measurement

Percent brain water content was determined using the wet/dry weight method. After animals were killed by decapitation, brains were quickly removed and the cerebellum discarded. The right and the left hemispheres along the anatomic midline were separated, and the wet weight of each hemisphere was measured. The tissue was then completely dried in a desiccating oven at 95°C for 5 days, and dry weight of each hemisphere was recorded. The percentage water content (%water) was calculated for each hemisphere as follows: %water=[(wet weight−dry weight)/wet weight] ×100. Water content was reported in the left uninjured hemisphere and the right injured hemisphere comparing vehicle and SR treated groups.

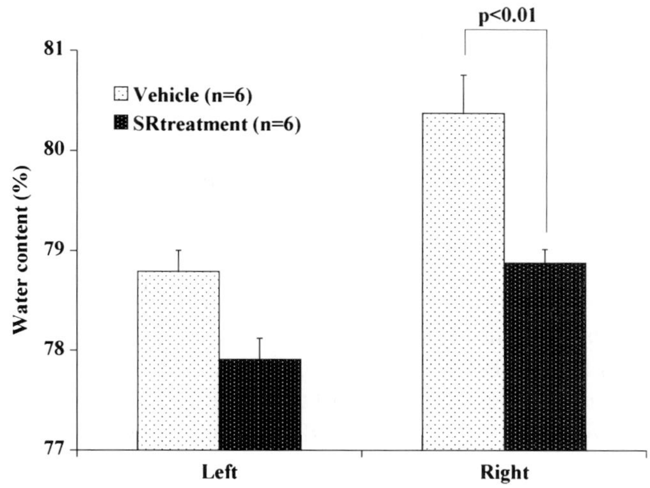

Fig. 2 Graph demonstrates that SR49059 treatment significantly reduced brain water content in the traumatized hemisphere. Both sections in the traumatized hemispheres (Right) showed increased water content compared to those in the non-traumatized hemispheres (Left)

Immunoblotting

Cerebral tissue was immediately cut into four consecutive 4 mm coronal sections excluding the most rostral and caudal sections from further analysis. The central 4 mm coronal section was then bisected into anterior and posterior sections and then bisected at the midline. This yielded 4 sections of tissue for analysis, two on the uninjured side designated left, anterior (LA) and right, posterior (RP) and two sections on the contused, or injured side designated, right anterior (RA) and right posterior (RP). Each of the four sections was homogenized on ice with a tissue homogenizer in 1200 μL of radioimmunoprecipitation buffer (50 mM Tris, 150 mM NaCl, 1% Igepal, 0.5% sodium *n*-dodecyl sulfate (SDS), 1%, pH 7.2) containing proteolysis inhibitors (Aprotinin 1.5 nM, E-64 Protease Inhibitor 0.01 μM, Leupeptin 0.01 μM). Homogenates were centrifuged at 13500 g at 4°C for 30 minutes to obtain supernatants and remove nuclei and mitochondria. Each protein concentration of supernatants was determined using protein assay kit (Bio-Rad Laboratories, Hercules, CA) and loading samples were adjusted to the same concentration (0.6 μg/μL) using sample buffer (Invitrogen, Carlsbad, CA). Protein (15 μg) from each sample was loaded for electrophoresis into 4–12% Bis-Tris polyacrylamide gels (Invitrogen, Carlsbad, CA), and subsequently transferred to a nitrocellulose membrane (Invitrogen, Carlsbad, CA). After the transfer, membranes were blocked for 45 minutes at room temperature in tris buffered saline plus Tween-20 (TBS-T)(10 mM Tris, 150 mM NaCl, 0.05% Tween-20, pH 7.5) with 3% milk powder, then incubated overnight at 4°C in AQP4 monoclonal antibody (Abcam, Inc., Cambridge, MA) diluted 1:750. The next morning, the membrane in primary antibody was incubated at room temperature for 20 minutes, washed three times for 10 minutes in TBS, blocked for 30 minutes, and subsequently incubated for 2 hours in secondary goat anti-mouse antibody (Rockland Gilbertsville, PA) diluted 1:5000. After two washes in TBS-T and three in TBS, immunodetection of AQP4 proteins was accomplished using an enhanced chemiluminescence system (Amersham, Buckinghamshire UK). Densitometric analysis was used to quantify AQP4 protein expression levels by determining intensity values for each band relative to cyclophilin-A (used as an internal control for lane loading).

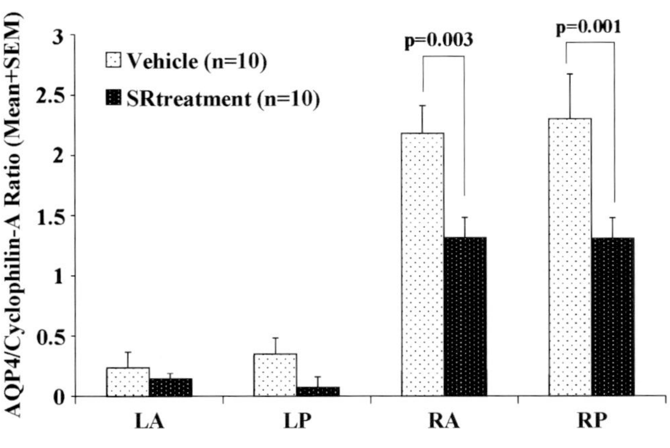

Fig. 1 Graph demonstrates that selective V1a receptor antagonist, SR49059, significantly reduced trauma-induced AQP4 up-regulation in the contusional areas. Both traumatized areas (RA, RP) showed the strong up-regulation of AQP4 compared to non-traumatize areas (LA, LP)

Statistical analysis

All data are given as means ± SEM. Statistical analyses were performed by Student's unpaired two-tail t-test. Values of $p<0.05$ were considered significant.

Result

All experiments were carried out without injury-induced morality. MABP and arterial blood gases were kept within physiological limits throughout the experimental procedure, requiring few adjustments in the isoflurane concentration and ventilation parameters. Elevated AQP4 expression levels and water content were observed on traumatized side (RA and RP) compared to the non-traumatized side (RA and RP). In brain sections bearing the contusion, the average AQP4 expression level in rats subjected to SR treatment (RA:1.313±0.172, RP:1.308±0.175) was significantly decreased from those treated with vehicle (RA:2.181±0.232, RP: 2.303±0.370, p=0.001, p=0.003) (Fig. 1). Water content levels on SR treatment (78.89± 0.14) was also significantly decreased from vehicle levels (80.38±0.38, $p<0.01$) in the traumatized hemisphere (Fig. 2).

Discussion

Arginine vasopressin (AVP) is a neuropeptide that is synthesized in the hypothalamus and transported via the axonal fiber system to the neurohypophysis, which releases the hormone into the blood stream [4]. AVP has been implicated in numerous central functions, including influence on brain water permeability [12, 26], regulation of intracranial pressure [24] and cerebrospinal fluid production [14]. The intraventricular injection of AVP aggravated cold injury edema and ischemic brain edema in mammals [10, 13, 27]. In vitro, AVP has also been shown to regulate glial cell volume [5, 7, 19]. Moreover, treatment with AVP receptor antagonist prevented cold-induced vasogenic edema [3] and hemorrhagic brain edema [28]. Accordingly, the increase of AVP is strongly associated with brain edema formation and facilitating transport of water from blood to brain.

It is generally assumed that V1a receptors mediate the central effects of AVP. The compound SR49059 has been characterized recently as the most potent and selective non-peptide AVP V1a antagonist described thus far, with marked affinity, selectivity and efficacy in both animal and human receptors [29, 32]. This compound inhibits AVP-induced vascular smooth cell contraction, blood pressure elevation and platelet aggregation [29, 31].

In this study, we investigated the effects of intravenously administered SR49059, a non-peptide selective AVP V1a receptor antagonist, on trauma-induced brain edema in rats. Our results showed that the increase in brain water content induced by CCI was significantly suppressed by treatment with SR49059. Recently, in our laboratory we have also found that treatment with SR49059 significantly reduced brain water content following middle cerebral artery occlusion model [17]. Collectively, these studies provide further experimental confirmation that the V1a receptor antagonist has protective effects on brain edema.

Water can cross cell membranes through different pathways: specific water channels (aquaporins), the lipid bilayer [15], or ion-water co-transport proteins [35, 36]. Specifically, in the central nervous system, AQP4 has been suggested to play a crucial role in cerebral water balance because of its anatomical and cellular localization. One possible explanation for the protective effects of SR49059 on brain edema formation found in this study is that AQP4 expression or activity may be regulated by AVP through V1a receptor. In these experiments, we found that trauma-induced AQP4 up-regulation was prevented by SR49059. To the best of our knowledge, this study is the first to demonstrate that following CCI, AQP4 expression is reduced by treatment with SR49059, the selective AVP V1a receptor antagonist. Although the precise molecular mechanisms of AQP4 regulation are not yet known, taken in concert, our study strengthens the basic concept that AQP4 expression can be regulated pharmacologically to effect a reduction in brain edema. Further experiments are needed to elucidate the fundamental mechanisms of AQP4 regulation and the relationship of AVP to V1a to provide novel treatments for reducing brain edema following TBI.

Acknowledgments This research was supported by grants NS 12587 and NS19235 from the National Institutes of Health, Bethesda, MD.

Conflict of interest statement We declare that we have no conflict of interest.

References

1. Agre P, Kozono D (2003) Aquaporin water channels: molecular mechanisms for human deseases. FEBS Lett 555:72–78
2. Amiry-Moghaddam M, Ottersen OP (2003) The molecular basis of water transport in the brain. Nat Rev Neurosci 4:991–1001
3. Bemana I, Nagao S (1999) Treatment of brain edema with a nonpeptide arginine vasopressin V1 receptor antagonist OPC-21268 in rats. Neurosurgery 44:148–155
4. Buijs RM (1978) Intra and extrahypothalamic vasopressin and oxytocin pathway in the rat. Pathways to limbic system, medulla oblongata and spinal cord. Cell Tissue Res 192:423–435
5. Chen Y, McNeill JR, Hajek I, Hertz L (1992) Effect of vasopressin on brain swelling at the cellular level: do astrocytes

exhibit a furosemide-vasopressin-sensitive mechanism for volume regulation? Can J Physiol Pharmacol 70:S367–S373
6. Cserr HF, Latzkovits L (1992) A role for centrally released vasopressin in brain ion and volume regulation: a hypothesis. Prog Brain Res 91:3–6
7. Del Bigio MR, Fedoroff S (1990) Swelling of astroglia in vitro and the effect of arginine vasopressin and atrial natriuretic peptide. Acta Neurochir Suppl 51:14–16
8. DePasquale M, Patlak CS, Cserr HF (1989) Brain ion and volume regulation during acute hypernatremia in Brattleboro rats. Am J Physiol 256:F1059–F1066
9. de Vries GJ, Miller MA (1998) Anatomy and function of extrahypothalamic vasopressin systems in the brain. Prog Brain Res 199:3–20
10. Dickinson LD, Betz AL (1992) Attenuated development of ischemic brain edema in vasopressin deficient rats. J Cereb Blood Flow Metab 12:681–690
11. Dixon CE, Clifton GL, Lighthall JW, Yaghmai AA, Hayes RL (1991) A controlled cortical impact model of traumatic brain injury in the rat. J Neurosci Methods 39:253–262
12. Doczi T, Szerdahelyi P, Gluya K, Kiss J (1982) Brain water accumulation after the central administration of vasopressin. Neurosurgery 11:402–407
13. Doczi T, Laszlo FA, Szerdahelyi P, Joo F (1984) Involvement of vasopressin in brain edema formation: further evidence obtained from the Brattleboro diabetes insipidus rat with experimental subarachnoid hemorrhage. Neurosurgery 14:436–441
14. Faraci FM, Mayhan WG, Farrrell WJ, Heistad DD (1988) Humoral regulation of blood flow to choroid plexus: role of arginine vasopressin. Circ Res 63:373–379
15. Finkelstein A (1987) Water movements through lipid bilayers, pores and plasma membranes: theory and reality. John Wiley & Sons, New York
16. Graham DI, Adams JH, Nicoll JA, Maxwell WL, Gennarelli TA (1995) The nature, distribution and causes of traumatic brain injury. Brain Pathol 5:397–406
17. Kleindienst A, Fazzina G, Dunbar JG, Glisson R, Marmarou A (2006) Protective effect of the V1a receptor antagonist SR49059 on brain edema formation following middle cerebral artery occlusion in the rat. Acta Neurochir Suppl 96:303–306
18. Landgraf R (1992) Central release of vasopressin: stimuli, dynamics, consequences. Prog Brain Res 91:29–39
19. Latzkovits L, Cserr HF, Park JT, Patlak CS, Pettigrew KD, Rimanoczy A (1993) Effects of arginine vasopressin and atriopeptin on glial cell volume measured as 3-MG space. Am J Physiol 264:C603–C608
20. Manley GT, Fujimura M, Ma T, Noshita N, Filiz F, Bollen AW, Chan P, Verkman AS (2000) Aquaporin4 deletion in mice reduces brain edema after acute water intoxication and ischemic stroke. Nat Med 6:159–163
21. Marmarou A (1994) Traumatic brain edema: an overview. Acta Neurochir Suppl 60:421–424
22. Nielsen S, Nagelhus EA, Amiry-Moghaddam M, Bourque C, Agre P, Ottersen OP (1997) Specialized membrane domains for water transport in glial cells: high- resolution immunogold cytochemistry of aquaporin4 in rat brain. J Neurosci 17:171–180
23. Nierman H, Amry-Moghaddam M, Holthoff K, Witte Ow, Ottersen OP (2001) A novel role of vasopressin in the brain: modulation of activity-dependent water flux in the neocortex. J Neurosci 21:3045–3051
24. Noto T, Nakajima T, Saji Y, Nagawa Y (1978) Effect of vasopressin on intracranial pressure of rabbit. Endocrinol Jpn 25:591–596
25. Papadopoulos MC, Krishna S, Verkman AS (2002) Aquaporin water channels and brain edema. Mt Sinai J Med 69:242–248
26. Raichle ME, Grubb RL (1978) Regulation of brain water permeability by centrally released vasopressin. Brain Res 143: 191–194
27. Reeder RF, Nattie EE, North WG (1986) Effect of vasopressin on cold-induced brain edema in cats. J Neurosurg 64:941–950
28. Rosenberg GA, Scremin O, Estrada E, Keyner WT (1992) Arginine vasopressin V1 antagonist and atrial natriuretic peptide reduce hemorrhagic brain edema in rats. Stroke 23:1767–1774
29. Serradeil-Le Gal C, Wagnon J, Garcia C, Lacour C, Guiraudou P, Christophe B, Villanova G, Nisato D, Maffrand JP, Fur GL, Guillon G, Cantau B, Barberies C, Trueba M, Ala Y, Jard S (1993) Biochemical and pharmacological properties of SR49059, a new, potent, nonpeptide antagonist of rat and human vasopressin V1a receptors. J Clin Invest 92:224–231
30. Shuaib A, Wang CX, Yang T, Noor R (2002) Effects of nonpeptide V1 vasopressin receptor antagonist SR49059 on infarction volume and recovery of function in a focal embolic stroke model. Stroke 33:3033–3037
31. Thibonnier M, Kilani A, Rahman M, DiBlasi TP, Warner K, Smith MC, Leenhardt AF, Brouard R (1999) Effects of the nonpeptide V1 vasopressin receptor antagonist SR49059 in hypertensive patients. Hypertension 34:1293–1300
32. Tribollet E, Raufaste D, Maffrand JP, Serradeil-Le Gal C (1999) Binding of the non-peptide vasopressin V1a receptor antagonist SR49059 in the rat brain: an in vitro and in vivo autoradiographic study. Neuroendocrinology 69:113–120
33. Venero JL, Vizuete ML, Machado A, Cano J (2001) Aquaporins in the central nervous system. Prog Neurobiol 63:321–336
34. Verkman AS (2002) Physiological importance of aquaporin water channels. Ann Med 34:192–200
35. Zeuthen T (1994) Cotransport of K^+,Cl^- and H_2O by membrane proteins from choroid plexus epithelium of Necturus maculosus. J Physiol 478:203–219
36. Zeuthen T, Meinild AK, Klaerke DA, Loo DD, Wright EM, Belhage B, Litman T (1997) Water transport by the Na^+/ glucose cotrans porter under isotonic conditions. Biol Cell 89:307–312

Acta Neurochir Suppl (2008) 102: 431–436

The modulation of aquaporin-4 by using PKC-activator (phorbol myristate acetate) and V1a receptor antagonist (SR49059) following middle cerebral artery occlusion/reperfusion in the rat

Kenji Okuno · Keisuke Taya · Christina R. Marmarou · Pinar Ozisik · Giovanna Fazzina · Andrea Kleindienst · Salih Gulsen · Anthony Marmarou

Abstract

Background We have pursued the concept that traumatic brain edema is predominantly cellular and that water entry is modulated in part by aquaporins. Aquaporin-4 (AQP4) has been shown to play a significant role in cellular edema formation. Phorbol myristate acetate (PMA) is a potent PKC activator; purportedly involved in modulation of AQP4 activity. Alternatively, AQP4 may be regulated by arginine-vasopressin. Administration of the vasopressin antagonist (SR49059) reduced brain water content and sodium shift following MCAo. To investigate if edema formation is affected by the reduction of AQP4 expression, we utilized PMA and SR49059 following middle cerebral artery occlusion model (MCAo), and measured AQP4 expression by Western-Blot (WB) techniques.

Methods Male Sprague Dawley rats were randomly assigned to sham (n=4) or MCAo groups (vehicle, PMA or SR49059 infusion; n=6 each). Each solution was infused for 5 hours, starting 1 hour before injury. After a two-hour period of ischemia and two-hour reperfusion, animals were sacrificed and brain regions of interest were processed by WB to quantify the effect of treatment on AQP4 expression.

This research was supported by grant NS 12587 and grant NS 19235 from the National Institutes of Health.

K. Okuno · K. Taya · C. R. Marmarou · P. Ozisik · G. Fazzina · A. Kleindienst · S. Gulsen · A. Marmarou
Department of Neurosurgery, Medical College of Virginia, Virginia Commonwealth University, Richmond, VA, USA

K. Taya · C. R. Marmarou · A. Marmarou
Department of Neurosurgery, Virginia Commonwealth University, 1001 East Broad Street, Old City Hall, Suite 235, Richmond, VA 23298-0508, USA

K. Okuno
Department of Emergency Medicine, Jikei University School of Medicine, 3-25-8 Nishi-shinbashi Minato-ku, Tokyo 105-8461, Japan

G. Fazzina
Liceo Scientifico Statale "O.M. Corbino" IX ISTITUTO DI ISTRUZIONE SECONDARIA SUPERIORE, Viale Regina Margherita, 16, 96100 SIRACUSA, Italy

A. Kleindienst
Department of Neurosurgery, Georg-August-Universitat Gottingen, Robert-Koch-Str. 40, 37075 Gottingen, Germany

S. Gulsen
Department of Neurosurgery, Baskent University, Fevzi Cakmak Caddesi, 10. sokak No:45 C Blok 1. katBahcelievler, Ankara 06490, Turkey

P. Ozisik
Kirkkonaklar Mah. 23. Cadde, Simkent Sitesi 2. Blok, No:6/13 GOP06610, Ankara, Turkey

A. Marmarou (✉)
Department of Neurosurgery, Virginia Commonwealth University Medical Center, 1101 East Marshall Street, Box 980508, Richmond, VA 23298-0508, USA
e-mail: marmarou@abic.vcu.edu

Results These studies demonstrate that MCAo results in a significant up-regulation of AQP4 on the ischemic zone when compared to the contralateral un-injured hemisphere ($p<0.05$) and that PMA and SR49059 treatment significantly down-regulated AQP4 expression compared to the vehicle group ($p<0.05$).
Conclusions These studies support the hypotheses that PMA and SR49059 may be useful in reducing cerebral water accumulation by modulating AQP4 expression and that pharmacological manipulation of AQP4 may emerge as a viable strategy for the reduction of fulminating edema following ischemic injury.

Keywords AQP water channel · Vasopressin · Middle cerebral artery occlusion · Brain edema

Introduction

Brain edema contributes to the high rate of secondary morbidity and mortality in injured patients and is a frequent complication of brain injuries such as cerebral ischemia, trauma, brain tumor, inflammation, intoxications and metabolic disorders [18, 19]. Although, the pathophysiological mechanisms of water and solute movement in traumatic brain edema are poorly understood, two major types of brain edema, vasogenic and cytotoxic can be discriminated. Vasogenic edema is due to a dramatic increase in blood brain barrier (BBB) permeability, which leads to a net influx of plasma components and fluid from the blood compartment into the extracellular space of the brain. Alternatively, cytotoxic edema results from subtle disturbance in BBB permeability associated with cellular disruptions in ionic homeostasis, and results in an intracellular retention of water. The main feature of cytotoxic edema is the swelling of brain cells, in particular the enlargement of astrocytic end-feet and concomitant contraction of the extracellular space that may be related to increased AQP4 expression [2, 3, 17]. AQP4 is the primary water channel proteins found in astrocytic end-feet surrounding the capillaries [23] In several studies, AQP4 has been shown to play a significant role in brain edema formation [1, 9, 12–14, 17, 21, 22, 24].

Recently, our laboratory has offered evidence suggesting a primarily cellular route for cerebral sodium and water inlet, supporting the role of AQP4 modulation as a potential therapeutic avenue for the treatment of fulminating edema [1, 13, 14]. AQP4 is strategically located on astrocytic end-feet that surround cerebral vessels [2, 3, 23]. Further, multiple investigations demonstrate that changes in AQP4 expression in reactive astrocytes following brain injury may be a key factor in subsequent edema formation [1, 9, 12–14, 17, 21, 22, 24]. We therefore tested the hypothesis that AQP4 may provide a potential target for the pharmacological intervention of edema using two modulators of AQP4 activity, phorbol myristate acetate (PMA), an upstream regulator of AQP4 activity and SR49059 a vasopressin antagonist that acts by blocking the vasopressin receptor (V1a). We utilized the middle cerebral artery occlusion/reperfusion (MCAo) model as it provides a predominantly cellular type of edema in the absence of BBB breakdown [4, 8, 22]. We used a MCAo model to measure the effect of PMA or SR49059 to reduce cellular swelling and concomitant water influx within the edematous brain

Phorbol myristate acetate (PMA) is a potent PKC activator that mimics the effect of the natural activator diacylglycerol (DAG) [9, 10]. In vitro studies suggest that the activity of AQP4 water channels can be regulated by PMA, through a PKC-dependent phosphorylation mechanism [27]. Alternatively, many authors have described that centrally released arginine-vasopressin (AVP) plays an important role in the brain capillary water permeability and ionic homeostasis [6, 13, 20]. In vitro, AQP4-mediated water flux is facilitated by vasopressin V1a receptor agonist. The non-peptide V1 receptor antagonist, OPC-21268 has been shown to significantly reduce brain edema after cold brain injury [5]. In our previous studies, we explored the effect of PMA to reduce of brain edema formation following MCAo and cortical contusion models and demonstrated a decrease in brain water content, sodium accumulation and AQP4 [1, 13]. In a comparison study, administration of the vasopressin antagonist (SR49059) reduced brain water content and sodium shift following MCAo suggesting the potential relationship of SR49059 to act on AQP4 [14]. These studies provided fundamental information, but the mechanism of AQP4 regulation and the potential for treatment of water accumulation in a cellular model of edema was unclear. In this study, we utilized two different types of drugs (PMA and SR49059) following MCAo in rats, and measured AQP4 expression by use of Western-Blot techniques.

Materials and methods

Animals and surgical procedure

The studies were conducted under approval of the Virginia Commonwealth University (VCU) Institutional Animal Care and Use Committee (IACUC) and in accordance with the recommendations provided in the National Institutes of Health (NIH) guide for the Care and Use of Laboratory Animals. Experiments were carried out using 350–400 g adult, male Sprague-Dawley rats (Harlan, Indianapolis IN), housed with a 12:12 hr light/dark cycle, and at 22±1°C with 60% humidity, pellet food and water ad libitum. Surgery was performed after intubation under isoflurane anesthesia and controlled ventilation (1.0–1.5% isoflurane in 70% nitrous oxide and 30% oxygen). Rectal temperature was maintained at 37.0±0.5°C using a heating lamp. The left femoral artery and vein were cannulated with polyethylene tubing (P.E. 50, Becton Dickinson and Company, Sparks, MD) for continuous monitoring of mean arterial blood pressure (MABP), blood sampling or drug infusion, respectively. Adequate ventilation was verified by an arterial blood gas measurement after one hour of anaesthesia.

Cerebral blood flow (CBF) over the supply territory of the right MCA was continuously monitored by Laser Doppler Flowmetry (LaserFlo vasamedics INC, St Paul, MN) through a burr hole located 1 mm posterior and 5 mm lateral to bregma leaving the dura intact. Animals were placed in a supine position over the laser doppler probe, and CBF as well as MABP were recorded continuously using a data acquisition system (AD Instruments, Colorado Springs, CO).

MCA occlusion was induced using the intra-luminal suture method described elsewhere [4] with slight modifications. Through a midline neck incision, the bifurcation of the right common carotid artery (CCA) was exposed, and branches of the external carotid artery (ECA) and internal carotid artery (ICA) including the occipital, lingual and maxillary arteries were separated and coagulated. The ECA was ligated with a 4-0 silk suture, and after temporary occlusion of ICA and CCA with vascular mini-clips, a 4-0 monofilament nylon suture (4-0 SN-644 MONOSOF nylon Polyamide) with a silicon tip of 0.30 to 0.35 mm diameter was inserted through the ECA stump and secured by a suture. The clips were removed and the filament was advanced through the ICA into the circle of Willis while occluding the pterygopalatine artery with forceps. A CBF reduction between 70 to 90% of baseline was observed when the suture was advanced at a distance of 22 to 24 mm from the carotid bifurcation thereby verifying proper occlusion of the MCA. Two hours after occlusion, and a 2 hr period of reperfusion in the MCA territory was performed by withdrawing the suture into the ECA stump under confirmation by a consecutive increase in CBF.

Study protocol and drug preparation

The objective of these experiments was to assess the effect of intravenous PMA and SR49059 infusion following MCA occlusion on AQP4 expression. Twenty-two animals were randomized into the following four groups; Group 1 Sham vehicle group (n=4), sham procedure with 1% DMSO infusion as vehicle solution (Sigma Chemical Co., St Louis, MO). Group2 MCAo-vehicle group (n=6), MCAo with 1% DMSO infusion. Group 3 MCAo-PMA group (n=6), MCAo with PMA (200 μg/kg; Sigma Chemical Co., St Louis, MO) infusion dissolved in 1% DMSO. Group 4 MCAo-SR group (n=6), MCAo with SR49059 (1 mg/Kg; Sanofi Recherche, Montpellier, France) infusion dissolved in 1% DMSO.

Each solution was infused at a uniform rate of 480 μL/hour for the 5 hours using a continuous infusion pump (Kd Scientific, sp210w syringe pump), starting 1 hour before injury. After a two-hour period of ischemia and two-hour reperfusion, animals were then sacrificed by an overdose of isoflurane, decapitated and the brains removed.

Tissue processing and immunoblot

Cerebral tissue was immediately cut into four consecutive 4 mm coronal sections excluding the most rostral and caudal sections from further analysis. After division into the right and left hemispheres along the anatomic midline, the four sections were homogenized on ice with a tissue homogenizer in 1.2 ml of radioimmunoprecipitation buffer (50 mM Tris, 150 mM NaCl, 1% NP40, 1% deoxycholic acid, 0.5% sodium *n*-dodecyl sulfate, pH 7.2) containing proteolysis inhibitors (Aprotinin 1.5 nM, E-64 Protease Inhibitor 0.01 μM, Leupeptin 0.01 μM). These homogenates were centrifuged in a Thermo MicroMax RF centrifuge (Thermo Electron Corporation, Needham Heights, MA) at 13500×g for 30 min at 4°C to remove nuclei and mitochondria. Each protein concentration of supernatants was determined using protein assay kit (Bio-Rad Laboratories, Hercules, CA) and samples were adjusted to same concentrations (0.6 μg/μL) using sample buffer (Invitrogen, Carlsbad, CA). Protein (15 μg) from each sample were loaded for electrophoresis into 12% Bis-Tris polyacrylamide gels (Invitrogen Nu-Page, Carlsbad, CA), and subsequently transferred to a nitrocellulose membrane (Invitrogen, Carlsbad, CA). After transfer by electro-elution to nitrocellulose membranes, membranes were blocked for 45 minutes at room temperature in tris buffered saline plus Tween-20 (TBS-T)(10 mM Tris, 150 mM NaCl, 0.05% Tween-20, pH 7.5) with 3% milk powder, then incubated overnight at 4°C in AQP4 monoclonal antibody (Abcam, Inc., Cambridge, MA) diluted 1:750. The next morning, the membrane in primary antibody was incubated at room temperature for 20 minutes, washed three times for 10 minutes in TBS, blocked for 30 minutes, and subsequently incubated for 2 hours in horseradish peroxidase-conjugated secondary goat anti-mouse antibody (Rockland Gilbertsville, PA) diluted 1:5000. After two washes in TBS-T and three in TBS, immunodetection of AQP4 proteins was accomplished using an enhanced chemiluminescence system (Amersham, Buckinghamshire UK). Controls were made by replacing primary antibody with non-immune IgG (Cyclophilin). Quantification was performed by densitometry.

Densitometry analysis was used to quantify AQP4 protein expression levels by determining intensity values

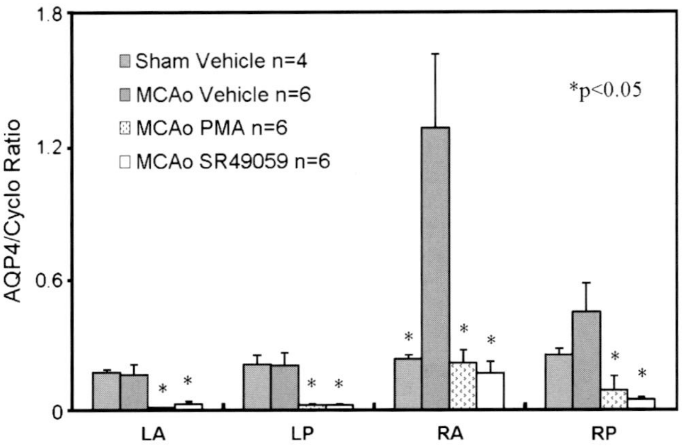

Fig. 1 Our results show that MCAo significantly up-regulated AQP4 on the ischemic side ($p<0.05$). We also found that PMA and SR49059 treatment significantly down-regulated AQP4 expression compared to the vehicle group in both hemisphere

for each band relative to cyclophilin-A (used as an internal control for lane loading). AQP4 expression was expressed as the ratio of AQP4 and cyclophilin-A densitometries.

Statistical analysis

All data are expressed as mean ± SD. Statistical analyses were performed by Student's unpaired two-tail t-test. Values of $p<0.05$ were considered significant.

Results

The experiments were performed without complication or procedure related mortality. Continuous recording of blood pressure and repeated blood gas were kept within physiological limits throughout the experimental procedure, requiring few adjustments in the isoflurane concentration and ventilation parameters. Statistical analysis showed no significant alterations in physiological parameters in Sham, MCA_O, Vehicle, PMA, or SR49059 groups.

MCA occlusion with subsequent reperfusion significantly up-regulated AQP4 expression on the ischemic region (Fig. 1). A significant difference in AQP4 expression was found in RA region, between the Sham Vehicle group (0.230±0.018) and the MCAo Vehicle group (1.277±0.332, $p<0.03$).

PMA treatment significantly down-regulated AQP4 expression compared to the vehicle group in both of the injured hemispheres. A significant difference in AQP4 expression was found in LA, LP, RA and RP region, between the MCAo Vehicle group (0.160±0.046, 0.201±0.061, 1.277±0.332, 0.445±0.129) and the MCAo PMA group (0.010±0.003, $p<0.03$, 0.022±0.010, $p<0.05$, 0.213±0.063, $p<0.03$, 0.089±0.063, $p<0.05$; Table 1).

We observed that the SR49059 treatment significantly down-regulated AQP4 expression compared to the vehicle group in both hemispheres. A significant difference in AQP4 expression was found in LA, LP, RA and RP region, between the MCAo Vehicle group and the MCAo SR group (0.030±0.013, $p<0.05$ 0.024±0.007, $p<0.05$, 0.163±0.054, $p<0.03$, 0.049±0.009, $p<0.03$; Table 1).

Discussion

Edema formation involves the infiltration and accumulation of water in the brain [18, 19] Current concepts regarding the passage of water flux across cell membranes support different pathways involved in maintaining fluid homeostasis that may include the interplay of water channels termed "aquaporins" and ion-water co-transport proteins [7, 16, 25, 27–29]. Presently, although the mechanisms are unclear, because of its specific anatomical and cellular localization in the central nervous system, AQP4 has been suggested to play a significant role in the regulation of water movement in the edematous brain [1–3, 5, 11–14, 17–19, 21, 24, 26]. We have put forth the notion that brain edema is a combination of vasogenic and cellular edema with the cellular component predominating [18, 19]. Vasogenic and cytotoxic edema often occur simultaneously and a pure form of either edema is unlikely to exist. In cerebral ischemia models, both types of edema are involved. However, most agree that during the early stages of ischemia, brain edema can be considered primarily cellular in nature whereas vasogenic edema is not found to occur within several hours after an ischemic insult [8, 18, 19, 21, 22].

Our findings demonstrate that MCAo significantly up-regulates AQP4 expression within the ischemic region. This data suggests that the up-regulation of AQP4 expression plays a significant role in cellular edema formation. In our previous

Table 1 The comparison of AQP4 expression between Vehicle vs PMA, and Vehicle vs SR49058

	Vehicle	PMA	p value Vehicle vs PMA	SR49059	p value Vehicle vs SR49059
LA	0.160±0.046	0.010±0.003	$p<0.03$	0.030±0.013	$p<0.05$
LP	0.201±0.061	0.022±0.010	$p<0.05$	0.024±0.007	$p<0.05$
RA (ischemic side)	1.277±0.332	0.213±0.063	$p<0.03$	0.163±0.054	$p<0.03$
RP (ischemic side)	0.445±0.129	0.089±0.063	$p<0.05$	0.049±0.009	$p<0.03$

studies, we examined the reduction of brain edema formation through the regulation of AQP4 by PMA and selective vasopressin V1a antagonist (SR49059) following MCAo models and showed that edema was reduced [1, 13, 14]. The current study demonstrates that PMA and SR49059 down-regulate AQP4 expression on the ischemic and non-ischemic region following MCAo model. We conclude that the concomitant reduction of edema and AQP4 by these drugs support our concept of water entry via AQP4 channels.

PMA is a potent protein kinase C (PKC) activator that mimics the effect of the natural activator diacylglycerol (DAG) [9, 10, 27]. PMA is able to inhibit AQP4 activity and to regulate AQP4 expression at the transcriptional level in vitro, through biochemical pathways that involve PKC activation [10, 27]. PKC is a second messenger ubiquitously found throughout the body; the inhibition of PKC by PMA may inhibit all second messenger pathways and may be toxic. In these studies, we experienced no toxic effects of PMA during experimental procedures.

Alternately, evidence indicates that AQP4 is regulated by vasopressin, a neuropeptide endogenous to the brain released via a vasopressin-containing fiber system [6, 15, 20]. V1a receptors are coupled via G-proteins to phospholipase C resulting in an IP3-dependent Ca^{2+} release from internal stores [9, 20]. V1a receptor stimulation also causes activation of PKC, and evidence has been provided by other investigators that vasopressin exerts part of its facilitator effect through this signalling pathway [20].

In these experiments, we found that ischemia- induced AQP4 up-regulation was prevented by PMA and SR49059. To the best of our knowledge, this study is the first to demonstrate that following MCAo, AQP4 expression is reduced by treatment with PMA and the selective AVP V1a receptor antagonist, SR49059, Although the precise molecular mechanisms of AQP4 regulation are not yet known, these studies provide further support to our notion that AQP4 can be pharmacologically manipulated and effectively reduce brain edema. Further experiments are needed to elucidate the precise mechanisms of AQP4 regulation to facilitate our understanding of the potential of new treatments for brain edema.

Acknowledgments This research was supported by grants NS 12587 and NS19235 from the National Institutes of Health, Bethesda, MD.

Conflict of interest statement We declare that we have no conflict of interest.

References

1. Amorini AM, Dunbar JG, Marmarou A (2003) Modulation of aquaporin-4 water transport in a model of TBI. Acta Neurochir Suppl 69:261–263
2. Badaut J, Verbavatz JM et al (2000) Presence of aquaporin-4 and muscarinic receptors in astrocytes and ependymal cells in rat brain: a clue to a common function? Neurosci Lett 292(2):75–78
3. Badaut J, Lasbennes F, Magistretti PJ, Regli L (2002) Aquaporins in brain: distribution, physiology and pathophysiology. J Cereb Blood Flow Metab 22(4):367–378
4. Belayev L, Alonso OF, Busto R, Zhao W, Ginsberg MD (1996) Middle cerebral artery occlusion in the rat by intraluminal suture. Neurological and pathological evaluation of an improved model. Stroke 27(9):1616–1622 discussion 1623
5. Bemana I, Nagao S (1999) Treatment of brain edema with a nonpeptide arginine vasopressin V1 receptor antagonist OPC-21268 in rats. Neurosurgery 44(1):148–154 discussion 154–5
6. de Vries GJ, Miller MA (1998) Anatomy and function of extrahypothalamic vasopressin systems in the brain. Prog Brain Res 1193–1120
7. Finkelstein A (1987) Water movements through lipid bilayers, pores and plasma membranes. Theory and reality. New York, John Wiley & Sons
8. Gotoh O, Asano T et al (1985) Ischemic brain edema following occlusion of the middle cerebral artery in the rat. I: The time courses of the brain water, sodium and potassium contents and blood-brain barrier permeability to 125I-albumin. Stroke 16(1):101–109
9. Gunnarson E, Zelenina M, Aperia A (2004) Regulation of brain aquaporins. Neuroscience 129(4):947–955
10. Han Z, Wax MB, Patil RV (1998) Regulation of aquaporin-4 water channels by phorbold ester-dependent protein phosphorylation. J Biol Chem 273(11):6001–6004
11. Ke C, Poon WS et al (2001) Heterogeneous responses of aquaporin-4 in oedema formation in a replicated severe traumatic brain injury model in rats. Neurosci Lett 301(1):21–24
12. Kiening KL, van Landeghem FK et al (2002) Decreased hemispheric Aquaporin-4 is linked to evolving brain edema following controlled cortical impact injury in rats. Neurosci Lett 324(2):105–108
13. Kleindienst A, Fazzini G, Amorin AM, Dunbar JG, Glisson R, Marmarou A (2006) Modulation of AQP4 expression by the protein kinase C activator, phorbol myristate acetate, decreses ischemia-induced brain edema. Acta Neurochir Suppl 96:393–397
14. Kleindienst A, Fazzina G, Amorini AM, Dunbar JG, Glisson R, Marmarou A (2006) Protective effect of the V1a receptor antagonist SR49059 on brain edema formation following middle cerebral artery occlusion in the rat. Acta Neurochir Suppl 96:303–306
15. Landgraf R (1992) Central release of vasopressin: stimuli, dynamics, consequences. Prog Brain Res 9129–9139
16. Loo DD, Zeuthen T et al (1996) Cotransport of water by the Na+/glucose co-transporter. Proc Natl Acad Sci U SA 93(23):13367–13370
17. Manley GT, Fujimura M, Ma T, Noshita N, Filiz F, Bollen AW, Chan P, Verkman AS (2000) Aquaporin-4 deletion in mice reduces brain edema after acute water intoxication and ischemic stroke. Nat Med 6(2):159–163
18. Marmarou A (2003) Pathophysiology of traumatic brain edema: current concepts. Acta Neurochir Supl 86:7–10
19. Marmarou A (2007) A review of progress in understanding the pathophysiology and treatment of brain edema. Neurosurg Focus 15; 22 (5) E1 Review
20. Niermann H, Amiry-Moghaddam M, Holthoff K, Witte OW, Otterson OP (2001) A novel role of vasopressin in the brain: modulation of activity-dependent water flux in the neocortex. J Neurosci 21(9):3045–3051
21. Sato S, Umenishi F, Inamasu G, Sato M, Ishikawa M, Nishizawa M, Oizumi T (2000) Expression of water channel mRNA following cerebral ischemia. Acta Neurochir Suppl 76239–76241
22. Taniguchi M, Yamashita T, Kumura E, Tamatani M, Kobayashi A, Yokawa T, Maruno M, Kat A, Ohnishi T, Kohmura E, Tohyama M, Yoshimine T (2000) Induction of aquaporin-4 water channel

mRNA after focal cerebral ischemia in rat. Brain Res Mol Brain Res 78(1–2):131–137
23. Venero JL, Vizuete ML, Machado A, Cano J (2001) Aquaporins in the central nervous system. Prog Neurobiol 63(3):321–336
24. Vizuete ML, Venero JL, Vargas C, Ilundain AA, Echevarria M, Machado A, Cano J (1999) Differential upregulation of aquaporin-4 mRNA expression in reactive astrocytes after brain injury: potential role in brain edema. Neurobiol Dis 6(4):245–258
25. Wright EM, Loo DD (2000) Coupling between Na+, sugar, and water transport across the intestine. Ann N Y Acad Sci 91554–91566
26. Yamamoto N, Yoneda K, Asai K, Sobue K, TAda T, Fujita Y, Katsuya H, Fujita M, Aihara N, Mase M, Yamada K, Miura Y, Kato T (2001) Alterations in the expression of the AQP family in cultured rat astrocytes during hypoxia and reoxygenation. Brain Res Mol Brain Res 90(1):26–38
27. Zelenina M, Zelenin S, Bondar AA, Brismar H, Aperia A (2002) Water permeability of aquaporin-4 is decreased by protein kinase C and dopamine. Am J Physiol Renal Physil 283(2):F309—F318
28. Zeuthen T (1994) Cotransport of K+, Cl- and H2O by membrane proteins from choroid plexus epithelium of Necturus maculosus. J Physiol 478(Pt 2):203–219
29. Zeuthen T, Meinild AK, Klaerke DA, Loo DD, Wright EM, Belhave B, Litman T (1997) Water transport by the Na+/glucose cotransporter under isotonic conditions. Biol Cell 89(5–6):307–312

Acta Neurochir Suppl (2008) 102: 437–439

Pro-inflammatory and pro-apoptotic elements of the neuroinflammatory response are activated in traumatic brain injury

J. C. Goodman · M. Van · S. P. Gopinath · C. S. Robertson

Abstract

Background The inflammatory response may contribute to cerebral edema, increased intracranial pressure and cellular loss in traumatic brain injury (TBI). Cytokines are biomarkers of this inflammatory response and new methods allow simultaneous measurement of multiple cytokines.

Methods We examined the IL-1β, IL-6, IL-8 and IL-12, TNFα, and IL-10 in arterial and jugular blood as well as cerebrospinal fluid in patients with severe traumatic brain injury.

Findings Multiple cytokines, particularly pro-inflammatory cytokines, are up-regulated following TBI. Cerebrospinal fluid and arteriovenous differences of some of the cytokines suggest production within the central nervous system. Anti-inflammatory cytokines are not up-regulated.

Conclusions Cytokine up-regulation may contribute to the neuroinflammatory reaction that follows traumatic brain injury and may contribute to secondary injury.

Keywords Traumatic brain injury · cytokines · inflammation · apoptosis

Introduction

The inflammatory response may contribute to cerebral edema, increased intracranial pressure and cellular loss in traumatic brain injury (TBI). Cytokines are biomarkers of this inflammatory response and new methods allow simultaneous measurement of multiple cytokines. We examined the IL-1β, IL-6, IL-8 and IL-12, TNFα, and IL-10 in arterial and jugular blood as well as cerebrospinal fluid in patients with severe traumatic brain injury.

Cytokines are signaling molecules that were initially described in the immune system were they modulate activation and inhibition of innate, cellular and humoral immunity [1]. They are expressed in a wide range of cells well beyond the classical immune system, and in the central nervous system the astrocytes and microglia appear to play a major role in cytokine release. These molecules participate in complex interlocking signaling pathways that ultimately up-regulate (pro-inflammatory) or down-regulate (anti-inflammatory) elements of the host response to injury. They also signal bias toward apoptosis (pro-apoptotic) or toward cell survival (anti-apoptotic) pathways in cellular targets. Interest in the role of these molecules in neurological disorders has been increasing, and it has become apparent that the central nervous system is capable of mounting a sustained and robust neuroinflammatory re-

J. C. Goodman · M. Van · S. P. Gopinath · C. S. Robertson
Department of Neurosurgery, Baylor College of Medicine,
One Baylor Plaza MS:BCM650,
Houston, TX 77030, USA

M. Van
e-mail: mvan@bcm.tmc.edu

S. P. Gopinath
e-mail: shankarg@bcm.tmc.edu

C. S. Robertson
e-mail: claudiar@bcm.tmc.edu

J. C. Goodman
Department of Neurology, Baylor College of Medicine,
Houston, TX 77030, USA

J. C. Goodman (✉)
Department of Pathology, Baylor College of Medicine,
One Baylor Plaza, MS BCM 315,
Houston, TX 77030, USA
e-mail: jgoodman@bcm.tmc.edu

sponse in conditions as diverse stroke, multiple sclerosis, tumors, infections and traumatic brain injury [1, 2, 4, 5]

Materials and methods

Samples of arterial and jugular blood as well as cerebrospinal fluid were collected acutely and daily for three days in 23 patients with severe TBI. The mean patient age was 32.7 year (15–57 years), the gender distribution was 17 males and 6 females, and the injuries resulted from motor vehicle accident in 19, falls in 2, and assaults in 2. The Glasgow Coma Scale (GCS) was less than 8 with a day 1 GCS of 4.5, day 2 of 5.2, and day 3 of 5.3. Detailed clinical and physiological monitoring data including temperature, intracranial pressure, mean arterial pressure, cerebral perfusion pressure, white blood count, concurrent injuries, and brain tissue pO_2 were collected and stored in a central data base along with the cytokine data.

Cytokine measurements were performed using multiplex flow cytometric bead array analysis (Beckton Dickinson Cytometric Bead Array, Human Inflammation Kit, San Jose, CA). This method permits the simultaneous measurement of 6 cytokines on a single sample.

Results

The pro-inflammatory and pro-apoptotic cytokines were elevated in blood and CSF in all patients while the anti-inflammatory response was not vigorous. Serum IL-6, IL-8 and IL-10 were consistently elevated and CSF levels were often extremely elevated indicating intracranial production. The mean values (pg/ml) with standard deviations (S.D.) for each of the cytokines for the first three days are shown below.

Tumor necrosis factor (TNF) is a pro-apoptotic cytokine had an arterial concentration of 6.35 (S.D. 6.36), jugular venous concentration of 7.37 (S.D. 7.68) and a cerebrospinal fluid concentration of 27.22 (S.D. 62.95).

Interleukin-12 (IL-12) is a pro-inflammatory cytokine had an arterial concentration of 9.4 (S.D. 13.44), jugular venous concentration of 10.07 (S.D. 14.69) and a cerebrospinal fluid concentration of 11.42 (14.24).

Interleukin-6 (IL-6) is a pro-inflammatory cytokine had an arterial concentration of 853.22 (2038.40), jugular venous concentration of 611.37 (S.D.1374.22) and a cerebrospinal fluid concentration of 2755.63 (S.D. 4828.18).

Interleukin-1 (IL-1) is a pro-inflammatory cytokine had an arterial concentration of 9.79 (S.D. 9.76), jugular venous concentration of 9.44 (S.D. 8.65) and a cerebrospinal fluid concentration of 88.55 (S.D. 217.2).

Interleukin-8 (IL-8) is a pro-inflammatory cytokine had an arterial concentration of 351.04 (S.D. 823.53), jugular venous concentration of 358.37 (S.D. 756.88) and a cerebrospinal fluid concentration of 3591.87 (S.D. 5668.12).

Interleukin-10 (IL-10) is an anti-inflammatory cytokine had an arterial concentration of 13.09 (S.D. 14.30), jugular venous concentration of 14.87 (S.D. 16.44) and a cerebrospinal fluid concentration of 10.32 (S.D. 35.48).

There was no clear relationship between arterial or CSF cytokines with core temperature, white blood count, ICP or CPP. There are currently insufficient clinical follow-up times to permit conclusions about the prognostic significance of the cytokine elevations.

Discussion

Tumor necrosis factor (TNF-α) is produced by monocytes/microglia and astrocytes, and it exerts pro-inflammatory actions including increased leukocyte adhesion, augmented coagulation, increased apoptosis and activation of other pro-inflammatory cytokine production. This cytokine has previously been studies and is up-regulated acutely in experimental and clinical TBI, and inhibitors of production are neuroprotective in experimental systems. In our study, plasma levels were elevated 1.5 times over normal while CSF levels were elevated 80-fold, indicating CNS production [1, 3].

IL-12 is produced by microglia and macrophages and exerts pro-inflammatory actions including increased production of other pro-inflammatory cytokines and proliferation of cytotoxic lymphocytes. Increased IL-12 has been described in CSF and serum in experimental and clinical TBI. In the present study, IL-12 serum levels were elevated 4 times over normal and CSF levels were elevated 10 times normal [1, 3].

IL-6 is produced by macrophages/microglia, astrocytes and neurons; and exerts pro-inflammatory effects. This cytokine is rapidly upregulated in experimental and clinical TBI. Experimental studies suggest that IL-6 may be deleterious at high concentrations but that it may be neuroprotective at low concentrations. The present study demonstrated a 45-fold increase over normal of plasma levels and a 120-fold increase over normal CSF levels [1, 3].

IL-1 is produced predominantly macrophages/microglial but also is produced by astrocytes. It exerts strong pro-inflammatory responses and is a well recognized endogenous pyrogen. IL-1 is elevated acutely in experimental and clinical TBI, it is reduced by hypothermia, and antagonists are neuroprotective in experimental TBI. We found that plasma levels were double normal values while CSF levels were elevated 12-fold [1, 3].

IL-8 is a chemokine (chemoattractant cytokine) derived from monocytes and endothelial cells and it exerts generally pro-inflammatory effects. IL-8 is Increased in CSF and serum in experimental and clinical TBI. We found plasma levels were increased 4-fold while CSF levels were elevated by 120-fold [1, 3].

IL-10 is produced by macrophages and microglia, and it exerts anti-inflammatory effects including Inhibition of pro-inflammatory cytokine synthesis and down regulation of cell-mediated immunity. It does have some local pro-inflammatory effects serving as a chemoattractant and it stimulates humoral activity at sites of inflammation. Previous studies have shown increased in plasma but not CSF in clinical TBI. We found that plasma levels were increased 1.5 times normal while CSF levels were increased 10-fold [1, 3].

Multiple pro-inflammatory and pro-apoptotic components of the inflammatory response are activated in severe TBI. In addition to the systemic response, there is a significant intracranial contribution. While elements of this response may be beneficial, it is likely that the neuroinflammatory response contributes to increased intracranial pressure, cerebral edema and neuroglial cell death following injury. Measurement and modulation of this response may have a role in the management of the head injured patient.

Acknowledgement This work was performed in compliance of the Baylor College of Medicine Animal Use and Care Regulations. Funding from NIH NINDS PO-1 NS38660–01 is gratefully acknowledged.

Conflict of interest statement We declare that we have no conflict of interest.

References

1. Brambilla R, Bracchi-Ricard V, Bethea JR (2006) Cytokines in brain trauma and spinal cord injury. In: Ransohoff RM, Beneviste EN (eds) Cytokines and the CNS, 2nd Edition. CRC Press, New York, pp 299–326
2. Lenzlinger PM, Morganti-Kossmann MC, Laurer HL, McIntosh TK (2001) The duality of the inflammatory response to traumatic brain injury. Mol Neurobiol 24(1–3):169–181
3. Maier B, Laurer HL, Rose S, Buurman WA, Marzi I (2005) Physiological levels of pro- and anti-inflammatory mediators in cerebrospinal fluid and plasma: a normative study. J Neurotrauma 22(7):822–835
4. Morganti-Kossmann MC, Rancan M, Otto VI, Stahel PF, Kossmann T (2001) Role of cerebral inflammation after traumatic brain injury: a revisited concept. Shock 16(3):165–177
5. Raghupathi R, McIntosh TK, Smith DH (1995) Cellular responses to experimental brain injinjury. Brain Pathol 5(4): 437–442

Acta Neurochir Suppl (2008) 102: 441–445

Protective effect of hyperbaric oxygen therapy on experimental brain contusions

Cornelia Voigt · Annette Förschler · Matthias Jaeger · Jürgen Meixensberger · Lea Küppers-Tiedt · Martin U. Schuhmann

Abstract
Background We evaluated the effect of hyperbaric oxygen therapy (HBO) on experimental brain contusions in rats using magnetic resonance imaging (MRI).
Materials and Methods Ten Sprague-Dawley rats were investigated at 24 h and 72 h after controlled cortical impact injury. One hour after trauma, 5 rats were treated for 60 min with 100% oxygen at 2.5 absolute atmosphere (ATA), 5 were kept at normobaric room air. MRI was performed longitudinally at 24 h and 72 h after injury. Lesion volume was determined in T2 weighted MRI scans. Relative apparent diffusion coefficient (ADC) changes were calculated in comparison to the contralateral side.
Results Following HBO, T2 lesion volume was smaller at 24 h versus controls (63.1 ± 16.5 mm^3 vs. 87.4 ± 13.8 mm^3, $p<0.05$), and decreased further at 72 h (46.8 ± 17.8 mm^3 vs. 92.5 ± 13.1 mm^3, $p<0.01$). At 24 h, the mean relative ADC change in the lesion area decreased from $+\ 26.8\pm2.3\%$ in controls to $+\ 2.3\pm12.2\%$ in HBO animals ($p<0.01$). At 72 h, the HBO effect on relative ADC values was less when compared to 24 h.
Discussion A 60-minute exposure to hyperbaric oxygen starting 1 h after impact injury significantly attenuated lesion growth and relative increase of ADC values within the contused area for up to 72 h. Thus, a "single-shot" HBO treatment seems to have long-lasting neuroprotective effects on the contused brain and its penumbra.

Keywords Hyperbaric oxygen · Controlled cortical impact · Traumatic brain injury · Magnetic resonance imaging

C. Voigt · J. Meixensberger
Department of Neurosurgery, University of Leipzig, 04103 Leipzig, Germany

C. Voigt
e-mail: dat_conny@web.de

J. Meixensberger
e-mail: meix@medizin.uni-leipzig.de

A. Förschler
Division of Neuroradiology, Klinikum rechts der Isar, Technical University Munich, 81675 Munich, Germany
e-mail: annette.foerschler@roe.med.tum.de

M. Jaeger
Department of Neurosurgery, Liverpool Health Service, University of New South Wales, Liverpool, NSW 2170, Australia
e-mail: ma.jaeger@gmx.net

L. Küppers-Tiedt
Department of Neurology, University of Leipzig, 04103 Leipzig, Germany
e-mail: lea-kueppers@medizin.uni-leipzig.de

M. U. Schuhmann (✉)
Department of Neurosurgery, Eberhard Karls University Tübingen, Hoppe-Seyler-Str. 3, D-72076 Tübingen, Germany
e-mail: martin.schuhmann@med.uni-tuebingen.de

Introduction

Hyperbaric oxygen therapy (HBO) following traumatic brain injury has been evaluated experimentally as early as 1970 [7] and was applied in several clinical series during the 1980's, culminating in a prospectively randomized single center study published in 1992 [11]. Although this study showed a decrease in the mortality rate, wide-spread use of HBO therapy did not occur as a result mainly due to two factors, 1) the rate of good outcomes was not increased and 2) the difficulties in transporting severely head injured patients into a hyperbaric treatment chamber. Recently, there has been renewed interest in HBO, with some encouraging clinical results [10, 12] and experimental data using fluid percussion injury (FPI) and a contusion model applying suction to the cortex. One FPI study showed that HBO significantly increased brain tissue PO_2 and restored mitochondrial redox potential within 4 hours post-injury [1]. Another investigation found that HBO therapy following severe FPI attenuated ICP increase and decreased mortality [13]. In the focal suction model, inflammatory mechanisms were found to be attenuated after HBO [19]. In a focal post-embolic stroke model, HBO was shown to be very effective in reducing ischemic lesion size as measured in MRI and histology [4].

We therefore investigated the therapeutic potential of HBO in the most widely used experimental focal traumatic brain injury model, the controlled cortical impact model. In this pilot study we focused on the growth of the contusion and the pericontusional "penumbra" zone in the first 72 h after injury. In human brain injury, contusion growth and development of pericontusional edema in the first 24–48 h is a clinically relevant phenomenon, causing secondary space occupying effects, elevations in ICP and neurological deterioration, necessitating prolonged intubation and sedation, surgical evacuation and in some cases, decompressive craniectomy.

Materials and methods

Subjects

Ten male Sprague Dawley rats weighting 250–300 g were studied. All animals were kept on a 12 h-light/dark cycle with free access to food and water. Five animals were assigned to HBO or control group, respectively. All received left parietal controlled cortical impact injury. Each animal was studied by MRI longitudinally at 24 h and 72 h after injury.

Procedures

Anesthesia was induced by breathing 5% Isoflurane in N_2O/O_2 (2:1) for two minutes. Animals were then injected with fentanyl (0.005 mg/kg), midazolam (2 mg/kg) and medetomidine (0.15 mg/kg) intramuscularly and breathed spontaneously. In a stereotactic frame the left parietal bone was exposed via a midline incision and a left-sided 7 mm craniotomy was centered 3.5 mm posterior and 4 mm lateral to the bregma. Animals underwent a controlled cortical impact injury with a device similar to that described by Lighthall [6]. A 5 mm round impactor tip was accelerated to 4 m/s with a vertical deformation depth of 2 mm and impact duration of 200 ms. Body temperature was monitored with a rectal probe and kept between 36.5 and 37.5 °C using a heating pad. Dural tears were not repaired but the bone flap was reinserted and fixed with dental cement. The skin was closed tightly. Anesthesia was then immediately reversed. One hour after impact HBO animals were pressurized within 10 minutes to 2.5 atmospheres absolute (ATA) of 100% O_2, which was maintained for 60 minutes, and then decompressed within 10 min. Controls were also transferred into the chamber for a comparable time interval, but breathed normobaric room air.

Magnetic resonance imaging

We employed a 1.5 T Philips Intera scanner and a microscopy coil. One mm slices were acquired with the

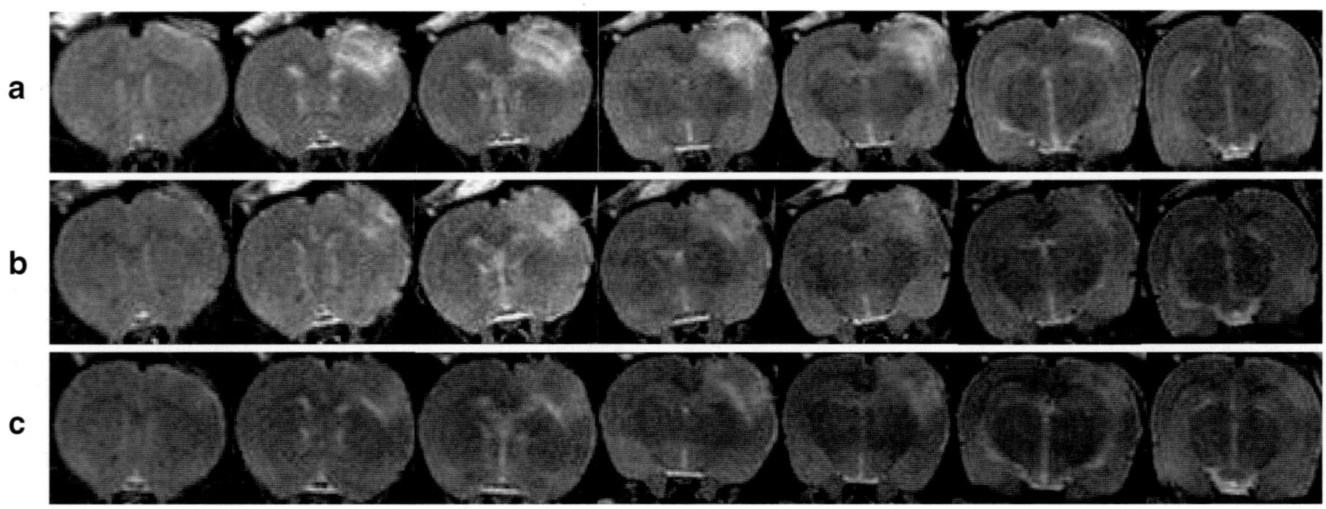

Fig. 1 Serial 1 mm coronal T_2 weighted scans from a control animal at 72 h (**a**) and a HBO treated animal at 24 h (**b**) and 72 h (**c**) post injury. Compared to control (A), the effect of HBO treatment is already apparent at 24 h (B) with a less prominent water signal in the lesion area, which is also smaller. At 72 h (C) there is less increase in tissue water resulting in a smaller lesion in the same animal

following sequences. T2: TE/TR (ms) 3200/100, Res 0.22× 0.28 mm, FOV 50 mm. DWI: TE/TR (ms) 4770/100, Res 0.55×0.60, FOV 50 mm. For MR investigations, the animals were anesthetized as they were for surgery.

Analysis of MRI

Lesion volume (comprising all hyperintense areas at the site of impact, i.e. contusion and surrounding edema) was determined from T2 weighted scans planimetrically using ImageJ software (NIH, Bethesda, USA). ADC values of the lesion were expressed as relative change compared to the corresponding contra-lateral area.

Data analysis

Lesion volumes were expressed in mm^3 and calculated as mean±SD per group. ADC values were expressed in % change to contralateral side. Differences between HBO treated animals and controls were analyzed using the Mann-Whitney U-test. A p value of <0.05 was required to reject the null hypothesis of equivalence.

Results

Animals in both groups were matched for body weight and age. At 24 h, the lesion appearance in the T2 weighted images was much less prominent in the HBO group as compared to controls (see Fig. 1). This was reflected in a smaller lesion volume of 63.1±16.5 mm^3 vs. 87.4±13.8 mm^3 in controls ($p<0.05$). Lesion volume at 72 h remained nearly unchanged in the normobaric group, however decreased further in the HBO group to 46.8±17.8 mm^3 vs. 92.5±13.1 mm^3 in controls ($p<0.01$, see Fig. 2a). At 24 h the ADC value of the lesion in control animals increased compared to the contralateral side by +26.8±2.3%. In HBO animals, the ADC, compared to the contralateral side, was only slightly elevated in 2 of 5 animals and decreased in 3 of 5 animals, resulting in a mean of +2.3±12.2%, $p<0.01$. At 72 h, there were no relative decreases of ADC in any HBO animal, however relative ADC was still less compared to control animals (10.1±14.8% vs. 32.8±11.2%, $p<0.02$, see Fig. 2b).

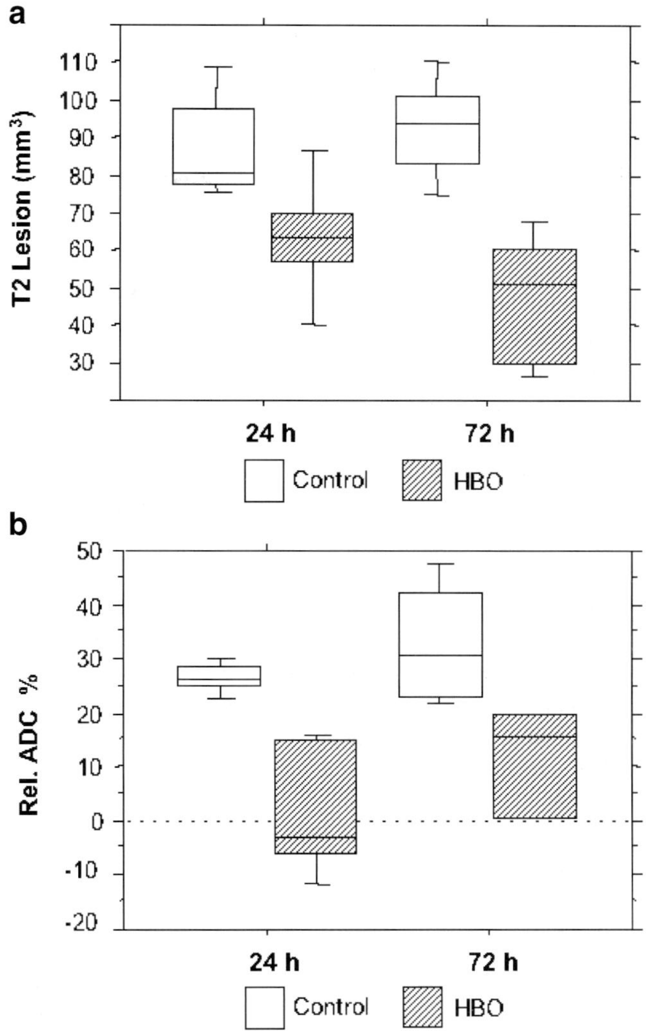

Fig. 2 Results of (**a**) planimetrical determination of lesion volume in mm^3 according to T_2 weighted MR scans (T2 Lesion) and (**b**) relative changes of ADC value in the lesion ROI compared to the corresponding contra-lateral ROI (rel. ADC) in diffusion weighted images

Discussion

The mechanisms responsible for brain contusion expansion are similar to those that result in secondary brain injury. The different molecular mechanisms that cause secondary brain injury are numerous. Therefore, it is not surprising that many studies utilizing the CCI model have shown that a variety of different pharmacological agents and therapeutic maneuvers were able to limit secondary brain injury and contusion growth, and result in improved neuronal survival or functional outcome. However, most of these therapies have not entered clinical use. Although oxygen is a universally available "drug" in the ICU, it is well understood that treating a heterogeneous patient population with HBO will likely result in a wide range of outcomes. Therefore possible beneficial effects of a chosen therapeutic approach, which might act only on certain aspects of the ongoing secondary injury cascades, may not be detected. This may be one reason Rockswold's randomized trial of HBO could not show an improvement in outcome, although the mortality rate in the treatment group, particularly among the more severely injured patients, was clearly lowered [11].

The evolution of brain contusions is similar in the human TBI and in the experimental set-up and can be primarily evaluated by various imaging modalities and secondarily be related to histological or functional outcome. Our pilot study, applying a one hour HBO treatment early after injury (1 h), showed that this "early single shot intervention" effectively halted contusion expansion, which is known to occur in this model within the first 24 to 48 h, with a maximum growth between 4 h and 24 h [14–16]. Interestingly, we not only noted cessation of contusion expansion, but a decrease in contusion size at 72 h compared with 24 h. Thus intra- and pericontusional edema had regressed between these time points, whereas it remained unchanged in controls.

The pericontusional penumbra is known to be an area of low cerebral blood flow [2, 3, 5] and cytotoxic brain edema [9, 16]. Hyperbaric oxygenation appears to interrupt secondary injury mechanisms early on after injury in the contusion and the pericontusional penumbra, which in turn seems to have a long-lasting effect on edema evolution. Importantly, hyperbaric oxygenation appears to speed up restoration of physiological conditions, as reflected by normalization of brain water content measured by T_2-weighted MR imaging.

The positive effects of the HBO therapy might be explained by the known increase of CaO_2 and CBF, both increasing the O_2 availability in the tissue at risk. This could be measured in the experimental setting by an increase of partial pressure of brain tissue O_2. In previous studies this was accompanied by a recovery of aerobic metabolism as expressed by restoration of the mitochondrial redox potential [1].

Others have investigated the role of normobaric oxygenation (NBO) and HBO on neuroinflammation following TBI, using a focal lesion model applying cortical suction. HBO resulted in a more pronounced reduction of inflammatory infiltration and of TUNEL positive cells around the focus of necrotic brain tissue than NBO (19). However, NBO still demonstrated positive effects compared to control.

In our study ADC mapping was not able to distinguish between different types of edema in the lesion, which is composed of the mechanically disintegrated contusion core and the pericontusional penumbra zone. This may be due to the fact that we used a 1.5 T machine, where spatial resolution is lower despite the use of a microscopy coil compared to investigations in dedicated animal scanners. The increase of relative ADC values in the lesion argues for a predominant increase of extracellular water in the region, which is in contrast to the findings of other investigators, showing an ADC decrease [9, 16]. However, in these studies the trauma intensity was more severe than in our study.

In the present study the HBO intervention seems to shift pathophysiology away from a predominance of extracellular free water towards a predominance of intracellular water accumulation, with a negative relative ADC in 3 of 5 animals at 24 h (-3% to -11%), and only slightly positive relative ADC in the remaining 2 animals (+14 to +16%). This suggests a protective effect of HBO on blood brain barrier integrity [8, 18] and attenuated vascular basal lamina degradation [17], which have been demonstrated in stroke models of permanent MCA occlusion.

This pilot research used magnetic resonance imaging to investigate whether an early "single shot" HBO therapy at 2.5 ATA has detectable beneficial effects on traumatic brain contusion development. This was found to be true regarding lesion size and changes in brain water distribution as detected by MRI. Future studies are necessary to investigate whether these effects translate into improved neuronal survival, attenuated inflammatory response, and, most importantly, improved outcomes. If so, HBO therapy, may hold promise as a "single shot early therapy" for TBI patients that can be administered in the ICU immediately following initial diagnosis in the CT scanner.

Conflict of interest statement We declare that we have no conflict of interest.

References

1. Daugherty WP, Levasseur JE, Sun D, Rockswold GL, Bullock RM (2004) Effects of hyperbaric oxygen therapy on cerebral oxygenation and mitochondrial function following moderate lateral fluid-percussion injury in rats. J Neurosurg 101:499–504
2. Forbes ML, Hendrich KS, Kochanek PM, Williams DS, Schiding JK, Wisniewski SR, Kelsey SF, DeKosky ST, Graham SH, Marion DW, Ho C (1997) Assessment of cerebral blood flow and CO2 reactivity after controlled cortical impact by perfusion magnetic resonance imaging using arterial spin-labeling in rats. J Cereb Blood Flow Metab 17:865–874
3. Hendrich KS, Kochanek PM, Williams DS, Schiding JK, Marion DW, Ho C (1999) Early perfusion after controlled cortical impact in rats: quantification by arterial spin-labeled MRI and the influence of spin-lattice relaxation time heterogeneity. Magn Reson Med 42:673–681
4. Henninger N, Kueppers-Tiedt L, Sicard KM, Guenther A, Schneider D, Schwab S (2006) Neuroprotective effect of hyperbaric oxygen therapy monitored by MR-imaging after embolic stroke in rats. Exp Neurol 201:316–323
5. Kroppenstedt SN, Stover JF, Unterberg AW (2000) Effects of dopamine on posttraumatic cerebral blood flow, brain edema, and cerebrospinal fluid glutamate and hypoxanthine concentrations. Crit Care Med 28:3792–3798
6. Lighthall JW (1988) Controlled cortical impact: a new experimental brain injury model. J Neurotrauma 5:1–15
7. Miller JD, Fitch W, Li M (1970) The effect of hyperbaric oxygen on experimentally increased intracranial pressure. J Neurosurg 33:287–296
8. Mink RB, Dutka AJ (1995) Hyperbaric oxygen after global cerebral ischemia in rabbits reduces brain vascular permeability and blood flow. Stroke 26:2307–2312
9. Portella G, Beaumont A, Corwin F, Fatouros P, Marmarou A (2000) Characterizing edema associated with cortical contusion

and secondary insult using magnetic resonance spectroscopy. Acta Neurochir Suppl 76:273–275
10. Ren H, Wang W, Ge Z (2000) Glasgow coma scale, brain electric activity mapping and Glasgow Outcome Scale after hyperbaric oxygen treatment of severe brain injury. Chin J Traumatol 4:239–241
11. Rockswold GL, Ford SE, Anderson DC, BErgman TA, Sherman RE (1992) Results of a prospective randomized trial for treatment of severely brain-injured patients with hyperbaric oxygen. J Neurosurg 76:929–934
12. Rockswold SB, Rockswold GL, Vargo JM, Erickson CA, Sutton RL, Bergman TA, Biros MH (2001) Effects of hyperbaric oxygenation therapy on cerebral metabolism and intracranial pressure in severely brain injured patients. J Neurosurg 94:403–411
13. Rogatsky GG, Kamenir Y, Mayevsky A (2005) Effect of hyperbaric oxygenation on intracranial pressure elevation rate in rats during the early phase of severe traumatic brain injury. Brain Research 1047:131–136
14. Schuhmann MU, Mokhtarzadeh M, Stichtenoth DO, Skardelly M, Klinge PM, Gutzki FM, Samii M, Brinker T (2003) Temporal profiles of cerebrospinal fluid leukotrienes, brain edema and inflammatory response following experimental brain injury. Neurol Res 25:481–491
15. Schuhmann MU, Stiller D, Skardelly M, Mokktarzadeh M, Thomas S, Brinker T, Samii M (2002) Determination of contusion and oedema volume by MRI corresponds to changes of brain water content following controlled cortical impact injury. Acta Neurochir 81(Suppl):213–215
16. Stroop R, Thomale UW, Pauser S, Bernarding J, Vollmann W, Wolf KJ, Lanksch WR, Unterberg AW (1998) Magnetic resonance imaging studies with cluster algorithm for characterization of brain edema after controlled cortical impact injury (CCII). Acta Neurochir Suppl (Wien) 71:303–305
17. Veltkamp R, Bieber K, Wagner S, Beynon C, Siebing DA, Veltkamp C, Schwaninger M, Marti HH (2006) Hyperbaric oxygen reduces basal lamina degradation after transient focal cerebral ischemia in rats. Brain Res 1076:231–237
18. Veltkamp R, Siebing DA, Sun L, Heiland S, Bieber K, Marti HH, Nagel S, Schwab S, Schwaninger M (2005) Hyperbaric oxygen reduces blood–brain barrier damage and edema after transient focal cerebral ischemia. Stroke 36:1679–1683
19. Vlodavsky E, Palzur E, Soustiel JF (2006) Hyperbaric oxygen therapy reduces neuroinflammation and expression of matrix metalloproteinase-9 in the rat model of traumatic brain injury. Neuropathol Appl Neurobiol 32:40–50

Author index

Index of keywords